Human Development
2007/2008

DATE DUE

e change:

ty in a divided world

			PRINTED IN U.S.A.

Published for the
United Nations

ISBN 978-0-230-54704-9

First Published in 2007 by
Palgrave Macmillan
Houndmills, Basingstoke, Hampshire RG21 6XS and
175 Fifth Avenue, New York, NY 10010

Companies and representatives throughout the world.

Palgrave Macmillan is the global academic imprint of the Palgrave Macmillan division of
St. Martin's Press, LLC and of Palgrave Macmillan Ltd.

Macmillan is a registered trademark in the United States, United Kingdom, and other countries.
Palgrave is a registered trademark in the European Union and other countries.

10 9 8 7 6 5 4 3 2 1
Printed in the U.S.A. by RR Donnelley/Hoechstetter (Pittsburgh). Cover is printed on International Paper's 15 pt Carolina
low-density coated-one-side paper that is chlorine free and meets the Sustainable Forest Initiative guidelines.
Text pages are printed on Cascades Mills' 60# Rolland Opaque30 Smooth text that is de-inked 30% post-industrial fibre,
Forest Stewardship Council certified and produced with Biogas energy. Both cover and text papers are printed with vegetable-based inks and
produced by means of environmentally-compatible technology.

Mixed Sources

**Product group from well-managed
forests, controlled sources and
recycled wood or fiber**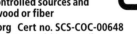

FSC www.fsc.org Cert no. SCS-COC-00648
© 1996 Forest Stewardship Council

Editing: Green Ink Inc.
Cover: talking-box
Information design: Mapping Worlds, Phoenix Design Aid and Zago
Layout: Phoenix Design Aid

For a list of any errors or omissions found subsequent to printing, please visit our
website at http://hdr.undp.org

Team for the preparation of Human Development Report 2007/2008

Director and lead author

Kevin Watkins

Research and statistics

Cecilia Ugaz (Deputy Director and chief editor), Liliana Carvajal, Daniel Coppard, Ricardo Fuentes Nieva, Amie Gaye, Wei Ha, Claes Johansson, Alison Kennedy (Chief of Statistics), Christopher Kuonqui, Isabel Medalho Pereira, Roshni Menon, Jonathan Morse and Papa Seck.

Production and translation

Carlotta Aiello and Marta Jaksona

Outreach and communications

Maritza Ascencios, Jean-Yves Hamel, Pedro Manuel Moreno and Marisol Sanjines (Head of Outreach)

The Human Development Report Office (HDRO): The Human Development Report is the product of a collective effort. Members of the National Human Development Report Unit (NHDR) provide detailed comments and advice throughout the research process. They also link the Report to a global research network in developing countries. The NHDR team comprises Sharmila Kurukulasuriya, Mary Ann Mwangi and Timothy Scott. The HDRO administrative team makes the office function and includes Oscar Bernal, Mamaye Gebretsadik, Melissa Hernandez and Fe Juarez-Shanahan. Operations are managed by Sarantuya Mend.

Foreword

What we do today about climate change has consequences that will last a century or more. The part of that change that is due to greenhouse gas emissions is not reversible in the foreseeable future. The heat trapping gases we send into the atmosphere in 2008 will stay there until 2108 and beyond. We are therefore making choices today that will affect our own lives, but even more so the lives of our children and grandchildren. This makes climate change different and more difficult than other policy challenges.

Climate change is now a scientifically established fact. The exact impact of greenhouse gas emission is not easy to forecast and there is a lot of uncertainty in the science when it comes to predictive capability. But we now know enough to recognize that there are large risks, potentially catastrophic ones, including the melting of ice-sheets on Greenland and the West Antarctic (which would place many countries under water) and changes in the course of the Gulf Stream that would bring about drastic climatic changes.

Prudence and care about the future of our children and their children requires that we act now. This is a form of insurance against possibly very large losses. The fact that we do not know the probability of such losses or their likely exact timing is not an argument for not taking insurance. We know the danger exists. We know the damage caused by greenhouse gas emissions is irreversible for a long time. We know it is growing with every day of inaction.

Even if we were living in a world where all people had the same standard of living and were impacted by climate change in the same way, we would still have to act. If the world were a single country, with its citizens all enjoying similar income levels and all exposed more or less to the same effects of climate change, the threat of global warming could still lead to substantial damage to human well-being and prosperity by the end of this century.

In reality, the world is a heterogeneous place: people have unequal incomes and wealth and climate change will affect regions very differently. This is, for us, the most compelling reason to act rapidly. Climate change is already starting to affect some of the poorest and most vulnerable communities around the world. A worldwide average 3° centigrade increase (compared to preindustrial temperatures) over the coming decades would result in a range of localized increases that could reach twice as high in some locations. The effect that increased droughts, extreme weather events, tropical storms and sea level rises will have on large parts of Africa, on many small island states and coastal zones will be inflicted in our lifetimes. In terms of aggregate world GDP, these short term effects may not be large. But for some of the world's poorest people, the consequences could be apocalyptic.

In the long run climate change is a massive threat to human development and in some places it is already undermining the international community's efforts to reduce extreme poverty.

Violent conflicts, insufficient resources, lack of coordination and weak policies continue to slow down development progress, particularly in Africa. Nonetheless in many countries there have been real advances. For instance, Viet Nam has been able to halve poverty and achieve universal primary education way ahead of the 2015 target. Mozambique has also managed to significantly reduce poverty and increase school enrollment as well as improving the rates of child and maternal mortality.

This development progress is increasingly going to be hindered by climate change. So we must see the fight against poverty and the fight against the effects of climate change as interrelated efforts. They must reinforce each other and success must be achieved on both fronts jointly. Success will have to involve a great deal of adaptation, because climate change is still going to affect the poorest countries significantly even if serious efforts to reduce emissions start immediately. Countries will need to develop their own adaptation plans but the international community will need to assist them.

Responding to that challenge and to the urgent request from leaders in developing countries, particularly in sub-Saharan Africa, UNEP and UNDP launched a partnership in Nairobi during the last climate convention in November 2006. The two agencies committed to provide assistance in reducing vulnerability and building the capacity of developing countries to more widely reap the benefits of the Clean Development Mechanism (CDM) in areas such as the development of cleaner and renewable energies, climate proofing and fuel-switching schemes.

This partnership, that will enable the UN system to act promptly in response to the needs of governments trying to factor in climate-change impacts into their investment decisions, constitutes a living proof of the United Nation's determination to 'deliver as One' on the climate change challenge. For example, we can help countries improve existing infrastructure to enable people to cope with increased flooding and more frequent and severe extreme weather events. More weather resistant crops could also be developed.

While we pursue adaptation we must start to reduce emissions and take other steps at mitigation so that the irreversible changes already underway are not further amplified over the next few decades. If mitigation does not start in earnest right now, the cost of adaptation twenty or thirty years from now will become prohibitive for the poorest countries.

Stabilizing greenhouse emissions to limit climate change is a worthwhile insurance strategy for the world as a whole, including the richest countries, and it is an essential part of our overall fight against poverty and for the Millennium Development Goals. This dual purpose of climate policies should make them a priority for leaders around the world.

But having established the need for limiting future climate change and for helping the most vulnerable adapt to what is unavoidable, one has to move on and identify the nature of the policies that will help us get the results we seek.

Several things can be said at the outset: First, non-marginal changes are needed, given the path the world is on. We need big changes and ambitious new policies.

Second, there will be significant short term costs. We have to invest in limiting climate change. There will be large net benefits over time, but at the beginning, like with every investment, we must be willing to incur the costs. This will be a challenge for democratic governance: political systems will have to agree to pay the early costs to reap the long term gains. Leadership will require looking beyond electoral cycles.

We are not too pessimistic. In the fight against the much higher inflation rates of the distant past, democracies did come up with the institutions such as more autonomous central banks and policy pre-commitments that allowed much lower inflation to be achieved despite the short term temptations of resorting to the printing press. The same has to happen with climate and the environment: societies will have to pre-commit and forego short term gratification for longer-term well being.

We would like to add that while the transition to climate protecting energy and life styles will have short term cost, there may be economic benefits beyond what is achieved by stabilizing temperatures. These benefits are likely to be realized through Keynesian and Schumpeterian mechanisms with new incentives for massive investment stimulating overall demand and creative destruction leading to innovation and productivity jumps in a wide array of sectors. It is impossible to quantitatively predict how large these effects will be but taking them into account could lead to higher benefit-cost ratios for good climate policies.

The design of good policies will have to be mindful of the danger of excessive reliance on bureaucratic controls. While government leadership is going to be essential in correcting the huge externality that is climate change, markets and prices will have to be put to work, so that private sector decisions can lead more naturally to optimal investment and production decisions.

Carbon and carbon equivalent gases have to be priced so that using them reflects their true social cost. This should be the essence of mitigation policy. The world has spent decades getting rid of quantity restrictions in many domains, not least foreign trade. This is not the time to come back to a system of massive quotas and bureaucratic controls because of climate change. Emission targets and energy efficiency targets have an important role to play but it is the price system that has to make it easier to achieve our goals. This will require a much deeper dialogue between economists and climate scientists as well as environmentalists than what we have seen so far. We do hope that this Human Development Report will contribute to such a dialogue.

The most difficult policy challenges will relate to distribution. While there is potential catastrophic risk for everyone, the short and medium-term distribution of the costs and benefits will be far from uniform. The distributional challenge is made particularly difficult because those who have largely caused the problem—the rich countries—are not going to be those who suffer the most in the short term. It is the poorest who did not and still are not contributing significantly to green house gas emissions that are the most vulnerable. In between, many middle income countries are becoming significant emitters in aggregate terms—but they do not have the carbon debt to the world that the rich countries have accumulated and they are still low emitters in per capita terms. We must find an ethically and politically acceptable path that allows us to start—to move forward even if there remains much disagreement on the long term sharing of the burdens and benefits. We should not allow distributional disagreements to block the way forward just as we cannot afford to wait for full certainty on the exact path climate change is likely to take before we start acting. Here too we hope this Human Development Report will facilitate the debate and allow the journey to start.

Kemal Derviş
Administrator
United Nations Development Programme

Achim Steiner
Executive Director
United Nations Environment Programme

The analysis and policy recommendations of the Report do not necessarily reflect the views of the United Nations Development Programme, its Executive Board or its Member States. The Report is an independent publication commissioned by UNDP. It is the fruit of a collaborative effort by a team of eminent consultants and advisers and the Human Development Report team. Kevin Watkins, Director of the Human Development Report Office, led the effort.

Acknowledgements

This Report could not have been prepared without the generous contribution of the many individuals and organizations listed below. Special mention must be made of Malte Meinshausen of the Potsdam Institute for Climate Impact Research, who provided constant and patient advice on a wide range of technical issues. Many other individuals contributed to the Report either directly through background papers, comments on draft text, and discussions, or indirectly through their research. The authors wish also to acknowledge their debt to the fourth assessment of the International Panel on Climate Change, which provides an unrivalled source of scientific evidence, and to the work of Sir Nicholas Stern and the team behind his Report on *The Economics of Climate Change*. Many colleagues in the United Nations system were extremely generous in sharing their time, expertise and ideas. The Human Development Report team received helpful advice from the UNDP Administrator, Kemal Derviş. We thank all of those involved directly or indirectly in guiding our efforts, while acknowledging sole responsibility for errors of commission and omission.

Contributors

Background studies, papers and notes were prepared on a wide range of thematic issues relating to the Report. Contributors were: Anu Adhikari, Mozaharul Alam, Sarder Shafiqul Alam, Juan Carlos Arredondo Brun, Vicki Arroyo, Albertina Bambaige, Romina Bandura, Terry Barker, Philip Beauvais, Suruchi Bhadwal, Preety Bhandari, Isobel Birch, Maxwell Boykoff, Karen O'Brien, Oli Brown, Odón de Buen, Peter Chaudhry, Pedro Conceição, Pilar Cornejo, Caridad Canales Dávila, Simon D. Donner, Lin Erda, Alejandro de la Fuente, Richard Grahn, Michael Grimm, Kenneth Harttgen, Dieter Helm, Caspar Henderson, Mario Herrero, Saleemul Huq, Ninh Nguyen Huu, Joseph D. Intsiful, Katie Jenkins, Richard Jones, Ulka Kelkar, Stephan Klasen, Arnoldo Matus Kramer, Kishan Khoday, Roman Krznaric, Robin Leichenko, Anthony Leiserowitz, Junfeng Li, Yan Li, Yue Li, Peter Linguiti, Gordon MacKerron, Andrew Marquard, Ritu Mathur, Malte Meinshausen, Mark Misselhorn, Sreeja Nair, Peter Newell, Anthony Nyong, David Ockwell, Marina Olshanskaya, Victor A. Orindi, James Painter, Peter D. Pederson, Serguey Pegov, Renat Perelet, Alberto Carillo Pineda, Vicky Pope, Golam Rabbani, Atiq Rahman, Mariam Rashid, Bimal R. Regmi, Hannah Reid, J. Timmons Roberts, Greet Ruysschaert, Boshra Salem, Jürgen Schmid, Dana Schüler, Rory Sullivan, Erika Trigoso Rubio, Md. Rabi Uzzaman, Giulio Volpi, Tao Wang, James Watson, Harald Winkler, Mikhail Yulkin and Yanchun Zhang.

Several organizations generously shared their data and other research materials: Agence Française de Développement; Amnesty International; Carbon Dioxide Information and Analysis Center; Caribbean Community

Secretariat; Center for International Comparisons of Production, Income and Prices at the University of Pennsylvania; Development Initiatives; Department for International Development; Environmental Change Institute at Oxford University; European Commission; Food and Agriculture Organization; Global Environment Facility; Global IDP Project; IGAD Climate Prediction and Applications Centre; Institute of Development Studies; International Centre for Prison Studies; Internally Displaced Monitoring Centre; International Research Institute for Climate and Society; International Energy Agency; International Institute for Environment and Development; International Institute for Strategic Studies; International Labour Organization; International Monetary Fund; International Organization for Migration; International Telecommunication Union; Inter-Parliamentary Union; Joint United Nations Programme on HIV/AIDS; Luxembourg Income Study; Macro International; Organisation for Economic Co-operation and Development; Overseas Development Institute; Oxfam; Pew Center for Climate Change; Practical Action Consulting; Stockholm International Peace Research Institute; Stockholm International Water Institute; Tata Energy Research Institute; Met Office; United Nations Children's Fund; United Nations Conference on Trade and Development; United Nations Department of Economic and Social Affairs, Statistics Division and Population Division; United Nations Development Fund for Women; United Nations Educational, Scientific and Cultural Organization, Institute for Statistics; United Nations High Commissioner for Refugees; United Nations Office on Drugs and Crime, Treaty Section; United Nations Office of Legal Affairs; University of East Anglia; WaterAid; World Bank; World Health Organization; World Meteorological Organization; World Trade Organization; World Intellectual Property Organization and the World Wildlife Fund.

Advisory Panel

The Report benefited greatly from intellectual advice and guidance provided by an external advisory panel of experts. The panel comprised Monique Barbut, Alicia Bárcena, Fatih Birol, Yvo de Boer, John R. Coomber, Mohammed T. El-Ashry, Paul Epstein, Peter T. Gilruth, José Goldemberg, HRH Crown Prince Haakon, Saleem Huq, Inge Kaul, Kivutha Kibwana, Akio Morishima, Rajendra Pachauri, Jiahua Pan, Achim Steiner, HRH Princess Basma Bint Talal, Colleen Vogel, Morris A. Ward, Robert Watson, Ngaire Woods and Stephen E. Zebiak. An advisory panel on statistics made an invaluable contribution, in particular Tom Griffin, the Report's Senior Statistical Advisor. The panel members are: Carla Abou-Zahr, Tony Atkinson, Haishan Fu, Gareth Jones, Ian D. Macredie, Anna N. Majelantle, John Male-Mukasa, Marion McEwin, Francesca Perucci, Tim Smeeding, Eric Swanson, Pervez Tahir and Michael Ward. The team is grateful to Partha Deb, Shea Rutstein and Michael Ward who reviewed and commented on an HDRO analysis of risk and vulnerability and lent their statistical expertise.

Consultations

Members of the Human Development Report team benefited individually and collectively from a wide-ranging process of consultation. Participants in a Human Development Network discussion provided some wide-ranging insights and observations on the linkages between climate change and human development. The Report team wishes also to thank Neil Adger, Keith Allott, Kristin Averyt, Armando Barrientos, Haresh Bhojwani, Paul Bledsoe, Thomas A. Boden, Keith Briffa, Nick Brooks, Katrina Brown, Miguel Ceara-Hatton, Fernando Calderón, Jacques Charmes, Lars Christiansen, Kirsty Clough, Stefan Dercon, Jaime de Melo, Stephen Devereux, Niky Fabiancic, Kimberley Fisher, Lawrence Flint, Claudio Forner, Jennifer Frankel-Reed, Ralph Friedlaender, Oscar Garcia, Stephen Gitonga, Heather Grady, Barbara Harris-White, Molly E. Hellmuth, John Hoddinott, Aminul Islam, Tarik-ul-Islam, Kareen Jabre, Fortunat Joos, Mamunul Khan, Karoly Kovacs, Diana Liverman, Lars Gunnar Marklund, Charles McKenzie, Gerald A. Meehl, Pierre Montagnier,

Jean-Robert Moret, Koos Neefjes, Iiris Niemi, Miroslav Ondras, Jonathan T. Overpeck, Vicky Pope, Will Prince, Kate Raworth, Andrew Revkin, Mary Robinson, Sherman Robinson, Rachel Slater, Leonardo Souza, Valentina Stoevska, Eric Swanson, Richard Tanner, Haiyan Teng, Jean Philippe Thomas, Steve Price Thomas, Sandy Tolan, Emma Tompkins, Emma Torres, Kevin E. Trenberth, Jessica Troni, Adriana Velasco, Marc Van Wynsberghe, Tessa Wardlaw and Richard Washington.

UNDP Readers

A Readers Group, made up of colleagues in UNDP, provided many useful comments, suggestions and inputs during the writing of the Report. Special mention must be made of the contribution and advice of Pedro Conceição, Charles Ian McNeil and Andrew Maskrey. All of them were generous with their time and made substantive contributions to the Report. Other inputs were received from: Randa Aboul-Hosn, Amat Al-Alim Alsoswa, Barbara Barungi, Winifred Byanyima, Suely Carvalho, Tim Clairs, Niamh Collier-Smith, Rosine Coulibaly, Maxx Dilley, Philip Dobie, Bjørn Førde, Tegegnework Gettu, Yannick Glemarec, Luis Gomez-Echeverri, Rebeca Grynspan, Raquel Herrera, Gilbert Fossoun Houngbo, Peter Hunnam, Ragnhild Imerslund, Andrey Ivanov, Bruce Jenks, Michael Keating, Douglas Keh, Olav Kjorven, Pradeep Kurukulasuriya, Oksana Leshchenko, Bo Lim, Xianfu Lu, Nora Lustig, Metsi Makhetha, Cécile Molinier, David Morrison, Tanni Mukhopadhyay, B. Murali, Simon Nhongo, Macleod Nyirongo, Hafiz Pasha, Stefano Pettinato, Selva Ramachandran, Marta Ruedas, Mounir Tabet, Jennifer Topping, Kori Udovicki, Louisa Vinton, Cassandra Waldon and Agostinho Zacarias.

Editing, Production and Translation

The Report benefited from the advice and contribution of an editorial team at Green Ink. Anne Moorhead provided advice on structure and presentation of the argument. Technical and production editing was carried out by Sue Hainsworth and Rebecca Mitchell. Cover and dividers were designed by Talking Box, with conceptual inputs by Martín Sánchez and Ruben Salinas, on the basis of a template designed by Grundy & Northedge in 2005. Information design was done by Phoenix Design Aid and Zago; one map (map 1.1) was designed by Mapping Worlds. Phoenix Design Aid, under the coordination of Lars Jørgensen, also did the Report's layout.

The production, translation, distribution and promotion of the Report benefited from the help and support of the UNDP Office of Communications, and particularly of Maureen Lynch and Boaz Paldi. Translations were reviewed by Iyad Abumoghli, Bill Bikales, Jean Fabre, Albéric Kacou, Madi Musa, Uladzimir Shcherbau and Oscar Yujnovsky.

The Report also benefited from the dedicated work of Jong Hyun Jeon, Isabelle Khayat, Caitlin Lu, Emily Morse and Lucio Severo. Swetlana Goobenkova and Emma Reed made valuable contributions to the statistical team. Margaret Chi and Juan Arbelaez of the UN Office of Project Services provided critical administrative support and management services.

Kevin Watkins
Director
Human Development Report 2007/2008

Contents

Chapter 3   Avoiding dangerous climate change: strategies for mitigation   109

Chapter 4   Adapting to the inevitable: national action and international cooperation   163

Boxes

Tables

Figures

Maps

Special contributions

Human development indicators

Fighting climate change: human solidarity in a divided world

"Human progress is neither automatic nor inevitable. We are faced now with the fact that tomorrow is today. We are confronted with the fierce urgency of now. In this unfolding conundrum of life and history there is such a thing as being too late...We may cry out desperately for time to pause in her passage, but time is deaf to every plea and rushes on. Over the bleached bones and jumbled residues of numerous civilizations are written the pathetic words: Too late."

Martin Luther King Jr. *'Where do we go from here: chaos or community'*

Delivered in a sermon on social justice four decades ago, Martin Luther King's words retain a powerful resonance. At the start of the 21st Century, we too are confronted with the "fierce urgency" of a crisis that links today and tomorrow. That crisis is climate change. It is still a preventable crisis—but only just. The world has less than a decade to change course. No issue merits more urgent attention—or more immediate action.

Climate change is the defining human development issue of our generation. All development is ultimately about expanding human potential and enlarging human freedom. It is about people developing the capabilities that empower them to make choices and to lead lives that they value. Climate change threatens to erode human freedoms and limit choice. It calls into question the Enlightenment principle that human progress will make the future look better than the past.

The early warning signs are already visible. Today, we are witnessing at first hand what could be the onset of major human development reversal in our lifetime. Across developing countries, millions of the world's poorest people are already being forced to cope with the impacts of climate change. These impacts do not register as apocalyptic events in the full glare of world media attention. They go unnoticed in financial markets and in the measurement of world gross domestic product (GDP). But increased exposure to drought, to more intense storms, to floods and environmental stress is holding back the efforts of the world's poor to build a better life for themselves and their children.

Climate change will undermine international efforts to combat poverty. Seven years ago, political leaders around the world gathered to set targets for accelerated progress in human development. The Millennium Development Goals (MDGs) defined a new ambition for 2015. Much has been achieved, though many countries remain off track. Climate change is hampering efforts to deliver the MDG promise. Looking to the future, the danger is that it will stall and then reverse progress built-up over generations not just in cutting extreme poverty, but in health, nutrition, education and other areas.

Climate change provides a potent reminder of the one thing that we share in common. It is called planet Earth. All nations and all people share the same atmosphere

How the world deals with climate change today will have a direct bearing on the human development prospects of a large section of humanity. Failure will consign the poorest 40 percent of the world's population—some 2.6 billion people—to a future of diminished opportunity. It will exacerbate deep inequalities within countries. And it will undermine efforts to build a more inclusive pattern of globalization, reinforcing the vast disparities between the 'haves' and the 'have nots'.

In today's world, it is the poor who are bearing the brunt of climate change. Tomorrow, it will be humanity as a whole that faces the risks that come with global warming. The rapid build-up of greenhouse gases in the Earth's atmosphere is fundamentally changing the climate forecast for future generations. We are edging towards 'tipping points'. These are unpredictable and non-linear events that could open the door to ecological catastrophes—accelerated collapse of the Earth's great ice sheets being a case in point—that will transform patterns of human settlement and undermine the viability of national economies. Our generation may not live to see the consequences. But our children and their grandchildren will have no alternative but to live with them. Aversion to poverty and inequality today, and to catastrophic risk in the future provides a strong rationale for urgent action.

Some commentators continue to cite uncertainty over future outcomes as grounds for a limited response to climate change. That starting point is flawed. There are indeed many unknowns: climate science deals in probability and risk, not in certainties. However, if we value the well-being of our children and grandchildren, even small risks of catastrophic events merit an insurance-based precautionary approach. And uncertainty cuts both ways: the risks could be greater than we currently understand.

Climate change demands urgent action now to address a threat to two constituencies with a little or no political voice: the world's poor and future generations. It raises profoundly important questions about social justice, equity and human rights across countries and generations. In the *Human Development Report 2007/2008*

we address these questions. Our starting point is that the battle against climate change can—and must—be won. The world lacks neither the financial resources nor the technological capabilities to act. If we fail to prevent climate change it will be because we were unable to foster the political will to cooperate.

Such an outcome would represent not just a failure of political imagination and leadership, but a moral failure on a scale unparalleled in history. During the 20th Century failures of political leadership led to two world wars. Millions of people paid a high price for what were avoidable catastrophes. Dangerous climate change is the avoidable catastrophe of the 21st Century and beyond. Future generations will pass a harsh judgement on a generation that looked at the evidence on climate change, understood the consequences and then continued on a path that consigned millions of the world's most vulnerable people to poverty and exposed future generations to the risk of ecological disaster.

Ecological interdependence

Climate change is different from other problems facing humanity—and it challenges us to think differently at many levels. Above all, it challenges us to think about what it means to live as part of an ecologically interdependent human community.

Ecological interdependence is not an abstract concept. We live today in a world that is divided at many levels. People are separated by vast gulfs in wealth and opportunity. In many regions, rival nationalisms are a source of conflict. All too often, religious, cultural and ethnic identity are treated as a source of division and difference from others. In the face of all these differences, climate change provides a potent reminder of the one thing that we share in common. It is called planet Earth. All nations and all people share the same atmosphere. And we only have one.

Global warming is evidence that we are overloading the carrying capacity of the Earth's atmosphere. Stocks of greenhouse gases that trap heat in the atmosphere are accumulating at an unprecedented rate. Current concentrations have reached 380 parts per million (ppm) of carbon dioxide equivalent (CO$_2$e) exceeding the natural range of the last 650,000 years. In the course of the 21st Century, average global temperatures could increase by more than 5°C.

To put that figure in context, it is equivalent to the change in temperature since the last ice age—an era in which much of Europe and North America was under more than one kilometre of ice. The threshold for dangerous climate change is an increase of around 2°C. This threshold broadly defines the point at which rapid reversals in human development and a drift towards irreversible ecological damage would become very difficult to avoid.

Behind the numbers and the measurement is a simple overwhelming fact. We are recklessly mismanaging our ecological interdependence. In effect, our generation is running up an unsustainable ecological debt that future generations will inherit. We are drawing down the stock of environmental capital of our children. Dangerous climate change will represent the adjustment to an unsustainable level of greenhouse gas emissions.

Future generations are not the only constituency that will have to cope with a problem they did not create. The world's poor will suffer the earliest and most damaging impacts. Rich nations and their citizens account for the overwhelming bulk of the greenhouse gases locked in the Earth's atmosphere. But, poor countries and their citizens will pay the highest price for climate change.

The inverse relationship between responsibility for climate change and vulnerability to its impacts is sometimes forgotten. Public debate in rich nations increasingly highlights the threat posed by rising greenhouse gas emissions from developing countries. That threat is real. But it should not obscure the underlying problem. Mahatma Gandhi once reflected on how many planets might be needed if India were to follow Britain's pattern of industrialization. We are unable to answer that question. However, we estimate in this Report that if all of the world's people generated greenhouse gases at the same rate as some developed countries, we would need nine planets.

While the world's poor walk the Earth with a light carbon footprint they are bearing the brunt of unsustainable management of our ecological interdependence. In rich countries, coping with climate change to date has largely been a matter of adjusting thermostats, dealing with longer, hotter summers, and observing seasonal shifts. Cities like London and Los Angeles may face flooding risks as sea levels rise, but their inhabitants are protected by elaborate flood defence systems. By contrast, when global warming changes weather patterns in the Horn of Africa, it means that crops fail and people go hungry, or that women and young girls spend more hours collecting water. And, whatever the future risks facing cities in the rich world, today the real climate change vulnerabilities linked to storms and floods are to be found in rural communities in the great river deltas of the Ganges, the Mekong and the Nile, and in sprawling urban slums across the developing world.

The emerging risks and vulnerabilities associated with climate change are the outcomes of physical processes. But they are also a consequence of human actions and choices. This is another aspect of ecological interdependence that is sometimes forgotten. When people in an American city turn on the air-conditioning or people in Europe drive their cars, their actions have consequences. Those consequences link them to rural communities in Bangladesh, farmers in Ethiopia and slum dwellers in Haiti. With these human connections come moral responsibilities, including a responsibility to reflect upon—and change—energy policies that inflict harm on other people or future generations.

The case for action

If the world acts now it will be possible—just possible—to keep 21st Century global temperature increases within a 2°C threshold above preindustrial levels. Achieving this future will require a high level of leadership and unparalleled international cooperation. Yet climate change is a threat that comes with an opportunity. Above all, it provides an opportunity for the world to

We are recklessly mismanaging our ecological interdependence. Our generation is running up an unsustainable ecological debt that future generations will inherit

The real choice facing political leaders and people today is between universal human values, on the one side, and participating in the widespread and systematic violation of human rights on the other

come together in forging a collective response to a crisis that threatens to halt progress.

The values that inspired the drafters of the Universal Declaration of Human Rights provide a powerful point of reference. That document was a response to the political failure that gave rise to extreme nationalism, fascism and world war. It established a set of entitlements and rights—civil, political, cultural, social and economic—for "all members of the human family". The values that inspired the Universal Declaration were seen as a code of conduct for human affairs that would prevent the "disregard and contempt for human rights that have resulted in barbarous acts which have outraged the conscience of mankind".

The drafters of the Universal Declaration of Human Rights were looking back at a human tragedy, the second world war, that had already happened. Climate change is different. It is a human tragedy in the making. Allowing that tragedy to evolve would be a political failure that merits the description of an "outrage to the conscience of mankind". It would represent a systematic violation of the human rights of the world's poor and future generations and a step back from universal values. Conversely, preventing dangerous climate change would hold out the hope for the development of multilateral solutions to the wider problems facing the international community. Climate change confronts us with enormously complex questions that span science, economics and international relations. These questions have to be addressed through practical strategies. Yet it is important not to lose sight of the wider issues that are at stake. The real choice facing political leaders and people today is between universal human values, on the one side, and participating in the widespread and systematic violation of human rights on the other.

The starting point for avoiding dangerous climate change is recognition of three distinctive features of the problem. The first feature is the combined force of inertia and cumulative outcomes of climate change. Once emitted, carbon dioxide (CO_2) and other greenhouse gases stay in the atmosphere for a long time. There are no rapid rewind buttons for running down stocks. People living at the start of the 22nd Century will live with the consequences of our emissions, just as we are living with the consequences of emissions since the industrial revolution. Time-lags are an important consequence of climate change inertia. Even stringent mitigation measures will not materially affect average temperatures changes until the mid-2030s—and temperatures will not peak until 2050. In other words, for the first half of the 21st Century the world in general, and the world's poor in particular, will have to live with climate change to which we are already committed.

The cumulative nature of the climate change has wide-ranging implications. Perhaps the most important is that carbon cycles do not follow political cycles. The current generation of political leaders cannot solve the climate change problem alone because a sustainable emissions pathway has to be followed over decades, not years. However, it has the power either to prise open the window of opportunity for future generations, or to close that window.

Urgency is the second feature of the climate change challenge—and a corollary of inertia. In many other areas of international relations, inaction or delayed agreements have limited costs. International trade is an example. This is an area in which negotiations can break down and resume without inflicting long-term damage on the underlying system—as witnessed by the unhappy history of the Doha Round. With climate change, every year of delay in reaching an agreement to cut emissions adds to greenhouse gas stocks, locking the future into a higher temperature. In the seven years since the Doha Round started, to continue the analogy, stocks of greenhouse gases have increased by around 12 ppm of CO_2e—and those stocks will still be there when the trade rounds of the 22nd Century get underway.

There are no obvious historical analogies for the urgency of the climate change problem. During the Cold War, large stockpiles of nuclear missiles pointed at cities posed a grave threat to human security. However, 'doing nothing' was a strategy for containment of the risks. Shared recognition of the reality of mutually assured destruction offered a perversely predictable stability. With climate change, by contrast, doing nothing offers a guaranteed route to a further build-up greenhouse gases, and to mutually assured destruction of human development potential.

The third important dimension of the climate change challenge is its global scale. The Earth's atmosphere does not differentiate greenhouse gases by country of origin. One tonne of greenhouse gases from China carries the same weight as one tonne of greenhouse gases from the United States—and one country's emissions are another country's climate change problem. It follows that no one country can win the battle against climate change acting alone. Collective action is not an option but an imperative. When Benjamin Franklin signed the American Declaration of Independence in 1776, he is said to have commented: "We must all hang together, or most assuredly, we shall all hang separately." In our unequal world, some people—notably poor people—might hang sooner than others in the event of a failure to develop collective solutions. But ultimately, this is a preventable crisis that threatens all people and all countries. We too have the choice between hanging together and forging collective solutions to a shared problem, or hanging separately.

Seizing the moment—2012 and beyond

Confronted with a problem as daunting as climate change, resigned pessimism might seem a justified response. However, resigned pessimism is a luxury that the world's poor and future generations cannot afford—and there is an alternative.

There is cause for optimism. Five years ago, the world was still engaged in debating whether or not climate change was taking place, and whether or not it was human-induced. Climate change scepticism was a flourishing industry. Today, the debate is over and climate scepticism is an increasingly fringe activity. The fourth assessment review of the International Panel on Climate Change has established an overwhelming scientific consensus that climate change is both real and man-made. Almost all governments are part of that consensus.

Following the publication of the Stern Review on *The Economics of Climate Change*, most governments also accept that solutions to climate change are affordable—more affordable than the costs of inaction.

Political momentum is also gathering pace. Many governments are setting bold targets for cutting greenhouse gas emissions. Climate change mitigation has now registered firmly on the agenda of the Group of Eight (G8) industrialized nations. And dialogue between developed and developing countries is strengthening.

All of this is positive news. Practical outcomes are less impressive. While governments may recognize the realities of global warming, political action continues to fall far short of the minimum needed to resolve the climate change problem. The gap between scientific evidence and political response remains large. In the developed world, some countries have yet to establish ambitious targets for cutting greenhouse gas emissions. Others have set ambitious targets without putting in place the energy policy reforms needed to achieve them. The deeper problem is that the world lacks a clear, credible and long-term multilateral framework that charts a course for avoiding dangerous climate change—a course that spans the divide between political cycles and carbon cycles.

With the expiry of the current commitment period of the Kyoto Protocol in 2012, the international community has an opportunity to put that framework in place. Seizing that opportunity will require bold leadership. Missing it will push the world further on the route to dangerous climate change.

Developed countries have to take the lead. They carry the burden of historic responsibility for the climate change problem. And they have the financial resources and technological capabilities to initiate deep and early cuts in emissions. Putting a price on carbon through taxation or cap-and-trade systems is the starting point. But market pricing alone will not be enough. The development of regulatory systems and public–private partnerships for a low-carbon transition are also priorities.

No one country can win the battle against climate change acting alone. Collective action is not an option but an imperative

The principle of "common but differentiated responsibility"—one of the foundations of the Kyoto framework—does not mean that developing countries should do nothing. The credibility of any multilateral agreement will hinge on the participation of major emitters in the developing world. However, basic principles of equity and the human development imperative of expanding access to energy demand that developing countries have the flexibility to make the transition to a low-carbon growth path at a rate consistent with their capabilities.

International cooperation has a critical role to play at many levels. The global mitigation effort would be dramatically enhanced if a post-2012 Kyoto framework incorporated mechanisms for finance and technology transfers. These mechanisms could help remove obstacles to the rapid disbursement of the low-carbon technologies needed to avoid dangerous climate change. Cooperation to support the conservation and sustainable management of rainforests would also strengthen the mitigation effort.

Adaptation priorities must also be addressed. For too long, climate change adaptation has been treated as a peripheral concern, rather than as a core part of the international poverty reduction agenda. Mitigation is an imperative because it will define prospects for avoiding dangerous climate change in the future. But the world's poor cannot be left to sink or swim with their own resources while rich countries protect their citizens behind climate-defence fortifications. Social justice and respect of human rights demand stronger international commitment on adaptation.

Our legacy

The post-2012 Kyoto framework will powerfully influence prospects for avoiding climate change—and for coping with the climate change that is now unavoidable. Negotiations on that framework will be shaped by governments with very different levels of negotiating leverage. Powerful vested interests in the corporate sector will also make their voices heard. As governments embark on the negotiations for a post-2012 Kyoto Protocol, it is important that they reflect on two constituencies with a limited

voice but a powerful claim to social justice and respect for human rights: the world's poor and future generations.

People engaged in a daily struggle to improve their lives in the face of grinding poverty and hunger ought to have first call on human solidarity. They certainly deserve something more than political leaders who gather at international summits, set high-sounding development targets and then undermine achievement of the very same targets by failing to act on climate change. And our children and their children's grandchildren have the right to hold us to a high standard of accountability when their future—and maybe their survival—is hanging in the balance. They too deserve something more than a generation of political leaders who look at the greatest challenge humankind has ever faced and then sit on their hands. Put bluntly, the world's poor and future generations cannot afford the complacency and prevarication that continues to characterize international negotiations on climate change. Nor can they afford the large gap between what leaders in the developed world say about climate change threats and what they do in their energy policies.

Twenty years ago Chico Mendes, the Brazilian environmentalist, died attempting to defend the Amazon rainforest against destruction. Before his death, he spoke of the ties that bound his local struggle to a global movement for social justice: "At first I thought I was fighting to save rubber trees, then I thought I was fighting to save the Amazon rainforest. Now I realise I am fighting for humanity."

The battle against dangerous climate change is part of the fight for humanity. Winning that battle will require far-reaching changes at many levels—in consumption, in how we produce and price energy, and in international cooperation. Above all, though, it will require far-reaching changes in how we think about our ecological interdependence, about social justice for the world's poor, and about the human rights and entitlements of future generations.

The 21st Century climate challenge

Global warming is already happening. World temperatures have increased by around 0.7°C since the advent of the industrial era—and the rate of increase is quickening. There is overwhelming scientific evidence linking the rise in temperature to increases in the concentration of greenhouse gases in the Earth's atmosphere.

There is no hard-and-fast line separating 'dangerous' from 'safe' climate change. Many of the world's poorest people and most fragile ecological systems are already being forced to adapt to dangerous climate change. However, beyond a threshold of 2°C the risk of large-scale human development setbacks and irreversible ecological catastrophes will increase sharply.

Business-as-usual trajectories will take the world well beyond that threshold. To have a 50:50 chance of limiting temperature increase to 2°C above preindustrial levels will require stabilization of greenhouse gases at concentrations of around 450ppm CO_2e. Stabilization at 550ppm CO_2e would raise the probability of breaching the threshold to 80 percent. In their personal lives, few people would knowingly undertake activities with a serious injury risk of this order of magnitude. Yet as a global community, we are taking far greater risks with planet Earth. Scenarios for the 21st Century point to potential stabilization points in excess of 750ppm CO_2e, with possible temperature changes in excess of 5°C.

Temperature scenarios do not capture the potential human development impacts. Average changes in temperature on the scale projected in business-as-usual scenarios will trigger large-scale reversals in human development, undermining livelihoods and causing mass displacement. By the end of the 21st Century, the spectre of catastrophic ecological impacts could have moved from the bounds of the possible to the probable. Recent evidence on the accelerated collapse of ice sheets in the Antarctic and Greenland, acidification of the oceans, the retreat of rainforest systems and melting of Arctic permafrost all have the potential—separately or in interaction—to lead to 'tipping points'.

Countries vary widely in their contribution to the emissions that are driving up atmospheric stocks of greenhouse gases. With 15 percent of world population, rich countries account for almost half of emissions of CO_2. High growth in China and India is leading to a gradual convergence in 'aggregate' emissions. However, per capita carbon footprint convergence is more limited. The carbon footprint of the United States is five times that of China and over 15 times that of India. In Ethiopia, the average per capita carbon footprint is 0.1 tonnes of CO_2 compared with 20 tonnes in Canada.

What does the world have to do to get on an emissions trajectory that avoids dangerous climate change? We address that question by drawing upon climate modeling simulations. These simulations define a carbon budget for the 21st Century.

If everything else were equal, the global carbon budget for energy-related emissions would amount to around 14.5 Gt CO_2 annually. Current emissions are running at twice this level. The bad news is that emissions are on a rising trend. The upshot: the carbon budget for the entire 21st Century could expire as early as 2032. In effect, we are running up unsustainable ecological debts that will lock future generations into dangerous climate change.

Carbon budget analysis casts a new light on concerns over the share of developing countries in global greenhouse gas emissions. While that share is set to rise, it should not divert attention from the underlying responsibilities of rich nations. If every person in the developing world had the same carbon footprint as the average person in Germany or the United Kingdom, current global emissions would be four times the limit defined by our sustainable emissions pathway, rising to nine times if the developing country per capita footprint were raised to Canadian or United States levels.

Changing this picture will require deep adjustments. If the world were a single country it would have to cut emissions of greenhouse gases by half to 2050 relative to 1990 levels, with sustained reductions to the end of the 21st Century. However, the world is not a single country. Using plausible assumptions, we estimate that avoiding dangerous climate change will require rich nations to cut emissions by at least 80 percent, with cuts of 30 percent by 2020. Emissions from developing countries would peak around 2020, with cuts of 20 percent by 2050.

By the end of the 21st Century, the spectre of catastrophic ecological impacts could have moved from the bounds of the possible to the probable

Our stabilization target is stringent but affordable. Between now and 2030, the average annual cost would amount to 1.6 percent of GDP. This is not an insignificant investment. But it represents less than two-thirds of global military spending. The costs of inaction could be much higher. According to the Stern Review, they could reach 5–20 percent of world GDP, depending upon how costs are measured.

Looking back at emission trends highlights the scale of the challenge ahead. Energy related CO_2 emissions have increased sharply since 1990, the reference years for the reductions agreed under the Kyoto Protocol. Not all developed countries ratified the Protocol's targets, which would have reduced their average emissions by around 5 percent. Most of those that did are off track for achieving their commitments. And few of those that are on track can claim to have reduced emissions as a result of a policy commitment to climate change mitigation. The Kyoto Protocol did not place any quantitative restrictions on emissions from developing countries. If the next 15 years of emissions follows the linear trend of the past 15, dangerous climate change will be unavoidable.

Projections for energy use point precisely in this direction, or worse. Current investment patterns are putting in place a carbon intensive energy infrastructure, with coal playing a dominant role. On the basis of current trends and present policies, energy-related CO_2 emissions could rise by more than 50 percent over 2005 levels by 2030. The US$20 trillion projected to be spent between 2004 and 2030 to meet energy demand could lock the world on to an unsustainable trajectory. Alternatively, new investments could help to decarbonize economic growth.

Climate shocks: risk and vulnerability in an unequal world

Climate shocks already figure prominently in the lives of the poor. Events such as droughts, floods and storms are often terrible experiences for those affected: they threaten lives and leave people feeling insecure. But climate shocks also erode long-term opportunities for human development, undermining productivity and eroding human capabilities. No single climate shock can be attributed to climate change. However, climate change is ratcheting up the risks and vulnerabilities facing the poor. It is placing further stress on already over-stretched coping mechanisms and trapping people in downward spirals of deprivation.

Vulnerability to climate shocks is unequally distributed. Hurricane Katrina provided a potent reminder of human frailty in the face of climate change even in the richest countries—especially when the impacts interact with institutionalized inequality. Across the developed world, public concern over exposure to extreme climate risks is mounting. With every flood, storm and heat wave, that concern is increasing. Yet climate disasters are heavily concentrated in poor countries. Some 262 million people were affected by climate disasters annually from 2000 to 2004, over 98 percent of them in the developing world. In the Organisation for Economic Co-operation and Development (OECD) countries one in 1,500 people was affected by climate disaster. The comparable figure for developing countries is one in 19—a risk differential of 79.

High levels of poverty and low levels of human development limit the capacity of poor households to manage climate risks. With limited access to formal insurance, low incomes and meagre assets, poor households have to deal with climate-related shocks under highly constrained conditions.

Strategies for coping with climate risks can reinforce deprivation. Producers in drought prone areas often forego production of crops that could raise income in order to minimize risk, preferring to produce crops with lower economic returns but resistant to drought. When climate disasters strike, the poor are often forced to sell productive assets, with attendant implications for recovery, in order to protect consumption. And when that is not enough households cope in other ways: for example, by cutting meals, reducing spending on health and taking children out of school. These are desperation measures that can create life-long cycles of disadvantage, locking vulnerable households into low human development traps.

Research carried out for this report underlines just how potent these traps can be. Using microlevel household data we examined some of the long-term impacts of climate-shocks in the lives of the poor. In Ethiopia and Kenya, two of the world's most drought-prone countries, children aged five or less are respectively 36 and 50 percent more likely to be malnourished if they were born during a drought. For Ethiopia, that translates into some 2 million additional malnourished children in 2005. In Niger, children aged two or less born in a drought year were 72 percent more likely to be stunted. And Indian women born during a flood in the 1970s were 19 percent less likely to have attended primary school.

The long-run damage to human development generated through climate shocks is insufficiently appreciated. Media reporting of climate-related disasters often plays an important role in informing opinion—and in capturing the human suffering that comes with climate shocks. However, it also gives rise to a perception that these are 'here-today-gone-tomorrow' experiences, diverting attention from the long-run human consequences of droughts and floods.

Climate change will not announce itself as an apocalyptic event in the lives of the poor. Direct attribution of any specific event to climate change will remain impossible. However, climate change will steadily increase the exposure of poor and vulnerable households to climate-shocks and place increased pressure on coping strategies, which, over time, could steadily erode human capabilities.

We identify five key transmission mechanisms through which climate change could stall and then reverse human development:

- *Agricultural production and food security.* Climate change will affect rainfall, temperature and water availability for agriculture in vulnerable areas. For example, drought-affected areas in sub-Saharan Africa could expand by 60–90 million hectares, with dry land zones suffering losses of US$26 billion by 2060 (2003 prices), a figure in excess of bilateral aid to the region in 2005. Other developing regions—including Latin America and South Asia—will also experience losses in agricultural production,

undermining efforts to cut rural poverty. The additional number affected by malnutrition could rise to 600 million by 2080.
- *Water stress and water insecurity.* Changed run-off patterns and glacial melt will add to ecological stress, compromising flows of water for irrigation and human settlements in the process. An additional 1.8 billion people could be living in a water scarce environment by 2080. Central Asia, Northern China and the northern part of South Asia face immense vulnerabilities associated with the retreat of glaciers—at a rate of 10–15 metres a year in the Himalayas. Seven of Asia's great river systems will experience an increase in flows over the short term, followed by a decline as glaciers melt. The Andean region also faces imminent water security threats with the collapse of tropical glaciers. Several countries in already highly water-stressed regions such as the Middle East could experience deep losses in water availability.
- *Rising sea levels and exposure to climate disasters.* Sea levels could rise rapidly with accelerated ice sheet disintegration. Global temperature increases of 3–4°C could result in 330 million people being permanently or temporarily displaced through flooding. Over 70 million people in Bangladesh, 6 million in Lower Egypt and 22 million in Viet Nam could be affected. Small island states in the Caribbean and Pacific could suffer catastrophic damage. Warming seas will also fuel more intense tropical storms. With over 344 million people currently exposed to tropical cyclones, more intensive storms could have devastating consequences for a large group of countries. The 1 billion people currently living in urban slums on fragile hillsides or flood-prone river banks face acute vulnerabilities.
- *Ecosystems and biodiversity.* Climate change is already transforming ecological systems. Around one-half of the world's coral reef systems have suffered 'bleaching' as a result of warming seas. Increasing acidity in the oceans is another long-term threat to marine ecosystems. Ice-based ecologies have also suffered devastating climate change

Global temperature increases of 3–4°C could result in 330 million people being permanently or temporarily displaced through flooding impacts, especially in the Arctic region. While some animal and plant species will adapt, for many species the pace of climate change is too rapid: climate systems are moving more rapidly than they can follow. With 3°C of warming, 20–30 percent of land species could face extinction.

- *Human health.* Rich countries are already preparing public health systems to deal with future climate shocks, such as the 2003 European heatwave and more extreme summer and winter conditions. However, the greatest health impacts will be felt in developing countries because of high levels of poverty and the limited capacity of public health systems to respond. Major killer diseases could expand their coverage. For example, an additional 220–400 million people could be exposed to malaria—a disease that already claims around 1 million lives annually. Dengue fever is already in evidence at higher levels of elevation than has previously been the case, especially in Latin America and parts of East Asia. Climate change could further expand the reach of the disease.

None of these five separate drivers will operate in isolation. They will interact with wider social, economic and ecological processes that shape opportunities for human development. Inevitably, the precise mix of transmission mechanisms from climate change to human development will vary across and within countries. Large areas of uncertainty remain. What is certain is that dangerous climate change has the potential to deliver powerful systemic shocks to human development across a large group of countries. In contrast to economic shocks that affect growth or inflation, many of the human development impacts—lost opportunities for health and education, diminished productive potential, loss of vital ecological systems, for example—are likely to prove irreversible.

Avoiding dangerous climate change: strategies for mitigation

Avoiding the unprecedented threats posed by dangerous climate change will require an unparalleled collective exercise in international cooperation. Negotiations on emis-

sion limits for the post-2012 Kyoto Protocol commitment period can—and must—frame the global carbon budget. However, a sustainable global emissions pathway will only be meaningful if it is translated into practical national strategies—and national carbon budgets. Climate change mitigation is about transforming the way that we produce and use energy. And it is about living within the bounds of ecological sustainability.

Setting credible targets linked to global mitigation goals is the starting point for the transition to a sustainable emissions pathway. These targets can provide a basis for carbon budgeting exercises that provide a link from the present to the future through a series of rolling plans. However, credible targets have to be backed by clear policies. The record to date in this area is not encouraging. Most developed countries are falling short of the targets set under the Kyoto Protocol: Canada is an extreme case in point. In some cases, ambitious 'Kyoto-plus' targets have been adopted. The European Union and the United Kingdom have both embraced such targets. For different reasons, they are both likely to fall far short of the goals set unless they move rapidly to put climate mitigation at the centre of energy policy reform.

Two major OECD countries are not bound by Kyoto targets. Australia has opted for a wide-ranging voluntary initiative, which has produced mixed results. The United States does not have a federal target for reducing emissions. Instead, it has a 'carbon-intensity' reduction goal which measures efficiency. The problem is that efficiency gains have failed to prevent large aggregate increases in emissions. In the absence of federal targets, several United States' states have set their own mitigation goals. California's Global Warming Solutions Act of 2006 is a bold attempt to align greenhouse gas reduction targets with reformed energy policies.

Setting ambitious targets for mitigation is an important first step. Translating targets into policies is politically more challenging. The starting point: putting a price on carbon emissions. Changed incentive structures are a vital condition for an accelerated transition to low-carbon growth. In an optimal scenario, the carbon price would be global. This is politically unrealistic in the short-run because the world lacks the required governance system. The more realistic option is for rich countries to develop carbon pricing structures. As these structures evolve, developing countries could be integrated over time as institutional conditions allow.

There are two ways of putting a price on carbon. The first is to directly tax CO_2 emissions. Importantly, carbon taxation does not imply an increase in the overall tax burden. The revenues can be used in a fiscally neutral way to support wider environmental tax reforms—for example, cutting taxes on labour and investment. Marginal taxation levels would require adjustment in the light of greenhouse gas emission trends. One approach, broadly consistent with our sustainable emissions pathway, would entail the introduction of taxation at a level of US\$10–20/t CO_2 in 2010, rising in annual increments of US\$5–10/t CO_2 towards a level of US\$60–100/t CO_2. Such an approach would provide investors and markets with a clear and predictable framework for planning future investments. And it would generate strong incentives for a low-carbon transition.

The second route to carbon pricing is cap-and-trade. Under a cap-and-trade system, the government sets an overall emissions cap and issues tradable allowances that grant business the right to emit a set amount. Those who can reduce emissions more cheaply are able to sell allowances. One potential disadvantage of cap-and-trade is energy price instability. The potential advantage is environmental certainty: the cap itself is a quantitative ceiling applied to emissions. Given the urgency of achieving deep and early quantitative cuts in greenhouse gas emissions, well-designed cap-and-trade programmes have the potential to play a key role in mitigation.

The European Union's Emissions Trading Scheme (ETS), is the world's largest cap-and-trade programme. While much has been achieved, there are serious problems to be addressed. The caps on emissions have been set far too high, primarily because of the failure of European Union member states to resist the lobbying efforts of powerful vested interests. Some sectors—notably power—have secured windfall gains at public expense. And only a small fraction of ETS permits—less than 10 percent in the second phase—can be auctioned, depriving governments of revenue for tax reform and opening the door to political manipulation and generating inefficiencies. Restricting ETS quota allocations in line with the European Union's commitment to a 20–30 percent cut in emissions by 2020 would help to align carbon markets with mitigation goals.

Carbon markets are a necessary condition for the transition to a low-carbon economy. They are not a sufficient condition. Governments have a critical role to play in setting regulatory standards and in supporting low-carbon research, development and deployment.

There is no shortage of positive examples. Renewable energy provision is expanding in part because of the creation of incentives through regulation. In Germany, the 'feed-in' tariff has boosted the share of renewable suppliers in the national grid. The United States has successfully used tax incentives to encourage the development of a vibrant wind power industry. However, while the rapid growth of renewable energy has been encouraging, overall progress falls far short of what is possible—and of what is required for climate change mitigation. Most OECD countries have the potential to raise the share of renewable energy in power generation to at least 20 percent.

Enhanced energy efficiency has the potential to deliver a 'double dividend'. It can reduce CO_2 emissions *and* cut energy costs. If all electrical appliances operating in OECD countries in 2005 had met the best efficiency standards, it would have saved some 322 Mt CO_2 of emissions by 2010—equivalent to taking over 100 million cars off the road. Household electricity consumption would fall by one-quarter.

Personal transportation is another area where regulatory standards can unlock double-dividends. The automobile sector accounts for about 30 percent of greenhouse gas emissions in developed countries—and the share is rising. Regulatory standards matter because they can influence fleet efficiency, or the average number of miles travelled per gallon (and hence CO_2 emissions). In the United States,

Carbon markets are a necessary condition for the transition to a low-carbon economy. They are not a sufficient condition

The rapid development and
deployment of low-carbon
technologies is vital to
climate change mitigation

fuel efficiency standards have slipped over time. They are now lower than in China. Raising standards by 20 miles per gallon would cut oil consumption by 3.5 million barrels a day and save 400 Mt CO_2 emissions a year—more than the total emissions from Thailand. Efforts to raise fuel efficiency standards are often countered by powerful vested interests. In Europe, for example, European Commission proposals to raise standards have been countered by a coalition of automobile manufacturers. Several member states have rejected the proposals, raising wider questions about the European Union's capacity to translate climate change goals into tangible policies.

International trade could play a much larger role in expanding markets for alternative fuels. Brazil is more efficient than either the European Union or the United States in producing ethanol. Moreover, sugar-based ethanol is more efficient at cutting carbon emissions. The problem is that imports of Brazilian ethanol are restricted by high import tariffs. Removing these tariffs would generate gains not just for Brazil, but for climate change mitigation.

The rapid development and deployment of low-carbon technologies is vital to climate change mitigation. Picking winners in technology is a hazardous affair. Governments have at best a mixed record. However, confronted with a national and global threat on the scale of climate change, governments cannot afford to stand back and wait for markets to deliver. Energy policy is an area in which the scale of upfront investments, time horizon, and uncertainty combine to guarantee that markets alone will fail to deliver technological change at the pace required by mitigation. In earlier periods, major technological breakthroughs have followed decisive government action: the Manhattan Project and the United States space programme are examples.

Carbon Capture and Storage (CSS) is a key breakthrough technology. Coal is the major source of power for electricity generation worldwide. Reserves are widely dispersed. Coupled with rising prices for oil and natural gas, this is one reason why coal figures prominently in the present and planned energy mix of major

emitters such as the China, India and the United States. CCS is important because it holds out the promise of coal-fired power generation with near-zero emissions. With a more active programme of public–private investment, aligned with carbon pricing, CCS technologies could be developed and deployed more rapidly. Both the European Union and the United States have the capacity to put in place at least 30 demonstration plants by 2015.

Low levels of energy efficiency in developing countries are currently a threat to climate change mitigation efforts. Raising efficiency levels through international cooperation could transform that threat into an opportunity, generating large gains for human development in the process. We demonstrate this by examining the impact on CO_2 emissions of an accelerated technology transfer programme for the coal sector in China. For China alone, emissions in 2030 would be 1.8 Gt CO_2 below the level projected by the International Energy Agency. That figure is equivalent to around one-half of current European Union emissions. Similar efficiency gains are attainable in other areas.

Enhanced energy efficiency is a win–win scenario. Developing countries stand to gain from improved energy efficiency and lower environmental pollution. All countries stand to gain from CO_2 mitigation. Unfortunately, the world currently lacks a credible mechanism for unlocking this win–win scenario. We propose the development, under the auspices of the post-2012 Kyoto framework, of a Climate Change Mitigation Facility (CCMF) to fill this gap. The CCMF would mobilize US$25–50 billion annually to finance low-carbon energy investments in developing countries. Financing provisions would be linked to the circumstances of individual countries, with a menu of grants, concessional support and risk guarantees available. Support would be programme-based. It would cover the incremental costs of achieving defined emission reduction targets by scaling-up nationally-owned energy policies in areas such as renewable energy, clean-coal and enhanced efficiency standards for transport and buildings.

Deforestation is another key area for international cooperation. Currently, the world is losing the carbon assets contained in rainforests at a fraction of the market value they would have even at low carbon prices. In Indonesia, every US$1 generated through deforestation to grow palm oil would translate into a US$50–100 loss if the reduced carbon capacity could be traded on the European Union's ETS. Beyond these market failures, the loss of rainforests represents the erosion of a resource that plays a vital role in the lives of the poor, in the provision of ecosystem services and in sustaining biodiversity.

There is scope for exploring the potential of carbon markets in the creation of incentives to avoid deforestation. More broadly, carbon finance could be mobilized to support the restoration of degraded grasslands, generating benefits for climate change mitigation, adaptation and environmental sustainability.

Adapting to the inevitable: national action and international cooperation

Without urgent mitigation action the world cannot avoid dangerous climate change. But even the most stringent mitigation will be insufficient to avoid major human development setbacks. The world is already committed to further warming because of the inertia built into climate systems and the delay between mitigation and outcome. For the first half of the 21st Century there is no alternative to adaptation to climate change.

Rich countries already recognize the imperative to adapt. Many are investing heavily in the development of climate defence infrastructures. National strategies are being drawn up to prepare for more extreme and less certain future weather patterns. The United Kingdom is spending US$1.2 billion annually on flood defences. In the Netherlands, people are investing in homes that can float on water. The Swiss alpine ski industry is investing in artificial snow-making machines.

Developing countries face far more severe adaptation challenges. Those challenges have to be met by governments operating under severe financing constraints, and by poor people themselves. In the Horn of Africa, 'adaptation' means that women and young girls walk further to collect water. In the Ganges Delta, people are erecting bamboo flood shelters on stilts. And in the Mekong Delta people are planting mangroves to protect themselves against storm surges, and women and children are being taught to swim.

Inequalities in capacity to adapt to climate change are becoming increasingly apparent. For one part of the world—the richer part—adaptation is a matter of erecting elaborate climate defence infrastructures, and of building homes that 'float on' water. In the other part adaptation means people themselves learning to 'float in' flood water. Unlike people living behind the flood defences of London and Los Angeles, young girls in the Horn of Africa and people in the Ganges Delta do not have a deep carbon footprint. As Desmond Tutu, the former Archbishop of Cape Town, has argued, we are drifting into a world of adaptation apartheid.

Planning for climate change adaptation confronts governments in developing countries with challenges at many levels. These challenges pose systemic threats. In Egypt, delta flooding could transform conditions for agricultural production. Changes to coastal currents in southern Africa could compromise the future of Namibia's fisheries sector. Hydroelectric power generation will be affected in many countries.

Responding to climate change will require the integration of adaptation into all aspects of policy development and planning for poverty reduction. However, planning and implementation capacity is limited:

- *Information.* Many of the world's poorest countries lack the capacity and the resources to assess climate risks. In sub-Saharan Africa, high levels of rural poverty and dependence on rainfed agriculture makes meteorological information an imperative for adaptation. However, the region has the world's lowest density of meteorological stations. In France, the meteorological budget amounts to US$388 million annually, compared with just US$2 million in Ethiopia. The 2005 G8 summit pledged action to strengthen Africa's meteorological monitoring capacity.

We are drifting into a world of adaptation apartheid

Support for the MDGs provides another powerful rationale for action: adaptation is a key requirement for achieving the 2015 targets and creating the conditions for sustained progress

Follow-up has fallen far short of the commitments made.

- *Infrastructure.* In climate change adaptation, as in other areas, prevention is better than cure. Every US$1 invested in pre-disaster risk management in developing countries can prevent losses of US$7. In Bangladesh, research among impoverished populations living on *char* islands shows that adaptation against flooding can strengthen livelihoods, even in extreme conditions. Many countries lack the financial resources required for infrastructural adaptation. Beyond disaster prevention, the development of community-based infrastructure for water harvesting can reduce vulnerability and empower people to cope with climate risks. Partnerships between communities and local governments in Indian states such as Andhra Pradesh and Gujarat provide examples of what can be achieved.

- *Insurance for social protection.* Climate change is generating incremental risks in the lives of the poor. Social protection programmes can help people cope with those risks while expanding opportunities for employment, nutrition and education. In Ethiopia the Productive Safety Net Programme is an attempt to strengthen the capacity of poor households to cope with droughts without having to sacrifice opportunities for health and education. In Latin America conditional cash transfers have been widely used to support a wide range of human development goals, including the protection of basic capabilities during a sudden crisis. In southern Africa cash transfers have been used during droughts to protect long-run productive capacity. While social protection figures only marginally in current climate change adaptation strategies, it has the potential to create large human development returns.

The case for international action on adaptation is rooted in past commitments, shared values, the global commitment to poverty reduction and the liability of rich nations for climate change problems. Under the terms of the United Nations Framework Convention on Climate Change (UNFCCC), northern governments are obliged to support adaptation capacity development. Support for the MDGs provides another powerful rationale for action: adaptation is a key requirement for achieving the 2015 targets and creating the conditions for sustained progress. Application of the legal principles of protection from harm and compensation for damage would constitute further grounds for action.

Expressed in diplomatic language, the international response on adaptation has fallen far short of what is required. Several dedicated multilateral financing mechanisms have been created, including the Least Developed Country Fund and the Special Climate Change Fund. Delivery through these mechanisms has been limited. Total financing to date has amounted to around US$26 million—a derisory response. For purposes of comparison, this is equivalent to one week's worth of spending under the United Kingdom flood defence programme. Current pledged funding amounts to US$279 million for disbursement over several years. This is an improvement over past delivery but still a fraction of what is required. It represents less than one-half of what the German state of Baden-Würtemberg will allocate to the strengthening of flood defences.

It is not just the lives and the livelihoods of the poor that require protection through adaptation. Aid programmes are also under threat. We estimate that around one-third of current development assistance is concentrated in areas facing varying degrees of climate change risk. Insulating aid budgets from that risk will require additional investment of around US$4.5 billion. At the same time, climate change is contributing to a diversion of aid into disaster relief. This has been one of the fastest-growing areas for aid flows, accounting for 7.5 percent of total commitments in 2005.

Estimating the aid financing requirements for adaptation is inherently difficult. In the absence of detailed national assessments of climate change risks and vulnerabilities, any assessment must remain a 'guesstimate'. Our 'guesstimate' is that by 2015 at least US$44 billion will be required annually for 'climate proofing' development investments (2005 prices). Building human resilience is another priority area.

Investments in social protection and wider human development strategies are needed to strengthen the capacity of vulnerable people to cope with risk. Our ballpark estimate is that at least US$40 billion will be needed by 2015 to strengthen national strategies for poverty reduction in the face of climate change risks. To put this figure in context, it represents around 0.5 percent of projected 2015 GDP for low income and lower middle income countries. Provision for disaster and post-disaster recovery will also have to be strengthened as droughts, floods, storms and landslides pose greater threats. Provision of an additional US$2 billion a year is implied by our estimates.

Adaptation financing requirements should be seen as 'new and additional' commitments. That is, they should supplement rather than divert existing aid commitments. Northern governments have pledged to double aid by 2010, though the record on delivery is mixed. Any shortfall in delivery will compromise progress towards the MDGs and compound problems in climate change adaptation.

The headline figure for new and additional adaptation financing appears large—but has to be placed in context. The total of around US$86 billion by 2015 may be required to prevent aid diversion. It would represent around 0.2 percent of developed country GDP, or around one-tenth of what they currently allocate to military expenditure. Measured in terms of returns for human security, adaptation financing is a highly cost-effective investment. There are a range of innovative financing mechanisms that could be explored to mobilize resources. These include carbon taxation, levies administered under cap-and-trade programmes and dedicated levies on air transport and vehicles.

International support for adaptation has to go beyond financing. Current international efforts suffer not just from chronic under-financing, but also a lack of coordination and coherence. The patchwork of multilateral mechanisms is delivering small amounts of finance with very high transaction costs, most of it through individual projects. While project-based support has an important role to play, the locus for adaptation planning has to be shifted towards national programmes and budgets.

The integration of adaptation planning into wider poverty reduction strategies is a priority. Successful adaptation policies cannot be grafted on to systems that are failing to address underlying causes of poverty, vulnerability and wider disparities based on wealth, gender and location. Dialogue over Poverty Reduction Strategy Papers (PRSPs) provides a possible framework for integrating adaptation in poverty reduction planning. Revision of PRSPs through nationally-owned processes to identify financing requirements and policy options for adaptation could provide a focal point for international cooperation.

Conclusion and summary of recommendations

Climate change confronts humanity with stark choices. We can avoid 21st Century reversals in human development and catastrophic risks for future generations, but only by choosing to act with a sense of urgency. That sense of urgency is currently missing. Governments may use the rhetoric of a 'global security crisis' when describing the climate change problem, but their actions—and inactions—on energy policy reform tell a different story. The starting point for action and political leadership is recognition on the part of governments that they are confronted by what may be the gravest threat ever to have faced humanity.

Facing up to that threat will create challenges at many levels. Perhaps most fundamentally of all, it challenges the way that we think about progress. There could be no clearer demonstration than climate that economic wealth creation is not the same thing as human progress. Under the current energy policies, rising economic prosperity will go hand-in-hand with mounting threats to human development today and the well-being of future generations. But carbon-intensive economic growth is symptomatic of a deeper problem. One of the hardest lessons taught by climate change is that the economic model which drives growth, and the profligate consumption in rich nations that goes with it, is ecologically unsustainable. There could be no greater challenge to our

There could be no clearer demonstration than climate that economic wealth creation is not the same thing as human progress

For the current generation, the challenge is to keep open the window of opportunity by bending greenhouse gas emissions in a downward direction

assumptions about progress than that of realigning economic activities and consumption with ecological realities.

Combating climate change demands that we place ecological imperatives at the heart of economics. That process has to start in the developed world—and it has to start today. The uncertainties have to be acknowledged. In this report we have argued that, with the right reforms, it is not too late to cut greenhouse gas emissions to sustainable levels without sacrificing economic growth: that rising prosperity and climate security are not conflicting objectives.

The current state of international cooperation and multilateralism on climate change is not fit for the purpose. As a priority, the world needs a binding international agreement to cut greenhouse gas emissions across a long time horizon, but with stringent near-term and medium-term targets. The major developing countries have to be party to that agreement and make commitments to reduce emissions. However, those commitments will need to reflect their circumstances and capabilities, and the overarching need to sustain progress in poverty reduction. Any multilateral agreement without quantitative commitments from developing countries will lack credibility in terms of climate change mitigation. At the same time, no such agreement will emerge unless it incorporates provisions for finance and technology transfer from the rich nations that bear historic responsibility for climate change.

International cooperation must also address the pressing issue of climate change adaptation. Even with stringent mitigation, the world is already committed to sustained global warming for the first half of the 21st Century. Having created the problem, the world's richest countries cannot stand aside and watch the hopes and the aspirations of the world's poor be undermined by increased exposure to the risks and vulnerabilities that will come with climate change.

Fighting climate change is a cross-generational exercise. For the current generation, the challenge is to keep open the window of opportunity by bending greenhouse gas emissions in a downward direction. The world has a historic opportunity to begin this task. In 2012, the current commitment period of the Kyoto Protocol expires. The successor agreement could set a new course, imposing stringent limits on future emissions and providing a framework for international collective action. Negotiations could be brought forward so that the quantitative targets are set by 2010, providing governments with goals for national carbon budgets. Carbon budgeting backed by radical energy policy reforms and government action to change incentive structures for consumers and investors is the foundation for effective climate change mitigation. There is no such thing as a last chance in human affairs. But the post-2012 Kyoto framework comes close.

Recommendations

1 Develop a multilateral framework for avoiding dangerous climate change under the post-2012 Kyoto Protocol

- Establish an agreed threshold for dangerous climate change at 2°C above preindustrial levels.
- Set a stabilization target for atmospheric concentrations of CO_2e at 450 ppm (the costs are estimated at 1.6 percent of average global GDP to 2030).
- Agree to a global sustainable emissions pathway aimed at 50 percent reductions of greenhouse gas emissions by 2050 from 1990 levels.
- Targets under the current Kyoto commitment period implemented by developed countries, with a further agreement to cut greenhouse gas emissions by at least 80 percent by 2050, with 20–30 percent cuts by 2020.
- Major emitters in developing countries to aim at an emissions trajectory that peaks in 2020, with 20 percent cuts by 2050.

2 Put in place policies for sustainable carbon budgeting—the agenda for mitigation

- Set a national carbon budget in all developed countries with targets for reducing overall emissions from a 1990 reference year incorporated into national legislation.
- Put a price on carbon through taxation or cap-and-trade programmes consistent with national carbon budget goals.
- Carbon taxation to be introduced at a level of US\$10–20/t CO_2 in 2010, rising in annual increments to US\$60–100/t CO_2.
- Adopt cap-and-trade programmes that aim at 20–30 percent cuts in CO_2 emissions by 2020 with 90–100 percent of allowances auctioned by 2015.

- Utilise revenues from carbon taxation and cap-and-trade to finance progressive tax reform, with reductions in taxes on labour and investments, and the development of incentives for low-carbon technology.
- Reform of the European Union's Emissions Trading Scheme to reduce quotas, increase auctioning and limit windfall gains for the private sector.
- Create an enabling environment for renewable energy through 'feed-in' tariffs and market regulation, with a 20 percent target by 2020 in renewable power generation.
- Increase energy efficiency through regulatory standards for appliances and buildings.
- Reduce CO_2 emissions from transport through stronger fuel efficiency standards in the European Union, with a target of 120g CO_2/km by 2012 and 80g CO_2/km by 2020, and more stringent Corporate Average Fuel Economy Standards (CAFE) in the United States with the introduction of taxation of aviation.
- Increase financing, incentives and regulatory support for the development of breakthrough technologies, with a focus on Carbon Capture and Storage (CCS)—the United States should aim at 30 demonstration plants by 2015, and the European Union should have a comparable level of ambition.

3 Strengthen the framework for international cooperation

- Develop international cooperation to enhance access to modern energy services and reduce dependence on biomass, the primary source of energy for about 2.5 billion people.
- Reduce the rate of increase in carbon emissions in developing countries through strengthened energy sector reforms, backed by finance and technology transfer.

- Create a Climate Change Mitigation Facility (CCMF) to mobilize the US$25–50 billion needed annually to support low-carbon transitions in developing countries through a mix of grants, concessional aid and risk guarantees for investment under nationally-owned energy sector reform programmes.
- Integrate project based carbon-financing through the Clean Development Mechanism and other Kyoto flexibility provisions into programme-based and sectoral national strategies for supporting low-carbon transition.
- Significantly strengthen international cooperation on coal, with the creation of incentives for the development and deployment on Integrated Gasification Combined Cycle (IGCC) technology and CCS.
- Develop international incentives for the conservation and sustainable management of rainforests.
- Extend carbon financing beyond industrial sector mitigation to land-use programmes—such as forest conservation and grasslands restoration—that offer benefits for the poor.

4 Put climate change adaptation at the centre of the post-2012 Kyoto framework and international partnerships for poverty reduction

- Recognize that the world is committed to significant climate change, that even stringent mitigation will not materially affect temperature change until the mid-2030s, and that average global temperatures will rise to 2050 even under a 'good case' scenario.
- Strengthen the capacity of developing countries to assess climate change risks and integrate adaptation into all aspects of national planning.

- Act on G8 commitments to strengthen meteorological monitoring capacity in sub-Saharan Africa through partnerships under the Global Climate Observing System.
- Empower and enable vulnerable people to adapt to climate change by building resilience through investments in social protection, health, education and other measures.
- Integrate adaptation into poverty reduction strategies that address vulnerabilities linked to inequalities based on wealth, gender, location and other markers for disadvantage.
- Provide at least US$86 billion in 'new and additional' finance for adaptation through transfers from rich to poor by 2016 to protect progress towards the MDGs and prevent post-2015 reversals in human development.
- Expand multilateral provisions for responding to climate-related humanitarian emergencies and supporting post-disaster recovery to build future resilience, with US$2 billion in financing by 2016 under arrangements such as the UN's Central Emergency Response Fund and the World Bank's Global Facility for Disaster Reduction and Recovery.
- Explore a range of innovative financing options beyond development assistance to mobilize support for adaptation, including carbon taxation, levies on quotas issued under cap-and-trade programmes, air transport taxes and wider measures.
- Streamline the current structure of dedicated multilateral funds which are delivering limited support (US$26 million to date and US$253 million in the pipeline, with high transition costs), and shift the locus of support from projects to programme-based financing.
- Use Poverty Reduction Strategy Papers (PRSPs) to conduct national estimates of the costs of scaling-up existing programmes, identifying priority areas for reducing vulnerability.

1

The 21ˢᵗ Century climate challenge

"One generation plants a tree; the next generation gets the shade."

Chinese Proverb

• •

"You already know enough. So do I. It is not knowledge we lack. What is missing is the courage to understand what we know and to draw conclusions."

Sven Lindqvist

1 The 21st Century climate challenge

The supreme reality of our time is the spectre of dangerous climate change

Easter Island in the Pacific Ocean is one of the most remote locations on Earth. The gigantic stone statues located in the Rono Raraku volcanic crater are all that remain of what was a complex civilization. That civilization disappeared because of the over-exploitation of environmental resources. Competition between rival clans led to rapid deforestation, soil erosion and the destruction of bird populations, undermining the food and agricultural systems that sustained human life.[1] The warning signs of impending destruction were picked up too late to avert collapse.

The Easter Island story is a case study in the consequences of failure to manage shared ecological resources. Climate change is becoming a 21st Century variant of that story on a global scale. There is, however, one important difference. The people of Easter Island were overtaken by a crisis that they could not anticipate—and over which they had little control. Today, ignorance is no defence. We have the evidence, we have the resources to avert crisis, and we know the consequences of carrying on with business-as-usual.

President John F. Kennedy once remarked that "the supreme reality of our time is our indivisibility and our common vulnerability on this planet".[2] He was speaking in 1963 in the aftermath of the Cuban missile crisis at the height of the Cold War. The world was living with the spectre of nuclear holocaust. Four decades on, the supreme reality of our time is the spectre of dangerous climate change.

That spectre confronts us with the threat of a twin catastrophe. The first is an immediate threat to human development. Climate change affects all people in all countries. However, the world's poorest people are on the front line. They stand most directly in harm's way—and they have the least resources to cope. This first catastrophe is not a distant future scenario. It is unfolding today, slowing progress towards the Millennium Development Goals (MDGs) and deepening inequalities within and across countries. Left unattended, it will lead to human development reversals throughout the 21st Century.

The second catastrophe is located in the future. Like the threat of nuclear confrontation during the Cold War, climate change poses risks not just for the world's poor, but for the entire planet—and for future generations. Our current path offers a one-way route to ecological disaster. There are uncertainties relating to the speed of warming, and to the exact timing and forms of the impacts. But the risks associated with accelerated disintegration of the Earth's great ice sheets, the warming of the oceans, the collapse of rainforest systems and other possible outcomes are real. They have the potential to set in train processes that could recast the human and physical geography of our planet.

Our generation has the means—and the responsibility—to avert that outcome. Immediate risks are heavily skewed towards the world's poorest countries and their most vulnerable citizens. However, there are no risk free havens over the long term. Rich countries and people not on the front line of the unfolding disaster will ultimately be affected. That is why precautionary climate change mitigation is

The Earth's capacity to absorb carbon dioxide and other greenhouse gases is being overwhelmed

an essential insurance against future catastrophe for humanity as a whole, including future generations in the developed world.

The heart of the climate change problem is that the Earth's capacity to absorb carbon dioxide (CO_2) and other greenhouse gases is being overwhelmed. Humanity is living beyond its environmental means and running up ecological debts that future generations will be unable to repay.

Climate change challenges us to think in a profoundly different way about human interdependence. Whatever else divides us, humanity shares a single planet just as surely as the people of Easter Island shared a single island. The ties that bind the human community on the planet stretch across countries and generations. No nation, large or small, can be indifferent to the fate of others, or oblivious to the consequences of today's actions for people living in the future.

Future generations will see our response to climate change as a measure of our ethical values. That response will provide a testimony on how political leaders today acted on their pledges to combat poverty and build a more inclusive world. Leaving large sections of humanity even more marginalized would signify a disregard for social justice and equity between countries. Climate change also asks tough questions about how we think about our links to people in the future. Our actions will serve as a barometer of our commitment to cross-generational social justice and equity—and as a record against which future generations will judge our actions.

There are encouraging signs. Five years ago, climate change scepticism was a flourishing industry. Liberally financed by large companies, widely cited in the media, and attentively listened to by some governments, climate sceptics exercised an undue influence on public understanding. Today, every credible climate scientist believes that climate change is real, that it is serious, and that it is linked to the release of CO_2. Governments across the world share that view. Scientific consensus does not mean that debates on the causes and consequences of global warming are over: the science of climate change deals in probabilities, not certainties.

But at least the political debate is now rooted in scientific evidence.

The problem is that there is a large gap between scientific evidence and political action. So far most governments have been failing the test on climate change mitigation. Most have responded to the recently released Fourth Assessment Report of the Intergovernmental Panel on Climate Change (IPCC) by recognizing that the evidence on climate change is "unequivocal" and that urgent action is needed. Successive meetings of the Group of Eight (G8) industrialized countries have reaffirmed the need for concrete measures to be put in place. They have acknowledged that the ship is heading for an object that looks ominously like an iceberg. Unfortunately, they have yet to initiate decisively evasive action by charting a new emissions trajectory for greenhouse gases.

There is a very real sense in which time is running out. Climate change is a challenge that has to be addressed throughout the 21st Century. No quick technological fixes are available. But the long-time horizon is not a window of opportunity for prevarication and indecision. In forging a solution, governments have to confront the problems of stocks and flows in the global carbon budget. Stocks of greenhouse gases are building up, driven by rising emissions. However, even if we stopped all emissions tomorrow the stocks would fall only very slowly. The reason: once emitted CO_2 stays in the atmosphere a long time and climate systems respond slowly. This inertia built into the system means that there is a long time-lag between today's carbon mitigation and tomorrow's climate outcomes.

The window of opportunity for successful mitigation is closing. There is a limit to the amount of carbon dioxide that the Earth's sinks can absorb without creating dangerous climate change effects—and we are nearing those limits. We have less than a decade to ensure that the window of opportunity is kept open. That does not mean we have a decade to decide on whether to act and to formulate a plan, but a decade in which to start the transition to low-carbon energy systems. One certainty in an area marked by high levels of uncertainty is this: if the next decade looks the same as the last one, then the world will be locked on course for the avoidable 'twin catastrophe' of near-term human development reversals and the risk of ecological disaster for future generations.

Like the catastrophe that struck Easter Island, that outcome is preventable. Expiry of the current commitment period of the Kyoto Protocol in 2012 provides an opportunity to develop a multilateral strategy that could redefine how we manage global ecological interdependence. The priority, as the world's governments negotiate that agreement, is to define a sustainable carbon budget for the 21st Century, and to develop a strategy for budget implementation that recognizes the "common but differentiated" responsibilities of countries.

Success will require the world's richest countries to demonstrate leadership: they have both the deepest carbon footprints, and the technological and financial capabilities to achieve deep and early cuts in emissions. However, a successful multilateral framework will require the active participation of all major emitters, including those in the developing world.

Establishing a framework for collective action that balances urgency with equity is the starting point for avoiding dangerous climate change.

This chapter sets out the scale of the challenge ahead. Section 1 looks at the interaction between climate change and human development. In section 2, we set out the evidence provided by climate science and scenarios for temperature changes. Section 3 provides a breakdown of the world's carbon footprint. Then in section 4, we contrast current emission trends with a sustainable emissions pathway for the 21st Century, drawing upon climate modelling work—and we

---

Special contribution | Climate change—together we can win the battle

The *Human Development Report 2007/2008* comes at a time when climate change—long on the international agenda—is starting to receive the very highest attention that it merits. The recent findings of the IPCC sounded a clarion call; they have unequivocally affirmed the warming of our climate system and linked it directly to human activity.

The effects of these changes are already grave, and they are growing. This year's Report is a powerful reminder of all that is at stake: climate change threatens a 'twin catastrophe', with early setbacks in human development for the world's poor being succeeded by longer term dangers for all of humanity.

We are already beginning to see these catastrophes unfold. As sea levels rise and tropical storms gather in intensity, millions of people face displacement. Dryland inhabitants, some of the most vulnerable on our planet, have to cope with more frequent and more sustained droughts. And as glaciers retreat, water supplies are being put at risk.

This early harvest of global warming is having a disproportionate effect on the world's poor, and is also hindering efforts to achieve the MDGs. Yet, in the longer run, no one—rich or poor—can remain immune from the dangers brought by climate change.

I am convinced that what we do about this challenge will define the era we live in as much as it defines us. I also believe that climate change is exactly the kind of global challenge that the United Nations is best suited to address. That is why I have made it my personal priority to work with Member States to ensure that the United Nations plays its role to the full.

Tackling climate change requires action on two fronts. First, the world urgently needs to step up action to mitigate greenhouse gas emissions. Industrialized countries need to make deeper emission reductions. There needs to be further engagement of developing countries, as well as incentives for them to limit their emissions while safeguarding economic growth and efforts to eradicate poverty.

Adaptation is the second global necessity. Many countries, especially the most vulnerable developing nations, need assistance in improving their capacity to adapt. There also needs to be a major push to generate new technologies for combating climate change, to make existing renewable technologies economically viable, and to promote a rapid diffusion of technology.

Climate change threatens the entire human family. Yet it also provides an opportunity to come together and forge a collective response to a global problem. It is my hope that we will rise as one to face this challenge, and leave a better world for future generations.

Ki Moon Ban

Ban Ki-moon
Secretary-General of the United Nations

Climate change will be one of the defining forces shaping prospects for human development during the 21st Century

look at the cost of making the transition to a more sustainable future. Section 5 contrasts our sustainable emissions pathway with the business-as-usual alternative. The chapter ends by setting out the ethical and economic case for urgent action on climate change mitigation and adaptation.

1.1 Climate change and human development

Human development is about people. It is about expanding people's real choices and the substantive freedoms—the capabilities—that enable them to lead lives that they value. Choice and freedom in human development mean something more than the absence of constraints.[3] People whose lives are blighted by poverty, ill-health or illiteracy are not in any meaningful sense free to lead the lives that they value. Neither are people who are denied the civil and political rights they need to influence decisions that affect their lives.

Climate change will be one of the defining forces shaping prospects for human development during the 21st Century. Through its impact on ecology, rainfall, temperature and weather systems, global warming will directly affect all countries. Nobody will be immune to its consequences. However, some countries and people are more vulnerable than others. In the long term, the whole of humanity faces risks but more immediately, the risks and vulnerabilities are skewed towards the world's poorest people.

Climate change will be superimposed upon a world marked by large human development deficits. While there are many uncertainties about the timing, nature and scale of future impacts, the forces unleashed by global warming can be expected to magnify existing disadvantages. Location and livelihood structures will emerge as powerful markers for disadvantage. Concentrated in fragile ecological areas, drought-prone arid lands, flood-prone coastal areas, and precarious urban slums, the poor are highly exposed to climate change risks—and they lack the resources to manage those risks.

The backdrop

The interface between climate change and human development outcomes will be shaped by differences in localized climate effects, by differences in social and economic coping capacities, and by public policy choices, among other factors. The starting point for any consideration of how climate change scenarios might play out is the human development backdrop.

That backdrop includes some good news stories that are often overlooked. Since the first *Human Development Report* was published in 1990 there have been spectacular—if spectacularly uneven—advances in human development. The share of the population living in developing countries on less than US$1 a day has fallen from 29 percent in 1990 to 18 percent in 2004. Over the same period, child mortality rates have fallen from 106 deaths per thousand live births to 83 and life expectancy has increased by 3 years. Progress in education has gathered pace. Globally, the primary school completion rate rose from 83 percent to 88 percent between 1999 and 2005.[4]

Economic growth, a condition for sustained progress in poverty reduction, has accelerated across a large group of countries. Based on this strong growth, numbers living in extreme poverty fell by 135 million between 1999 and 2004. Much of this progress has been driven by East Asia in general and China in particular. More recently, the emergence of India as a high-growth economy, with per capita incomes rising at an average of 4–5 percent since the mid-1990s, has created enormous opportunities for accelerated human development. While sub-Saharan Africa lags behind on many dimensions of human development, here too there are signs of progress. Economic growth has picked up since 2000 and the share of people in the region living in extreme poverty has finally started to fall, although the absolute number of poor has not declined.[5]

The bad news is that forces generated by climate change will be superimposed on a world marked by deep and pervasive human development deficits, and by disparities that divide the 'haves' and the 'have-nots'. While globalization has created unprecedented opportunities for some, others have been left behind. In some countries—India is an example—rapid economic growth has produced modest progress in poverty reduction and in nutrition. In others—including most of sub-Saharan Africa—economic growth is too slow and uneven to sustain rapid progress in poverty reduction. Despite high growth across much of Asia, on current trends most countries are off track for achieving the MDG targets for reducing extreme poverty and deprivation in other areas by 2015.

The state of human development is presented in more detail elsewhere in this Report. What is important in the context of climate change is that emerging risks will fall disproportionally on countries already characterized by high levels of poverty and vulnerability:

- *Income poverty*. There are still around 1 billion people living at the margins of survival on less than US$1 a day, with 2.6 billion—40 percent of the world's population—living on less than US$2 a day. Outside East Asia, most developing regions are reducing poverty at a slow pace—too slowly to achieve the MDG target of halving extreme poverty by 2015. Unless there is an acceleration of poverty reduction from 2008 onwards, the target looks likely to be missed by around 380 million people.[6]
- *Nutrition*. Around 28 percent of all children in developing countries are estimated to be underweight or stunted. The two regions that account for the bulk of the deficit are South Asia and sub-Saharan Africa—and both are off track in terms of achieving the MDG target of halving under-nutrition by 2015. If India's high economic growth is unequivocal good news, the bad news is that this has not been translated into accelerated progress in cutting under-nutrition.

One-half of all rural children are underweight for their age—roughly the same proportion as in 1992.[7]

- *Child mortality*. Progress on child mortality lags behind progress in other areas. Around 10 million children die each year before the age of 5, the vast majority from poverty and malnutrition. Only around 32 countries out of 147 monitored by the World Bank are on track to achieve the MDG of a two-thirds reduction in child mortality by 2015.[8] South Asia and sub-Saharan Africa are comprehensively off track. On current trends the MDG target will be missed by a margin that will represent 4.4 million additional deaths in 2015.[9]
- *Health*. Infectious diseases continue to blight the lives of the poor across the world. An estimated 40 million people are living with HIV/AIDS, with 3 million deaths in 2004. Every year there are 350–500 million cases of malaria, with 1 million fatalities: Africa accounts for 90 percent of malarial deaths and African children account for over 80 percent of malaria victims worldwide.[10]

These deficits in human development draw attention to deep inequalities across the world. The 40 percent of the world's population living on less than US$2 a day accounts for 5 percent of global income. The richest 20 percent accounts for three-quarters of world income. In the case of sub-Saharan Africa, a whole region has been left behind: it will account for almost one-third of world poverty in 2015, up from one-fifth in 1990.

Income inequality is also rising within countries. Income distribution influences the rate at which economic growth translates into poverty reduction. More than 80 percent of the world's population lives in countries where income differentials are widening. One consequence is that more growth is needed to achieve an equivalent poverty reduction outcome. According to one analysis, developing countries have to grow at over three times the pre-1990 rate to achieve the same reduction in poverty incidence.[11]

Skewed income distribution intersects with wider inequalities. Child death rates among the poorest one-fifth in the developing world

With the global rise in temperature, local rainfall patterns are changing, ecological zones are shifting, the seas are warming and ice caps are melting

are falling at half the average rate for the richest, reflecting deep disparities in nutrition and access to health provision.[12] In an increasingly urbanized world, disparities between rural and urban populations remain substantial. Rural areas account for three in every four people living on less than US$1 a day and a similar share of the world population suffering from malnutrition.[13] However, urbanization is not synonymous with human progress. Urban slum growth is outpacing urban growth by a wide margin.

The state of the world's environment is a vital link between climate change and human development. In 2005, the United Nations' (UN) *Millennium Ecosystem Assessment* drew attention to the global deterioration of vital ecosystems, including mangrove swamps, wetlands and forests. These ecosystems are highly vulnerable to climate change—as are the people who depend on the services they provide.

At a time when climate change concerns are mounting across the world, it is important that complex future scenarios are considered in the context of initial human development conditions. Climate change is a global phenomenon. However, the human development impacts of climate change cannot automatically be inferred from global scenarios, or from predicted movements in average global temperatures. People (and countries) vary in their resilience and capacity to manage the incremental risks associated with climate change. They vary in their capacity to adapt.

Inequalities in capacity to cope with these risks will fuel wider inequalities in opportunity. As the incremental risks created by climate change intensify over time, they will interact with existing structures of disadvantage. Prospects for sustained human development in the years and decades after the 2015 target date for the MDGs are directly threatened.

Dangerous climate change—five human development 'tipping points'

Average global temperature has become a popular metric for the state of the global climate.[14] That metric tells us something

important. We know that the world is warming and that the average global temperature has increased by around 0.7°C (1.3°F) since the advent of the industrial era. We know also that the trend is accelerating: average global mean temperature is rising at 0.2°C every decade. With the global rise in temperature, local rainfall patterns are changing, ecological zones are shifting, the seas are warming and ice caps are melting. Forced adaptation to climate change is already happening across the world. In the Horn of Africa, adaptation means that women have to walk further to find water in the dry season. In Bangladesh and Viet Nam, it means that small-scale farmers have to cope with losses caused by more intense storms, floods and sea surges.

Fifteen years have now passed since the UN Framework Convention on Climate Change (UNFCCC) set out the broad objectives for multilateral action. Those objectives include stabilizing greenhouse gas concentrations in the atmosphere at "a level that would prevent dangerous anthropogenic interference with the climate system". Indicators for the prevention of danger include stabilization within a time frame that allows ecosystems to adapt naturally, the avoidance of disruption to food systems, and the maintenance of conditions for sustainable economic development.

Defining dangerous

At what point does climate change become dangerous? That question invites another: Dangerous for whom?[15] What is dangerous for a small-scale farmer living in Malawi might not appear very dangerous for a large, mechanized farm in the Midwest of the United States. Climate change scenarios for rising sea levels that might be viewed with equanimity from behind the flood defence systems of London or lower Manhattan might reasonably be regarded with alarm in Bangladesh, or in Viet Nam's Mekong Delta.

Such considerations caution against the drawing of hard and fast lines separating 'safe' from 'dangerous' climate change. Dangerous climate change cannot be inferred from a set of scientific observations alone. The threshold for what is dangerous depends on value judgements over what is an unacceptable cost in social, economic and ecological terms at any given level of warming. For millions of people and for many ecosystems the world has already passed the danger threshold. Determining what is an acceptable upper-limit target for future global temperature increases raises fundamental questions about power and responsibility. The extent to which those facing the greatest risks are able to articulate their concerns, and the weight attached to their voice, matters a great deal.

Yet with all of these caveats any successful climate change mitigation effort has to start by establishing a target. Our starting point is the growing consensus among climate scientists on the threshold marker for dangerous climate change. That consensus identifies 2°C (3.6°F) as a reasonable upper-bound.[16]

Beyond this point, the future risks of catastrophic climate change rise sharply. Accelerated melting of the Greenland and West Antarctic ice sheets could set in train irreversible processes, leading eventually to sea levels rising by several metres—an outcome that would cause forced human resettlement on a vast scale. Large areas of rainforest could be transformed into savannah. The world's already shrinking glaciers would be set on course for rapid decline. Above the 2°C threshold, the pressure on ecological systems such as coral reefs and biodiversity would intensify. Complex carbon on biodiversity feedback effects linked to the warming of the oceans, the loss of rainforests and melting ice sheets would accelerate the pace of climate change.

Crossing the 2°C threshold would be a step across the boundary that marks significant risk of catastrophic outcomes for future generations. More immediately, it would trigger setbacks in human development. Developing countries are at a double disadvantage in this area: they are located in tropical areas that stand to experience some of the most severe early impacts from climate change; and agriculture—the sector most immediately affected—plays a far greater social and economic role. Above all, they are characterized by high levels of poverty, malnutrition and disadvantage in health. The combination of acute deprivation on the one side, with weak social insurance provision and limited infrastructural capacity to contain climate risks on the other, points to a high potential for human development reversals.

From climate change to stalled human progress—the transmission mechanisms

Climate change is global but the effects will be local. Physical impacts will be determined by geography and microlevel interactions between global warming and existing weather patterns. The immense scope of these impacts makes generalization difficult: drought-prone areas in sub-Saharan Africa will face different problems from flood-prone areas in South Asia. Human development impacts will also vary as changes in climate patterns interact with pre-existing social and economic vulnerabilities. However, five specific risk-multipliers for human development reversals can be identified:

- *Reduced agricultural productivity.* Around three-quarters of the world's population living on less than US$1 a day depend directly on agriculture. Climate change scenarios point to large losses in productivity for food staples linked to drought and rainfall variation in parts of sub-Saharan Africa and South and East Asia. Projected revenue losses for dryland areas in sub-Saharan Africa amount to 26 percent by 2060, with total revenue losses of US$26 billion (in constant 2003 terms)—in excess of bilateral aid transfers to the region. Through its impact on agriculture and food security, climate change could leave an additional 600 million facing acute malnutrition by the 2080s over and above the level in a no-climate change scenario.[17]
- *Heightened water insecurity.* Exceeding the 2°C threshold will fundamentally change the distribution of the world's water resources. Accelerated glacial melt in the Himalayas will compound already severe ecological problems across northern China, India and Pakistan, initially increasing floods before reducing the flow of water to major river systems vital for irrigation. In Latin America, accelerated melting of tropical glaciers will threaten water supplies for urban populations, agriculture and hydroelectricity,

Through its impact on agriculture and food security, climate change could leave an additional 600 million facing acute malnutrition by the 2080s

How does human development relate to our environmental concerns in general and to climate change in particular? There are well established traditions in policy discussions to make us think of the demands of development and the preservation of the environment in rather antagonistic terms. Attention is often concentrated on the fact that many of the deteriorating environmental trends in the world, including global warming and other disturbing evidence of climate change, are linked with heightened economic activity, such as industrial growth, increased energy consumption, more intensive irrigation, commercial felling of trees, and other activities that tend to correlate with economic expansion. At a superficial level, it may well appear that the process of development is responsible for environmental damage.

On the other side, environmental protagonists are frequently accused by development enthusiasts of being 'anti-development' since their activism often takes the form of being rather unwelcoming to processes that can raise incomes and reduce poverty—because of their allegedly unfavourable environmental impact. The battle lines may or may not be very sharply drawn, but it is hard to escape the sense of tension that does exist, in varying degrees, between the champions of poverty reduction and development, on one side, and the advocates of ecology and environmental preservation, on the other.

Does the human development approach have something to offer to make us understand whether this apparent conflict between development and environmental sustainability is real or imaginary? There is a huge contribution that the human development approach can make by invoking the central perspective of seeing development as the expansion of substantive human freedom, which is indeed the point of departure of the human development approach. In this broader perspective, assessment of development cannot be divorced from considering the lives that people can lead and the real freedoms that they can enjoy. Development cannot be seen merely in terms of enhancement of inanimate objects of convenience, such as a rise in the GNP (or in personal incomes). This is the basic insight that the human development approach brought to the development literature right from the outset of that approach, and this insight is critically important today for clarity regarding environmental sustainability.

Once we appreciate the necessity of seeing the world in the broader perspective of the substantive freedoms of human beings, it immediately becomes clear that development cannot be divorced from ecological and environmental concerns. Indeed, important components of human freedoms—and crucial ingredients of our quality of life—are thoroughly dependent on the integrity of the environment, involving the air we breathe, the water we drink, the epidemiological surroundings in which we live, and so on. Development has to be environment-inclusive, and the belief that development and environment must be on a collision course is not compatible with the central tenets of the human development approach.

The environment is sometimes misleadingly seen as the state of 'nature', reflected by such measures as the extent of forest cover, the depth of the groundwater table, and so on. This understanding, however, is seriously incomplete for two important reasons.

First, the value of the environment cannot be just a matter of what there is, but also of the opportunities it actually offers. The impact of the environment on human lives must *inter alia* be among the relevant considerations in assessing the richness of the environment. Indeed, the visionary report of the World Commission on Environment and Development chaired by Gro Brundtland, *Our Common Future* (1987), made this clear by focusing on sustaining the fulfilment of human 'needs'. We can, in fact, go beyond the Brundtland Report's focus on human needs and bring in the larger domain of human freedoms, since the human development approach requires us to see people not merely as 'needy', but as people whose freedom to do what they have reason to do is important and demands sustaining (and if possible expansion).

People have reason to satisfy their needs, of course, and the elementary applications of the human development approach (for example what we get from the simple Human Development Index, the HDI) do indeed focus exactly on that. But the domain of freedom can go well beyond that, and the use of the fuller human development perspective can take into account the freedom of people to do things that are not governed exclusively by their own needs. Human beings may not, for example, 'need' spotted owls in any obvious sense, and yet if they have reason to object to the extinction of such species, then the value of their freedom to achieve this deliberated goal can be the basis of a reasoned judgement. Prevention of the extinction of animal species that we human beings want to preserve (not so much because we 'need' these animals in any specific way, but because we judge that it is a bad idea to let existing species disappear forever) can be an integral part of the human development approach. In fact, the preservation of biodiversity is likely to be among the concerns in our responsible thinking about climate change.

Second, the environment is not only a matter of passive preservation, but also one of active pursuit. We must not think of the environment exclusively in terms of pre-existing natural conditions, since the environment can also include the results of human creation. For example, purification of water is a part of improving the environment in which we live. The elimination of epidemics, such as smallpox (which has already occurred) and malaria (which ought to occur very soon if we can get our acts together), is a good illustration of an environmental improvement that we can bring about.

This positive recognition does not, of course, change the significant fact that the process of economic and social development can, in many circumstances, also have strongly destructive consequences. Those unfavourable effects have to be clearly identified and firmly resisted, along with strengthening the positive and constructive contributions of development. Even though many human activities that accompany the process of development may have destructive consequences, it is also within human power to resist and reverse many of these bad consequences if timely action is taken.

In thinking about the steps that may be taken to halt environmental destruction we have to search for constructive human intervention. For example, greater levels of female education and women's employment can help to reduce fertility rates, which in the long run can reduce the pressure on global warming and the increasing destruction of natural habitats. Similarly, the spread of school education and improvements in its quality can make us more environmentally conscious. Better communication and a richer media can make us more aware of the need for environment-oriented thinking.

Indeed, the need for public participation in ensuring environmental sustainability is critically important. It is also crucial not to reduce important issues of human evaluation, which demand reflection and deliberative social assessment, into narrowly technocratic matters of formulaic calculation. For example, consider the ongoing debate on what 'discount rate' to use in balancing present sacrifices against future security. A central aspect of such discounting is social evaluation of gains and losses over time. This is at bottom a deeply reflective exercise and a matter for public deliberation, rather than one for some kind of a mechanical resolution on the basis of some simple formula.

Perhaps the most telling concern here comes from the uncertainty that is inescapably associated with any future prediction. One reason for being cautious about the 'best guess' regarding the future is the possibility that if we get it wrong, the world we end up with may be extremely precarious. There are even fears that what can be prevented now may become close to irreversible if no preventive action is taken without delay, no matter how much the future generations might be ready to spend to reverse the catastrophe. Some of these predicaments may be particularly damaging for the developing world (for example, the submerging of parts of Bangladesh or the whole of the Maldives due to rising sea levels).

These are critically important matters for public consideration and discussion, and the development of such public dialogue is an important part of the human development approach. The need for such public deliberation is as important in dealing with climate change and environmental dangers as it is in tackling more traditional problems of deprivation and continuing poverty. What characterizes human beings—perhaps more than anything else—is our ability to think and to talk to each other, and to decide what to do and then to do it. We need to make good use of this quintessential human capability as much for reasoned sustaining of the environment as we do for coordinated eradication of old-fashioned poverty and deprivation. Human development is involved in both.

Amartya Sen

By 2080, climate change could increase the number of people facing water scarcity around the world by 1.8 billion

especially in the Andean region. By 2080, climate change could increase the number of people facing water scarcity around the world by 1.8 billion.[18]

- *Increased exposure to coastal flooding and extreme weather events.* The IPCC forecasts an increase in extreme weather events.[19] Droughts and floods are already the main drivers of a steady increase in climate-related disasters. On average around 262 million people were affected each year between 2000 and 2004, over 98 percent of them living in developing countries. With an increase in temperatures above 2°C, warmer seas will fuel more violent tropical cyclones. Drought-affected areas will increase in extent, jeopardizing livelihoods and compromising progress in health and nutrition. The world is already committed to rising sea levels in the 21ˢᵗ Century because of past emissions. Temperature increases in excess of 2°C would accelerate the rise, causing the widespread displacement of people in countries such as Bangladesh, Egypt and Viet Nam and the inundation of several small-island states. Rising sea levels and more intense tropical storm activity could increase the number of people experiencing coastal flooding by between 180 million and 230 million.[20]

- *The collapse of ecosystems.* All predicted species extinction rates accelerate beyond the 2°C threshold, with 3°C marking the point at which 20–30 percent of species would be at 'high risk' of extinction.[21] Coral reef systems, already in decline, would suffer extensive 'bleaching' leading to the transformation of marine ecologies, with large losses of biodiversity and ecosystem services. This would adversely affect hundreds of millions of people dependent upon fish for their livelihoods and nutrition.

- *Increased health risks.* Climate change will impact on human health at many levels. Globally an additional 220–400 million people could be at increased risk of malaria. Exposure rates for sub-Saharan Africa, which accounts for around 90 percent of deaths, are projected to increase by 16–28 percent.[22]

These five drivers for major human development reversal cannot be viewed in isolation. They will interact with each other, and with pre-existing human development problems, creating powerful downward spirals. While the processes are already apparent in many countries, breaching the 2°C threshold would mark a qualitative shift: it would mark a transition to far greater ecological, social and economic damage.

That transition will have important implications for long term human development prospects. Climate change scenarios provide a snapshot of a plausible future. They enable us not to predict when or where a specific climate event might happen, but the average probabilities associated with emerging climate patterns.

From a human development perspective, these are outcomes that can set in train dynamic and cumulative processes of disadvantage. In chapter 2 we set out a model that captures this process through detailed analysis of household survey data. The results powerfully illustrate a hidden dimension of human costs associated with climate change. To give one example, Ethiopian children who were born in a drought year in their district are 41 percent more likely than their counterparts born in a non-drought year to be stunted. For 2 million Ethiopian children this translates into diminished opportunities for the development of human capabilities. The important implication is that even a small incremental risk of more droughts can lead to large human development setbacks. Climate change will create large incremental risks.

Not all of the human development costs associated with climate change can be measured in terms of quantitative outcomes. At a fundamental level, human development is also about people having a say in the decisions that affect their lives. In articulating a vision of development as freedom, the Nobel Laureate Amartya Sen draws attention to the role of human beings as agents of social change, emphasizing both "the processes that allow freedoms of actions and decisions, and actual opportunities that people have, given their personal and social circumstances".[23] Climate change is a profound denier of freedom of action and a source of disempowerment. One section of humanity—broadly the poorest

2.6 billion—will have to respond to climate change forces over which they have no control, manufactured through political choices in countries, where they have no voice.

The world is now at or near the warmest level on record in the current interglacial period, which began around 12,000 years ago

1

The 21st Century climate challenge

1.2 Climate science and future scenarios

Understanding the scientific evidence on climate change is a starting point for understanding the human development challenges of the 21st Century. There is a vast amount of scientific literature on the subject. Here we focus on the consensus set out by the IPCC, while drawing attention to the large areas of uncertainty over future outcomes. In looking at the future under climate change there are many 'known unknowns'—events that can be predicted but without any certainty as to their timing or magnitude. It should come as no surprise that scientists cannot be certain about precisely how the Earth's ecological systems will respond to human-induced greenhouse gas emissions: we are living with an experiment that has never been conducted before.

One of the 'knowns' is that we are on a trajectory that, if uncorrected, will lead to a very high probability of dangerous climate change outcomes. Those outcomes would provide a continuum from near-term human development setbacks to long term ecological disaster.

Human-induced climate change

Throughout its history, the earth has experienced oscillations between warm and cool periods. These shifts in climate have been traced to a wide variety of 'climate forcings', including orbital variations, solar fluctuations, volcanic activity, water vapour, and the atmospheric concentration of greenhouse gases, such as CO_2. The changes that we see happening today are occurring at a more rapid rate, with stronger magnitudes and patterns that cannot be explained by natural cycles.

Average global surface temperature is the most fundamental measure of climate change. Temperatures in the past half-century have probably been the highest of any 50-year period for the past 1,300 years. The world is now at or near the warmest level on record in the current interglacial period, which began around 12,000 years ago. There is strong evidence that the process is accelerating. Eleven of the twelve warmest years since 1850 occurred between 1995 and 2006. Over the past 100 years the Earth has warmed by 0.7°C. There are large interannual variations. However, on a decade-by-decade basis, the linear warming trend for the past 50 years is nearly twice that for the past 100 years (figure 1.1).[24]

There is an overwhelming body of scientific evidence linking rising temperatures to increased atmospheric concentrations of CO_2 and other greenhouse gases. The effect of these gases in the atmosphere is to retain part of the outgoing solar radiation, thereby raising the temperature of the Earth. This natural 'greenhouse effect' is what keeps our planet habitable: without it, the planet would be 30°C colder. Throughout the Earth's four previous glacial and warming cycles, there has been a high correlation between atmospheric concentrations of CO_2 and temperature.[25]

What is different about the current warming cycle is the rapid rate at which CO_2 concentrations are increasing. Since preindustrial times, atmospheric CO_2 stocks have increased by one-third—a rate of increase unprecedented during at least the last 20,000 years. Evidence from ice cores shows that current atmospheric concentrations exceed the natural range of the last 650,000 years. The increase in stocks of CO_2 has been accompanied by rising concentrations of other greenhouse gases.

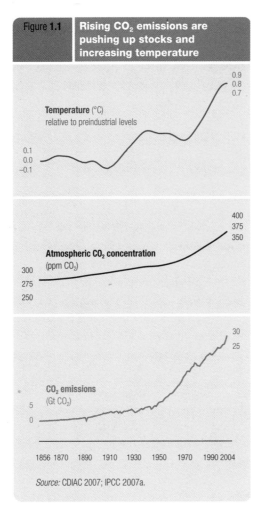

Figure 1.1 Rising CO₂ emissions are pushing up stocks and increasing temperature

Temperature (°C) relative to preindustrial levels

0.9
0.8
0.7

0.1
0.0
−0.1

Atmospheric CO₂ concentration (ppm CO₂)

400
375
350

300
275
250

CO₂ emissions (Gt CO₂)

30
25

5
0

1856 1870   1890   1910   1930   1950   1970   1990 2004

Source: CDIAC 2007; IPCC 2007a.

While the current warming cycle is not unique in terms of temperature change, it is unique in one important respect: it is the first time that humanity has decisively changed a cycle. Mankind has been releasing CO_2 into the atmosphere through burning and land-use changes for over 500,000 years. But climate change can be traced back to two great transformations in energy use. In the first, water power was replaced by coal—a source of energy condensed by nature over millions of years. It was coal harnessed to new technologies that fuelled the industrial revolution, unleashing unprecedented increases in productivity.

The second great transformation happened 150 years later. Oil had been a source of human energy for millennia: China had oil wells in the 4th Century. However, the harnessing of oil to the internal combustion engine in the early 20th Century marked the start of a revolution in transport. The burning of coal and oil,

supplemented by natural gas, has transformed human societies, providing the energy that has driven vast increases in wealth and productivity. It has also fuelled climate change.

In recent years there has been a protracted debate over the attribution of global temperature changes to human activities. Some scientists have argued that natural cycles and other forces are more important. However, while natural factors such as volcanic activity and solar intensity can explain much of the global temperature trend in the early 19th Century, they do not explain the rise since then. Other candidates for explaining global warming have also been rejected. For example, it has been argued that recent temperature changes can be traced not to greenhouse gases but to increases in the sun's output and cosmic rays. Detailed research investigating this claim showed that, for the past two decades, the sun's output has in fact declined while temperatures on Earth have risen.[26]

Debates on attribution may continue. But the scientific jury came in with the verdict on the core issues some time ago. That verdict was confirmed in the IPCC's most recent assessment, which concluded that "it is extremely unlikely that global climate change can be explained without external forcing".[27] Put differently, there is greater than 90 percent likelihood that most of the observed warming is due to human-generated greenhouse gases.

Global carbon accounting—stocks, flows and sinks

Climate change has provided an important reminder of a sometimes forgotten fact. Human activities take place in ecological systems that are not marked by national borders. Unsustainable management of these systems has consequences for the environment and for the well-being of people today and in the future. Reduced to its essentials, the threat of dangerous climate change is a symptom of unsustainable ecological resource management on a global scale.

Human energy systems interact with global ecological systems in complex ways. The burning of fossil fuels, land-use changes and other activities release greenhouse gases, which are continuously recycled between the atmosphere, oceans and land biosphere. Current concentrations of greenhouse gases are the net results of past emissions, offset by chemical and physical removal processes. The Earth's soils, vegetation and oceans act as large 'carbon sinks'. Emissions of CO_2 are the primary source of increased concentrations. Other long-lived greenhouse gases like methane and nitrous dioxide generated from agricultural activities and industry, mix with CO_2 in the atmosphere. The total warming or 'radiative forcing' effect is measured in terms of CO_2 equivalence, or CO_2e.[28] The sustained rate of increase in radiative forcing from greenhouse gases over the past four decades is at least six times faster than at any time before the industrial revolution.

The global carbon cycle can be expressed in terms of a simple system of positive and negative flows. Between 2000 and 2005 an average of 26 Gt CO_2 was released into the atmosphere each year. Of this flow, around 8 Gt CO_2 was absorbed into the oceans and another 3 Gt CO_2 was removed by oceans, land and vegetation. The net effect: an annual increase of 15 Gt CO_2 in the Earth's atmospheric stocks of greenhouse gases.

Global mean concentration of CO_2 in 2005 was around 379 ppm. Other long-lived greenhouse gases add about 75 ppm to this stock measured in terms of radiative forcing effects. However, the net effect of all human-induced greenhouse gas emissions is reduced by the cooling effect of aerosols. [29] There are large degrees of uncertainty associated with these cooling effects. According to the IPCC, they are roughly equivalent to the warming generated by non-CO_2 greenhouse gases.[30]

Atmospheric concentrations of CO_2 are on a sharply rising trend.[31] They are increasing at around 1.9 ppm each year. For CO_2 alone the annual concentration growth rate over the past 10 years has been around 30 percent faster than the average for the past 40 years.[32] In fact, in the 8,000 years prior to industrialization, atmospheric CO_2 increased by only 20 ppm.

Current rates of absorption by carbon sinks are sometimes confused with the 'natural' rate. In reality, carbon sinks are being overwhelmed.

Take the world's largest sink—its oceans. These naturally absorb just 0.1 Gt more CO_2 per year than they release. Now they are soaking up an extra 2 Gt a year—more than 20 times the natural rate.[33] The result is serious ecological damage. Oceans are becoming warmer and increasingly acidic. Rising acidity attacks carbonate, one of the essential building blocks for coral and small organisms at the start of the marine food chain. Based on current trends, future carbon dioxide releases could produce chemical conditions in the oceans that have not been witnessed in the past 300 million years, except during brief catastrophic events.[34]

The future rate of accumulation in greenhouse gas stocks will be determined by the relationship between emissions and carbon sinks. There is bad news on both fronts. By 2030 greenhouse gas emissions are set to increase by between 50 and 100 percent above 2000 levels.[35] Meanwhile, the capacity of the Earth's ecological systems to absorb these emissions could shrink. This is because feedbacks between the climate and the carbon cycle may be weakening the absorptive capacity of the world's oceans and forests. For example, warmer oceans absorb less CO_2 and rainforests could shrink with higher temperatures and reduced rainfall.

Even without taking into account uncertainties over future carbon absorption we are heading for a rapid increase in greenhouse gas stock accumulation. In effect, we are opening the taps to increase the flow of water into an already overflowing bath. The overflow is reflected in the rate at which CO_2 is entering and being locked into the Earth's atmosphere.

Climate change scenarios—the known, the known unknowns, and the uncertain

The world is already committed to future climate change. Atmospheric stocks of greenhouse gases are rising with increases in emissions. Total emissions of all greenhouse gases amounted to around 48 Gt CO_2e in 2004—an increase of one-fifth since 1990. Rising concentrations of greenhouse gases mean that global temperatures will continue to increase over time. The rate of increase and the ultimate level of temperature change will be determined by concentrations of CO_2 and other greenhouse gases.

Climate models cannot predict specific events associated with global warming. What they can do is simulate ranges of probability for average temperature change. While the modelling exercises themselves are enormously complex, one simple conclusion emerges: following current trends concentrations of greenhouse gases could commit the world to climate change at levels far above the 2°C threshold.

The world is warming

One of the early pioneers of climate science, the Swedish physicist Svante Arrenhuis, predicted with surprising accuracy that a doubling of CO_2 stocks in the Earth's atmosphere would raise average global temperatures between 4 and 5°C—a marginal overestimate according to recent IPCC models.[36] Less accurately, Arrhenuis assumed that it would take around 3,000 years for atmospheric concentrations to double over preindustrial levels. On current trends that point, around 550 ppm, could be reached by the mid-2030s.

Future temperature increases will depend on the point at which stocks of greenhouse gases stabilize. At whatever level, stabilization requires that emissions must be reduced to the point at which they are equivalent to the rate at which CO_2 can be absorbed through natural processes, without damaging the ecological systems of the carbon sinks. The longer that emissions remain above this level, the higher the point at which accumulated stocks will stabilize. Over the long term, the Earth's natural capacity to remove greenhouse gases without sustaining damage to the ecological systems of carbon sinks is probably between 1 and 5 Gt CO_2e. With emissions running at around 48 Gt CO_2e, we are currently overloading the Earth's carrying capacity by a factor of between 10 and 50.

If emissions continue to rise following current trends then stocks will be increasing at 4–5 ppm a year by 2035—almost double the current rate. Accumulated stocks will have risen to 550 ppm. Even without further increases in the rate of emissions, stocks of greenhouse gases would reach over 600 ppm by 2050 and 800 ppm by the end of the 21ˢᵗ Century.[37]

The IPCC has developed a family of six scenarios identifying plausible emissions pathways for the 21ˢᵗ Century. These scenarios are differentiated by assumptions about population change, economic growth, energy use patterns and mitigation. None of the scenarios points to stabilization below 600 ppm and three are associated with greenhouse gas concentrations of 850 ppm or above.

The relationship between stabilization point and temperature change is uncertain. The IPCC scenarios have been used to identify a set of possible ranges for 21ˢᵗ Century temperature change, with a 'best-estimate' indicator within each range (table 1.1 and figure 1.2). That best estimate is between 2.3°C and 4.5°C (factoring in the 0.5°C increase from the start of the industrial era to 1990).[38] With the doubling of atmospheric concentrations, the IPCC projects a temperature increase of 3°C as the most likely outcome with the rider that "values substantially higher than 4.5°C cannot be excluded."[39] In other words, none of the IPCC scenarios point to a future below the 2°C threshold for dangerous climate change.

Table **1.1**	**Temperature ranges rise with CO₂ stocks— projections for 2080**	
IPCC scenarios	**Relative to 1980–1999 average temperature (°C)**	**Relative to preindustrial temperature (°C)**
Constant year 2000 concentrations	0.6 (0.3–0.9)	1.1
B1 scenario	1.8 (1.1–2.9)	2.3
A1T scenario	2.4 (1.4–3.8)	2.9
B2 scenario	2.4 (1.4–3.8)	2.9
A1B scenario	2.8 (1.7–4.4)	3.3
A2 scenario	3.4 (2.0–5.4)	3.9
A1FI scenario	4.0 (2.4–6.4)	4.5

Note: **IPCC scenarios** describe plausible future patterns of population growth, economic growth, technological change and associated CO_2 emissions. The **A1 scenarios** assume rapid economic and population growth combined with reliance on fossil fuels (A1FI), non-fossil energy (A1T) or a combination (A1B). The **A2 scenario** assumes lower economic growth, less globalization and continued high population growth. The **B1** and **B2 scenarios** contain some mitigation of emissions, through increased resource efficiency and technology improvement (B1) and through more localized solutions (B2).
Source: IPCC 2007a.

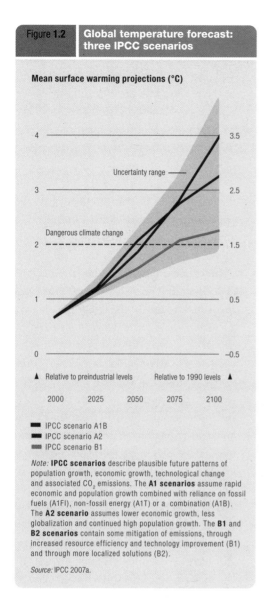

Figure 1.2 Global temperature forecast: three IPCC scenarios

Mean surface warming projections (°C)

Uncertainty range ——

Dangerous climate change

▲ Relative to preindustrial levels   Relative to 1990 levels ▲

2000   2025   2050   2075   2100

■ IPCC scenario A1B
■ IPCC scenario A2
■ IPCC scenario B1

Note: **IPCC scenarios** describe plausible future patterns of population growth, economic growth, technological change and associated CO_2 emissions. The **A1 scenarios** assume rapid economic and population growth combined with reliance on fossil fuels (A1FI), non-fossil energy (A1T) or a combination (A1B). The **A2 scenario** assumes lower economic growth, less globalization and continued high population growth. The **B1** and **B2 scenarios** contain some mitigation of emissions, through increased resource efficiency and technology improvement (B1) and through more localized solutions (B2).

Source: IPCC 2007a.

Heading for dangerous climate change

In two important respects the IPCC's best-estimate range for the 21st Century might understate the problem. First, climate change is not just a 21st Century phenomenon. Temperature adjustments to rising concentrations of CO_2 and other greenhouse gases will continue to take place in the 22nd Century. Second, IPCC best-estimates do not rule out the possibility of higher levels of climate change. At any given level of stabilization, there is a probability range for exceeding a specified temperature. Illustrative probability ranges identified in modelling work include the following:

- Stabilization at 550 ppm, which is below the lowest point on the IPCC scenarios, would carry an 80 percent probability of overshooting the 2°C dangerous climate change threshold.[40]

- Stabilization at 650 ppm carries a probability of between 60 and 95 percent of exceeding 3°C. Some studies predict a 35–68 percent likelihood of overshooting 4°C.[41]

- At around 883 ppm, well within the IPCC non-mitigation scenario range, there would be a 50 percent chance of exceeding a 5°C temperature increase.[42]

Probability ranges are a complex device for capturing something of great importance for the future of our planet. An increase in average global temperature in excess of 2–3°C would bring with it enormously damaging ecological, social and economic impacts. It would also create a heightened risk of catastrophic impacts, acting as a trigger for powerful feedback effects from temperature change to the carbon cycle. Temperature increases above 4–5°C would amplify the effects, markedly increasing the probability of catastrophic outcomes in the process. In at least three of the IPCC scenarios, the chances of exceeding a 5°C increase are greater than 50 percent. Put differently, under current scenarios, there is a far stronger likelihood that the world will overshoot a 5°C threshold than keep within the 2°C climate change threshold.

One way of understanding these risks is to reflect on what they might mean in the lives of ordinary people. We all live with risks. Anybody who drives a car or walks down a street faces a very small risk of an accident that will create serious injury. If the risk of such an accident increased above 10 percent most people would think twice about driving or taking a stroll: a one in ten chance of serious injury is not a negligible risk. If the odds on a serious accident increased to 50:50, the case for embarking upon serious risk reduction measures would become overwhelming. Yet we are on a greenhouse gas emission course that makes dangerous climate change a virtual certainty, with a very high risk of crossing a threshold for ecological catastrophe. This is an overwhelming case for risk reduction, but the world is not acting.

In the course of one century or slightly more, there is a very real prospect that current

Today, we are living with the consequences of the greenhouses gases emitted by earlier generations—and future generations will live with the consequences of our emissions

trends will see global temperatures increase by more than 5°C. That figure approximates the increase in average temperature that has taken place since the end of the last ice age some 10,000 years ago. During that age, most of Canada and large areas of the United States were under ice. The giant Laurentide glacier covered much of the north-east and north-central United States with ice several miles deep. The retreat of that ice created the Great Lakes and scoured-out new land formations, including Long Island. Much of northern Europe and north-west Asia were also covered in ice.

Comparisons between 21ˢᵗ Century climate change and the transition from the last ice age should not be overstated. There is no direct analogy for the warming processes now underway. However, geological evidence strongly suggests that temperature changes on the scale and at the pace of those now underway could culminate in transformations of the Earth's geography, along with marked changes in the distribution of species and human geography.

Probability ranges for temperature change associated with greenhouse gas concentrations help to identify targets for mitigation. By changing the flow of emissions we can alter the rate at which stocks of greenhouse gases accumulate and hence the probabilities of overshooting specific temperature targets. However, the relationship between greenhouse gas flows, accumulated stocks and future temperature scenarios is not simple. Long time-lags between today's actions and tomorrow's outcomes are built into the system. Policies for climate change mitigation have to deal with powerful forces of inertia that have an important bearing on the timing of mitigation.

- *Current emissions define future stocks.* Basic chemistry is one force of inertia. When CO_2 is released into the atmosphere it stays there a long time. Half of every tonne emitted remains in the atmosphere for a period of between several centuries and several thousand years. What this means is that traces of the CO_2 released when the first coal-powered steam engines designed by John Newcomen were operating in the early 18ᵗʰ Century are still in the atmosphere. So are traces of the emissions generated by the world's first coal-fired

power station, designed by Thomas Edison and opened in lower Manhattan in 1882. Today, we are living with the consequences of the greenhouses gases emitted by earlier generations—and future generations will live with the consequences of our emissions.

- *Stocks, flows and stabilization.* There are no rapid rewind buttons for running down stocks of greenhouse gases. People living at the end of the 21ˢᵗ Century will not have the opportunity to return in their lifetime to a world of 450 ppm if we continue on a business-as-usual path. The accumulated stock of greenhouse gases that they inherit will depend on the emissions pathway that links the present to the future. Keeping emissions at current levels would not reduce stocks because they exceed the absorptive capacity of the Earth's carbon sinks. Stabilizing emissions at 2000 levels would increase stocks by over 200 ppm by the end of the 21ˢᵗ Century. Because of cumulative processes, the rate of emissions reduction required to meet any stabilization goal is very sensitive to the timing and the level of the peak in global emissions. The later and the higher the peak, the deeper and the more rapid the cuts needed to achieve a specified stabilization target.

- *Climate systems respond slowly.* By the late 21ˢᵗ Century, actions taken today will be the major factor affecting climate change. However, mitigation efforts today will not produce significant effects until after 2030.[43] The reason: changing emission pathways does not produce a simultaneous response in climate systems. The oceans, which have absorbed about 80 percent of the increase in global warming, would continue to rise, and ice sheets would continue melting under any medium-term scenario.

Uncertain future and 'nasty surprises'—catastrophic risk under climate change

Rising global average temperature is a predictable climate change outcome. It is one of the 'knowns' that emerge from climate modelling exercises. There is also a wide range of 'known unknowns'. These are predictable events with large areas of uncertainty attached to their timing and magnitude. Uncertain but significant risks of catastrophic outcomes are part of the emerging climate change scenario.

The IPCC's fourth assessment draws attention to a wide range of uncertainties linked to potentially catastrophic events. Two such events have figured prominently in debates on climate change. The first is a reversal of the meridional overturning circulation (MOC), the vast conveyor of warm water in the Atlantic Ocean. The heat transported by the Gulf Stream is equivalent to around 1 percent of humanity's current energy use.[44] As a result of this heat transport, Europe is up to 8°C warmer, with the largest effects apparent in winter. It is the threat to the comparatively mild European climate, as well as climate concerns elsewhere, that has given rise to worries about the future of the MOC.

Additional fresh water flowing into the North Atlantic as a result of glacial melting has been identified as a potential force for shutting down or slowing the MOC. Switching off the Gulf Stream would put northern Europe on course for an early ice age. While the IPCC concludes that a large abrupt transition is very unlikely in the 21st Century, it warns that "longer-term changes in the MOC cannot be assessed with confidence". Moreover, the likelihood range for an abrupt transition is still 5–10 percent. While this may be "very unlikely" in terms of the IPCC's statistical accounting, the magnitude of the threat and the considerable uncertainty that surrounds it make a powerful case for precautionary behaviour in the interests of future generations.

The same applies to rising sea levels. The IPCC scenarios point to rises of between 20 and 60 centimetres by the end of the 21st Century. That is more than a marginal change. Moreover, the fourth assessment acknowledges that "larger values cannot be excluded." Outcomes will depend upon complex ice formation and melting processes, and on wider carbon cycle effects. The IPCC anticipates the continuing contraction of the great ice sheet in Greenland as a source of rising sea levels, with uncertainty over the future of the ice sheets of Antarctica. However, in the case of Antarctica the IPCC acknowledges that recent models provide evidence pointing to processes that could "increase the vulnerability of the ice-sheets to warming".[45]

These uncertainties are of more than passing academic concern. Consider first the evidence on the melting of ice sheets and rising sea levels. So far, the rise in sea level has been dominated by thermal expansion due to increased temperatures rather than glacial melt—but this could change. For humanity as a whole, the accelerated disintegration and eventual demise of the Greenland and West Antarctic ice sheets are perhaps the greatest of all the threats linked to climate change. Recent evidence suggests that warming ocean waters are now thinning some West Antarctic ice shelves by several metres a year. The area of Greenland on which summer melting of ice took place has increased by more than 50 percent during the past 25 years. Concern over the fate of Antarctic ice shelves has been gathering since the enormous Larsen B ice shelf collapsed in 2002. Several more ice shelves have broken up rapidly in recent years.[46]

One of the reasons for uncertainty about the future is that ice sheet disintegration, unlike ice sheet formation, can happen very rapidly. According to one of the world's most prominent climate scientists working at the North American Space Agency (NASA), a business-as-usual scenario for ice sheet disintegration in the 21st Century could yield sea level rises in the order of 5 metres this century. Note that this does not take into account accelerated melting of the Greenland ice sheet, the complete elimination of which would add around 7 metres to sea levels.[47] The IPCC sets out what can be thought of as a lowest common denominator consensus. However, its assessment of the risks and uncertainties does not include recent evidence of accelerated melting, nor does it factor in the possibility of large-scale, but imperfectly understood, carbon cycle effects. The upshot is that the headline risk numbers may err on the side of understatement.

The 'known unknowns' surrounding rising sea levels are a particularly striking example of threats facing the whole of humanity. The one certainty is that current trends and past evidence are a weak guide to the future. Climate change could trigger a range of 'surprises':

rapid, non-linear responses of the climate system to human-induced forcing (box 1.1).

Climate scientists have drawn a distinction between 'imaginable surprises', which are currently seen as possible but unlikely (deglaciation of polar ice sheets or MOC reversals are examples) and 'true surprises', or risks that have not been identified because of the complexity of climate systems.[48] Feedback effects between climate change and the carbon cycle, with changes in temperature giving rise to unpredictable outcomes, are the source of these potential surprises.

There is growing evidence that natural carbon absorption will weaken as temperatures rise. Modelling by the Hadley Centre suggests that climate change feedback effects could reduce the absorptive capacity consistent with stabilization at 450 ppm by 500 Gt CO_2, or 17 years of global emissions at current levels.[49] The practical consequence of carbon cycle feedback effects is that emissions may need to peak at lower levels or be cut more rapidly, especially at higher levels of greenhouse gas concentrations.

The focus on potentially catastrophic outcomes should not divert attention from the more immediate risks. There is a large section of humanity that would not have to await the advanced disintegration of ice sheets to experience catastrophe under these conditions. Precise numbers can be debated, but for the poorest 40 percent of the world's population—around 2.6 billion people—we are on the brink of climate change events that will jeopardize prospects for human development. We will develop this point further in chapter 2.

Risk and uncertainty as a case for action

How should the world respond to the uncertainties associated with climate change? Some commentators argue for a 'wait-and-see' approach, with the mitigation effort to be scaled up in light of developments. The fact that the IPCC's assessment and wider climate science point to uncertain risks with low probabilities of global catastrophe in the medium term is cited as grounds for delayed action.

Such responses fail a number of public policy tests for the development of climate change mitigation strategies. Consider first the response to the range of possibilities identified by climate science. These ranges are not a

---

Box 1.1   Feedback effects could accelerate climate change

There are many positive feedback effects that could transform climate change scenarios for the 21st Century. High levels of uncertainty about positive feedback effects are reflected in IPCC scenario projections.

Multiple feedbacks have been observed in ice sheet disintegration. One example is the 'albedo flip'—a process that occurs when snow and ice begin to melt. Snow-covered ice reflects back to space most of the sunlight that strikes it. When surface ice melts, darker wet ice absorbs more solar energy. The meltwater produced burrows through the ice sheet, lubricating its base, and speeding the discharge of icebergs into the ocean. As an ice sheet discharges more icebergs into the ocean, it loses mass and its surface sinks to a lower altitude, where the temperature is warmer, causing it to melt even faster. Meanwhile, warming oceans add yet another positive feedback to this process, melting the offshore accumulation of ice—ice shelves—that often form a barrier between ice sheets and the ocean.

The accelerated melting of permafrost in Siberia with global warming is another concern. This could release vast amounts of methane—a highly potent greenhouse gas—into the atmosphere, which would increase warming and the rate at which permafrost melts.

The interaction between climate change and the carbon sink capacity of rainforests provides another example of positive feedback uncertainties. Rainforests can be thought of as vast 'carbon banks'. Trees in the Amazon region of Brazil alone store 49 billion tonnes of carbon. Another 6 billion tonnes is stored in Indonesia's forests. As global temperatures rise, changing climate patterns could generate processes that will lead to the release of large amounts of carbon from these reservoirs.

Rainforests are already contracting at an alarming rate in the face of commercial pressures, illegal logging and other activities. Under a business-as-usual scenario, climate models forecast temperatures in most of the Amazon region rising by 4–6°C by 2100. This could convert up to 30 percent of the Amazon rain forest into a type of dry savannah, according to research carried out under the auspices of Brazil's National Space Research Institute. Such an outcome would in turn drive up net global emissions of CO_2. Because rainforests recycle at least half of rainfall back into the atmosphere, accelerated deforestation would also increase drought and fuel the spread of savannah areas.

Source: FAO 2007b; Hansen 2007a, 2007b; Houghton 2005; Nobre 2007; Volpi 2007.

justification for inaction. They are an invitation to assess the nature of identified risks and to develop strategies for risk mitigation. As a group of eminent United States military leaders has argued, no commander in the field would look at risks comparable to those posed by climate change and decide not to act because of uncertainty: "We cannot wait for certainty. Failing to act because a warning is not precise enough is not acceptable." [50]

The nature of the risks associated with climate change uncertainties reinforces that assessment on three counts. First, these are risks that threaten the whole of future generations of humanity with catastrophic outcomes. The sea level rises that would accompany the collapse of the ice sheets on Greenland and the West Antarctic would overwhelm the flood defences of even the richest countries, submerging large areas of Florida and much of the Netherlands, as well as inundating the Ganges Delta, Lagos and Shanghai. Second, the outcomes associated with the risks are irreversible: the West Antarctic ice sheet cannot be restored by future generations. Third, uncertainty cuts both ways: there is as much chance of outcomes being more malign as there is of them being more benign.

In a one-country world inhabited by citizens who shared a concern for the well-being of future generations, climate change mitigation would be an urgent priority. It would be viewed as an insurance policy against catastrophic risk and as an imperative rooted in considerations of cross-generational equity. Uncertainty in this one-country world would be viewed not as grounds for inaction but as evidence of the case for acting with resolve to reduce the risks.

In a world of many countries at vastly different levels of development there is a complementary case for urgent action. That case is first of all rooted in considerations of social justice, human rights and ethical concern for the world's poorest and most vulnerable people. Millions of these people are already dealing with the early impacts of climate change. These impacts are already slowing human progress and all plausible scenarios point to more of the same, and worse. Because mitigation will have a limited influence on climate change for several decades, investment in adaptation should be seen as part of the insurance policy for the world's poor.

Both mitigation and adaptation should be seen as human security imperatives in a broader sense. Dangerous climate change, and the ecological damage that will follow in its wake, threatens to cause massive human displacement and the collapse of livelihoods on a vast scale. The ripple effects would extend far beyond the localities of those most immediately affected. Associated outcomes will extend from the movement of displaced people across national borders to the potential collapse of fragile states. In an interdependent world, no country would be immune to the consequences. Of course, many rich countries might seek to protect their citizens against climate insecurity through investment in flood defences and other actions. However, the anger and resentment that would be felt by those most immediately affected would create wider insecurities.

> In a one-country world inhabited by citizens who shared a concern for the well-being of future generations, climate change mitigation would be an urgent priority

1.3 From global to local—measuring carbon footprints in an unequal world

For global carbon accounting purposes the world is a single country. The Earth's atmosphere is a common resource without borders. Emissions of greenhouse gases mix freely in the atmosphere over time and space. It makes no difference for climate change whether the marginal tonne of CO_2 comes from a coal-fired power plant, from a car, or from a loss of carbon sinks in tropical rainforests. Similarly, when greenhouse gases enter the Earth's atmosphere they are not segmented by country of origin: a tonne of

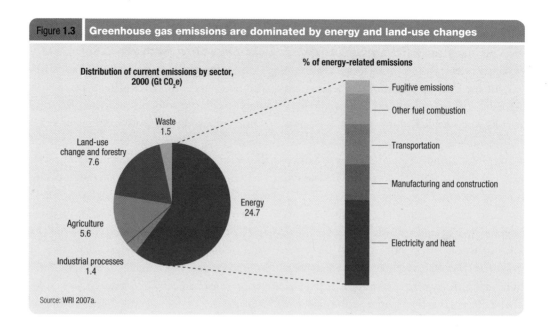

Figure **1.3** Greenhouse gas emissions are dominated by energy and land-use changes

Distribution of current emissions by sector, 2000 (Gt CO_2e)

Waste 1.5
Land-use change and forestry 7.6
Agriculture 5.6
Industrial processes 1.4
Energy 24.7

% of energy-related emissions

Fugitive emissions
Other fuel combustion
Transportation
Manufacturing and construction
Electricity and heat

Source: WRI 2007a.

CO_2 from Mozambique is the same weight as a tonne of CO_2 from the United States.

While each tonne of carbon dioxide carries equal weight, the global account masks large variations in contributions to overall emissions from different sources. All activities, all countries and all people register in the global carbon account—but some register far more heavily than others. In this section we look at the carbon footprint left by emissions of CO_2. Differences in the depth of carbon footprints can help to identify important issues of equity and distribution in approaches to mitigation and adaptation.

National and regional footprints— the limits to convergence

Most human activities—fossil fuel combustion for power generation, transport, land-use changes and industrial processes—generate emissions of greenhouse gases. That is one of the reasons why mitigation poses such daunting challenges.

The breakdown of the distribution of greenhouse gas emissions underlines the scope of the problem (figure 1.3). In 2000, just over half of all emissions came from the burning of fossil fuels. Power generation accounted for around 10 Gt CO_2e, or around one-quarter of the total. Transport is the second largest source of energy-related CO_2e emissions. Over the past three decades, energy supply and transport have increased their greenhouse gas emissions by 145 and 120 percent respectively. The critical role of the power sector in global emissions is not fully captured by its current share. Power generation is dominated by capital-intensive infrastructural investments. Those investments create assets that have a long lifetime: power plants opening today will still be emitting CO_2 in 50 years time.

Land-use change also plays an important role. Deforestation is by far the largest source of CO_2 emissions in this context, releasing sequestered carbon into the atmosphere as a

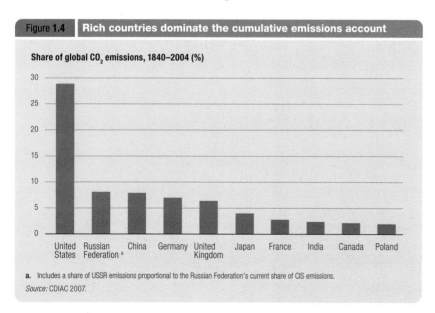

Figure **1.4** Rich countries dominate the cumulative emissions account

Share of global CO_2 emissions, 1840–2004 (%)

United States
Russian Federation [a]
China
Germany
United Kingdom
Japan
France
India
Canada
Poland

a. Includes a share of USSR emissions proportional to the Russian Federation's current share of CIS emissions.
Source: CDIAC 2007.

result of burning and loss of biomass. Data in this area are more uncertain than for other sectors. However, best estimates suggest that around 6 Gt CO_2 are released annually.[51] According to the IPCC, the share of CO_2 originating from deforestation ranges between 11 and 28 percent of total emissions.[52]

One of the conclusions to emerge from the sectoral analysis of carbon footprints is that mitigation aimed at reducing CO_2 emissions from power generation, transport and deforestation is likely to generate high returns.

National carbon footprints can be measured in terms of stocks and flows. National footprint depth is closely related to historic and current energy use patterns. While the aggregate footprint of the developing world is becoming deeper, historic responsibility for emissions rests heavily with the developed world.

Rich countries dominate the overall emissions account (figure 1.4). Collectively, they account for about 7 out of every 10 tonnes of CO_2 that have been emitted since the start of the industrial era. Historic emissions amount to around 1,100 tonnes of CO_2 per capita for Britain and America, compared with 66 tonnes per capita for China and 23 tonnes per capita for India.[53] These historic emissions matter on two counts. First, as noted earlier, cumulative past emissions drive today's climate change. Second, the envelope for absorbing future emissions is a residual function of past emissions. In effect, the ecological 'space' available for future emissions is determined by past action.

Turning from stocks to flows produces a different picture. One striking feature of that picture is that emissions are highly concentrated in a small group of countries (figure 1.5). The United States is the largest emitter, accounting for around one-fifth of the total. Collectively, the top five—China, India, Japan, the Russian Federation and the United States —account for more than half; the top ten for over 60 percent. While climate change is a global problem, national and multilateral action involving a relatively small group of countries or groupings—such as the G8, the

European Union (EU), China and India—would encompass a large share of the total flow of emissions.

Much has been made of the convergence in emissions between developed and developing

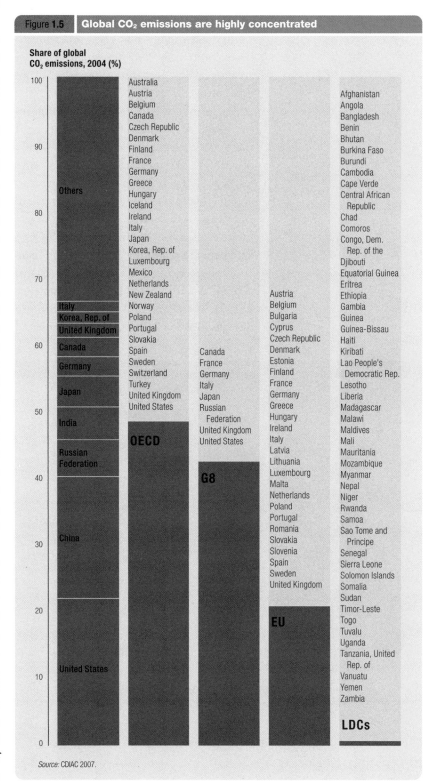

Figure 1.5 | Global CO_2 emissions are highly concentrated

Share of global CO_2 emissions, 2004 (%)

Source: CDIAC 2007.

countries. At one level, the process of convergence is real. Developing countries account for a rising share of global emissions. In 2004, they accounted for 42 percent of energy-related CO_2 emissions, compared to around 20 percent in 1990 (appendix table). China may be about to overtake the United States as the world's largest emitter and India is now the world's fourth largest emitter. By 2030 developing countries are projected to account for just over half of total emissions.[54]

Factoring in deforestation reconfigures the global CO_2 emissions league table. If the world's rainforests were a country, that country would stand at the top of the world's league table for CO_2 emissions. Taking into account just emissions from deforestation, Indonesia, would rank as the third largest source of annual CO_2 emissions (2.3 Gt CO_2) with Brazil ranking fifth (1.1 Gt CO_2).[55] There are large interannual variations in emissions, making it difficult to compare countries. In 1998, when El Niño events triggered severe droughts in South-east Asia, an estimated 0.8–2.5 billion tonnes of carbon were released to the atmosphere through fires in peat forests.[56] In Indonesia, land-use change and forestry are estimated to release about 2.5 Gt CO_2e annually—around six times the emissions from energy and agriculture combined.[57] For Brazil, emissions linked to land use changes account for 70 percent of the national total.

Convergence in aggregate emissions is sometimes cited as evidence that developing countries as a group need to embark on rapid mitigation. That assessment overlooks some important considerations. Developing country participation will be required if global mitigation is to succeed. However, the extent of convergence has been heavily overstated.

With just 15 percent of the world population, rich countries account for 45 percent of CO_2 emissions. Sub-Saharan Africa also accounts for around 11 percent of the world population, but represents 2 percent of global emissions. Low income countries as a group account for one-third of the world's population but for just 7 percent of emissions.

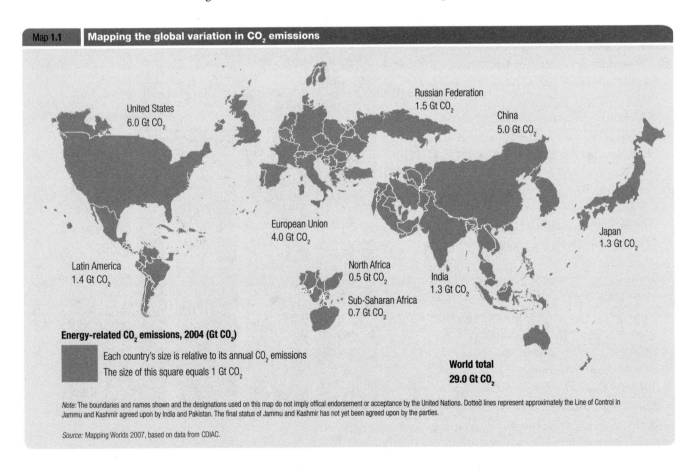

Map 1.1 **Mapping the global variation in CO_2 emissions**

United States
6.0 Gt CO_2

Russian Federation
1.5 Gt CO_2

China
5.0 Gt CO_2

European Union
4.0 Gt CO_2

Japan
1.3 Gt CO_2

Latin America
1.4 Gt CO_2

North Africa
0.5 Gt CO_2

Sub-Saharan Africa
0.7 Gt CO_2

India
1.3 Gt CO_2

Energy-related CO_2 emissions, 2004 (Gt CO_2)

Each country's size is relative to its annual CO_2 emissions
The size of this square equals 1 Gt CO_2

**World total
29.0 Gt CO_2**

Note: The boundaries and names shown and the designations used on this map do not imply offical endorsement or acceptance by the United Nations. Dotted lines represent approximately the Line of Control in Jammu and Kashmir agreed upon by India and Pakistan. The final status of Jammu and Kashmir has not yet been agreed upon by the parties.

Source: Mapping Worlds 2007, based on data from CDIAC.

Inequalities in carbon footprinting—some people walk more lightly than others

Differences in the depth of carbon footprints are linked to the history of industrial development. But, they also reflect the large 'carbon debt' accumulated by rich countries—a debt rooted in the over-exploitation of the Earth's atmosphere. People in the rich world are increasingly concerned about emissions of greenhouse gases from developing countries. They tend to be less aware of their own place in the global distribution of CO_2 emissions (map 1.1). Consider the following examples:

- The United Kingdom (population 60 million) emits more CO_2 than Egypt, Nigeria, Pakistan, and Viet Nam combined (total population 472 million).
- The Netherlands emits more CO_2 than Bolivia, Colombia, Peru, Uruguay and the seven countries of Central America combined.
- The state of Texas (population 23 million) in the United States registers CO_2 emissions of around 700 Mt CO_2 or 12 percent of the United States' total emissions. That figure is greater than the total CO_2 footprint left by sub-Saharan Africa—a region of 720 million people.
- The state of New South Wales in Australia (population 6.9 million) has a carbon footprint of 116 Mt CO_2. This figure is comparable to the combined total for Bangladesh, Cambodia, Ethiopia, Kenya, Morocco, Nepal and Sri Lanka.
- The 19 million people living in New York State have a higher carbon footprint than the 146 Mt CO_2 left by the 766 million people living in the 50 least developed countries.

Extreme inequalities in national carbon footprints reflect disparities in per capita emissions. Adjusting CO_2 emission accounts to factor in these disparities demonstrates the very marked limits to carbon convergence (figure 1.6).

Carbon footprint convergence has been a limited and partial process that has started from different emission levels. While China may be about to overtake the United States as the world's largest emitter of CO_2, per capita emissions are just one-fifth of the size. Emissions from India are on a rising trend. Even so, its per capita carbon footprint is less than one-tenth of that in high-income countries. In Ethiopia, the average per capita carbon footprint is 0.1 tonnes, compared with 20 tonnes in Canada. The per capita increase in emissions since 1990 for the United States (1.6 tonnes) is higher than the total per capita emissions for India in 2004 (1.2 tonnes). The overall increase in emissions from the United States exceeds sub-Saharan Africa's total emissions. The per capita increase for Canada since 1990 (5 tonnes) is higher than per capita emissions for China in 2004 (3.8 tonnes).

The distribution of current emissions points to an inverse relationship between climate change risk and responsibility. The world's poorest people walk the Earth with a very light carbon footprint. We estimate the carbon footprint of the poorest 1 billion people on the planet at around 3 percent of the world's total footprint. Living in vulnerable rural areas and urban slums, the poorest billion people are highly exposed to climate change threats for which they carry negligible responsibility.

The global energy divide

Inequalities in aggregate and per capita carbon footprints are intimately related to wider inequalities. They mirror the relationship between economic growth, industrial development and access to modern energy services. That relationship draws attention to an important human development concern. Climate change and the curtailment of excessive fossil fuel use may be the greatest challenge of the 21st Century, but an equally urgent and more immediate challenge is the expanded provision of affordable energy services to the world's poor.

Living without electricity affects many dimensions of human development. Energy services play a critical role not just in supporting economic growth and generating employment, but also in enhancing the quality of people's lives. Around 1.6 billion people in the world lack access to such services (figure 1.7). Most

Figure 1.6   Rich countries—deep carbon footprints

CO_2 emissions (t CO_2 per capita)

2004 ●
1990 ○

United States
20.6
19.3

Canada
20.0
15.0

Russian Federation
10.6
13.4 (1992)

United Kingdom
9.8
10.0

France
6.0
6.4

China
3.8
2.1

Egypt 2.3 1.5

Brazil 1.8 1.4

Viet Nam 1.2 0.3

India 1.2 0.8

Nigeria 0.9 0.5

Bangladesh 0.3 0.1

Tanzania 0.1 0.1

Ethiopia 0.1 0.1

Source: CDIAC 2007.

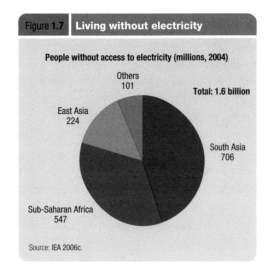

Figure 1.7 Living without electricity

People without access to electricity (millions, 2004)

Total: 1.6 billion

Others 101

East Asia 224

South Asia 706

Sub-Saharan Africa 547

Source: IEA 2006c.

modern energy services in 2030 if current trends continue (box 1.2).[60] Currently some 2.5 billion people depend on biomass (figure 1.8).

Changing this picture is vital for human development. The challenge is to expand access to basic energy services while limiting increases in the depth of the developing world's per capita carbon footprint. Enhanced efficiency in energy use and the development of low-carbon technologies hold the keys, as we show in chapter 3.

There are overwhelming practical and equitable grounds for an approach that reflects past responsibility and current capabilities. Mitigation responsibilities and capabilities cannot be derived from the arithmetic of carbon footprinting. Even so, that arithmetic does provide some obvious insights. For example, if everything else were equal, a cut of 50 percent in CO_2 emissions for South Asia and sub-Saharan Africa would reduce global emissions by 4 percent. Similar reductions in high-income countries would reduce emissions by 20 percent. The equity arguments are equally compelling. An average air-conditioning unit in Florida emits more CO_2 in a year than a person in Afghanistan or Cambodia during their lifetime. And an average dishwasher in Europe emits as much CO_2 in a year as three Ethiopians. While climate change mitigation is a global challenge, the starting place for mitigation is with the countries that carry the bulk of historic responsibility and the people that leave the deepest footprints.

live in sub-Saharan Africa,[58] where only around one-quarter of people use modern energy services, and South Asia.

The vast global deficit in access to basic energy services has to be considered alongside concerns over the rise in CO_2 emissions from developing countries. Emissions of CO_2 from India may have become a matter of global concern for climate security. That perspective is very partial. The number of people in India living without access to modern electricity is around 500 million—more than the total population of the enlarged European Union. These are people who live without so much as a light bulb in their homes and rely on firewood or animal dung for cooking.[59] While access to energy is increasing across the developing world, progress remains slow and uneven, holding back advances in poverty reduction. Worldwide, there will still be 1.4 billion people without access to

1.4 Avoiding dangerous climate change—a sustainable emissions pathway

Climate change is a global problem that demands an international solution. The starting point must be an international agreement on the limitation of greenhouse gas emissions. Strategies for limitation have to be developed at a national level. What is required at the international level is a framework that sets limits on overall emissions. That framework

has to chart an emissions pathway consistent with the objective of avoiding dangerous climate change.

In this section we set out such a pathway. We start by identifying a global carbon budget for the 21st Century. The concept of a carbon budget is not new. It was developed by the architects of the Kyoto Protocol and has been taken

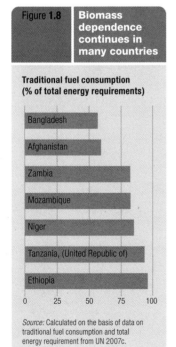

Figure 1.8 Biomass dependence continues in many countries

Traditional fuel consumption (% of total energy requirements)

Bangladesh

Afghanistan

Zambia

Mozambique

Niger

Tanzania, (United Republic of)

Ethiopia

0   25   50   75   100

Source: Calculated on the basis of data on traditional fuel consumption and total energy requirement from UN 2007c.

| Box 1.2 | Millions are denied access to modern energy services |

"Our day starts before five in the morning as we need to collect water, prepare breakfast for the family and get our children ready for school. At around eight, we start collecting wood. The journey is several kilometres long. When we cannot get wood we use animal dung for cooking—but it is bad for the eyes and for the children."
Elisabeth Faye, farmer, aged 32, Mbour, Senegal

In most rich countries access to electricity is taken for granted. At the flick of a switch the lights come on, water is heated and food is cooked. Employment and prosperity are supported by the energy systems that sustain modern industry, drive computers and power transport networks.

For people like Elisabeth Faye access to energy has a very different meaning. Collecting wood for fuel is an arduous and time-consuming activity. It takes 2–3 hours a day. When she is unable to collect wood, she has no choice but to use animal dung for cooking—a serious health hazard.

In developing countries there are some 2.5 billion people like Elisabeth Faye who are forced to rely on biomass—fuelwood, charcoal and animal dung—to meet their energy needs for cooking (figure 1.8). In sub-Saharan Africa, over 80 percent of the population depends on traditional biomass for cooking, as do over half of the populations of India and China.

Unequal access to modern energy is closely correlated with wider inequalities in opportunities for human development. Countries with low levels of access to modern energy systems figure prominently in the low human development group. Within countries, inequalities in access to modern energy services between rich and poor and urban and rural areas interact with wider inequalities in opportunity.

Poor people and poor countries pay a high price for deficits in modern energy provision:

- *Health.* Indoor air pollution resulting from the use of solid fuels is a major killer. It claims the lives of 1.5 million people each year, more than half of them below the age of five: that is 4000 deaths a day. To put this number in context, it exceeds total deaths from malaria and rivals the number of deaths from tuberculosis. Most of the victims are women, children and the rural poor. Indoor air pollution is also one of the main causes of lower respiratory tract infections and pneumonia in children. In Uganda, children under the age of five are reported to suffer 1–3 episodes of acute respiratory tract infection annually. In India, where three in every four households in rural areas depend on firewood and dung for cooking and heat, pollution from unprocessed biofuels accounts for some 17 percent of child deaths. Electrification is often associated with wider advances in health status. For example, in Bangladesh, rural electrification is estimated to increase income by 11 percent—and to avert 25 child deaths for every 1000 households connected.

- *Gender.* Women and young girls have to allocate large amounts of time to the collection of firewood, compounding gender inequalities in livelihood opportunities and education. Collecting fuelwood and animal dung is a time-consuming and exhausting task, with average loads often in excess of 20kg. Research in rural Tanzania has found that women in some areas walk 5–10 kilometres a day collecting and carrying firewood, with loads averaging 20kg to 38kg. In rural India, average collection times can amount to over 3 hours a day. Beyond the immediate burden on time and body, fuelwood collection often results in young girls being kept out of school.

- *Economic costs.* Poor households often spend a large share of their income on fuelwood or charcoal. In Guatemala and Nepal, wood expenditure represents 10–15 percent of total household expenditure in the poorest quintile. Collection time for fuelwood has significant opportunity costs, limiting opportunities for women to engage in income generating activities. More broadly, inadequate access to modern energy services restricts productivity and helps keep people poor.

- *Environment.* Deficits in access to modern energy can create a vicious circle of environmental, economic and social reversal. Unsustainable production of charcoal in response to rising urban demand has placed a huge strain on areas surrounding major cities such as Luanda in Angola and Addis Ababa in Ethiopia. In some cases, charcoal production and wood collection has contributed to local deforestation. As resources shrink, dung and residues are diverted to fuel use instead of being ploughed back into fields, undermining soil productivity.

Expanded access to affordable electricity for the poor remains an overarching development priority. Current projections show that the number of people relying on biomass will increase over the next decade and beyond, especially in sub-Saharan Africa. This will compromise progress towards several MDGs, including those relating to child and maternal survival, education, poverty reduction and environmental sustainability.

Source: IEA 2006c; Kelkar and Bhadwal 2007; Modi et al. 2005; Seck 2007b; WHO 2006; World Bank 2007b.

up by some governments (chapter 3). In effect, the carbon budget is akin to a financial budget. Just as financial budgets have to balance spending against resources, so carbon budgets have to balance greenhouse gas emissions against ecological capacity. However, carbon budgets have to operate over a very long time-horizon.

Because the emissions that drive the accumulation of greenhouse gas stocks are cumulative and long-lived, we have to set an expenditure framework that spans decades rather than years.

There are further parallels between financial budgeting and carbon budgeting. When

1

Our carbon budget has a single goal: keeping average global temperature increases (over preindustrial levels) below 2°C

households or governments set budgets they target a range of objectives. Households have to avoid unsustainable spending patterns or face the prospect of debt. Government budgets are geared towards a range of public policy goals in areas such as employment, inflation and economic growth. If public spending exceeds revenues by large margins, the consequences are reflected in large fiscal deficits, inflation and the accumulation of debt. Ultimately, budgets are about living within the bounds of financial sustainability.

Carbon budgeting for a fragile planet

Carbon budgets define the bounds of ecological sustainability. Our carbon budget has a single goal: keeping average global temperature increases (over preindustrial levels) below 2°C. The rationale for this goal is, as we have seen, rooted in climate science and human development imperatives. Climate science identifies 2°C as a potential 'tipping point' for long-run catastrophic outcomes. More immediately, it represents a 'tipping point' for large scale human development reversals during the 21st Century. Remaining within the 2°C threshold should be seen as a reasonable and prudent long term objective for avoiding dangerous climate change. Many governments have adopted that objective. Sustainable carbon budget management should be seen as a means to that end.

What is the upper limit on greenhouse gas emissions for a world committed to avoiding dangerous climate change? We address that question by using simulations carried out at the Potsdam Institute for Climate Impact Research (PIK).

Stabilization of greenhouse gas stocks requires a balance between current emissions and absorption. A specific stabilization target can be achieved through a number of possible emission trajectories. In broad terms, emissions can peak early and decline gradually, or they can peak later and decline more rapidly. If the aim is to avoid dangerous climate change, the starting point is to identify a stabilization target consistent with the world staying within the 2°C dangerous climate change threshold.

Keeping within 2°C—the 'fifty–fifty' point

In our simulation we set the bar at the lowest reasonable level. That is, we identify the level of greenhouse gas stocks consistent with an approximately even chance of avoiding dangerous climate change. This level is around 450 ppm CO_2e. It might be argued that this is insufficiently ambitious: most people would not stake their future well-being on the toss of a coin. However, stabilizing at 450 ppm CO_2e will entail a sustained global effort.

Setting the bar above our target would lengthen the odds on avoiding dangerous climate change. At greenhouse gas stock levels of 550 ppm CO_2e the likelihood of overshooting the dangerous climate change threshold of 2°C increases to around 80 percent (figure 1.9). Opting for a 550 ppm CO_2e target would be taking a gamble at very long odds on the future of the planet and 21st Century human development prospects. In fact, there would be a one-in-three chance of overshooting 3°C.

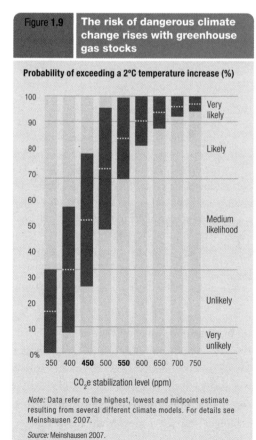

Figure 1.9 **The risk of dangerous climate change rises with greenhouse gas stocks**

Probability of exceeding a 2°C temperature increase (%)

CO$_2$e stabilization level (ppm)

Note: Data refer to the highest, lowest and midpoint estimate resulting from several different climate models. For details see Meinshausen 2007.

Source: Meinshausen 2007.

The emerging consensus that climate change must be limited to a 2°C ceiling sets an ambitious but achievable goal. Realising that goal will require concerted strategies to limit the accumulation of greenhouse gas stocks to 450 ppm. While there is uncertainty at the margin, this remains the most plausible best-estimate for a sustainable carbon budget.

If the world were a single country, it would be implementing a recklessly extravagant and unsustainable carbon budget. If that budget were a financial budget the government of that country would be running a large fiscal deficit, exposing its citizens to hyperinflation and unsustainable debt. The lack of prudence in carbon budgeting can best be described by looking across the whole century.

We use the PIK simulations to address this task. Our approach focuses on fossil fuel-related CO_2 emissions because these are of the most direct relevance to policy debates on climate change mitigation. It identifies a level of emissions consistent with avoiding dangerous climate change. Briefly summarized, the 21st Century budget amounts to 1,456 Gt CO_2, or around 14.5 Gt CO_2 on a simple annual average basis.[61] Current emissions are running at twice this level. Put in financial budget terms, expenditure is outstripping income by a factor of two.

The bad news is that things are worse than they look because emissions are rising with population growth and economic growth. Using IPCC scenarios, the 21st Century budget consistent with avoiding dangerous climate change could expire as early as 2032, or in 2042 under more benign assumptions (figure 1.10).

Scenarios for climate security— time is running out

These projections tell an important story in two parts. The first part relates to basic budget management. As a global community, we are failing the most basic tests of sound budget practice. In effect, we are spending our monthly pay cheque in 10 days. Today's energy use and emission patterns are running down the Earth's ecological assets, and running up unsustainable ecological debts. Those

debts will be inherited by future generations, who will have to compensate at great human and financial cost for our actions and also face the threats posed by dangerous climate change.

The second part of the budget story is equally stark. It is that time is running out. The fact that the carbon budget is set to expire between 2032 and 2042 does not mean we have two or three decades to act. Once the critical threshold has been reached, there is no way back to a more secure climate option. Moreover, emissions pathways cannot be changed overnight. They require extensive reforms in energy policies and behaviour implemented over several years.

How many planets?

On the eve of India's independence, Mahatma Gandhi was asked whether he thought the country could follow the British model of industrial development. His response retains a powerful resonance in a world that has to redefine its relation to the earth's ecology: "It took Britain half the resources of this planet to

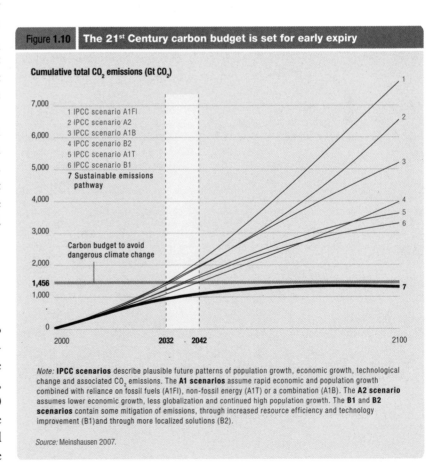

Figure 1.10 The 21st Century carbon budget is set for early expiry

Cumulative total CO_2 emissions (Gt CO_2)

1 IPCC scenario A1FI
2 IPCC scenario A2
3 IPCC scenario A1B
4 IPCC scenario B2
5 IPCC scenario A1T
6 IPCC scenario B1
7 Sustainable emissions pathway

Carbon budget to avoid dangerous climate change

Note: **IPCC scenarios** describe plausible future patterns of population growth, economic growth, technological change and associated CO_2 emissions. The **A1 scenarios** assume rapid economic and population growth combined with reliance on fossil fuels (A1FI), non-fossil energy (A1T) or a combination (A1B). The **A2 scenario** assumes lower economic growth, less globalization and continued high population growth. The **B1** and **B2 scenarios** contain some mitigation of emissions, through increased resource efficiency and technology improvement (B1) and through more localized solutions (B2).

Source: Meinshausen 2007.

achieve its prosperity. How many planets will India require for development?"

We ask the same question for a world edging towards the brink of dangerous climate change. Using the annual ceiling of 14.5 Gt CO_2, if emissions were frozen at the current level of 29 Gt CO_2 we would need two planets. However, some countries are running a less sustainable account than others. With 15 percent of the world population, rich countries are using 90 percent of the sustainable budget. How many planets would we need if developing countries were to follow the example of these countries?

If every person living in the developing world had the same carbon footprint as the average for high income countries, global CO_2 emissions would rise to 85 Gt CO_2—a level that would require six planets. With a global per capita footprint at Australian levels, we would need seven planets, rising to nine for a world with Canada and United States levels of per capita emissions (table 1.2).

The answer to Gandhi's question raises some wider questions about social justice in climate change mitigation. As a global community, we are running up a large and unsustainable carbon debt, but the bulk of that debt has been accumulated by the world's richest countries.

The challenge is to develop a global carbon budget that charts an equitable and sustainable course away from dangerous climate change.

Charting a course away from dangerous climate change

We use the PIK model to identify plausible pathways for keeping within the 2°C threshold. One pathway treats the world as a single country, which for carbon accounting purposes it is, then identifies targets for rationing or 'burden sharing'. However, the viability of any system of burden sharing depends on participants in the system perceiving the distribution of rations to be fair. The UNFCCC itself acknowledges this through an injunction to "protect the climate system...on the basis of equity and in accordance with...common but differentiated responsibilities and respective capabilities."

While interpretation of that injunction is a matter for negotiation, we have distinguished between industrialized countries and developing countries, charting separate pathways for the two groups. The results are summarized in figure 1.11. The cuts from a 1990 base-year on our sustainable emissions pathway are as follows:

- *The world.* Emissions for the world will have to be reduced by around 50 percent by 2050, with a peak around 2020. Emissions would fall towards zero in net terms by the end of the 21st Century.
- *Developed countries.* High-income countries would have to target an emissions peak between 2012 and 2015, with 30 percent cuts by 2020 and at least 80 percent cuts by 2050.
- *Developing countries.* While there would be large variations, major emitters in the developing world would maintain a trajectory of rising emissions to 2020, peaking at around 80 percent above current levels, with cuts of 20 percent against 1990 levels by 2050.

Contraction and convergence— sustainability with equity

We emphasize that these are feasible pathways. They are not specific proposals for individual countries. Yet the pathways do serve an important purpose. Governments are embarking

Table **1.2**	**Global carbon footprints at OECD levels would require more than one planet** [a]		
	CO_2 emissions per capita (t CO_2) 2004	Equivalent global CO_2 emissions (Gt CO_2) 2004 [b]	Equivalent number of sustainable carbon budgets [c]
World [d]	4.5	29	2
Australia	16.2	104	7
Canada	20.0	129	9
France	6.0	39	3
Germany	9.8	63	4
Italy	7.8	50	3
Japan	9.9	63	4
Netherlands	8.7	56	4
Spain	7.6	49	3
United Kingdom	9.8	63	4
United States	20.6	132	9

a. As measured in sustainable carbon budgets.
b. Refers to global emissions if every country in the world emitted at the same per capita level as the specified country.
c. Based on a sustainable emissions pathway of 14.5 Gt CO_2 per year.
d. Current global carbon footprint.

Source: HDRO calculations based on Indicator Table 24.

Figure 1.11 | **Halving emissions by 2050 could avoid dangerous climate change**

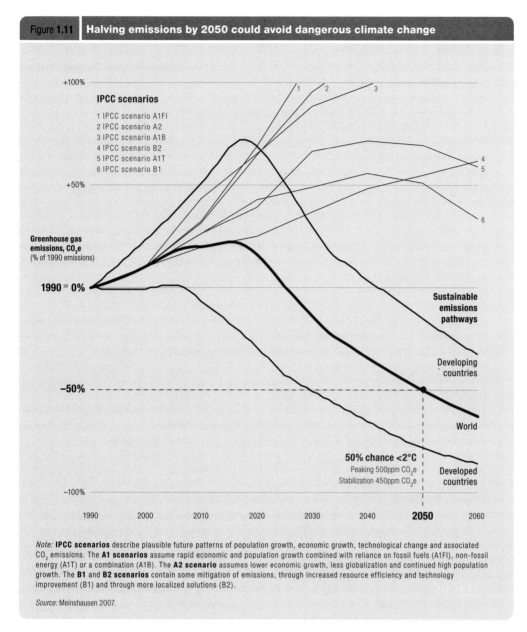

IPCC scenarios

1 IPCC scenario A1FI
2 IPCC scenario A2
3 IPCC scenario A1B
4 IPCC scenario B2
5 IPCC scenario A1T
6 IPCC scenario B1

Greenhouse gas
emissions, CO_2e
(% of 1990 emissions)

Sustainable
emissions
pathways

Developing
countries

World

50% chance <2°C
Peaking 500ppm CO_2e
Stabilization 450ppm CO_2e

Developed
countries

Note: **IPCC scenarios** describe plausible future patterns of population growth, economic growth, technological change and associated CO_2 emissions. The **A1 scenarios** assume rapid economic and population growth combined with reliance on fossil fuels (A1FI), non-fossil energy (A1T) or a combination (A1B). The **A2 scenario** assumes lower economic growth, less globalization and continued high population growth. The **B1** and **B2 scenarios** contain some mitigation of emissions, through increased resource efficiency and technology improvement (B1) and through more localized solutions (B2).

Source: Meinshausen 2007.

on negotiations for the multilateral framework to succeed the current Kyoto Protocol following the expiry of the current commitment period in 2012. The PIK simulations identify the scale of emission reductions that will be required to put the world on a pathway that avoids dangerous climate change. There are various trajectories that could be adopted to achieve the 2050 targets. What our sustainable emissions pathway does is to emphasize the importance of linking near-term and long term goals.

The emissions pathways also serve to highlight the importance of early and concerted action. In theory starting points for carbon emission reductions could be pushed back.

But the corollary would be far deeper cuts required over a reduced time horizon. In our view that would be a prescription for failure because costs would rise and adjustments would become even more difficult. Another scenario could be drawn up in which some major Organisation for Economic Co-operation and Development (OECD) countries do not participate in quantitative carbon budgeting. Such an approach would all but guarantee failure. Given the magnitude of emission reductions required in the OECD countries, it is unlikely that participating countries would be able to compensate for the non-participation of major emitters. Even if they did, it is unlikely that they would embrace an agreement that allowed 'free riding'.

Participation of the developing world in quantitative reductions is equally vital. In some respects, our 'two-country' model oversimplifies the issues to be addressed in negotiations. The developing world is not homogenous: the United Republic of Tanzania is not in the same position as China, for example. Moreover, what matters is the overall volume of emission reductions. From a global carbon budget perspective, deep reductions in sub-Saharan Africa carry negligible weight relative to reductions in major emitting countries.

However, with developing countries accounting for nearly half of worldwide emissions, their participation in any international agreement is increasingly important. At the same time, even high growth developing countries have pressing human development needs that must be taken into account. So too must the very large 'carbon debt' that the rich countries owe the world. Repayment of that debt and recognition of human development imperatives demand that rich countries cut emissions more deeply and support low-carbon transitions in the developing world.

We acknowledge that many other emissions' pathways are possible. One school of thought argues that every person in the world ought to enjoy an equivalent right to emit greenhouse gases, with countries that exceed their quota compensating those that underutilize their entitlement. Although proposals in this framework are often couched in terms of rights and equity, it is not clear that they have a rights-based foundation: the presumed 'right to emit' is clearly something different than the right to vote, the right to receive an education or the right to enjoy basic civil liberties.[62] At a practical level, attempts to negotiate a 'pollution rights' approach is unlikely to gain broad support. Our pathway is rooted in a commitment to achieve a practical goal: namely, the avoidance of dangerous climate change. The route taken requires a process of overall contraction in greenhouse gas flows and convergence in per capita emissions (figure 1.12).

Urgent action and delayed response—the case for adaptation

Deep and early mitigation does not offer a short-cut for avoiding dangerous climate change. Our sustainable emissions pathway demonstrates the importance of the time lag between mitigation actions and outcomes. Figure 1.13 captures the lag. It compares the degree of warming above preindustrial levels associated with the IPCC's non-mitigation scenarios, with the anticipated warming if the world stabilizes greenhouse gas stocks at 450 ppm CO_2e. Temperature divergence begins between 2030 and 2040, becoming more emphatically marked after 2050, by which time all but one of the IPCC scenarios breach the 2°C dangerous climate change threshold.

The timing of the temperature divergence draws attention to two important public policy issues. First, even the stringent mitigation implied by our sustainable emissions pathway will not make a difference to world temperature trends until after 2030. Until then, the world in general and the world's poor in particular will have to live with the consequences of past emissions. Dealing with these consequences while maintaining progress towards the MDGs and building on that progress after 2015 is a matter not for mitigation but for adaptation. Second, the real benefits of mitigation will build cumulatively across the second half of the 21st Century and beyond.

| Figure **1.12** | **Contracting and converging to a sustainable future** |

Emissions per capita for stabilization at 450 ppm CO_2e (t CO_2 per capita)

Developed and transition countries
Developing countries
World

Note: **IPCC scenarios** describe plausible future patterns of population growth, economic growth, technological change and associated CO_2 emissions. The **A1 scenarios** assume rapid economic and population growth combined with reliance on fossil fuels (A1FI), non-fossil energy (A1T) or a combination (A1B). The **A2 scenario** assumes lower economic growth, less globalization and continued high population growth. The **B1** and **B2 scenarios** contain some mitigation of emissions, through increased resource efficiency and technology improvement (B1) and through more localized solutions (B2).

Source: Meinshausen 2007.

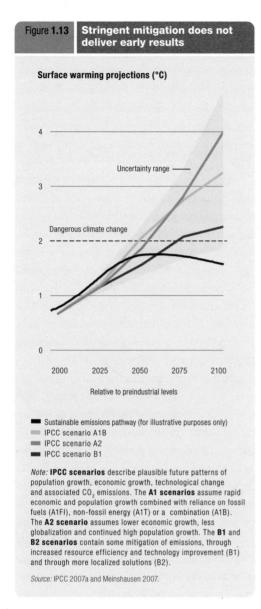

Figure 1.13 | **Stringent mitigation does not deliver early results**

Surface warming projections (°C)

Uncertainty range ——

Dangerous climate change

4

3

2

1

0

2000   2025   2050   2075   2100

Relative to preindustrial levels

■ Sustainable emissions pathway (for illustrative purposes only)
■ IPCC scenario A1B
■ IPCC scenario A2
■ IPCC scenario B1

Note: **IPCC scenarios** describe plausible future patterns of population growth, economic growth, technological change and associated CO_2 emissions. The **A1 scenarios** assume rapid economic and population growth combined with reliance on fossil fuels (A1FI), non-fossil energy (A1T) or a combination (A1B). The **A2 scenario** assumes lower economic growth, less globalization and continued high population growth. The **B1** and **B2 scenarios** contain some mitigation of emissions, through increased resource efficiency and technology improvement (B1) and through more localized solutions (B2).

Source: IPCC 2007a and Meinshausen 2007.

One important implication is that the motivation for urgent mitigation has to be informed by a concern for future generations. The world's poor will face the most immediate adverse impacts of temperature divergence. By the end of the 21st Century, with some of the IPCC scenarios pointing to temperature increases of 4–6°C (and rising), humanity as a whole will be facing potentially catastrophic threats.

The cost of a low-carbon transition—is mitigation affordable?

Setting carbon budgets is an exercise that has implications for financial budgets. While there have been many studies looking at the cost of achieving specific mitigation goals, our 2°C threshold is a far more stringent target than those assessed in most of these studies. While our sustainable climate pathway may be desirable, is it affordable?

We address that question by drawing on an approach that combines quantitative results from a large number of models in order to investigate the costs of achieving specified stabilization outcomes.[63] These models incorporate dynamic interactions between technology and investment, exploring a range of scenarios for achieving specified mitigation targets.[64] We use them to identify global costs for achieving a target of 450 ppm CO_2e.

Emissions of CO_2 can be cut in several ways. Increased energy efficiency, reduced demand for carbon-intensive products, changes in the energy mix—all have a role to play. Mitigation costs will vary according to how reductions are achieved and the time frame for achieving them. They arise from financing the development and deployment of new technologies and from the cost to consumers of switching to lower-emissions goods and services. In some cases, major reductions can be achieved at low cost: increased energy efficiency is an example. In others initial costs can generate benefits over the longer term. Deployment of a new generation of efficient, low-emission coal-fired power stations might fit in this category. Gradually reducing the flow of greenhouse gases over time is a lower-cost option than abrupt change.

Modelling work carried out for this Report estimates the costs of stabilization at 450 ppm CO_2e under various scenarios. Expressed in terms of headline dollars, the figures are very large. However, the costs of action are spread over many years. In a simple reference scenario, averaging out these costs produces a figure of around 1.6 percent of annual world GDP between now and 2030.[65]

That is not an insignificant investment. It would be wrong to underestimate the massive effort required to stabilize CO_2e emissions close to 450 ppm. However, the costs have to be put in perspective. As the Stern Review powerfully reminded the world's governments, they have to be evaluated against the costs of inaction.

Measured in economic terms the case for stringent mitigation makes good business sense

The 1.6 percent of global GDP required to achieve the 450 ppm targets for CO_2e represents less than two-thirds of global military expenditures. In the context of OECD countries, where government expenditure typically represents 30 to 50 percent of GDP, the stringent mitigation goals hardly appear unaffordable, especially if expenditures in other areas—such as military budget and agricultural subsidies—can be reduced.

The human and ecological costs of dangerous climate change cannot readily be captured in simple cost–benefit analysis. However, measured in economic terms the case for stringent mitigation makes good business sense. Over the long term the costs of inaction will be larger than the costs of mitigation. Estimating the costs of climate change impacts is intrinsically difficult. With warming of 5–6°C economic models that include the risk of abrupt and large-scale climate change point to losses of 5 to 10 percent of global GDP. Poor countries could suffer losses in excess of 10 percent.[66] Catastrophic climate change impacts could push the losses above this level. Reducing the risk of catastrophic outcomes is one of the most powerful arguments for early investment in mitigation to achieve the 450 ppm target.

It has to be emphasized that there are large margins of uncertainty in any assessment of mitigation costs. Most obviously, the cost structures for future low-carbon technologies, the timing of their introduction, and other factors are unknown. Higher costs than those indicated above are perfectly plausible—and political leaders need to communicate the uncertainties of financing for a 2°C climate change threshold. At the same time, it is also possible that costs could be lower. International emissions trading and the integration of carbon taxation into wider environmental tax reforms have the potential to drive down mitigation costs.[67]

All governments have to assess the financial implications of achieving climate change mitigation targets. Multilateral climate protection architecture will be left on an insecure foundation if it is not rooted in financial commitments. The 1.6 percent of average global GDP required for stringent mitigation implies a claim on scarce resources. But the alternatives are not cost-free. Political debate on financing must also address the question of whether dangerous climate change is an affordable option.

That question goes to the heart of the twin case for urgent action set out in this chapter. Given the momentous nature of the catastrophic ecological risks that will accompany dangerous climate change, 1.6 percent of global GDP might be seen as a small price to pay on an insurance policy to protect the well-being of future generations. Given that the same investment has the potential to prevent large-scale and very immediate reversals in human development for millions of the more vulnerable people across the world, the cross-generational and the cross-country social justice imperatives are mutually reinforcing.

1.5 Business-as-usual—pathways to an unsustainable climate future

Trend is not destiny and past performance can be a weak guide to future outcomes. In the case of climate change that is unequivocally a good thing. If the next 20 years look like the past 20 the battle against dangerous climate change will be lost.

Looking back—the world since 1990

Experience under the Kyoto Protocol provides some important lessons for the development of a 21st Century carbon budget. The Protocol provides a multilateral framework that sets limits on greenhouse gas emissions. Negotiated under the auspices of the UNFCCC, it took 5 years to reach an agreement—and another 8 years before that agreement was ratified by enough countries to become operational.[68] The headline target for greenhouse gas emissions cuts was 5 percent from 1990 levels.

Measured in terms of aggregate global emissions the Kyoto protocol did not set particularly ambitious targets. Moreover, quantitative ceilings were not applied to developing countries. The decisions of Australia and the United States not to ratify the protocol further limited the size of the intended cuts. The implication of these exceptions can be illustrated by reference to energy-related CO_2 emissions. From the 1990 base year the commitment made under the Kyoto protocol translates into a 2.5 percent reduction of energy-related CO_2 emissions in real terms by the 2010/2012 target date.[69]

Delivery against the targets has been disappointing so far. In 2004, overall greenhouse emissions for Annex I countries were 3 percent below 1990 levels.[70] However, the headline figure masks two major problems. First, since 1999 overall emissions have been on a rising trend, raising questions about whether the overall target will be achieved. Second, there are large variations in country performance (figure 1.14). Much of the overall decline can be traced to deep reductions in emissions in the Russian Federation and other transition economies—in some cases in excess of 30 percent. This outcome owes less to energy policy reform than to the effects of deep economic recession in the 1990s. Emissions are now rising with economic recovery. As a group, non-transition Annex I parties—broadly the OECD—have increased emissions by 11 percent from 1990 to 2004 (box 1.3).

Looking ahead—locked on a rising trajectory

Looking back, trends since the 1990 reference-point for the Kyoto Protocol are cause for concern. Looking ahead, the scenarios for future energy use and emissions point unmistakably towards a dangerous climate future, unless the world changes course.

Changing course will require a shift in energy use patterns as far-reaching as the

energy revolution that shaped the industrial revolution. Even without climate change, the future of fossil-fuel energy systems would be the subject of intense debate. Energy security—broadly defined as access to reliable and affordable supplies—is an increasingly prominent theme on the international agenda.

Since 2000, oil prices have increased by a factor of five in real terms, to around US$70

Looking ahead, the scenarios for future energy use and emissions point unmistakably towards a dangerous climate future, unless the world changes course

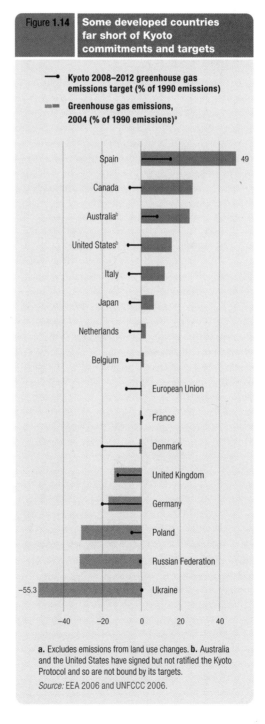

Figure **1.14** **Some developed countries far short of Kyoto commitments and targets**

— Kyoto 2008–2012 greenhouse gas emissions target (% of 1990 emissions)

▬ Greenhouse gas emissions, 2004 (% of 1990 emissions)[a]

Spain 49
Canada
Australia[b]
United States[b]
Italy
Japan
Netherlands
Belgium
European Union
France
Denmark
United Kingdom
Germany
Poland
Russian Federation
Ukraine −55.3

−40 −20 0 20 40

a. Excludes emissions from land use changes. **b.** Australia and the United States have signed but not ratified the Kyoto Protocol and so are not bound by its targets.

Source: EEA 2006 and UNFCCC 2006.

Box **1.3**   **Developed countries have fallen short of their Kyoto commitments**

The Kyoto Protocol was a first step in the multilateral response to climate change. It set targets for cutting greenhouse gas emissions against 1990 levels by 2010–2012. With governments embarking on negotiations for the post-2012 multilateral framework that will build on the current commitment period, it is important that lessons are learned.

There are three particularly important lessons. The first is that the level of ambition matters. Targets adopted under the first commitment period were modest, averaging around 5 percent for developed countries. The second lesson is that binding targets matter. Most countries are off track for delivering on their Kyoto commitments. The third lesson is that the multilateral framework has to cover all major emitting nations. Under the current Protocol, two major developed countries—Australia and the United States—signed the agreement but did not ratify it, creating an exemption for the targets. There are also no quantitative targets for developing countries.

While it is too early to deliver a final verdict on outcomes under the Kyoto protocol, the summary record to date on emissions without land-use changes is not encouraging. Most 68 countries are off track. Moreover, emissions' growth has strengthen since 2000.

Among the preliminary outcomes:

- The European Union made average emission reduction commitments of 8 percent under Kyoto. Actual cuts have amounted to around 2 percent and European Environment Agency projections suggest that current policies will leave this picture unchanged by 2010. Emissions from the transport sector increased by one-quarter. Emissions from electricity and heat generation increased by 6 percent. Large increases in renewable energy supply will be required to meet the Kyoto targets, but the European Union is falling short of the investments needed to meet its own target of 20 percent provision by 2020.

- The United Kingdom has surpassed its Kyoto target of a 12 percent emissions reduction, but is off track to meet a national target to reduce emissions by 20 percent against 1990 levels. Most of the reduction was achieved before 2000 as a result of industrial restructuring and market liberalization measures that led to a switch from carbon-intensive coal to natural gas. Emissions increased in 2005 and 2006 as a result of switching from natural gas and nuclear to coal (chapter 3).

- Germany's emissions were 17 percent lower in 2004 than in 1990. Reductions reflect deep cuts from 1990 to 1995 following reunification and industrial restructuring in East Germany (over 80 percent of the total reduction), supplemented by a decline in emissions from the residential sector.

- Italy and Spain are far off track for their Kyoto targets. In Spain emissions have increased by almost 50 percent since 1990, with strong economic growth and increased use of coal power following droughts. In Italy, the primary driver of increased emissions has been the transport sector.

- Canada agreed under the Kyoto Protocol to target a 6 percent cut in emissions. In the event, emissions have increased by 27 percent and the country is now around 35 percent above its Kyoto target range. While greenhouse gas intensity has fallen, efficiency gains have been swamped by an increase in emissions from an expansion in oil and gas production. Net emissions associated with oil and gas exports have more than doubled since 1990.

- Japan's emissions in 2005 were 8 percent above 1990 levels. The Kyoto target was for a 6 percent reduction. On current trends it is projected that the country will miss its target by around 14 percent. While emissions from industry have fallen marginally since 1990, large increases have been registered in emissions from transportation (50 percent for passenger vehicles) and the residential sector. Household emissions have grown more rapidly than the number of households.

- The United States is a signatory to the Kyoto Protocol but it has not ratified the treaty. If it had, it would have been required to reduce its emissions to 7 percent below 1990 levels by 2010. Overall emissions have increased by 16 percent. By 2010 projected emissions are 1.8 Gt above 1990 levels on a rising trend. Emissions have grown across all major sectors despite a 21 percent decline in greenhouse gas intensity of the United States' economy, as measured by the ratio of greenhouse gas emissions to GDP.

- Like the United States, Australia did not ratify the Kyoto Protocol. Overall emissions have grown at around twice the rate that would have been required had the country participated, with emissions rising by 21 percent since 1990. High levels of dependence on coal-fired power generation contributed to large increases in the energy sector, with CO_2 emissions rising by over 40 percent.

Looking to the post-2012 period, the challenge is to forge an international agreement that engages all major emitting countries in a long term effort to achieve a sustainable carbon budget for the 21ˢᵗ Century. There is little that governments can do today that will have significant effects on emissions between 2010 and 2012: like oil tankers, energy systems have large turning circles.

What is needed now is a framework for beating dangerous climate change. That framework will have to provide a far longer time-horizon for policymakers, with short term commitment periods linked to medium-term and long term goals. For developed countries, those goals have to include emission reductions of around 30 percent by 2020 and at least 80 percent by 2050—consistent with our sustainable emissions pathway. Reductions by developing countries could be facilitated through financial and technology transfer provisions (chapter 3).

Source: EEA 2006; EIA 2006; Government of Canada 2006; IEA 2006c; Government of the United Kingdom 2007c; Ikkatai 2007; Pembina Institute 2007a.

per barrel. While prices may retreat, a return to the low levels of the late 1990s is unlikely. Some commentators interpret these market trends as evidence to support the 'peak oil' thesis—the idea that production is in long-run decline towards the exhaustion of known reserves.[71] In parallel to these market developments, political concern over the security of energy supplies has mounted in the face of growing terrorist threats, political instability in major exporting regions, high-profile disruptions in supply, and disputes between importers and exporters.[72]

Energy security and climate security —pulling in different directions?

The energy security background is important for climate change mitigation strategies. However, hopes that rising prices for fossil fuels will automatically trigger an early transition to a low-carbon future are likely to prove misplaced. Proponents of the 'peak oil' argument overstate their case. New supplies are almost certainly going to be more costly and more difficult to extract and deliver, raising the marginal price of a barrel of oil over time. Yet the world will not run out of oil any time soon: proven reserves could cover four decades of current consumption and much more may be discovered.[73] The bottom line is that there is more than enough affordable fossil fuel available to take the world over the threshold of dangerous climate change.

With current technologies, exploitation of even a small fraction of the Earth's vast reservoir of fossil fuels would guarantee such an outcome. Whatever the pressure on conventional oil sources, proven reserves of oil slightly exceed the volume used since 1750. In the case of coal, known reserves are around 12 times post-1750 use. Using just half of the world's known coal reserves during the 21st Century would add around 400 ppm to atmospheric stocks of greenhouse gases, guaranteeing dangerous climate change in the process.[74] The availability of fossil fuel reserves underlines the case for prudent carbon budget management.

Current market trends reinforce that case. One possible response to the rise in prices for oil and natural gas is a 'dash for coal'. This is the world's cheapest, most widely dispersed and most CO_2-intensive fossil fuel: for each unit of energy generated, coal generates around 40 percent more CO_2 than oil and almost 100 percent more than natural gas. Moreover, coal figures very prominently in the current and future energy profiles of major CO_2 emitters such as China, Germany, India and the United States. Experience in the transition economies points to wider problems. Consider the direction of energy policy in the Ukraine. Over the past 10 to 15 years coal has been steadily replaced by cheaper (and less polluting) imported natural gas. However, with the interruption of supplies from the Russian Federation in early 2006 and the doubling of import prices, the Ukrainian government is considering a shift back towards coal.[75] The case demonstrates the way in which national energy security may conflict with global climate security goals.

Energy demand scenarios confirm that rising fossil fuel prices are not pushing the world towards a sustainable emissions pathway. Demand is projected to increase by half between now and 2030, with over 70 percent of the increase coming from developing countries.[76] These projections suggest that the world will spend around US$20 trillion between 2005 and 2030 in meeting those demands. Much of that investment is still being directed towards carbon-intensive infrastructures that will still be generating energy—and emitting CO_2—in the second half of the 21st Century. The consequences can be assessed by comparing energy-related CO_2 emission scenarios developed by the International Energy Agency (IEA) and the IPCC with our sustainable emissions pathway simulations:

- Our sustainable emissions pathway points to a trajectory that requires a 50 percent cut in greenhouse gas emissions worldwide by 2050 against 1990 levels. The IEA scenario, in contrast, points to an increase of around 100 percent. Between 2004 and 2030 alone, energy-related emissions are projected to increase by 14 Gt CO_2, or 55 percent.

There is more than enough affordable fossil fuel available to take the world over the threshold of dangerous climate change

- While our sustainable emissions pathway points to an indicative target of cuts in the range of at least 80 percent for OECD countries, the IEA reference scenario indicates a 40 percent increase—an aggregate expansion of 4.4 Gt CO_2. The United States will account for around half the increase, taking emissions 48 percent above 1990 levels (figure 1.15).

- According to the IEA, developing countries will account for three-quarters of the increase in global CO_2 emissions, whereas our sustainable emissions pathway points to the need for cuts of around 20 percent by 2050 against 1990 levels. The projected expansion would represent a fourfold increase over 1990 levels.

- While per capita emissions will increase most rapidly in developing countries, convergence will be limited. By 2030, OECD emissions are projected at 12 tonnes of CO_2 per capita, compared with 5 tonnes CO_2 for developing countries. In 2015, per capita emissions from China and India are projected at 5.2 and 1.1 tonnes, compared with 19.3 tonnes for the United States.

- IPCC scenarios are more comprehensive than those developed by the IEA because they incorporate other sources of emissions,

including agriculture, changes in land use, and waste, and a wider range of greenhouse gases. These scenarios point to emission levels of 60–79 Gt CO_2e by 2030, on a sharply rising trend. The lower end of this range is 50 percent above the 1990 baseline. One of the IPCC's non-mitigation scenarios has emissions doubling in the three decades to 2030.[77]

Drivers for increased emissions

As with any future scenario, these figures have to be treated with caution. They represent a best-estimate based on underlying assumptions about economic growth, population change, energy markets, technology and current policies. The scenarios do not chart a predetermined trajectory. What they draw attention to is the hard fact that the world is currently on an emissions trajectory that guarantees a collision between people and planet.

Changing trajectories will be difficult. There are three powerful drivers of rising emissions that will interact with technology, changes in energy markets and public policy choices.

- *Demographic trends.* Current projections point to an increase in world population from 6.5 billion today to 8.5 billion by 2030. At a global level, just standing still in terms of overall emissions will require 30 percent reductions in average per capita emissions—and standing still will not be enough to avoid dangerous climate change. Almost all the increase in population will take place in developing countries, where there are currently large unmet energy needs and lower levels of energy efficiency.

- *Economic growth.* Economic growth and the carbon intensity of growth—a function of energy mix and sectoral composition—are two of the most powerful drivers of emission trends. Any projections in this area are subject to uncertainty. Climate change itself could act as a brake on future growth, especially in the event of catastrophic sea-level

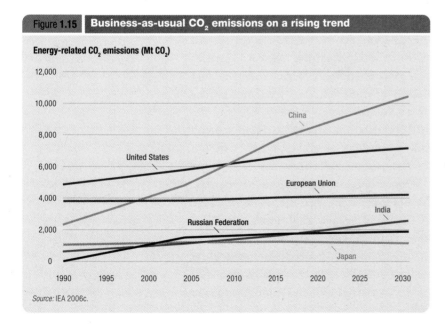

Figure 1.15 | **Business-as-usual CO_2 emissions on a rising trend**

Energy-related CO_2 emissions (Mt CO_2)

China
United States
European Union
Russian Federation
India
Japan

Source: IEA 2006c.

rises or unanticipated 'nasty surprises'. However, that brake may not be applied in the next few decades: most models do not expect climate to have significant effects on the drivers of world growth until towards the end of the 21st Century.[78] More immediately, the global economy is experiencing one of the longest periods of sustained growth in history. World GDP growth has averaged over 4 percent per annum for the past decade.[79] At this rate, output doubles every 18 years, pushing up demand for energy and emissions of CO_2 in the process. The amount of CO_2 generated by every dollar of growth in the world economy—the 'carbon-intensity' of world GDP—has been falling over the past two-and-a-half decades, weakening the link between GDP and carbon emissions. That trend reflects improvements in energy efficiency, changes in economic structure—with the share of carbon-intensive manufacturing falling relative to service sectors in many countries—and changes in the energy mix. However, the decline in carbon intensity has stalled since 2000, creating further upward pressure on emissions (figure 1.16).

- *Energy mix.* For the past quarter of a century, energy-related CO_2 emissions have grown less rapidly than primary energy demand. However, under the IEA scenario, the period to 2030 could see CO_2 emissions rise more rapidly than primary energy demand. The reason: an increase in the share of coal in primary energy demand. Emissions of CO_2 from coal are projected to increase by 2.7 percent a year in the decade to 2015—a rate that is 50 percent higher than for oil.

Achieving climate change mitigation on the scale required in the face of these pressures will require a sustained public policy effort backed by international cooperation. Current trends in energy markets alone are not going to push the world on to a low-carbon trajectory. However, recent market trends and concerns over energy security could provide an impetus towards a low-carbon future. With prices for oil and natural gas set to remain at high levels, the incentives for developing low-carbon energy capacity have moved in a favourable direction. Similarly, governments concerned about 'addiction to oil' and the security of energy supply have strong grounds for advancing programmes aimed at enhancing energy efficiency, creating incentives for the development and deployment of low-carbon technologies, and promoting greater self-reliance through renewable energy. We look in more detail at the mitigation framework in chapter 3. But the four building blocks for success are:

- Putting a price on carbon emissions through taxation and cap-and-trade systems.
- Creating a regulatory framework that enhances energy efficiency, sets standards for reducing emissions and creates market opportunities for low-carbon energy suppliers.
- Agreeing on multilateral international cooperation to finance technology transfers to developing countries supporting a transition to low-carbon energy sources.
- Developing a post-2012 multilateral framework to build on the first phase of the Kyoto Protocol, with far more ambitious targets for cutting greenhouse gas emissions.

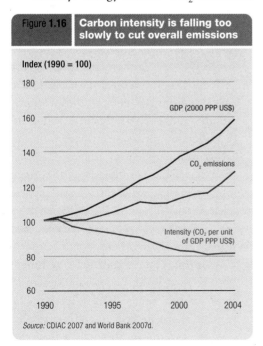

Figure 1.16 Carbon intensity is falling too slowly to cut overall emissions

Index (1990 = 100)

GDP (2000 PPP US$)

CO_2 emissions

Intensity (CO_2 per unit of GDP PPP US$)

Source: CDIAC 2007 and World Bank 2007d.

Policies for mitigating greenhouse gas emissions will require far-reaching changes in energy policy and behaviour

1.6 Why we should act to avoid dangerous climate change

We live in a deeply divided world. Extremes of poverty and prosperity retain the power to shock. Differences in religious and cultural identification are a source of tension between countries and people. Competing nationalisms pose threats to collective security. Against this backdrop, climate change provides a hard lesson in a basic fact of human life: we share the same planet.

Wherever people live and whatever their belief systems, they are part of an ecologically interdependent world. Just as flows of trade and finance are linking people together in an integrated global economy, so climate change draws our attention to the environmental ties that bind us in a shared future.

Climate change is evidence that we are mismanaging that future. Climate security is the ultimate public good: the world's atmosphere is shared by all in the obvious sense that nobody can be 'excluded' from it. By contrast, dangerous climate change is the ultimate public bad. While some people (the world's poor) and some countries stand to lose faster than others, everybody stands to lose in the long run, with future generations facing increased catastrophic risks.

Writing in the 4th Century BC, Aristotle observed that "what is common to the greatest number has the least care bestowed upon it". He could have been commenting on the Earth's atmosphere and the absence of care bestowed on our planet's capacity to absorb carbon. Creating the conditions for change will require new ways of thinking about human interdependence in a world heading for dangerous climate change outcomes.

Climate stewardship in an interdependent world

Tackling climate change confronts governments with difficult choices. Complex issues involving ethics, distributional equity across generations and countries, economics, technology and personal behaviour are at stake. Policies for mitigating greenhouse gas emissions will require far-reaching changes in energy policy and behaviour.

In this chapter we have looked at a range of issues that are important in framing the response to climate change. Four themes merit special emphasis because they go to the heart of the ethics and economics of any public policy framework for mitigation:

- *Irreversibility*. Emissions of CO_2 and other greenhouse gases are, for all practical purposes, irreversible. The duration of their residence in the Earth's atmosphere is measured in centuries. Similar logic applies to climate system impacts. Unlike many other environmental issues, where damage can be cleaned up relatively swiftly, the damage wrought by climate change has the potential to extend from vulnerable populations today across generations to the whole of humanity in the distant future.

- *Global scale*. The climate forcing generated through a build-up of greenhouse gases does not distinguish between nations, even if the effects differ. When a country emits CO_2 the gas flows into a stock that affects the whole world. Greenhouse gas emissions are not the only form of transboundary environmental pollution: acid rain, oil spillages and river pollution also create externalities that cross national borders. What is different with climate change is the scale and the consequence: that no nation state acting alone can solve the problem (even though some countries can do more than others).

- *Uncertainty and catastrophe*. Climate change models deal in probabilities—and probabilities imply uncertainties. The combination of uncertainty and catastrophic risk for future generations is a powerful rationale for investment in risk insurance through mitigation.

- *Near-term human development reversals.* Long before catastrophic events due to climate change impact on humanity, many millions of people will be profoundly affected. It might be possible to protect Amsterdam, Copenhagen and Manhattan from rising sea levels in the 21ˢᵗ Century, albeit at high cost. But coastal flood defences will not save the livelihoods or the homes of hundreds of millions of people living in Bangladesh and Viet Nam or the Niger or Nile deltas. Urgent climate change mitigation would reduce the risks of human development setbacks over the course of the 21ˢᵗ Century, though most of the benefits will occur after 2030. Reducing human costs prior to that date will require support for adaptation.

Social justice and ecological interdependence

There are many theories of social justice and approaches to efficiency that can be brought to bear on climate change debates. Perhaps the most apposite was crafted by the Enlightenment philosopher and economist Adam Smith. In considering how to determine a just and ethical course of action, he suggested a simple test: "to examine our own conduct as we imagine any other fair and impartial spectator would examine it".[80]

Such a "fair and impartial spectator" would take a dim view of a generation that failed to act on climate change. Exposing future generations to potentially catastrophic risks might be considered inconsistent with a commitment to core human values. Article Three of the Universal

---

Special contribution | **Our common future and climate change**

Sustainable development is about meeting the needs of present generations without compromising the ability of future generations to meet their own needs. More than that, it is about social justice, equity and respect for the human rights of future generations.

Two decades have now passed since I had the privilege of chairing the World Commission on the Environment. The Report that emerged from our proceeding had a simple message that was captured in its title, *Our Common Future*. We argued that humanity was overstepping the limits of sustainability and running down the world's ecological assets in a way that would compromise the well-being of future generations. It was also clear that the vast majority of the world's population only had a small share in the overuse of our finite resources. Unequal opportunities and unequal distribution were at the heart of the problems we identified.

Today we need to reflect in detail on climate change. But is there any more powerful demonstration of what it means to live unsustainably?

The *Human Development Report 2007/2008* sets out what it describes as a 'carbon budget' for the 21ˢᵗ Century. Drawing upon the best climate science, that budget establishes the volume of greenhouse gases that can be emitted without causing dangerous climate change. If we continue on our current emissions trajectory, the carbon budget for the 21ˢᵗ Century will expire in the 2030s. Our energy consumption patterns are running up vast ecological debts that will be inherited by future generations—debts that they will be unable to repay.

Climate change is an unprecedented threat. Most immediately, it is a threat to the world's poorest and most vulnerable people: they are already living with the consequences of global warming.

In our already deeply divided world, global warming is magnifying disparities between rich and poor, denying people an opportunity to improve their lives. Looking to the future, climate change poses risks of an ecological catastrophe.

We owe it to the world's poor and to future generations to act with resolve and urgency to stop dangerous climate change. The good news is that it is not too late. There is still a window of opportunity, but let's be clear: the clock is ticking, and time is running out.

Rich nations must show leadership and acknowledge their historic responsibility. Their citizens leave the biggest carbon footprint in the Earth's atmosphere. Moreover, they have the financial and technological capabilities needed to make deep and early cuts in carbon emissions. None of this means that mitigation has to be left to the rich world. Indeed, one of the most urgent priorities is international cooperation on technology transfer to enable developing countries to make the transition to low-carbon energy systems.

Today, climate change is teaching us the hard way some of the lessons that we attempted to communicate in *Our Common Future*. Sustainability is not an abstract idea. It is about finding a balance between people and planet—a balance that addresses the great challenges of poverty today, while protecting the interests of future generations.

Gro H. Brundtland

Gro Harlem Brundtland
Chair of the World Commission on Sustainable Development
Former Prime Minister of Norway

The challenge is to sustain human progress today while facing the incremental risks created by climate change in the lives of a significant section of humanity

Declaration on Human Rights establishes that "everyone has a right to life, liberty and personal security." Inaction in the face of the threat posed by climate change would represent a very immediate violation of that universal right.

The principle of cross-generational equity is at the heart of the idea of sustainability. Two decades have now passed since the World Commission on Environment and Development brought the idea of sustainable development to the centre of the international agenda. The core principle is worth restating, if only to highlight how comprehensively it will be violated by a continued failure to prioritize climate change mitigation: "Sustainable development seeks to meet the needs and aspirations of the present without compromising the ability to meet those of the future."[81]

That vision retains a powerful resonance and an application to public policy debates on climate change. Of course, sustainable development cannot mean that every generation leaves the world's environment exactly as it found it. What need to be conserved are the opportunities for future generations to enjoy substantive freedoms, make choices and lead lives that they value.[82] Climate change will eventually limit those freedoms and choices. It will deny people control over their destinies.

Thinking about the future does not mean that we should think less about social justice in our lifetime. An impartial observer might also reflect on what inaction in the face of climate change might say about attitudes to social justice, poverty and inequality today. The ethical foundation of any society has to be measured partly on the basis of how it treats its most vulnerable members. Allowing the world's poor to bear the brunt of a climate change problem that they did not create would point to a high level of tolerance for inequality and injustice.

In human development terms, the present and the future are connected. There is no long term trade-off between climate change mitigation and the development of human capabilities. As Amartya Sen argues in his special contribution to this Report, human development and environmental sustainability are integral elements in the substantive freedom of human beings.

Tackling climate change with well-designed policies will reflect a commitment to expand the substantive freedoms that people enjoy today without compromising the ability of future generations to build on those freedoms.[83] The challenge is to sustain human progress today while facing the incremental risks created by climate change in the lives of a significant section of humanity.

There is a fundamental sense in which climate change challenges us to think differently about human interdependence. Greek philosophers argued that human affinity could be understood in terms of concentric circles stretching out from family, to locality, country and the world—and weakening with every remove from the centre. Enlightenment economists such as Adam Smith and philosophers such as David Hume sometimes used this framework to explain human motivation. In today's economically and ecologically more interdependent world, the concentric circles have become closer to each other. As the philosopher Kwame Appiah has written: "Each person you know about and affect is someone to whom you have responsibilities: to say this is just to affirm the very idea of morality."[84] Today we "know about" people in far-distant places—and we know about how our use of energy "affects" their lives through climate change.

Viewed from this perspective, climate change poses some tough moral questions. Energy use and the associated emissions of greenhouse gases are not abstract concepts. They are aspects of human interdependence. When a person switches on a light in Europe or an air-conditioning unit in America, they are linked through the global climate system to some of the world's most vulnerable people—to small-scale farmers eking out a living in Ethiopia, to slum dwellers in Manila, and to people living in the Ganges Delta. They are also linked to future generations, not only their own children and grandchildren but also to the children and grandchildren of people across the world. Given the evidence about the implications of dangerous climate change for poverty and future catastrophic risks, it would be a denial of morality to disregard the responsibilities that come with the ecological interdependence that is driving climate change.

The moral imperative to tackle climate change is rooted above all in ideas about stewardship, social justice and ethical responsibility. In a world where people are often divided by their beliefs, these are ideas that cross religious and cultural divides. They provide a potential foundation for collective action by faith group leaders and others (box 1.4).

The economic case for urgent action

Ambitious climate change mitigation requires spending today on a low-carbon transition. The costs will fall predominantly on today's generation, with the rich world facing the biggest bill. Benefits will be distributed across countries and

Box 1.4	Stewardship, ethics and religion—common ground on climate change

"We do not inherit the Earth from our ancestors, we borrow it from our children"

American Indian proverb

Sustainability was not a concept invented at the Earth Summit in 1992. Belief in the values of stewardship, cross-generational justice and shared responsibility for a shared environment underpin a wide range of religious and ethical systems. Religions have a major role to play in highlighting the issues raised by climate change.

They also have the potential to act as agents of change, mobilizing millions of people on the basis of shared values to take action on an issue of fundamental moral concern. While religions vary in their theological or spiritual interpretation of stewardship, they share a common commitment to the core principles of cross-generational justice and concern for the vulnerable.

At a time when the world focuses too often on religious difference as a source of conflict, climate change offers opportunities for inter-faith dialogue and action. With some notable exceptions, religious leaders could do more in the public sphere. One result is that there has been insufficient moral reflection on the issues raised by climate change. The foundations for inter-faith action are rooted in basic scriptures and current teaching:

- *Buddhism.* The Buddhist term for individual is *Santana*, or stream. It is intended to capture the idea of interconnectedness between people and their environment, and between generations. Buddhist teaching places an emphasis on personal responsibility to achieve change in the world through change in personal behaviour.
- *Christianity.* Theologians from a wide range of Christian traditions have taken up the issue of climate change. From a Catholic perspective, the Holy See's Permanent Observer to the UN has called for an "ecological conversion" and "precise commitments that will effectively confront the problem of climate change." The World Council of Churches has issued a powerful and compelling call to action rooted in theological concerns: "The poor and vulnerable communities in the world and future generations will suffer the most from climate change...The rich nations use far more than their fair share of the global commons. They must pay that ecological debt to other peoples by fully compensating them for the costs of adaptation to climate change. Drastic emission reductions by the rich are required

to ensure that the legitimate development needs of the world's poor can be met."

- *Hinduism.* The idea of nature as a sacred construction is deeply rooted in Hinduism. Mahatma Gandhi drew on traditional Hindu values to emphasize the importance of non-violence, respect for all forms of life and harmony between people and nature. Ideas of stewardship are reflected in statements of Hindu faith on ecology. As the spiritual leader Swami Vibudhesha has written: "This generation has no right to use up all the fertility of the soil and leave behind an unproductive land for future generations."
- *Islam.* The primary sources of Islamic teaching about the natural environment are the *Quaran*, the collections of *hadiths*—discrete anecdotes about the Prophet's sayings and actions—and Islamic Law (*al-Sharia*). Because humans are seen as part of nature, a recurrent theme in these sources is opposition to wastefulness and environmental destruction. Islamic Law has numerous injunctions to protect and guard common environmental resources on a shared basis. The Quaranic concept of 'tawheed' or oneness captures the idea of the unity of creation across generations. There is also an injunction that the Earth and its natural resources must be preserved for future generations, with human beings acting as custodians of the natural world. Drawing on these teachings, the Australian Council of Islamic Councils has commented: "God entrusts humans to enjoy the bounty of nature on the strict condition that they take care of it...Time is running out. People of religion must forget their theological differences and work together to save the world from climatic ruin."
- *Judaism.* Many of Judaism's deepest beliefs are consistent with environmental protection. As one theologian puts it, while the Torah may give humanity a privileged place in the order of creation, this is not "the dominion of a tyrant"—and many commandments concern the preservation of the natural environment. Applying Judaic philosophy to climate change, the Central Conference of American Rabbis has commented: "We have a solemn obligation to do whatever we can within reason to prevent harm to current and future generations and to preserve the integrity of creation... Not to do so when we have the technological capacity—as in the case of non-fossil fuel energy and transport technologies—is an unforgivable abdication of our responsibilities."

Source: Climate Institute 2006; IFEES 2006; Krznaric 2007.

1

Do the costs and benefits
of climate change
mitigation support the
case for urgent action?

time. Future generations will gain from lower risks and the world's poor will benefit from enhanced prospects for human development within our own lifetime. Do the costs and benefits of climate change mitigation support the case for urgent action?

That question was addressed by the Stern Review on *The Economics of Climate Change*. Commissioned by the United Kingdom Government, the Review provided a strong response. Using cost–benefit analysis based on long-run economic modelling it concluded that the future costs of global warming would be likely to fall between 5 and 20 percent of annual world GDP. These future losses could be avoided, according to the review analysis, by incurring relatively modest annual mitigation costs of around 1 percent of GDP to achieve greenhouse gas stabilization at 550 ppm CO_2e (rather than the more ambitious 450 ppm advocated in this Report). The conclusion: an overwhelming case for urgent, immediate, and rapid reductions in emissions of greenhouse gases on the grounds that prevention is better, and cheaper, than inaction.

Some critics of the Stern Review have reached different conclusions. They maintain that cost–benefit analysis does not support the case for early and deep mitigation. The counterarguments are wide-ranging. The Stern Review and its critics start from a similar proposition: namely, that the real global damages from climate change, whatever their level, will be incurred far into the future. Where they differ is in their evaluation of these damages. The Stern review's critics argue that the welfare of people living in the future should be discounted at a higher rate. That is, it should receive less weight than allowed for in the Stern Review compared to costs incurred in the present.

Policy prescriptions emerging from these opposing positions are different.[85] Unlike the Stern review, the critics argue for a modest rate of emission reductions in the near future, followed by sharper reductions in the longer term as the world economy grows richer—and as technological capacities develop over time.[86]

The ongoing debate following the Stern review matters at many levels. It matters most

immediately because it goes to the heart of the central question facing policymakers today: namely, should we act with urgency now to mitigate climate change? And it matters because it raises questions about the interface of economics and ethics—questions that have a bearing on how we think about human interdependence in the face of the threats posed by dangerous climate change.

Discounting the future—ethics and economics

Much of the controversy has centred on the concept of social discounting. Because climate change mitigation implies current costs to generate future benefits, one critical aspect of the analysis is about how to treat future outcome relative to present outcome. At what rate should future impacts be discounted to the present? The discount rate is the tool used to address that question. Determining the rate involves placing a value on future welfare simply because it is in the future (the rate of pure time preference). It also involves a decision on the social value of an extra dollar in consumption. This second element captures the idea of diminishing marginal utility as incomes rise.[87]

The argument between the Stern review and its critics over the costs and benefits of mitigation—and the timing of action—can be attributed in large measure to the discount rate. To understand why the different approaches matter for climate change mitigation, consider the following example. At a discount rate of 5 percent, it would be worth spending only US$9 today to prevent an income loss of US$100 caused by climate change in 2057. Without any discounting, it would be worth spending up to US$100 today. So, as the discount rate goes up from zero, the future damages from warming evaluated today shrink. Applied over the long time-horizon necessary for considering climate change impacts, the magic of compound interest in reverse can generate a strong cost–benefit case for deferred action on mitigation, if discount rates are high.

From a human development perspective, we believe that the Stern review is right in its central choice for a low value for the rate of pure time preference—the component of the discount rate that weighs the welfare of future generations in comparison with ours.[88] Discounting the well-being of those that will live in the future just because they live in the future is unjustified.[89] How we think about the well-being of future generations is an ethical judgement. Indeed, the founding father of discounting described a positive rate of pure time preference as a practice which is "ethically indefensible and arises merely from the weakness of the imagination".[90] Just as we do not discount the human rights of future generations because they are equivalent to ours, so we should accept a 'stewardship of the earth' responsibility to accord future generations the same ethical weight as the current generation. Selecting a 2 percent rate of pure time preference would halve the ethical weight given to somebody born in 2043 relative to somebody born in 2008.[91]

Denying the case for action today on the grounds that future generations with a lower weight should be expected to shoulder a greater burden of mitigation costs is not an ethically defensible proposition—and it is inconsistent with the moral responsibilities that come with membership of a human community linked across generations. Ethical principles are the primary vehicle through which the interests of people not represented in the market place (future generations) or lacking a voice (the very young) are brought into policy formulation. That is why the issue of ethics has to be addressed explicitly and transparently in determining approaches to mitigation.[92]

Uncertainty, risk and irreversibility— the case for catastrophic risk insurance

Any consideration of the case for and against urgent action on climate change has to start from an assessment of the nature and timing of the risks involved. Uncertainty is critical to the argument.

As shown earlier in this chapter, uncertainty under climate change is closely associated with the possibility of catastrophic outcomes. In a world that has more chance of going over

5°C than staying under 2°C, 'nasty surprises' of a catastrophic nature will become more probable over time. The impact of those surprises is uncertain. However, they include possible disintegration of the West Antarctic ice sheet, with attendant implications for human settlements and economic activity. Ambitious mitigation can be justified as a down payment on catastrophic risk insurance for future generations.[93]

Catastrophic risks of the order posed by climate change provide grounds for early action. The idea that costly actions today should be deferred until more is known is not applied to other areas. In dealing with national defence and protection against terrorism, governments do not refuse to put in place investments today because they are uncertain about the future benefits of those investments, or the precise nature of future risks. Rather, they assess risks and determine on the balance of probabilities whether there is sufficient likelihood of severe future damage to take anticipatory action aimed at risk reduction.[94] That is, they weigh-up the costs, the benefits and the risks, and try to insure their citizens against uncertain but potentially catastrophic outcomes.

The case against urgent action on climate change suffers from wider shortcomings. There are many areas of public policy in which a 'wait-and-see' approach might make sense—but climate change is not one of them. Because the accumulation of greenhouse gases is cumulative and irreversible, policy errors cannot be readily corrected. Once CO_2e emissions have reached, say, 750 ppm, future generations will not enjoy the option of expressing a preference for a world that stabilized at 450 ppm. Waiting to see whether the collapse of the West Antarctic ice sheet produces catastrophic outcomes is a one-way option: ice sheets cannot be reconnected to the bottom of the sea. The irreversibility of climate change places a high premium on the application of the precautionary principle. And the potential for genuinely catastrophic outcomes in an area marked by large areas of uncertainty makes the use of marginal analysis a restrictive framework for the formulation of

In dealing with national defence and protection against terrorism, governments do not refuse to put in place investments today because they are uncertain about the future benefits of those investments, or the precise nature of future risks

The costs of delayed mitigation will not be equally spread across countries and people

responses to the challenge of climate change mitigation. To put it differently: a small probability of an infinite loss can still represent a very big risk.

Beyond one world—why distribution matters

There has also been a debate on the second aspect of the discount rate. How should we weight the value of an extra dollar of consumption in the future if the overall amount of consumption is different from today's? Most people who would accord the same ethical weight to future generations would agree that, if those generations were going to be more prosperous, an increase in their consumption should be worth less than it is today. As income increases over time, the question arises as to the value of an additional dollar. How much we discount increasing consumption in the future depends on social preference: the value attached to the additional dollar. The critics of the Stern review have argued that its choice of parameter was too low, leading in turn to what is, in their eyes, an unrealistically low overall discount rate. The issues relating to this part of the debate are different from those relating to pure time preference and involve projected growth scenarios under conditions of great uncertainty.

If the world were a single country with an ethical concern for the future of its citizens, it should be investing heavily in catastrophic risk insurance through climate change mitigation. In the real world, the costs of delayed mitigation will not be equally spread across countries and people. The social and economic impacts of climate change will fall far more heavily on the poorest countries and their most vulnerable citizens. Distributional concerns linked to human development greatly reinforce the case for urgent action. In fact, these concerns represent one of the most critical parts of that case. This point is widely ignored by those arguing about discount rates in 'one world' models.

Global cost–benefit analysis without distribution weights can obscure the issues in thinking about climate change. Small impacts on the economies of rich countries (or rich

people) register more strongly on the cost–benefit balance sheet precisely because they are richer. The point can be illustrated by a simple example. If the 2.6 billion poorest people in the world saw their incomes cut by 20 percent, per capita world GDP would fall by less than 1 percent. Similarly, if climate change led to a drought that halved the income of the poorest 28 million people in Ethiopia, it would barely register on the global balance sheet: world GDP would fall by just 0.003 percent. There are also problems in what cost–benefit analysis does not measure. The value that we attach to things which are intrinsically important are not easily captured by market prices (box 1.5).

Distributional imperatives are often overlooked in the case for action on climate change mitigation. As with the wider debate on discounting, the weighting of consumption gains and losses for people and countries with different levels of income must be explicitly considered. There is, however, a key difference between the distribution issues relating to intergeneration distribution and those relating to distribution between current populations. In the former, the case for ambitious mitigation rests on the need to insure against uncertain but potentially catastrophic risk. In the latter case of distribution of income in our lifetimes, it rests in the 'certain' costs of climate change for the livelihoods of the poorest people in the world.[95]

Concern for distributional outcomes between countries and people at very different levels of development is not restricted to mitigation. Mitigation today will create a steady flow of human development benefits that strengthen in the second half of the 21ˢᵗ Century. In the absence of urgent mitigation, poverty reduction efforts will suffer and many millions of people will face catastrophic outcomes. Mass displacement due to flooding in countries like Bangladesh and mass hunger linked to drought in sub-Saharan Africa are two examples.

However, there is no neat dividing line between present and future. Climate change is already impacting on the lives of the poor and the world is committed to further climate change irrespective of mitigation efforts.

What this means is that mitigation alone will not provide a safeguard against adverse distributional outcomes linked to climate change—and that, for the first half of the 21ˢᵗ Century, adaptation to climate change must be a priority, alongside ambitious mitigation efforts.

Mobilizing public action

Through the work of the IPCC and others, climate science has improved our understanding of global warming. Debates on the economics of climate change have helped to identify choices over resource allocation. In the end though, it is public concern that will drive policy change.

Public opinion—a force for change

Public opinion matters at many levels. An informed public understanding of why climate change is such an urgent priority can create the political space for governments to introduce radical energy reforms. As in many other areas,

public scrutiny of government policies is also critical. In the absence of scrutiny, there is a danger that high-sounding declarations of intent will substitute for meaningful policy action—a perennial problem with G8 commitments on aid to developing countries. Climate change poses a distinctive challenge because, perhaps more than in any other sphere of public policy, the reform process has to be sustained over a long time-horizon.

Powerful new coalitions for change are emerging. In the United States, the Climate Change Coalition has brought together non-government organizations (NGOs), business leaders and bipartisan research institutions. Across Europe, NGOs and church-based groups are building powerful campaigns for urgent action. 'Stop Climate Chaos' has become a statement of intent and a rallying point for mobilization. At an international level, the Global Climate Campaign is building a network that mobilizes across national borders, bringing pressure to bear on governments before,

---

Box 1.5   Cost–benefit analysis and climate change

Much of the debate over the case for and against urgent mitigation has been conducted in terms of cost–benefit analysis. Important issues have been raised. At the same time, the limitations of cost–benefit approaches have to be acknowledged. The framework is essential as an aide to rational decision making. But it has severe limitations in the context of climate change analysis and cannot by itself resolve fundamental ethical questions.

One of the difficulties with the application of cost–benefit analysis to climate change is the time-horizon. Any cost–benefit analysis is a study in uncertainty. Applied to climate change mitigation, the range of uncertainty is very large. Projecting costs and benefits over a 10- or 20-year period can be challenging even for simple investment projects such as building a road. Projecting them over 100 years and more is a largely speculative exercise. As one commentator puts it: "Trying to forecast costs and benefits of climate change scenarios a hundred years from now is more the art of inspired guesstimating by analogy than a science."

The more fundamental problem concerns what is being measured. Changes in GDP provide a yardstick for measuring an important aspect of the economic health of nations. Even here there are limitations. National income accounts record changes in wealth and the depreciation of the capital stock used in its creation. They do not capture the costs of environmental damage or the depreciation of ecological assets such as forests or water resources. Applied to

climate change, the wealth generated through energy use shows up in national income, the damage associated with the depletion of the Earth's carbon sinks does not.

Abraham Maslow, the great psychologist, once said: "If the only tool you have is a hammer, every problem begins to look like a nail." In the same way, if the only tool used to measure cost is a market price, things that lack a price tag—the survival of species, a clean river, standing forests, wilderness—might look like they have no value. Items not in the balance sheet can become invisible, even though they have great intrinsic value for present and future generations. There are some things that, once lost, no amount of money can bring back. And there are some things that do not lend themselves to market pricing. For these things asking questions just through cost–benefit analysis can produce the wrong answers.

Climate change touches in a fundamental way on the relationship between people and ecological systems. Oscar Wilde once defined a cynic as "someone who knows the price of everything and the value of nothing". Many of the impacts that will come with unmitigated climate change will touch upon aspects of human life and the environment that are intrinsically valuable—and that cannot be reduced to the economics of the ledger sheet. That, ultimately, is why investment decisions on climate change mitigation cannot be treated in the same way as investment decisions (or discount rates) applied to cars, industrial machines or dishwashers.

Source: Broome 2006b; Monbiot 2006; Singer 2002; Weitzman 2007.

For all the progress that has been achieved, the battle for public hearts and minds is not yet won

during and after high-level intergovernmental meetings. As little as five years ago, most large multinational companies were either indifferent or hostile to advocacy on climate change. Now an increasing number are pressing for action and calling for clear government signals to support mitigation. Many business leaders have realized that current trends are unsustainable and that they need to steer their investment decisions in a more sustainable direction.

Throughout history public campaigns have been a formidable force for change. From the abolition of slavery, through struggles for democracy, civil rights, gender equity and human rights, to the *Make Poverty History* campaign, public mobilization has created new opportunities for human development. The specific challenge facing campaigners on climate change is rooted in the nature of the problem. Time is running out, failure will lead to irreversible setbacks in human development, and policy change has to be sustained across many countries over a long period of time. There is no 'quick fix' scenario.

Opinion surveys tell a worrying story

For all the progress that has been achieved, the battle for public hearts and minds is not yet won. Assessing the state of that battle is difficult. Yet opinion surveys tell a worrying story—especially in the world's richest nations.

Climate change now figures prominently in public debates across the developed world. Media coverage has climbed to unprecedented levels. The film *An Inconvenient Truth* has reached an audience of millions. Successive reports—the Stern review being an outstanding example—have narrowed the space between popular understanding and rigorous economic analysis. The planet health warnings set out by the IPCC provide a clear basis for understanding the evidence on climate change. In the face of all of this, public attitudes continue to be dominated by a mindset that combines apathy and pessimism.

Headline numbers from recent surveys demonstrate the point. One major cross-country survey found that people in the developed world see climate change as a far less pressing

threat than people in the developing world. For example, only 22 percent of Britons saw climate change as "one of the biggest issues" facing the world, compared with almost one-half in China and two-thirds in India. Developing countries dominated the ranking for countries whose citizens see climate change as the world's most worrying concern, with Brazil, China and Mexico topping the league table. The same survey found a far higher level of fatalism in rich countries, with a high level of scepticism about the prospects for avoiding climate change.[96]

Detailed national level surveys confirm these broad global findings. In the United States, climate change mitigation is now a subject of intense debate in Congress. However, the current state of public opinion does not provide a secure foundation for urgent action:

- Roughly four in ten Americans believe that human activity is responsible for global warming, but just as many believe that warming can be traced to natural patterns in the Earth's climate systems alone (21 percent) or that there is no evidence of global warming (20 percent).[97]

- While 41 percent of Americans see climate change as a "serious problem", 33 percent see it as only "somewhat serious" and 24 percent as "not serious". Only 19 percent expressed a great deal of personal concern—a far lower level than in other G8 countries and dramatically lower than in many developing countries.[98]

- Concern remains divided along party-political lines. Democrat voters register higher levels of concern than Republican voters, though neither locates climate change near the top of their list of priorities. On a ranking scale of 19 electoral issues, climate change registered 13ᵗʰ for Democrats and 19ᵗʰ for Republicans.

- Moderate levels of public concern are linked to perceptions of where risks and vulnerabilities are located. In a ranking of public concerns, only 13 percent of people covered were most concerned about impacts on their family or community, while half saw the most immediate impacts as affecting people in other countries, or nature.[99]

Caution has to be exercised in interpreting opinion survey evidence. Public opinion is not static and it may be changing. There is some positive news. Some 90 percent of Americans who have heard of global warming think that the country should reduce its greenhouse gas emissions, regardless of what other countries do.[100] Even so, if "all politics is local", then current public risk assessments are unlikely to provide a powerful political impetus. Climate change is still perceived overwhelmingly as a moderate and distant risk that will primarily impact people and places far away in space and time.[101]

Evidence that European opinion is far ahead of American opinion is not corroborated by opinion survey evidence. More than eight in every ten European Union citizens are aware that the way they consume and produce energy has a negative impact on climate.[102] Yet only half say that they are "to some degree concerned"—a far higher share express concern about the need for Europe to have greater diversity in energy supply.

In some European countries, public attitudes are marked by an extraordinary degree of pessimism. For example, in France, Germany and the United Kingdom the share of people agreeing with the statement that "we will stop climate change" ranges from 5 to 11 percent. Alarmingly, four in every ten people in Germany thought that it was not even worth trying to do anything, most of them on the grounds that nothing can be done.[103] All of this suggests a strong case for a greater emphasis on public education and campaigning.

The evidence from opinion surveys is worrying at several levels. It raises questions first of all about the understanding of people in rich nations about the consequences of their actions. If the public had a clearer understanding of the consequences of their actions for future generations, and for vulnerable people in developing countries, the imperative to act might be expected to register far more strongly. The fact that so many people see climate change as an intractable problem is another barrier to action because it creates a sense of powerlessness.

The role of the media

The media have a critical role to play in informing and changing public opinion. Apart from their role in scrutinizing government actions and holding policymakers to account, the media are the main source of information for the general public on climate change science. Given the immense importance of the issues at stake for people and planet, this is a role that carries great responsibilities.

The development of new technologies and globalized networks has enhanced the power of the media across the world. No government in a democracy can ignore the media. But power and responsibility have not always gone together. Speaking in 1998, Carl Bernstein said: "The reality is that the media are probably the most powerful of all our institutions today and they, or rather we [journalists], too often are squandering our power and ignoring our obligations."[104] That observation has a powerful resonance for the debate on climate change.

There are very large variations in the way that the media within and across countries have responded to climate change. Many journalists and many media organs have performed an extraordinary service in keeping public debates alive and deepening knowledge. However, the flip side has to be acknowledged. Until recently, the principle of 'editorial balance' has been applied in ways that have served to hold back informed debate. One study in the United States[105] found that the balance norm resulted in over half of articles in the country's most prestigious newspapers between 1990 and 2002 giving equal weight to the findings of the IPCC and of the climate science community, and the views of climate sceptics—many of them funded by vested interest groups. Continued confusion in public opinion is one consequence.[106]

Editorial balance is a laudable and essential objective in any free press. But balance between what? If there is a strong and overwhelming 'majority' view among the world's top scientists dealing with climate change, citizens have a right to expect to be informed about that view. Of course, they also have a right to be informed about minority views that do not reflect a scientific consensus. However, informed judgement

Dangerous climate change is a predictable crisis that comes with an opportunity

is not helped when editorial selection treats the two views as equivalent.

Media coverage of climate change has suffered from wider problems. Many of the issues that have to be addressed are enormously complex and inherently difficult to communicate. Some media reporting has clouded public understanding. For example, there has been a far stronger focus on catastrophic risk, than on more immediate human development threats—and in many cases the two dimensions are confused.

Over the past two years the quantity of climate change coverage has increased and the quality has improved. But in some areas media treatment continues to hold back informed debate. Sharp peaks in attention during weather-related disasters or around the launch of key reports are often followed by lengthy troughs in coverage. The tendency to focus on emergencies today and apocalyptic future events obscures an important fact: that the most damaging medium-term effects of climate change will take the form of gradually intensifying pressures on highly vulnerable people. Meanwhile, the responsibility of people and governments in rich countries for these pressures is a heavily under-represented theme. One consequence is that public awareness of the importance of support for adaptation measures to build resilience remains limited—as does international development assistance for adaptation.

Conclusion

The science of climate change has established a clear and reasonable target for international action. That target is a threshold for average temperature increases of 2°C. The Stern review has provided a powerful economic rationale for action. The proposition that the battle against climate change is affordable and winnable is one that has achieved powerful traction with policymakers.

The argument for long-run insurance against catastrophic risk and the human development imperative provide powerful rationales for action. Mitigation of climate change poses real financial, technological and political challenges. But it also asks profound moral and ethical questions of our generation. In the face of clear evidence that inaction will hurt millions of people and consign them to lives of poverty and vulnerability, can we justify inaction? No civilized community adhering to even the most rudimentary ethical standards would answer that question in the affirmative, especially one that lacked neither the technology nor the financial resources to act decisively.

Dangerous climate change is a predictable crisis that comes with an opportunity. That opportunity is provided by negotiations on the Kyoto Protocol. Under a revitalized post-2012 multilateral framework, the Protocol could provide a focal point for deep cuts in emissions, allied to a plan of action on adaptation that deals with the consequences of past emissions.

| Appendix table 1.1 | Measuring the global carbon footprint—selected countries and regions | 1 |

| | Carbon dioxide emissions[a] | | | | | | | | |
| Top 30 CO₂ emitters | Total emissions (Mt CO₂) | | Growth rate (%) | Share of world total (%) | | Population share (%) | CO₂ emissions per capita (t CO₂) | | CO₂ emissions or sequestration from forests[b] (Mt CO₂ / year) |
	1990	2004	1990–2004	1990	2004	2004	1990	2004	1990–2005
1 United States	4,818	6,046	25	21.2	20.9	4.6	19.3	20.6	-500
2 China[c]	2,399	5,007	109	10.6	17.3	20.0	2.1	3.8	-335
3 Russian Federation	1,984[d]	1,524	-23[d]	8.7[d]	5.3	2.2	13.4[d]	10.6	72
4 India	682	1,342	97	3.0	4.6	17.1	0.8	1.2	-41
5 Japan	1,071	1,257	17	4.7	4.3	2.0	8.7	9.9	-118
6 Germany	980	808	-18	4.3	2.8	1.3	12.3	9.8	-75
7 Canada	416	639	54	1.8	2.2	0.5	15.0	20.0	..
8 United Kingdom	579	587	1	2.6	2.0	0.9	10.0	9.8	-4
9 Korea (Republic of)	241	465	93	1.1	1.6	0.7	5.6	9.7	-32
10 Italy	390	450	15	1.7	1.6	0.9	6.9	7.8	-52
11 Mexico	413	438	6	1.8	1.5	1.6	5.0	4.2	..
12 South Africa	332	437	32	1.5	1.5	0.7	9.1	9.8	(.)
13 Iran (Islamic Republic of)	218	433	99	1.0	1.5	1.1	4.0	6.4	-2
14 Indonesia	214	378	77	0.9	1.3	3.4	1.2	1.7	2,271
15 France	364	373	3	1.6	1.3	0.9	6.4	6.0	-44
16 Brazil	210	332	58	0.9	1.1	2.8	1.4	1.8	1,111
17 Spain	212	330	56	0.9	1.1	0.7	5.5	7.6	-28
18 Ukraine	600[d]	330	-45[d]	2.6[d]	1.1	0.7	11.5[d]	7.0	-60
19 Australia	278	327	17	1.2	1.1	0.3	16.3	16.2	..
20 Saudi Arabia	255	308	21	1.1	1.1	0.4	15.9	13.6	(.)
21 Poland	348	307	-12	1.5	1.1	0.6	9.1	8.0	-44
22 Thailand	96	268	180	0.4	0.9	1.0	1.7	4.2	18
23 Turkey	146	226	55	0.6	0.8	1.1	2.6	3.2	-18
24 Kazakhstan	259[d]	200	-23[d]	1.1[d]	0.7	0.2	15.7[d]	13.3	(.)
25 Algeria	77	194	152	0.3	0.7	0.5	3.0	5.5	-6
26 Malaysia	55	177	221	0.2	0.6	0.4	3.0	7.5	3
27 Venezuela (Bolivarian Republic of)	117	173	47	0.5	0.6	0.4	6.0	6.6	..
28 Egypt	75	158	110	0.3	0.5	1.1	1.5	2.3	-1
29 United Arab Emirates	55	149	173	0.2	0.5	0.1	27.2	34.1	-1
30 Netherlands	141	142	1	0.6	0.5	0.2	9.4	8.7	-1
World aggregates									
OECD[e]	11,205	13,319	19	49	46	18	10.8	11.5	-1,000
Central & Eastern Europe & CIS	4,182	3,168	-24	18	11	6	10.3	7.9	-166
Developing countries	6,833	12,303	80	30	42	79	1.7	2.4	5,092
East Asia and the Pacific	3,414	6,682	96	15	23	30	2.1	3.5	2,294
South Asia	991	1,955	97	4	7	24	0.8	1.3	-49
Latin America & the Caribbean	1,088	1,423	31	5	5	8	2.5	2.6	1,667
Arab States	734	1,348	84	3	5	5	3.3	4.5	44
Sub-Saharan Africa	456	663	45	2	2	11	1.0	1.0	1,154
Least developed countries	74	146	97	(.)	1	11	0.2	0.2	1,098
High human development	14,495	16,616	15	64	57	25	9.8	10.1	90
Medium human development	5,946	10,215	72	26	35	64	1.8	2.5	3,027
Low human development	78	162	108	(.)	1	8	0.3	0.3	858
High income	10,572	12,975	23	47	45	15	12.1	13.3	-937
Middle income	8,971	12,163	36	40	42	47	3.4	4.0	3,693
Low income	1,325	2,084	57	6	7	37	0.8	0.9	1,275
World	22,703[f]	28,983[f]	28	100[f]	100[f]	100	4.3	4.5	4,038

NOTES

a Data refer to carbon dioxide emissions stemming from the consumption of solid, liquid and gaseous fossil fuels and from gas flaring and production of cement.

b Data refer only to living biomass—above and below ground, carbon in deadwood, soil and litter are not included. Refer to annual average net emissions or sequestration due to changes in carbon stock of forest biomass. A positive number suggests carbon emissions while a negative number suggests carbon sequestration.

c CO₂ emissions for China do not include emissions for Taiwan, Province of China, which were 124 Mt CO₂ in 1990 and 241 Mt CO₂ in 2004.

d Data refer to 1992 and growth rate values refer to the 1992–2004 period .

e OECD as a region includes the following countries that are also included in other subregions listed here: Czech Republic, Hungary, Mexico, Poland, Republic of Korea and Slovakia. Therefore, in some instances, the sum of individual regions may be greater than the world total.

f The world total includes carbon dioxide emissions not included in national totals, such as those from bunker fuels and oxidation of non-fuel hydrocarbon products (e.g., asphalt), and emissions by countries not shown in the main indicator tables. These emissions amount to approximately 5% of the world total.

Source: Indicator Table 24.

2

Climate shocks:
risk and vulnerability
in an unequal world

"The countries most vulnerable are least able to protect themselves. They also contribute least to the global emissions of greenhouse gases. Without action they will pay a high price for the actions of others."

Kofi Annan

"Like slavery and apartheid, poverty is not natural. It is man-made and it can be overcome and eradicated by the actions of human beings."

Nelson Mandela

It is easy to lose sight of the human face of the people who are most vulnerable to climate change

"Hurricane Jeanne took all that I had...my job and my home are gone. I used to have food. Now I beg in the market."

Rosy-Claire Zepherin, Gonaives, Haiti, 2005[1]

"We are eating only a little once a day to make the maize last longer, but even then it will last only a short time. Then we are in trouble."

Margaret Mpondi, Mphako, Malawi, 2002[2]

"If the rains fail like they did last year we will go hungry. The rich have savings. They have stocks of food. They can sell their oxen for cash. But what do I have? If I sell my ox how will I plant next year? If my crop fails we have nothing. It is always like that. Everything depends on rain."

Kaseyitu Agumas, Lat Gayin, southern Gonda, Ethiopia, 2007[3]

"We had never seen such floods before. Lots of houses were destroyed, lots of people died, our agricultural land was submerged, crops stored in houses were lost. Many livestock were lost too. We were just not prepared to face such big flooding. So we didn't have any savings of money or food."

Pulnima Ghosh Mahishura Gram Panchayat, Nadia District, West Bengal, India, 2007[4]

"There are more floods now and the river banks are being washed away faster. There's nowhere to go. My land is in the river, I have nothing now."

Intsar Husain, Antar Para, north-western Bangladesh, 2007.[5]

Climate science deals in measurement. Emissions of carbon dioxide (CO_2) are weighed in tonnes and gigatonnes. Concentrations of greenhouse gases in the Earth's atmosphere are monitored in parts per million (ppm). Confronted with the data, it is easy to lose sight of the human face of the people who are most vulnerable to climate change—people such as those quoted above.

The human face of climate change cannot be captured and packaged in statistics. Many of the current impacts are impossible to separate from wider pressures. Others will happen in the future. There is uncertainty about the location,

What the world's poor are facing is a relentless increase in the risks and vulnerabilities associated with climate

timing and magnitude of these impacts. However, uncertainty is not a cause for complacency. We know that climate-related risks are a major cause of human suffering, poverty and diminished opportunity. We know that climate change is implicated. And we know that the threat will intensify over time. In chapter 1 we identify catastrophic future risks for the whole of humanity as one of the most powerful grounds for urgent action in tackling climate change. In this chapter we focus on a more immediate potential catastrophe: the prospect of large-scale human development reversals in the world's poorest countries.

That catastrophe will not announce itself as a 'big bang' apocalyptic event. What the world's poor are facing is a relentless increase in the risks and vulnerabilities associated with climate. The source of these incremental risks can be traced through climate change to energy consumption patterns and political choices in the rich world.

The climate already figures as a powerful force in shaping the life chances of poor people. In many countries, poverty is intimately related to repeated exposure to climate risks. For people whose livelihoods depend on agriculture, variable and uncertain rainfall is a potent source of vulnerability. For urban slum dwellers, floods pose a constant threat. Across the world, the lives of the poor are punctuated by the risks and vulnerabilities that come with an uncertain climate. Climate change will gradually ratchet up these risks and vulnerabilities, putting pressure on already over-stretched coping strategies and magnifying inequalities based on gender and other markers for disadvantage.

The scale of the potential human development reversals that climate change will bring has been heavily underestimated. Extreme climate events such as droughts, floods and cyclones are terrible occurrences in their own right. They bring suffering, distress and misery to the lives of those affected, subjecting whole communities to forces beyond their control and providing a constant reminder of human frailty. When climate shocks strike, people must first deal with the immediate consequences: threats to health and nutrition, the loss of savings and assets, damage to property, or the destruction of

crops. The short-term costs can have devastating and highly visible consequences for human development.

The long-term impacts are less visible but no less devastating. For the 2.6 billion people who live on less than US$2 a day climate shocks can trigger powerful downward spirals in human development. Whereas the rich can cope with shocks through private insurance, by selling off assets or by drawing on their savings, the poor face a different set of choices. They may have no alternative but to reduce consumption, cut nutrition, take children out of school, or sell the productive assets on which their recovery depends. These are choices that limit human capabilities and reinforce inequalities.

As Amartya Sen has written: "The enhancement of human capabilities also tends to go with an expansion of productivities and earning power."[6] The erosion of human capabilities has the opposite effect. Setbacks in nutrition, health and education are intrinsically damaging, reducing the prospects for employment and economic advancement. When children are withdrawn from school to help their parents make up income losses, or suffer malnutrition because of reduced food availability, the consequences can stay with them for their whole lives. And when poor people suddenly lose the assets they have built up over years, this reinforces their poverty and holds back efforts to reduce vulnerability and extreme deprivation in the medium to longer term. Single climate shocks can thus create cumulative cycles of disadvantage that are transmitted across generations.

Climate change matters because it can be expected to increase the intensity and frequency of climate shocks. Over the medium and long term, outcomes will be influenced by the international mitigation effort. Deep and early cuts in carbon emissions would diminish the incremental risks associated with climate change from the 2030s onwards. Until then, the world in general, and the world's poor in particular, will have to live with the consequences of past emissions. That is why, as argued in chapter 4, adaptation strategies are so critical for human development prospects.

In this chapter we look at the past impacts of climate shocks on human development in order to cast a light on future threats. We draw a critical distinction between risk and vulnerability. Climate risk is an external fact of life for the entire world. Vulnerability is something very different. It describes an inability to manage risk without being forced to make choices that compromise human well-being over time. Climate change will strengthen the transmission mechanisms that convert risk into vulnerability, militating against the efforts of the poor to advance human development.

The first section of this chapter sets out the evidence on a range of climate impacts. It examines the distribution of exposure to climate disasters and the long-run consequences of these disasters on human development. In the second section we use climate scenarios developed by the IPCC and others to examine the mechanisms through which the incremental risks generated by climate change might impact on human development during the 21st Century.

Climate risk is an external fact of life for the entire world. Vulnerability is something very different

2.1 Climate shocks and low human development traps

Climate disasters have been a recurrent theme in human history. Plato's Atlantis myth captures the destructive power of floods. The collapse of the Mayan civilization was triggered by a succession of droughts. The 21st Century has already provided some potent reminders of the frailty of people in the face of extreme climate.

Climate disasters are increasing in frequency and touching the lives of more people. The immediate consequences are horrific. But climate shocks are also reinforcing wider risks and vulnerabilities, leading to long-term setbacks for human development.

Climate disasters—the rising trend

Extreme climate events are a source of mounting concern across the world. In recent decades, the number of people affected by climate disasters such as droughts, floods and storms has been rising. Almost every disaster is accompanied by speculation about possible links to climate change. As climate science develops it will provide clearer insights into the relationship between global warming and weather system outcomes. However, current evidence points very clearly in one direction: namely, that climate change will increase the risk of exposure to climate disaster.

Reported climate disasters are on a rising trend. Between 2000 and 2004 an average of 326 climate disasters was reported each year.

Some 262 million people were affected annually from 2000 to 2004, more than double the level in the first half of the 1980s (figure 2.1).[7]

Rich countries have registered a mounting roll-call of climate disasters. During 2003, Europe was hit by the most intense heat wave in more than 50 years—an event that caused thousands of deaths among the elderly and other vulnerable people. A year later, Japan was hit by

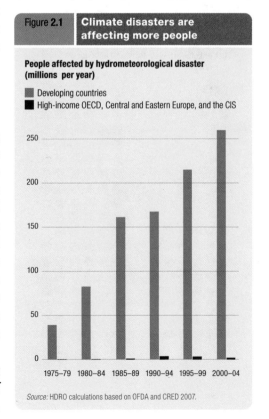

Figure **2.1** | **Climate disasters are affecting more people**

People affected by hydrometeorological disaster (millions per year)

■ Developing countries
■ High-income OECD, Central and Eastern Europe, and the CIS

1975–79 1980–84 1985–89 1990–94 1995–99 2000–04

Source: HDRO calculations based on OFDA and CRED 2007.

For the period 2000–2004, on an average annual basis one in 19 people living in the developing world was affected by a climate disaster

more tropical cyclones than in any other year over the previous century.[8] In 2005, Hurricane Katrina, one event in the worst Atlantic hurricane season on record, provided a devastating reminder that even the world's richest nations are not immune to climate disaster.[9]

The intensive media coverage that accompanies climate disasters in rich countries ensures widespread public awareness of the impacts. It also creates a distorting prism. While climate disasters are affecting more and more people across the world, the overwhelming majority lives in developing countries (figure 2.2). For the period 2000–2004, on an average annual basis one in 19 people living in the developing world was affected by a climate disaster. The comparable figure for OECD countries was one in 1,500 affected—a risk differential of 79.[10] Flooding affected the lives of some 68 million people in East Asia and 40 million in South Asia. In sub-Saharan Africa 10 million were affected by drought and 2 million by flooding, in many cases with near simultaneous episodes. Here are some examples of events behind the reported headline numbers:[11]

- The 2007 monsoon period in East Asia displaced 3 million people in China, with large tracts of the country registering the heaviest rainfall since records began. According to the China Meteorological Association, the floods and typhoons of the previous year caused the second deadliest toll on record in terms of lives lost.
- Monsoon floods and storms in South Asia during the 2007 season displaced more than 14 million people in India and 7 million in Bangladesh. Over 1,000 people lost their lives across Bangladesh, India, southern Nepal and Pakistan.
- The 2006/2007 cyclone season in East Asia, which saw large areas of Jakarta flooded, displaced 430,000 people, with Hurricane Durian causing mudslides and extensive loss of life in the Philippines, followed by widespread storm damage in Viet Nam.
- In terms of overall activity, the 2005 Atlantic hurricane season was the most active on record. Hurricane Katrina made most of the headlines, causing widespread devastation in New Orleans. However, the 27 named storms of the season—including Stan, Wilma and Beta—affected communities across Central America and the Caribbean. Hurricane Stan caused the deaths of more than 1,600 mainly Mayan people in the Central Highlands of Guatemala—a greater human toll than Hurricane Katrina.[12]
- Droughts in the Horn of Africa and southern Africa during 2005 threatened the lives of over 14 million people across a swathe of countries from Ethiopia and Kenya to Malawi and Zimbabwe. In the following year, drought gave way to extensive flooding across many of the same countries.[13]

Reported data on the numbers affected by climate disasters provide important insights. However, the data captures only the tip of the iceberg. Many local climate disasters go unreported, or under-reported—and many more do not figure at all, because they do not meet the criteria for a humanitarian disaster (box 2.1).

Gender bias in the impact of disasters is also under-reported. When disasters strike, they hurt whole communities—but women often bear the

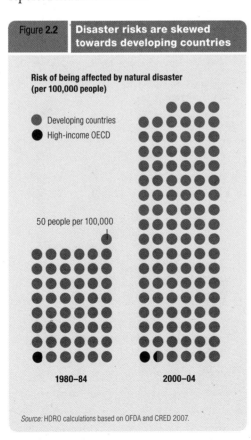

Figure 2.2 Disaster risks are skewed towards developing countries

Risk of being affected by natural disaster (per 100,000 people)

- Developing countries
- High-income OECD

50 people per 100,000

1980–84   2000–04

Source: HDRO calculations based on OFDA and CRED 2007.

Box 2.1 | **Under-reporting climate disasters**

Figures on climate-related disasters come from the EM-DAT *International Disasters Database* maintained by the Centre for Research on the Epidemiology of Disasters (CRED). The database has played a valuable role in improving the flow of information on disasters over time. However, it has certain limitations.

Sources for EM-DAT range from government agencies and the UN system to NGOs, insurance companies and press agencies. Some events are more reported than others: high-profile disasters like Hurricane Katrina attract more media attention than local droughts. Similarly, some groups are almost certainly under-reported: slum dwellers and people living in remote or marginal rural areas are examples.

The criteria for an event being categorized as a disaster are restrictive. Eligibility requirements include numbers killed or affected (at least 10 and 100 respectively), the declaration of a national emergency, or a call for international assistance. Some climate disasters do not meet these criteria. For example, during 2007, just over 1 million people in Ethiopia were receiving drought relief under international aid programmes that registered on the climate disasters database. Seven times this number were receiving support under a national programme to protect nutrition levels in drought-prone areas. That programme did not figure in the database because it was not counted as humanitarian aid.

There are wider sources of under-reporting. During 2006 a crisis caused by late rains in Tanzania did not figure in the CRED database. However, a national food security vulnerability assessment found that the event and rising food prices had left 3.7 million people at risk of hunger, with 600,000 destitute. Disaster statistics also fail to expose the imminent risks faced by the poor. In Burkina Faso, for example, a good harvest in 2007 meant that the country did not make an emergency food aid appeal. Even so, the United States Agency for International Development (USAID) food security assessment warned that over 2 million people were at risk of food insecurity in the event of any disruption to rainfall.

Finally, the disasters database provides a snapshot of numbers affected immediately after the event, but not subsequently. When Hurricane Stan struck Guatemala in October 2005, it affected half a million people, the majority of them from poor, indigenous households in the Western Highlands. They figured in the database for that year. During 2006, food security assessments showed that many of those affected had been unable to restore their assets and that production by subsistence farmers had not recovered. Meanwhile, food prices had increased sharply. The result was an increase in chronic malnutrition in areas affected by Hurricane Stan. That outcome represented a local disaster that was not recorded in the database.

Source: Hoyois et al. 2007; Maskrey et al. 2007; USAID FEWS NET 2006.

brunt. Floods frequently claim far more female victims because their mobility is restricted and they have not been taught to swim. When Bangladesh was hit by a devastating cyclone and flood in 1991, the death rate was reportedly five times higher among women. In the aftermath of a disaster, restrictions on the legal rights and entitlements of women to land and property can limit access to credit needed for recovery.[14]

Reported economic losses also paint a distorted picture. While over 98 percent of people affected by climate disasters live in developing countries, economic impacts are skewed towards rich countries. The reason for this is that costs are assessed on the basis of property values and insured losses, which have been rising steeply (figure 2.3). All eight of the climate disasters registering more than US$10 billion in damages reported since 2000 took place in rich countries, six of them in the United States.

Insurance markets under-report losses in developing countries, especially those sustained by the poor. This is because loss claims reflect the value of the assets and the wealth of those affected. When tropical cyclones sweep across Florida, they hit one of the world's prime real estate locations, with properties protected by high levels of insurance coverage. When the same cyclones hit slums in Haiti or Guatemala, the market value is lower and the real estate of the poor is largely uninsured.

Is climate change implicated in the increase in climate disasters? Direct attribution is impossible. Every weather event is the product of random forces and systemic factors. If Hurricane Katrina had stayed out at sea it would have been just another powerful tropical cyclone. However, climate change is creating systemic conditions for more extreme weather events. All hurricanes gather their strength from the heat of the oceans—and the world's oceans are warming as a result of climate change. More intense storms with higher peak wind speeds and heavier precipitation are a predictable outcome. Similarly, while individual droughts in sub-Saharan Africa cannot be directly attributed

Figure 2.3 **Climate disasters are driving up insured losses**

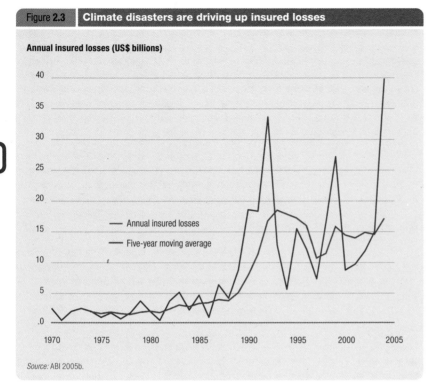

Annual insured losses (US$ billions)

— Annual insured losses
— Five-year moving average

Source: ABI 2005b.

to climate change, climate models predict systemic decreases in rainfall in sub-tropical areas—over 20 percent in some regions.

The precise role of climate change in driving up the number of people affected by climate disaster is also open to debate. Social factors have clearly contributed. Population growth, the expansion of human settlements in hazardous areas—for example, urban slums perched on fragile hillsides and villages located in flood zones—and ecological stress have all played a role in adding to risk exposure. However, climate hazards have also increased. The record shows that droughts in sub-Saharan Africa have become more frequent and protracted. Tropical storms have increased in intensity. Climate change may not provide a full explanation—but it is heavily implicated.[15]

Debates over attribution will continue. As shown in chapter 1, climate science does not provide certainties. However, uncertainty does not constitute a case for inaction. The global insurance industry has been forced into a radical reappraisal of the implications of climate risk for its business models (box 2.2). Across the world, people are being forced to adapt to emerging climate risks in their everyday lives. For small-scale farmers, urban slum dwellers and people living in low-lying coastal areas these

risks threaten to become a powerful obstacle to human development.

Risk and vulnerability

Climate change scenarios provide a framework for identifying structural shifts in weather systems. How those shifts are transmitted through to human development outcomes is conditioned by the interplay of risk and vulnerability.

Risk affects everyone. Individuals, families and communities are constantly exposed to risks that can threaten their well-being. Ill-health, unemployment, violent crime, or a sudden change in market conditions can, in principle, affect anyone. Climate generates a distinctive set of risks. Droughts, floods, storms and other events have the potential to disrupt people's lives, leading to losses of income, assets and opportunities. Climate risks are not equally distributed, but they are widely disbursed.

Vulnerability is different from risk. The etymological root of the word is the Latin verb 'to wound'. Whereas risk is about exposure to external hazards over which people have limited control, vulnerability is a measure of capacity to manage such hazards without suffering a long-term, potentially irreversible loss of well-being.[16] The broad idea can be reduced to "some sense of insecurity, of potential harm people must feel wary of—'something bad' can happen and 'spell ruin'."[17]

Climate change threats illustrate the distinction between risk and vulnerability.[18] People living in the Ganges Delta and lower Manhattan share the flood risks associated with rising sea levels. They do not share the same vulnerabilities. The reason: the Ganges Delta is marked by high levels of poverty and low levels of infrastructural protection. When tropical cyclones and floods strike Manila in the Philippines, they expose the whole city to risks. However, the vulnerabilities are concentrated in the over-crowded, makeshift homes of the slums along the banks of the Pasig River, not in Manila's wealthier areas.[19]

The processes by which risk is converted into vulnerability in any country are shaped by the underlying state of human development,

Climate-related insurance claims have increased rapidly over the past two decades or more. While climate sceptics and some governments continue to question the links between climate change and climate disasters, many global insurance companies are drawing the opposite conclusion.

In the five years to 2004, insured losses from climate events averaged US$17 billion a year—a fivefold increase (in 2004 terms) over the four years to 1990. Climate-related insurance claims are rising more rapidly than population, income and insurance premiums, prompting the industry to reassess the viability of current business models.

That reassessment has taken different forms in different countries. In some cases the industry has emerged as a forceful advocate for the development of infrastructure aimed at reducing insured losses. In Canada and the United Kingdom, for example, insurance companies have led demands for increased public investment in storm and flood-defence systems, while also calling on Government to underwrite losses as an insurer of last resort.

In the United States, insurance companies were actively reviewing their exposure to climate risks even before Hurricane Katrina rewrote the history books in terms of storm damage costs. They have been putting caps on paid losses, shifting a greater part of the risk on to consumers, and withdrawing from high-risk areas.

One of the side-effects of Hurricane Katrina has been to fuel the rise of catastrophic risk bonds, which transfer risk from insurers to capital markets: payments to bond holders cease in the event of a climate catastrophe. The market in 2006 stood at US$3.6 billion, compared with US$1 billion two years earlier.

Federal and state government insurance programmes have not been immune to climate-related pressures. The exposure of two major programmes—the National Flood Insurance Programme (exposure nearing US$1 trillion) and the Federal Crop Insurance Programme (exposure US$44 billion)—has prompted the Government Accountability Office to warn that "Climate change has implications for the fiscal health of the Federal Government."

Experience in developed country insurance markets highlights a wider problem. Climate change creates large uncertainties. Risk is a feature of all insurance markets. Premiums are calculated on the basis of risk assessment. With climate change, insurance claims are likely to rise over time. Based on one estimate from the Association of British Insurers, a doubling of CO_2 could increase insured losses from extreme storm events alone for the global industry by US$66 billion annually (at 2004 prices). The difficulty for the industry is that this trend will be punctuated by catastrophic events that will undermine pooled risk arrangements.

Source: ABI 2004, 2005b; Brieger, Fleck and Macdonald 2001; CEI 2005; GAO 2007; Mills 2006; Mills, Roth and Leomte 2005; Thorpe 2007.

including the inequalities in income, opportunity and political power that marginalize the poor. Developing countries and their poorest citizens are most vulnerable to climate change. High levels of economic dependence on agriculture, lower average incomes, already fragile ecological conditions, and location in tropical areas that face more extreme weather patterns, are all vulnerability factors. The following are among the factors that create a predisposition for the conversion of risk into vulnerability:

- *Poverty and low human development.* High concentrations of poverty among populations exposed to climate risk are a source of vulnerability. The 2.6 billion people—40 percent of the world's population—living on less than US$2 a day are intrinsically vulnerable because they have fewer resources with which to manage risks. Similarly, for the 22 countries with a combined population of 509 million people in the low human development category of the Human Development Index (HDI),

even small increases in climate risk can lead to mass vulnerability. Across much of the developing world (including countries in the medium human development category) there is a two-way interaction between climate-related vulnerability, poverty and human development. Poor people are often malnourished partly because they live in areas marked by drought and low productivity; and they are vulnerable to climate risks because they are poor and malnourished. In some cases, that vulnerability is directly linked to climate shocks. Disaggregated HDI data for Kenya, for example, show a close fit between food emergencies linked to drought and districts where human development is low (table 2.1). In Ghana, half of children in the drought-prone northern region are malnourished, compared with 13 percent in Accra.[20]

- *Disparities in human development.* Inequalities within countries are another marker for vulnerability to climate shocks.

Table 2.1	Drought-related food emergencies and human development are closely linked in Kenya

Kenyan districts	Human Development Index value 2005
Districts suffering food emergency (November 2005–October 2006)	
Garissa	0.267
Isiolo	0.580
Mandera	0.310
Masrabit	0.411
Mwingi	0.501
Samburu	0.347
Turkana	0.172
Wajir	0.256
Others	
Mombassa	0.769
Nairobi	0.773
Kenya national average	0.532

Source: UNDP 2006a; USAID FEWS NET 2007.

One recent quantitative assessment of the human impacts of disasters has found that "countries with high levels of income inequality experience the effects of climate disasters more profoundly than more equal societies".[21] Average levels of human development can obscure high levels of deprivation. Guatemala, for example, is a medium human development country marked by large social disparities between indigenous and non-indigenous people. Malnutrition among indigenous people is twice as high as for non-indigenous people. When Hurricane Stan swept across the Western Highlands of Guatemala in 2005 its impact was felt most heavily by indigenous people, the majority of them subsistence farmers or agricultural labourers. Losses of basic grains, the depletion of food reserves and the collapse of employment opportunities magnified already severe levels of deprivation, with inequality acting as a barrier to early recovery.[22] Disparities in human development also expose vulnerable populations to climate risks in some of the world's richest countries. When Hurricane Katrina hit New Orleans, some of America's poorest communities were affected. Recovery was hampered by deep underlying inequalities (box 2.3).

- *Lack of climate-defence infrastructure.* Infrastructural disparities help to explain why similar climate impacts produce very different outcomes. The elaborate system of dykes in the Netherlands acts as a powerful buffer between risk and vulnerability. Flood defence systems, water infrastructure and early warning systems all reduce vulnerability. Japan faces a higher exposure to risks associated with cyclones and flooding than the Philippines. Yet between 2000 and 2004, average fatalities amounted to 711 in the Philippines and only 66 in Japan.[23]

- *Limited access to insurance.* Insurance can play an important role in enabling people to manage climate risks without having to reduce consumption or run down their assets. Private markets and public policy can play a role. Households in rich countries have access to private insurance to protect themselves against climate-related losses. Most poor households in developing countries do not. Social insurance is another buffer against vulnerability. It enables people to cope with risks without eroding long-term opportunities for human development. It can provide for people in old age, afford

Figure 2.4	Social insurance provision is far greater in rich countries

Source: World Bank 2006g.

Box **2.3** | **Hurricane Katrina—the social demographics of a disaster**

When Hurricane Katrina breached the levees of New Orleans it caused human suffering and physical damage on a vast scale. As the flood waters receded, they revealed the acute vulnerabilities associated with high levels of pre-existing social inequality. Flood damage was superimposed on a divided city, just as climate change damage will be superimposed on a divided world. Two years after the tragedy, inequalities continue to hamper recovery.

Located on the Gulf Coast of the United States, New Orleans is in one of the world's high-risk hurricane zones. In August 2005 the flood defences mitigating that risk were overwhelmed, with tragic consequences. Hurricane Katrina claimed over 1,500 lives, displaced 780,000 people, destroyed or damaged 200,000 homes, crippled the city's infrastructure and traumatized its population.

The hurricane impacted on the lives of some of the poorest and most vulnerable people in the world's richest nation. Pre-Katrina child poverty rates in New Orleans were among the highest in the United States, with one in three living below the poverty line. Health provision was limited, with some 750,000 people lacking insurance coverage.

Hurricane Katrina selected its victims overwhelmingly from the most disadvantaged areas of the city. Poorer districts dominated by black communities bore the brunt. Flood damage interacted with deep racial inequalities (poverty rates among blacks three times higher than for whites). An estimated 75 percent of the population living in flooded neighbourhoods was black. The Lower Ninth Ward and the Desire/Florida communities, two of the poorest and most vulnerable in the city, were both totally devastated by Katrina.

Images of the human suffering in New Orleans were beamed around the world as the city became a magnet for international media attention. Yet as people sought to rebuild their lives after the cameras had departed, pre-hurricane inequalities emerged as a barrier to recovery.

The health sector provides a striking example. Many of the health facilities in the safety net system serving the poor were damaged by Hurricane Katrina, with the Charity Hospital, which provided most of the medical care for this group—emergency, acute and basic—still closed. While a special Medicaid waiver was introduced to provide temporary coverage for uninsured evacuees, eligibility rules limited entitlements for low-income households without children, leading to a large number of rejected claims. It took Congress and the Administration 6 months to authorize a US$2 billion provision for Medicaid to cover uninsured health costs.

Research conducted by the Kaiser Family Foundation 6 months after the storm revealed that many people had been unable to maintain pre-existing treatment or to access the care needed to deal with new conditions. In household interviews, over 88 percent of respondents identified the need for expanded and improved health provision as a vital challenge for the city. Two years on, that challenge remains.

Of the many factors blocking the social and economic recovery of New Orleans, the health care system may be the most important. Only one of the city's seven general hospitals is operating at its pre-hurricane level; two more are partially open, and four remain closed. The number of hospital beds in New Orleans has dropped by two-thirds. There are now 16,800 fewer medical jobs than before the storm, down 27 percent, in part because nurses and other workers are in short supply.

Two important lessons emerge from Hurricane Katrina that have a wider bearing on climate change strategies. The first is that high levels of poverty, marginalization and inequality create a predisposition for risk to convert into mass vulnerability. The second is that public policy matters. Policies that provide people with entitlements to health and housing provision can facilitate early recovery, while weak entitlements can have the opposite effect.

Poverty in New Orleans

People living in poverty, 2000 (%)	New Orleans	United States
Total population	28	12
Children 18 years and younger	38	18
Whites	12	9
African–Americans	35	25

Source: Perry et al. 2006.

Source: Perry et al. 2006; Rowland 2007; Turner and Zedlewski 2006; Urban Institute 2005.

protection during periods of sickness or unemployment, assist child development and protect basic nutrition. Countries vary widely in their support for social insurance (figure 2.4). Rich countries spend a greater share of their far higher average incomes on social insurance. In terms of global climate change risk management this means that there is an inverse relationship between vulnerability (which is concentrated in poor countries) and insurance (which is concentrated in rich countries).

Gender inequalities intersect with climate risks and vulnerabilities. Women's historic disadvantages—their limited access to resources, restricted rights, and a muted voice in shaping decisions—make them highly vulnerable to climate change. The nature of that vulnerability varies widely, cautioning against generalization. But climate change is likely to magnify existing

Climate shocks: risk and vulnerability in an unequal world

For many generations, Inuit have closely observed the environment, accurately predicting the weather so as to allow safe travel on the sea ice. However our ability to read and predict weather patterns and conditions around us is now greatly challenged as a result of climate change. For decades, our hunters have reported melting permafrost, thinning ice, receding glaciers, new invasive species, rapid coastal erosion and dangerously unpredictable weather. From our far Northern perspective, we have observed that the global climate change debate too often focuses on economic and technical matters rather than on the human impacts and consequences of climate change. Inuit are already experiencing these impacts and will soon face dramatic social and cultural dislocation.

Climate change is our greatest challenge: overarching, complex and requiring immediate action. It also presents an opportunity to reconnect with each other as a shared humanity, despite our differences. With this in mind I decided to look at the international human rights regimes that are in place to protect peoples from cultural extinction—the very situation we Inuit could be facing. The question was always how can we bring some clarity of purpose and focus to a debate that seems always to be caught up in technical arguments and competing short term ideologies? I believe it is significant internationally for global climate change to be debated and examined in the arena of human rights. As Mary Robinson said "human rights and the environment are interdependent and interrelated". That is why, together with 61 other Inuit, I worked to launch the Climate Change Human Rights Petition in December 2005.

In essence the petition states that governments should develop their economies using appropriate technologies that significantly limit greenhouse gas emissions. But we have also achieved much more than that.

Through this work we have made human faces—and our fates—the centre of attention. We have changed the international discourse from dry technical discussions to debates about human values, human development and human rights. We have given United Nations conferences a heartbeat, a renewed sense of urgency. We did this by reminding people far away from the Arctic that we are all connected: that the Inuit hunters falling through the thinning ice are connected to the people facing the melting glaciers of the Himalayas and the flooding of the small island states; but that this is also connected to the way the world goes about its daily life in terms of the cars we drive, the industries we support and the policies we choose to make and enforce.

A brief window of opportunity still remains to save the Arctic and, ultimately, the planet. Coordinated action can still forestall the future projected in the Arctic Climate Impact Assessment. Nations can again come together, as we did in Montreal in 1987 and Stockholm in 2001. Already our ozone is mending; already the toxic chemicals that poisoned the Arctic are decreasing. Now the world's greatest emitters must make binding commitments to act. I only hope that nations take this opportunity to once more come together through the understanding of our connectivity and our shared atmosphere, ultimately our shared humanity.

Sheila Watt-Cloutier

Sheila Watt-Cloutier
Advocate for Arctic climate change

---

patterns of gender disadvantage. In the agricultural sector, rural women in developing countries are the primary producers of staple food, a sector that is highly exposed to the risks that come with drought and uncertain rainfall. In many countries, climate change means that women and young girls have to walk further to collect water, especially in the dry season. Moreover, women can be expected to contribute much of the labour that will go into coping with climate risks through soil and water conservation, the building of anti-flood embankments and increased off-farm employment. One corollary of gender vulnerability is the importance of women's participation in any planning process for adaptation to climate change.[24]

Climate change is also providing a reminder of the symbiotic relationship between human culture and ecological systems. This relationship is very evident in the Arctic, where some of the world's most fragile ecosystems are being affected by rapid warming. Indigenous people in the Arctic have become sentinels for a world undergoing climate change. As one of the leaders of the Inuit community has commented: "The Arctic is the world's climate change barometer. Inuit are the mercury in that barometer."[25] For Inuit people, business-as-usual warming will disrupt or even destroy a culture based on hunting and food sharing, as reduced sea ice causes the animals on which they depend to become less accessible, and possibly decline towards extinction. In December 2005, representatives of Inuit organizations submitted a petition to the Inter-American Commission on Human Rights, claiming that unrestricted emissions from the United States were violating the human rights of the Inuit. The aim was not to seek damages but rather redress, in the form of leadership in mitigating dangerous climate change.

Low human development traps

Human development is about expanding freedom and choice. Climate-related risks force people into trade-offs that limit substantive freedom and erode choice. These trade-offs can constitute a one-way ticket into low human development traps—downward spirals of disadvantage that undermine opportunities.

Climate shocks affect livelihoods in many ways. They wipe out crops, reduce opportunities for employment, push up food prices and destroy property, confronting people with stark choices. Wealthy households can manage shocks by drawing upon private insurance, using their savings, or trading in some of their assets. They are able to protect their current consumption—'consumption smoothing'—without running down their productive capacities or eroding their human capabilities. The poor have fewer options.

With limited access to formal insurance, low income and meagre assets, poor households have to adapt to climate shocks under more constrained conditions. In an effort to protect current consumption, they are often forced to sell productive assets, compromising future income generation. When incomes fall from already low levels, they may have no choice but to reduce the number of meals they eat, cut spending on health, or withdraw their children from school to increase labour supply. The coping strategies vary. However, the forced trade-offs that follow climate shocks can rapidly erode human capabilities, setting in train cycles of deprivation.

Poor households are not passive in the face of climate risks. Lacking access to formal insurance, they develop self-insurance mechanisms. One of these mechanisms is to build up assets—such as livestock—during 'normal' times for sale in the event of a crisis. Another is to invest household resources in disaster prevention. Household surveys in flood-prone urban slums in El Salvador record families spending up to 9 percent of their income on strengthening their homes against floods, while also using family labour to build retaining walls and maintain drainage channels.[26] Diversification of production and income sources is another form of self-insurance. For example, rural households seek to reduce their risk exposure by inter-cropping food staples and cash crops, and by engaging in petty trade. The problem is that self-insurance mechanisms often break down in the face of severe and recurrent climate shocks.

Research points to four broad channels or 'risk multipliers' through which climate shocks can undermine human development: 'before-the-event' losses in productivity, early coping costs, asset erosion of physical capital and asset erosion of human opportunities.

'Before-the-event' losses in productivity

Not all of the human development costs of climate shocks happen after the event. For people with precarious livelihoods in areas of climate variability, uninsured risk is a powerful impediment to increased productivity. With less capacity to manage risk, the poor face barriers to engage in higher-return but higher-risk investment. In effect, they are excluded from opportunities to produce their way out of poverty.

It is sometimes argued that the poor are poor because they are less 'entrepreneurial' and choose to avoid risky investments. The fallacy in this view lies in confusion between risk aversion and innovative capacity. As households move closer to extreme poverty they become risk averse for a very good reason: adverse outcomes can affect life chances at many levels. Operating without formal insurance in areas of high risk exposure—such as floodplains, drought-prone regions or fragile hillsides—poor households rationally choose to forego potentially higher return investments in the interests of household security. Farmers may be forced to make production decisions that are less sensitive to rainfall variation, but also less profitable.

Research in Indian villages in the 1990s found that even slight variations in rainfall timing could reduce farm profits for the poorest quartile of respondents by one-third, while having a negligible impact on profitability for the richest quartile. Faced with high risk, poor farmers tended to over-insure: production decisions led to average profits that were lower than they could have been in an insured risk environment.[27] In Tanzania, village-level research found poor farmers specializing in the production of drought-resistant crops—like sorghum and cassava—which provide more food security but a lower financial return. The crop portfolio of the wealthiest quintiles yielded 25 percent more than that of the poorest quintile.[28]

This is part of a far wider pattern of de facto risk insurance that, interacting with other factors, increases inequality and locks poor households into low-return systems of production.[29] As climate change gathers pace, agricultural production in many developing countries will become riskier and less profitable (see section on Agriculture and food security below). With three-quarters of the world's poor dependent on agriculture, this has important implications for global poverty reduction efforts.

It is not just the world's poor that will have to adjust to new climate patterns. Agricultural producers in rich countries will also have to deal with the consequences, however, the risks are less severe, and they are heavily mitigated through large-scale subsidies—around US$225 billion in OECD countries in 2005—and public support for private insurance.[30] In the United States, Federal Government insurance payments for crop damage averaged US$4 billion a year from 2002 to 2005. The combination of subsidies and insurance enables producers in developed countries to undertake higher-risk investments to obtain higher returns than would occur under market conditions.[31]

The human costs of 'coping'

The inability of poor households to cope with climate shocks is reflected in the immediate human impacts, and in increasing poverty. Droughts provide a potent example.

When rains fail the ripple effects are transmitted across many areas. Losses in production can create food shortages, push up prices, undermine employment, and depress agricultural wages. The impacts are reflected in coping strategies that range from reduced nutrition to the sale of assets (table 2.2). In Malawi, the 2002 drought left nearly 5 million people in need of emergency food aid. Long before the aid arrived, households had been forced to resort to extreme survival measures, including such activities as theft and prostitution.[32] The acute vulnerabilities that can be triggered by climate shocks in countries at low levels of human development were powerfully demonstrated in the 2005 food security crisis in Niger (box 2.4).

Droughts are often reported as short term, single events. That practice obscures some important impacts in countries where multiple or sequential droughts create repeated shocks over several years. Research in Ethiopia illustrates the point. The country has experienced at least five major national droughts since 1980, along with literally dozens of local droughts. Cycles of drought create poverty traps for many households, constantly thwarting efforts to build up assets and increase income. Survey data show that between 1999 and 2004 more than half of all households in the country experienced at least one major drought shock.[33] These shocks are a major cause of transient poverty: had households been able to smooth consumption, then poverty in 2004 would

Table 2.2	Drought in Malawi—how the poor cope

Behaviours adopted to cope with drought, 1999 (% of people)	Blantyre Town (%)	Rural Zomba (%)
Dietary adjustments		
• Substituted meat for vegetables	73	93
• Ate smaller portions to make meals last longer	47	91
• Reduced number of meals per day	46	91
• Ate different foods, such as cassava instead of maize	41	89
Expenditure reduction		
• Bought less firewood or paraffin	63	83
• Bought less fertilizer	38	33
Cash generation for food		
• Depleted savings	35	0
• Borrowed money	36	7
• Searched for casual labour (ganyu) for cash and food	19	59
• Sold livestock and poultry	17	15
• Sold household items and clothes	11	6
• Sent children to look for money	10	0

Source: Devereux 1999.

have been at least 14% lower (table 2.3)—a figure that translates into 11 million fewer people below the poverty line.[34]

The human impacts of current climate shocks provide a widely ignored backdrop for understanding the human development implications of climate change. Malnutrition levels rise and people get locked into poverty traps. If climate change scenarios predicting more frequent and more intense droughts and floods are correct, the consequences could be large and rapid reversals in human development in the countries affected.

Asset erosion—physical capital

Climate shocks can have devastating consequences for household assets and savings. Assets such as live animals represent something more than a safety net for coping with climate shocks. They provide people with a productive resource, nutrition, collateral for credit, and a source of income to meet health and education costs, while also providing

Table 2.3	The impact of drought shocks in Ethiopia

	People in poverty (%)
Observed poverty	47.3
Predicted poverty with no drought shocks	33.1
Predicted poverty with no shocks of any kind	29.4

Source: Dercon 2004.

security in the event of crop failure. Their loss increases future vulnerability.

Climate shocks create a distinctive threat to coping strategies. Unlike, say, ill-health, many climate shocks are covariate: that is, they affect entire communities. If all affected households sell their assets at the same time in order to protect consumption, asset prices can be expected to fall. The resulting loss of value can rapidly and severely undermine coping strategies, reinforcing wider inequalities in the process.

Research on the 1999/2000 drought in Ethiopia illustrates this point. The disaster began with a failure of the short or *belg* rains,

---

Box 2.4	Drought and food insecurity in Niger

Niger is one of the poorest countries in the world. It ranks close to bottom of the HDI, with a life expectancy of nearly 56 years, 40 percent of children having low weight for their age in an average year, and more than one in five children dying before their fifth birthday. Vulnerability to climate shocks in Niger is linked to several factors, including widespread poverty, high levels of malnutrition, precarious food security in 'normal' years, limited health coverage and agricultural production systems that have to cope with uncertain rainfall. During 2004 and 2005 the implications of these underlying vulnerabilities were powerfully demonstrated by a climate shock, with an early end to rains and widespread locust damage.

Agricultural production was immediately affected. Output fell sharply, creating a cereals deficit of 223,000 tonnes. Prices of sorghum and millet rose 80 percent above the 5-year average. In addition to high cereal prices, deteriorating livestock conditions deprived household of a key source of income and risk insurance. The loss of pasture and nearly 40 percent of the fodder crop, along with rising animal feed prices and 'distress sales', pushed down livestock prices, depriving households of a key source of income and risk insurance. With vulnerable households trying to sell under-nourished animals for income to buy cereals, the drop in prices adversely affected their food security and terms of trade.

By the middle of 2005 around 56 zones across the country were facing food security risks. Some 2.5 million people—around a fifth of the country's population—required emergency food assistance. Twelve zones in regions such as Maradi, Tahou and Zinder were categorized as 'extremely critical', meaning that people were reducing the number of meals eaten each day, consuming wild roots and berries, and selling female cattle and production equipment. The crisis in agriculture led to severe human costs, including:

- Migration to neighbouring countries and less critically affected zones.
- In 2005 Médecins Sans Frontières (MSF) re-reported an acute malnutrition rate of 19 percent among children aged 6–59 months in Maradi and Tahoua, representing a significant deterioration over average levels. MSF also reported a fourfold increase in the number of children suffering from severe malnutrition in therapeutic feeding centres.
- USAID survey team reported women spending entire days collecting *anza*, a wild food.

In some respects, Niger's low level of human development makes the country an extreme case. However, developments during 2005 demonstrated in stark fashion the mechanisms through which increased climate-related risk can disrupt coping strategies and create extensive vulnerabilities.

Source: Chen and Meisel 2006; Mousseau and Mittal 2006; MSF 2005; Seck 2007a.

The trade-offs forced upon people by climate shocks reinforce and perpetuate wider inequalities based on income, gender and other disparities

which can fall between February and April. This frustrated farmers' attempts to plough and sow crops. Reduced rainfall during the long rainy season (the June–September *meher* rains) caused widespread crop failure. When the subsequent *belg* season in early 2000 also saw poor rainfall, the result was a major food security crisis. Distress sales of assets—mainly livestock—began early and continued for 30 months. By the end of 1999, livestock sellers were receiving less than half the pre-drought price, constituting a huge loss of capital. However, not all farmers adopted the same coping strategy. The top two quartiles, with far more cattle, sold animals early in a classic 'consumption smoothing' pattern, trading in their insurance risk premium in order to maintain access to food. In contrast, the lowest two quartiles stubbornly held on to their small number of animals, with only small decreases in livestock ownership until the end of the drought period. The reason: their animals were a vital productive resource for ploughing. In effect, the rich were able to smooth consumption without detrimentally eroding their productive assets, whereas the poor were forced to choose between the two.[35]

Agropastoral and pastoral households, which are even more reliant on livestock for their livelihoods, also suffer severe asset losses during droughts. As experience in Ethiopia has repeatedly shown, the consequences are likely to include adverse impacts for their terms of trade, with livestock prices falling sharply relative to cereal prices.

Another example comes from Honduras. In 1998 Hurricane Mitch cut a wide path of destruction across the country. In this case, the poor were forced to sell a far greater share of their assets than wealthier households in order to cope with a steep increase in poverty. By running down the productive assets of the poor, the climate shock in this case created conditions for an increase in future inequalities (box 2.5).

Asset erosion—human opportunities

Media images of human suffering during climate shocks do not capture the damaging trade-offs into which poor households are forced. When

droughts, floods, storms and other climate events disrupt production, cut income and erode assets, the poor face a stark choice: they must make up income losses or cut spending. Whatever the choice, the consequences are long-term costs that can jeopardize human development prospects. The trade-offs forced upon people by climate shocks reinforce and perpetuate wider inequalities based on income, gender and other disparities. Some examples:

- *Nutrition.* Climate shocks such as drought and floods can cause grave setbacks in nutritional status as food availability declines, prices rise and employment opportunities shrink. Deteriorating nutrition provides the most telling evidence that coping strategies are failing. The drought that swept across large areas of eastern Africa in 2005 illustrates the point. In Kenya, it put the lives of an estimated 3.3 million people in 26 districts at risk of starvation. In Kajiado, the worst affected district, the cumulative effect of the two poor rainy seasons in 2003 and the total failure of rains in 2004 almost completely wiped out production. Particularly, decline in the production of rainfed crops such as maize and beans harmed both people's diet and their purchasing power. Health centres in the district reported an increase in malnutrition, with 30 percent of children seeking medical assistance found to be underweight compared to 6 percent in normal years.[36] In some cases, the trade-offs between consumption and survival can exacerbate gender bias in nutrition. Research in India has found that girls' nutrition suffers most during periods of low consumption and rising food prices, and that rainfall shortages are more strongly associated with deaths among girls than boys.[37]

- *Education.* For the poorest households, increasing labour supply can mean transferring children from classrooms into the labour market. Even in 'normal' years, poor households are often forced to resort to child labour, for example during the lean season before harvests. Droughts and floods intensify these pressures. In Ethiopia and Malawi, children are routinely taken out of school to engage in income-generating activities. In Bangladesh and India, children in poor households work on farms, tend cattle or engage in other tasks in exchange for food during periods of stress. In Nicaragua in the aftermath of Hurricane Mitch, the proportion of children working rather than attending school increased from 7.5 to 15.6 percent in affected households.[38] It is not only low-income countries that are affected. Household research in Mexico covering the period 1998–2000 shows an increase in child labour in response to drought.

- *Health.* Climate shocks are a potent threat to the poor's most valuable assets—their health and their labour. Deteriorating nutrition and falling incomes generate a twin threat: increased vulnerability to illness and fewer resources for medical treatment. Droughts and floods are often catalysts for wide-ranging health problems, including an increase in diarrhoea among children, cholera, skin problems and acute under-nutrition. Meanwhile, capacity to treat old problems and cope with new ones is hampered by increased poverty. Research for this Report shows that in Central Mexico during the period 1998 to 2000, children under five saw their chances of falling sick increasing when they suffered a weather shock: the probability of illness increased by 16 percent with droughts and by 41 percent with floods.[39] During the 2002 food crisis in southern Africa, over half of households in Lesotho and Swaziland reported reduced health spending.[40] Reduced or delayed treatment of diseases is an enforced choice that can have fatal consequences.

Forced trade-offs in areas such as nutrition, education and health have consequences that

Climate shocks are a potent threat to the poor's most valuable assets—their health and their labour

---

Box 2.5 Distress sales in Honduras

Climate change will bring with it more intense tropical storms as sea temperatures rise. The incremental risks will be borne across societies. However, poor households with limited risk management capacity will suffer the most. Evidence from Central America, which will be one of the worst affected regions, shows how storms can erode assets and exacerbate inequality.

In contrast to droughts, which emerge as 'slow-fuse' crises over months, storms create instantaneous effects. When Hurricane Mitch tore into Honduras in 1998 it had an immediate and devastating impact. Data collected shortly after the hurricane showed that poor rural households lost 30–40 percent of their income from crop production. Poverty increased by 8 percent, from 69 to 77 percent at a national level. Low-income households also lost on average 15–20 percent of their productive assets, compromising their prospects for recovery.

Some 30 months after Hurricane Mitch a household survey provided insights into asset management strategies in a distress coping environment. Almost half of all households reported a loss of productive assets. Not surprisingly, especially in a highly unequal country like Honduras, the value of the loss increased with wealth: the average pre-Mitch asset value reported by the wealthiest quartile was 11 times greater than for the poorest quartile. However, the poorest quartile lost around one-third of the value of their assets, compared with 7 percent for the wealthiest quartile (see table).

In the reconstruction effort, average aid to the richest 25 percent amounted to US$320 per household—slightly more than double the level for the poorest quartile.

Detailed analysis of post-shock asset recovery has drawn attention to the way in which Hurricane Mitch has reinforced asset-based inequality. When asset value growth rates over the two-and-a-half years after Mitch were compared with the predicted trend based on pre-Mitch data, it emerged that, while both rich and poor were rebuilding their asset base, the net growth rate for the poorest quartile was 48 percent below the predicted pre-Mitch trend, whereas for the richest quartile it was only 14 percent below.

The rise in asset inequality has important implications. Honduras is one of the most unequal countries in the world, with a Gini index for income distribution of 54. The poorest 20 percent account for 3 percent of national income. Asset loss among the poor will translate into diminished opportunities for investment, increased vulnerability and rising income inequality in the future.

Hurricane Mitch devastated the assets of the poor

	Poorest 25%	Second 25%	Third 25%	Wealthiest 25%
Share of assets lost as a result of Hurricane Mitch (%)	31.1	13.9	12.2	7.5

Source: Carter et al. 2005.

Source: Carter et al. 2005; Morris et al. 2001.

extend far into the future. Detailed household survey analysis in Zimbabwe demonstrates the longevity of human development impacts linked to climate shocks. Taking a group of children that were aged 1–2 years during a series of droughts between 1982 and 1984, researchers interviewed the same children 13–16 years later. They found that the drought had reduced average stature by 2.3 centimetres, delayed the start of school and resulted in a loss of 0.4 years of schooling. The education losses translated into a 14 percent loss of lifetime earnings. Impacts in Zimbabwe were most severe among children in households with few livestock—the main self-insurance asset for smoothing consumption.[41]

Caution must be exercised in interpreting results from one specific case. But the Zimbabwe experience demonstrates the transmission

mechanisms from climate shocks through nutrition, stunting and educational deprivation into long-run human development losses. Evidence from other countries confirms the presence and the durability of these mechanisms. When Bangladesh was hit by a devastating flood in 1998, the poorest households were forced into coping strategies that led to long-term losses in nutrition and health. Today many adults are living with the consequences of the deprivation they suffered as children in the immediate aftermath of the flood (box 2.6).

From climate shocks today to deprivation tomorrow—low human development traps in operation

The idea that a single external shock can have permanent effects provides a link from climate shocks—and climate change—to the relationship between risk and vulnerability set out in this chapter. The direct and immediate impact of droughts, hurricanes, floods and other climate shocks can be ghastly. But the after-shocks interact with wider forces that hold back the development of human capabilities.

These after-shocks can be understood through a poverty trap analogy. Economists have long recognized the presence of poverty traps in the lives of the poor. While there are many versions of the poverty trap, they tend to focus on income and investment. In some accounts, poverty is seen as the self-sustaining outcome of credit constraints that limit the capacity of the poor to invest.[42] Other accounts point to a self-reinforcing cycle of low productivity, low income, low savings and low investments. Linked to these are poor health and limited opportunities for education, which in turn restrict opportunities for raising income and productivity.

When climate disasters strike, some households are rapidly able to restore their livelihoods and rebuild their assets. For other households, the recovery process is slower. For some—especially the poorest—rebuilding may not be possible at all. Poverty traps can be thought of as a minimum threshold for assets or income, below which people are

---

Box 2.6	The 'flood of the century' in Bangladesh

Flooding is a normal part of the ecology of Bangladesh. With climate change, 'abnormal' flooding is likely to become a standing feature of the future ecology. Experience following the flood event of 1998—dubbed the 'flood of the century'—highlights the danger that increased flooding will give rise to long term human development setbacks.

The 1998 flood was an extreme event. In a normal year, around a quarter of the country experiences inundation. At its peak, the 1998 flood covered two-thirds of the country. Over 1,000 people died and 30 million were made homeless. Around 10 percent of the country's total rice crop was lost. With the duration of the flood preventing replanting, tens of millions of households faced a food security crisis.

Large-scale food imports and government food aid transfers averted a humanitarian catastrophe. However, they failed to avert some major human development setbacks. The proportion of children suffering malnutrition doubled after the flood. Fifteen months after the flood, 40 percent of the children with poor nutritional status at the time of the flood had still not regained even the poor level of nutrition they had prior to the flood.

Households adjusted to the floods in several ways. Reduced spending, asset sales and increased borrowing all featured. Poor households were more likely both to sell assets and to take on debts. Fifteen months after the floods had receded, household debt for the poorest 40 percent averaged 150 percent of monthly expenditure—twice the pre-flood level.

Management of the 1998 floods is sometimes seen as a success story in disaster management. To the extent that an even larger loss of life was averted, that perception is partially justified. However, the flood had long term negative impacts, notably on the nutritional status of already malnourished children. The affected children may never be in a position to recover from the consequences. Poor households suffered in the short term through reduced consumption and increased illness, and through having to take on high levels of household debt—a strategy that may have added to vulnerability.

Source: del Ninno and Smith 2003; Mallick et al. 2005.

unable to build productive assets, educate their children, improve their health and nutrition and increase income over time.[43] People above that threshold are able to manage risks in ways that do not lead to downward cycles of poverty and vulnerability. People below it are unable to reach the critical point beyond which they can escape the gravitational pull of poverty.

Analysis of income poverty traps has drawn attention to the processes by which deprivation is transmitted through time. By the same token, it has underplayed the importance of human capabilities—the wider set of attributes that determine the choices open to people. Shifting the focus towards capability does not mean ignoring the role of income. Low income is clearly a major cause of human deprivation. However, limited income is not the only thing that holds back the development of capabilities. Exclusion from opportunities for basic education, health and nutrition are sources of capability deprivation. In turn, these are linked to lack of progress in other dimensions, including the ability of people to participate in decision-making and to assert their human rights.

Like poverty traps, low human development traps occur when people are unable to pass a threshold beyond which they can engineer a virtuous circle of capability expansion. Climate shocks are among the many external factors that sustain such traps over time. They interact with other events—ill-health, unemployment, conflict and disruptions in markets. While these are important, climate shocks are among the most potent forces sustaining low human development traps.

Research carried out for this Report provides evidence of low human development traps in operation. In order to track the impact of climate shocks across time in the lives of those affected, we developed an econometric model to explore microlevel household survey data (*Technical Note 2*). We looked at specific human development outcomes associated with an identified climate shock. What difference does it make to the nutritional status of children if they were born during a drought? Using our model we addressed that question for several countries that face recurrent droughts. The results

demonstrate the damaging impact of drought on the life chances of affected children:

- In Ethiopia, children aged five or less are 36 percent more likely to be malnourished and 41 percent more likely to be stunted if they were born during a drought year and affected by it. This translates into some 2 million 'additional' malnourished children.
- For Kenya, being born in a drought year increases the likelihood of children being malnourished by 50 percent.
- In Niger, children aged two or under who were born during a drought year and were affected by it are 72 percent more likely to be stunted, pointing to the rapid conversion of droughts into severe nutritional deficits.

These findings have important implications in the context of climate change. Most obviously, they demonstrate that the inability of poor households to cope with 'current' climate shocks is already a major source of human capability erosion. Malnutrition is not an affliction that is shaken off when the rains return or the flood waters recede. It creates cycles of disadvantage that children will carry with them throughout their lives. Indian women born during a drought or a flood in the 1970s were 19 percent less likely to ever attend primary school, when compared with women the same age who were not affected by natural disasters. The incremental risks associated with climate change have the potential to reinforce these cycles of disadvantage.

We stress the word 'potential'. Not every drought is the prelude to famine, malnutrition or educational privation. And not every climate shock gives rise to the distress sale of assets, long-run increases in vulnerability or the spread of low human development traps. This is an area in which public policies and public institutions make a difference. Governments can play a critical role in creating mechanisms that build resilience, support pro-poor risk management and reduce vulnerability. Policies in these areas can create an enabling environment for human development. With climate change, international cooperation on adaptation is a key condition for scaling-up these policies to meet incremental risks—an issue to which we return in chapter 4.

> Governments can play a critical role in creating mechanisms that build resilience, support pro-poor risk management and reduce vulnerability

Developing countries are likely to become more dependent on imports from the rich world, with their farmers losing market shares in agricultural trade

2.2 Looking ahead—old problems and new climate change risks

"Prediction is very difficult, especially if it's about the future," commented the Danish physicist and Nobel laureate Niels Bohr. The observation applies with special force to climate. However, while specific events are uncertain, changes in average conditions associated with climate change can be predicted.

The IPCC's Fourth Assessment Report provides a best-estimate set of projections for future climate. These projections are not weather forecasts for individual countries. What they offer is a range of probabilities for broad changes in climate patterns. The underlying story has important implications for human development. Over the decades ahead there will be a steady increase in human exposure to such events as droughts, floods and storms. Extreme weather events will become more frequent and more intense, with less certainty and predictability in the timing of monsoons and rainfall.

In this section we provide an overview of the links from the IPCC's projections to human development outcomes.[44] We focus on 'likely' and 'very likely' outcomes for climate, defined respectively as results with an occurrence probability in excess of 66 and 90 percent.[45] While these outcomes relate only to average global and regional conditions, they help to identify emerging sources of risk and vulnerability.

Agricultural production and food security

IPCC projection: Increases in precipitation in high latitudes and decreases in sub-tropical latitudes, continuing the current pattern of drying in some regions. Warming is likely to be above the global average throughout sub-Saharan Africa, eastern Asia and South Asia. In many water-scarce regions, climate change is expected to further reduce water availability through increased frequency of droughts, increased evaporation and changes in patterns of rainfall and runoff.[46]

Human development projection: Major losses in agricultural production leading to increased malnutrition and reduced opportunities for poverty reduction. Overall, climate change will lower the incomes and reduce the opportunities of vulnerable populations. By 2080, the number of additional people at risk of hunger could reach 600 million—twice the number of people living in poverty in sub-Saharan Africa today.[47]

Global assessments of the impact of climate change on agriculture obscure very large variations across and even within countries. In broad terms, climate change will increase the risks to and reduce the productivity of developing country agriculture. In contrast, production could be boosted in developed countries, so that the distribution of world food production may shift. Developing countries are likely to become more dependent on imports from the rich world, with their farmers losing market shares in agricultural trade.[48]

Emerging patterns of climate change risk in agriculture will have important implications for human development. Around three in every four people in the world living on less than US$1 a day reside in rural areas. Their livelihoods depend on smallholder agriculture, farm employment, or pastoralism.[49] The same constituency also accounts for most of the 800 million people in the world who are malnourished. Climate change impacts on agriculture will thus have important multiplier effects. Agricultural production and employment underpin many national economies (table 2.4). The agricultural sector accounts for over one-third of export earnings in around 50 developing countries and for almost half of employment in the developing world.[50] In sub-Saharan Africa in particular, economic growth rates are closely tied to rainfall, as demonstrated by the experience of Ethiopia (figure 2.5). Moreover, every US$1 generated in agriculture in sub-Saharan Africa is estimated to generate up to US$3 in the non-agricultural sector.[51]

Climate modelling exercises point to very large changes in production patterns. One study has averaged out the findings of six such exercises, identifying changes in output potential for the 2080s.[52] The results paint a worrying picture. At global level, aggregate agricultural output potential will be relatively little affected by climate change. However, the average masks significant variations. By the 2080s, agricultural potential could increase by 8 percent in developed countries, primarily as a result of longer growing seasons, while in the developing world it could fall by 9 percent, with sub-Saharan Africa and Latin America projected to experience the greatest losses (figure 2.6).

Sub-Saharan Africa—a region at risk

As the world's poorest and most rainfall-dependent region, sub-Saharan Africa is a cause for special concern. Across the region, agricultural producers are operating with limited resources in fragile environments sensitive to even minor shifts in temperature and rainfall patterns. In dryland areas sophisticated intercropping systems—maize and beans, cowpea and sorghum, and millet and groundnut, for example—have been developed to manage risk and sustain livelihoods. Climate change poses a direct threat to these systems and to the livelihoods that they sustain.

Part of that threat comes from expansion of the area vulnerable to drought, as projected by the Hadley Centre for Climate Change (map 2.1). Arid and semi-arid areas are projected to increase by 60–90 million hectares. By 2090, in some regions, climate change has the potential to cause extreme damage. Southern Africa faces especially acute threats: yields from rainfed agriculture could be reduced by up to 50 percent between 2000 and 2020, according to the IPCC.[53]

Dryland agricultural systems will register some of the most damaging impacts from climate change. One study has looked at the potential implications for dryland areas in

Table 2.4	Agriculture plays a key role in developing regions	
	Agricultural value added (% of GDP) 2005	Agricultural labour force (% of total labour force) 2004
Arab States	7	29
East Asia and the Pacific	10	58
Latin America and the Caribbean	7	18
South Asia	17	55
Sub-Saharan Africa	16	58

Source: Column 1: World Bank 2007d; column 2: WRI 2007b.

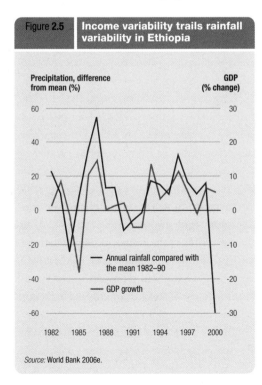

Figure 2.5 Income variability trails rainfall variability in Ethiopia

- Annual rainfall compared with the mean 1982–90
- GDP growth

Source: World Bank 2006e.

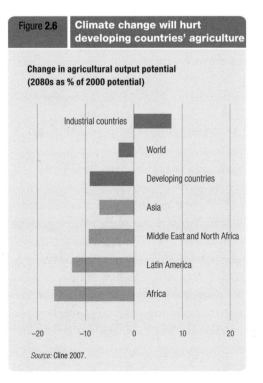

Figure 2.6 Climate change will hurt developing countries' agriculture

Change in agricultural output potential (2080s as % of 2000 potential)

Source: Cline 2007.

2

Map 2.1   Drying out: Africa's drought area is expanding

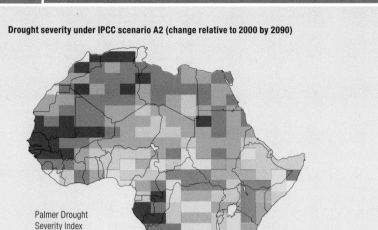

Drought severity under IPCC scenario A2 (change relative to 2000 by 2090)

Palmer Drought
Severity Index

- −5
- −3
- −1
- 0
- 1
- 3
- 5

Note: The boundaries shown and the designations used on this map do not imply official endorsement or acceptance by the United Nations.
IPCC scenarios describe plausible future patterns of population growth, economic growth, technological change and associated CO_2 emissions. The **A1 scenarios** assume rapid economic and population growth combined with reliance on fossil fuels (A1FI), non-fossil energy (A1T) or a combination (A1B). The **A2 scenario**, used here, assumes lower economic growth, less globalization and continued high population growth. A negative change in the Palmer Drought Severity Index, calculated based on precipitation and evaporation projections, implies more severe droughts.

Source: Met Office 2006.

sub-Saharan Africa of a 2.9°C increase in temperature, coupled with a 4 percent reduction in rainfall by 2060. The result: a reduction in revenue per hectare of about 25 percent by 2060. In 2003 prices, overall revenue losses would represent around US$26 billion in 2060 [54]—a figure in excess of bilateral aid to the region in 2005. More broadly, the danger is that extreme food insecurity episodes, such as those that have frequently affected countries like Malawi, will become more common (box 2.7).

Cash crop production in many countries could be compromised by climate change. With an increase of 2°C in average temperatures, it is projected that the land area available for growing coffee in Uganda will shrink.[55] This is a sector that accounts for a large share of cash income in rural areas and figures prominently in export earnings. In some cases, modelling exercises produce optimistic results that mask pessimistic processes. For example, in Kenya it would be

possible to maintain tea production—but not in current locations. Production on Mount Kenya would have to move up to higher slopes currently occupied by forests, suggesting that environmental damage could be a corollary of sustained production.[56]

Climate change on the scale projected for sub-Saharan Africa will have consequences that extend far beyond agriculture. In some countries, there are very real dangers that changed climate patterns will become drivers for conflict. For example, climate models for Northern Kordofan in Sudan indicate that temperatures will rise by 1.5°C between 2030 and 2060, with rainfall declining by 5 percent. Possible impacts on agriculture include a 70 percent drop in yields of sorghum. This is against the backdrop of a long-term decline in rainfall that, coupled with overgrazing, has seen deserts encroach in some regions of Sudan by 100 kilometres over the past 40 years. The interaction of climate change with ongoing environmental degradation has the potential to exacerbate a wide range of conflicts, undermining efforts to build a basis for long-term peace and human security.[57]

The wider threats

These extreme threats facing sub-Saharan Africa should not distract from wider risks for human development. Climate change will have important but uncertain consequences for rainfall patterns across the developing world.

Large uncertainties surround the El Niño/Southern Oscillation (ENSO)—an ocean–atmosphere cycle that spans a third of the globe. In broad terms, El Niño increases the risk of drought across southern Africa and large areas of South and East Asia, while increasing hurricane activity in the Atlantic. Research in India has found evidence of links between El Niño and the timing of the monsoon, on which the viability of entire agricultural systems depends.[58] Even small changes in monsoon intensity and variability could have dramatic consequences for food security in South Asia.

Global projections of climate change can obscure important local effects. Consider the case of India. Some projections point to

Box **2.7** | **Climate change in Malawi—more of the same, and worse**

Climate change models paint a bleak picture for Malawi. Global warming is projected to increase temperatures by 2–3°C by 2050, with a decline in rainfall and reduced water availability. The combination of higher temperatures and less rain will translate into a marked reduction in soil moisture, affecting the 90 percent of smallholder farmers who depend on rainfed production. Production potential for maize, the main smallholder food crop, which in a normal year is the source of three-quarters of calorie consumption, is projected to fall by over 10 percent.

It is hard to overstate the implications for human development. Climate change impacts will be superimposed on a country marked by high levels of vulnerability, including poor nutrition and among the world's most intense HIV/AIDS crisis: almost one million people are living with the disease. Poverty is endemic. Two in every three Malawians live below the national poverty line. The country ranks 164 out of the 177 countries measured in the HDI. Life expectancy has fallen to about 46 years.

Successive droughts and floods in recent years have demonstrated the added pressures that climate change could generate. In 2001/2002, the country suffered one of the worst famines in recent living memory as localized floods cut maize output by one-third. Between 500 and 1,000 people in the central and southern part of the country died during the disaster or in the immediate aftermath. Up to 20,000 are estimated to have died as an indirect result of associated malnutrition and disease. As maize prices rose, malnutrition increased: from 9 percent to 19 percent between December 2001 and March 2002 in the district of Salima.

The 2001/2002 drought undermined coping strategies. People were forced not just to cut back on meals, withdraw children from school, sell household goods and increase casual labour, but also to eat seeds that would have been planted and exchange productive assets for food. As a result, many farmers had no seed to plant in 2002. In 2005, the country was again in the grip of a crisis caused by drought, with more than 4.7 million people out of a population of over 13 million experiencing food shortages.

Climate change threatens to reinforce the already powerful cycles of deprivation created by drought and flood. Incremental risks will be superimposed upon a society marked by deep vulnerabilities. In a 'normal' year, two-thirds of households are unable to produce enough maize to cover household needs. Declining soil fertility, associated with limited access to fertilizer, credit and other inputs, has reduced maize production from 2.0 tonnes per hectare to 0.8 tonnes over the past two decades. Productivity losses linked to reduced rainfall will make a bad situation far worse.

Apart from its immediate consequences for health, HIV/AIDS has created new categories of vulnerable groups. These include households lacking adult labour or headed by elderly people or children, and households with sick family members unable to maintain production. Women are faced with the triple burden of agricultural production, caring for HIV/AID victims and orphans, and collecting water and firewood. Almost all HIV/AIDS-affected households covered in a survey of the Central region reported reduced agricultural production. HIV/AIDS-affected groups will be in the front line facing incremental climate change risks.

For a country like Malawi climate change has the potential to produce extreme setbacks for human development. Even very small increments to risk through climate change can be expected to create rapid downwards spirals. Some of the risks can be mitigated through better information, flood management infrastructure and drought-response measures. Social resilience has to be developed through social provision, welfare transfers and safety nets that raise the productivity of the most vulnerable households, empowering them to manage risk more effectively.

Source: Devereux 2002, 2006c; Menon 2007a; Phiri 2006; Republic of Malawi 2006.

substantial aggregate increases in rainfall for the country as a whole. However, more rain is likely to fall during intense monsoon periods in already rain-abundant parts of the country (creating increased risk of flooding), while other large areas will receive less rainfall. These include drought-prone areas in Andhra Pradesh, Gujarat, Madhya Pradesh and Rajasthan. Microlevel climate research for Andhra Pradesh shows temperatures rising by 3.5°C by 2050, leading to a decline of 8–9 percent in yields for water-intensive crops such as rice.[59]

Losses on this scale would represent a source of greatly increased vulnerability in rural livelihoods. Falling production would reduce the amount of food grown by households for their own consumption, cut supplies to local markets and diminish opportunities for employment. This is another area in which evidence from the past can cast light on future threats. In Andhra Pradesh, one survey covering eight districts in dryland areas found that droughts occurred on average once every 3–4 years, leading to losses in output value of 5–10 percent. This is enough to push many farmers below the poverty line. Models for farm income in India as a whole suggest that a 2–3.5°C temperature increase could be associated with a net farm revenue reduction of 9–25 percent.[60]

Climate shocks: risk and vulnerability in an unequal world

Losses of productivity linked to climate change will increase inequalities between rainfed and commercial producers, undermine livelihoods and add to pressures that are leading to forced migration

The implications of this projection should not be underestimated. While India is a high-growth economy, the benefits have been unequally shared and there is a large human development backlog. Around 28 percent of the population, some 320 million people, live below the poverty line, with three-quarters of the poor in rural areas. Unemployment among rural labourers, one of the poorest groups, is increasing, and almost half of rural children are underweight for their age.[61] Superimposing incremental climate change risks on this large human development deficit would compromise the ambition of 'inclusive growth' set out in India's Eleventh Five–Year Plan.

Projections for other countries in South Asia are no more encouraging:

- Climate scenario exercises for Bangladesh suggest that a 4°C temperature increase could reduce rice production by 30 percent and wheat production by 50 percent.[62]
- In Pakistan, climate models simulate agricultural yield losses of 6–9 percent for wheat with a 1°C increase in temperature.[63]

National projections for climate change in other regions confirm potentially large-scale economic losses and damage to livelihoods. In Indonesia, climate models simulating the impact of temperature changes, soil moisture content and rainfall on agricultural productivity show a wide dispersion of results, with yields falling by 4 percent for rice and 50 percent for maize. Losses will be especially marked in coastal areas where agriculture is vulnerable to salt water incursion.[64]

In Latin America, smallholder agriculture is particularly vulnerable, partly because of limited access to irrigation and partly because maize, a staple across much of the region, is highly sensitive to climate. There is considerable uncertainty in climate model projections for crop production. However, recent models point to the following as plausible outcomes:

- Smallholder losses for maize yields averaging around 10 percent across the region, but rising to 25 percent for Brazil.[65]
- Losses for rainfed maize production will be far higher than for irrigated production

with some models predicting losses of up to 60 percent for Mexico.[66]

- Increased soil erosion and desertification caused by increased rainfall and higher temperatures in southern Argentina, with heavy precipitation and increased exposure to flooding damaging production of soya in the central humid Pampas.[67]

Changes in agricultural production linked to climate change will have important implications for human development in Latin America. While agriculture accounts for a shrinking share of regional employment and GDP, it remains the source of livelihood for a large section of the poor. In Mexico, for example, around 2 million low-income producers depend on rainfed maize cultivation. Maize is the main food staple for producers in the 'poverty-belt' states of southern Mexico, such as Chiapas. Productivity in these states is currently around a third of the level in irrigated commercial agriculture, holding back poverty reduction efforts. Losses of productivity linked to climate change will increase inequalities between rainfed and commercial producers, undermine livelihoods and add to pressures that are leading to forced migration.

Water stress and scarcity

IPCC projection: Changing climate patterns will have important implications for water availability. It is very likely that mountain glaciers and snow cover will continue to retreat. With rising temperatures, changes in runoff patterns and increased water evaporation, climate change will have a marked impact on the distribution of the world's water—and on the timing of flows.

Human development projection: Large areas of the developing world face the imminent prospect of increased water stress. Flows of water for human settlements and agriculture are likely to decrease, adding to already acute pressures in water-stressed areas. Glacial melting poses distinctive human development threats. In the course of the 21st Century water supply stored in glaciers and snow cover will decline, posing immense risks for agriculture, the environment and human settlements. Water stress will figure prominently in low human development traps, eroding the ecological resources on which the poor depend, and restricting options for employment and production.

Water is a source of life and livelihoods. As we showed in the *Human Development Report 2006,* it is vital to the health and well-being of households and an essential input into agriculture and other productive activities. Secure and sustainable access to water—water security in its broadest sense—is a condition for human development.

Climate change will be superimposed on wider pressures on water systems. Many river basins and other water sources are already being unsustainably 'mined'. Today, around 1.4 billion people live in 'closed' river basins where water use exceeds discharge levels, creating severe ecological damage. Symptoms of water stress include the collapse of river systems in northern China, rapidly falling groundwater levels in South Asia and the Middle East, and mounting conflicts over access to water.

Dangerous climate change will intensify many of these symptoms. Over the course of the 21st Century, it could transform the flows of water that sustain ecological systems, irrigated agriculture and supplies of household water. In a world that is already facing mounting pressure on water resources, climate change could add around 1.8 billion people to the population living in a water-scarce environment—defined in terms of a threshold of 1000 cubic metres per capita per annum—by 2080.[68]

Scenarios for the Middle East, already the world's most water-stressed region, point in the direction of increasing pressure. Nine out of fourteen countries in the region already have average per capita water availability below the water scarcity threshold. Decreased precipitation is projected for Egypt, Israel, Jordan, Lebanon and Palestine. Meanwhile, rising temperatures and changes in runoff patterns will influence the flow of rivers upon which countries in the region depend. The following are among the findings to emerge from national climate modelling exercises:

- In Lebanon, a 1.2°C increase in temperature is projected to decrease water availability by 15 percent because of changed runoff patterns and evaporation.[69]
- In North Africa even modest temperature increases could dramatically change water availability. For example, a 1°C increase could reduce water runoff in Morocco's Ouergha watershed by 10 percent by 2020. If the same results hold for other watersheds, the result would be equivalent to losing the water contained by one large dam each year.[70]
- Projections for Syria point to even deeper reductions: a 50 percent decline in renewable water availability by 2025 (based on 1997 levels).[71]

Climate change scenarios for water in the Middle East cannot be viewed in isolation. Rapid population growth, industrial development, urbanization and the need for irrigation water to feed a growing population are already placing immense pressure on water resources. The incremental effects of climate change will add to that pressure within countries, potentially giving rise to tensions over water flowing between countries. Access to the waters of the River Jordan, cross-border aquifers, and the River Nile could become flashpoints for political tensions in the absence of strengthened water-management systems.

Glaciers in retreat

Glacial melting poses threats to more than 40 percent of the world's population.[72] The precise timing and magnitude of these threats remains uncertain. However, they are not a distant prospect. Glaciers are already melting at an accelerating rate. That trend is unlikely to be reversed over the next two to three decades, even with urgent mitigation. Climate change scenarios point to increased flows in the short term, followed by long term drying.

The thousands of glaciers located across the 2,400 kilometres of the Himalayan range are at the epicentre of an emerging crisis. These glaciers form vast water banks. They store water and snow in the form of ice, building up stores during the winter and releasing them during the summer. The flow sustains river systems that are

Climate change will be superimposed on wider pressures on water systems. Many river basins and other water sources are already being unsustainably 'mined'

The past 25 years have seen some glacier systems in the tropics transformed. Their impending disappearance has potentially disastrous implications for economic growth and human development

the lifeblood of vast ecological and agricultural systems.

Himalayas is a Sanskrit word that translates as 'abode of snow'. Today the glacial abode, the largest mass of ice outside of the polar caps, is shrinking at a rate of 10–15 metres a year.[73] The evidence shows the pace of melting to be uneven. But the direction of change is clear.

At current rates two-thirds of China's glaciers—including Tien Shan—will disappear by 2060, with total melting by 2100.[74] The Gangotri glacier, one of the main water reservoirs for the 500 million people living in the Ganges basin, is shrinking by 23 metres a year. One recent study by the Indian Space Research Organisation, using satellite images and covering 466 glaciers, found a 20 percent reduction in size. Glaciers on the Qinghai–Tibet plateau, a barometer of world climate conditions and the source of the Yellow and Yangtze rivers, have been melting by 7 percent a year.[75] With any climate change scenario in excess of the 2°C dangerous climate change threshold, the rate of glacial retreat will accelerate.

Accelerated glacial melt creates some immediate human development risks. Avalanches and floods pose special risks to densely populated mountain regions. One of the countries facing severe risks today is Nepal, where glaciers are retreating at a rate of several metres each year. Lakes formed by melting glacier waters are expanding at an alarming rate—the Tsho Rolpa Lake being a case in point, having increased more than sevenfold in the last 50 years. A comprehensive assessment completed in 2001 identified 20 glacial lakes that could potentially burst their banks, with catastrophic consequences for people, agriculture and hydropower infrastructure, unless urgent action is taken.[76]

As glacial water banks are run down, water flows will diminish. Seven of Asia's great river systems—the Brahmaputra, the Ganges, the Huang He, the Indus, the Mekong, the Salween and the Yangtze—will be affected. These river systems provide water and sustain food supplies for over 2 billion people.[77]

- The flow of the Indus, which receives nearly 90 percent of its water from upper mountain catchments, could decline by as much as 70 percent by 2080.
- The Ganges could lose two-thirds of its July–September flow, causing water shortages for over 500 million people and one-third of India's irrigated land area.
- Projections for the Brahmaputra point to reduced flows of between 14 and 20 percent by 2050.
- In Central Asia, losses of glacial melt into the Amu Darya and Syr Darya rivers could restrict the flow of water for irrigation into Uzbekistan and Kazakhstan, and compromise plans to develop hydroelectric power in Kyrgyzstan.

Climate change scenarios for glacial melting will interact with already severe ecological problems and put pressure on water resources. In India, competition between industry and agriculture is creating tensions over the allocation of water between states. Reduced glacial flows will intensify those tensions. Northern China is already one of the world's most water-stressed regions. In parts of the Huai, Hai and Huang (Yellow) basins (the '3-H' river basins) current water extraction is 140 percent of renewable supply—a fact that explains the rapid shrinkage of major river systems and falling groundwater tables. Over the medium term, changed glacial melting patterns will add to that stress. In an area that is home to around half of China's 128 million rural poor, contains about 40 percent of the country's agricultural land area and accounts for one-third of GDP, this has serious implications for human development (box 2.8).[78]

Tropical glaciers are also shrinking

Tropical glaciers are retreating even more rapidly than those in the Himalayas. In the lifetime of a glacier, a quarter of a century represents the blink of an eye. But the past 25 years have seen some glacier systems in the tropics transformed. Their impending disappearance has potentially disastrous implications for economic growth and human development.

Surveys by geologists suggest that the rate at which Latin America's glaciers are retreating is increasing. There are 2,500 square kilometres of glaciers in the tropical

Box 2.8   Climate change and China's water crisis

Over the past two decades China has emerged as the manufacturing workshop of the world. Rapid economic growth has gone hand-in-hand with a steep decline in poverty and improving human development indicators. Yet China is highly vulnerable to climate change.

By 2020 average temperatures in China are projected to be between 1.1 and 2°C above 1961–1990 levels. In a country as vast as China, spanning several climatic zones, the effects will be complex and diverse. However, a National Climate Change Assessment predicts more droughts, spreading deserts and reduced water supplies. Projections for agriculture suggest that the production of rice, maize and wheat could fall by 10 percent by 2030, and by up to 37 percent during the second half of the century because of climate-related factors.

As in other countries, climate change in China will interact with underlying stresses. The river systems of northern China provide a powerful demonstration of the ecological pressures generated by rapid economic growth. The Hai, Huai and Huang (Yellow) River Basins (the 3-H river basins) supply just under half of China's population with water. With the growing demands of industry, urban centres and agriculture, water is being withdrawn from the basins at twice the rate of replenishment. The result: rivers that no longer reach the sea and sinking groundwater tables.

Any reduction in water flows through the 3-H basins could rapidly turn an ecological crisis into an outright social and economic disaster. Around one-third of China's GDP originates in the basins, along with a large share of its grain production. One in every two of the rural poor lives here—most of them directly dependent on agriculture. As drought, rising temperatures and reduced runoff under climate change take effect, an obvious danger is that the adjustment costs will be borne first by the poor.

In western China entire ecological systems are under threat. Projected temperature increases for this region are 1–2.5°C by 2050. The Qinghai–Tibet plateau covers a landmass the size of Western Europe and contains more than 45,000 glaciers. These glaciers are retreating at the dramatic rate of 131.4 square kilometres annually. On current trends, most will disappear altogether by the end of the century.

What is happening to China's glaciers constitutes a national ecological security crisis of the first order. In the short term, increased flows of water from ice melt are likely to lead to more flooding. In the long term, the retreat of the glaciers will deprive communities living in the mountains of their water and transform large swathes of China's environment. Desertification will gather pace as rising temperatures and unsustainable land-use practices continue to accelerate to soil erosion. Events such as the 13 major dust-storms recorded in 2005, one of which deposited 330,000 tonnes of sand in Beijing, will become more common. Meanwhile, flows into the Yangtze, the Yellow and other rivers that originate on the Qinghai–Tibet plateau will decline, adding to the stress on water-based ecological systems.

It is not only rural environments that stand to suffer. The city of Shanghai is particularly vulnerable to climate-related hazards. Located at the mouth of the Yangtze river, at an elevation of only 4 metres above sea level, the city faces acute flood risks. Summer typhoons, storm surges and excessive river runoff contribute to extreme flooding.

All of Shanghai's 18 million residents face the risk of flooding. Rising sea levels and increased storm surges have put the coastal city on the danger list. However, vulnerability is heavily concentrated among the estimated 3 million temporary residents who have migrated from rural areas. Living in transient encampments around construction sites or in flood-prone areas, and with limited rights and entitlements, this population is faced with a high exposure to risk with extreme vulnerability.

Source: Cai 2006; O'Brien and Leichenko 2007; People's Republic of China 2007; Shen and Liang 2003.

Andes, of which 70 percent are located in Peru and 20 percent in Bolivia. The remaining mass is accounted for by Colombia and Ecuador. Since the early 1970s, it is estimated that, the surface area of glaciers in Peru has declined between 20 and 30 percent, with the Quelcayya ice cap in the vast Cordillera Blanca range losing almost a third of its area. Some of the smaller glaciers in Bolivia have already disappeared (figure 2.7). Research by the World Bank predicts that many of the lower glaciers in the Andes will be a matter for the history books within a decade.[79]

One immediate danger is that melting ice will lead to the formation of larger glacial lakes, leading to increased risk of flooding, avalanches, mudslides and the bursting of dams. The warning signs are already evident: for example, the surface area of Lake Safuna Alta, in the Cordillera Blanca in Peru, has increased by a factor of five since 1975.[80] Many basins fed by glaciers have experienced an increase in runoff in recent years. However, models predict a rapid fall-off in flows after 2050, especially in the dry season.

This is a particular concern for Peru. Populations living in arid coastal areas, including the capital Lima, depend critically on water supplies from melting glaciers in the Andes. In a country that is already struggling to provide basic water services to urban populations, glacial

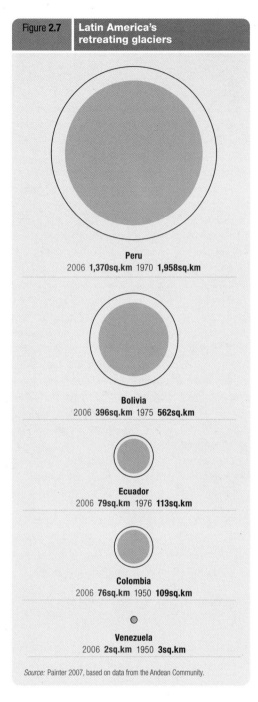

Figure 2.7 Latin America's retreating glaciers

Peru
2006 **1,370sq.km** 1970 **1,958sq.km**

Bolivia
2006 **396sq.km** 1975 **562sq.km**

Ecuador
2006 **79sq.km** 1976 **113sq.km**

Colombia
2006 **76sq.km** 1950 **109sq.km**

Venezuela
2006 **2sq.km** 1950 **3sq.km**

Source: Painter 2007, based on data from the Andean Community.

melting poses a real and imminent threat to human development (box 2.9).

Rising seas and exposure to extreme weather risks

The IPCC projection: It is likely that tropical cyclones—typhoons and hurricanes—will become more intense as oceans warm, with higher peak speeds and heavier precipitation. All typhoons and hurricanes are driven by energy released from the sea—and energy levels will rise. One study has found a doubling of power dissipation in tropical cyclones over the past three decades.[81] Sea levels will continue to rise, though there is uncertainty about by how much. Oceans have absorbed over 80 percent of the increased heat generated by global warming, locking the world into continued thermal expansion.[82] Drought and floods will become more frequent and widespread across much of the world.

The human development projection: Emerging risk scenarios threaten many dimensions of human development. Extreme and unpredictable weather events are already a major source of poverty. They bring near-term human insecurity and destroy long-term efforts aimed at raising productivity, improving health and developing education, perpetuating the low human development traps described earlier in this chapter. Many countries have large and highly vulnerable populations that will face a steep increase in climate-related risks, with people living in coastal areas, river deltas, urban slums and drought-prone regions facing immediate threats.

Climate change is only one of the forces that will influence the profile of risk exposure in the decades ahead. Other global processes—ecological stress, urbanization and population growth among them—will also be important. However, climate change will reconfigure patterns of risk and vulnerability across many regions. The combination of increasing climate hazards and declining resilience is likely to prove a lethal mix for human development.

Any increase to climate-related risk exposure has to be assessed against the backdrop of current exposure. That backdrop includes the following numbers of people facing climate-related hazards:[83]

- 344 million exposed to tropical cyclones;
- 521 million exposed to floods;
- 130 million exposed to droughts;
- 2.3 million exposed to landslides.

As these figures indicate, even small increases to risk over time will affect very large numbers of people. Like climate change itself, the potential linkages between changing weather patterns and evolving trends in risk

For centuries, the runoff from glaciers in the Andean range has watered agricultural lands and provided human settlements with a predictable flow of water. Today, the glaciers are among the early casualties of climate change. They are melting fast—and their impending disappearance has potentially negative implications for human development in the Andean region.

Peru and Bolivia are the location for the world's largest expanse of tropical glaciers—around 70 percent of the total for Latin America is in Peru and 20 percent in Bolivia. These countries are also home to some of the largest concentrations of poverty and social and economic inequalities in Latin America—the world's most unequal region. Glacial melt threatens not just to diminish water availability, but to exacerbate these inequalities.

Geography is part of the explanation for the risks now facing countries like Peru. Eastern Peru has 98 percent of the country's water resources, but two in every three Peruvians live on the western desert coast—one of the world's most arid regions. Urban water supplies and economic activity are sustained by some 50 rivers flowing from the Andes, with around 80 percent of the fresh water resources originating from snow or glacial melt. Glacier-fed surface waters constitute the source of water, not only for many rural areas but also for major cities and hydroelectric power generation.

Peru has registered some of the most rapid rates of glacial retreat in the world. Between 20 and 30 percent of the glacial surface area has been lost in the last three decades. That area is equivalent to the total glacial surface in Ecuador.

The capital city Lima, with a population of nearly 8 million, is on the coast. Lima gets its water from the Rio Rimac and other rivers in the Cordillera Central, all of which depend to varying degrees on glacial melt. There is already a large gap between supply and demand for water. Overall population is growing at 100,000 a year, driving up demand for water. Rationing is already common in the summer. With limited reservoir storage and exposure to drought increasing, the city would face more rationing in the short term.

Rapid glacial recession in the vast Cordillera Blanca, in the northern Andes, would call into question the future of agriculture, mining, power generation and water supplies across large areas. One of the rivers nourished by the Cordillera Blanca is the Rio Santa. The river sustains a wide array of livelihoods and economic activity. At altitudes of between 2,000 and 4,000 metres, the river delivers the water that irrigates mostly small-scale agriculture. In the lower valleys it irrigates large-scale commercial agriculture, including two large irrigation projects for export crops. Its flow generates hydroelectric power and delivers drinking water to two major urban areas on the Pacific coast—Chimbote and Trujillo—with a combined population of more than one million people.

The problem is that up to 40 percent of the dry season discharge from the Rio Santa originates in melting ice that is not being replenished through annual precipitation. Major economic losses and damage to livelihoods could result. The Chavimochic irrigation scheme on the Rio Santa has contributed to a remarkable national boom in non-traditional agriculture. Total exports from the sector increased from US$302 million in 1998 to US$1 billion in 2005. The boom has been sustained by water-intensive products such as artichokes, asparagus, tomatoes and other vegetables. Glacial melting threatens to erode the viability of investments in irrigation, undermining employment and economic growth in the process.

Monitoring the retreat of tropical glaciers in the Peruvian Andes is relatively straightforward. Developing a response is more challenging. Compensating for the loss of glacial flows in the medium term will require billions of dollars of investment in the construction of tunnels beneath the Andes. Compensating for power losses will require investments in thermal power generation estimated by the World Bank at US$1.5 billion. The price tag points to tough questions about cost sharing at both the domestic and international levels. People in Peru are not responsible for glacial melting: they account for 0.1 percent of the world's carbon emissions. Yet they face the prospect of paying a high financial and human price for the far higher carbon emissions of other countries.

Source: Carvajal 2007; CONAM 2004; Coudrain, Francou and Kundzewicz 2005; Painter 2007.

2

Climate shocks: risk and vulnerability in an unequal world

and vulnerability are complex. They are also non-linear. There is no ready-made calculus for assessing the human development impact of a 2-metre sea-level rise coupled with an increase in tropical storm intensity. But it is possible to identify some of the linkages and transmission mechanisms.

Drought

Increased exposure to drought is of particular concern in sub-Saharan Africa, though other regions, including South Asia and Latin America, could also be affected. Agricultural production is likely to suffer in these regions, especially those dominated by rainfed production. In sub-Saharan Africa, the areas suitable for agriculture, the length of growing seasons and the yield potential of food staples are all projected to decline (see section on Agricultural production and food security above). By 2020, between 75 million and 250 million more people in sub-Saharan Africa could have their livelihoods and human development prospects compromised by a combination of drought, rising temperature and increased water stress.[84]

Floods and tropical storms

There are large margins of uncertainty in projections for populations exposed to risk from flooding.[85] Accelerated disintegration of the West Antarctic ice sheet could multiply sea-level rises by a factor of five over and above the ceiling predicted by the IPCC. However, even more benign scenarios are a source of concern.

One model using an IPCC scenario for high population growth estimates the number of additional people experiencing coastal flooding at 134–332 million for a 3–4°C rise in temperature.[86] Factoring in tropical storm activity could increase the numbers affected to 371 million by the end of the 21st Century.[87] Among the consequences of a 1-metre rise in sea levels:

- In Lower Egypt, possible displacement of 6 million people and flooding of 4,500km² of farmland. This is a region marked by high levels of deprivation in many rural areas, with 17 percent of the population—some 4 million people—living below the poverty line.[88]

- The displacement of up to 22 million people in Viet Nam, with losses of up to 10 percent of GDP. Flooding and more intensive storms could slow human development progress in major population areas, including the Mekong Delta (box 2.10).

- In Bangladesh, one metre rise in sea level would inundate 18 percent of land area, directly threatening 11 percent of the population. The impact on river levels from sea rises could affect over 70 million people.[89]

While most of the people affected by rising sea levels live in a small number of countries with large populations, the impacts will be far more widely distributed (table 2.5). For many low-lying small-island states, rising sea levels and storms point to a highly predictable social, economic and ecological crisis. For the Maldives, where 80 percent of the land area is less than 1 metre above sea level, even the most

---

Box 2.10 Climate change and human development in the Mekong Delta

Over the past 15 years, Viet Nam has made spectacular progress in human development. Poverty levels have fallen and social indicators have improved, putting the country ahead of schedule on almost all of the MDGs. Climate change poses a real and imminent danger to these achievements—and nowhere more so than in the Mekong Delta.

Viet Nam has a long history of dealing with extreme weather. Located in a typhoon zone, with a long coastline and extensive river deltas, the country is close to the top of the natural disasters league table. On average, there are six to eight typhoons each year. Many leave an extensive trail of destruction, killing and injuring people, damaging homes and fishing boats, and destroying crops. The country's 8,000 kilometres of sea and river dykes, some of which have been developed through communal labour over centuries, testify to the scale of national investment in risk management.

The Mekong Delta is an area of special concern. One of the most densely populated parts of Viet Nam, it is home to 17.2 million people. It is also the 'rice basket' of the country, playing a critical role in national food security. The Mekong Delta produces half of Viet Nam's rice and an even larger share of fisheries and fruit products.

The development of agriculture has played a pivotal role in poverty reduction in the Mekong Delta. Investment in irrigation and support for marketing and extension services has enabled farmers to intensify production, growing two or even three crops a year. Farmers have also constructed dykes and embankments to protect their fields from the flooding that can accompany typhoons and heavy rains.

Climate change poses threats at several levels. Rainfall is predicted to increase and the country will face more intensive tropical storms. Sea levels are expected to rise by 33 cm by 2050 and 1 metre by 2100.

For the low-lying Mekong Delta this is a particularly grim forecast. The sea-level rise projected for 2030 would expose around 45 percent of the Delta's land area to extreme salinization and crop damage through flooding. Crop productivity for rice is forecast to fall by 9 percent. If sea levels rise by 1 metre, much of the Delta would be completely inundated for some periods of the year.

How might these changes impact on human development in the Mekong Delta? While poverty levels have been falling, inequality has been increasing, driven partly by high levels of landlessness. There are still 4 million people living in poverty in the Delta. Many of these people lack basic health protection and school drop-out rates for their children are high. For this group, even a small decline in income or loss of employment opportunities linked to flooding would have adverse consequences for nutrition, health and education. The poor face a double risk. They are far more likely to live in areas vulnerable to flooding—and they are less likely to live in more robust permanent homes.

Source: Chaudhry and Ruysschaert 2007; Nguyen 2007; UNDP and AusAID 2004.

benign climate change scenarios point to deep vulnerabilities.

Small-island developing states are on the front line of climate change. They are already highly vulnerable to climate disasters. Annual damages for the Pacific islands of Fiji, Samoa and Vanuatu are estimated at 2–7 percent of GDP. In Kiribati, one estimate of the combined annual damage bill from climate change and sea-level rises in the absence of adaptation puts the figure at a level equivalent to 17–34 percent of GDP. [90]

Islands in the Caribbean are also at risk. With a 50 centimetre increase in sea levels, over one-third of the Caribbean's beaches would be lost, with damaging implications for the region's tourist industry. An increase of 1 metre would permanently submerge about 11 percent of the land area in the Bahamas. Meanwhile, the intrusion of salt water would compromise freshwater supplies, forcing governments to undertake costly investments in desalination.[91]

More intense tropical storm activity is one of the givens of climate change. Warming seas will fuel more powerful cyclones. At the same time, higher sea temperatures and wider climate change may also alter the course of cyclone tracks and the distribution of storm activity. The first-ever hurricane in the South Atlantic struck Brazil in 2004, and 2005 marked the first hurricane to hit the Iberian peninsula since the 1820s.

Scenarios for tropical storm activity demonstrate the importance of interactions with social factors. In particular, rapid urbanization is placing a growing population in harm's way. Approximately 1 billion people already live in informal urban settlements, and numbers are rising. UN-HABITAT estimates that if current trends continue there will be 1.4 billion people living in slums by 2020 and 2 billion by 2030: one in every three urban dwellers. While more than half the world's slum population today lives in Asia, sub-Saharan Africa has some of the world's fastest growing slums.[92]

Living in makeshift homes often located on hillsides vulnerable to flooding and landslides, slum dwellers are both highly exposed and highly vulnerable to climate change impacts.

| Table 2.5 | Rising sea levels would have large social and economic impacts |

Magnitude of sea level rise (m)	Impact (% of global total)					
	Land area	Population	GDP	Urban area	Agricultural area	Wetland area
1	0.3	1.3	1.3	1.0	0.4	1.9
2	0.5	2.0	2.1	1.6	0.7	3.0
3	0.7	3.0	3.2	2.5	1.1	4.3
4	1.0	4.2	4.7	3.5	1.6	6.0
5	1.2	5.6	6.1	4.7	2.1	7.3

Source: Dasgupta et al. 2007.

These impacts will not be determined purely through physical processes. Public policies can improve resilience in many areas, ranging from flood control to infrastructural protection against landslides and the provision of formal settlement rights to urban slum dwellers. In many cases the absence of formal rights is a deterrent to investment in more robust building materials.

Climate change will create mounting threats. Even robust mitigation will do little to lessen those threats until 2030. Until then, the urban poor will have to adapt to climate change. Supportive public policies could help that adaptation. The starting points: creating more secure tenure rights, investing in slum upgrading and providing clean water and sanitation to the urban poor.

Ecosystems and biodiversity

IPCC projection: There is a high confidence probability that the resilience of many ecosystems will be undermined by climate change, with rising CO_2 levels reducing biodiversity, damaging ecosystems and compromising the services that they provide.

Human development projection: The world is heading towards unprecedented losses of biodiversity and the collapse of ecological systems during the 21st Century. At temperature increases in excess of 2°C, rates of extinction will start to increase. Environmental degradation will gather pace, with coral, wetland and forest systems suffering rapid losses. The processes are already under

Losses of biodiversity are mounting in many regions. Climate change is one of the forces driving these trends. Over time it will become a more powerful force

way. Losses of ecosystems and biodiversity are intrinsically bad for human development. The environment matters in its own right for current and for future generations. However, vital ecosystems that provide wide ranging services will also be lost. The poor, who depend most heavily on these services, will bear the brunt of the cost.

As in other areas, the processes of climate change will interact with wider pressures on ecosystems and biodiversity. Many of the world's great ecosystems are already under threat. Losses of biodiversity are mounting in many regions. Climate change is one of the forces driving these trends. Over time it will become a more powerful force.

The rapidly deteriorating state of the global environment provides the context for assessing the impact of future climate change. In 2005, the *Millennium Ecosystem Assessment* found that 60 percent of all ecosystem services were either degraded or being used unsustainably.[93] The loss of mangrove swamps, coral reef systems, forests and wetlands was highlighted as a major concern, with agriculture, population growth and industrial development acting together to degrade the environmental resource base. Nearly one in four mammal species is in serious decline.[94]

Losses of environmental resources will compromise human resilience in the face of climate change. Wetlands are an example. The world's wetlands provide an astonishing range of ecological services. They harbour biodiversity, provide agricultural, timber and medicinal products, and sustain fish stocks. More than that, they buffer coastal and riverside areas from storms and floods, protecting human settlements from sea surges. During the 20th Century, the world lost half its wetlands through drainage, conversion to agriculture and pollution. Today, the destruction continues apace at a time when climate change threatens to generate more intensive storms and sea surges.[95] In Bangladesh, the steady erosion of the mangrove areas in the Sundabarns and other regions has undermined livelihoods while increasing exposure to rising sea levels.

Climate change is transforming the relationship between people and nature. Many ecosystems and most species are highly susceptible to shifts in climate. Animals and plants are adapted to specific climate zones. Only one species has the ability to adjust the climate through thermostats attached to heating or cooling devices—and that is the species responsible for global warming. Plants and animals have to adapt by moving.

Ecological maps are being redrawn. Over the past three decades, the lines marking regions in which average temperatures prevail—'isotherms'—have been moving towards the North and South Poles at a rate of about 56 kilometres per decade.[96] Species are attempting to follow their climate zones. Changes in flowering seasons, migratory patterns and the distribution of flora and fauna have been detected across the world. Alpine plants are being pushed towards higher altitudes, for example. But when the pace of climate change is too rapid, or when natural barriers such as oceans block migration routes, extinction looms. The species most at risk are those in polar climates, because they have nowhere to go. Climate change is literally pushing them off the planet.

Climate change has already contributed to a loss of species—and global warming in the pipeline will add to that loss. But far greater impacts will take off at 2°C over preindustrial levels. This is the threshold at which predicted extinction rates start to rise. According to the IPCC, 20–30 percent of plant and animal species are likely to be at increased risk of extinction if global average temperature increases exceed 1.5–2.5°C, including polar bears and fish species that feed on coral reefs. Some 277 medium or large mammals in Africa would be at risk in the event of 3°C warming.[97]

The Arctic under threat

The Arctic region provides an antidote to the view that climate change is an uncertain future threat. Here, fragile ecological systems have come into contact with rapid and extreme temperature increases. Over the past 50 years, mean annual surface temperature in areas from

Alaska to Siberia has increased by 3.6°C—more than twice the global average. Snow cover has declined by 10 percent in the past 30 years, and average sea ice cover by 15–20 percent. Permafrost is melting and the tree line is shifting northwards.

Climate change scenarios point in a worrying direction. Mean surface temperatures are projected to increase by another 3°C by 2050, with dramatic reductions in summer sea ice, the encroachment of forests into tundra regions, and extensive loss of ecosystems and wildlife. Entire species are at risk. As the Arctic Climate Impact Assessment puts it: "Marine species dependent on sea ice, including polar bears, ice-living seals, walrus and some marine birds, are very likely to decline, with some facing extinction."[98]

The United States has acknowledged the impact of climate change on the Arctic. In December 2006, the US Department of the Interior proposed, on the basis of "the best scientific evidence", placing the polar bear on the Endangered Species list. That act effectively acknowledges the role played by climate change in increasing its vulnerability—and it requires government agencies to protect the species. More recently, polar bears have been joined on the list by 10 species of penguin which are also under threat. Unfortunately, the "best scientific evidence" points in a worrying direction: within a couple of generations, the only polar bears on the planet could be those on display in the world's zoos. The late summer Arctic sea ice, on which they depend for hunting, has been shrinking at over 7 percent a decade since the late 1970s. Recent scientific studies of adult polar bears in Canada and Alaska have shown weight loss, reduced cub survival, and an increase in the number of bears drowning as they are forced to swim further in search of prey. In western Hudson Bay, populations have fallen by 22 percent.[99]

The United States Department of the Interior's actions establish an important principle of shared responsibility across borders. That principle has wider ramifications. Polar bears cannot be treated in isolation. They are part of a wider social and ecological system. And if the impact of climate change and associated responsibilities of governments

are recognized for the Arctic the principle should be more widely applied. People living in drought-prone areas of Africa and flood-prone regions of Asia are also affected. Applying one set of rules for polar bears and another for vulnerable people in approaches to climate change mitigation and adaptation would be inconsistent.

The sheer pace of climate change across the Arctic is creating challenges at many levels. Loss of permafrost could unlock vast amounts of methane—a potent greenhouse gas that could undermine mitigation efforts by acting as a driver for 'positive feedbacks'. The rapid melting of Arctic ice has opened up new areas to exploration for oil and natural gas, giving rise to tensions between states over the interpretation of the 1982 Convention on the Law of the Sea.[100] Within countries, climate change could lead to immense social and economic harm, damaging infrastructure and threatening human settlements.

Scenarios for Russia illustrate the point. With climate change, Russia will experience warming effects that could raise agricultural production, though increased exposure to drought may negate any benefits. One of the more predictable consequences of climate change for Russia is increased thawing of the permafrost which covers approximately 60 percent of the country. Thawing has already led to increases in winter flows of major rivers. Accelerated melting will affect coastal and river bank human settlements, exposing many to flood risks. It will also require heavy investments in infrastructural adaptation, with roads, electrical transmission lines and the Baikal Amur railway potentially affected. Plans are already being drawn up to protect the planned East Siberia–Pacific export oil pipeline through extensive trenching to combat coastal erosion linked to permafrost melting—a further demonstration that ecological change carries real economic costs.[101]

The coral reef—a climate change barometer

Arctic regions provide the world with a highly visible early warning system for climate change.

The "best scientific evidence" points in a worrying direction: within a couple of generations, the only polar bears on the planet could be those on display in the world's zoos

2

Coral reefs are not just havens of exceptional biodiversity, but also a source of livelihoods, nutrition and economic growth for over 60 countries

Other ecosystems provide an equally sensitive though less immediately visible barometer. Coral reefs are an example. During the 21st Century, warming oceans and rising acidification could destroy much of the world's coral, with devastating social, ecological and economic consequences.

Warming seas have contributed to the destruction of coral reefs on an extensive scale, with half of all systems in decline.[102] Even fairly short periods of abnormally high temperature—as little as 1°C higher than the long term average—can cause corals to expel the algae that supply most of their food, resulting in 'bleaching' and sudden death of the reef.[103]

The world's coral reef systems already bear scars from climate change. Around half these systems have already been affected by bleaching. The 50,000 km² of coral reef in Indonesia, 18 percent of the world's total, is deteriorating rapidly. One survey in Bali Barat National Park in 2000 found that the majority of the reef had been degraded, most of it by bleaching.[104] Aerial views of the Great Barrier Reef in Australia also capture the extent of bleaching.

There could be far worse to come. With average temperature increases above 2°C, annual bleaching would be a regular event. The major bleaching events that accompanied the 1998 El Niño, when 16 percent of the world's coral was destroyed in 9 months, would become the rule, rather than the exception. Localized bleaching episodes are becoming more frequent in many regions, providing a worrying pointer for the future. For example, in 2005, the eastern Caribbean suffered one of the worst bleaching episodes on record.[105]

Bleaching is just one of the threats posed by climate change. Many marine organisms, including coral, make their shells and skeletons out of calcium carbonate. The upper ocean is super-saturated with these minerals. However, the increases in ocean acidity caused by the 10 billion tonnes of CO_2 being absorbed by the oceans each year attacks carbonate, removing one of the essential building blocks needed by coral.[106]

Marine scientists have pointed to a worrying parallel. Ocean systems respond slowly and over very long time horizons to changes in the atmospheric environment. Business-as-usual climate change in the 21st Century could make the oceans more acidic over the next few centuries than they have been at any time for 300 million years, with one exception: a single catastrophic episode that occurred 55 million years ago. That episode was the result of the rapid ocean acidification caused by the release of 4,500 gigatonnes of carbon.[107] It took over 100,000 years for the oceans to return to their previous acidity levels. Meanwhile, geological records show a mass extinction of sea creatures. As one of the world's leading oceanographers puts it: "Nearly every marine organism that made a shell or a skeleton out of calcium carbonate disappeared from the geologic record … if CO_2 emissions are unabated, we may make the oceans more corrosive to carbonate minerals than at any time since the extinction of the dinosaurs. I personally believe that this will cause the extinction of corals."[108]

The collapse of coral systems would represent a catastrophic event for human development in many countries. Coral reefs are not just havens of exceptional biodiversity, but also a source of livelihoods, nutrition and economic growth for over 60 countries. Most of the 30 million small-scale fishers in the developing world are dependent in some form on coral reefs for maintaining feeding and breeding grounds. More than half of the protein and essential nutrients in the diets of 400 million poor people living in tropical coastal areas is supplied by fish.

Coral reefs are a vital part of the marine ecosystems that sustain fish stocks, though warming oceans pose wider threats. In Namibia, anomalously warm water currents in 1995—the Benguela Niño current—resulted in fish stocks moving 4–5° of latitude south—an outcome that destroyed a small-scale fisheries industry for pilchards.[109]

Beyond their value in the lives and nutrition of the poor, corals have a wider economic value. They generate income, exports and, in regions such as the Indian Ocean and the Caribbean, support tourism. Recognition of the important role of coral in economic, ecological and social life has prompted many governments and aid donors to invest in rehabilitation. The problem is that climate change is a powerful force pulling in the other direction.

Human health and extreme weather events

IPCC projection: Climate change will affect human health through complex systems involving changes in temperature, exposure to extreme events, access to nutrition, air quality and other vectors. Currently small health effects can be expected with very high confidence to progressively increase in all countries and regions, with the most adverse effects in low-income countries.

Human development projection: Climate will interact with human health in diverse ways. Those least equipped to respond to changing health threats—predominantly poor people in poor countries—will bear the brunt of health setbacks. Ill-health is one of the most powerful forces holding back the human development potential of poor households. Climate change will intensify the problem.

Climate change is likely to have major implications for human health in the 21st Century. Large areas of uncertainty surround assessments, reflecting the complex interaction between disease, environment and people. However, in health, as in other areas, recognition of uncertainty is not a case for inaction. The World Health Organization (WHO) predicts that the overall impact will be negative.[110]

Public health outcomes linked to climate change will be shaped by many factors. Pre-existing epidemiology and local processes will be important. So, too, will pre-existing levels of human development and the capacities of public health systems. Many of the emerging risks for public health will be concentrated in developing countries where poor health is already a major source of human suffering and poverty—and where public health systems lack the resources (human and financial) to manage new threats. An obvious danger is that climate change under these conditions will exacerbate already extreme global inequalities in public health.

Malaria gives rise to some of the greatest causes for concern. This is a disease that currently claims around 1 million lives annually, over 90 percent of them in Africa. Some 800,000 children under the age of 5 in sub-Saharan Africa die as a result of malaria each year, making it the third largest killer of children worldwide.[111] Beyond these headline figures, malaria causes immense suffering, robs people of opportunities in education, employment and production, and forces people to spend their limited resources on palliative treatment. Rainfall, temperature and humidity are three variables that most influence transmission of malaria—and climate change will affect all three.

Increased rain, even in short downpours, warmer temperatures and humidity create a 'perfect storm' for the spread of the *Plasmodium* parasite that causes malaria. Rising temperatures can extend the range and elevation of mosquito populations, as well as halving incubation periods. For sub-Saharan Africa in particular, any extension of the malaria range would pose grave risks to public health. Some four in five people in the region already live in malarial areas. Future projections are uncertain, though there are concerns that the malarial range could expand in upland areas. More disconcerting still, the seasonal transmission period may also increase, effectively increasing average per capita exposure to malarial infection by 16–28 percent.[112] Worldwide it is estimated that an additional 220–400 million people could be exposed to malaria.[113]

Changing weather patterns are already producing new disease profiles in many regions. In eastern Africa, flooding in 2007 created new breeding sites for disease vectors such as mosquitoes, triggering epidemics of Rift Valley Fever and increasing levels of malaria. In Ethiopia, an epidemic of cholera following the extreme floods in 2006 led to widespread loss of life and illness. Unusually dry and warm conditions in eastern Africa have been linked to the spread of *chikungunya* fever, a viral disease that has proliferated across the region.[114]

Climate change could also increase the population exposed to dengue fever. This is a highly climate-sensitive disease that is currently

Changing weather patterns are already producing new disease profiles in many regions

Urgent action is needed to conduct assessments of the risks posed by climate change to public health in the developing world, followed by a mobilization of resources to create an enabling environment for risk management

largely confined to urban areas. Latitudinal expansion linked to climate change could increase the population at risk from 1.5 billion people to 3.5 billion by 2080.[115] Dengue fever is already in evidence at higher elevations in previously dengue-free areas of Latin America. In Indonesia, warmer temperatures have led to the mutation of the dengue virus, leading to an increase in fatalities in the rainy season. While there is no proven evidence that climate change is implicated, in the late 1990s El Niño and La Niña events in the country were associated with severe outbreaks of both dengue and malaria, with malaria spreading to high elevations in the highlands of Irian Jaya.[116]

Extreme climate events provide another set of threats. Floods, droughts and storms bring in their wake increased health risks, such as cholera and diarrhoea among children. There is already evidence in developing countries of the impacts of rising temperatures. During 2005, Bangladesh, India and Pakistan faced temperatures 5–6°C above the regional average. There were 400 reported deaths in India alone, though unreported deaths would multiply this figure many times over.[117] Public health in developed countries has not been immune. The heat-wave that hit Europe in 2003 claimed between 22,000 and 35,000 lives, most of them elderly. In Paris, the worst affected city, 81 percent of the victims were aged over 75 years.[118] More events of this nature are likely. For example, the incidence of heat waves in most United States' cities is expected to approximately double by 2050.[119]

Public health authorities in rich nations are being forced to confront the challenges posed by climate change. The city of

New York provides an example of a wider process. Climate impact assessments have pointed to higher summer-season temperatures, with increasing frequency and duration of heat waves. The prognosis: a projected increase in summer-season heat stress morbidity, particularly among the elderly poor. Summer heat-related mortality could increase 55 percent by the 2020s, more than double by the 2050s and more than triple by the 2080s.[120] Climate change could also contribute indirectly to at least three classes of wider health problems: incidence of certain vector-borne diseases such as West Nile Virus, Lyme disease and malaria may rise; water-borne disease organisms may become more prevalent; and photochemical air pollution may increase.[121] Strategies are being developed to address the risks.

Governments in the developed world have to respond to the public health threats posed by climate change. Many authorities— as in New York—acknowledge the special problems faced by poor and vulnerable populations. Yet it would be wrong for countries with first class health systems and the financial resources needed to counteract climate change threats at home, to turn a blind eye to the risks and vulnerabilities faced by the poor in the developing world. Urgent action is needed to conduct assessments of the risks posed by climate change to public health in the developing world, followed by a mobilization of resources to create an enabling environment for risk management. The starting point for action is the recognition that rich countries themselves carry much of the historic responsibility for the threats now facing the developing world.

Conclusion

"We are made wise not by the recollection of our past" wrote George Bernard Shaw, "but by the responsibility for our future." Viewed from the perspective of human development, climate change brings the past and the future together.

In this chapter we have looked at the 'early harvest' climate change catastrophe. That harvest, which has already begun, will initially slow progress in human development. As climate change develops, large-scale reversals will become more likely. Evidence from the past provides us with insights into the processes that will drive these reversals, but the future under climate change will not look like the past. Setbacks for human development will be non-linear, with powerful mutually reinforcing feedback effects. Losses in agricultural productivity will reduce income, diminishing access to health and education. In turn, reduced opportunities in health and education will restrict market opportunities and reinforce poverty. At a more fundamental level, climate change will erode the ability of the world's most vulnerable people to shape decisions and processes that impact on their lives.

Catastrophic human development setbacks are avoidable. There are two requirements for changing the 21st Century scenario to a more favourable direction. The first is climate change mitigation. Without early and deep cuts in emissions of CO_2, dangerous climate change will happen—and it will destroy human potential on a vast scale. The consequences will be reflected in surging inequalities within and across countries and rising poverty. Rich countries may escape the immediate effects. They will not escape the consequences of the anger, resentment and transformation of human settlement patterns that will accompany dangerous climate change in poor countries.

The second requirement for averting the threats set out in this chapter is adaptation. No amount of mitigation will protect vulnerable people in developing countries from the incremental climate change risks that they face today, or from the global warming to which the world is already committed. Increased risk exposure is inevitable—human development reversals are not. Adaptation is ultimately about building the resilience of the world's poor to a problem largely created by the world's richest nations.

Catastrophic human development setbacks are avoidable

3

Avoiding dangerous climate change: strategies for mitigation

"We shall require a substantially new manner of thinking if mankind is to survive."

Albert Einstein

• •

"Speed is irrelevant if you are going in the wrong direction."

Mahatma Gandhi

• •

"Alone we can do so little; together we can do so much."

Helen Keller

CHAPTER 3

Avoiding dangerous climate change: strategies for mitigation

Living within a sustainable 21st Century carbon budget requires that rich countries cut emissions of greenhouse gases by at least 80 percent by 2050, with 30 percent cuts by 2020

Climate change is an immense, long-term and global challenge that raises difficult questions about justice and human rights, both within and across generations. Humanity's ability to address these questions is a test of our capacity to manage the consequences of our own actions. Dangerous climate change is a threat, not a pre-ordained fact of life. We can choose to confront and eliminate that threat, or we can choose to let it evolve into a fully fledged crisis for poverty reduction and for future generations.

Approaches to mitigation will determine the outcome. The more we delay action, the more atmospheric concentrations of greenhouse gases will rise, the more difficult it will be to stabilize below the 450 ppm CO_2e target—and the more likely the 21st Century will experience dangerous climate change.

On our sustainable emissions pathway set out in chapter 1, mitigation would start to make a difference after 2030 and world temperatures would peak around 2050. These outcomes highlight the lag between action and results in tackling climate change. They also draw attention to the importance of thinking beyond the time-horizon defined by political cycles. Dangerous climate change is not a short term emergency amenable to a quick fix. The current generation of political leaders cannot solve the problem. What they can do is to keep open and then widen the window of opportunity for future generations to take up the battle. The 21st Century carbon budget set out in chapter 1 provides a roadmap for achieving this objective.

Keeping the window open will require early and radical shifts in energy policy. Since the industrial revolution, economic growth and human prosperity have been fuelled by carbon-based energy systems. Over the next few decades, the world needs an energy revolution

that enables all countries to become low-carbon economies. That revolution has to start in the developed world. Living within a sustainable 21st Century carbon budget requires that rich countries cut emissions of greenhouse gases by at least 80 percent by 2050, with 30 percent cuts by 2020. If the targets are to be achieved, the collective emissions curve will have to peak and start bending in a downwards direction between 2012 and 2015. Developing countries will also have to chart a low-carbon transition pathway, albeit at a pace that reflects their more limited resources and the imperative of sustaining economic growth and cutting poverty.

This chapter looks at the strategies needed to achieve a rapid transition to a low-carbon future. The 21st Century carbon budget provides a roadmap for reaching the agreed destination— a world free of dangerous climate change. But targets and roadmaps are not a substitute for policies. They will only contribute to the battle against climate change if they are backed by effective mitigation strategies.

There are three foundations for success. The first is putting a price on carbon emissions. Market-based instruments have a critical role to play in creating incentives that signal to business and consumers that there is a value in reducing emissions—and that the Earth's capacity for

Successful mitigation ultimately requires that consumers and investors shift demand to low-carbon energy sources

absorbing CO_2 is marked by scarcity. The two broad options for pricing emissions are taxation and cap-and-trade.

The second foundation for mitigation is behavioural change in the broadest sense. Successful mitigation ultimately requires that consumers and investors shift demand to low-carbon energy sources. Price incentives can encourage behavioural change—but prices alone will not deliver reductions on the scale or at the pace required. Governments have a critical role to play in encouraging behavioural change to support the transition to a low-carbon economy. Setting standards, providing information, encouraging research and development, and—where necessary—restricting choices that compromise efforts to tackle climate change are all key parts of the regulatory toolkit.

International cooperation represents the third leg of the mitigation tripod. Rich countries have to take the lead in tackling dangerous climate change: they have to make the deepest and earliest cuts. However, any international framework that does not establish targets for all major greenhouse gas emitting countries will fail. Avoiding dangerous climate change requires a low-carbon transition in developing countries too. International cooperation can help to facilitate that transition, ensuring that reduced emission pathways do not compromise human development and economic growth.

This chapter provides an overview of the mitigation challenge. It starts out by looking from global to national carbon budgeting. Converting the global 21st Century carbon budget

into national budgets is the first step towards mitigation of dangerous climate change. It is also a precondition for the successful implementation of any multilateral agreement. With governments negotiating the post-2012 framework for the Kyoto Protocol, it is important that national targets are aligned with credible global targets. Currently, many target-setting exercises suffer from a lack of clarity and consistency, compounded in some cases by a divergence between stated goals and energy policy frameworks.

In section 3.2 we then turn to the role of market-based instruments in the transition to sustainable carbon budgeting. We set out the case for carbon taxation and cap-and-trade schemes, while highlighting the problems that have reduced the effectiveness of the world's largest such scheme—the European Union Emissions Trading Scheme (EU ETS). Section 3.3 looks beyond taxation and cap-and-trade to the critical role of wider regulation and standards and public–private partnerships in research and development.

The chapter concludes by highlighting the underexploited potential of international cooperation. In section 3.4 we show how financial support and technology transfer could raise the energy efficiency of developing countries, providing a win–win scenario for human development and climate change: extending access to affordable energy while cutting emissions. Deforestation and land-use change, currently the source of about 20 percent of world greenhouse gas emissions, is another area of unexploited opportunity in international cooperation.

3.1 Setting mitigation targets

Expiry of the current commitment period of the Kyoto Protocol in 2012 creates an opportunity for early progress in climate change mitigation. In chapter 1, we argued for a multilateral framework geared towards well-defined global carbon budget goals. Such a framework has to combine long-term goals (a 50 percent reduction on 1990 levels

in emissions of greenhouse gases by 2050), with medium-term benchmarks set out in rolling commitment periods. The multilateral framework also has to provide a practical guide for implementing the principle of "common but differentiated responsibility", identifying broad pathways for developed and developing countries.

Without a credible multilateral framework the world will not avoid dangerous climate change. However, no multilateral framework will deliver results unless it is underpinned by national targets, and by policies that are aligned with those targets. The corollary of a meaningful global carbon budget for the 21st Century is the development of national carbon budgets that operate within the global resource envelope.

Carbon budgeting—living within our ecological means

National carbon budgeting is a necessary foundation for the post-2012 multilateral framework. At their most basic level, carbon budgets set a limit on the total quantity of CO_2e emissions over a specified period of time. By setting a rolling budget period of, say, 3–7 years, governments can strike a balance between the certainty needed to meet national and global emission reduction targets, and the annual variation that will accompany fluctuations in economic growth, fuel prices or the weather. From a carbon mitigation perspective, what matters is the trend in emissions over time rather than annual variations.

There are parallels between global and national carbon budgeting. Just as the global carbon budget discussed in chapter 1 establishes a bridge between current and future generations, national carbon budgets provide for continuity across political cycles. In money markets, uncertainties over the future direction of policies on interest rates, money supply or price level can all fuel instability. That is why many governments use independent central banks to address the problem. In the case of climate change, uncertainty is an obstacle to successful mitigation. In any democracy, it is difficult for a government to irrevocably commit its successors to specific mitigation policies. However, fixing multilateral commitments into national legislation aimed at achieving long-run mitigation goals is vital for policy continuity.

National carbon budgeting is also a foundation for international agreements. Effective multilateral agreements have to be based on shared commitments and transparency. For countries participating in international agreements aimed at rationing global greenhouse gas emissions, it is important that partners are seen to stick to their side of the bargain. Perceived free-riding is guaranteed to weaken agreements by eroding confidence. Ensuring that multilateral commitments are enshrined in transparent national carbon budgets can counteract this problem.

At a national level, carbon budgets can reduce the threat of economic disruption by sending clear signals to investors and consumers on the future direction of policy. Beyond the market, carbon budgets can also play an important role in increasing public awareness and holding governments to account, with citizens using carbon budget outcomes to assess the contribution of their governments to multilateral mitigation efforts.

Emission reduction targets are proliferating

Recent years have witnessed an increase in target-setting exercises on climate change. National governments have adopted a wide range of goals. Within countries, state and regional governments have also been active in setting emission reduction targets (table 3.1).

The growth of target setting has produced some impressive results. The Kyoto Protocol itself was an exercise in setting national limits linked to global mitigation goals. Most OECD countries—Australia and the United States are the major exceptions—are committed to achieving reductions by 2008–2012 against a 1990 base year. Many have even embraced additional targets. The European Union is an example. Under the Kyoto Protocol, the European Union is required to achieve an 8 percent reduction in emissions. However, in 2007 it committed itself to cutting greenhouse gas emissions by "at least" 20 percent by 2020 and by 30 percent if an international agreement is reached, with a reduction of 60–80 percent by 2050. Several member states have adopted national targets for reductions against 1990 levels, among them:

- The United Kingdom has set itself a 'Kyoto–plus' target in the form of a 20 percent cut on 1990 levels by 2010. Legislation

3

Avoiding dangerous climate change: strategies for mitigation

| Table **3.1** | **Emission reduction targets vary in ambition** |

Greenhouse gas reduction targets and proposals	Near term (2012–2015)	Medium term (2020)	Long term (2050)
HDR sustainable emissions pathway (for developed countries)	**Emissions peaking**	**30%**	**at least 80%**
Selected countries			
	Kyoto targets [a] **(2008–2012)**	**Post-Kyoto**	
European Union [b]	8%	20% (individually) or 30% (with international agreement)	60–80% (with international agreements)
France	0%	–	75%
Germany	21%	40%	–
Italy	6.5%	–	–
Sweden	4% increase (4% reduction national target) (by 2010)	25%	–
United Kingdom	12.5% (20% national target)	26–32%	60%
Australia [c]	8% increase	–	–
Canada	6%	20% relative to 2006	60–70% relative to 2006
Japan	6%	–	50%
Norway	1% increase (10% reduction national target)	30% (by 2030)	100%
United States [c]	7%	–	–
Selected United States state-level proposals			
Arizona	–	2000 levels	50% below 2000 (by 2040)
California	2000 levels (by 2010)	1990 levels	80% below 1990 levels
New Mexico	2000 levels (by 2012)	10% below 2000 levels	75% below 2000 levels
New York	5% below 1990 (by 2010)	10% below 1990 levels	–
Regional Greenhouse Gas Initiative (RGGI) [d]	Stabilization at 2002–2004 levels (by 2015)	10% below 2002–2004 levels (by 2019)	–
Selected United States Congress proposals			
Climate Stewardship and Innovation Act	2004 levels (by 2012)	1990 levels	60% below 1990 levels
Global Warming Pollution Reduction Act	–	2% per year reduction from 2010–2020	80% below 1990 levels
Climate Stewardship Act	2006 level (by 2012)	1990 levels	70% below 1990 levels
Safe Climate Act of 2007	2009 level (by 2010)	2% per year reduction from 2011–2020	80% below 1990 levels
United States non-governmental proposals			
United States Climate Action Partnership	0–5% increase of current level (by 2012)	0–10% below "current level" (by 2017)	60–80% below "current level"

a. Kyoto reduction targets are generally against 1990 emission levels for each country, by 2008–2012, except that for some greenhouse gases (hydrofluorocarbons, perfluorocarbons and sulphur hexafluoride) some countries chose 1995 as their base year.
b. Kyoto targets only refer to 15 countries which were members of the European Union in 1997 at the time of signing.
c. Signed but did not ratify the Kyoto Protocol, therefore commitment is not binding.
d. Participating states include Connecticut, Delaware, Maine, Maryland, Massachusetts, New Hampshire, New Jersey, New York, Rhode Island and Vermont.

Source: Council of the European Union 2007; Government of Australia 2007; Government of Canada 2007; Government of France 2007; Government of Germany 2007; Government of Norway 2007; Government of Sweden 2006; Pew Center on Climate Change 2007c; RGGI 2005; State of California 2005; The Japan Times 2007; UNFCCC 1998; USCAP 2007.

under preparation would establish a statutory obligation on Government to achieve reductions of 26–32 percent by 2020, and 60 percent by 2050.[1]

- France has a national target of a 75 percent cut in emissions by 2050.[2]

- In 2005, Germany updated its National Climate Change Programme to include the target of a 40 percent reduction by 2020 (subject to the European Union subscribing to a 30 percent reduction).[3] In August 2007, the German Federal Government reaffirmed this commitment by adopting a policy package to achieve the target.[4]

Target setting has also emerged as an issue on the agenda of the G8. At their 2007 summit, the G8 leaders accepted in principle the need for urgent and concerted action to avoid dangerous climate change. No formal targets were adopted. However, the summit agreed to "consider seriously" decisions made by Canada, the European Union and Japan to set a level of ambition aimed at halving global emissions by 2050.[5]

Target setting from below in the United States

The United States currently lacks a national target for overall emission reductions. Under the 2002 Global Climate Change Initiative (GCCI), the Federal Government set a national goal for reducing greenhouse gas emissions intensity, as measured by the ratio of greenhouse gas emissions to GDP. However, the absence of a national emission reduction goal has not prevented the emergence of a range of target-setting initiatives, with states and cities setting out quantitative goals of their own. Prominent examples include:

- *State initiatives.* With the passage of the 2006 Global Warming Solutions Act, California has set an enforceable target of achieving 1990 levels of greenhouse gas emissions by 2020, with an 80 percent reduction on 1990 levels by 2050 (box 3.1). Concerns that these targets will necessarily compromise competitiveness and employment are not well supported by the evidence. Modelling work has found that new incentives created by the state's cap on emissions could create an additional US$59 billion in income and 20,000 new jobs by 2020.[6] In total, there are now 17 states across the United States with emissions targets.[7]
- *Regional initiatives.* The Regional Greenhouse Gas Initiative (RGGI) established in 2005 is the first mandatory cap-and-trade programme in the United States, setting limits on emissions from power plants. It now extends to 10 states.[8] The target is to cap emissions at current levels from 2009 to 2015 and then to reduce them by 10 percent

by 2019. In 2007, the creation of the Western Regional Climate Action Initiative—involving Arizona, California, New Mexico, Oregon, Utah and Washington—expanded the reach of regional initiatives. The Canadian provinces of British Columbia and Manitoba joined in 2007, turning it into an international partnership. By 2009, these states will set a regional emissions target and devise market-based programmes to achieve them.[9]

- *City initiatives.* Cities are also setting emission reduction targets. In total, around 522 mayors, representing 65 million Americans, are aiming to reach what would have been the United States Kyoto target of a 7 percent reduction below 1990 levels by 2012.[10] New York has introduced caps on emissions from the city's power stations. The New York City Government has also passed legislation that requires a city-wide inventory of greenhouse gas emissions and a city-wide goal of 7 percent reductions below 1990 levels by 2020. While the reductions are voluntary for the private sector, the City Government is committed to 30 percent emissions cuts.[11]

These initiatives have to be placed in context. If California were a country, it would be the world's fourteenth largest source of CO_2 emissions—that is why its leadership is of global importance. However, the bulk of emissions still originate in states with no planned caps on emissions: California and the RGGI states together account for around 20 percent of United States' greenhouse gas emissions. Just as greenhouse gases from India and the United States mix in the Earth's atmosphere, so a tonne of CO_2 from San Francisco has the same impact as a tonne from Houston. In the absence of binding Federal targets, emission reductions in some states can be swamped by increases in others. Even so, state-level and regional government initiatives have created a political impetus towards the establishment of emission ceilings at the Federal level.

That impetus is reflected in the United States Congress. Recent years have witnessed a steady proliferation in proposed legislation aimed at setting targets for future emissions of

At their 2007 summit, the G8 leaders accepted in principle the need for urgent and concerted action to avoid dangerous climate change

| Box 3.1 | Leadership by example in carbon budgeting—California |

The world's sixth largest economy, California has long been a national and international leader on energy conservation and environmental stewardship. Today, it is setting the standard for global action on climate change mitigation.

The 2006 Global Warming Solutions Act requires California to cap greenhouse gas emissions by 2020 at 1990 levels, with a long-term reduction goal of 80 percent by 2050. This legislation represents the first enforceable state-wide programme to cap emissions from all major industries, with in-built penalties for non-compliance.

Legislation is rooted in strong institutional provisions. The state plan grants the State Air Resources Board (SARB) authority to establish how much industry groups contribute to emission reductions, assigning emission targets and setting non-compliance penalties. It sets a 2010 deadline for establishing how the system will work, allowing industries three years to prepare for implementation. The SARB is also required to develop a strategy "for achieving the maximum technologically feasible and cost-effective reductions in greenhouse gas emissions by 2020". That strategy, to be enforceable by 2010, includes a cap-and-trade programme based on quantitative targets.

California's targets are backed by substantive policies. Among the most important:

- *Vehicle emission standards.* Over the past four years California has pioneered higher emission standards. Current vehicle standards legislation will require a 30 percent reduction in greenhouse gas emissions from new vehicles by 2016. The state is also developing a low Carbon Fuel Standard aimed at reducing fuel emissions intensity by 10 percent to 2020. This is expected to create incentives for emissions cuts in petroleum processing, biofuels and electricity-driven vehicles.

- *Performance standards for electricity.* Public policy action in this area has received less public attention than the Global Warming Solutions Act, but it has important implications. Under the relevant legislation, the California Energy Commission is required to set stringent emission standards for electricity procured under long-term contracts, whether the power is produced within the state or imported from plants in other states. The standards will drive low-carbon electricity generation, including research and development of power plants that capture and store CO_2.

- *Renewable energy.* California is one of twenty-one states with a 'renewable portfolio standard' setting a target for renewable energy. By 2020 California aims to generate 20 percent of its power from renewable sources. The state will pay an estimated US$2.9 billion in rebates over 10 years to households and businesses that install solar panels, with further tax credits to cover 30 percent of the cost of installation. These subsidies are part of the 'One Million Solar Roofs' initiative.

- *Setting conservation standards.* During 2004 California announced a stringent energy conservation target aimed at saving the equivalent of 30,000 GWh by 2013. In order to achieve this goal, new appliance and building standards have been introduced.

Three important features of the California case have wider lessons for carbon budgeting. First, the legislation establishes a credible target. Applied by all developed countries, the 80 percent reduction by 2050 would put the world on to a potentially sustainable emissions trajectory. Second, compliance and monitoring are overseen through strong institutional mechanisms that provide a basis for transparency and accountability. Third, the legislation establishes a balance between mandated targets, incentives and regulatory measures aimed at cutting emissions and spurring innovation.

Source: Arroyo and Linguiti 2007.

greenhouse gases. In the first half of 2007, seven separate bills aimed at setting economy-wide quantitative ceilings were under consideration in Congress.[12] One of these—the Climate Stewardship and Innovation Act—envisages an emissions pathway with 20 percent cuts below 1990 levels by 2030, deepening to 60 percent by 2050, for the electricity generation, transportation, industrial and commercial sectors.

Beyond Congress, there has been a surge of multi-constituency initiatives bringing together industry, environmentalists and others. The United States Climate Action Partnership (USCAP) is an example. An alliance of 28 major companies—including BP America, Caterpillar, Duke Energy,

DuPont and General Electric—and six leading NGOs (with a membership of over one million), USCAP has called for a combination of mandatory approaches, technological incentives and other actions to achieve a peak of emissions by 2012, with reductions up to 10 percent by 2017, and 80 percent by 2050 with respect to 'current' levels.[13] Many of the companies involved have set voluntary targets for reducing emissions, anticipating the future development of mandatory targets.

USCAP's proposals are instructive. Beyond the targets themselves, they reflect important changes in approaches to climate change mitigation. Five years ago, many of America's largest companies were hostile in principle to the

In the battle against climate change, it's easy to talk about lofty, far-away goals, but the question is: What are you doing today to achieve them? In New York City, we recently unveiled an ambitious yet achievable plan to combat global warming and create the first truly sustainable 21st Century city. The plan, which we call *PlaNYC,* includes 127 specific initiatives designed to reduce air and water pollution, clean-up polluted land, modernize our infrastructure and energy network, and significantly reduce the city's carbon footprint. In short, it's about leaving our children a greener, greater city.

Gone are the days when public and private sector leaders could act as though environmental sustainability and economic competitiveness work against one another. In fact, the very opposite has proven true. Fighting global warming begins, in many ways, with learning how to become more efficient. Investing in energy-saving technology allows governments, businesses and families to save significant amounts of money over the long term. As part of *PlaNYC,* for instance, New York City has committed to reducing its energy use by 30 percent over the next 10 years. We're also incentivizing private sector 'green' construction. And we're in the process of upgrading all 13,000 of our famous yellow taxi cabs, doubling their fuel efficiency to match or beat today's hybrid cars. This will not only mean less CO_2 and air pollution, but also lower gas bills for drivers—and that means more money in their pockets.

PlaNYC will help us to maintain our economic growth and protect our environment. But it will also allow us to fulfill our broader responsibilities as global citizens. The *Human Development Report 2007/2008* states plainly that climate change is one of the greatest challenges facing humanity, and it is the world's most vulnerable populations who are most immediately at risk. The actions of the wealthiest nations—those generating the vast majority of greenhouse gases—have tangible consequences for people in the rest of the world, especially in the poorest nations.

We can't sit back and wait for others to act—and that's why cities around the world are leading the charge. Leaders of cities focus on results, not politics—on taking action, not toeing the party line. Although international climate accords have been difficult to reach and harder to enforce, city leaders have been driving new innovations and sharing best practices. In February 2007, the United States Conference of Mayors launched the Climate Protection Center to provide mayors with the guidance and assistance they need to lead their cities' efforts to reduce greenhouse gas emissions. And in May of this year, New York City hosted the C40 Large Cities Climate Summit, which brought together more than 30 mayors from the world's largest cities to exchange ideas and best practices for combating climate change.

The leading role that cities have played against climate change is evidenced by the fact that many of the initiatives in *PlaNYC* were inspired by other cities. We drew on the experiences of London, Stockholm and Singapore in formulating our traffic-reducing congestion pricing plan; on Berlin for our renewable energy and green roof policies; on Delhi, Hong Kong and Shanghai for our innovative transit improvements; on Copenhagen for our pedestrian and cycling upgrades; on Chicago and Los Angeles for our plan to plant one million more trees; on Amsterdam and Tokyo for our transit-oriented development policies; and on Bogota for our plans for Bus Rapid Transit. By taking a global approach to a global problem, we were able to formulate a distinctly local plan that will allow us to do our part in the fight against climate change—and, we hope, to be a model for others to follow.

As the *Human Development Report 2007/2008* makes clear, it is no longer acceptable for the world's governments to ignore the threat of climate change, or for elected officials to announce distant goals without putting forth substantive plans to achieve them, including interim targets that allow the public to hold those officials and their successors accountable for making steady progress. As public leaders, we have a responsibility to take bold action that will lead to real change—starting today.

Michael R. Bloomberg

Michael R. Bloomberg
Mayor of the City of New York

idea of mandatory quantitative restrictions on greenhouse gas emissions. That is now changing. Increasingly, companies see quantitative targets not as a threat but as an opportunity that will create incentives and prospects for low-carbon investments.

Ironically, the absence of a national framework setting mandatory ceilings on greenhouse gas emissions is now regarded by many major companies as a problem, partly because it creates market uncertainty, and partly because the surge of state-level and regional-level initiatives is creating a complex patchwork of regulatory systems. The Alliance of Automobile Manufacturers, which includes General Motors and Ford Motor Company, has called for "a national, federal, economy-wide approach to addressing greenhouse gases".[14]

Many of the targets set are, at best, only weakly related to sustainable carbon budget requirements

The Electric Power Supply Association also announced its support for "comprehensive, mandatory federal legislation to minimize the impact of greenhouse gases".[15]

Four targeting problems in carbon budgeting

Is the new trend towards target setting in developed countries providing a foundation for carbon budgets that will enable the world to avoid dangerous climate change?

The answer to that question is a qualified 'no'. While the adoption of targets is an encouraging indication that public concern is registering on the political radar screen, many of the targets set are, at best, only weakly related to sustainable carbon budget requirements. Insufficient ambition is a common problem. Another is the confusion associated with a proliferation of targets, especially when those targets are inadequately reflected in energy policies. There are four broad potential sources of error in carbon budget targeting that need to be addressed:

- *Insufficient ambition*. Our sustainable emissions pathway establishes two plausible benchmarks for assessing where emissions ceilings need to be set by developed countries. The broad trajectory: peaking in the period 2012 to 2015, cuts of 30 percent by 2020 and cuts of at least 80 percent by 2050, against a 1990 baseline. There are two problems. First, some targets—the United Kingdom's and several proposals in the United States are examples—fall short of these benchmarks (table 3.1). Second, the selection of reference years can obscure under-ambition in target setting. For example, some governments interpret the commitment made at the G8 to "seriously consider" halving emissions by 2050 as an implied reduction from 'current' levels. Simple carbon arithmetic demonstrates why changes in reference years matters. Shifting the United States reference year from 1990 to 2004, for example, would increase the permitted emissions base by

over 900 Mt CO_2e—roughly equivalent to total German emissions in 2004.[16] For Canada, the same shift in reference years would raise the baseline for emissions by 27 percent over 1990 levels. From a carbon budgeting perspective, any change in base year should include adjustments in reduction targets to compensate for any increase in emissions from 1990.

- *Inaccurate indicators*. Some governments present targets for reduced carbon intensity as equivalent to climate change mitigation goals. This confuses means and ends. Reducing the amount of CO_2 emitted for every dollar in wealth created (the carbon intensity of growth), or for every unit of power generated (the carbon intensity of energy), is an important goal. No mitigation strategy is likely to succeed without progress in these areas. However, what ultimately matters is the 'overall reduction' in emissions. From a sustainable carbon budget perspective, carbon intensity targets in isolation are a mitigation red-herring. Many countries have an impressive record in cutting carbon intensity but still have an overall increase in emissions (figure 3.1). The United States has reduced greenhouse gas intensity by around 25 percent since 1990 but its overall emissions have gone up by an equivalent amount. The GCCI targets a further reduction in greenhouse gas intensity of 18 percent between 2002 and 2012—broadly consistent with the trend since 1980. However, the Energy Information Administration projects an increase in CO_2 emissions over the same period of around 25 percent.[17]

- *Inadequate sectoral coverage*. Effective carbon accounting requires that all emissions are reflected in the budget. Unfortunately, current reporting systems keep some sectors 'off-budget'. For example, aviation is excluded from international inventories of greenhouse gases for the Kyoto Protocol. The Earth's atmosphere is less discriminating. Since 1990, emissions of CO_2 from aviation fuel have increased from 331 Mt CO_2 annually to 480 Mt CO_2. The latter figure represents around 2 percent of global emissions. However, because the emissions are released directly into the high atmosphere, the radiative forcing effects are far stronger, accounting for 3 percent (2–8 percent range) of global warming.[18] For several OECD countries, aviation represents a significant and growing share of the national contribution to global warming. In the United Kingdom, annual emissions from aviation are projected to grow by between 62 and 161 Mt CO_2 by 2050. In order to offset emissions from the aviation sector and achieve the national target of a 60 percent reduction in overall emissions by 2050, other sectors would have to reduce their emissions by 71–87 percent.[19] This is not a plausible option, suggesting that aviation will have to be subject to cuts in emissions.

- *Insufficient urgency.* Sometimes decisions in public policy can be postponed without great cost. That is not the case with climate change. Because emissions are long-lived, delaying the decision to reduce them adds to the stock of greenhouse gases and cuts the time frame for reducing it. Several legislative proposals for the United States envisage limited cuts to 2020 against 1990 levels, followed by steeper declines thereafter. That approach may be ill-advised. One study for the United States shows that a pathway for contributing to a global stabilization level at 450 ppm CO_2e can be achieved with annualized reductions of 3 percent a year by 2050. However, delaying action until 2020 would require reductions of 8.2 percent a year—which would require stringent adjustments and an implausible rate of technological innovation.[20]

Targets matter, but so do outcomes

Setting targets is not the same as delivering results. Experience under the Kyoto Protocol provides a constant reminder of the limited progress made in aligning climate security goals with energy policies.

The experience of two countries at different ends of the Kyoto Protocol performance league is instructive. In Canada, energy-intensive economic growth has comprehensively undermined the prospects for delivery against the country's Kyoto commitments (box 3.2). Unlike Canada, the United Kingdom is on-track to meet its Kyoto targets, though not primarily as a result of energy policy reform: a shift in energy mix from coal to natural gas has been more important. The country has now defined an ambitious carbon budget that sets a pathway for reduced emissions through to 2050. However, CO_2 emissions from the United Kingdom have not fallen over the past decade—and there are serious questions over whether or not the country will achieve national targets for reduced emissions (box 3.3).

Institutional arrangements play an important role in determining the credibility of emissions reduction targets. In carbon budgeting, as in financial budgeting, governance matters a great deal, not least in ensuring that targets are translated into outcomes. This is another area in which California has provided leader-

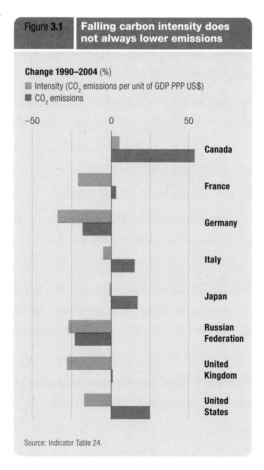

| Figure **3.1** | **Falling carbon intensity does not always lower emissions** |

Change 1990–2004 (%)
- Intensity (CO_2 emissions per unit of GDP PPP US$)
- CO_2 emissions

Source: Indicator Table 24.

Experience under the Kyoto Protocol provides a constant reminder of the limited progress made in aligning climate security goals with energy policies

3

Avoiding dangerous climate change: strategies for mitigation

Box 3.2   Targets and outcomes diverge in Canada

Carbon-intensive economic growth has pushed Canada well off track from its Kyoto commitments. The country's experience powerfully demonstrates the difficulties in aligning domestic economic policies with international commitments.

In 2004, Canadians contributed around 639 million tonnes of CO_2 to the Earth's atmosphere. While this is only 2 percent of the world total, Canada has one of the highest levels of per capita emissions in the world—and the carbon footprint is deepening. Since 1990, CO_2 emissions from fossil fuel have increased by 54 percent, or 5 tonnes per capita. That increase is greater than the total per capita CO_2 emissions from China.

Canada is far from meeting its Kyoto Protocol commitments. Emissions have increased by 159 million tonnes of CO_2e since 1990—a 27 percent overall increase and 33 percent above Kyoto target levels.

Why has Canada missed its Kyoto targets by such a wide margin? Rapid economic growth has been one factor. Another has been the carbon intensity of growth, driven by a surge in investments in natural gas and oil production. Greenhouse gas emissions associated with exports from this sector have increased from 21 million to 48 million tonnes per annum since 1990.

Developments in oil and natural gas markets have contributed to Canada's Kyoto deficit. With rising oil prices, it has become commercially viable to exploit tar sands in Alberta. Unlike conventional oil extracted through wells, oil is extracted from tar sands by stripping away upper layers of soils, or by using high-pressure steam to heat the underlying sands and make the bitumen less viscous. The energy requirements and the greenhouse gas intensity per barrel of oil extracted from tar sands are almost double that for conventional oil.

Oil sands exploration has important implications for Canada's greenhouse gas emissions trajectory. The Canadian Association of Petroleum Producers and the Canadian National Energy Board estimate that C\$95 billion (US\$108 billion) will be spent on oil sands operations from 2006 to 2016. Output is expected to triple, to over three million barrels a day. Translated into carbon footprint terms, greenhouse gas emissions from oil sands could increase by a factor of five to 2020, rising to over 40 percent of national emissions by 2010.

Changing this trajectory will be difficult given the high levels of investment already in place. In 2006, new targets were set under a Clean Air Act that specifies reductions of 45–65 percent below 2003 levels by 2050. However, the targets are not binding—and they are not linked to specific policies. Initiatives at a provincial and municipal level have established more concrete provisions, producing some impressive results. For example, Toronto has achieved deep cuts in emissions (40 percent below 1990 levels in 2005) through energy efficiency initiatives, retro-fitting of old buildings and land fill policy.

Canada has a long history of global leadership on global atmospheric environmental issues, from acid rain to ozone depletion and climate change. Maintaining this tradition will require tough decisions. The David Suzuki Foundation has called for a 25 percent cut in emissions by 2020, with an 80 percent cut by 2050. Those targets are attainable, but not with current policies. Among the options:

- Accelerated deployment of low-carbon technologies and increased investment in carbon sequestration to reduce long-term emissions;
- A requirement on exporters that the purchase of Canadian oil and natural gas is linked to the purchase of verifiable emissions reductions through carbon market trading;
- The introduction of a carbon tax on investors in oil sands production to finance technological innovation and the purchase of emissions credits;
- Strict regulation of production standards and price incentives for low-emission production of oil sands and natural gas.

Source: Bramley 2005; Government of Canada 2005; Henderson 2007; Pembina Institute 2007a, 2007b.

ship. In order to implement the state's cap on emissions, a strong agency—the California Air Resources Board—has been directed to develop regulations, establish a mandatory reporting system and monitor emission levels. While the targets are set by elected political leaders, implementation and administration are conducted through public agencies with a strong technical capacity. At the same time, the targets have been backed by far-reaching reforms in energy policy (see box 3.1). By contrast, the European Union has set ambitious targets for cutting emissions, without having either an institutional framework for implementation or a coher-

ent agenda for energy reform: energy policy is overwhelmingly a national responsibility (box 3.4). Transition economies have also adopted targets under the Kyoto Protocol. While most are on track for achieving the targets, this owes more to the economic recession of the 1990s than to energy reform—an area in which progress has been mixed (box 3.5).

The limits to voluntarism

Some countries have relied primarily on voluntary programmes to achieve climate change mitigation goals. Results have been mixed. In some cases, voluntary action has made a difference.

The United Kingdom's Climate Change Bill is a bold and innovative proposal to create a national carbon budget that supports global mitigation efforts. Legislation would commit Government to mandatory cuts in emissions over time. Applied more widely across the developed world, the broad approach could underpin a strengthened post-2012 Kyoto system. However, there are serious questions about the level of ambition—and about the United Kingdom's capacity to meet its own carbon reduction targets.

The Climate Change Bill charts a pathway for emissions reductions to 2050. An expressed aim is to contribute to international efforts to avoid dangerous climate change, which the United Kingdom Government identifies as a global mean temperature increase in excess of 2°C. The roadmap sets the 2050 target for greenhouse gas emissions reductions at 60 percent, with an interim target of 26–32 percent reductions by 2020 against levels in 1990.

These targets would be fixed in a system of 'carbon budgets'—rolling 5-year limits on CO_2 emissions. Three budgets would be set in advance, helping to create a long-term horizon for business and investment decisions. Legislation would create enabling powers that make future policies for controlling emissions quicker and easier to introduce. However, two issues will have to be addressed if the Climate Bill is to provide the framework for a sustainable carbon budget.

The first problem is one of overall ambition. Emission targets in the Climate Bill are not consistent with the objective of avoiding dangerous climate change. Our sustainable emissions pathway suggests that developed countries need to cut emissions of greenhouse gases by at least 80 percent by 2050 against 1990 levels, not 60 percent. Moreover, the current framework excludes aviation and shipping. Factoring them in would raise the cumulative United Kingdom carbon budget to 2050 by around 5.5 Gt CO_2, or 27 percent.

If the rest of the developed world followed the pathway envisaged in the United Kingdom's Climate Change Bill, dangerous climate change would be inevitable. It would lead to approximate atmospheric concentrations of greenhouse gases in excess of 660 ppm CO_2e, and possibly 750 ppm CO_2e. These are outcomes that would correspond to a rise in average global temperatures of 4–5°C, well beyond the dangerous climate change threshold. The overarching requirement for keeping within the 2°C threshold is a stabilization of greenhouse gas stocks at 450 ppm CO_2e.

The second problem to be addressed is the direction of current greenhouse gas emissions (see figure). On a positive note, the United Kingdom is one of a small group of European Union countries that is on-track for achieving its Kyoto Protocol target. While the economy has expanded by 47 percent since the 1990 base year for Kyoto, emissions of CO_2 are 5 percent lower. The less positive news is that all the reduction took place prior to 1995. Since 2000, emission levels have increased by 9 Mt CO_2 (to 567 Mt CO_2 in 2006). The upshot is that the national target of reducing CO_2 emissions to 20 percent below 1990 levels by 2010 is now unattainable: the likely outcome is a reduction less than one-half this target.

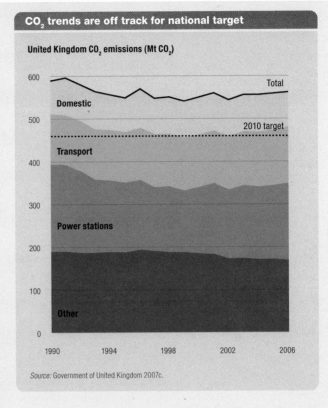

CO_2 trends are off track for national target

United Kingdom CO_2 emissions (Mt CO_2)

Source: Government of United Kingdom 2007c.

Breaking down emission sources for CO_2 by sector helps to identify some of the challenges facing the United Kingdom. Emissions from power stations, which represent around one-third of the total, have increased in five of the last seven years. The transport sector, now the second largest source of emissions, is on a sharply rising trajectory, while emissions from industry and the residential sector have not moved significantly. Changing these CO_2 emission trajectories to make possible a reduction of 26–32 percent by 2020 will require radical new policies that align energy policy with climate change mitigation goals. Among the options:

- *Carbon taxation and strengthened cap-and-trade.* Carbon pricing is critical to sustainable carbon budgeting. Signalling a commitment to carbon taxation in the range outlined in this chapter offers one route for aligning energy markets with sustainable carbon budget goals. Working through the European Union's cap-and-trade scheme is another option (section 3.2), provided that the ceiling on emissions is set at a level consistent with 26–32 percent cut in emissions by 2020.

- *Power generation.* The future energy mix in power generation will shape the United Kingdom's emissions trajectory. Since early 2000, increased use of coal, the most polluting fossil fuel, has been instrumental in driving up emissions. Regulatory mechanisms could be deployed to initiate the rapid retirement of highly polluting plants, with a commitment to the accelerated introduction of zero-emission coal plants. Britain also lags far behind best European Union practice on renewable energy: it currently produces only 2 percent of its overall energy from renewables.

The Renewables Obligation, a regulatory instrument, stipulates the amount of electricity that power suppliers have to access from renewable sources. It has achieved mixed results. The current target is for the share of renewables to reach 10 percent by 2010, rising to 15 percent by 2015. However, current trends fall far short of these targets, and shorter still of the European Union's 20 percent target by 2020. If Britain is to achieve its own stated goals, it will need to accelerate the development of wind and tidal power. One option would be a system of renewables support modelled on the German feed-in tariff system, with stronger price incentives backed by public investment.

- *Cutting emissions from transport.* Taxation and regulation are mutually reinforcing instruments for cutting transport emissions. Increased taxation on petrol is one demand management mechanism. More broadly, vehicle excise duties could be adjusted, with a steeper graduation to reflect the higher CO_2 emissions associated with low fuel-efficiency vehicles, especially sports utility vehicles. The national carbon budget could establish 'carbon pricing' in vehicle taxation as

a source of revenue for investment in renewable energy, with vehicle tax registration for all new cars after 2010 graduated to reflect more stringent pricing on CO_2 emissions. Rising emissions from transport also reflect weaknesses in the public transport infrastructure and a decline in the cost of private transport relative to public transport.

- *The residential sector.* Energy use in the residential sector remains highly inefficient. An average existing home requires four times as much energy to heat as a new home. Around one-third of the homes that will be occupied in 2050 are yet to be built. With adoption and implementation of the best European Union standards, this represents an opportunity for deep cuts in emissions.

Setting the right targets is the starting point for sustainable carbon budgeting. Ultimately though, governments have to be judged on policies and outcomes. Impressive inflation targets count for little in the face of uncontrolled money supply. The same applies to climate change targets. The challenge for the United Kingdom is to align a more stringent target with wide-ranging energy policy reform.

Source: Anderson and Bowes 2007; Government of the United Kingdom 2006b, 2006c, 2007b, 2007c, 2007e; Seager and Milner 2007.

However, faced with a threat on the scale posed by climate change, voluntarism cannot substitute for effective state action.

Developed countries that have not ratified the Kyoto Protocol have relied on voluntary targets. The only Federal target in the United States is the (non-binding) emissions intensity target. Other flagship programmes—such as the Combined Heat and Power Partnership and the Clean Energy–Environment State Partnership—attempt to encourage voluntary reductions by the corporate sector. In Australia, the national climate change strategy does have a non-binding target: emission cuts of 87 Mt CO_2 by 2010.[21] Voluntary measures, such as consumer education and engagement with the private sector, are the primary mechanism for achieving the objective.

Outcomes have not been encouraging. The centrepiece of the voluntary programme in Australia is the Greenhouse Challenge Plus (GCP) initiative. Participating companies are required to develop and publish company-level greenhouse gas inventories and strategies for cutting emissions. The GCP has played an important role in informing public debate and many participating companies have adopted innovative strategies for cutting

emissions. However, Australia's overall greenhouse gas emissions in 2004, not including land-use change, were 25 percent above 1990 levels.[22] Emissions of CO_2 from energy were up by one-third and by 16 percent for industrial processes.[23] Voluntarism is clearly not delivering the required outcome.

Recognition of this fact has prompted several state and territory governments to argue for a national programme for mandatory emissions cuts to supplement voluntary efforts. One prominent example is New South Wales, which has set a target of reducing greenhouse gas emissions by 60 percent by 2050.[24] More immediately, state legislation passed in 2002 aims to cut emissions per capita from the production and use of electricity from 8.6 tonnes to 7.3 tonnes between 2003 and 2007—a reduction of 5 percent against the Kyoto Protocol threshold.[25] The Greenhouse Gas Abatement Scheme sets annual statewide greenhouse gas reduction targets, and then requires individual electricity retailers to meet mandatory benchmarks based on the size of their share of the electricity market.[26] As in the United States, this is an example of political leadership on climate change from below.

"The aim is that the European Union leads the world in accelerating the shift to a low-carbon economy."

José Manuel Barroso, President of the European Commission, January 2007

What the European Union does in energy policy matters for the world. Its 27 countries account for around 15 percent of CO_2 emissions worldwide and Europe has a strong voice in international negotiations. Making that voice count depends critically on the demonstration of leadership by practical example.

Ambitious targets have been set. In 2006, European governments agreed to aim at cuts of 20 percent in greenhouse gas emissions against 1990 levels by 2020, rising to 30 percent in the event of an international agreement. At the heart of the strategy for achieving the target is a commitment to a 20 percent increase in energy efficiency.

Translating targets into concrete policies is proving more difficult. Proposals from the European Union to achieve greater efficiency through market liberalization, including the 'unbundling' of energy production, are contested by several member states. More broadly, there is no European Union-wide strategy for translating the 20 percent reduction commitment into national carbon budgets through taxation, strengthened efficiency standards or a more stringent cap-and-trade system. The European Union Emission Trading Scheme (EU ETS) is the world's largest cap-and-trade programme but it is not geared towards attainment of the 20–30 percent cuts in emissions (section 3.2).

Prospects for the European Union meeting its Kyoto Protocol reduction commitments remain uncertain. For the pre-2004 member states, it is estimated that current policies will achieve a reduction of 0.6 percent from the 1990 baseline. This means that the member states are less than one-tenth of the way to achieving the target of an 8 percent reduction. More stringent enforcement of existing energy efficiency regulations would go a long way towards closing the gap.

The European Union has taken one step towards leadership in global carbon mitigation: it has set ambitious targets. Translating these targets into a coherent set of policies will require greater coherence and bold reforms of the EU ETS, including far more stringent cuts in quota.

Source: CEC 2006b, 2007a; EC 2006c, 2007b; High-Level Task Force on UK Energy Security, Climate Change and Development Assistance 2007.

Governments in countries that ratified the Kyoto Protocol have also engaged with the private sector in voluntary initiatives. In Japan, the Voluntary Action Plan (VAP) was drawn up by Government in consultation with the Japanese Business Federation. It covers seven major industrial sectors. The problem is that companies are free to set their own targets. In 2005, the Japanese Government set out a new plan aimed at getting the country back on-track to meet its Kyoto commitments by achieving a 9 percent cut in emissions of the industrial sector by 2010. The target under the VAP is for the industrial and energy converting sectors is to achieve emissions levels in 2010 that are below those in 1990.[27]

None of this is to downplay the importance of corporate sector voluntary action. In the United States, many companies are not waiting for mandatory government targets to change business practices. They are acting now.[28] In 2003, 35 investors with US$4.5 trillion in assets signed up to the Carbon Disclosure Project—a voluntary arrangement for reporting corporate emissions. There are now 155 institutional investors with combined assets of US$21 trillion represented.[29] Many are participating in a voluntary programme—'Energy Star'—that sets standards for energy efficiency. Companies in the power sector are investing in the development of renewable energy capacity. Meanwhile, one of the world's largest energy supply companies—American Electric Power—has set itself the ambitious target of building one or more Integrated Gas Combine Cycle power-plants by 2010. Pollution-intensive industries—such as steel and cement—have also developed technologies to cut emissions.

Box 3.5 | Reducing carbon intensity in transition economies

The experience of countries in Central and Eastern Europe (CEE) and the Commonwealth of Independent States (CIS) serves to highlight the important role of markets—and the consequences of sending the wrong price signals.

When these countries moved from communist rule some 18 years ago, they exhibited some of the highest levels of energy intensity in the world. Heavy subsidies for coal-based energy generation and low prices for energy users created strong disincentives for efficiency, and high levels of CO_2 pollution.

The transition from centrally planned economies has taken the region through a painful restructuring process. During the first half of the 1990s, energy demand and CO_2 emissions tracked the economy in a dramatic decline—a fact that explains why transition economies 'over-achieved' against their Kyoto targets. Since then, energy policy reforms have produced a mixed picture.

Energy intensity (energy consumption per unit of GDP) and the carbon intensity of GDP have fallen in all countries, albeit at very different rates—and for different reasons (see table). In the Czech Republic, Hungary and Poland advances have been driven by economic reforms and privatization. Poland has almost halved energy intensity against 1990 levels. Deep reforms in the energy sector, including sharp increases in real prices, and the transition from an economy based on large state enterprises to private sector firms, have spurred rapid technological change. Ten years ago, Poland used 2.5 times more energy per unit of cement production than the European Union average. That differential has now been eliminated. The energy intensity of GDP has fallen by half since 1990.

Ukraine has achieved far lower reductions in energy and carbon intensity. Moreover, the reductions owe less to reform than to a change in energy mix: imports of natural gas from the Russian Federation have halved the share of coal. The energy reform process has yet to take off. Energy prices remain heavily subsidized, creating disincentives for efficiency gains in industry. An influential commission created by the Government—the Blue Ribbon Commission—has called for far-reaching reforms. The proposals range from cost-recovery pricing to the creation of an independent energy regulator and the withdrawal of subsidies. Progress towards implementation has been slow, but has gathered pace following an interruption of gas supplies from the Russian Federation in 2006.

Developments in the Russian Federation's energy sector are a matter of global concern for climate change. The country is the world's third largest emitter of CO_2, with a per capita carbon footprint close to the OECD average.

The Russian Federation ratified the Kyoto Protocol in 2004. When it did so, greenhouse gas emissions were 32 percent below 1990 levels—a fact that bears testimony to the depth of the recession that accompanied transition. Compared with 1990 levels, there has been considerable progress. However, the Russian Federation remains an energy intensive economy—twice as intensive as Poland. One reason for this can be traced to the partial nature of economic reforms. While many of the most inefficient state enterprises have been dismantled, economic recovery has been driven by energy-intensive sectors, such as minerals and natural gas.

Energy reform has also been partial. The natural gas sector illustrates the problem. In 2004, it is estimated that Gazprom, the state energy company, lost nearly 10 percent of its total production through leaks and inefficient compressors. Inefficient flaring of gas is another problem. Independent estimates suggest that around 60 billion cubic metres of natural gas—another 8 percent of production—is lost through flaring, suggesting that the Russian Federation may be responsible for around one-third of global emissions from this source.

Countries such as the Russian Federation demonstrate the immense potential for achieving win–win outcomes for national energy efficiency and climate change mitigation. Emissions trading through carbon markets such as the EU ETS could play a role in supporting low-carbon investment. However, unlocking the win–win potential will require the creation of new incentive structures through energy reform. Higher energy prices, the scaling down of subsidies, the introduction of a more competitive energy sector with strengthened independent regulation, and wider governance reforms are among the priorities.

Carbon and energy intensity is reducing in transition economies

	Total CO$_2$ emissions (Mt CO$_2$)			CO$_2$ emissions per capita (t CO$_2$)		Energy intensity (Energy use per unit of GDP PPP US$)		Carbon intensity (CO$_2$ per unit of GDP PPP US$)	
	1990	2000	2004	1990	2004	1990	2004	1990	2004
Russian Federation [a]	1,984	1,470	1,524	13.4	10.6	0.63	0.49	1.61	1.17
Poland	348	301	307	9.1	8.0	0.36	0.20	1.24	0.68
Ukraine [a]	600	307	330	11.5	7.0	0.56	0.50	1.59	1.18
Hungary	60	55	57	5.8	5.6	0.24	0.17	0.50	0.37
Czech Republic [a]	138	119	117	13.4	11.4	0.32	0.26	1.03	0.66
Slovakia [a]	44	35	36	8.4	6.7	0.37	0.26	0.96	0.51
CEE and the CIS	4,182	2,981	3,168	10.3	7.9	0.61	0.47	1.49	0.97
OECD	11,205	12,886	13,319	10.8	11.5	0.23	0.20	0.53	0.45

a. 1990 data refer to 1992.

Source: HDRO calculations based on Indicator Tables 22 and 24.

Source: GUS 2006; High-Level Task Force on UK Energy Security, Climate Change and Development Assistance 2007; Olshanskaya 2007; Perelet, Pegov and Yulkin 2007; Stern 2006; UNDP, Ukraine 2005; Ürge-Vorsatz, Miladinova and Paizs 2006.

As these positive examples suggest, voluntary initiatives for climate change mitigation have an important role to play. They can inform consumer choice, create incentives for companies and establish best practice models. But voluntary action is not enough. It has not been enough to push emission trends in a downward direction in Australia or in the United States. In other areas of public policy—national security, nuclear safety or the regulation of environmental pollution, for example—governments would not consider reliance on voluntary action alone. Yet when it comes to climate change, there is a damaging tendency to overstate the role of 'choice' and understate the importance of government action. Ultimately, failure to recognize the limits to voluntarism will compromise climate change mitigation.

> The monetary and wider social costs of carbon emissions are large but uncertain—and they are spread across countries and generations

3.2 Putting a price on carbon—the role of markets and governments

The debate on climate change has shifted in recent years. The argument is no longer about whether or not the world is warming, or whether or not human-induced climate change is responsible. Today, the debate is about how to tackle the problem.

In an ideal world, the marginal cost of carbon would be aligned with the damage—or externalities—caused by additional emissions, leaving the actors responsible for those emissions to pay the full social cost of their actions. In the real world, putting the full-cost price on carbon is a tricky business. The monetary and wider social costs of carbon emissions are large but uncertain—and they are spread across countries and generations. One important outcome is that emitters do not face the consequences of their own pollution.

None of this represents an insurmountable obstacle to the development of carbon pricing. We may not be able to calculate the precise social costs of emissions. However, we know the order of magnitude for emission reductions required to avoid dangerous climate change. Our sustainable emissions pathway provides a first approximation. The immediate challenge is to push the price of carbon to a level consistent with this pathway, either through taxation or quota, or both.

Taxation versus 'cap-and-trade'

The case for putting a price on carbon as part of a climate change mitigation strategy is increasingly widely accepted. But where should the price be set? And how should it be generated? These questions are at the heart of a somewhat polarized debate over the relative merits of carbon taxation and 'cap-and-trade' programmes. The polarization is unhelpful—and unnecessary.

Both carbon taxation and cap-and-trade systems would create economic incentives to drive emission reductions. Under a carbon tax, emitters are required to pay a price for every tonne of emissions they generate. Using a tax to achieve a specified reduction in emissions requires decisions on the level of tax, who should pay and what to do with the revenue. Under a cap-and-trade programme, the government sets an overall emissions cap. It then issues tradable allowances—in effect, 'permits to pollute'—that allow business the right to emit a set amount. Those who can reduce their emissions more cheaply are able to sell their allowances to others who would otherwise be unable to comply. Using a cap-and-trade programme means taking decisions on where to set the pollution ceiling, who should be issued with allowances and how many of the allowances should be sold rather than given away free.

The case for carbon taxation

Proponents of carbon taxation claim a broad range of advantages over cap-and-trade systems.[30] These can be clustered into four categories:

There are strong grounds for introducing cap-and-trade, especially to meet the short term and medium-term goals upon which success in avoiding dangerous climate change ultimately depends

- *Administration.* Advocates of tax-based approaches maintain that they offer wider administrative advantages. In principle, duties on CO_2 emissions can be introduced through the standard tax system, with opportunities for evasion limited by enforcement at key points in the economy. One estimate for the United States suggests that a carbon tax applied to 2000 entities could cover virtually all fossil fuel consumption, limiting opportunities for evasion.[31]

- *Limiting distortions caused by vested interests.* As in any system of quota allocation, cap-and-trade schemes are open to manipulation by vested interests. As one commentator has written, issuing allowances is "in essence printing money for those in control of the permits".[32] Who gets how many permits and at what price are issues that have to be determined through political processes. Inevitably those processes are open to influence by powerful actors—power companies, oil companies, industry and retailing, to name a few. Pandemic cheating has been highlighted as the Achilles' heel of cap-and-trade approaches.

- *Price predictability.* While both taxation and cap-and-trade raise the cost of CO_2 emissions, they do so in very different ways. Carbon taxes directly influence price in a predictable fashion. By contrast, cap-and-trade schemes control quantity. By fixing the quantity of emissions, such schemes will drive prices through whatever adjustment corresponds to the quota ceiling. Critics of cap-and-trade argue that quotas will accentuate energy price fluctuations, affecting business investment and household consumption decisions.

- *Revenue mobilization.* Carbon taxation has the potential to generate large streams of revenue. Because the tax base for carbon levies is so large, even a modest tax could deliver considerable amounts. For the OECD, a tax on energy-related CO_2 emissions set at US$20/t CO_2 would release up to US$265 billion annually.[33] Revenues derived from carbon taxation can provide a source of finance for the reform of taxation systems, while maintaining fiscal neutrality (leaving the tax-to-GDP ratio unchanged).

Carbon tax revenue can be used to reduce taxation on employment and investment, or to create new incentives for the development of low-carbon technologies. For example, in the early 1990s Norway introduced a carbon tax on energy which now generates almost 2 percent of GDP in revenue. The revenue flows from carbon taxation have supported technological innovation and financed reductions in labour taxes.[34] In Denmark, carbon taxation has played an important role in reducing carbon intensity and promoting the development of renewable energy. Since 1990, the share of coal in primary energy use has fallen from 34 to 19 percent, while the share of renewables has more than doubled to 16 percent.

Taxes and quotas: the difference can be exaggerated

Carbon taxation does offer an effective route for cutting emissions. Many of the claimed advantages are real—as are many of the problems highlighted with cap-and-trade systems. Yet there are strong grounds for introducing cap-and-trade, especially to meet the short term and medium-term goals upon which success in avoiding dangerous climate change ultimately depends. Moreover, differences between cap-and-trade and taxation can be overstated. In practice, neither approach is inherently more complex than the other. Both require monitoring, enforcement and effective governance systems—and both have to address the question of how to distribute costs and benefits across society.

Administrative complexity is one area in which the differences have been overstated. Quota-based systems in any economic sector can create formidably difficult administrative problems.[35] However, the concentration of CO_2 emissions in large-scale power plants and carbon-intensive industries makes it possible to operate cap-and-trade schemes through a relatively small number of enterprises. The EU ETS, considered in more detail below, operates through less than 11,000 enterprises.

Administration of carbon levies through the tax system may have some operational advantages. Even so, tax systems can also be highly complex, especially when, as would be the case with carbon taxation, they incorporate exemptions and special provisions. Moreover, the design and implementation of taxation systems is no less open to lobbying by vested interests than permit allocations under cap-and-trade programmes.

Price volatility is a challenge in cap-and-trade systems. Here too, however, it is important not to over emphasize the differences. If the policy aim is to achieve quantitative goals in the form of reduced emissions, carbon taxation will have to be constantly amended in the light of quantitative outcomes. Marginal tax rates would have to be adjusted to reflect undershooting or overshooting, and uncertainties over marginal tax rates could become a source of instability in energy prices.

What about the argument that carbon taxation offers a predictable revenue stream to finance wider tax reform? This is an important potential benefit. However, cap-and-trade programmes can also generate revenues, provided that they auction permits. Transparent auctioning offers several advantages apart from revenue mobilization. It enhances efficiency and reduces the potential for lobbying by vested interest groups, addressing two of the major drawbacks with quota systems. Signalling the gradual introduction and scaling up of auctioning to cover 100 percent of permit allocation should be an integral part of cap-and-trade design. Unfortunately, this is not happening under the EU ETS, though several states of the United States have proposed the development of auction-based cap-and-trade systems.

From a climate change mitigation perspective, cap-and-trade offers several advantages. In effect, taxes offer greater price certainty, while cap-and-trade offers greater environmental certainty. Strict enforcement of the quota guarantees a quantitative limit on emissions, leaving markets to adjust to the consequences. The United States acid-rain programme provides an example of a cap-and-trade scheme that has delivered tangible environmental benefits. Introduced in 1995, the programme targeted a 50 percent reduction in emissions of sulphur dioxide (SO_2). Tradable permits were distributed in two phases to power plants and other SO_2-intensive units, creating incentives for rapid technological change. Today, the targets are close to attainment—and sensitive ecosystems are already recovering.[36]

In the context of climate change, quotas may be the most effective option for achieving the stringent near-term goals for emission reductions. Put simply, cap-and-trade offers a quantitative mechanism for achieving quantitative targets. Getting the price right on marginal tax would produce an equivalent effect over time. But getting the price wrong in the early stages would compromise mitigation efforts because it would lead to higher emissions requiring more stringent future adjustments.

What is important in the context of any debate over the relative merits of carbon taxation and cap-and-trade is clarity of purpose. The ambition has to be aligned with the carbon emissions trajectory for avoiding dangerous climate change. For developed countries, that trajectory requires 30 percent cuts by 2020 and at least 80 percent cuts by 2050 against 1990 levels. The credibility of any cap-and-trade scheme as a mechanism for avoiding dangerous climate change rests on its alignment with these targets—a test that the EU ETS currently fails (see below).

Estimating carbon taxation levels consistent with our sustainable emissions pathway is difficult. There is no blueprint for estimating the marginal taxation rate consistent with that pathway. One reason for this is uncertainty about the relationship between changed market incentives and technological innovation. Economic modelling exercises suggest that a carbon price in the range of US$60–100/t CO_2 would be broadly consistent with the mitigation efforts required. The introduction of the tax would have to be carefully sequenced to achieve the twin goal of signalling the long-term direction of policy, without disrupting markets. One possible option is a graduated approach along the following lines:

- A tax of US$10–20/t CO_2 introduced in 2010;

Economic modelling exercises suggest that a carbon price in the range of US$60–100/t CO_2 would be broadly consistent with the mitigation efforts required

The climate change benefits of carbon taxation or cap-and-trade systems will be limited if governments do not complement reforms in these areas with a curtailment of fossil-fuel subsidies

• An annualized increase in taxation of US$5–10/t CO$_2$ adjusted on a rolling basis to take into account the national emissions trajectory.[37]

It should be emphasized that the aim of introducing carbon taxation is climate change mitigation—not revenue raising. Taxes on CO$_2$ can be increased without raising the overall tax burden. Indeed, fiscally neutral carbon tax reform offers a potential to finance wider reforms of the taxation system. As seen before, lowering taxes on employment or investment can create incentives for the development of low-carbon technologies. Because carbon taxation has the potential to feed through into higher prices for energy, overcoming the regressive effects by using revenues to support low income groups is also important.

Where should carbon taxes or cap-and-trade programmes be applied? The optimal approach would be to create a single global price for carbon, with the distributional consequences addressed through international transfers (just as national transfers are used to compensate for the effects of taxation). In theory, it is possible to design a transitional route to this goal, with taxes or cap-and-trade quotas graduated to reflect the circumstances of rich and poor countries. In practice, the world lacks the political, administrative and financial governance structures to oversee taxation or cap-and-trade systems covering both developed and developing countries.

That does not mean that the world cannot move towards a global carbon price regime. The issue is one of sequencing. For developed countries, the priority is to build upon current cap-and-trade schemes or to introduce carbon taxation consistent with the emission reduction targets set out in our sustainable emissions pathway. Integrating emerging carbon markets in Australia, Europe, Japan and the United States provides a skeletal structure for global carbon trading. Developing countries could gradually integrate into international systems by establishing their own cap-and-trade schemes, or by introducing carbon taxation as they seek to reduce their emissions over a longer-term time horizon.

Eliminating perverse subsidies

Whatever their respective merits, the climate change benefits of carbon taxation or cap-and-trade systems will be limited if governments do not complement reforms in these areas with a curtailment of fossil-fuel subsidies. While OECD countries as a group have been reducing these subsidies over time, they continue to distort markets and create incentives for carbon-intensive investments. Overall, OECD subsidies for fossil-fuel energy are estimated at US$20–22 billion annually. From a climate change mitigation perspective, these subsidies are sending precisely the wrong market signals by encouraging investments in carbon-intensive infrastructure. Among the examples:

• In the United States, the congressional Joint Committee on Taxation estimates tax concessions for exploration and development of fossil fuels at US$2 billion annually for 2006–2010.[38] Old coal power plants in the United States are also subject to weaker pollution controls under the Clean Air Act than newer plants—in effect providing them with an indirect subsidy for pollution.[39]

• In 2004, the European Environment Agency estimated on-budget state subsidies for coal production to total €6.5 billion (US$8.1 billion), dominated by Germany (€3.5 billion, some US$4.4 billion) and Spain (€1 billion, some US$1.2 billion), with off-budget support generating a similar amount.[40] In 2005, the European Commission approved a €12 billion (US$15 billion) grant for 10 coal mines in Germany.[41]

• Aviation fuel used in domestic and international flights is exempt from fuel duty in many countries. This is an obvious contrast to the position for petrol used in cars, where fuel duties figure prominently in final prices paid by consumers. The tax advantage enjoyed by aviation fuel represents an implicit subsidy on air transport, though the level of subsidy varies across countries.[42]

Subsidy elimination and taxation on flights and fuel, or the application of cap-and-trade to the aviation industry are priorities.

Cap-and-trade—lessons from the EU Emission Trading Scheme

Climate change *realpolitik* presents a powerful case for cap-and-trade. Whatever the theoretical and practical merits of carbon taxation, the political momentum behind cap-and-trade is gathering pace. The next few years are likely to witness the emergence of mandatory emissions controls in the United States with an expansion of institutionalized carbon trading. More broadly, there is a prospect that the post-2012 Kyoto framework will witness a process of integration between carbon markets in the developed world, with strengthened carbon financing links to developing countries. None of this precludes an expanded role for carbon taxation. However, cap-and-trade programmes are emerging as the primary vehicle for market-based mitigation—and it is vital that they are implemented to achieve the central objective of avoiding dangerous climate change. These are important lessons to be learnt from the European Union.

The EU Emission Trading Scheme—a big scheme with a short history

The EU ETS is by far the world's largest cap-and-trade scheme. For the European Union it represents a landmark contribution to climate change mitigation. To its critics, the EU ETS is a design-flawed confirmation of all that is wrong with cap-and-trade schemes. Reality is more prosaic.

The first phase of the EU ETS ran from 2005 to 2007. Phase II will run for a 5-year period to the end of 2012.[43] Writing off an experiment on the scale of the EU ETS before the end of its pilot phase might be considered a case-study in premature judgement. However, the scheme has undoubtedly suffered from a number of flaws in design and implementation.

The origins of the EU ETS can be traced to the 'flexibility mechanisms' introduced under the Kyoto Protocol.[44] Through these mechanisms, the Protocol aimed to create a mechanism for achieving emission reductions at lower cost. The EU ETS operates through the allocation and trading of greenhouse gas emission permits.

The permits are allocated to member states and distributed to identified emitters, which in turn have the flexibility to buy additional allowances or to sell surplus allowances. In the first phase of the EU ETS, 95 percent of allowances had to be distributed free of charge, severely restricting the scope for auctioning.

Other Kyoto flexibility mechanisms have been linked to the EU ETS. The Clean Development Mechanism (CDM) is an example. This allows countries with a Kyoto target to invest in projects that abate emissions in developing countries. The rules governing the generation of mitigation credits through the CDM are based on the twin principles of 'supplementarity' and 'additionality'. The former requires that domestic action on mitigation should be the primary source of emission reductions (though there are no quantitative guidelines); the latter requires evidence that the abatement would not have occurred in the absence of the CDM investment. Between the end of 2004 and 2007, there were 771 registered projects with a declared reduction commitment of 162.5 Mt CO_2e. Just four countries—Brazil, China, India and Mexico—accounted for three-quarters of all projects, with sub-Saharan Africa representing less than 2 percent.[45]

Rapid institutional development is one of the positive lessons to emerge from the EU ETS. During the first phase, the scheme covered around one-half of the European Union's total greenhouse gas emissions, spanning 25 countries and over 10,000 installations in a wide range of sectors (including power, metals, minerals and paper). It has spawned a large market. In 2006, transactions involving 1.1 billion tonnes of CO_2e worth €18.7 billion (US$24.4 billion) took place in a global carbon market worth €23 billion (US$30 billion).[46]

Three systematic problems

The EU ETS provides an institutional structure that has the potential to play a key role in an ambitious European Union climate change mitigation strategy. That potential has yet to be realized, however. During the first phase, three systemic problems emerged:

- *Overallocation of permits, creating the wrong price signals.* In the initial stages of

Rapid institutional development is one of the positive lessons to emerge from the EU ETS

allowance trading, prices climbed to €30/t CO_2 (US$38/t CO_2) in April 2006, before collapsing and stabilizing at prices below €1/t CO_2 (US$1.3/t CO_2) in 2007.[47] The reason for the collapse: publication of data showing that the cap had been set *above* emission levels.[48] Overallocation, the short time-horizon for the first phase, and uncertainty about allocations in the second phase have fuelled price volatility and kept prices depressed though there are signs of recovery (figure 3.2).

- *Windfall profits for the few.* Carbon trading during the first 3 years of the EU ETS did little to reduce overall emissions, but it did generate very large profits for some. In the power sector in particular, companies were able to cover their emissions through free quotas, pass on costs to consumers and benefit from market opportunities to trade excess quotas.[49] The United Kingdom Government estimates that large electricity generators gained £1.2 billion (US$2.2 billion) in 2005.[50] Estimates for the power sectors in France, Germany and the Netherlands put the windfall profit generated through emissions trading at around €6 billion (US$7.5 billion) for 2005.[51]

- *Lost opportunities for revenue mobilization.* CO_2 emissions permits have a real market value. For their holders they are the same as cash-in-hand. Selling quotas through auction can enable governments to mobilize resources, avoid political manipulation and achieve efficiency goals. This has not happened under the EU ETS. In the first phase, a ceiling of 5 percent was set on the share of allowances that could be auctioned. In the event, just one country—Denmark—took advantage of this limited opportunity. Allowances have been distributed on the basis of historic emissions, rather than efficiency—an arrangement known as 'grandfathering'. The result is that governments have foregone opportunities for revenue mobilization and/or tax reductions, with the 'rents' from emissions trading privatized.

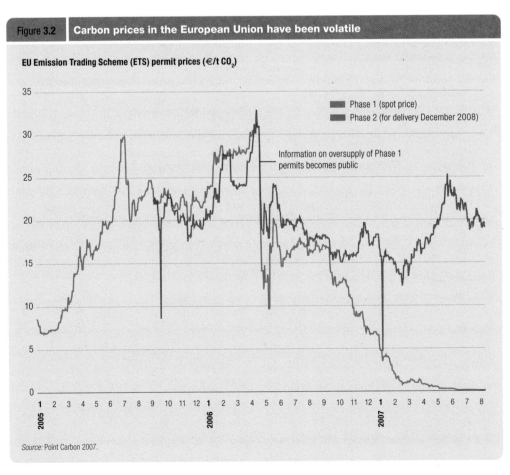

| Figure 3.2 | Carbon prices in the European Union have been volatile |

EU Emission Trading Scheme (ETS) permit prices (€/t CO_2)

Phase 1 (spot price)
Phase 2 (for delivery December 2008)

Information on oversupply of Phase 1 permits becomes public

Source: Point Carbon 2007.

Prospects for the second phase

Will these problems in the EU ETS be corrected in the second phase, which runs from 2008 to 2012? While the scheme has been strengthened in some areas, serious problems remain. Governments have not seized the opportunity to use the EU ETS to institutionalize deep cuts in emissions. Most seriously, the scheme remains de-linked from the European Union's own emissions reduction targets for 2020.

Allowances have so far been approved for 22 member states.[52] The cap for these countries has been lowered: it is around 10 percent below the level set for the first phase and marginally below verified 2005 emissions. There is already evidence that markets are responding to stronger political signals. Prices for Phase II allowances on futures markets have recovered. Market forecasts by Point Carbon anticipate a price range of €15–30/t CO_2 (US$19–37/t CO_2), depending on the costs of abatement.

These are positive developments. Even so, when measured against the yardstick of sustainable carbon budget management the design of the second phase of the EU ETS has to be judged quite harshly. The cap set for 2008 to 2012 is just 2 percent below verified emissions for 2005. This is not compatible with a sustainable emissions pathway that would lead to a 30 percent cut in emissions by 2020 based on 1990 levels. For most countries, the EU ETS second phase will not require major adjustments (table 3.2). An underlying problem is that the EU ETS has been interpreted by European Union governments as a vehicle for delivering on the very limited Kyoto commitments, rather than as an opportunity to act on the 2020 commitments. This is despite of the fact that the mandate for the EU ETS extends to "emissions development and reduction potential".[53] Another element of continuity with the first phase is auctioning. While the bar has been raised, there is still a limit of 10 percent on the share of permits that can be distributed through auctioning, perpetuating losses for public finance and efficiency.[54]

Negotiations on the second phase of the EU ETS have highlighted a number of wider challenges for the European Union. As long as cap-setting remains the remit of individual member states, the battle to set more robust targets will continue. Most governments sought Phase II allowances above 2005 emission levels. The underlying problem is that cap setting at a national level is a highly political exercise that opens the door to intensive, and highly effective, lobbying by national industries and 'energy champions.' So far, European governments have shown a tendency to succumb to pressure from highly polluting industries, with the result that very weak limits have been placed on overall emissions.[55] Bluntly stated, European Union governments have been bolder in setting aspirational targets for 2020 than they have been in setting concrete emission caps under the actually functioning EU ETS.

Against this backdrop, there is a strong case for empowering the European Commission to set—and enforce—more robust targets aligned with the European Union's 2020 emission reduction goals. Another priority is to rapidly increase the share of quotas that are auctioned in order to generate the incentives for efficiency gains and finance wider environmental tax reforms. Aiming at 100 percent auctioning by

Table **3.2**	**Proposals for the European Union Emissions Trading Scheme**			
		Emissions cap for 2008–2012 period		
	2005 verified emissions under Phase II of ETS (Mt CO_2)	Proposed by government (Mt CO_2)	Allowed by European Commission (Mt CO_2)	Allowed by European Commission as % of 2005 emissions
Austria	33	33	31	94
Belgium	56	63	59	105
Czech Republic	83	102	87	105
Finland	33	40	38	115
France	131	133	133	102
Hungary	26	31	27	104
Germany	474	482	453	96
Greece	71	76	69	97
Ireland	22	23	21	95
Italy	226	209	196	87
Netherlands	80	90	86	108
Spain	183	153	152	83
Sweden	19	25	23	121
United Kingdom	242 a	246	246	101
Total	1,943 a	2,095	1,897	98

a. Does not include the United Kingdom's installations which were temporarily excluded from the scheme in 2005 but will be covered in 2008 to 2012, estimated to amount to 30 Mt CO_2.

Source: European Union 2007c.

Effective public policies can help create win–win outcomes for global climate security, national energy security and living standards

2015 is a realistic goal. For sectors—such as power generation— facing limited competition, rules could be revised to allow for one-half of permits to be auctioned by 2012.

There are two CDM-related dangers that the European Union also has to address. The first is the danger of overuse. Opportunities for generating emission trading credits overseas should not totally displace mitigation in the European Union. If companies are able to meet their EU ETS obligations primarily by 'buying in' mitigation in developing countries while putting in place carbon-intensive investments at home, that is evidence for insufficiently ambitious targets. One detailed study of national allocation plans for nine countries estimates that between 88 and 100 percent of emissions reductions under the second phase of the EU

ETS could take place outside of the European Union.[56] Against this backdrop, it is important that emission credits play a supplementary role, as envisaged under the Kyoto Protocol.

The second danger concerns the authenticity of CDM emission reductions. Rules governing the arrangement require that emission reductions are 'additional'—that is, they would not have happened in the absence of CDM investments. In practice, this is difficult to verify. There is evidence that some CDM credits have been acquired for investments that would have taken place anyway.[57] Far more stringent independent monitoring is required to ensure that carbon trading does not act to dilute real mitigation. The need for such stringent monitoring raises questions about the further expansion of the CDM based on the current model.

3.3 The critical role of regulation and government action

Putting a price on carbon either through taxation or cap-and-trade schemes is a necessary condition for avoiding dangerous climate change. But carbon pricing alone will not be sufficient to drive investments and change behaviour at the scale or speed required. There are other barriers to a breakthrough in climate change mitigation—barriers that can only be removed through government action. Public policies on regulation, energy subsidies and information have a central role to play.

There are no blueprints for identifying in advance the appropriate policies to create an enabling environment for low-carbon transition. However, the problems to be addressed are well-known. Changing the energy mix in favour of low-carbon energy requires large up-front investments and a long-term planning horizon. Markets alone will not deliver. Government regulatory mechanisms backed by subsidies and incentives have a key role in guiding investment decisions. Energy efficiency standards for buildings, electrical appliances and vehicles can dramatically curtail emissions at low

cost. Meanwhile, policy support for research and development can create conditions for a technological breakthrough.

Effective public policies can help create win–win outcomes for global climate security, national energy security and living standards. Improvements in end-use efficiency illustrate the potential. Scenarios developed by the International Energy Agency (IEA) point to the potential for efficiency savings to cut emissions by 16 percent in OECD countries by 2030. Every US$1 invested in securing these reductions through more efficient electrical appliances could save US$2.2 in investment in power plants. Similarly, every US$1 invested in more efficient fuel standards for vehicles could save US$2.4 in oil imports.[58]

While estimates of the cost–benefit ratios for efficiency gains vary, as these figures demonstrate, there are large gains on offer. Those gains can be measured in terms of consumer savings, reduced dependence on oil imports and reduced costs for industry. They can also be measured in terms of cut-price climate change mitigation. Viewed differently, the failure to unlock efficiency gains is a route to 'lose–lose' outcomes for global climate security, national energy security and consumers. In this section we look at the place of regulatory provision and public policy in four key areas:

- Power generation;
- Residential sector;
- Vehicle emission standards;
- Research, development and deployment of low-carbon technologies.

Power generation—changing the emissions trajectory

Power generation is the main source of CO_2 emissions. It accounts for four in every ten tonnes of CO_2 dispatched to the Earth's atmosphere. How countries generate electricity, how much they generate and how much CO_2 gets emitted with each unit of energy produced are critical in shaping the prospects for stringent climate change mitigation.

Current scenarios point in some worrying directions. World electricity demand is projected to double by 2030. Cumulative investments for meeting this demand are projected by the IEA at US$11 trillion from 2005 to 2030.[59] Over half of this investment will happen in developing countries characterized by low levels of energy efficiency. China alone will account for around one-quarter of projected global investments. Projected investments for the United States are estimated at US$1.6 trillion, reflecting a large-scale replacement of existing power generation stock.

Emerging power generation investment patterns point in a worrying direction. They suggest that the world is being too locked into the growth of highly carbon-intensive infrastructures. Coal figures with growing prominence in planned power supply. The largest increases in investment are planned in China, India and the United States—three of the four largest current sources of CO_2 emissions. In each of these countries, rapid expansion in coal-fired power generation capacity is already under way or in the pipeline. In 2006, China was building an estimated two new coal-fired power stations every week.

Authorities in the United States are considering proposals for building over 150 coal-fired power plants, with planned investment of US$145 billion to 2030.[60] Over the next 10 years, India is planning to increase its coal-fired electricity generation capacity by over 75 percent.[61] In each case, the expansion in capacity is one of the major drivers of a large projected increase in national CO_2 emissions (figure 3.3).

What are the prospects for achieving deep cuts in CO_2 emissions linked to power generation? The answer to that question will depend partly on the rate at which new low-carbon technologies are developed and deployed, partly on the rate at which major developing countries adopt these technologies, and partly on demand-side factors such as savings through efficiency gains—issues that we consider in later sections of this chapter. Public policies that shape the energy mix will be important in each of these areas.

Power generation is the main source of CO_2 emissions. It accounts for four in every ten tonnes of CO_2 dispatched to the Earth's atmosphere

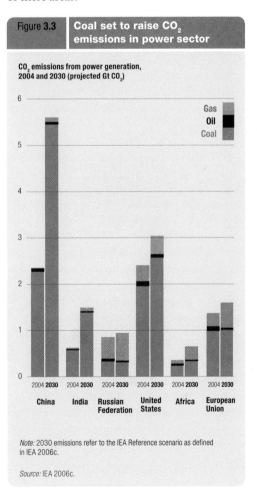

Figure **3.3** **Coal set to raise CO_2 emissions in power sector**

CO_2 emissions from power generation, 2004 and 2030 (projected Gt CO_2)

Gas
Oil
Coal

China · India · Russian Federation · United States · Africa · European Union

Note: 2030 emissions refer to the IEA Reference scenario as defined in IEA 2006c.

Source: IEA 2006c.

The energy mix

Current energy mix in the OECD countries is heavily dominated by fossil fuels. Changing this mix in favour of low-carbon or zero-carbon energy could lead to cuts in emissions. However, energy systems cannot be transformed overnight.

Nuclear power is one low-carbon option. However, it is an option that raises some difficult questions for policymakers. On the one hand, nuclear power offers a source of electricity with a near-zero carbon footprint. It has the additional advantages of reducing dependence on imported fossil fuels and providing a source of energy that is less subject to price volatility than fossil fuel. On the other hand, nuclear energy raises concerns about safety, the environmental repercussions and the proliferation of nuclear weapons—concerns that are reflected in widespread public opposition to expansion. On balance, nuclear energy is likely to remain an important part of overall supply. However, in terms of long-run climate mitigation potential, it is unlikely to play a prominent role and its market share could shrink (box 3.6).[62]

Renewable energy from the sun, wind and sea tides remains substantially underexploited. Discounting hydroelectricity, the renewables sector currently accounts for only around 3 percent of power generation in OECD countries. Achieving a target of 20 percent by 2020, as envisaged by the European Union, is a practical goal. With current technologies, renewable energy is not competitive with coal-fired power. However, scaling up a tax on carbon emissions to US$60–100/t CO_2 would radically change incentive structures for investment, eroding the advantage currently enjoyed by carbon-intensive power suppliers. At the same time, a range of supportive policies are required to stimulate

---

Box 3.6   Nuclear power—some thorny questions

Does nuclear power provide a cost-effective route for aligning energy security and climate security? Proponents point to potential benefits for carbon mitigation, price stability and reduced dependence on oil and gas imports. Critics of nuclear energy contest the economic arguments and claim that the environmental and military risks outweigh the benefits. The real answer probably lies somewhere in between these positions.

Nuclear energy reduces the global carbon footprint. It currently accounts for around 17 percent of the world's electricity generation. Some four-fifths of this capacity is located in 346 reactors in OECD countries. The share of nuclear in the national energy mix for electricity production ranges from over 20 percent for the United Kingdom and the United States to 80 percent in France. Phasing out nuclear energy without phasing in an equivalent supply of non-nuclear, zero-carbon energy from an alternative source is a prescription for increased emissions of CO_2.

That does not make nuclear power a panacea for climate change. In 2006, one reactor was started up—in Japan—while six were shut down in other OECD countries. Just to keep pace with retirements, eight new plants a year will be needed to 2017. While some countries (such as Canada and France) have announced plans for expanding nuclear energy, in others (including Germany and Sweden) a phase-out is under active consideration. In the United States, no nuclear plants have been ordered for over three decades. Medium-term projections point to a static or shrinking nuclear share in global energy supply.

These projections could change—but there are big economic questions to be addressed. Nuclear plants are highly capital-intensive. Capital costs range from US$2–3.5 billion per reactor, even before decommissioning and the disposal of nuclear waste are factored in. In the absence of government action to provide guaranteed markets, reduce risks and dispose of nuclear waste, there would be little private sector interest in nuclear power. The question for governments is whether nuclear is more cost-effective over the long term than low-carbon alternatives, such as wind power and solar power.

Non-economic questions relating to governance and regulation also loom large in nuclear energy debates. In many countries, public concerns over safety remain deeply entrenched. At an international level, there is a danger that nuclear technologies can be used to generate weapons-grade fissile material, irrespective of whether the material is designated for military purposes. Without an international agreement to strengthen the Non-Proliferation of Nuclear Weapons Treaty, the rapid expansion of nuclear energy would pose grave risks to all countries. Institutional mechanisms to restrict the crossover between civilian and military applications of nuclear energy have to include enhanced verification and inspection. Greater transparency, allied to clearly defined, monitorable and enforceable rules on the use and disposal of weapons-usable material (highly enriched uranium and plutonium) in civilian nuclear programmes, is also required. Developed countries could do far more to meet the governance challenge, notably by reducing their own nuclear arsenals and promoting more active diplomacy to advance non-proliferation.

Source: Burke 2007; IEA 2006c; NEA 2006.

investment through the creation of predictable and stable markets for renewable energy.

Current trends underline the potential for rapid growth in renewable energy provision. Both wind power and solar power are expanding sources of energy. Global investment in renewable energy has increased rapidly, from US$27 billion in 2004 to US$71 billion in 2006 alone.[63] Remarkable efficiency gains have been registered. Modern wind turbines produce 180 times more energy at less than half the cost per unit than turbines 20 years ago.[64] Investments in the United States have increased wind capacity by a factor of six in the intervening period (figure 3.4).[65] Much the same has happened in solar power. The efficiency with which photovoltaic cells convert sunlight into electricity has climbed from 6 percent in the early 1990s to 15 percent now, while their cost has fallen by 80 percent.[66]

Public policies have the potential to support a rapid expansion in renewable energy. Regulatory intervention is one instrument for the creation of incentives. In the United States, around 21 states have renewable portfolio standards requiring a certain proportion of power sold to come from renewable energy suppliers: in California, the proportion is 20 percent by 2017.[67] By providing guaranteed markets and setting favourable tariffs over several years, governments can provide renewable suppliers with a secure market in which to plan investments.

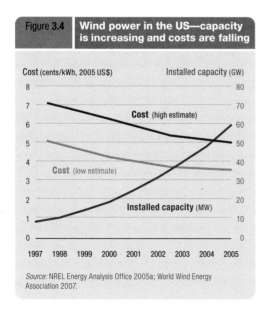

Figure 3.4 | **Wind power in the US—capacity is increasing and costs are falling**

Cost (cents/kWh, 2005 US$)   Installed capacity (GW)

Cost (high estimate)

Cost (low estimate)

Installed capacity (MW)

1997 1998 1999 2000 2001 2002 2003 2004 2005

Source: NREL Energy Analysis Office 2005a; World Wind Energy Association 2007.

Germany's Renewable Sources Act is an example. This has been used to fix the price of renewable power for 20 years on a sliding scale. The aim has been to create a long-term market while at the same time creating competitive pressures that create incentives for efficiency gains (box 3.7). In Spain, the Government has used a national premium tariff to increase the contribution of wind power. This now meets around 8 percent of the country's electricity demand, rising to more than 20 percent in the densely populated provinces of Castilla-La Mancha and Galicia. In 2005 alone, the increase in wind turbine capacity in Spain saved around 19 million tonnes of CO_2 emissions.[68]

Fiscal policy also has an important role to play in supporting renewable energy development. The United States has emerged as one of the world's most dynamic markets for renewable energy, with states such as California and Texas now established as global leaders in wind power generation. Market support has been provided through a three-year Production Tax Credit programme. However, uncertainty over the renewal of tax credits has given rise in the past to large fluctuations in investment and demand.[69] Many countries have combined a wide range of instruments to promote renewable energy. In Denmark, the wind power sector has been encouraged through tax breaks on capital investment, preferential pricing and a mandated target. The result: in the space of two decades, wind power has increased its share of electricity generation from less than 3 percent to 20 percent.[70]

The development of renewable energy is not a panacea for climate change. Because supplies are contingent on natural forces, there are problems with intermittent output. The initial capital costs of connecting to national grids can also be high, which is why the rapid expansion of the industry in recent years has been linked to the provision of subsidies. However, fossil fuel based energy has also been heavily subsidized over many decades—and in contrast to fossil fuels, renewable energy provides important returns for climate change mitigation.

Many countries have combined a wide range of instruments to promote renewable energy

The residential sector—low-cost mitigation

Some ways of cutting CO_2 emissions are cheaper than others. And some ways cost nothing at all over the long run. The residential and services sector provides a particularly striking example. Current practices across the world forcefully demonstrate the scope for measures that will save electricity, reduce emissions and cut costs for households and national economies.

Energy use patterns in the residential sector have an important bearing on the global carbon footprint. In the OECD countries, around one-third of the electricity produced ends up in heating and cooling systems, domestic refrigerators, ovens, lamps and other household devices. The residential sector accounts for around 35–40 percent of national CO_2 emissions from all fossil fuels, with appliances alone producing roughly 12 percent.[71]

There is an enormous untapped potential for energy savings in the residential sector. Realizing that potential would generate a double benefit: international climate change mitigation efforts would gain with a fall in CO_2 emissions, and the public would save money. Recent studies have highlighted the scale of this potential. One detailed exercise for OECD countries examines

a wide range of policies on building standards, procurement regulations, appliance standards and energy-efficiency obligations to assess the potential costs and benefits of achieving emission reductions.[72] The results point to a 29 percent saving in emissions by 2020, representing a reduction of 3.2 Gt CO_2—a figure equivalent to around three-times current emissions from India. The resulting energy savings would counterbalance the costs. Another study estimates that the average European Union household could save €200–1000 (US$250–1243) annually through improved energy efficiency (2004 prices).[73]

Electrical appliances are another major potential source of efficiency gains. Some appliances use energy more efficiently, and produce a lower carbon footprint, than others. If all electrical appliances operating in OECD countries from 2005 onwards met the best efficiency standards, it would save some 322 million tonnes of CO_2 emissions by 2010.[74] This would be equivalent to taking 100 million cars off the road—a figure that represents all vehicles in Canada, France and Germany combined.[75] By 2030, these higher standards would avoid emissions of 572 Mt CO_2 a year, which would be equivalent to removing 200 million cars from the road or closing 400 gas-fired power stations.

Would these efficiency gains deal a devastating blow to household budgets? On the contrary, they would reduce residential electricity consumption by around one-quarter by 2010. For North America, where households consume 2.4 times more electricity per household than in Europe, that reduction would save consumers an estimated US$33 billion for the period. By 2020, for every tonne of CO_2 emissions avoided, each household in the United States would save around US$65. "In Europe, each tonne of CO_2 avoided would save consumers some €169"[76] (reflecting Europe's higher electricity cost and lower efficiency standards).

Lighting provides another example. World lighting represents around 10 percent of global electricity demand and generates 1.9 Gt CO_2 per year—7 percent of total CO_2 emissions. As a glance around any developed country city day or night will confirm, much of this electricity is wasted. Light is routinely cast on spaces where nobody is present and delivered through inefficient sources. Simple installation of low-cost sources—such as compact fluorescent lamps—could reduce total lighting energy use by 38 percent.[77] The payback period for investment in more efficient lighting? Around 2 years on average for OECD countries.

Regulation and information are two of the keys for unlocking energy efficiency gains in the building and residential sector. Public policy has a key role to play not just in enhancing consumer awareness but in prohibiting or creating strong disincentives for practices that drive down efficiency and drive up carbon emissions. While there are costs associated with regulation and information provision, there are substantial climate change mitigation benefits. There are also large consumer costs associated with regulatory standards that allow inefficient energy use. Enhanced energy efficiency in this area can achieve emission savings with a net benefit. Among the public policy instruments:

- *Appliance standards.* These are among the most cost-effective mitigation measures. One example comes from Japan's 'Top Runner' scheme. Introduced in 1998 to support national efforts to comply with Kyoto reduction commitments, this scheme requires that all new products meet specified efficiency standards. Energy efficiency gains of over 50 percent have been recorded for some products, including cars, fridges, freezers and televisions. Research in a wide group of countries points to large benefits from reducing CO_2 through improved energy standards. This is an area in which effective demand management can cut carbon and energy costs, creating win–win benefits for the economy and the environment. Research in the European Union and the United States points to estimated benefits in a range from US$65/t CO_2 to 190/t CO_2.[78]

- *Information.* This is one of the keys to unlocking efficiency gains. In the United States, the Energy Star programme, a voluntary endorsement labelling scheme, provides consumers with extensive information on the energy efficiency of over 30 products. It is estimated to have delivered annual savings of US$5 billion in 2002.[79] In Australia, mandatory labelling of certain appliances—including freezers and dishwashers—has contributed in savings of CO_2 with benefits estimated at around US$30/t CO_2.[80]

- *Building codes.* Building standard regulations can generate very large savings in CO_2 emissions linked to energy use. Enforcement matters as much as the rules. In Japan, where the implementation of energy efficiency standards in buildings is voluntary, energy savings have been moderate. Far greater savings have been registered in countries such as in Germany and the United States, where compliance is enforced more stringently. The European Union estimates that efficiency gains in energy consumption could be increased by one-fifth, with potential savings of €60 billion (US$75 billion).[81] One-half of the gains would result from simple implementation of existing regulatory standards, most of them in the building sector.

Vehicle emission standards

Personal transportation is the world's largest consumer of oil—and its fastest growing source

Regulation and information are two of the keys for unlocking energy efficiency gains in the building and residential sector

3

Avoiding dangerous climate change: strategies for mitigation

of CO_2 emissions. In 2004, the transport sector produced 6.3 Gt CO_2. While the share of developing countries is rising, OECD countries account for two-thirds of the total.[82] The automobile sector in these countries accounts for about 30 percent of total greenhouse gas emissions, and the share is rising over time.[83]

The regulatory environment for transport is a critical part of the international carbon mitigation effort. Aggregate greenhouse gas emissions from any vehicle is a function of three factors: miles travelled, amount of fuel used for each mile travelled, and the carbon content of the fuel. Emissions are rising in many countries because the distances travelled are growing faster than fuel-use efficiency, and because fuel economy gains have been reduced by a trend towards bigger and more powerful vehicles.

Setting the standard

Countries vary widely in their fuel efficiency standards. The European Union and Japan have the highest standards, while the United States has the lowest in the developed world—lower, in fact, than in China (figure 3.5).[84]

Efficiency standards in the United States relative to the rest of the world have slipped over time. One reason for this is that they have changed only marginally over the past two decades, whereas other countries have been setting higher standards. Another is the prevalence of regulatory gaps favouring low-efficiency sports utility vehicles.

These gaps have reduced fleet efficiency and driven up emissions. Since 1990, emissions from transport have increased at an annual average rate of 1.8 percent, almost double the rate for all other sources. The primary driver of the emissions upsurge is vehicle miles travelled (which has climbed by 34 percent) and an increase in the use of light-duty trucks (box 3.8).[85]

Improvements in United States regulatory standards could make a global difference in climate change mitigation, with large associated benefits for national energy security. According to the National Commission for Energy, increasing the fuel efficiency requirement for cars in the United States by 20 miles per gallon (equivalent to 8.5 kilometres per litre) would reduce projected oil consumption by 3.5 million barrels a day, diminishing CO_2 emissions by 400 million tonnes per year in the process.[86] The savings from that regulatory shift would be equivalent to France's total CO_2 emissions. Apart from the benefits for climate change mitigation, the associated reduction in oil imports would achieve one of the central goals of United States energy security policy.

While the European Union has attained relatively higher fuel efficiency than the United States, it faces problems in aligning standards with its stated climate change goals. Since 1990,

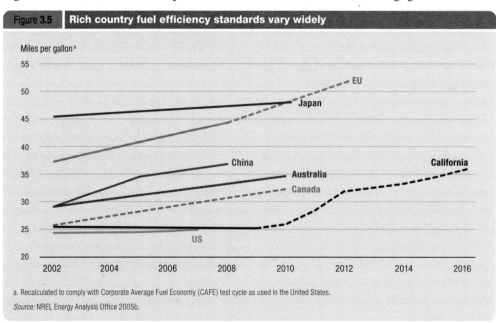

Figure 3.5 | **Rich country fuel efficiency standards vary widely**

Miles per gallon[a]

a. Recalculated to comply with Corporate Average Fuel Economy (CAFE) test cycle as used in the United States.

Source: NREL Energy Analysis Office 2005b.

Box 3.8   **Vehicle emissions standards in the United States**

Established in 1975, the United States' Corporate Average Fuel Economy (CAFE) programme is one of the world's oldest regulatory regimes on fuel efficiency. It is also one of the most important: the United States accounts for around 40 percent of oil-based CO_2 emissions from transport.

Where the United States sets its vehicle fuel efficiency standards registers in the world's carbon footprint. In the 1970s, CAFE rules were instrumental in doubling vehicle fuel economy, spurring investment in new technologies. However, fuel economy standards have not been increased for passenger cars over the past 20 years, and they have increased only slightly for light trucks.

As a result, the fuel efficiency standard divide between the United States and the rest of the world has widened. Today, the United States' standard is just over one half of the level in Japan. The 136 million passenger cars on United States' roads contribute 35 percent of national transport-based greenhouse gas emissions, and the 87 million light trucks another 27 percent.

The design of CAFE standards has had an important bearing on transport-related emissions. Average fuel standards for cars (27.5 miles per gallon or 11.7 kilometres per litre) are higher than for light trucks (20.7 mpg or 8.8 km/L). Rising demand for light trucks has led to an overall decrease in the fuel economy of new light-duty vehicles. In 2002, the number of light trucks sold exceeded new passenger cars sold for the first time. The upshot: fuel efficiency today is lower than in 1987.

CAFE standards are at the centre of an active national debate. The 2007 State of the Union Address proposed CAFE standard reforms to achieve a 5 percent reduction in gasoline consumption, based on projected future demand (rather than current levels). No numerical target for fuel efficiency was identified.

Would more stringent targets undermine employment and competitiveness? That question is at the centre of debates over CAFE standards. Research indicates that light-duty fuel efficiency could be increased by one-quarter to one-third at less than the cost of the fuel saved—and without compromising vehicle safety. Over the medium term, more stringent standards would create incentives for investment in advanced diesel engines, hybrid vehicles and hydrogen-powered fuel-cell vehicles.

With oil prices and concerns over CO_2 emissions rising, weak efficiency standards could send the wrong signals to the automobile industry. While recent years have seen significant improvements in engine technologies and vehicle design, such improvements have been used to increase power, performance and safety rather than to enhance fuel economy. One result is that firms in the United States have lost out to Japanese competitors in markets for more fuel-efficient models.

More stringent CAFE standards in the United States could create a triple benefit. They would demonstrate United States leadership in international climate change mitigation efforts, advance national energy security goals by reducing dependence on imported oil and open up new opportunities for investment in the automobile industry.

Source: Arroyo and Linguiti 2007; Merrill Lynch and WRI 2005; NCEP 2004b; Sperling and Cannon 2007.

---

the European Union has reduced overall emissions of greenhouse gases by around 1 percent. However, emissions from road transport have increased by 26 percent. As a result, the share of transport in overall emissions has climbed from around one-sixth to over one-fifth in little more than a decade.[87] Road transport is the biggest source of rising emissions, with passenger vehicles accounting for around one-half of the total. If domestic transport greenhouse gas emissions continue to rise with economic growth, they could be 30 percent above 1990 levels by 2010 and 50 percent by 2020.[88] Thus current trends in the transport sector are not consistent with the European Union's commitment to achieving 20–30 percent reductions in overall greenhouse gas emissions by 2020.

Aligning regulatory policies with more stringent climate change mitigation goals has been difficult. Current approaches are based on three pillars: voluntary commitments by the automobile industry, fuel-economy labelling and promotion of efficiency through fiscal measures. The long-standing aim has been to achieve a fuel-efficiency goal of 120g CO_2/km. However, the target date for achieving this goal has repeatedly been pushed back, initially from 2005 to 2010 and now to 2012, in the face of lobbying by the automobile industry and opposition in some member states. The interim target is now 140g CO_2/km by 2008–09.

As for the United States, where the European Union sets the fuel-efficiency bar matters for international climate change mitigation. It matters in a very immediate sense because more stringent standards will cut emissions of CO_2. Over the 10-year period to 2020, a 120g CO_2/km target would reduce emissions by about 400 Mt CO_2—more than the total emissions from France or Spain in 2004. That figure represents around 45 percent of total current European Union

Avoiding dangerous climate change: strategies for mitigation

3

Many governments now see biofuels as a technology that kills two birds with one stone, helping to fight global warming while reducing dependence on oil imports

emissions from transport. More broadly, because the European Union is the world's largest automobile market, tighter emission standards would signal an important change in direction to the global automobile industry, creating incentives for components suppliers to develop low carbon technologies. However, the European Union is not on track for achieving its long-standing target. As an assessment by the European Commission puts it: "In the absence of additional measures, the European Union objective of 120g CO_2/km will not be met at a 2012 time horizon."[89]

Efforts to change this picture have produced a political deadlock. The European Commission has proposed regulatory measures to raise fleet average efficiency standards to achieve the long-standing 120g CO_2/km goal by 2020. As in the past, the proposal has attracted opposition from the European Automobile Manufacturers Association—a coalition of 12 global automobile companies. Some European governments have supported that opposition, arguing that more stringent regulation could undermine the competitiveness of the industry.

This is a position that is difficult to square with a commitment to the European Union's 2020 targets. Arguments on economic competitiveness are also not well supported by the evidence. Several companies in the global automobile industry have lost out in fast-expanding markets for low-emission vehicles precisely because they have failed to raise efficiency standards. With supporting policies, it would be possible for the European Union to sustain progressive improvements in efficiency standards consistent with its climate goals, with fleet average standards improving to 80g CO_2/km by 2020.[90]

Regulatory standards cannot be viewed in isolation. Car taxation is a powerful instrument through which governments can influence the behaviour of consumers. Graduated taxation that rises with the level of CO_2 emissions could help to align energy policies in transport with climate change mitigation goals. Annual vehicle excise taxes and registration taxes on new vehicles would be means to this end. Such measures would support the efforts of car manufacturers

to meet improved efficiency standards, along with the efforts of governments to achieve their stated climate change goals.

The role of alternative fuels

Changing the fuel mix within the transport sector can play an important role in aligning energy policies with carbon budgets. The CO_2 emissions profile of an average car journey can be transformed by using less petroleum and more ethanol produced from plants. Many governments now see biofuels as a technology that kills two birds with one stone, helping to fight global warming while reducing dependence on oil imports.

Developing countries have demonstrated what can be achieved through a judicious mix of incentives and regulation in the transport sector. One of the most impressive examples comes from Brazil. Over the past three decades, the country has used a mix of regulation and direct government investment to develop a highly efficient industry. Subsidies for alcohol-based fuel, regulatory standards requiring automobile manufacturers to produce hybrid vehicles, preferential duties and government support for a biofuel delivery infrastructure have all played a role. Today, biofuels account for around one-third of Brazil's total transport fuel, creating wide-ranging environmental benefits and reducing dependence on imported oil.[91]

Several countries have successfully changed the national transport sector fuel-mix by using a mixture of regulation and market incentives to promote compressed natural gas (CNG). Prompted partly by concerns over air quality in major urban centres, and partly by a concern to reduce dependence on imported oil, both India and Pakistan have seen a major expansion of CNG use. In India, several cities have used regulatory mechanisms to prohibit a range of vehicles from using non-CNG fuel. For example, Delhi requires all public transport vehicles to use CNG. In Pakistan, price incentives have supplemented regulatory measures. Prices for CNG have been held at around 50–60 percent of the price of petroleum, with Government supporting the development of an infrastructure for

Climate change is the defining challenge facing political leaders across the world today. Future generations will judge us on how we respond to that challenge. There are no easy solutions—and no blueprints. But I believe that we can win the battle against climate change by acting nationally and working together globally.

If we are to succeed in tackling climate change we have to start by setting out the ground rules. Any international strategy has to be built on the foundations of fairness, social justice and equity. These are not abstract ideas. They are guides to action.

The *Human Development Report 2007/2008* should be mandatory reading for all governments, especially those in the world's richest nations. It reminds us that historic responsibility for the rapid build-up of greenhouse gases in the Earth's atmosphere rests not with the world's poor, but with the developed world. It is people in the richest countries that leave the deepest footprint. The average Brazilian has a CO_2 footprint of 1.8 tonnes a year compared with an average for developed countries of 13.2 tonnes a year. As the Report reminds us, if every person in the developing world left the same carbon footprint as the average North American we would need the atmospheres of nine planets to deal with the consequences.

We only have one planet—and we need a one-planet solution for climate change. That solution cannot come at the expense of the world's poorest countries and poorest people, many of whom do not have so much as a light in their home. Developed countries have to demonstrate that they are serious by cutting their emissions. After all, they have the financial and the technological resources needed to act.

Every country faces different challenges, but I believe the experience of Brazil is instructive. One of the reasons that Brazil has such a low per capita footprint is that we have developed our renewable energy resources and now have one of the world's cleanest energy systems. Hydro-power accounts for 92 percent of our electricity generation, for example. The upshot is that Brazil not only has a lighter carbon footprint than rich nations, but that we generate less than half as much CO_2 for every dollar in wealth that we generate. Put differently, we have lowered our emissions by reducing the carbon intensity and the energy intensity of our economy.

The transport sector provides a striking example of how clean energy policies can generate national and global benefits. Brazil's experience with the development of ethanol from sugar cane as a motor fuel goes back to the 1970s. Today, ethanol-based fuels reduce our overall emissions by about 25.8 million tonnes of CO_2e every year. Contrary to the claims made by some commentators lacking familiarity with Brazilian geography, the sugar production

that sustains our ethanol industry is concentrated in São Paulo, far from the Amazon region.

Today, we are expanding our ethanol programme. In 2004, we launched the National Program of Biodiesel Production and Use (PNPB). The aim is to raise the share of biodiesel in every litre of diesel sold in Brazil to 5 percent by 2013. At the same time, PNPB has introduced fiscal incentives and subsidies aimed at expanding market opportunities for biofuel production for small family farms in the North and the North-East region.

Brazil's experience with biofuels can help to support the development of win–win scenarios for energy security and climate change mitigation. Oil dominates the transport fuels sector. However, concerns over high prices, reserve levels, and security of supply are prompting many countries—rich and poor—to develop policies for reducing oil-dependency. Those policies are good for energy efficiency and good for climate change.

As a developing country Brazil can play an important role in supporting the transition to low-carbon energy. South–South cooperation has a vital role to play—and Brazil is already supporting the efforts of developing countries to identify viable alternative energy sources. However, we should not downplay the potential for international trade. North America and the European Union are both scaling-up heavily subsidized biofuel programmes. Measured against Brazil's ethanol programme these score badly both in terms of costs and in terms of efficiency in cutting CO_2 emissions. Lowering import barriers against Brazilian ethanol would reduce the costs of carbon abatement and enhance economic efficiency in the development of alternative fuels. After all, there is no inherent virtue in self-reliance.

Finally, a brief comment on rainforests. The Amazon region is a treasured national ecological resource. We recognize that this resource has to be managed sustainably. That is why we introduced in 2004 an Action Plan for Preventing and Controlling Deforestation in the Amazon. Encompassing 14 ministries, the plan provides a legal framework for land use management, establishes monitoring arrangements, and creates incentives for sustainable practices. The decline since 2004 in the rate of deforestation recorded in states such as Mato Grosso demonstrates that it is possible to reconcile economic growth with sustainable environmental management.

Luiz Inácio Lula da Silva
President of the Federative Republic of Brazil

production and distribution. Some 0.8 million vehicles now use CNG and the market share is rising fast (figure 3.6). Apart from cutting emissions of CO_2 by around 20 percent, using natural gas creates wide-ranging benefits for air quality and public health.

In the developed world biofuel development is one of the energy-based growth industries of the past 5 years. The United States has set particularly far-reaching goals. In his 2007 State of the Union Address, President Bush set a target of increasing the use of biofuels to 35 billion gallons in 2017—five times current levels. The ambition is to replace around 15 percent of imported oil with domestically produced ethanol.[92] The European Union is also actively promoting biofuels. Targets include raising to 10 percent the share of biofuels in all road-transport fuel consumption by 2020. That figure is double the target for 2010—and around 10 times the current share.[93]

Impressive targets have been backed with impressive subsidies for the development of the biofuels sector. In the United States, tax credits for maize-based ethanol production were estimated at US$2.5 billion in 2006.[94] Overall subsidies to ethanol and biodiesel, currently estimated at US$5.5–7.5 billion discounting direct payments to maize farmers, are expected to rise with production.[95] With the share of maize production directed towards ethanol mills growing, prices are rising sharply. In 2007 they reached a 10-year high, even though the crop of the previous year was the third highest on record.[96] Because the United States is the world's largest exporter of maize, the diversion of supply to the bioethanol industry has been instrumental in pushing up world prices. In Mexico and other countries in Central America, rising prices for imported maize could create food security problems for poor households.[97]

'Biofuel mania' has not so far left such a deep mark on the European Union. However, this is likely to change. Projections by the European Commission point to increasing prices for oilseeds and cereals. The arable area for producing biofuels will rise from an estimated 3 million hectares in 2006 to 17 million hectares in 2020.[98] Most of the increase in supply of biofuel in the European Union will come from domestic production of cereals and oilseeds, though imports are projected to account for 15–20 percent of total demand by 2020. For European agriculture, the prospective bio-diesel boom offers lucrative new markets. As the Commission puts it: "The targets for renewable energy can be seen as good news for European

agriculture: they [...] promise new outlets and a positive development of demand and prices at a time when farmers are increasingly faced with international competition."[99] Under the reformed Common Agricultural Policy, a special premium is payable to farmers for the production of energy crops.[100]

Unfortunately, what is good for subsidized agriculture and the biofuels industry in the European Union and the United States is not inherently good for climate change mitigation. Biofuels do represent a serious alternative to oil for use in transport. However, the cost of production of those fuels relative to the real amount of CO_2 abatement is also important. This is an area in which the United States and the European Union do not score very well. For example, sugarcane-based ethanol can be produced in Brazil at half the unit price of maize-based ethanol in the United States and whereas sugar-based ethanol in Brazil cuts emissions by some 70 percent, the comparable figure for the maize-based ethanol used in the United States is 13 percent.[101] The European Union is at an even greater cost disadvantage (figure 3.7).

Comparative advantage explains an important part of the price differentials. Production costs in Brazil are far lower because of climatic factors, land availability and the greater efficiency of sugar in converting the sun's energy into cellulosic ethanol. These differences point to a case for less reliance on domestic production and an expanded role for international trade in the European Union and the United States.

There is no inherent virtue in self-reliance. From a climate change mitigation perspective, the priority is to achieve carbon abatement at the lowest marginal cost. The problem is that trade barriers and subsidies are driving up the cost of carbon mitigation, while simultaneously adding to the cost of reducing oil dependency.

Most developed countries apply import restrictions on alternative fuels such as bio-ethanol. The structure of protection varies widely—but the net effect is to substantially lower consumer demand. The European Union allows duty free market access for ethanol for around 100 developing countries, most of which do not export ethanol. In the case of

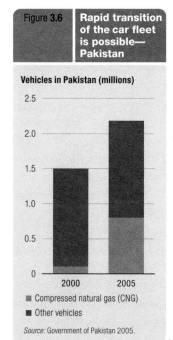

Figure **3.6** | **Rapid transition of the car fleet is possible—Pakistan**

Vehicles in Pakistan (millions)

■ Compressed natural gas (CNG)
■ Other vehicles

Source: Government of Pakistan 2005.

Brazil, an import duty of €0.73 (US$1) per gallon is applied by the European Union—a tariff equivalent in excess of 60 percent.[102] In the United States, Brazilian ethanol faces an import duty of US$0.54 a gallon.[103] While lower than in the European Union, this still represents a tariff of around 25 percent at 2007 domestic market prices for ethanol.

Trade policies applied to ethanol conflict with a wide range of climate change goals. Ethanol from Brazil is disadvantaged even though it is cheaper to produce, generates lower CO_2 emissions in production, and is more efficient in reducing the carbon-intensity of vehicle transport. More broadly, the high levels of tariff applied to Brazilian ethanol raise serious questions for economic efficiency in the energy sector. The bottom-line is that abolishing ethanol tariffs would benefit the environment, climate change mitigation, and developing countries which—like Brazil—enjoy favourable production conditions. In the European Union, Sweden has argued strongly for a reduced emphasis on protectionism and stronger policies for the development of 'second-generation' biofuels in areas such as forest biomass.[104]

Not all international trade opportunities linked to biofuels offer benign outcomes. As in other areas, the social and environmental impacts of trade are conditioned by wider factors—and benefits are not automatic. In Brazil, the sugar production that sustains the ethanol industry is concentrated in the southern State of São Paulo. Less than 1 percent originates from the Amazonia. As a result, the development of biofuels has had a limited environmental impact, and has not contributed to rainforest destruction. The picture in other countries and for other crops is mixed. One potential source of agricultural inputs for biodiesel is oil palm. Expansion of cultivation of that crop in East Asia has been associated with widespread deforestation and violation of human rights of indigenous people. There is now a danger that the European Union's ambitious biofuel targets will encourage the rapid expansion of oil palm estates in countries that have failed to address these problems (box 3.9). Since 1999, European Union imports of palm oil (primarily from Malaysia and Indonesia)

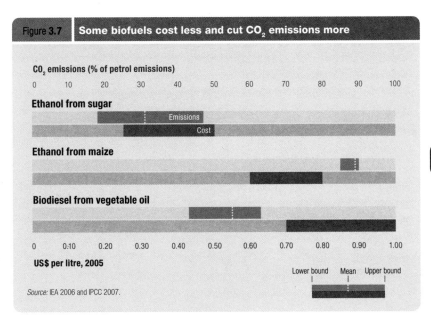

Figure 3.7 **Some biofuels cost less and cut CO₂ emissions more**

CO₂ emissions (% of petrol emissions)

Ethanol from sugar — Emissions / Cost

Ethanol from maize

Biodiesel from vegetable oil

US$ per litre, 2005

Lower bound / Mean / Upper bound

Source: IEA 2006 and IPCC 2007.

have more than doubled to 4.5 million tonnes, or almost one-fifth of world imports.[105] Rapid expansion of the market has gone hand-in-hand with an erosion of the rights of small farmers and indigenous people.

R&D and deployment of low-carbon technologies

Joseph Schumpeter coined the phrase 'creative destruction' to describe a "process of industrial mutation that incessantly revolutionizes the economic structure from within, incessantly destroying the old one, incessantly creating a new one". He identified three phases in the process of innovation: invention, application and diffusion.

Successful climate change mitigation will require a process of accelerated 'creative destruction', with the gap between these phases shrinking as rapidly as possible. Carbon pricing will help to create incentives for the emergence of these technologies—but it will not be enough. Faced with very large capital costs, uncertain market conditions and high risks, the private sector alone will not develop and deploy technologies at the required pace, even with appropriate carbon price signals. Governments will have to play a central role in removing obstacles to the emergence of breakthrough technologies.

The case for public policy action is rooted in the immediacy and the scale of the

Avoiding dangerous climate change: strategies for mitigation

Box 3.9   Palm oil and biofuel development—a cautionary tale

The European Union's ambitious targets for expanding the market share of biofuels have created strong incentives for the production of cereals and oils, including palm oil. Opportunities for supplying an expanding European Union market have been reflected in a surge of investment in palm oil production in East Asia. Is this good news for human development?

Not under current conditions. Oil palm can be grown and harvested in environmentally sustainable and socially responsible ways, especially through small-scale agroforestry. Much of the production in West Africa fits into this category. However, large-scale mono-cropping plantations in many countries do not have a good record. And much of the recent surge in palm oil production has taken place on such plantations.

Even before the European Union's renewable energy targets generated a new set of market incentives, oil palm cultivation was expanding at a prolific rate. By 2005, global cultivation had reached 12 million hectares—almost double the area in 1997. Production is dominated by Indonesia and Malaysia, with the former registering the fastest rate of increase in terms of forests converted into oil palm plantations. The estimated annual net release emissions of CO_2 from forest biomass in Indonesia since 1990 is 2.3 Gt. European Union markets for biofuel materials can be expected to create a further impetus for oil palm plantations. Projections by the European Commission suggest that imports will account for around one-quarter of the supply of biodiesel fuels in 2020, with palm oil representing 3.6 million out of a total of 11 million tonnes of imports.

Palm oil exports represent an important source of foreign exchange. However, the expansion of plantation production has come at a high social and environmental price. Large areas of forest land traditionally used by indigenous people have been expropriated and logging companies have often used oil palm plantations as a justification for harvesting timber.

With palm oil prices surging, ambitious plans have been developed to expand cultivation. One example is the Kalimantan Border Oil Palm Project in Indonesia, which aims at converting 3 million hectares of forest in Borneo. Concessions have already been given to companies. While national legislation and voluntary guidelines for industry stipulate protection for indigenous people, enforcement has been erratic at best and—in some cases—ignored. Areas deemed suitable for oil palm concessions include forest areas used by indigenous people—and there are extensively documented reports of people losing land and access to forests.

In Indonesia, as in many other countries, the judicial process is slow, the legal costs are beyond the capacities of indigenous people, and links between powerful investors and political elites make it difficult to enforce the rights of forest dwellers. Against this backdrop, the European Union has to carefully consider the implications of internal directives on energy policy for external human development prospects.

Source: Colchester et al. 2006a, 2006b; Tauli-Corpuz and Tamang 2007.

---

threat posed by climate change. As shown in previous chapters of this Report, dangerous climate change will lead to rising poverty in poor countries, followed by catastrophic risks for humanity as a whole. Avoiding these outcomes is a human development challenge. More than that, it is a global and national security imperative.

In earlier periods of history, governments have responded to perceived security threats by launching bold and innovative programmes. Waiting for markets to generate and deploy the technologies to reduce vulnerability was not considered an option. In 1932, Albert Einstein famously concluded: "There is not the slightest indication that nuclear energy will ever be obtainable." Just over a decade later, the Allied powers had created the Manhattan Project. Driven by perceived national security imperatives, this was a research effort that brought together the world's top scientists in a US$20 billion (in 2004 terms) programme that pushed back technological frontiers. The same thing happened under President Eisenhower and President Kennedy, when Cold War rivalries and national security concerns led to government leadership of ambitious research and development drives, culminating in the creation of the Apollo space programme.[106]

Contrasts with the R&D effort to achieve a low-carbon transition are strikingly evident. R&D spending in the energy sectors of OECD countries today is around one-half of the level in the early 1980s in real terms (2004 prices).[107] Measured as a share of turnover in the respective sectors, the R&D expenditure of the power industry is less than one-sixth of that for the automobile industry and one-thirtieth of that for the electronics industry. The distribution of research spending is equally problematic. Public spending on R&D has been dominated by nuclear energy, which still accounts for just under half of the total.

3

These R&D patterns can be traced to a variety of factors. The power sector, in particular, is characterized by large central power plants dominated by a small number of suppliers, with restricted competition for market share. Heavy subsidies to fossil fuel-based power and nuclear energy have created powerful disincentives for investment in other areas such as renewable energy. The end result is that the energy sector has been characterized by a slow pace of innovation, with many of the core technologies for coal and gas power generation now over three decades old.

'Picking winners' in coal

Developments in the coal sector demonstrate both the potential for technological breakthroughs in climate change mitigation and the slow pace of progress. There is currently around 1200 GigaWatts (GW) of coal-fired power capacity worldwide accounting for 40 percent of the world's electricity generation and CO_2 emissions. With natural gas prices rising and coal reserves widely disbursed across the world, the share of coal in world energy generation is likely to rise over time. Coal-fired power generation could be the driver that takes the world beyond the threshold of dangerous climate change. However, it also provides an opportunity.

Coal-fired power plants vary widely in their thermal efficiency.[108] Increased efficiency, which is largely a function of technology, means that plants generate more power with less coal—and with fewer emissions. The most efficient plants today use super-critical technologies that have attained efficiency levels of around 45 percent. During the 1990s, new Integrated Gasification Combined Cycle (IGCC) technologies emerged. These are able to burn synthetic gas produced from coal or another fuel and to clean gas emissions. Supported by public funding in the European Union and the United States, five demonstration plants were constructed in the 1990s. These plants have attained levels of thermal efficiency comparable to the best conventional plants, with high levels of environmental performance.[109]

What is the link between IGCC plants and climate change mitigation? The real potential breakthrough technology for coal is a process known as Carbon Capture and Storage (CCS). Using CCS technology, it is possible to separate the gas emitted when fossil fuels are burned, process it into liquefied or solid form, and transport it by ship or pipeline to a location—below the sea-bed, into disused coal mines, depleted oil wells, or other locations—where it can be stored. Applied to coal plants, CCS technology offers the potential for near-zero CO_2 emissions. In theory, any conventional coal plant can be retrofitted with CCS technology. In practice, IGCC plants are technologically the most adaptable to CCS, and by far the lowest cost option.[110]

No single technology offers a magic bullet for climate change mitigation, and 'picking winners' is a hazardous affair. Even so, CCS is widely acknowledged to be the best-bet for stringent mitigation in coal-fired power generation. Large-scale development and deployment of CCS could reconcile the expanding use of coal with a sustainable carbon budget. If successful, it could take the carbon out of electricity generation, not just in power stations but also from other carbon-intensive sites of production such as cement factories and petrochemical facilities.

Demonstration plants operated through private–public partnerships in the European Union and the United States have shown the feasibility of CCS technology, though some challenges and uncertainties remain.[111] For example, the storage of CO_2 beneath sea-beds is the subject of international conventions and there are safety concerns about the potential for leaks. Encouraging as the demonstration project results have been, the current effort falls far short of what is needed. CCS technology is projected to come on-stream very slowly in the years ahead. With planned rates of deployment, there will be just 11 CCS plants in operation by 2015. The upshot of this late arrival is that the plants will collectively save only around 15 Mt CO_2 in emissions, or 0.2 percent of total coal-fired power emissions.[112] At this rate, one of the key technologies in the battle against global warming will arrive on the battlefield far too late to help the world avoid dangerous climate change.

The real potential breakthrough technology for coal is a process known as Carbon Capture and Storage

At present, conventional coal-fired power plants enjoy a commercial advantage for one simple reason: their prices do not reflect the costs of their contribution to climate change

Barriers to accelerated development and disbursement of CCS technologies are rooted in markets. Power generation technologies that can facilitate rapid deployment of CCS are still not widely available. In particular, IGCC plants are not fully commercialized, partly because there has been insufficient R&D. Even if full-scale CCS systems were available today, cost would be a major obstacle to deployment. For new plants, capital costs are estimated to be up to US$1 billion higher than conventional plants, though there are large variations: retrofitting old plants is far more costly than applying CCS technology to new IGCC plants. Carbon capture is also estimated to increase the operational costs of electricity generation in coal plants by 35–60 percent.[113] Without government action, these cost barriers will continue to hold back deployment.

Coal partnerships—too few and too limited

Some of the obstacles to the technological transformation of coal-fired power generation could be removed through carbon pricing. At present, conventional coal-fired power plants enjoy a commercial advantage for one simple reason: their prices do not reflect the costs of their contribution to climate change. Imposing a tax of US$60–100/t CO_2 or introducing a stringent cap-and-trade scheme, would transform incentive structures in the coal industry, putting more highly polluting power generators at a disadvantage. Creating the market conditions for increased capital investment through tax incentives is one of the conditions for a low-carbon transition in energy policy.

Policies in the United States are starting to push in this direction. The 2005 Energy Act has already boosted planning applications for IGCC plants by putting in place a US$2 billion Clean Coal Power Initiative (CCPI) that includes subsidies for coal gasification.[114] Tax credits have been provided for private investment in nine advanced clean coal facilities. Public–private partnerships have also emerged. One example is the seven Carbon Sequestration Regional Partnerships that bring together the Department of Environment, state

governments and private companies. The total value of the projects is around US$145 million over the next four years. Another example is FutureGen, a public–private partnership that is scheduled to produce the United States' first near-zero power plant in 2012.[115]

The European Union has also moved to create an enabling environment for the development of CCS. The formation of the European Technology Platform for Zero Emissions Fossil Fuel has provided a framework that brings together governments, industry, research institutes and the European Commission. The aim: to stimulate the construction and operation by 2015 of up to 12 demonstration plants, with all coal-fired power plants built after 2020 fitted with CCS.[116] Total estimated funding for CO_2 capture and storage technologies for 2002 to 2006 was around €70 million (US$88 million).[117] However, under the current European Union research framework, up to €400 million (US$500 million) will be provided towards clean fossil-fuel technologies between 2007 and 2012, with CCS a priority.[118] As in the United States, a range of demonstration projects are under way, including collaboration between Norway and the United Kingdom on the storage of carbon in North Sea oil fields.[119]

Emerging private–public partnerships have achieved important results. However, far more ambitious approaches are needed to accelerate technological change in the coal industry. The Pew Center on Global Climate Change has argued for the development of a 30-plant programme over 10 years in the United States to demonstrate technical feasibility and create the conditions for rapid commercialization. Incremental costs are estimated at around US$23–30 billion.[120] The Pew Center has proposed the establishment of a trust fund created by a modest fee on electricity generation to cover these costs. While there are a range of financing and incentive structures that could be considered, the target of a 30-plant programme by 2015 is attainable for the United States. With political leadership, the European Union could aim for a comparable level of ambition.

The danger is that public policy failures will create another obstacle to CCS development and deployment. Higher costs associated with CCS-equipped plants could give rise to a 'non-CCS lock-in' as a result of investment decisions on the replacement of current coal-fired capacity. In the absence of long-term carbon price signals and incentive structures to reward low-carbon electricity, power generators might take decisions that would make it more difficult to make the transition to CCS.

This would signal another lost opportunity. Around one-third of existing coal-fired capacity in the European Union is expected to reach the end of its technical lifetime in the next 10–15 years.[122] In the United States, where coal is resurgent, applications or proposals have been made for the development of over 150 new coal-fired power plants to 2030, with a projected investment of around US$145 billion.[123]

Both the European Union and the United States have an opportunity to use the retirement of old coal-fired power stock to create an enabling environment for an early transition to CCS. Seizing that opportunity will require bold steps in energy policy. Increasing investment in demonstration projects, signalling a clear intent to tax carbon emissions and/or introducing more stringent cap-and-trade provisions, and using regulatory authority to limit the construction of non-IGCC plants are among the policy requirements.

Increased financial and technological support for low-carbon power generation in developing countries is one priority area

3.4 The key role of international cooperation

International cooperation could open the door to wide-ranging win–win scenarios for human development and climate change mitigation. Increased financial and technological support for low-carbon power generation in developing countries is one priority area. Cooperation here could expand access to energy and improve efficiency, lowering carbon emissions and supporting poverty reduction efforts in the process. Deforestation is another problem that offers an opportunity. International action to slow the pace of rainforest destruction would reduce the global carbon footprint while generating a range of social, economic and environmental benefits.

Current approaches are failing to unlock the potential in international cooperation. Under the terms of the UNFCCC, international cooperation was identified as a key element in climate change mitigation. Developed countries pledged to "take all practicable steps to promote, facilitate and finance, as appropriate, the transfer of, or access to, environmentally sound technologies".[124] In 2001, an agreement was drawn up—the Marrakesh Accords—aimed at giving greater substance to the commitment on technology transfer. Yet delivery has fallen far short of the pledges made, and even

further short of the level of ambition required. Progress in tackling deforestation is similarly discouraging.

Negotiations on the next commitment period for the Kyoto Protocol provide an opportunity to change this picture. There are two urgent priorities. First, the world needs a strategy to support low-carbon energy transitions in developing countries. Developed countries should see this not as an act of charity but as a form of insurance against global warming and as an investment in human development.

In the absence of a coherent international strategy for finance and technology transfer to facilitate the spread of low-carbon energy, developing countries will have little incentive to join a multilateral agreement that sets emission ceilings. There are 1.6 billion people in the world lacking access to electricity—often women who walk many miles to fetch wood and/or collect cow dung to use as fuel. Expecting governments that represent them to accept medium-term ceilings on emissions that compromise progress in access to energy is unrealistic and unethical. It is also inconsistent with international commitments on poverty reduction.

The second priority is the development of a strategy on deforestation. Carbon markets

One unit of electricity produced in a developing country emits 20 percent more CO₂ than an average unit in developed countries

and financial transfers alone do not provide an answer to the problem. However, they can help to reduce the perverse incentives that currently act to promote deforestation, with negative consequences for people and the planet.

An expanded role for technology transfer and finance

Low levels of energy efficiency hold back human development and economic growth in many countries. Enhanced efficiency is a means to generate more power with less fuel—and fewer emissions. Rapidly narrowing the efficiency gap between rich and poor countries would act as a powerful force for climate change mitigation, and it could act as a force for human development.

Coal provides a powerful demonstration of the point. The average thermal efficiency for coal plants in developing countries is around 30 percent, compared with 36 percent in OECD countries.[124] This means that one unit of electricity produced in a developing country emits 20 percent more CO_2 than an average unit in developed countries. The most efficient supercritical plants in OECD countries, so called because they burn coal at higher temperatures with less waste, have achieved efficiency levels of 45 percent.[125] Projections for future emissions from coal-fired power generation are highly sensitive to the tech-nological choices that will influence overall efficiency. Closing the efficiency gap between these plants and the average in developing countries, would halve CO_2 emissions from coal-fired power generation in developing countries.[126]

The potential mitigation impact of efficiency gains can be illustrated by reference to China and India. Both countries are diversifying energy sources and expanding renewable energy provision. However, coal is set to remain the main source of power generation: the two countries will account for around 80 percent of the increase in global demand for coal to 2030. Average thermal efficiency in coal-fired power plants is increasing for both countries, but is still only around 29–30 percent.[127] Rapid expansion of coal-fired power generation built on this level of efficiency would represent a climate change

disaster. With large investments going into new plants, there is an opportunity to avert that disaster by raising efficiency levels (table 3.3). Getting more energy from less coal would unlock wide ranging benefits for national economies, the environment and climate change mitigation.

China and India highlight the tension between national energy security and global climate security goals. Coal is at the heart of these tensions. Over the next decade, China will become the world's largest source of CO_2 emissions.[128] By 2015, power generation capacity will increase by around 518 GW, double current levels. It will increase again by around 60 percent, according to IEA projections, by 2030. To put the figures in context, the increase in power generation to 2015 is equivalent to current capacity in Germany, Japan and the United Kingdom combined. Coal will account for roughly three-quarters of the total increase by 2030.

Coal-fired power capacity is also expanding rapidly in India. In the decade to 2015, India will add almost 100 GW in power generation capacity—roughly double current power generation in California. The bulk of the increase will come from coal. Between 2015 and 2030, coal-fired power capacity is projected to double again, according to the IEA. While both China and India will continue to have far smaller per capita footprints than OECD countries, the current pattern of carbon-intensive energy growth clearly has worrying implications for climate change mitigation efforts.

Enhanced energy efficiency has the potential to convert a considerable climate change threat into a mitigation opportunity. We demonstrate this potential by comparing IEA scenarios for China and India covering the period 2004 to 2030, with more ambitious scenarios based on strengthened international cooperation. While any scenario is sensitive to assumptions, the results graphically illustrate both the benefits of multilateral action in supporting national energy policy reform and the implied costs of inaction.

Even modest reforms to enhance energy efficiency can deliver significant mitigation. The IEA compares a business-as-usual 'reference scenario' for future emissions with an 'alternative scenario' in which governments deepen energy sector reforms. Under these reforms, it is assumed that overall coal-fired efficiency in China and India increases from current levels of around 30 percent to 38 percent by 2030. Most of the reforms would build incrementally on existing measures aimed at reducing demand.

It is possible to imagine a more ambitious scenario. Energy efficiency standards could be strengthened. Inefficient old plants could be retired more rapidly and be replaced by new supercritical plants and IGCC technologies, paving the way for an early transition to carbon capture and storage. Of course, these options would require additional financing and the development of technological capabilities. But, they would also deliver results.

Looking beyond the IEA scenario, we consider a more rapid transition to low-carbon, high-efficiency coal-fired power generation. That transition would see average efficiency levels raised to 45 percent by 2030—the level of the best-performing OECD plants today. We also factor in an additional element: early introduction of CCS technology. We assume that 20 percent of the additional capacity introduced between 2015 and 2030 takes the form of CCS.

These assumptions may be bold—but they are hardly beyond the realm of technological feasibility. Measured in terms of climate change mitigation, the emission reductions that would result are considerable:

- *China.* By 2030, emissions in China would be 1.8 Gt CO_2 below the IEA reference scenario level. That figure represents about one-half of current energy-related CO_2 emissions from the European Union. Put differently, it would reduce overall projected CO_2 emissions from all developing countries by 10 percent against the IEA reference scenario.
- *India.* Efficiency gains would also generate large mitigation effects in India. These amount to 530 Mt CO_2 in 2030 against the IEA reference level—a figure that exceeds current emissions from Italy.

Both of these illustrations underline the potential for rapid mitigation through efficiency gains in the power sector

(figure 3.8). In important respects, the headline figures understate the potential gains for climate change mitigation through enhanced energy efficiency. One reason for this is that our alternative scenario focuses just on coal. It does not consider the potential for very large energy efficiency gains and CO_2 reductions through wider technological innovations in natural gas and renewable energy, for example. Nor do we factor in the large potential for achieving efficiency gains through technological breakthroughs in carbon-intensive industrial sectors, such as cement and heavy industry (table 3.4). Moreover, we present the gains in terms of a static one-year snapshot for 2030,

Table 3.3	Carbon emissions are linked to coal plant technology		
	Approx. CO_2 emissions (g/kWh)	Reduction from Chinese average (%)	Lifetime CO_2 saving (Mt CO_2)[a]
Coal-fired plants:			
Chinese coal-fired fleet average, 2006	1140	–	–
Global standard	892	22	73.3
Advanced cleaner coal	733	36	120.5
Supercritical coal with carbon capture	94	92	310.8

a. Lifetime savings assume a 1GW plant running for 40 years at an average capacity factor of 85 percent in comparison with a similar plant with Chinese average efficiency (currently 29 percent).

Source: Watson et al. 2007.

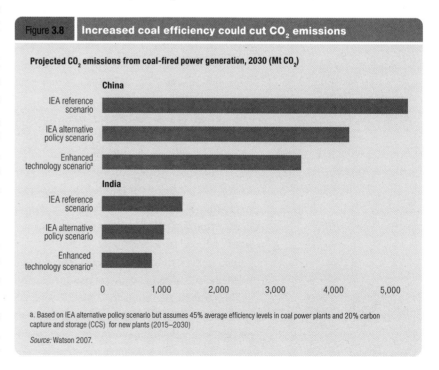

Figure 3.8 Increased coal efficiency could cut CO_2 emissions

Projected CO_2 emissions from coal-fired power generation, 2030 (Mt CO_2)

a. Based on IEA alternative policy scenario but assumes 45% average efficiency levels in coal power plants and 20% carbon capture and storage (CCS) for new plants (2015–2030)

Source: Watson 2007.

Table 3.4	Industrial energy efficiency varies widely			
Energy consumption per unit produced (100=most efficient country)		**Steel**	**Cement**	**Ammonia**
Japan		100	100	–
Europe		110	120	100
United States		120	145	105
China		150	160	133
India		150	135	120
Best available technology		75	90	60

Source: Watson et al. 2007.

whereas the benefits of emission reductions, like the costs of rising emissions, are cumulative. Accelerated introduction of CCS technologies in particular could produce very large cumulative gains in the post-2030 era.

Our focus on China and India also understates the wider potential benefits. We apply our alternative energy scenario to these countries because of their weight in global emissions. However, the exercise has broader relevance.

Consider the case of South Africa. With an energy-sector dominated by low-efficiency coal-fired power generation (which accounts for over 90 percent of electricity generation) and an economy in which mining and minerals production figure prominently, South Africa is the only country in sub-Saharan Africa with a carbon footprint to rival that of some OECD countries. The country has a deeper footprint than countries such as France and Spain—and it accounts for two-thirds of all CO_2 emissions from sub-Saharan Africa.[129] Raising average efficiency levels for coal-fired power generation in South Africa to 45 percent would reduce emissions by 130 Mt CO_2 by 2030. That figure is small by comparison with China and India. But it still represents over one-half of all energy-related CO_2 emissions from sub-Saharan Africa (excluding South Africa).[130] In South Africa itself enhanced efficiency in the coal sector would help address one of the country's most pressing environmental concerns: the serious problems caused by emissions of nitrous dioxide and sulphur dioxide from coal combustion.[131]

For the world as a whole, enhanced energy efficiency in developing countries offers some obvious advantages. If climate security is a global public good, then enhanced efficiency is an investment in that good. There are also potentially large national benefits. For example, China is attempting to reduce emissions from coal plants to address pressing public health concerns (box 3.10). About 600 million people are exposed to sulphur dioxide levels above WHO guidelines and respiratory illness is the fourth most common cause of death in urban areas. In India, inefficiencies in the power sector have been identified by the Planning Commission as a constraint on employment creation and poverty reduction (box 3.11).[132] As these examples demonstrate, both countries stand to gain from enhanced energy efficiency and reduced pollution—and the entire world stands to gain from the CO_2 mitigation that would come with improved efficiency. Conversely, all parties stand to lose if the gaps in coal-fired energy efficiency are not closed.

If the potential for win–win outcomes is so strong why are the investments in unlocking those outcomes failing to materialize? For two fundamental reasons. First, developing countries themselves face constraints in financing and capacity. In the energy sector, setting a course for low-carbon transition requires large front-loaded investments in new technologies, some of which are still in the early stages of commercial application. The combination of large capital cost, higher risk and increased demands on technological capabilities represents an obstacle to early deployment. Achieving a breakthrough towards a low-carbon transition will impose substantial incremental costs on developing countries, many of which are struggling to finance current energy reforms.

Failures in international cooperation represent the second barrier. While the international climate security benefits of a low-carbon transition in the developing world may be substantial, the international financing and capacity-building mechanisms needed to unlock those benefits remain underdeveloped. In energy, as in other areas, the international community has not succeeded in developing a strategy for investing in global public goods.

This is not to understate the importance of a range of programmes that are now underway.

Box 3.10 Coal and energy policy reform in China

With the world's fastest growing economy, one-fifth of its population, and a highly coal-intensive energy system, China occupies a critical place in efforts to tackle climate change. It is the world's second largest source of CO_2 emissions after the United States and is on the verge of becoming the largest emitter. At the same time, China has a small per capita carbon footprint by international standards, just one-fifth of that in the United States and a third of the average for developed countries.

Climate change confronts China with two distinctive but related challenges. The first challenge is one of adaptation. China is already registering highly damaging climate change impacts. Extreme weather events have become more common. Droughts in north-eastern China, flooding in the middle and lower reaches of the Yangtze River and coastal flooding in major urban centres such as Shanghai are all examples. Looking to the future, it would be no exaggeration to say that China faces the prospect of a climate change emergency. Yields of the three major grains—wheat, rice and maize—are projected to decline with rising temperatures and changed rainfall patterns. Glaciers in western China are projected to thin by 27 percent to 2050. Large reductions in water availability are projected across several river systems, including those in northern China—already one of the world's most ecologically stressed regions.

As these scenarios suggest, China has a strong national interest in supporting global mitigation efforts. The challenge is to change the emissions trajectory in a high-growth economy without compromising human development. Currently, emissions are on a sharply rising trend. They are projected by the IEA to double to 10.4 Gt CO_2 by 2030. Under its 11th Five-Year Plan, the Chinese Government has set a wide range of goals for lowering future emissions:

- *Energy intensity.* The current targets include a goal of reducing energy intensity by 20 percent below 2005 levels by 2010. Achieving that goal would reduce business-as-usual CO_2 emissions by 1.5 Gt by 2020. Progress to date has been slower than anticipated, at around one-quarter of the required level.
- *Large enterprises.* In 2006 the National Development and Reform Commission (NDRC) launched a major programme—the Top 1000 Enterprises Programme—to improve energy efficiency in the country's largest enterprises through monitored energy efficiency improvement plans.
- *Advanced technology initiatives.* China is now becoming active in the development of IGCC technologies that could enhance energy efficiency and set the scene for an early transition

to CCS. However, while a demonstration project has been authorized, implementation has been delayed by financing constraints and uncertainties over commercial risks.
- *Retiring inefficient power plants and industrial enterprises.* In 2005, only 333 of China's 6,911 coal-fired power units had capacities in excess of 300 MW. Many of the remainder have a capacity of less than 100 MW. These smaller units tend to use outmoded turbine designs that combine low efficiency with high levels of emissions. An NDRC plan envisages the accelerated closure of small, inefficient plants with a capacity of less than 50 MW by 2010. Targets have also been set for closing inefficient plants in areas such as steel and cement production, with stipulated reduction quotas for regional and provincial governments. In 2004, large and medium-sized steel mills consumed 705 kg of coal per tonne of steel, while smaller mills consumed 1045 kg/tonne.
- *Renewable energy.* Under a 2005 renewable energy law, China has set a national target of producing 17 percent of primary energy from renewable sources by 2020—more than twice the level today. While hydropower is envisaged as the main source, ambitious goals have been set for wind power and biomass, backed by financial incentives and subsidies.

These are ambitious targets. Translating them into measures that shape energy market outcomes will be difficult. For example, very small and highly inefficient units (less than 200 MW) accounted for over one-third of the new capacity installed from 2002 to 2004. That outcome points to a governance challenge in energy policy. In effect, a significant proportion of Chinese coal-fired power plant development is out of central government control, with local government not enforcing national standards. Similarly, there are very large gaps in efficiency between small enterprises and the larger enterprises subject to government regulatory authority.

Enhancing energy efficiency and reducing carbon intensity will require sustained reforms in China. At the same time, the current direction of energy reform, with a growing emphasis on efficiency, renewables and carbon mitigation, opens up opportunities for international cooperation and dialogue on climate change. The entire world has an interest in China deploying coal technologies that will facilitate the earliest and most rapid cuts in CO_2 emissions—and the earliest transition to CCS. Multilateral financing and technology transfer could play a critical role by meeting the incremental costs of a low-carbon transition, creating incentives and supporting the development of capacity.

Sources: CASS 2006; Li 2007; Watson et al. 2007; World Bank 2006d.

Yet the experience of coal again provides a powerful demonstration of current failures in international cooperation. While there has been a proliferation of exercises in cooperation, delivery has been largely limited to dialogue. One example is the Asia-Pacific Partnership on Clean

Development. This brings together a large group of countries—including China, India, Japan and the United States—committed to expanding the development and deployment of low-carbon technology. However, the partnership is not based on binding commitments and has so far

3

Avoiding dangerous climate change: strategies for mitigation

Box 3.11   Decarbonizing growth in India

Rapid economic growth over the past two decades has created unprecedented opportunities for poverty reduction in India. Sustained growth, allied to policies that tackle deep social disparities, is a basic requirement for overcoming the country's large human development deficit. But is there a tension between the national energy security policies needed to support economic growth and global climate security?

From a global climate change mitigation perspective, rapid economic growth fuelled by coal in the world's second most populous country poses an obvious challenge. Yet it also provides an opportunity for international cooperation.

India is now the world's fourth largest emitter of CO_2. Between 1990 and 2004, emissions increased by 97 percent—one of the highest rates of increase in the world. However, per capita energy use is rising from a low base. The average Indian uses 439 kg of oil-equivalent energy (kgoe), less than one-half of the average for China. The comparable figure for the United States is 7,835 kgoe. India's per capita carbon footprint places the country 128th in the world league table.

The energy shortfalls behind these figures have implications for human development. Around half of India's population—some 500 million people—do not have access to electricity. At a household level, low levels of energy use are reflected in high levels of dependence on biofuels (see figure). Meanwhile, persistent power shortages and unreliable supply act as a constraint on economic growth, productivity and employment. The all-India average for peak power shortages is 12 percent.

Energy occupies a critical place in India's development planning. The ambition set out in its Eleventh Five-Year Plan is to sustain economic growth rates in excess of 8–9 percent a year. At this level, energy generation will also have to double. Over the longer term, sustaining growth at current levels through to 2030 will require a fivefold increase in energy generation.

Coal is likely to provide most of the increase. With abundant domestic supplies—India accounts for around 10 percent of the world's known reserves—and concerns over the security of imported energy supplies, coal will remain the preferred fuel. Business-as-usual scenarios point to an increase in the share of coal in power supply and CO_2 emissions. Coal–based emissions are projected to rise from 734 Mt CO_2 in 2004, to 1,078 Mt CO_2 in 2015 and 1,741 Mt CO_2 by 2030.

Radical changes to this emissions trajectory are possible. Low levels of energy efficiency are holding back India's efforts to increase energy supply and expand access to electricity, while driving up emissions. Research carried out by the Planning Commission estimates that India could generate the same amount of power with one-third less fuel. As shown in this chapter, efficiency gains have the potential to generate deep cuts in emissions.

Technology provides part of the explanation for the low levels of efficiency in the coal sector. Over 90 percent of India's coal generation capacity is subcritical, much of it concentrated in small-scale plants. Improving the efficiency of these plants would generate large energy sector benefits for India, along with global climate change mitigation benefits.

Domestic policy reform is one requirement for unlocking efficiency gains. The power sector in India is dominated by large monopolies that control both power supply and distribution. Most state power utilities are in a financially weak condition, with average annual losses running at 40 percent. Uncollected bills, the provision of heavily subsidized electricity to agriculture (where most benefits are captured by high income farmers) and wider inefficiencies all contribute to these losses. The upshot is that utilities lack the financial resources needed to upgrade technology.

Current reforms are addressing these problems. The 2003 Electricity Act provides a framework for more efficient and equitable tariffs. New regulatory structures have been created, and some states—such as Andhra Pradesh and Tamil Nadu—have started to break electricity boards up into more competitive units for generation, transmission and distribution.

Energy reform in India provides the international community with an opportunity to support national policies that will also advance global climate change mitigation goals. Early adoption of clean coal technologies and best-practice international standards would enable India to change its emissions trajectory while meeting rising energy demand.

Research carried out for this Report by the Tata Energy Research Institute estimates that an annualized increase in investment of around US$5 billion is needed for the period 2012–2017 to support a rapid transition to low-carbon energy generation, over and above current investment plans. Mobilizing these resources through the type of multilateral mechanisms proposed in this chapter could create a win–win outcome for energy efficiency in India and global climate change mitigation.

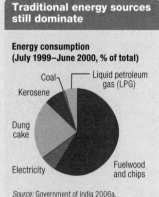

Traditional energy sources still dominate

Energy consumption
(July 1999–June 2000, % of total)

Coal
Liquid petroleum gas (LPG)
Kerosene
Dung cake
Electricity
Fuelwood and chips

Source: Government of India 2006a.

Source: Government of India 2006a, 2006b; Mathur and Bhandari 2007; MIT 2007; Watson et al. 2007.

produced little more than information exchange. Much the same is true of the G8's Plan of Action for Climate Change, Clean Energy and Sustainable Development.

The failure to develop substantive cooperation on CCS is particularly worrisome. From a global public goods perspective, there is an overwhelming interest in developed

Avoiding dangerous climate change: strategies for mitigation

3

countries speeding-up the deployment of CCS technologies at home, and then ensuring that they are available to developing countries as soon as possible and at the lowest price. Perhaps the most concrete example of cooperation in this area to date is the Near-Zero Emissions Coal Project, which is part of the European Union–China Partnership on Climate Change. The project is planned in three phases, starting with a three-year feasibility study (2005–2008) to explore technological options. The ultimate target is a single demonstration plant in 2020. However, progress in implementation has been slow—and details for implementing later phases have yet to be revealed.[133] Collaboration between the United States' FutureGen 'clean coal' project and Huaneng, China's third largest coal-power generation company, has been beset by similar uncertainties.

The missing link—a framework for finance and technology transfer

What is missing from the current patchwork of fragmented initiatives is an integrated international framework for finance and technology transfer. Developing that framework is a matter of urgency.

There are several areas in which international cooperation could help strengthen climate change mitigation efforts through support for national energy policy reforms. Under the UNFCCC, developed countries undertook to "meet the agreed full incremental costs" of a range of measures undertaken by developing countries in the three core areas of finance, technology and capacity building.[134] National resource mobilization will remain the primary financing vehicle for energy policy reform. Meanwhile, the focal point for international cooperation is the incremental financial cost and the enhanced technological capabilities required to achieve a low-carbon transition. For example, international cooperation would mobilize the resources to cover the 'price gap' between low-carbon options such as renewable energy and enhanced coal-efficiency options on the one side, and existing fossil-fuel based options on the other side.

The underlying problem is that developing countries already face deep financing constraints in energy policy. Estimates by the IEA suggest that an annual investment for electricity supply alone of US$165 billion annually is needed through to 2010, rising at 3 percent a year to 2030. Less than half of this financing is available under current policies.[135] Financing deficits have very real implications for human development. On current trends there will still be 1.4 billion people lacking access to electricity in 2030, and one-third of the world's population—2.7 billion people—will still be using biomass.[136]

Developing countries themselves have to address a wide range of energy sector reform problems. In many countries, heavily-subsidized energy prices and low levels of revenue collection represent a barrier to sustainable financing. Electricity subsidies are often directed overwhelmingly towards higher-income groups partly because they are distributed through large centralized grids to which the poor have limited access. Greater equity in energy financing and the development of decentralized grid systems that meet the needs of the poor are two of the foundations for meaningful reform. However, it is neither realistic nor equitable to expect the world's poorest countries to finance both the energy investments vital for poverty reduction at home and the incremental costs of a low-carbon transition to support international climate change mitigation.

These costs are linked to the capital requirements for new technologies, the increase in recurrent costs in power generation and the risks associated with the deployment of new technologies. As with any new technology, the risks and uncertainties associated with low-carbon technologies that have yet to be widely deployed even in the developed world represent a large barrier to deployment in developing countries.[137]

The multilateral framework for the post-2012 era will have to include mechanisms that finance these incremental costs, while at the same time facilitating technology transfer. Putting a figure on costs is difficult. One ballpark estimate for the investment costs to facilitate access to low-carbon technology broadly consistent with

On current trends there will still be 1.4 billion people lacking access to electricity in 2030

3

The Kyoto Protocol and the framework provided by the UNFCCC provide the primary platform for addressing global cooperation on climate change under United Nations leadership

our sustainable emissions pathway suggests that an additional US$25–50 billion per annum would be required for developing countries.[138] However, this is at best an approximation. One of the most urgent requirements for international cooperation is the development of detailed national financing estimates based on national energy policy plans.

Whatever the precise figure, financial transfers in the absence of cooperation on technology and capacity-building will be insufficient. The massive new investments required in developing countries' energy sectors over the next 30 years provide a window of opportunity for technological transformation. However, technological upgrading cannot be achieved through a simple process of technological transfer. New technologies have to be accompanied by the development of knowledge, capabilities in areas such as maintenance, and the development of national capacities to climb the technology-ladder. This is an area in which international cooperation—including South–South cooperation— has an important role to play.

Strengthened cooperation on financing, technology and capacity-building is vital for the credibility of the post-2012 Kyoto Protocol framework. Without that cooperation, the world will not get on to an emissions trajectory that avoids dangerous climate change. Moreover, developing countries will have little incentive to join a multilateral agreement that requires significant energy policy reforms on their part, without providing financial support.

History offers some important lessons. Perhaps the most successful of all international environmental treaties is the 1987 Montreal Protocol—the agreement forged to cut back emissions of ozone-depleting substances. Prompted by alarm over the expansion of the ozone hole above Antarctica, the treaty set stringent time-bound targets for phasing out these substances. Developing countries' participation was secured through a multilateral fund under which the incremental costs of achieving the targets were met by developed countries. Today, no countries are significantly off track for achieving the Montreal Protocol targets—and technology transfer is one of the

primary reasons for this outcome.[139] The benefits of international cooperation are reflected in the fact that the ozone hole is shrinking.

Experience under the Montreal Protocol has informed the multilateral response to climate change. Under the UNFCCC, the Global Environment Facility (GEF) became a financial instrument to mobilize resources for climate change activities in mitigation and adaptation. While overall financing has been limited, especially in the case of adaptation (see chapter 4), funds controlled under the GEF have demonstrated a capacity to leverage larger investments. Since its inception in 1991, the GEF has allocated US$3 billion, with co-financing of US$14 billion. Current resource mobilization is insufficient to finance low-carbon transition at the pace required. Moreover, the GEF continues to rely principally on voluntary contributions—an arrangement that reduces the predictability of finance. If the GEF is to play a more central role in mitigation in support of nationally-owned energy sector reforms, financing provisions may have to be placed on a non-voluntary basis.[140]

Building international cooperation on climate change is a formidable task. The good news is that the international community does not have to start by reinventing the wheel. Many of the individual elements for successful cooperation are already in place. The Kyoto Protocol and the framework provided by the UNFCCC provide the primary platform for addressing global cooperation on climate change under United Nations leadership. The CDM has provided a mechanism linking the mitigation agenda to financing for sustainable development in developing countries. This is done through greenhouse gas reducing projects that generate emission credits in developing countries which can be used by developed countries to offset their own domestic emissions. In 2006, CDM financing amounted to US$5.2 billion.[141] At one level, the CDM is potentially an important source of carbon financing for mitigation in developing countries. At another level, the CDM suffers from a number of shortcomings. Because it is project-based, transaction costs are high. Establishing that CDM emission reductions are 'additional', and monitoring outcomes, is also problematic. There are legitimate concerns that many of the emissions reductions under the CDM have been illusory. Moreover, carbon abatement has often been purchased at prices far higher than costs (box 3.12). Even without these problems, scaling-up the CDM in its current form to achieve emission reductions and financing transfers on the scale required would be enormously complex. It would require the establishment of thousands of projects, all of which would have to be validated and registered, with subsequent emission outcomes subject to verification and certification.

Box 3.12   **Linking carbon markets to the MDGs and sustainable development**

With cap-and-trade programmes set to play an increasingly prominent role in the mitigation efforts of rich countries, carbon markets are set to take-off on a global scale. Firms and governments will continue to seek low cost abatement opportunities in developing countries. Could flows of carbon finance help to expand opportunities for sustainable development and a low-carbon transition in the poorest countries?

Flexible mechanisms that have emerged from the Kyoto Protocol have created opportunities for developing countries to participate in carbon markets. The CDM market is set to grow from its current level of around US$5 billion. However, CDM projects are heavily concentrated in a small number of large developing countries. These countries have developed a strong capacity to market mitigation in large industrial enterprises. So far, the poorest developing countries have been bypassed—and there have been limited benefits for broad-based sustainable development (see figure).

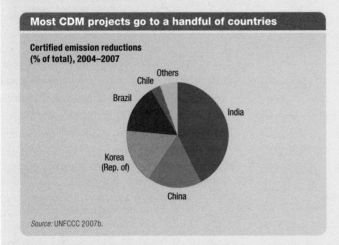

Most CDM projects go to a handful of countries

Certified emission reductions (% of total), 2004–2007

Others
Chile
Brazil
Korea (Rep. of)
China
India

Source: UNFCCC 2007b.

Perhaps unsurprisingly, carbon markets have concentrated finance in countries offering to reduce carbon emissions at the lowest abatement price. Sub-Saharan Africa represents less than 2 percent of credits, with only one country figuring in the 2007 project pipeline. Moreover, carbon finance flows have been heavily skewed towards greenhouse gases (other than CO_2) known as HFCs, especially in countries such as China and India. Because the cost of destroying these gases, which account for over one-third of all emission credits, is much lower than the price that credits can make on the open market, carbon trading has generated large profits for chemical companies and carbon brokers. Benefits for the world's poor have been less evident.

Market barriers provide one explanation for the limited participation of developing countries. Current rules for the flexibility mechanisms in the Kyoto Protocol restrict the scope of carbon financing linked to land use (section 3.4). The more serious structural problem is that groups such as small-scale farmers and forest dwellers do not have opportunities to engage in carbon markets, partly because the markets themselves are remote; and partly because they lack marketable rights in land and environmental resources. Marginal women farmers in Burkina Faso or Ethiopia are not well placed to negotiate with carbon brokers in the City of London—and carbon brokers seeking to minimize transaction costs have an inbuilt preference for large suppliers of mitigation credits.

Social organization is one of the keys to tapping the potential of carbon markets for sustainable development. In 2006, Kenya's Greenbelt Movement successfully marketed a programme to reforest two mountain areas in Kenya as part of an emissions reduction agreement. Women's groups will plant thousands of trees, with revenues coming from a carbon trade for the reduction of 350,000 tonnes of CO_2. The aim is to generate wide-ranging social and environmental benefits, including the restoration of eroded soils.

Innovative new approaches are being developed to address barriers to market entry. One example is the MDG Carbon Facility launched by the UNDP. In an effort to link carbon financing to sustainable development goals, UNDP 'bundled' a portfolio of projects sourced over 2 years, generating up to 15 Mt CO_2e within the first Kyoto commitment period (2008–2012). The credits will be marketed by Fortis Bank. One cluster of projects aims at renewable energy programmes to bring electricity to remote areas. Another will support the use of animal dung to generate biogas, freeing up women and children from fuelwood collection. Stringent processes have been established to ensure that the projects deliver mitigation and benefits for the poor.

The MDG Carbon Facility is an attempt to achieve a wider distribution of benefits from carbon markets. It involves the development of new operational and financing mechanisms. If successful, it will give some of the world's poorest countries the opportunity to participate in these markets. And it will link climate change mitigation to pro-poor sustainable development.

Source: UNDP 2007; UNFCCC 2007d; Zeitlin 2007.

Under a programme-based approach, developing countries could pledge to achieve a specified level of emission reduction, either in a specific sector (such as electricity generation) or for the country as a whole

Shifting the focus towards programme-based approaches could yield far more positive outcomes. Under a programme-based approach, developing countries could pledge to achieve a specified level of emission reduction, either in a specific sector (such as electricity generation) or for the country as a whole. The target could be set against a specific benchmark either in terms of reductions from a business-as-usual reference scenario or in terms of absolute cuts. Developed countries could support achievement of the targets by agreeing to meet the incremental costs of new technologies and capacity building. For example, current energy plans in China and India could be revisited to explore the potential and the costs for reductions in CO_2 emissions through the introduction of expanded programmes for renewable energy and accelerated introduction of clean coal technologies.

Negotiations on the post-2012 Kyoto Protocol framework provide an opportunity to put in place an architecture for international cooperation that links climate change mitigation to sustainable energy financing. One option would be the creation of an integrated Climate Change Mitigation Facility (CCMF). The CCMF would play a wide-ranging role. Its overarching objective would be to facilitate the development of low-carbon energy systems in developing countries. To that end, the aim would be to provide through multilateral channels support in key areas, including financing, technology transfer and capacity-building. Operations would be geared towards the attainment of emission reduction targets agreed under the post-2012 framework, with dialogue based on nationally-owned energy strategies. Rules and governance mechanisms would have to be developed to ensure that all parties deliver on commitments, with CCMF support geared towards well-defined quantitative goals and delivered in a predictable fashion. The following would be among the core priorities:

- *The mobilization of finance.* The CCMF would mobilize the US$25–50 billion needed annually to cover the estimated incremental costs of facilitating access to low-carbon technologies. Financing provisions would be linked to the circumstances of countries.

In middle-income countries—such as China and South Africa—concessionary finance might be sufficient, whereas low-income countries might require grants. The development of a programme-based CDM approach linking carbon markets in rich countries to mitigation in developing countries would be another instrument in the CCMF toolkit. One of the broad objectives of the CCMF would be to leverage private investment, domestic and foreign. Public finance could be partly or wholly generated through carbon taxation or levies on cap-and-trade permits.

- *Mitigating risks.* Commercial risks associated with the introduction of new, low-carbon technologies can act as a significant barrier to market entry. CCMF financing could be used to reduce risks through concessional loans, along with partial or full risk guarantees on loans for new technology—extending an approach developed under the World Bank's International Finance Corporation (IFC).

- *Building technological capabilities.* The CCMF could act as a focal point for wide-ranging cooperation on technology transfer. The agenda would extend from support for developing countries seeking financing for technology development, to the strengthening of capacity in state and non-state enterprises, strategies for sharing new technologies, and support for the development of specialized training agencies and centres of excellence in low-carbon technology development.

- *Buying out intellectual property.* It is not clear that intellectual property rights are a major barrier to low-carbon technology transfer. In the event that transfers of breakthrough technology were constrained by intellectual property provisions, the CCMF could be used to finance a structured buy-out of intellectual property rights, making climate-friendly technologies more widely accessible.

- *Expanding access to energy.* Meeting the needs of populations lacking access to modern energy services without fuelling dangerous climate change is one of the greatest challenges in international cooperation. There are strong efficiency and equity grounds for developing decentralized, renewable energy systems. Here too, however, there are large financing gaps. Under an Action Plan for Energy Access in Africa drawn up by the World Bank and others, strategies have been identified aimed at increasing access to modern energy from 23 percent today to 47 percent by 2030.[142] Implementation of these strategies will require an additional US$2 billion in concessional financing each year—roughly double current levels. The CCMF could provide a focal point for international efforts to mobilize these resources.

Creating a CCMF would not entail the development of vast new institutional structures. Large international bureaucracies that duplicate existing mechanisms will not help advance climate change mitigation. Neither will a 'more-of-the-same' model. If the world is to unite around a common mitigation agenda, it cannot afford to continue the current patchwork of fragmented initiatives. What is needed is a multilateral framework that links ambitious targets with ambitious and practical strategies for transferring low-carbon technologies. That framework should be developed under the auspices of the UNFCCC as part of the post-2012 Kyoto Protocol. And it should be designed and implemented through a process that gives developing countries, including the poorest countries, a real voice.

The starting point is political leadership. Stringent climate change mitigation will not happen through discrete technological fixes and bilateral dialogue. Government leaders need to send a clear signal that the battle against climate change has been joined—and that the future will look different to the past. That signal has to include a commitment on the part of developed countries to technology transfer and financing for a low-carbon transition. More broadly, what is needed is a partnership on mitigation. That partnership would be a two-way contract. Developing countries would draw on international support to strengthen current efforts to reduce emissions, setting quantitative targets that go beyond current

plans. Developed countries would underwrite attainment of incremental elements in these targets, supporting nationally-owned energy strategies that deliver tangible outcomes.

Developed through a CCMF framework, this approach could provide a focal point for a broad-based effort. Because a low-carbon transition is about far more than technology and finance, specialized agencies of the United Nations—such as UNDP and UNEP—could focus on an enhanced capacity-building effort, building the human resource base for deep energy reforms. The World Bank would be well-placed to oversee the financing provisions of the proposed CCMF. Its role could entail management of the subsidy element in the CCMF, the blending of concessional and non-concessional finance, oversight of subsidized credits to reduce risk, and the leveraging of private sector support. At a time when the future role of the World Bank in much of the developing world is uncertain, the CCMF could provide the institution with a clear mission that links improved access to energy and energy efficiency to climate change mitigation. Substantive engagement with the private sector would be imperative given its critical role in finance and technological innovation.

Reducing deforestation

The world's forests are vast repositories for carbon. The erosion of those repositories through deforestation accounts for about one-fifth of the global carbon footprint. It follows that preventing deforestation can mitigate climate change. But forests are more than a carbon bank. They play a crucial role in the lives of millions of poor people who rely on them for food, fuel and income. And tropical forests are sites of rich biodiversity. The challenge for international cooperation is to find ways of unlocking the triple benefits for climate mitigation, people and biodiversity that could be generated through the conservation of forests.

Governments are not currently meeting the challenge. The facts on deforestation tell their own story (figure 3.9). Between 2000 and 2005, net forest loss worldwide averaged 73 thousand

If the world is to unite around a common mitigation agenda, it cannot afford to continue the current patchwork of fragmented initiatives

Across the developing world, rainforests are being felled for gains which, in a functioning carbon market, would be dwarfed by the benefits of conservation

square kilometres a year—an area the size of a country like Chile.[143] Rainforests are currently shrinking at about 5 percent a year. Every hectare lost adds to greenhouse gas emissions. While forests vary in the amount of carbon that they store, pristine rainforest can store around 500 tonnes of CO_2 per hectare.

Between 1990 and 2005, shrinkage of the global forest estate is estimated to have added around 4 Gt CO_2 to the Earth's atmosphere each year.[144] If the world's forests were a country, that country would be one of the top emitters. On one estimate, deforestation, peat land degradation and forest fires have made Indonesia the third largest source of greenhouse gas emissions in the world.[145] Deforestation in the Amazon region is another of the great sources of global emissions. Data from the Instituto de Pesquisa Ambiental da Amazônia, a research institute in northern Brazil, suggest that deforestation is responsible for emissions of an estimated 730 Mt CO_2 each year.[146]

The many drivers of deforestation
Deforestation is driven by many forces. In some cases, poverty is the driver, with agricultural populations collecting fuelwood or expanding the frontier for subsistence agriculture. In others, opportunities for wealth generation are the main engine of destruction.

The expansion of national and international markets for products such as beef, soybeans, palm oil and cocoa can create strong incentives for deforestation. In Brazil, devaluation and a

30 percent increase in prices for soy exports from 1999 to 2004 gave a boost to forest clearance. In the 5 years to 2005, the states of Goias, Mato Grosso and Mato Grosso do Sul planted an additional 54,000 square kilometres of soy—an area slightly larger than Costa Rica. At the same time, forests are under pressure from commercial logging, much of it illegal. In Cambodia, to take one example, illegal logging of hardwood timbers for export was responsible for much of the 30 percent reduction in primary rainforest cover since 2000—one of the most rapid losses recorded by the FAO.[147]

Commercial pressures on rainforests are unlikely to dissipate in the near future. Croplands, pastures, plantations and logging are expanding into natural forests across the world. Population growth, rising incomes and opportunities for trade create incentives for deforestation—as does market failure on a global scale.

The scale of market failure is revealed in the basic economics of rainforest conversion. Across the developing world, rainforests are being felled for gains which, in a functioning carbon market, would be dwarfed by the benefits of conservation. Consider the following example. In Indonesia, oil palm cultivation generates an estimated value of US$114 per hectare. As the trees that stood on that hectare burn and rot, they release CO_2 into the atmosphere—perhaps 500 tonnes a hectare in dense rainforests. At a carbon price of US$20–30 a tonne, a plausible future range on the EU ETS, the carbon market value of that release would amount to US$10,000–15,000 a hectare. Put differently, farmers in Indonesia are trading a carbon bank asset worth at least US$10,000 in terms of climate change mitigation, for one worth US$114, or around 2% of its value.[148] Even commercial logging, which generates a higher market return, represents less than one-tenth of the value of the carbon bank. And these figures do not include the market and non-market values of environmental services and biodiversity.

Perverse incentives are at the heart of a 'lose–lose' scenario. The world is losing

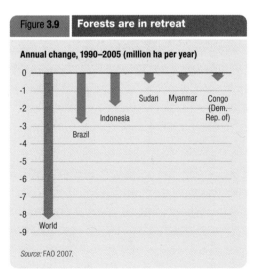

Figure 3.9 **Forests are in retreat**

Annual change, 1990–2005 (million ha per year)

Sudan Myanmar Congo (Dem. Rep. of)

Indonesia

Brazil

World

Source: FAO 2007.

immense opportunities for carbon mitigation through forest conversion. Countries are losing assets that could have a real value in terms of carbon finance. And people depending on forests for their livelihoods are losing out to economic activities operating on the basis of a false economy. Viewed in narrow commercial terms, deforestation makes sense only because markets attach no value to carbon repositories. In effect, standing trees are obstacles to the collection of money lying on the ground. While national circumstances vary, in many countries most of that money is appropriated by large-scale farmers, ranchers and illegal loggers. The upshot is that market failures are creating incentives that are bad for climate change, bad for national environmental sustainability and bad for equity.

What would it take to change the current incentive structure? Economic analysis can provide a very partial insight. The World Bank estimates that a price of US$27/t CO_2 would induce conservation of 5 million km^2 of rainforest by 2050, preventing the release of 172Gt CO_2.[149] However, markets cannot be considered independently of institutions and power relationships. Translating market incentives into rainforest conservation will require wide-ranging measures to distribute the benefits to poor farmers, thereby reducing poverty-related deforestation pressures, and to regulate the activities of large commercial farmers and illegal actors.

Carbon markets alone will not provide an automatic corrective for the wider forces driving deforestation. This is because forests are far more than carbon banks. Many of their ecological functions are unmarketed. Markets do not attach a price to the 400 plant species in Indonesia's Kerinci-Sebat National Park in Sumatra, nor to the immense biodiversity in Brazil's *cerrado* or savannah woodland. This generates an illusion that a zero price is associated with a zero economic value. As one commentator has written: "When conservation competes with conversion, conversion wins because its values have markets, whereas conservation values appear to be low. Prices and values should not be confused."[150]

Inequalities in political power are another source of deforestation not easily amenable to correction through the market. In Brazil, the incursion of commercial agriculture into rainforest areas has been associated with violations of the human rights of indigenous people and recourse to violence.[151] In Papua New Guinea, forest rights reside with indigenous communities in legislative theory. However, formal legal tenure has not prevented logging companies from operating without the consent of indigenous people.[152] In Indonesia, laws have been passed which recognize the rights of indigenous forest dwellers.[153] However, the eviction of indigenous people with the expansion of illegal logging and commercial plantations continues unabated. Living in remote areas, lacking economic power and with a weak voice in policy design and enforcement, forest dwellers carry less weight than powerful vested interests in forest management.

Governance of forests has to reflect their diverse functions. Forests are ecological resources that generate wide-ranging public and private benefits. They are the home and basis of livelihoods for many poor people and a source of potential profit for large commercial interests. They are a productive asset, but also a source of biodiversity. One of the challenges in forest governance is to balance the demands of competing interests with very different levels of power.

Some countries are developing institutional structures to address that challenge. In 2004, Brazil started implementing an Action Plan for Preventing and Controlling Deforestation. That plan integrates the work of 14 separate ministries. It establishes a legal framework for land-use decisions, strengthens monitoring and creates a legal framework for sustainable forest management. Outcomes will depend upon implementation and enforcement through state governments—an area where the record to date has been mixed. However, preliminary data for 2005 and 2006 suggests that the rate of deforestation has slowed by around 40 percent in the state of Mato Grosso.[154] Government commitment and the active engagement of civil society have been critical to this step in a positive direction.

Translating market incentives into rainforest conservation will require wide-ranging measures to distribute the benefits to poor farmers

Avoiding dangerous climate change: strategies for mitigation

The rehabilitation of severely degraded grasslands, and the conversion of degraded croplands to forests and agroforestry systems, can also build carbon storage capacity

International cooperation on climate change alone cannot resolve the wider problems driving deforestation. Respect for the human rights of indigenous people, the protection of biodiversity and conservation are issues for national political debate. However, the world is losing an opportunity to join up the climate change mitigation agenda with a range of wider human development benefits. International cooperation in the context of the post-2012 Kyoto commitment period could help to create incentives to unlock these benefits.

Filling the gaps

The current Kyoto Protocol suffers from a number of shortcomings as a framework for addressing the greenhouse gas emissions associated with land-use changes. There is significant potential for creating triple benefits from climate change mitigation, to adaptation and sustainable development. However, existing mechanisms limit the possibility of harnessing carbon finance as a mechanism for sustainable development.

Deforestation does not figure in the current Kyoto Protocol beyond a very limited provision to support 'afforestation' through the CDM. The rules of the CDM place a 1 percent cap on the share of carbon credits that can be generated through land use, land-use change and forestry, effectively de-linking activities in this sector from the climate change mitigation agenda. The Protocol does not allow developing countries to create emission reductions from avoided deforestation, limiting opportunities for transfers of carbon finance. Nor does it establish any financing mechanisms through which developed countries might provide incentives against deforestation.

Forests are the most visible ecological resource written out of the script for international cooperation on mitigation. But, they are not the only such resource. Carbon is also stored in soil and biomass. The rehabilitation of severely degraded grasslands, and the conversion of degraded croplands to forests and agroforestry systems, can also build carbon storage capacity. Because the environmental degradation of soils is both a cause and an effect of poverty, tapping

into carbon finance for these purposes could unlock multiple benefits. These include an increased flow of finance into environmental sustainability, support for more resilient livelihood systems in the face of climate change, and benefits for climate change mitigation.

Several innovative proposals have been developed to address the gaps in the current Kyoto approach. The Coalition of Rainforest Nations, led by Costa Rica and Papua New Guinea, has argued for 'avoided deforestation' to be brought into the Kyoto framework, opening the door to the use of CDM credits. Broadly, the idea is that every hectare of forest that would have been cut down but is left standing is a contribution to climate change mitigation. If incorporated into a CDM-type arrangement, this would open the door to potentially large flows of finance to countries with standing forests. A proposal tabled by Brazil sets out an alternative approach. This calls for the provision of new and additional resources for developing countries that voluntarily reduce their greenhouse gas emissions through reduced deforestation. However, under the Brazilian proposal the reductions would not register as developed country mitigation credits. Others have called for a revision of CDM rules to allow for an increased flow of carbon finance into soil regeneration and grassland restoration (box 3.12).

Proposals such as these merit serious consideration. The limitations of carbon markets as a vehicle for avoiding deforestation have to be recognized. Serious governance issues are at stake. 'Avoided deforestation' is clearly a source of mitigation. However, any standing rainforest is a potential candidate for classification as 'avoided deforestation'. Using trend rates for deforestation activity does not help resolve the problem of quantifying commitments, partly because information on trends is imperfect; and partly because changes in reference years can produce very big shifts in results. Other concerns, widely voiced during the last round of Kyoto negotiations, also have to be addressed. If avoided deforestation were integrated into the CDM without clear quantified limits, the sheer volume of CO_2 credits could swamp carbon markets, leading to a collapse in prices. Moreover, the permanence of mitigation through 'avoided deforestation' is difficult to establish.

Serious as the governance challenges are, none of these problems represents a case against the use of well-designed market instruments to create incentives for conservation, reforestation or the restoration of carbon-absorbing grasslands. There may be limits to what carbon markets can achieve. However, there are also vast and currently untapped opportunities for mitigation through reduced deforestation and wider land-use changes. Any action that keeps a tonne of carbon out of the atmosphere has the same climate impact, no matter where it occurs. Linking that action to the protection of ecosystems could create wide-ranging human development benefits.

Cooperation beyond carbon markets will be needed to tackle the wider forces driving deforestation. The world's forests provide a wide range of global public goods, of which climate change mitigation is one. By paying for the protection and upkeep of these goods through financial transfers, developed countries could create strong incentives for conservation.

International financial transfers, as advocated by Brazil, could play a key role in sustainable forest management. Multilateral mechanisms for such transfers should be developed as part of a broad-based strategy for human development. Without such arrangements international cooperation is unlikely to slow deforestation. However, successful outcomes will not be achieved just through unconditional financial transfers. Institutional mechanisms and governance structures for overseeing shared goals have to extend beyond conservation and emission targets to a far wider set of environmental and human development concerns, including respect for the human rights of indigenous people.

There are vast and currently untapped opportunities for mitigation through reduced deforestation and wider land-use changes

Avoiding dangerous climate change: strategies for mitigation

Conclusion

Stringent climate change mitigation will require fundamental changes in energy policy—and in international cooperation. In the case of energy policy, there is no alternative to putting a price on carbon through taxation and/or cap-and-trade. Sustainable carbon budgeting requires the management of scarcity—in this case the scarcity of the Earth's capacity to absorb greenhouse gases. In the absence of markets that reflect the scarcity implied by the stabilization target of 450 ppm CO_2e energy systems will continue to be governed by the perverse incentive to overuse carbon-intensive energy.

Without fundamental market-based reform the world will not avoid dangerous climate change. But pricing alone will not be enough. Supportive regulation and international cooperation represent the other two legs of the policy tripod for climate change mitigation. As we have shown in this chapter, there has been progress on all three fronts. However, that progress falls far short of what is required. Negotiations on the post-2012 framework for the Kyoto Protocol provide an opportunity to correct this picture. Incorporating an ambitious agenda for finance and technology transfer to developing countries is one urgent requirement. Another is international cooperation to slow the pace of deforestation.

4

Adapting to the inevitable: national action and international cooperation

"If you are neutral in a situation of injustice, you have chosen the side of the oppressor."

Archbishop Desmond Tutu

. . . .

"An injustice committed against anyone is a threat to everyone."

Montesquieu

4 Adapting to the inevitable: national action and international cooperation

All countries will have to adapt to climate change

The village of Maasbommel on the banks of the River Maas in Zeeland, southern Netherlands, is preparing for climate change. Like most of the Netherlands, this is a low-lying area at risk from rising sea levels and rivers swollen by rain. The landscape is dominated by water—and by the networks of dykes that regulate its flow. Located on the Maasbommel waterfront are 37 homes with a distinctive feature: they can float on water. Fixed to large steel stilts that are sunk into the river bed, the hollow foundations of the homes act like the hull of a ship, buoying the structure above water in the event of a flood. The floating homes of Maasbommel offer a case study in how one part of the developed world is adapting to the increased risks of flooding that will come with climate change.

People in the developing world are also adapting. In Hoa Thanh Hamlet in Viet Nam's Mekong Delta, people understand what it means to live with the risk of flooding. The greatest risks occur during the typhoon season, when storms that develop in the South China Sea produce sudden sea surges at a time when the Mekong is in flood. Vast networks of earth dykes maintained through the labour of farmers are an attempt to keep the flood waters at bay. Here too, people are dealing with climate change risks. Dykes are being strengthened, mangroves are being planted to protect villages from storm surges, and homes are being constructed on bamboo stilts. Meanwhile, part of an innovative 'living with floods' programme supported by donor agencies is providing vulnerable communities with swimming lessons and issuing life-jackets.

The contrasting experiences of Maasbommel and Hoa Thanh Hamlet illustrate how climate change adaptation is reinforcing wider global inequalities. In the Netherlands, public investment in an elaborate flood defence infrastructure provides a higher level of protection against risk. At a household level, technological capacity and financial resources offer people the choice of dealing with the threat of flooding by purchasing homes that enable them to float 'on' the water. In Viet Nam, a country that faces some of the world's most extreme threats from climate change, a fragile flood defence infrastructure provides limited protection. And in villages across the Mekong Delta, adaptation to climate change is a matter of learning to float 'in' the water.

All countries will have to adapt to climate change. In rich countries governments are putting in place public investments and wider strategies to protect their citizens. In developing countries adaptation takes a different form. Some of the world's most vulnerable people living with the risks of drought, floods and exposure to tropical storms are being left to cope using only their own very limited resources. Inequality in capacity to adapt to climate change is emerging as a potential driver of wider disparities in wealth, security and opportunities for human development. As Desmond Tutu, the former Archbishop of Cape Town, warns in his special contribution to this Report, we are drifting into a situation of global adaptation apartheid.

International cooperation on climate change demands a twin-track approach. The priority is to mitigate the effects that we can control and to support adaptation to those that we cannot. Adaptation is partly about investment in the 'climate-proofing' of basic infrastructure. But it is also about enabling people to manage climate-related risks without suffering reversals in human development.

If left uncorrected the lack of attention to adaptation will undermine prospects for human development for a large section of the world's most vulnerable people. Urgent action on mitigation is vital because no amount of adaptation planning, however well financed or well designed, will protect the world's poor from business-as-usual climate change. By the same token, no amount of mitigation will protect people from the climate change that is already inevitable. In a best case scenario, mitigation will start to make a difference from around 2030 onwards, but temperatures will increase to around 2050. Until then, adaptation is a 'no-choice' option. The bad news is that we are a very long way from a best-case scenario because mitigation has yet to take off.

Special contribution   **We do not need climate change apartheid in adaptation**

In a world that is so divided by inequalities in wealth and opportunity, it is easy to forget that we are part of one human community. As we see the early impacts of climate change registering across the world, each of us has to reflect on what it means to be part of that family.

Perhaps the starting point is to reflect on the inadequacy of language. The word 'adaptation' has become part of the standard climate change vocabulary. But what does adaptation mean? The answer to that question is different things in different places.

For most people in rich countries adaptation has so far been a relatively painfree process. Cushioned by heating and cooling systems, they can adapt to extreme weather with the flick of a thermostat. Confronted with the threat of floods, governments can protect the residents of London, Los Angeles and Tokyo with elaborate climate defence systems. In some countries, climate change has even brought benign effects, such as longer growing seasons for farmers.

Now consider what adaptation means for the world's poorest and most vulnerable people—the 2.6 billion living on less than US$2 a day. How does an impoverished woman farmer in Malawi adapt when more frequent droughts and less rainfall cut production? Perhaps by cutting already inadequate household nutrition, or by taking her children out of school. How does a slum dweller living beneath plastic sheets and corrugated tin in a slum in Manila or Port-au-Prince adapt to the threat posed by more intense cyclones? And how are people living in the great deltas of the Ganges and the Mekong supposed to adapt to the inundation of their homes and lands?

Adaptation is becoming a euphemism for social injustice on a global scale. While the citizens of the rich world are protected from harm, the poor, the vulnerable and the hungry are exposed to the harsh reality of climate change in their everyday lives. Put bluntly, the world's poor are being harmed through a problem that is not of their making. The footprint of the Malawian farmer or the Haitian slum dweller barely registers in the Earth's atmosphere.

No community with a sense of justice, compassion or respect for basic human rights should accept the current pattern of adaptation. Leaving the world's poor to sink or swim with their own meagre resources in the face of the threat posed by climate change is morally wrong. Unfortunately, as the *Human Development Report 2007/2008* powerfully demonstrates, this is precisely what is happening. We are drifting into a world of 'adaptation apartheid'.

Allowing that drift to continue would be short-sighted. Of course, rich countries can use their vast financial and technological resources to protect themselves against climate change, at least in the short-term—that is one of the privileges of wealth. But as climate change destroys livelihoods, displaces people and undermines entire social and economic systems, no country—however rich or powerful—will be immune to the consequences. In the long-run, the problems of the poor will arrive at the doorstep of the wealthy, as the climate crisis gives way to despair, anger and collective security threats.

None of this has to happen. In the end the only solution to climate change is urgent mitigation. But we can—and must—work together to ensure that the climate change happening now does not throw human development into reverse gear. That is why I call on the leaders of the rich world to bring adaptation to climate change to the heart of the international poverty agenda—and to do it now, before it is too late.

Desmond Tutu
Archbishop Emeritus of Cape Town

Adapting to the inevitable: national action and international cooperation

Mitigation is one part of a twin strategy for insurance under climate change. Investment in mitigation will provide high returns for human development in the second half of the 21st Century, reducing exposure to climate risks for vulnerable populations. It also offers insurance against catastrophic risks for future generations of humanity, regardless of their wealth and location. International cooperation on adaptation is the second part of the climate change insurance strategy. It represents an investment in risk reduction for millions of the world's most vulnerable people.

While the world's poor cannot adapt their way out of dangerous climate change, the impacts of global warming can be diminished through good policies. Adaptation actions taken in advance can reduce the risks and limit the human development damage caused by climate change.

Northern governments have a critical role to play. When they signed the United Nations Framework Convention on Climate Change (UNFCCC) in 1992, these governments agreed to help "the developing countr(ies) that are particularly vulnerable to the adverse effects of climate change in meeting costs of adaptation to those adverse effects". Fifteen years on that pledge has yet to be translated into action. To date, international cooperation on adaptation has been characterized by chronic under-financing, weak coordination and a failure to look beyond project-based responses. In short, the current framework provides the equivalent of an aid sponge for mopping up during a flood.

Effective adaptation poses many challenges. Policies have to be developed in the face of uncertainties on the timing, location and severity of climate change impacts. Looking to the future, the scale of these impacts will be contingent on mitigation efforts undertaken today: delayed or limited mitigation will drive up the costs of adaptation. These uncertainties have to be considered in the development of adaptation strategies and financing plans. However, they do not provide a justification for inaction. We know that climate change is impacting on the lives of vulnerable people

today—and we know that things will get worse before they get better.

In one respect, the developed world has shown the way. Here, no less than in the developing world, governments and people have to deal with climate change uncertainty. But that uncertainty has not acted as a barrier to large-scale investment in infrastructure, or to the development of broader adaptation capacities. As the primary architects of the dangerous climate change problem, the governments and citizens of the rich world cannot apply one rule at home and another to the vulnerable communities that are the prospective victims of their actions. Watching the consequences of dangerous climate change unfold in developing countries from behind elaborate climate defence systems is not just ethically indefensible. It is also a prescription for a widening gap between the world's haves and have-nots, and for mass resentment and anger—outcomes that will have security implications for all countries.

This chapter is divided into two parts. In the first section we focus on the national adaptation challenge, looking at how people and countries are responding to the challenge and at the strategies that can make a difference. Climate change poses such a threat because it is exposing vulnerable people to incremental risks. Enabling people to manage these risks requires public policies that build resilience through investment in infrastructure, social insurance and improved disaster management. It requires also a strengthened commitment to broader policies that bolster human development and reduce extreme inequalities.

In the second section we turn to the role of international cooperation. There is an overwhelming case for rich countries to play a greater role in supporting adaptation. Historic responsibility for the climate change problem, moral obligation, respect for human rights and enlightened self-interest combine to make this case. Increased financing for the integration of adaptation into national poverty reduction planning is one requirement. Another is the early development of a coherent multilateral structure for delivering support.

International cooperation on adaptation has been characterized by chronic under-financing, weak coordination and a failure to look beyond project-based responses

4

Adapting to the inevitable: national action and international cooperation

Planning for adaptation
to climate change is a
fast-growing industry in
developed countries

4.1 The national challenge

All countries will have to adapt to climate change. How they adapt, and the choices open to people and governments, will be determined by many factors. The nature of the risks associated with climate change varies across regions and countries. So too does the capacity to adapt. The state of human development, technological and institutional capabilities and financial resources all play a role in defining that capacity.

In some respects, the incremental risk posed by climate change is one of degree. The policies and institutions that can enable countries and people to adapt to climate risks today—social and economic policies that build capabilities and resilience against 'climate shocks', investment in infrastructural defences against flooding and cyclones, institutions for regulating watershed management—are the same as those that will be needed to address future threats. However, the scale of these threats poses both quantitative and qualitative challenges. Some countries—and some people—are far better equipped than others to respond.

Adaptation in the developed world

Planning for adaptation to climate change is a fast-growing industry in developed countries. National governments, regional planning bodies, local governments, city authorities and insurance companies are all drawing up adaptation strategies with a common goal: protecting people, property and economic infrastructure from emerging climate change risks.

Mounting public concern has been one factor shaping the adaptation agenda. In many developed countries there is a widespread perception that climate change is adding to weather-related risks. The 2003 European heatwave, the 2004 Japanese typhoon season, Hurricane Katrina and the devastation of New Orleans, and episodes of drought, flooding and extreme temperature across the developed world have been among the headline events fuelling public concern. Uncertainty over the future direction of climate change has done little to deter public calls for more proactive government responses.

The insurance industry has been a powerful force for change. Insurance provides an important mechanism through which markets signal changes in risk. By pricing risk, markets provide incentives for individuals, companies and governments to undertake risk reduction measures, including adaptation. In both Europe and the United States, the insurance industry has shown growing concern over the implications of climate change for risk-related losses (see chapter 2). Projections pointing to the increased frequency of extreme flood and storm events are one source of that concern. In several countries, the insurance industry has emerged as a forceful advocate of increased public investment in 'climate-proofing' infrastructure to limit private losses. For example, the Association of British Insurers is calling for a 50 percent increase in national flood defence spending by 2011.[1]

Adaptation in the developed world has taken many forms. The 'floating home' owners of Maasbommel provide a household-level illustration of behavioural shift. In other cases, business is being forced to adapt. One example comes from the European ski industry. Snow cover in European alpine areas is already in retreat, and the IPCC has warned that, at middle elevations, the duration of snow cover is expected to decrease by several weeks for each 1°C of temperature increase.[2] The Swiss ski industry has 'adapted' by investing heavily in artificial snow-making machines. Covering one hectare of ski slopes requires about 3,300 litres of water, and helicopters are used to ferry in the raw material, which is converted into snow through energy-intensive freezing.[3]

Many developed countries have conducted detailed studies on climate change impacts. Several are moving towards the implementation of adaptation strategies. In Europe, countries such as France, Germany and the United Kingdom have created national institutional structures for adaptation planning. The European Commission has urged member states to integrate adaptation into infrastructure programmes and for a good reason.[4] With a lifetime of 80–100 years, infrastructure such as bridges, ports and motorways have to take into account future climate change conditions. Sectors such as agriculture and forestry will have to deal with far earlier impacts, as will the public at large.

The scale of defensive climate change adaptation efforts in rich countries is not widely appreciated. While the record varies, the overall picture is one of rising investments in preventative action. Among the examples:

- *The Netherlands.* As a densely populated, low-lying country with more than one-quarter of its land area below sea level, the Netherlands faces acute climate change risks. The risks are contained through a vast network of canals, pumps and dykes. The dykes are constructed to withstand weather events that might happen only once in every 10,000 years. It is not only the sea that poses threats. The River Rhine, which forms a large delta with the Maas, is a constant flood threat. With sea levels rising, more intense storms occurring, and climate models predicting that precipitation could increase by 25 percent, adaptation planning in the Netherlands is viewed as a matter of national security. Dutch water policy recognizes that current infrastructure may be insufficient to deal with increased water levels in rivers and rising sea levels. In 2000 the national policy document—*Room for the River*—set out a detailed framework for adaptation. The framework includes more stringent planning controls on human settlements, Catchment Area Strategies implemented by regional authorities to develop flood-retention areas, and a budget of US$3 billion for investment to protect against flooding. The policy aims at protecting the Netherlands from discharges from the River Rhine of up to 18,000m³/s from 2015—around 50 percent above the highest recorded level to date.[5]
- *United Kingdom.* The United Kingdom Climate Impacts Programme (UKCIP)

has drawn up detailed region-by-region and sectoral studies looking at adaptation challenges. Management strategies for flooding are being developed in the light of risk assessments of rising sea levels and increased rainfall. Forecasted changes in climate, storms and rainfall patterns are expected to lead to an increased risk of flooding. In contrast to the Netherlands, Britain's flood defence systems are designed to cope with the biggest floods expected every 100–200 years. With sea levels rising and more storms and rain in prospect, flood defence strategies are under revision. Estimates by the insurance industry suggest that the number of homes at risk of flooding could rise from 2 million in 2004 to 3.5 million over the long term if flood defence infrastructure is not strengthened. Only around one-half of the national flood defence infrastructure is in good condition. The Environment Agency, a government body, has called for at least US$8 billion to be spent strengthening the Thames Barrier—a mechanized flood defence structure that protects London. Current spending on flood management and coastal erosion is around US$1.2 billion annually.[6] Major floods in 2007 led to renewed calls for increased spending.

- *Japan.* Concern over adaptation in Japan was heightened in 2004 when the country was hit by 10 tropical cyclones. This was more than in any other year over the previous century. Total losses amounted to US$14 billion, of which roughly one-half was covered by insurance. Rising temperatures and rising sea levels are also increasing risk: average sea levels are rising at 4–8mm a year. While Japan has one of the world's most highly developed flood defence infrastructures, ports and harbours are seen as sites of great vulnerability. More intensive tropical storm activity could lead to large-scale economic disruption. Plans developed by the Japanese Government to provide more effective defences in the face of a 21st Century sea level rise of 1 metre estimate costs at US$93 billion.[7]

The European Commission has urged member states to integrate adaptation into infrastructure programmes

Adapting to the inevitable: national action and international cooperation

- *Germany*. Large areas of Germany face increased risk of flooding with climate change. Research in the Neckar catchment area in Baden-Württemberg and Bavaria predicts an increase of 40–50 percent in small and medium-sized flood events by the 2050s, with a 15 percent increase in 'hundred year' floods. The Baden-Württemberg Ministry for Environment estimates the additional cost of long-term flood defence infrastructure at US$685 million. Following large-scale flooding in 2002 and 2003, Germany adopted a Flood Control Articles Act which integrates climate change assessment into national planning, imposing strict requirements on the designation of flood areas and human settlements.[8]

- *California*. Climate change will have serious implications for water supply in parts of California. Rising winter temperatures are expected to reduce the accumulation of snow in the Sierra Nevada, which functions as a large water storage system for the state. Reductions in snow cover in the Sacramento, San Joaquim and Trinity drainage areas (relative to the average for 1961–1990) are projected to amount to 37 percent for the period 2035–2064, rising to 79 percent for the period 2070–2090. As an already highly water-stressed state, California has developed an extensive system of reservoirs and water-transfer channels to maintain flows to dry areas. In its 2005 Water Plan Update, the Department of Water Resources (DWR) sets out a wide-ranging strategy to deal with reduced water flows, including efficiency measures to reduce water use in urban areas and agriculture. Increased investment in recycled water, with a target of 930 million cubic meters by 2020, or roughly twice current levels, also figures. California also faces increased flood threats from two directions: rising sea levels and accelerating snow melt. The DWR estimates the costs of upgrading the Central Valley flood control system and levees in the Delta alone at over US$3 billion. Climate change could redraw California's coastal map, with beachfront

real estate ending up under water, sea walls collapsing and cliffs eroding.[9]

These examples demonstrate that policy-makers in rich countries do not see climate change uncertainty as a cause for delaying adaptation. Public investments today are seen as an insurance against future costs. In the United Kingdom, government agencies estimate that every US$1 spent on flood defences saves around US$5 in flood damage.[10] The returns on early adaptation investments are likely to increase over time as climate change impacts strengthen. Estimates by the European Commission suggest that the damage caused by rising sea levels in 2020 will be up to four times higher than damage incurred if preventative measures are taken. By the 2080s, they could be over eight times higher.[11] Further, the costs of such defence measures are only a fraction of the damages they avoid (figure 4.1).

Not all adaptation is defensive. In the short term at least, climate change is likely to create winners as well as losers—and most of the winners will be in rich countries. Agriculture provides an illustration. While small-scale farmers in developing countries stand to lose under climate change, the medium-term impacts could create opportunities in much of the developed world. In the United States,

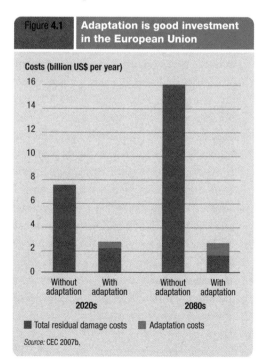

Figure 4.1 **Adaptation is good investment in the European Union**

Costs (billion US$ per year)

| | Without adaptation | With adaptation | Without adaptation | With adaptation |
| | 2020s | | 2080s | |

■ Total residual damage costs ■ Adaptation costs

Source: CEC 2007b.

national climate change projections show that near-term agricultural food production may increase, albeit with southern states lagging behind and the Great Plains facing more droughts as production centres move north.[12] Northern Europe also stands to gain from longer and warmer growing seasons, creating scope for improved competitiveness in a range of fruit and vegetables.[13] Displacement of imports from developing countries therefore remains a threat to human development in some product areas.

Living with climate change— adaptation in developing countries

While rich countries are preparing to adapt to climate change, it is developing countries that will be faced with the greatest and earliest burden in terms of adverse impacts on living standards, livelihoods, economic growth and human vulnerability. As in the developed world, people in the poorest countries will have to deal with the consequences of a changing climate. However, there are two important differences. First, developing countries in tropical and subtropical regions will register some of the strongest climate change effects. Second, the incremental risks that come with climate change will be superimposed on societies marked by mass poverty and acute vulnerability. While northern governments have the financial, technological and human capabilities to respond to the climate change risks facing their citizens, developing countries are far more constrained.

Adaptation to climate change is not a future scenario for the developing world. It is already happening—just as it is in rich countries. But the contrasts with adaptation in the developed world are striking. In London and New York, people are being protected against the risks associated with rising sea levels through public investment in infrastructure. In the poorest countries, adaptation is largely a matter of self-help. Millions of people with barely enough resources to feed, clothe and shelter their families are being forced to direct money and labour to adaptation. Among the examples of that struggle:

- In northern Kenya the increased frequency of droughts means that women are walking greater distances to collect water, often ranging from 10 to 15 km a day. This confronts women with personal security risks, keeps young girls out of school and imposes an immense physical burden—a plastic container filled with 20 litres of water weighs around 20 kg.[14]
- In West Bengal in India, women living in villages in the Ganges Delta are constructing elevated bamboo platforms known as *machan* on which to take refuge above monsoon floodwaters. In neighbouring Bangladesh, donor agencies and NGOs are working with people living on *chars*—highly flood-prone islands that are cut off during the monsoon—to raise their homes above flood levels by placing them on stilts or raised embankments.[15]
- Communities in Viet Nam are strengthening age-old systems of dykes and embankments to protect themselves against more powerful sea surges. In the Mekong Delta, agricultural collectives now levy a tax for coastal protection and are supporting the rehabilitation of mangrove areas as a barrier against storm surges.[16]
- Investments in small-scale water harvesting are increasing. Farmers in Ecuador are building traditional U-shaped detention ponds, or *albarradas*, to capture water during wetter years and recharge aquifers during drought years.[17] In Maharashtra, India, farmers are coping with increased exposure to drought by investing in watershed development and small-scale water-harvesting facilities to collect and conserve rainwater.[18]
- In Nepal, communities in flood-prone areas are building early warning systems—such as raised watchtowers—and providing labour and material to shore up embankments to prevent glacial lakes from bursting their banks.
- Farmers across the developing world are responding to emerging climate threats by drawing on traditional cultivation technology. In Bangladesh, women farmers are building 'floating gardens'—hyacinth rafts on which to grow vegetables in flood-prone areas. In Sri Lanka, farmers are experimenting with rice varieties that can withstand saline intrusion and cope with reduced water.[19]

It is developing countries that will be faced with the greatest and earliest burden in terms of adverse impacts on living standards, livelihoods, economic growth and human vulnerability

4

Adapting to the inevitable: national action and international cooperation

None of these cases provides evidence of adaptation directly attributable to climate change. It is impossible to establish causality between specific climate events and global warming. What has been established is an overwhelmingly probable link between climate change and the type of events—droughts, water shortages, storms and weather variability—that force adaptation. Attempting to quantify the climate change components of the increment to risk in any one case is an exercise in futility. But ignoring evidence of mounting systemic risks is a study in myopia.

Human development itself is the most secure foundation for adaptation to climate change. Policies that promote equitable growth and the diversification of livelihoods, expand opportunities in health and education, provide social insurance for vulnerable populations, improve disaster management and support post-emergency recovery all enhance the resilience of poor people facing climate risks. That is why climate change adaptation planning should be seen not as a new branch of public policy but as an integral part of wider strategies for poverty reduction and human development.

Good climate change adaptation planning will not override problems linked to inequality and marginalization. Experience in Kenya is instructive. For Kenya's 2 million pastoralists, increased exposure to future drought is a real threat. However, that threat is magnified by wider forces that are weakening pastoral livelihoods today, including a policy bias in favour of settled agriculture, the privatization of water rights and disregard for the customary rights of pastoralists. In the Wajir district of northern Kenya, to take one example, the encroachment of crop production into pastoral areas has restricted access to grazing lands, blocked migration corridors and undermined traditional water-sharing arrangements, leading to increased overgrazing and reduced milk production.[20]

Framing national adaptation policies

There are no blueprints for successful climate change adaptation. Countries face different types and degrees of risk, start from different levels of human development and vary widely in their technological and financial capabilities.

While policies for human development are the most secure foundation for adaptation, even the best human development practice will have to take into account emerging climate change risks. These risks will magnify the costs of past policy failure and will demand a reassessment of current human development practice, placing a premium on the integration of climate change scenarios into wider national programmes.

So far adaptation planning has been a fringe activity in most developing countries. To the extent that strategies for adaptation are emerging, the focus is on climate-proofing infrastructure. This is a critical area. But adaptation is about far more than infrastructure. The starting point is to build climate change risk assessment into all aspects of policy planning. In turn, risk management requires that strategies for building resilience are embedded in public policies. For countries with limited government capacity this is an immense task.

The magnitude of that task is insufficiently appreciated. In Egypt, a 0.5 metre increase in sea levels could lead to economic losses in excess of US$35 billion and the displacement of 2 million people.[21] The country is developing an institutional response through a high-level ministerial dialogue led by the Ministry of the Environment. But the sheer magnitude of the climate risks will require far-reaching policy reforms across the entire economy.

Another illustration comes from Namibia.[22] Here too climate change poses threats across many sectors. Fisheries provide an example. Commercial fish processing is now one of the mainstays of the Namibian economy: it represents almost one-third of total exports. One of the sources of Namibia's rich fishery revenues is the Benguela current—a cold water current that runs along its coast. With water temperatures warming, there is growing concern that key fish species will migrate southwards. This creates a major adaptation challenge for the fisheries sector. Given the uncertainties, should Namibia be increasing investments in fish processing? Or, should it be seeking diversity?

Adjusted for country context, these are the type of questions being asked of governments across the developing world. Providing answers requires vastly strengthened capacity in risk assessment and resilience planning. While an international response is emerging through mechanisms such as the Global Environmental Facility (GEF), that response remains under–financed, poorly coordinated and weakly managed.

Successful adaptation planning will require a transformational change in government practices. Reactive measures are guaranteed to prove insufficient, as are responses that fail to address transboundary climate change impacts through regional cooperation. But, the greatest transformation is required in planning for human development and poverty reduction. Building the resilience and coping capacity of the poorest and more vulnerable sections of the society will require something more than rhetoric pledges to the MDGs and pro-poor growth. It will require a fundamental reappraisal of poverty reduction strategies backed by a commitment to enhanced equity in tackling social disparities.

As in other areas, adaptation policies are likely to be more successful and responsive to the needs of the poor when the voice of the poor identifies priorities and shapes the design of policies. Accountable and responsive government and the empowerment of people to improve their own lives are necessary conditions for successful adaptation, just as they are for human development. The foundations for successful adaptation planning can be summarized under four 'i's:

- *Information* for effective planning;
- *Infrastructure* for climate-proofing;
- *Insurance* for social risk management and poverty reduction;
- *Institutions* for disaster risk management.

Information on climate risks

In planning for adaptation to climate change, information is power. Countries lacking the capacity and resources to track meteorological patterns, forecast impacts and assess risk cannot provide their citizens with good quality information—and are less able to target the public investments and policies that can reduce vulnerability.

At a global level there is an inverse relationship between climate change risk exposure and information. The IPCC acknowledges that current climate models for Africa provide insufficient information to downscale data on rainfall, the spatial distribution of tropical cyclones and the occurrence of droughts. One reason for this is that the region has the world's lowest density of meteorological stations, with one site for every 25,460 km²—one-eighth of the minimum level recommended by the World Meteorological Organization (WMO).[23] The Netherlands, by way of contrast, has one site for every 716 km²—four times above the WMO minimum (figure 4.2).

Inequalities in climate monitoring infrastructure are intimately linked to wider disparities. Opportunities in education and training are critical for the development of meteorological infrastructure and the conduct of relevant research. In countries with restricted

Adaptation policies are likely to be more successful and responsive to the needs of the poor when the voice of the poor identifies priorities and shapes the design of policies

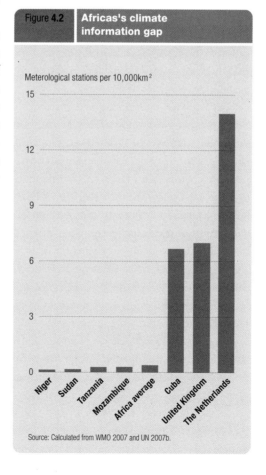

Figure 4.2 Africas's climate information gap

Meterological stations per 10,000km²

Source: Calculated from WMO 2007 and UN 2007b.

4

Adapting to the inevitable: national action and international cooperation

4

access to secondary and tertiary education, the human capital for these activities is often lacking. Evidence for this can be seen in the distribution of published international research. Whereas Europe and North America account for over two-thirds of all papers published in two major climate journals, Africa accounts for just 4 percent.[24]

Financing constraints widen the disparities in access to information. Developed countries are able to invest far more heavily than poorer countries in meteorological data collection and analysis, providing climate-sensitive sectors with a steady flow of information. Farmers in France, to take one example, benefit from a meteorological network that invests US$388 million annually in climate monitoring and analysis, using some of the world's most sophisticated forecasting systems.[25] By contrast, in Ethiopia, where over 90 percent of people depend on agriculture for their livelihoods, the national meteorological budget for 2005 was around US$2 million. By sub-Saharan African standards, Ethiopia is well endowed: in Malawi, the meteorological budget for 2005 was less than US$1 million.[26] Indeed, the French meteorological budget exceeds expenditure on climate monitoring and analysis for the whole of sub-Saharan Africa.[27]

Capacity for monitoring and forecasting climate can have an important bearing on livelihood security. For agricultural producers, advance warning of abrupt changes in rainfall patterns or temperature can mean the difference between a successful harvest and crop failure. Seasonal forecasting systems and effective dissemination of the information they generate can enable farmers to monitor potential hazards and respond by adjusting planting decisions or changing the mix of crops.

One successful example comes from Mali. Here the national meteorological service—the Direction Nationale de la Météorologie (DNM)—has developed a programme for transmitting rainfall and soil moisture information through a network of representative farmers' organizations, NGOs and local governments. Information is collected from diverse sources, including the WMO, regional

monitoring systems and a national network of simple rain gauges. Throughout the growing season, farmers receive regular bulletins, enabling them to adjust production practices. Evaluation of results in the 2003–2004 cropping season show that crop yields and incomes were higher in areas where agro-meteorological information was used, notably for maize.[28]

The Mali experience demonstrates that low income does not have to be a barrier to successful action. In this case, government, farmers and climatologists have worked together to generate and disseminate information in a way that empowers vulnerable producers, reducing the risks and uncertainties associated with erratic rainfall. In other countries, information is less available, and what is available is often unequally distributed, or presented in ways that are not useful to farmers or other users. All too often, large-scale commercial growers have access to good-quality meteorological information while smallholders in the marginal areas facing the greatest climate risks are in 'information-free' zones.

Building meteorological monitoring capacity will require international cooperation. Many developing countries lack both the financial and technological capabilities to scale up monitoring activities. Yet without improved access to information, governments and people across the developing world will be denied opportunities to develop effective climate adaptation strategies.

There have been some encouraging developments. At their summit in Gleneagles in 2005, G8 leaders recognized the importance of building capacity to monitor climate. They pledged to strengthen existing climate institutions in Africa and to help the region obtain the benefits of cooperation through the Global Climate Observation System (GCOS) with "a view to developing fully operational regional climate centres in Africa".[29] The Government of Finland has actively supported the development of meteorological infrastructure in eastern Africa. In the United Kingdom, the Meteorological Office's Hadley Centre has developed a low-cost, high-resolution climate monitoring model that has been made freely available, together with training and support, to 11 regional centres in the developing world.[30]

Encouraging as these initiatives have been, the international response has fallen far short of what is needed. Based on the commitments made by the G8, the Economic Commission for Africa and the WMO have drawn up plans requiring a modest US$200 million of expenditure over 10 years to expand the region's observation and infrastructure capacity.[31] However, donor support thus far has been limited. Resources have been mobilized only for initial scoping exercises—and the G8 has failed to monitor progress at subsequent summits. In a review of progress to date, the Africa Partnership Forum has concluded: "Despite the G8 commitment and strong support by key African institutions...the funding of the programme has yet to be realized."[32]

Infrastructure for climate-proofing

Throughout history, communities have attempted to protect themselves against the vagaries of climate by building infrastructure. Flood defence and drainage systems, reservoirs, wells and irrigation channels are all examples. No infrastructure provides immunity from climatic forces. What infrastructural investments can do is to provide partial protection, enabling countries and people to manage the risks and limit vulnerability.

Climate change has important implications for the planning of infrastructural investments. Rising sea levels, higher temperatures and increased exposure to floods and storms all affect the viability of such investments. Current approaches to adaptation planning in many developing countries focus on the 'climate-proofing' of existing investments against incremental risk. The following examples, drawn from National Adaptation Programmes of Action (NAPAs), illustrate these approaches:

- Cambodia estimates that US$10 million of investment will be required to construct water gates and culverts for newly rehabilitated road networks developed without factoring in increased risks of flooding.
- In Bangladesh, projects worth US$23 million have been identified by government

to create a coastal buffer zone in regions vulnerable to storm surges, with an additional US$6.5 million to counter the effects of increasing salinity in coastal soils. In the transport sector, the Government estimates that raising an 800 kilometre network of roads by between 0.5 and 1 metre to counter sea level rises will cost US$128 million over a 25-year period.

- In Haiti the national adaptation plan estimates that a budget of US$11 million is needed for investment in projects to counter water shortages and the threat of flooding through measures to tackle soil erosion.

The project-based approach to adaptation planning set out in NAPAs, which detail only immediate and urgent needs, provides a limited perspective on the scale of financing required for effective 'climate-proofing'. In Viet Nam, UN agencies and the Ministry of Agriculture and Rural Development have drawn up a comprehensive strategy for reducing disaster risk in the Mekong Delta. The strategy builds on assessments of communities and ecologies vulnerable to climate change, with adaptation planning integrated into a wider programme for coastal zone management. It includes investments aimed at strengthening drainage systems, reinforcing dykes and trenches around human settlements and agricultural areas, and supporting the restoration of mangrove areas. Capital investment costs are estimated at US$1.6 billion between 2006 and 2010 and at US$1.3 billion from 2010 to 2020.[33]

Viet Nam's strategy for disaster risk reduction in the Mekong Delta illustrates three important points of wider relevance in approaches to adaptation. The first is that effective adaptation planning in high-risk environments requires investments that are beyond the financing capacities of most governments acting alone. The second is that adaptation planning requires a long time-horizon—in the case of the Mekong it is 15 years. Third, adaptation planning is unlikely to succeed if it is approached as a stand-alone exercise. In Viet Nam, the Mekong strategy is integrated into the country's national poverty reduction strategy and medium-term expenditure framework,

Current approaches to adaptation planning in many developing countries focus on the 'climate-proofing' of existing investments against incremental risk

4

Adapting to the inevitable: national action and international cooperation

linking it to public policies aimed at overcoming hunger and reducing vulnerability—and to wider partnerships with donors.

Infrastructural development can be a cost-effective route to improved disaster risk management. In rich countries, recognition that disaster prevention is more cost-effective than cure has been an important factor in shaping government infrastructure investment. Similar cost–benefit principles apply in the developing world. One recent global study estimates that US$1 invested in pre-disaster risk management activities in developing countries can prevent US$7 in losses.[34] National research confirms this broad cost–benefit story. In China, the US$3 billion spent on flood defences in the four decades up to 2000 is estimated to have averted losses of US$12 billion.[35] Evidence from a mangrove-planting project designed to protect coastal populations from storm surges in Viet Nam estimated economic benefits that were 52 times higher than costs.[36]

Successful adaptation planning has the potential to avert economy-wide losses. Disaster risk analysis in Bangladesh provides an insight into returns to adaptation investments. Using risk analysis methods analogous to those deployed by the insurance industry, researchers assessed the economic asset losses associated with flooding risks today, in 2020 and in 2050, under a range of plausible climate change scenarios. If no adaptation was assumed, the costs associated with more extreme '50-year events' amounted to 7 percent of GDP in 2050. With adaptation they fell to around 2 percent.[37] The differential translates into potentially large setbacks in agricultural production, employment and investment, with negative implications for human development.

Consideration of distributional factors is critical to adaptation planning. Governments have to make tough decisions about where to allocate limited public investment resources. An obvious danger is that the adaptation needs of marginalized communities will be overlooked in the face of demands from more powerful groups with a stronger political voice.

Pro-poor adaptation strategies cannot be developed in isolation from wider policies aimed at reducing poverty and overcoming inequality. In Bangladesh, government and donors have started to identify adaptation strategies that reach some of the country's most marginalized people, such as those living on highly flood-prone *char* islands. As in other areas, there are strong cost–benefit grounds for undertaking pro-poor adaptation: the estimated return on investment in *char* islands is around 3:1 (box 4.1). The cost–benefit case is powerfully reinforced by basic equity considerations: US$1 in the household income of some of Bangladesh's poorest people has to be attached a higher weight than, say, US$1 saved by high-income groups.

Infrastructure for water management can play an important role in enhancing—or diminishing—the opportunities for human development. Some of the world's poorest agricultural producers will face some of the toughest climate change adaptation challenges. With their livelihoods dependent on the timing and duration of rainfall, temperature and water run-off patterns, the rural poor face immediate risks with very limited resources. This is especially true for producers dependent on rainfed rather than irrigated agriculture. Over 90 percent of sub-Saharan African agriculture is in this category. Moreover, the region has one of the lowest rates of conversion of precipitation into water flows, partly because of high evaporation and partly because of the lack of an irrigation tradition.[38] Although South Asia has wider access to irrigation, two in every three rural people still depend on rainfed agriculture.

Agricultural producers operating in water-stressed, rainfed environments already invest labour in developing water harvesting systems that conserve rainfall. As climate change increases the risks, one of the challenges in adaptation planning is to support these efforts. In many countries, the development of irrigation systems also has a role to play. In 2005 the Economic Commission for Africa called for a doubling of the arable area under irrigation by 2015. Improved access to irrigation could help simultaneously to raise productivity and reduce climate risks. However, proposals in this area must take into account the impact of future climate change on water availability.

Beyond irrigation there are wider opportunities to develop water harvesting, especially in countries—such as Ethiopia, Kenya and Tanzania—with relatively abundant, but concentrated rainfall.[39] Ethiopia spans 12 major river basins and has relatively abundant water, but one of the lowest reservoir storage capacities in the world: 50 cubic metres per person compared with 4,700 in Australia. In countries lacking water storage capacity, even increased rainfall may not enhance water availability. High levels of runoff and increased risks of flooding are more likely outcomes.

Experience from India is instructive. Here, as elsewhere, climate change will place additional pressures on already highly stressed water systems. While overall rainfall is projected to increase on average, much of the country will receive less rain. Local communities are already developing innovative responses to water stress.

| Box **4.1** | **Adaptation on the *char* islands of Bangladesh** |

River deltas in Bangladesh are on the front line of climate change. Located in the Ganges–Brahmaputra Delta, islands and other low-lying delta lands—known as *chars*—are home to over 2.5 million highly vulnerable people living under risk of frequent flooding. The human development imperative to help such communities adapt to the increased threats brought about by climate change has long been recognized. But innovative cost–benefit exercises are showing that it makes economic sense too.

The lives of *char* people are closely bound up with the flow of rivers—and with flooding. *Chars* themselves undergo constant erosion and reformation, as rivers wash away soil and deposit silt. Entire islands are vulnerable to erosion and flooding, though people living by unprotected river channels face special risks.

Coping capacity is limited by poverty. The riverine areas of Bangladesh are marked by high levels of human deprivation. Over 80 percent live in extreme poverty (see table). Indicators for nutrition, child mortality and public health are among the worst in the country. Flooding poses a constant threat. People cope by building embankments and ditches around agricultural lands—and by rebuilding their homes when they are destroyed. Even minor floods cause high levels of damage. Major events—such as the 1998 and 2004 floods—destroy agricultural production and homes on a large scale, isolating communities from crucial health and other public services in the process.

Government, donors and local communities have developed a range of approaches for reducing vulnerability. Protecting homes has been identified as a priority. Under the Chars Livelihood Programme, one pilot project aims at 'flood-proofing' homesteads against floods with a one in twenty years likelihood of occurrence (most homes are currently vulnerable to two-year events). The objective is to construct earth platforms to accommodate homes for four households, with trees and grass planted as a protection against soil erosion. Hand pumps and basic latrines are provided to secure access to clean water and sanitation. So far, around 56,000 *char* people have participated in this re-housing programme.

The benefits for those involved are revealed in reduced exposure to flooding. But does it make economic sense to scale up the initiative for all 2.5 million *char* people? Using information from local people to estimate the appropriate height for raised earth platforms, to identify the most appropriate material for limiting soil erosion and to project future damages under different climate change scenarios, researchers have conducted cost–benefit analysis to assess potential returns.

The results point to a strong economic case for investment. Creating the 125,000 raised platforms needed to protect all *char* people from 20-year floods would cost US$117 million. However, every US$1 of this is estimated to protect US$2–3 in assets and production that would otherwise be lost during floods. These figures understate the wider human development benefits. *Char* people are among the poorest in Bangladesh. It follows that losses sustained during floods have highly damaging implications for their nutrition, health and education. As shown in chapter 2, losses in these areas can trap people in long-term cycles of destitution, undermining lifelong opportunities and transmitting poverty across generations. There is, therefore, an urgent need to support in-country assessments of the costs and benefits of identified adaptation options, and to scale up such assessments to national budgetary planning exercises directed towards the needs of those most vulnerable to climate change.

Human deprivation on the *char* islands

2005	Char Island	Bangladesh average
Extreme poverty (%)	80	23
Literacy rate (males 10 years and older, %)	29	57
Literacy rate (females 10 years and older, %)	21	46
Share of households suffering food insecurity (%)		
1 month or more	95	..
2 months or more	84	..
3 months or more	24	..
4 months or more	9	..

Source: Dasgupta et al. 2005.

Source: Dasgupta et al. 2005; DFID 2002; Tanner et al. 2007.

In Gujarat, where persistent drought and problems in irrigation management have led to the depletion of groundwater, community initiatives have restored 10,000 check dams to store monsoon rains and recharge groundwater. National and state programmes are supporting community initiatives. In Andhra Pradesh, the Drought-Prone Areas Programme covers over 3,000 watershed areas, incorporating a wide range of 'drought-proofing' measures, including soil conservation, water harvesting and afforestation.[40]

Top-down planning, large-scale irrigation and huge water harvesting systems are not a panacea for the emerging risks facing agricultural producers as a result of climate change. The challenge is to support local initiatives through national and subnational strategies that mobilize resources and create incentives. Successful adaptation is not just about physical infrastructure. It is about where that infrastructure is created, who controls it and who has access to the water it conserves.

Insurance for social protection

Climate change will create incremental risks in the lives and livelihoods of the poor. Since many millions of poor people cannot fully manage current climate risks with their own resources, any adaptation strategy needs to strengthen risk management capabilities. Empowering people to cope with climate shocks—especially catastrophic shocks—without suffering the long-term setbacks analysed in chapter 2 is a condition for sustained progress in human development.

Prospects for successful adaptation to climate change will be shaped by wider human development conditions. Public policies in areas such as health, education, employment and economic planning can enhance or diminish the capacity for risk management. Ultimately, the first line of public policy defence against climate change risk is an effective strategy for overcoming poverty and extreme inequality. Social protection is an integral part of any such strategy.

Programmes for social protection encompass a wide range of interventions. They include

contributory schemes through which people can pool risks (old-age pensions and unemployment insurance are examples) and tax-funded transfers providing a variety of benefits to target populations. One of the overarching aims is to prevent temporary shocks from becoming a source of long-term destitution. In the context of climate change, social protection programmes implemented as part of a wider adaptation strategy can play a vital role in helping poor people to manage risks and avoid long-term human development reversals.

As we saw in chapter 2, climate shocks can rapidly erode the entitlements of vulnerable people through their impact on income, nutrition, employment, health and education. Well designed social protection measures can protect entitlements in these areas, while at the same time expanding opportunity. Incremental climate change risks, and adaptation to those risks, are not the sole motivation for an increased emphasis on social protection. Well designed policies in this area are critical in any national strategy for accelerating poverty reduction, reducing vulnerability and overcoming marginalization. However, climate change provides a strong rationale for strengthening social protection safety nets for the poor, especially in the following four areas:

- Employment programmes;
- Cash transfers;
- Crisis-related transfers;
- Insurance related transfers.

Employment programmes. Public work programmes can provide a measure for protecting nutrition and health, creating employment and generating income when climate shocks lead to a loss of agricultural employment or reduced food availability. Employment-based programmes to support cash-transfer or food-transfer schemes can also provide a longer-term safety net. One of the best known examples of such programmes is the Employment Guarantee Scheme in Maharashtra, India. The success of this programme in stabilizing household incomes and preventing food crises gave rise to a national campaign to secure 'the right to work'—and to all-India legislation. The 2005 National Rural Employment Guarantee

4

Act guarantees 100 days of employment at the minimum wage rate for every rural household in India.[41] The costs are estimated at US$10 billion annually, or around 1 percent of GDP.[42]

Even relatively small cash transfers can make a difference. In Ethiopia, the Productive Safety Net Programme (PSNP) provides people with transfers of up to US$4 a month in cash or food. Designed to overcome the uncertainties associated with annual food aid appeals, the programme provides some 5 million people with a predictable source of income and employment (box 4.2). Apart from reducing vulnerability to poor nutrition during episodes of drought, the transfers have enabled poor households to build up their productive assets and invest in health and education.

Cash transfers. Floods, droughts and other climate shocks can force poor households to withdraw children from school to increase labour supply, or to cut spending on health and nutrition. Such coping strategies narrow future opportunities, locking households into low human development traps. Cash transfers linked to clear human development goals can weaken the transmission mechanisms that convert risk into vulnerability. They can also create incentives for the development of human capabilities. Here are some examples:

- In Mexico the Oportunidades programme targets the poorest municipalities for transfers conditional on parents keeping their children in school and attending periodic health checks. In 2003 Progresa supported 4 million families at an annual cost of US$2.2 billion. Coverage under the programme has been found to reduce by 23 percent the probability that children aged 12–14 will leave school and enter the labour market in the event of drought, unemployment among parents or other shocks.[43]
- In Brazil a number of cash transfer programmes have been integrated into a single umbrella scheme—the Bolsa Família Programme (BFP)—which now covers about 46 million people, around one-quarter of the population. The BFP, which represents a legal entitlement for eligible households, has reduced vulnerability and

supported advances in human development across a broad front, enabling households to manage shocks without withdrawing children from school (box 4.3).
- Programmes in Central America have also built resilience against shocks. Since 2000, Nicaragua's Red de Protección Social (RPS) has provided cash transfers conditional on children attending school and health clinic checks. Randomized evaluation studies have shown that the RPS has successfully protected households from a range of shocks, including a slump in coffee prices. Expenditure levels in beneficiary households stayed constant in 2001 while a slump in coffee prices reduced income in non-beneficiary households by 22 percent. In Honduras, there is evidence that cash transfers have protected school attendance and child health during agricultural shocks through its Programa de Asignación Familiar (PRAF).[44]
- In Zambia the Kalomo pilot project provides US$6 a month (US$8 for those with children) to the poorest 10 percent of households, sufficient to meet the costs of a daily meal and preclude absolute poverty. Increased household investment and improved child nutrition and school attendance have already been observed among beneficiaries. Additionally, some households have saved some of the cash and have invested in seed and small animals. The project aims to reach over 9,000 households (58,000 people) by the end of 2007 and is being considered for national upscaling at a projected cost of US$16 million (0.2 percent of GDP or 1.6 percent of current aid flows) per year.[45]

Crisis-related transfers. Climate shocks have the potential to lock smallholder agriculture into downward spirals that undermine the prospects for human development. When a drought or a flood wipes out a crop, people are left facing immediate nutritional threats. But farmers are also left without the seeds, or the cash to purchase seeds and other inputs, for next season's crop. This increases the prospect of reduced income and employment, and hence of continuing dependence on food aid. This

Cash transfers linked to clear human development goals can weaken the transmission mechanisms that convert risk into vulnerability

Box 4.2   The Productive Safety Net Programme in Ethiopia

"Before this programme we could only eat twice. In the hungry time before the harvest perhaps we would only have one meal. The children suffered. Sometimes I could not keep them in school or pay for medicines when they were ill. Of course life is difficult—but at least now I have something to get us through the hard times. Now we eat better food, I can keep my nine-year-old in school, and I am saving to buy a calf."

These are the words of Debre Wondimi, a 28-year-old woman with four children living in Lay Gant *woreda* (district) of South Gondar, Ethiopia. Like millions of people across the country, her life is a struggle to cope with the lethal interaction of drought and poverty. Today, she is a participant in Ethiopia's Productive Safety Net Programme (PSNP), a bold attempt to tackle the food security threats posed by an uncertain climate. That programme could provide important lessons for countries addressing the risk management challenges posed by climate change.

When the rains fail in Ethiopia the well-being and even the lives of people like Debre Wondimi and her children are put at risk. Droughts and famines have recurred throughout the country's history. Since 2000 alone, there have been three major droughts, including a devastating episode in 2002–2003. These emergencies are superimposed on high levels of chronic deprivation. Ethiopia ranks 169 out of the 177 countries covered by the HDI. 23 percent of its population survives on less than US$1 a day, and nearly two in five (38 percent) of its children are underweight for their age.

Food insecurity is thus an integral part of poverty in Ethiopia. Traditionally, the response to food insecurity has been food aid. Every year, donors and government have estimated the amount of food aid needed to cover chronic deficits, topping up that amount through emergency appeals.

The PSNP is an attempt to break with this humanitarian model. It is an employment-based social transfer programme. Targeting people facing predictable food insecurity as a result of poverty rather than temporary shocks, it offers guaranteed employment for 5 days a month in return for transfers of either food or cash—US$4 per month for each household member. The aim is to extend coverage from 5 million people in 2005 to 8 million by 2009. Unlike the food aid model, the PSNP is a multi-year arrangement. Financed by government and donors it will operate for 5 years, shifting the mode of support away from sporadic emergency aid towards more predictable resource transfers.

Predictability is one of the foundations of the PSNP. The programme was prompted partly by concerns in the Ethiopian Government and donor community that emergency appeals were regularly falling short of their targets, or providing late and erratic support. For poor households, delayed support during a prolonged drought can have devastating consequences in both the short and longer term. In 1983–1984 it led to the death of thousands of vulnerable people.

Another distinction between the PSNP and humanitarian food aid is in its level of ambition. The objectives include not just smoothing household consumption by bridging production deficits, but also protecting household assets. Cash transfers are seen as a vehicle for building assets, increasing investment and stimulating rural markets, as well as for preventing the distress sales that push people into destitution.

How successful has the programme been? Independent evaluations give grounds for optimism on several counts. There is strong evidence that the transfers are reaching large numbers of poor people and making a difference to their lives (see table). The following are among the findings of a household survey on the impacts of PSNP transfers during the programme's first year:

- Three-quarters of households reported consuming more or better food than in the previous year; 60 percent also reported that they had been able to retain more of their own food to eat rather than selling for other needs;
- Three in five beneficiaries avoided having to sell assets to buy food—a common 'distress' response—with over 90 percent attributing this directly to the PSNP;
- Almost one-half of beneficiaries stated that they used healthcare facilities more than in the previous year; over one-third of households enrolled more of their children in school and almost a half kept children in school for longer;
- Around one-quarter of beneficiaries acquired new assets, with 55 percent directly attributing this to the PSNP.

The PSNP faces a number of challenges. Around 35 million of Ethiopia's people live below the national poverty line, suggesting many potential beneficiaries are currently excluded. The 'graduation' targets—the percentage of recipients 'passing out' of the programme after 3 years—may also be over-ambitious. It is not clear that the PSNP will equip people with the assets and resources needed to escape deprivation and poverty for good. However, the programme's early implementation phase does demonstrate the potential of well targeted interventions to support household coping strategies.

The human impact of safety nets

	Outcome of productive safety net programme (PSNP)	Beneficiary households (%)	Households directly attributing outcome to PSNP (% of beneficiary households)
Food security	Consumed more or better food than last year	74.8	93.5
	Retained food production for consumption	62.4	89.7
Asset protection	Avoided having to sell assets to buy food	62.0	91.3
	Avoided having to use savings to buy food	35.6	89.7
Access to services	Used healthcare facilities more than last year	46.1	75.9
	Kept children in school longer than last year	49.7	86.5
Asset creation	Acquired new household assets	23.4	55.3
	Acquired new skills or knowledge	28.6	85.5

Source: Devereux et al 2006.

Source: Devereux et al. 2006; Government of the Federal Republic of Ethiopia 2006; Menon 2007b; Sharp, Brown and Teshome 2006; Slater et al. 2006.

Adapting to the inevitable: national action and international cooperation

4

self-reinforcing downward spiral can be broken, or at least weakened, through the transfer of a range of productive inputs, for example:

- In Malawi, the subsidized transfer of a 'productive package' of seeds and fertilizers played an important role in facilitating recovery from the 2005 drought (box 4.4).
- Following a severe drought in the Gao region of Mali in 2005–2006, the international NGO Oxfam initiated a combined cash and credit work programme, acting through local government and community-based organizations. People were employed in creating small-scale water conservation structures, with half their income paid in cash and the other half as credit for the purchase of essential items, such as seeds, other inputs, livestock and schooling.[46]

- In Kenya, drought in pastoral areas is associated with the 'distress sale' of livestock as animal feed supplies decline—a coping strategy that pushes livestock prices down just as food grain prices are rising. An innovative government programme has provided transport subsidies to traders, enabling them to move their animals to markets outside drought areas, effectively putting a floor under prices.[47]

Insurance-related transfers. Coping with climate risk is an intrinsic part of life, especially for poor rural households. Formal insurance markets play a limited role in mitigating that risk. The barriers to market development are well-known. In any functioning insurance market, the price of premiums rises with risk. For poor households in high-risk marginal areas, insurance premiums are likely to prove

---

Box 4.3   Conditional cash transfers—Brazil's Bolsa Família Programme

Conditional cash transfers (CCTs) can play an important role in breaking the link between risk and vulnerability. By setting minimum guaranteed levels for income and wider entitlements to health, education and nutrition, CCTs can empower poor people by creating a legal basis for their entitlements. Brazil's *Bolsa Família* programme (BFP), one of the world's largest CCT schemes, demonstrates what is possible.

Developed initially to deter child labour during crises, Brazil's CCT was dramatically scaled up between 2001 and 2003. The original *Bolsa Escola* programme (a financial transfer contingent on parents keeping their children in school) was supplemented by three additional programmes. *Bolsa Alimentação* was designed as a cash or food transfer to reduce malnutrition among poor households. *Auxilio Gas* was a compensatory measure for poor households following the phasing out of cooking gas subsidies, and *Fome Zero* was introduced in 2003 in order to combat the worst forms of hunger in Brazil. Starting in 2003, efforts to consolidate these various CCTs into a single umbrella programme—the BFP—intensified.

Beneficiaries of the BFP are selected through various targeting methods, including geographic and household assessments based on per capita income. In 2006, eligibility requirements were set at monthly household income levels of Cr$60 (US$28) and Cr$120 (US$55) respectively for poor and moderately poor families.

As of June 2006, the BFP covered 11.1 million families or about 46 million people—a quarter of Brazil's population and almost all of its poor. Total projected costs are estimated at US$4 billion, or 0.5 percent of Brazilian GDP. This is a modest transfer that has produced impressive outcomes. Among the results:

- The programme reaches 100 percent of families living below the official poverty threshold of Cr$120 per month; 73 percent of all transfers go to the poorest families and 94 percent reach families living in the bottom two quintiles.
- BFP accounts for almost one-quarter of Brazil's recent precipitous drop in inequality and 16 percent of its decline in extreme poverty.
- BFP is also improving school enrolment rates. Studies have found that 60 percent of poor children aged 10–15 years currently not in school are expected to enrol in response to BFP and its predecessor. Drop-out rates have been reduced by around 8 percent.
- Some of the most pronounced impacts of the BFP have been on nutrition. The incidence of malnutrition among children aged 6–11 months was found to be 60 percent lower in poor households covered by the nutrition programme.
- Administration of the BFP has supported gender empowerment, with women established as beneficiaries with legal entitlements.

Each country faces different financial, institutional and political constraints in tackling vulnerability. One of the reasons why the BFP has worked in Brazil is that it has been implemented through a decentralized political system but with strong federal support in terms of setting rules, building capacity and holding providers to account. The Brazil case, like others cited in this chapter, demonstrates the potential for CCTs not only to reduce vulnerability but to go beyond this, enabling poor people to claim entitlements that facilitate human development breakthroughs.

Source: de Janvry et al. 2006c; Lindert et al. 2007; Vakis 2006.

Adapting to the inevitable: national action and international cooperation

One of the ways in which climate shocks create cycles of disadvantage is through their impact on agricultural production. When a drought or flood destroys a harvest, the resulting loss of income and assets can leave households unable to afford the seed, fertilizer and other inputs needed to restore production the following year. Well framed public policy interventions can break the cycle, as demonstrated by recent experience in Malawi.

The 2005 maize harvest in Malawi was one of the worst on record. Following successive droughts and floods, production fell from 1.6 million tonnes in the previous year to 1.2 million tonnes—a decline of 29 percent. Over 5 million people faced food shortages. With rural incomes in free fall, households lacked the resources to invest in inputs for the 2006 cropping season, raising the spectre of a famine on the scale of that experienced in 2002.

Supported by a group of donors, the Government of Malawi put in place a strategy for getting productive inputs into the hands of small-scale farmers. Around 311,000 tonnes of fertilizer and 11,000 tonnes of maize seed were sold at subsidized prices. Over 2 million households purchased fertilizer at US$7 for 50 kg—less than one-third of the world price. For distribution, the government used private sector outlets as well as state agencies, enabling farmers to choose their source of supply.

Subsequent harvests showed that this productive inputs programme was a moderate success. Good rains and an increase in the area planted to improved crop varieties raised productivity and overall output. It is estimated that the programme generated an additional 600,000–700,000 tonnes of maize in 2007, independent of rainfall variation. The value of this extra production has been estimated at between US$100 million and US$160 million, compared with the US$70 million cost of the programme. The Malawian economy has also benefited from a reduction in food import requirements. And the increased production has generated household income and employment opportunities.

The productive inputs programme is not a stand-alone strategy for human development. Nor is it a panacea for rural poverty. Far more needs to be done to strengthen the accountability of government, tackle deep-seated inequalities and increase the level of investment in basic service provision for the poor. The programme will have to be retained for several years if it is to break the cycle of low productivity that afflicts Malawian agriculture. Nevertheless, the country's experience underlines the role that public policies can play in reducing vulnerability to climate risk by creating an enabling environment for poverty reduction.

Source: Denning and Sachs 2007; DFID 2007.

unaffordable. Risk pooling and insurance arrangements also suffer from a range of agency problems. The verification of loss, especially in remote rural areas, and the creation of perverse incentives (such as declaring a loss rather than harvesting if crop prices are low) are two examples. To some degree, these problems can be addressed through weather-indexing (box 4.5). Public policies can also help vulnerable people create and manage their own schemes for coping with potentially catastrophic risks. When the 2001 Gujarat earthquake hit India, only 2 percent of those affected had insurance. Low insurance coverage increased vulnerability and hindered economic recovery. One positive outcome was the creation of a micro-insurance scheme for the poor supported by NGOs and the business community. The Afat Vimo scheme under the Regional Risk Transfer Initiative now covers 5,000 low-income families against 19 different types of disasters, with premiums of around US$5 a year. This exercise demonstrates the potential for risk-spreading across

geographic locations even in areas marked by high levels of poverty and vulnerability.[48]

Institutions for disaster risk management

Disaster risk management is an integral part of adaptation planning. Exposure to risk is a function not only of past human development but also of current public policy and institutional capacity. Not every flood or storm produces a climate disaster—and the same event can produce very different outcomes in different countries.

In 2004, the Dominican Republic and Haiti were simultaneously struck by Hurricane Jeanne. In the Dominican Republic, some 2 million people were affected and a major town was almost destroyed, but there were just 23 deaths and recovery was relatively swift. In Haiti, over 2,000 people were killed in the town of Gonaives alone. And tens of thousands were left trapped in a downward spiral of poverty.

The contrasting impacts were not the product of meteorology. In Haiti, a cycle of poverty and environmental destruction has

Can farm insurance schemes be scaled up as part of an integrated strategy for climate change adaptation and human development? Climate change has given an impetus to a range of initiatives aimed at extending access to micro-insurance and weather derivatives in the developing world. But there are difficulties in developing schemes that are accessible to the poor.

Attempts to expand market-based insurance have met with some success. In the Caribbean, for example, the Windward Island's Crop Insurance Programme has covered around 20 percent of the losses experienced by its members—caused by some 267 storm events between 1998 and 2004 alone—providing a safety net sufficient to get growers back on their feet.

However, as climate change increases the frequency and severity of droughts it will drive up the costs of insurance, pricing the most vulnerable people out of the market. The fact that the most vulnerable households are often poor precisely because they operate in high-risk environments adds to the problem, because insurance providers will attach a risk premium to proposals from people living in such environments.

A further problem is that the commonest form of farm insurance—traditional crop insurance—can create perverse incentives, including the incentive to let crops fail during periods of low prices. Weather-indexing can address this problem. In India, the Comprehensive Crop Insurance Scheme (CCI) insures farmers who use official credit systems, charging a small premium and using weather-indexes (rather than farm production) to determine claims.

Premium holders are paid in response to 'trigger events' such as delayed monsoons or abnormal rainfall. However, India's CCI currently has only 25,000 members, mainly wealthier producers.

The participation of small-scale-farmers' groups in the design of insurance packages and the provision of collateral through 'social capital' have produced some promising results. In Malawi, the World Bank and other donors have developed an insurance programme involving private sector companies and the National Smallholder Farmers Association. The programme offers insurance for groundnut and maize, with payments triggered when rainfall falls below a specified threshold determined by records at meteorological stations. This 'drought index insurance' is provided as part of an input loans package to groups of 20–30 farmers, with payouts triggered if there is insufficient rain during the planting season (a 'no-sow' provision) or during three key periods for crop development. The scheme has been successful in its first 2 years, motivating farmers to take the risk of using inputs to raise yields, but its spread is limited by Malawi's sparse network of meteorological stations.

The World Bank and a number of donors are exploring mechanisms for scaling up schemes of this kind, with additional pilot programmes in Ethiopia, Morocco, Nicaragua and Tunisia. While there is undoubtedly scope for enhanced insurance coverage using weather-indexing, there are limits to what private insurance markets can achieve for large vulnerable populations facing covariate risks linked to climate change.

Source: DFID 2004; IRI 2007; Mechler, Linnerooth-Bayer and Peppiatt 2006; Mosley 2000; World Bank 2006f.

4

denuded hillsides of trees and left millions of people in vulnerable slums. Governance problems, low levels of finance and a limited disaster response capacity left public agencies unable to initiate rescue and recovery operations on the scale required. In the Dominican Republic, national laws have limited deforestation and the civil defence force has a staff 10 times larger than its counterpart in Haiti to cater for a population of similar size.[49]

Institutional and infrastructural capacity for disaster risk management is not automatically linked to national wealth. Some countries have demonstrated that much can be achieved even at low levels of average income. Mozambique used the chastening experience of the 2000 floods to strengthen institutional capacity in disaster management, putting in place more effective early warning and response systems (box 4.6). Cuba provides another striking example of a country that has successfully built infrastructure that protects lives. Located at the centre of one of the world's most extreme tropical cyclone zones, the island is hit by several major storms every year. These cause extensive damage to property. However, loss of life and long-term development impacts are limited. The reason: an effective early warning system and a highly developed civil defense infrastructure based on community mobilization. Local authorities play a vital role in relaying early warning information and working with communities at risk. When Hurricane Wilma, then the most intense hurricane ever recorded in the Atlantic Basin, hit the island in 2005, over 640,000 people were evacuated—and there was just one fatality.[50]

Simple comparisons across countries provide only a crude indicator of the effectiveness of disaster risk management measures. The impact of storms and floods is conditioned not just by their intensity, but

Adapting to the inevitable: national action and international cooperation

Countries cannot escape from the accidents of geography that put them in harm's way and increase their exposure to climate risks. What they can do is reduce these risks through policies and institutions that minimize impacts and maximize resilience. The experience of Mozambique powerfully demonstrates that public policies can make a difference.

One of the poorest countries in the world, Mozambique is ranked 172 out of 177 on the HDI and has more than one-third of its people living on less than US$1 a day. Progress in human development has gathered pace over the past decade, but extreme climate events are a constant source of vulnerability. Tropical cyclones that gather in the Indian Ocean are a major cause of storms and flooding. The flooding is aggravated by the fact that Mozambique straddles the lowland basins of nine major rivers—including the *Limpopo* and *Zambezi*—that drain vast areas of south-eastern Africa before crossing the country on their way to the ocean.

In 2000 Mozambique was hit on two fronts. Heavy rains at the end of 1999 swelled river systems to near record levels. Then, in February 2000, cyclone Eline made landfall, causing extensive flooding in the centre and south of the country. Another cyclone—Gloria—arrived in March to make a bad situation worse. Emergency services were overwhelmed and donors were slow to respond. At least 700 people died and 650,000 people were displaced.

During 2007 Mozambique was revisited by a similar climate event. A powerful cyclone, accompanied by high rains, destroyed 227,000 hectares of cropland and affected almost half a million people in the Zambezi basin. Yet on this occasion 'only' 80 people died and recovery was more rapid. What made the difference?

The experience of the 2000 flood gave rise to intensive dialogue within Mozambique and between Mozambique and its aid donors. Detailed flood risk analysis was carried out across the country's river basins, identifying 40 districts with a population of 5.7 million that were highly vulnerable to flooding. Community-based disaster risk management strategies and disaster simulation exercises were conducted in a number of high-risk basins. Meanwhile, the meteorological network was strengthened: in flood-prone Sofala province, for example, the number of stations was increased from 6 to 14. In addition, Mozambique has developed a tropical cyclone early warning system.

Mozambique's policymakers also recognized the importance of the mass media in disaster preparedness. Radio is particularly important. The local language network of Radio Mozambique now provides regular updates on climate risks, communicating information from the National Institute of Meteorology. During 2007, early warning systems and the media enabled government and local communities to identify the most at-risk areas in advance. Mass evacuations were carried out in the most threatened low-lying districts. Elsewhere, emergency food supplies and medical equipment were put in place before the floods arrived.

While much remains to be done, Mozambique's experience demonstrates how countries can learn to live with the threat of floods, reducing vulnerability in at-risk communities.

Source: Bambaige 2007; Chhibber and Laajaj 2006; IRI 2007; World Bank 2005b; WFP 2007.

4

by the topography and pattern of human settlements in the countries that they strike. Even with this caveat, cross-country data say something important: well-developed risk management institutions work. Average income in Cuba is lower than in the Dominican Republic—a country that faces comparable climate risks. Yet in the decade to 2005 the international disasters database records that Cuba had around 10 times as many people affected by disaster but less than one-seventh of the deaths.[51] Much of the difference can be traced to Cuba's highly developed infrastructure and policies for managing climate risks. With tropical storms set to increase in intensity, there is considerable scope for cross-country learning from best practices in climate-related disaster risk management. The conclusion: considerable benefits can be gained from awareness-raising and institutional organization—measures that do not have to entail high capital investment.

4.2 International cooperation on climate change adaptation

The UNFCCC sets out a bold agenda for action on adaptation. It calls for international cooperation to prepare for the impacts of climate change in areas that range from agriculture, through coastal defence management, to lowland cities at risk of flooding. Under this broad umbrella, rich countries are required to support developing countries that are particularly vulnerable to the adverse effects of climate change, building their adaptive capacity and providing financial assistance.[52]

Northern governments have not honoured the spirit of the UNFCCC commitment. While investing heavily in adaptation at home they have failed to support parallel investments in developing countries. Increasingly, the world is divided between countries that are developing a capacity to adapt to climate change, and those that are not.

Inequalities in climate change adaptation cannot be viewed in isolation. They will interact with wider inequalities in income, health, education and basic human security. At any given level of climate change risk, countries with the most limited adaptation capacity will suffer the most adverse impacts on human development and economic growth. The danger is that inequalities in adaptation will reinforce wider drivers of marginalization, holding back efforts to forge a more inclusive model of globalization.

Enhanced international cooperation cannot guarantee effective adaptation or substitute for national political leadership. What it can do is help create an environment that enables developing countries to act and empowers vulnerable people, building the resilience needed to prevent increased risk from translating into greater vulnerability.

The case for international action

Why should the world's richest countries support the efforts of its poorest countries to adapt to climate change? The human development case for urgent international action is rooted in the ethical, social and economic implications of our ecological interdependence. Four considerations merit special emphasis.

Shared values
'Think of the poorest person you have ever seen,' said Gandhi, 'and ask if your next act will be of any use to him.' That injunction captures a basic idea: namely, that the true ethical test of any community lies not in its wealth but in how it treats its most vulnerable members. Turning a blind eye to the adaptation needs of the world's poor would not meet the criterion for ethical behaviour set by Gandhi, or any other ethical criteria. Whatever the motivation for action—a concern for the environment, religious values, secular humanism or human rights—action on climate change adaptation by developed countries is an ethical imperative.

The Millennium Development Goals
The MDGs have galvanized unprecedented efforts to address the needs of the world's poorest people. The time-bound targets for 2015—ranging from halving extreme poverty and hunger to providing universal education, cutting child deaths and promoting greater gender equity—have been embraced by governments, civil society and major development institutions. While the MDGs are not a complete human development agenda, they reflect a sense of urgency and define a set of shared priorities. With climate change already impacting on the lives of the poor, enhanced adaptation is a requirement for supporting progress to the 2015 targets. In the world beyond 2015, climate change will act as a brake on human development, holding back or even reversing human progress until mitigation starts to take effect. Scaling up adaptation to counter that threat should be seen as a part of the post-2015 strategy for building on the achievements of the MDG process. Failure to act on adaptation would rapidly erode what will have been achieved by then. It would be inconsistent with a commitment to the MDGs.

Common interest
While the most immediate victims of climate change and failed adaptation will be the world's poor, the fall-out will not respect the neat divides of national borders. Climate change has the potential to create humanitarian disasters, ecological collapse and economic dislocation on a far greater scale than we see today. Rich countries will not be immune to the consequences. Mass environmental

The human development case for urgent international action is rooted in the ethical, social and economic implications of our ecological interdependence

Adapting to the inevitable: national action and international cooperation

The starting point is that
donors have to deliver
on past commitments

displacement, the loss of livelihoods, rising hunger and water shortages have the potential to unleash national, regional and global security threats. Already fragile states could collapse under the weight of growing poverty and social tensions. Pressures to migrate will intensify. Conflicts over water could become more severe and widespread.

In an interdependent world, climate change impacts will inevitably flow across national borders. Meanwhile, if the countries that carry primary responsibility for the problem are perceived to turn a blind-eye to the consequences, the resentment and anger that will surely follow could foster the conditions for political extremism.

Responsibility and liability

Historic responsibility for climate change and continuing high current per capita emissions of CO_2 raise important questions for the citizens of rich countries. The principle of protection from harm by others is enshrined in the legal codes of almost all countries. One clear example is smoking. In 1998, Attorneys General representing five American states and eighteen cities prosecuted a group of tobacco companies in a class action lawsuit for causing a range of diseases. Punitive damages of US$206 billion were awarded, along with legal injunctions to change marketing behaviour.[53] Harm to the environment is also subject to the force of law. In 1989 the ship *Exxon Valdez* ran aground in Alaska, pouring 42 million litres of oil into a wilderness area of outstanding environmental importance. The United States National Transportation Safety Board claimed that negligence had contributed, leading to legal action that resulted in criminal damage and civil lawsuits worth over US$2 billion.[54] More widely, when factories pollute rivers or the air, the 'polluter pays' principle is applied to cover the costs of cleaning up. If the environmental damages generated by climate change were neatly contained within one legal jurisdiction, those who had created the damage would be faced with a legal obligation to compensate the victims. That would place an obligation on rich countries not just to stop harmful practices (mitigation) but to compensate for damage (adaptation).

Current adaptation financing— too little, too late, too fragmented

International cooperation on adaptation can be thought of as an insurance mechanism for the world's poor. Climate change mitigation will make a small difference to the human development prospects of vulnerable populations in the first half of the 21st Century—but a big difference in the second half. Conversely, adaptation policies can make a big difference over the next 50 years—and they will remain important thereafter. For governments concerned with achieving progress towards the MDGs over the next decade, and building on that progress afterwards, adaptation is the only option for limiting the damage caused by existing climate change.

National governments in developing countries have primary responsibility for developing the strategies needed to build resilience against climate change. Nonetheless, successful adaptation will require coordinated action on many fronts. Aid donors and development agencies will have to work with national governments to integrate adaptation into wider poverty reduction strategies and planning processes. Given that many of the most affected countries are among the poorest, international aid has a pivotal role to play in creating the conditions for adaptation.

Delivering on commitments

The starting point is that donors have to deliver on past commitments. Recent years have witnessed a remarkable change in the provision of aid. During the 1990s, development assistance flows went into steep decline, holding back global poverty reduction efforts. The 2000 UN Millennium Summit, then the largest gathering of world leaders in history, marked a turning point. It resulted in an unprecedented commitment to achieving shared goals—the MDGs—through a partnership between rich and poor countries. Commitments made at Monterrey in 2002, by the European Union in 2005 and by the G8 at Gleneagles backed that partnership with commitments on aid. The Monterrey Consensus reaffirmed a long-standing development assistance target of

0.7 percent of Gross National Income (GNI) for rich countries. Commitments made by the European Union and G8 in 2005 included a pledge to double aid flows by 2010—a US$50 billion increase, with around one-half earmarked for Africa. These are resources that could help countries meet the challenge of scaling up adaptation efforts.

Early signs on delivery are not encouraging. International aid has been increasing since the late 1990s. However, in 2006, development assistance fell by 5 percent—the first recorded fall since 1997. This figure partially exaggerates the decline because of exceptional debt relief provided for Iraq and Nigeria in 2005. But even excluding these operations, aid levels fell by 2 percent.[55] Headline numbers on aid also obscure some wider concerns. For example, much of the increase since 2004 can be traced to debt relief and humanitarian aid. Debt relief inflates the figure for real resource transfers for reasons of financial accounting: aid data record reductions in debt stock as increased aid flows. Humanitarian aid is heavily concentrated and—by definition—geared towards disaster response rather than long-term development.

Analysis by the OECD has raised important questions as to whether, on current trends, aid donors can meet their own commitments. Discounting debt reduction and humanitarian aid, the rate of increase will have to triple over the next four years if the 2005 commitment to double aid by 2010 is to be met (figure 4.3).[56] Of special concern is the stagnation since 2002 in aid flows for core development programmes in sub-Saharan Africa (figure 4.4). These trends are not compatible with the financing requirements for adaptation to climate change.

Limited delivery through dedicated adaptation mechanisms

In stark contrast to adaptation planning in developed countries, the multilateral aid response to adaptation financing in developing

Special contribution | **No choice is our choice**

The changing climate is changing our world for all times to come and for the worse—much worse. This much we know.

What we must now learn is how we can 'cope' with this changing climate and how indeed we can (and must) avert catastrophe by reducing our emissions. The fact is that even with the change in global temperature we've seen so far—some 0.7°C from the mid-1800s to now—we are beginning to see devastation all around us. We know that we are witnessing an increase in extreme weather events. We know that floods have ravaged millions across Asia; that cyclones and typhoons have destroyed entire settlements in coastal areas; that heatwaves have killed people even in the rich world. The list goes on.

But what we must remember is that this is limited damage. That we are living on borrowed time. If this is the level of devastation with just that seemingly small rise in temperature, then think what will happen when the world warms up another 0.7°C, which scientists now tell us is inevitable—the result of emissions we have already pumped into the atmosphere. Then think what happens if we are even more climate-irresponsible and temperatures increase, as predicted in all business-as-usual models, by 5°C. Just think: this is the difference in temperature between the last ice age and the world we know now. Think and act.

It is now clear that coping with changing climate is not new rocket science. It is about doing development. The poor already live on the margins of subsistence. Their ability to withstand the next drought, the next flood or the next natural disaster is already stretched to the limits. Adaptation is about investment in everything that will make societies, particularly the poorest and most climate-vulnerable, more resilient. Adaptation is about development for all. But it needs much more investment and much more speed.

This is one part of what is needed. The other, more difficult, is to reduce our current emissions, and drastically. There is no other truth. We also know emissions are linked to growth and that growth is linked to lifestyles. Because of this our efforts to reduce emissions have been high on rhetoric and low on action. This will have to change.

It will have to change even as we learn another truth: we live on one planet Earth and to live together we will have to share its resources. The fact is that even as the rich world must reduce its carbon footprint, the poor world must get ecological space to increase its wealth. It is about the right to development.

The only question is can we learn new ways to build wealth and well-being? The only answer is we have no choice.

Sunita Narain
Director Centre for Science and Environment

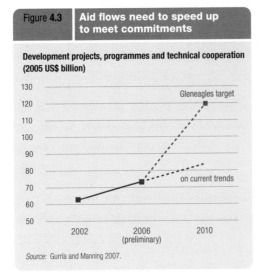

Figure 4.3 | Aid flows need to speed up to meet commitments

Development projects, programmes and technical cooperation (2005 US$ billion)

Gleneagles target

on current trends

2002   2006 (preliminary)   2010

Source: Gurría and Manning 2007.

countries has been slow to take off. Indeed, the response has been characterized by chronic underfinancing, fragmentation and weak leadership. To make matters worse, international cooperation on adaptation has not been developed as part of the wider international aid partnership on poverty reduction. The end result is that multilateral financing mechanisms are delivering small flows of finance with high transaction costs, yielding very limited results.

Multilateral mechanisms for adaptation have been developed under a range of initiatives (table 4.1). Two UNFCCC funds—the Least Developed Country Fund (LDCF) and the Special Climate Change Fund (SCCF)—have

been established under the auspices of the GEF. Both are financed through voluntary pledges by donors. In 2004, another mechanism, the Strategic Priority on Adaptation (SPA), was created to fund pilot projects from GEF's own resources over a 3-year period. The stated objective of the GEF funds is to reduce countries' vulnerability by supporting projects that enhance adaptive capacity. With the entry into force of the Kyoto Protocol in 2005, another potential source of financing was created in the form of the Adaptation Fund—a facility to be funded through Clean Development Mechanism (CDM) transactions (see chapter 3).

The record of delivery to date is not impressive. It can be summarized as follows:

- *The Least Developed Country Fund.* Created in 2001, the LDCF to date has received pledges from 17 donors amounting to just under US$157 million. Less than one-half of this amount has been delivered to GEF accounts. Actual spending in terms of delivery through projects amounts to US$9.8 million.[57] The most tangible output of the LDCF to date has been 20 completed NAPAs. Many of these plans include useful analytical work, providing important insights on priorities. However, they suffer from two basic shortcomings. First, they provide a very limited response to the adaptation challenge, focussing primarily on 'climate-proofing' through small-scale projects: the average country financing proposal generated in the plans amounts to US$24 million.[58] Second, the NAPAs have, in most countries, been developed outside the institutional framework for national planning on poverty reduction. The upshot is a project-based response that fails to integrate adaptation planning into the development of wider policies for overcoming vulnerability and marginalization (box 4.7).

- *The Special Climate Change Fund.* Operational since 2005, the SCCF has received pledges of US$67.3 million, of which US$56.7 million is specifically earmarked for adaptation.[59] The SCCF was created to address the special long-term adaptation needs of developing countries, with a remit covering health, agriculture, water and

Figure 4.4 | Core aid to sub-Saharan Africa is flat

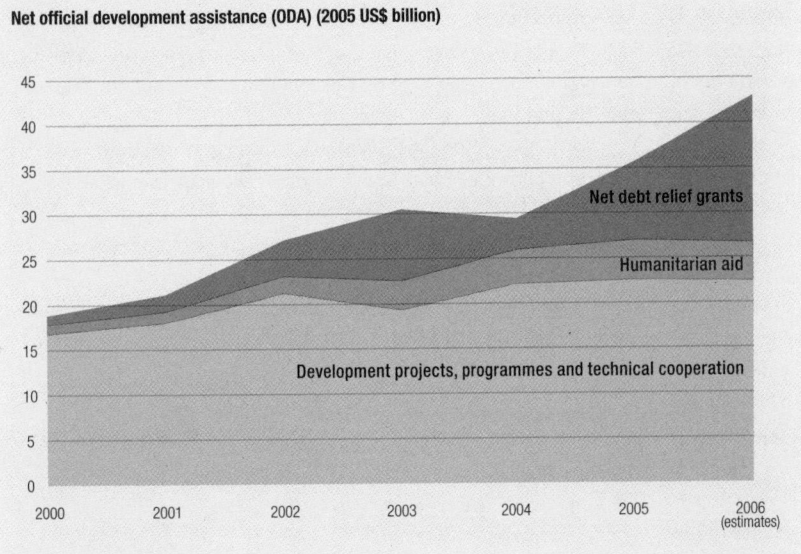

Net official development assistance (ODA) (2005 US$ billion)

Net debt relief grants

Humanitarian aid

Development projects, programmes and technical cooperation

2000   2001   2002   2003   2004   2005   2006 (estimates)

Source: Gurría and Manning 2007.

vulnerable ecosystems. Actual spending under projects to date amounts to US$1.4 million.[60]

- *The Strategic Priority on Adaptation*. This became operational in 2004. It earmarks US$50 million over a 3-year period for pilot projects in a wide range of areas, notably ecosystem management. To date, US$28 million has been committed, of which US$14.8 million has been disbursed.[61]

- *The Adaptation Fund*. This was created to support "concrete activities", to be financed through a 2 percent levy on credits generated through CDM projects. If implemented, the levy could generate a total income in

the range of US$160–950 million by 2012, depending on trade volumes and prices.[62] However, the Adaptation Fund has yet to support any activities because of disagreements over governance.

To reduce a complex story to a simple balance sheet, the record is as follows. By mid-2007, actual multilateral financing delivered under the broad umbrella of initiatives set up under the UNFCCC had reached a total of US$26 million. This is equivalent to one week's worth of spending on flood defence in the United Kingdom. Looking to the future, total committed financing for adaptation through dedicated multilateral funds amounts

---

Box 4.7   National Adaptation Programmes of Action (NAPAs)—a limited approach

National Adaptation Programmes of Action (NAPAs) are among the few tangible products of multilateral cooperation on adaptation. Funded through the GEF's Least Developed Countries (LDC) Fund, NAPAs are intended to identify urgent and immediate needs while at the same time developing a framework for bringing adaptation into the mainstream of national planning. Have they succeeded?

On balance the answer to that question is 'no'. Twenty NAPAs have been produced to date. While many include excellent analytical work, the overall exercise suffers from four inter-related shortcomings:

- *Inadequate financing.* Under the LDC Fund each country is initially allocated up to US$200,000 to fund the formulation of a NAPA. That figure represents a small fraction of what some districts and cities in Europe have spent on analytical risk and vulnerability assessments. Financial constraints have limited the scope of governments to consult with at-risk communities or conduct national research.

- *Underestimation of adaptation costs.* While NAPAs are not intended as stand-alone exercises, their financing provisions are unrealistically low. The proposed average financing envelope for the first 16 NAPAs is US$24 million, stretched over a budget cycle of 3–5 years. Countries in an advanced state of project preparation under the LDC Fund will receive an average of US$3–3.5 million each to start implementing the first priorities identified by their NAPAs. Even for countries at the higher end of this range, the headline figures are difficult to square with the urgent and immediate needs facing poor households. For example, the US$74 million proposed for Bangladesh and the US$128 million for Cambodia fall far short of requirements.

- *Project-based bias.* Most NAPAs focus entirely on small-scale, project-based interventions to be cofinanced by donors. For example, Niger identifies 14 projects in areas such as watershed management and livestock fodder development. Bangladesh

identifies a range of projects for coastal defence. While well designed projects are necessary to address the urgent needs of the most vulnerable, they cannot provide the basis for an effective adaptation strategy. As in other areas of aid, project-based support tends to come with high transaction costs, with an in-built bias towards donor preferences and priorities. Effective adaptation planning has to be developed through national programmes and national budgets, with governments setting the priorities through political structures that are responsive to the needs of those most affected. There is little evidence to suggest that this has been achieved on anything like the necessary scale.

- *Weak links to human development.* Some NAPAs provide important insights into the impact of emerging climate change risks on vulnerable groups. However, they do not provide a basis for integrating adaptation into national poverty reduction strategies. The focus is almost entirely on 'climate-proofing', to the exclusion of social protection and wider strategies for empowering poor households. The political disconnect between adaptation planning and poverty-reduction planning is evident in Poverty Reduction Strategy Papers (PRSPs), the documents that set out national development goals and priorities supported through aid partnerships. In a review of 19 PRSPs carried out for this report most identified climate events and weather variability as important drivers of poverty and constraints on human development. Yet only four countries—Bangladesh, India, Malawi and Yemen—identified specific links between climate change and future vulnerability. In many cases, adaptation planning is happening on an entirely separate track from poverty-reduction planning. For example, Mauritania did not include the findings of its 2004 NAPA in its 2006 PRSP—an outcome suggesting that climate change adaptation does not figure prominently in defining aid partnership priorities.

Source: Government of the People's Republic of Bangladesh 2005b; Matus Kramer 2007; Reid and Huq 2007; Republic of Niger 2006; Royal Government of Cambodia 2006.

4

Adapting to the inevitable: national action and international cooperation

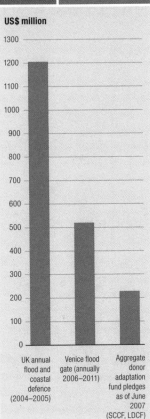

US$ million

UK annual flood and coastal defence (2004–2005)	Venice flood gate (annually 2006–2011)	Aggregate donor adaptation fund pledges as of June 2007 (SCCF, LDCF)

Source: Abbott 2004; DEFRA 2007 and GEF 2007.

to a total of US$279 million. These funds will be disbursed over several years. Contrasts with the adaptation effort in rich countries are striking. The German state of Baden-Würtemberg is planning to spend more than twice as much as the entire multilateral adaptation effort on strengthening flood defences. Meanwhile, the Venice Mose plan, which aims to protect the city against rising sea levels, will spend US$3.8 billion over five years (figure 4.5).[63]

The concern of rich countries to invest in their own climate change adaptation is, of course, entirely legitimate. The sustained and chronic under-financing of adaptation in developing countries is less legitimate, not least given the role of rich countries in creating climate change risks.

Aid portfolios under threat

Have other donors compensated for the shortfall in aid delivery through dedicated climate change adaptation funds? There are problems in assessing the wider aid effort, not least because there is no common definition of what represents an adaptation activity. However, detailed analysis suggests that the integration of adaptation planning into aid policies remains at an early stage.

Bilateral and multilateral donors are gradually increasing support for adaptation, from a low base. One review of 10 bilateral agencies accounting for almost two-thirds of international development assistance attempted to identify projects in which climate change adaptation was an explicit consideration. It documented total commitments of US$94

million over a 5-year period from 2001 to 2005—less than 0.2 percent of average development assistance flows.[64] Of course, this figure captures only what has happened in the past. There are signs that donors are starting to respond to climate change adaptation needs. Between 2005 and 2007 the World Bank's adaptation-related activity increased from around 10 to 40 projects, for example.[65] However, planning and financing for climate change adaptation remain marginal activities in most donor agencies.

Failure to change this picture will have consequences not just for poverty and vulnerability in developing countries but also for aid effectiveness. While most donors have been slow to respond to the challenge of adaptation, their aid programmes will be directly affected by climate change. Rural development programmes, to take an obvious example, will not be immune to the consequences of changed rainfall patterns. An increase in the frequency of droughts in sub-Saharan Africa will impact very directly on programmes for health, nutrition and education. And an increase in the severity and frequency of storms and flooding will compromise aid programmes in many areas. Media images of schools and health clinics being swept away during the 2007 floods in Bangladesh graphically capture the way in which social sector investments can be compromised by climate-related disasters.

Across the developing world large amounts of aid investment are tied up in projects and programmes that are vulnerable to climate change. The OECD's Development Assistance Committee (DAC) has developed a framework for identifying aid activities that are sensitive to climate change. It has applied that framework to a number of developing countries. In the cases of Bangladesh and Nepal the DAC estimates that over one-half of all aid is concentrated in activities that will be negatively affected by climate change.[66]

Using the DAC's reporting system, we have developed an 'aid-sensitivity' analysis for donor portfolios averaged across the period 2001–2005. Broadly, we identify development assistance activities that might be considered vulnerable to various levels of climate change risk. The

Table **4.1** The multilateral adaptation financing account			
Adaptation fund	**Total pledged** (US$ million)	**Total received** (US$ million)	**Total disbursed (less fees)** (US$ million)
Least Developed Countries Fund	156.7	52.1	9.8
Special Climate Change Fund	67.3	53.3	1.4
Adaptation Fund	5	5	–
Sub-total	229	110.4	11.2
Strategic Priority on Adaptation	50	50	14.8[a]
Total	279	160.4	26

a. Includes fees.
Note: data are as of 30th April 2007.

Source: GEF 2007a, 2007b, 2007c.

range for that risk extends from a narrow band of activities that are highly sensitive—such as agriculture and water supply—to a wider band of affected projects and programmes in sectors such as transport.[67]

The results are striking. Our analysis suggests that 17 percent of all development assistance falls into the narrow band of intensive risk, rising to 33 percent for the wider band. Expressed in financial terms, between US$16 billion and US$32 billion are at immediate risk. These figures suggest that 'climate-proofing' aid should be viewed as an important part of the adaptation challenge. Approximate costs for such 'climate-proofing' aid are around US$4.5 billion, or 4 percent of 2005 aid flows.[68] Bear in mind that this represents just the cost of protecting existing investments against climate change, not the incremental cost of using aid programmes to build resilience.

Beneath these headline numbers, there are variations between donors. Some major bilateral donors—including Canada, Germany, Japan and the United Kingdom—face high levels of risk exposure (figure 4.6). Multilateral agencies such as the African Development Bank (ADB) and the World Bank's International Development Association (IDA) portfolios are in a similar position.

Adapting disaster relief to climate change

Climate-related disasters pose a wider set of challenges for the donor community. Climate change will increase the frequency and severity of natural disasters. Increased investment in disaster risk reduction is an essential requirement for dealing with this challenge. However, the reality is that disasters will happen—and that the international community will have to respond through humanitarian relief. Increased aid provision and a strengthening capacity for supporting disaster recovery are two of the requirements.

Disaster relief is already one of the fastest growing areas of international aid, with bilateral spending reaching US$8.4 billion—or 7.5 percent of total aid—in 2005.[69] Climate-related disaster is among the strongest engines driving the increase in humanitarian aid, and climate

change will strengthen it still further. Exposure to the risk of climate disasters can be expected to rise with urbanization, the expansion of unplanned human settlements in slum areas, environmental degradation and the marginalization of rural populations. As shown in chapter 2, climate-related catastrophes can slow or stall progress in human development. But responding to the rising tide of disaster has the potential to divert aid from long-term development programmes in other areas—a prospect which points to the importance of new and additional aid resources to cope with future demands.

Aid quantity is not the only problem. Timing and fulfilment of pledges present further limitations. In 2004, for example, only 40 percent of the US$3.4 billion in emergency funds requested by the UN was delivered, much of it too late to avert human development setbacks.[70] An increase in climate-related disasters poses wider threats to development that will have to be addressed through improvements in aid quality. One danger is that low-profile 'silent emergencies' linked to climate change will not receive the attention that they demand. Persistent local droughts in sub-Saharan Africa generate less media attention than earthquakes or tsunami-type events, even though their long-term effects can be even more devastating. Unfortunately, less media attention has a tendency to translate into less donor interest and the underfinancing of humanitarian appeals.

Post-disaster recovery is another area of aid management that has important implications for adaptation. When vulnerable communities are hit by droughts, floods or landslides, immediate humanitarian suffering can swiftly transmute into long-term human development setbacks. Support for early recovery is vital to avert that outcome. However, while aid flows for disaster relief have been rising, recovery has been systematically underfinanced. As a result, the transition from relief to recovery is regularly compromised by insufficient funds and the non-disbursement of committed resources. Farmers are left without the seeds and credit they need to rebuild productive capacities, slum dwellers are left to rebuild their assets by their own efforts,

Figure 4.6 Aid is vulnerable to climate change

Official development assistance vulnerable to climate change, selected donors 2001–2005 (%)

United States
France
Italy
Norway
United Kingdom
Netherlands
Canada
Germany
Japan
Denmark

OECD–DAC Average

High estimate
34%, **US$ 32 billion**
Low estimate
17%, **US$ 16 billion**

Source: HDRO calculations based on OECD 2007b and Agrawala 2005.

4

and infrastructures for health and education are left devastated.

The foundations for a multilateral system equipped to deal with climate emergencies are just beginning to emerge. The Central Emergency Response Fund (CERF), managed under UN auspices, is an attempt to ensure that the international community has the resources available to initiate early action and to tackle 'silent emergencies'. Its aim is to provide urgent and effective humanitarian relief within the first 72 hours of a crisis. Since its launch in 2006, the CERF has received pledges from 77 countries. The current proposal is to have in place an annual revolving budget of US$450 million by 2008. The wider multilateral system is also reforming. The World Bank's Global Facility for Disaster Reduction and Recovery (GFDRR) also includes a mechanism—the Standby Recovery Financing Facility—a multi-donor trust fund aimed at supporting the transition to recovery through rapid, sustained and predictable financing. Both the CERF and the GFDRR directly address failings in the current emergency response system. However, the risk remains that the growing costs associated with emergency responses will divert assistance from long-term development assistance in other areas.

Rising to the adaptation challenge— strengthening international cooperation on adaptation

Climate change adaptation has to be brought to the top of the international agenda for poverty reduction. There are no blueprints to be followed—but there are two conditions for success.

The first is that developed countries have to move beyond the current system of underfinanced, poorly coordinated initiatives to put in place mechanisms that deliver on the scale and with the efficiency required. Faced with the threat to human development posed by climate change, the world needs a global adaptation financing strategy. That strategy should be seen not as an act of charity on the part of the rich but as an investment in climate change insurance for the world's poor. The

aim of the insurance is to empower vulnerable people to deal with a threat that is not of their making.

The second condition for successful adaptation is institutional. The risks and vulnerabilities that come with climate change cannot be dealt with through microlevel projects and 'special initiatives'. They have to be brought into the mainstream of poverty reduction strategies and budget planning. One possible framework for action is revision of the Poverty Reduction Strategy Papers (PRSPs) that provide the framework for nationally owned policies and partnerships with donors.

Financing adaptation insurance

Estimating the financing requirements for climate change adaptation poses some obvious problems. By definition, the precise costs of interventions cannot be known in advance. The timing and intensity of local impacts remain uncertain. Moreover, because interventions have to cover a wide range of activities, including physical infrastructure, livelihood support, the environment and social policy, it is difficult to assign costs to specific climate change risks. These are all important caveats. But they do not constitute a justification for business-as-usual approaches.

Several attempts have been made to provide ballpark estimates of the financing required for adaptation. Most have focused on 'climate-proofing'. That is, they have looked principally at the cost of adapting current investments and infrastructure to protect them against climate change risks. The World Bank has provided one set of estimates based on a range of current investments and 'guesstimates' of adaptation costs. Updating the World Bank's figures for 2005 points to a mid-range cost estimate of around US$30 billion (table 4.2). Importantly, these costs estimates are based on national economic indicators. Another valuable source of information comes from 'bottom-up' analysis. Extrapolating from current NAPA cost estimates, one study puts the financing needed for immediate 'climate-proofing' at between US$1.1 billion and US$2.2 billion for LDCs, rising to US$7.7–33 billion for all developing countries.[71] The figures are based on project costs contained in the NAPA.

Using a different approach, Oxfam has attempted to estimate the broad financing requirements for community-based adaptation. Drawing upon a range of project-based per capita estimates, it reaches an indicative figure of around US$7.5 billion in adaptation financing requirements for people living on less than US$2 a day.[72] Exercises such as this draw attention to some of the adaptation costs that fall directly on the poor—costs that are not captured in many national planning exercises.

All of these cost estimates provide an insight into plausible orders of magnitude for adaptation financing. Understanding the financial costs of 'climate-proofing' is critical for national economic planning. Governments cannot build credible plans in the absence of information on national financing requirements. At the same time, it is important for human development that community-based investments, many of which are not monetized, are also taken into account. Further research in these areas is critical to the integration of adaptation planning into long-term budget planning and poverty reduction strategies.

Consideration also has to be given to adaptation beyond 'climate-proofing'. Protecting infrastructure against climate risks is one critical element in adaptation. Another element is the financing of recovery from climate-related disasters. However, building resilience against incremental risks is about more than investment in physical infrastructure and post-emergency recovery. It is also about empowering people to cope with climate shocks through public policy investments that reduce vulnerability. One of the most serious problems in current approaches to adaptation is the overwhelming focus on 'climate-proofing' infrastructure, to the exclusion of strategies for empowering—and hence climate-proofing—people. The latter is more difficult to put a price on, but no less critical to successful adaptation.

Increased financing for human development should be viewed as a central element in international cooperation on adaptation: uncertainties over costs cannot be allowed to obscure the fact that climate change will diminish the benefits of aid flows and hold back the international poverty reduction effort. In effect, the incremental risks associated with climate change are pushing up the costs of achieving human development goals, especially the MDGs. That is why increased adaptation financing should be seen in part as a response to the increased financing requirements for delivering on the MDG targets, in 2015 and thereafter.

The critical starting point is that adaptation financing has to take the form of new and additional resources. That means that the international effort should be supplementary to the aid targets agreed at Gleneagles and supplementary to the wider aspiration of achieving an aid-to-GNI level of 0.7 percent by 2015. Estimates of the financing requirements for adaptation cannot be developed through the application of mechanistic formulae. Provisions have to be calibrated against human development impact assessments and the experience of the poor. Adjustments will have to be made in the light of new scientific evidence and national assessments. Over the longer term, the scale of the adaptation challenge will be determined in part by the mitigation effort. All of these considerations point to the importance of flexibility. But recognition of the case for flexibility is neither a reason for delaying action, nor a justification for what is clearly an inadequate international effort. Climate change is a real and present danger for the MDGs—and for post-2015 progress in human development.

Addressing that danger will require an enhanced resource mobilization effort that

Increased adaptation financing should be seen in part as a response to the increased financing requirements for delivering on the MDG targets

| Table **4.2** | The cost of climate-proofing development | | | | |

	Developing countries (US$ billion) 2005	Estimated portion sensitive to climate change (%)	Estimated costs of climate adaptation (%)	Estimated cost (US$ billion) 2005	Mid range of estimated cost (US$ billion) 2005
Investment (US$ billion)	2,724	2–10	5–20	3–54	~30
Foreign direct investment (US$ billion)	281	10	5–20	1–6	~3
Net official development assistance	107	17–33	5–20	1–7	~4

Source: Data on investment from IMF 2007; data on foreign direct investment from World Bank 2007d data on ODA from Indicator Table 18; assumptions on climate sensitivity and cost from Stern 2006.

includes, but goes beyond, climate-proofing. Our rough estimate for financing requirements in 2015 is as follows:

- *Climate-proofing development investment.* Carrying out detailed costing exercises for the protection of existing infrastructure is a priority. Building on the World Bank's methodology outlined above and updating for 2005 data, we estimate costs for climate-proofing development investments and infrastructure to be at least US$44 billion annually by 2015. [73]

- *Adapting poverty reduction programmes to climate change.* Poverty reduction programmes cannot be fully climate-proofed. However, they can be strengthened in ways that build resilience and reduce vulnerability. National poverty reduction plans and budgets are the most effective channel for achieving these goals. Social protection programmes of the kind described earlier in this chapter provide one cost-effective strategy. At their 2007 summit, the G8 leaders identified social protection as an area for future cooperation on development. At the same time, the incremental risks created by climate change require a broader response, including, for example, support for public health, rural development and community-based environmental protection. These investments will have to be scaled up over time. The 2015 target should be a commitment of at least US$40 billion per year—a figure that represents around 0.5% of GDP for low income and lower-middle income countries—for strengthening social protection programmes and scaling up aid in other key areas. [74]

- *Strengthening the disaster response system.* Disaster risk reduction investments through aid will deliver higher returns than post-disaster relief. However, climate disasters will happen—and climate change will add to wider pressures on international systems for dealing with humanitarian emergencies. How these systems respond will have a critical bearing on human development prospects for affected communities across the world. One of the greatest challenges is to ensure that resources are mobilized swiftly to deal with climate-related emergencies. Another is to finance the transition from relief to recovery. Provisions should be made for an increase in climate-related disaster response of US$2 billion a year in bilateral and multilateral assistance by 2015 to prevent the diversion of development aid.

The lower bound ballpark figures that emerge appear large. In total they amount to new additional adaptation finance of around US$86 billion a year by 2015 (table 4.3). Mobilizing resources on this scale will require a sustained effort. However, the figures have to be put in context. In total, developed countries would have to mobilize around 0.2 percent of GDP in 2015—roughly one tenth of what they currently mobilize for military expenditure.[75]

Rich countries' responsibility weighs heavily in the case for adaptation financing. The impact of climate change in the lives of the poor is not the result of natural forces. It is the consequence of human actions. More specifically, it is the product of energy use patterns and decisions taken by people and governments in the rich world. The case for enhanced financing of adaptation in developing countries is rooted partly in a simple ethical principle: namely that countries which are responsible for causing harm are also responsible for helping those affected deal with the consequences. International cooperation on adaptation should be viewed not as an act of charity, but as an expression of social justice, equity and human solidarity.

None of this is to understate the scale of the challenge facing donors. Mobilizing resources on the scale required for climate change

Table **4.3**	Investing in adaptation up to 2015		
		Estimated cost	
		% of OECD GDP **2015**	**US$ billion** **2015**
Estimated donor country cost			
Climate-proofing development investment		0.1	44
Adapting poverty reduction to climate change		0.1	40
Strengthening disaster response		(.)	2
Total		**0.2**	**86**

Source: HDRO estimates based on GDP projections from World Bank 2007d.

adaptation will require a high level of political commitment. Aid donors will need to work with developing country governments in identifying incremental climate change risks, assessing the financing requirements for responding to those risks, and engaging in dialogue on adaptation policies. At the same time, donors themselves will have to forge a far stronger consensus on the case for international action on adaptation, going beyond statements of principle to practical action. Given the scale of resource mobilization required, donors may also need to consider the urgent development of innovative financing proposals. There are several options:

- *Resource mobilization through carbon markets.* The Kyoto Protocol Adaptation Fund already establishes the principle that adaptation financing could be linked to carbon markets. That principle should be acted on. Mobilizing resources for adaptation through markets for mitigation offers two broad advantages: a predictable flow of finance and a link from the source of the problem to a partial solution. Carbon taxation provides one avenue for resource mobilization (see chapter 3). For example, a tax of just US$3/tonne CO_2 on OECD energy-related emissions would mobilize around US$40 billion per year (at 2005 emissions levels). Cap-and-trade schemes provide another market-based route for mobilizing adaptation finance. For example, the European Union's ETS will allocate around 1.9 Gt in emission allowances annually in the second phase to 2012. Under current rules up to 10 percent of these allowances can be auctioned. For illustrative purposes, an adaptation levy set at US$3/tonne CO_2 on this volume would raise US$570 million. With an increase in auctioning after 2012, the EU ETS auctioning could provide a more secure foundation for adaptation financing.

- *Wider levies.* In principle, adaptation financing can be mobilized through a range of levies. Applying levies to carbon emissions has the twin benefit of generating revenues for adaptation while at the same time improving the incentives to promote

mitigation. One example is an air-ticket levy. In 2006, France began collecting an 'international solidarity contribution' on all European and international flights.[76] The aim is to generate revenues of US$275 million to finance treatment for HIV/AIDS and other epidemics. An international drugs purchase facility has been created to disburse revenues from the scheme. The United Kingdom uses part of its Air Passenger Duty tax to fund immunization investments in developing countries. Establishing a levy of US$7 per flight would be unlikely to deter air transport on any scale, but it would yield around US$14 billion in revenues that could be allocated to adaptation.[77] Levies could be extended through taxation in other areas, including petrol, commercial electricity supply and CO_2 emissions from industry. An adaptation levy graduated to reflect the high level of CO_2 emissions of sports utility vehicles and other low fuel-efficiency vehicles could also be considered.

- *Financing linked to income and capabilities.* A number of commentators have argued for adaptation commitments to be linked to developed country wealth. One proposal is for all Annex I Parties under the Kyoto Protocol to set aside a fixed share of their GDP to finance adaptation.[78] Another advocates the development of a formula for contributions to adaptation financing that links responsibility for carbon emissions (as reflected in historic shares) and financing capabilities (measured by reference to the HDI and national income).[79]

Proposals in all of these areas merit serious consideration. One obvious requirement is that revenue mobilization to support adaptation should be transparent and efficient. There are potential pitfalls with the creation of special financing mechanisms and dedicated funding sources. Over-reliance on supplementary levies has the potential to introduce an element of unpredictability into revenue flows. Given the far-reaching and long-term nature of the adaptation financing challenge, there is a strong case for rooting it in normal budgetary processes. However, this does not rule out an expanded

Donors may also need to consider the urgent development of innovative financing proposals

4

The best PRSPs link well-defined targets to an analysis of poverty and to systems of financial allocation under annual budgets and rolling medium-term expenditure frameworks

role for supplementary financing, whether in the direct financing of adaptation or in mobilizing additional budgetary resources.

'Mainstreaming' adaptation

Financing is not the only constraint on the development of successful adaptation strategies. In most countries adaptation is not treated as an integral part of national programmes. Both donors and national governments are responding to the adaptation challenge principally through project-based institutional structures operating outside planning systems for budgets and poverty reduction strategies.

This backdrop helps to explain the low priority attached to adaptation in current aid partnerships. While arrangements vary, in many developing countries adaptation planning is located in environment ministries which have a limited influence on other ministries, notably finance. Most PRSPs—the documents that set out national priorities and define the terms for aid partnerships—provide a cursory treatment of climate change adaptation (see box 4.7). One result is that much of the aid financing for adaptation happens though project-based assistance. Current multilateral delivery mechanisms and the approach followed under NAPA point to more of the same.

Some projects on climate change adaptation are delivering results. Looking to the future, projects will continue to play an important role. However, project-based assistance cannot provide a foundation for scaling up adaptation partnerships at the pace or at the scale required. Project-based aid tends to increase transaction costs because of in-built donor preferences for their own reporting systems, weak coordination and strains on administrative capacity. Aid transaction costs in these areas already impose a heavy burden on capacity. In 34 aid-recipient countries covered by one OECD review in 2005, there were 10,507 donor missions in the course of the year.[80]

There is a danger that current approaches to adaptation could push up aid transaction costs. Developing countries already face constraints in integrating climate change adaptation into national planning processes. They are also

responding to pressing demands in many other areas—HIV/AIDS, nutrition, education and rural development, to name but a few—where they are often engaging with multiple donors. If the route to increased financing for adaptation to climate change is through several multilateral initiatives, each with its own reporting system, it can be confidently predicted that transaction costs will rise. Making the transition to a programme-based framework that is integrated into wider national planning exercises is the starting point for scaling up adaptation planning.

Small-island developing states have already demonstrated leadership in this area. Faced with climate change risks that touch all aspects of social, economic and ecological life, their governments have developed an integrated response linking national and regional planning. In the Caribbean, to take one example, the Mainstreaming Adaptation to Climate Change programme was initiated in 2002 to promote integration of adaptation and climate risk management strategies into water resource management, tourism, fisheries, agriculture and other areas. Another example is in Kiribati in the Pacific, where the Government has worked with donors to integrate climate change risk assessments into national planning, working through high-level ministerial committees. The 2-year preparation phase (2003–2005) is to be followed by a 3-year implementation period, during which donors are cofinancing incremental climate change adaptation spending in key areas.

Working through PRSPs

For low-income countries, dialogue on PRSPs provides an obvious vehicle for the transition to a stronger emphasis on programmes. The best PRSPs link well-defined targets to an analysis of poverty and to systems of financial allocation under annual budgets and rolling medium-term expenditure frameworks. Whereas projects operate on short-term cycles, adaptation planning and financing provisions have to operate over a longer time horizon. In countries with a proven capacity for delivery, channelling donor support through national budgets that finance national and subnational programmes is likely to prove more effective than funding dozens of small-scale projects. The PRSP provides a link from poverty reduction goals to national budgets and is thus the best tool for rolling out public spending programmes geared to the MDGs and wider macroeconomic goals.

In many countries, increased programme-level support could deliver an early harvest of benefits from adaptation that bolster wider poverty reduction efforts. Bangladesh provides an example. Many donors in the country are engaged in a wide range of projects and programmes aimed at reducing climate risks. However, far more could be done to expand programme support in key areas. Two examples:

- *Social safety net programmes (SSNPs).* Through the PRSP, poor people themselves have identified strengthened safety net programmes as a vital requirement for reducing vulnerability. Currently, Bangladesh has a large portfolio of such programmes, with spending estimated at around 0.8 percent of GDP. These include an old-age allowance scheme, allowances for distressed groups, a Rural Maintenance Programme and a Rural Infrastructure Development Programme—respectively providing cash for work and food for work—and conditional cash transfers that provide food for education and stipends for girls.[81] Apart from providing immediate relief, these programmes have offered a ladder for people to climb out of poverty. However, there are a number of problems. First, coverage is inadequate: there are around 24 million people in Bangladesh in the category of 'extremely poor', whereas safety nets only currently reach about 10 million. Second, there is no integrated national SSNP based on comprehensive and updated risk and vulnerability mapping. Each separate SSNP is funded by a range of donors and there are problems with unclear and overlapping mandates. Strengthened capacity and scaled up national programmes in these areas could provide millions of people facing immediate climate change risks with support for adaptation.[82]

- *Comprehensive disaster management.* Working with donors through a range of innovative programmes, Bangladesh has developed an increasingly effective disaster management system. Linked explicitly to the MDGs, it brings together a range of previously fragmented activities, including the development of early warning systems, community-based flood defence and post-flood recovery.[83] However, current funding—US$14.5 million over four years—is inconsistent with the ambitious goal of reducing the vulnerability of the poor to 'manageable and acceptable levels'.

While every country is different, these examples illustrate the wider potential for integrating strategies for adaptation into national planning. Dialogue on PRSPs provides a framework through which developed countries can support the efforts of developing country governments. It could also provide them with a mechanism through which to strengthen disaster risk management strategies.

Initial progress has already been made on multilateral assistance mechanisms. Under the Hyogo Framework for Action, an international disaster risk reduction framework signed by 168 countries in 2005, clear guidelines have been set out for the incorporation of disaster risk reduction into national planning processes. Elements of the architecture for turning guidelines into outcomes have started to emerge.[84] Similarly, the World Bank's GFDRR supports the Hyogo Framework. One of its core objectives is to build the capacity of low-income countries to integrate disaster risk reduction analysis and action (including that brought about by climate change) into PRSPs and wider strategic planning processes.[85] Total programme financing requirements to 2016 are estimated at US$2 billion.[86]

Key lessons emerge from the adaptation experience of developing countries related to requirements for developing such strategies:

- *Reforming dedicated multilateral funds.* The major multilateral funds should be unified into a single fund with simplified uptake procedures and a shift in emphasis towards programme-based adaptation.

Increased programme-level support could deliver an early harvest of benefits from adaptation that bolster wider poverty reduction efforts

- *Revising PRSPs.* All PRSPs should be updated over the next two years to incorporate a systematic analysis of climate change risks and vulnerabilities, identify priority policies for reducing vulnerability and provide indicative estimates for the financing requirements of such policies.

- *Putting adaptation at the centre of aid partnerships.* Donors need to mainstream adaptation across their aid programmes, so that the effects of climate change can be addressed in all sectors. By the same token, national governments need to mainstream adaptation across ministries, with the coordination of planning taking place at a high political level.

Conclusion

The limitations of adaptation strategies have to be recognized. Ultimately, adaptation is an exercise in damage limitation. It deals with the symptoms of a problem that can be cured only through mitigation. However, failure to deal with the symptoms will lead to large-scale human development losses.

The world's poorest and most vulnerable people are already adapting to climate change. For the next few decades, they have no choice but to continue adapting. In a good-case scenario, average global temperatures will peak around 2050 before they reach the 2°C dangerous climate change threshold. In a bad-case scenario, with limited mitigation, the world will breach the 2°C threshold before 2050 and be set on course for still further rises. Hoping—and working—for the best while preparing for the worst, serves as a useful first principle for adaptation planning.

Successful adaptation coupled with stringent mitigation holds the key to human development prospects for the 21st Century and beyond. The climate change that the world is already locked into has the potential to result in large-scale human development setbacks, first slowing, then stalling and reversing progress in poverty reduction, nutrition, health, education and other areas.

Developing countries and the world's poor cannot avert these setbacks by acting alone—nor should they have to. As shown in chapter 1 of this Report, the world's poor walk the earth with a light carbon footprint. With their historic responsibility for the energy emissions that are driving climate change and their far deeper current carbon footprints, rich countries have a moral obligation to support adaptation in developing countries. They also have the financial resources to act on that obligation. The business-as-usual model for adaptation is indefensible and unsustainable. Putting in place large-scale adaptation investments in rich countries while leaving the world's poor to sink or swim is not just a prescription for human development reversals. It is a prescription for a more divided, less prosperous and more insecure 21st Century.

Notes

Chapter 1

1 Diamond 2005.
2 Kennedy 1963.
3 Sen 1999.
4 UN 2007b.
5 World Bank 2007c.
6 UNDP 2006b.
7 Government of India 2007.
8 World Bank 2007c.
9 UNDP 2006b.
10 WHO 2006; WHO and UNICEF 2005.
11 Lopez 2006.
12 Wagstaff and Claeson 2004.
13 World Bank 2003.
14 Hansen et al. 2006.
15 ISSC 2005.
16 ISSC 2005; European Union 2007b; den Elzen and Meinshausen 2005; Schellnhuber 2006; Government of France 2006.
17 Warren et al. 2006.
18 Warren et al. 2006.
19 OFDA and CRED 2007.
20 Anthoff et al. 2006; Dasgupta et al. 2007.
21 IPCC 2007b, Chapter 4: Ecosystems, their Properties, Goods, and Services; Warren et al. 2006.
22 IPCC 2007b, Chapter 8: Human Health, Summary Table 8.2.
23 Sen 1999.
24 IPCC 2007d.
25 This correlation highlights carbon cycle feedbacks, with the biosphere losing carbon to the atmosphere in response to higher temperatures, which in turn drives temperatures upwards.
26 Lockwood and Fröhlich 2007.
27 IPCC 2007d.
28 The total radiative forcing effect of greenhouse gases is measured in terms of the equivalent concentration (in parts per million, or ppm) of CO_2. There are six greenhouse gases recognized under the Kyoto Protocol. These are carbon dioxide, methane, nitrous dioxide, perfluorocarbons (PFCs), hydrofluorocarbons (HFCs) and sulphur hexafluoride (SF_6).
29 Anthropogenic contributions to aerosols (mainly sulphate, organic carbon, black carbon, nitrates and dust) produce a cooling effect by blocking solar radiation.

30 The radiative forcing value for non-CO_2 long-lived greenhouse gases is 0.98 (Wm^{-2}) and the cooling effect of aerosols is 1.2 (Wm^{-2}) (IPCC 2007d).
31 ppm stands for parts per million and in this instance is the number of greenhouse gas molecules per million molecules of dry air.
32 IPCC 2007d.
33 Henderson 2006a.
34 Caldeira 2007; Caldeira, Jain and Hoffert 2003; Henderson 2006a.
35 IPCC 2007f.
36 Flannery 2005.
37 Stern 2006.
38 Preindustrial temperature refers to the average temperature for the period 1861-1890.
39 IPCC 2007a, Chapter 10: Global Climate Projections.
40 Meinshausen 2005.
41 Meinshausen 2005.
42 Personal correspondence with Dr Malte Meinshausen, Potsdam Institute for Climate Impact Research.
43 Personal correspondence with Dr Malte Meinshausen. The reference year period for the temperature increase is 1980 to 1999.
44 Schlesinger et al. 2005.
45 IPCC 2007d.
46 Hansen et al. 2007; Pritchard and Vaughn 2007.
47 Hansen 2007a, 2007b.
48 Schellnhuber and Lane 2006; Schellnhuber 2006.
49 Jones, Cox and Huntingford 2005.
50 CNA Corporation 2007.
51 Gullison et al. 2007.
52 IPCC 2007e.
53 WRI 2007a.
54 IEA 2006c.
55 Volpi 2007.
56 Volpi 2007.
57 PEACE 2007.
58 Modi et al. 2005.
59 IEA 2006c.
60 IEA 2006c.
61 The equivalent figure for a carbon equivalent budget covering all Kyoto greenhouse gases is around 600 Gt CO_2e, or 6Gt CO_2e annually. This translates into around 22 Gt CO_2e. Current emissions are

around double this level. In 2004, total greenhouse gas emissions were estimated by the IPCC at around 49 Gt CO_2e annually (IPCC 2007c).
62 Stern 2006.
63 Barker and Jenkins 2007.
64 For example, the Stern Review examined a stabilization scenario set at 550 ppm. Research carried out for this year's HDR extrapolates from these models to derive the cost implications of keeping within a 2°C threshold, or around 450 CO_2e.
65 HDRO calculations based on the annual cost expressed as percentage of GDP in Barker and Jenkins 2007. The calculation is the average yearly cost in the period 2000–2030 weighted by the size of the global economy over that period. Barker and Jenkins 2007 also present other scenarios with lower mitigation costs.
66 Stern (2006), on which these figures are based, discusses a wide range of estimates.
67 Barker and Jenkins (2007) project the cost of stabilization at 450ppm CO_2e at 2–3% of GDP, falling to 1–2% with permit trading. If the policy framework also allows for the revenues from auctioning permits and carbon taxes to be recycled, these would entail tax reform. National and global economies could benefit by as much as 5% of GDP above the 2030 baseline.
68 The Kyoto Protocol was negotiated in Japan in 1997 within the framework of the UNFCCC. Under the terms of the Protocol, Annex I parties representing 55 percent of 1990 emissions were required to accept binding limits on emissions. Ratification by the Russian Federation in 2004 provided the critical mass to meet this condition.
69 Calculation based on data from IEA 2006c.
70 Annex I parties include the industrialized countries that were members of the OECD in 1992, plus countries with economies in transition (the EIT Parties), including the Baltic States, several Central and Eastern European States and the Russian Federation. Non-Annex I parties are mostly developing countries.
71 Roberts 2005.

72 Council on Foreign Relations 2006.

73 IEA 2006c.

74 Hansen 2007c.

75 UNDP 2006b; UNDP Ukraine 2006.

76 IEA 2006c.

77 IPCC 2007f.

78 Stern 2006; Nordhaus 2007.

79 IMF 2006.

80 Smith 1854.

81 World Commission on Environment and Development 1987.

82 Anand and Sen 1996.

83 Sen 2004.

84 Appiah 2006.

85 Nordhaus 2007.

86 Nordhaus 2006.

87 The discount rate that emerges from a very simple standard economic model considering only one infinitely lived representative agent and other simplifying assumptions can be expressed by: $\rho = \delta + \eta g$, where δ is the social rate of time preference, g is the projected growth rate of consumption per capita, and η is the elasticity of the social weight—or marginal utility— attributed to a change in consumption. It is standard assumption that utility will decrease when consumption increases, making always positive. In this simplifying framework, it is also considered constant.

88 In fact, the only justifiable reason for discounting the welfare of future generations, according to Stern (2006), was the possibility of extinction. He therefore allows for a very small rate of pure time preference of 0.1 percent.

89 Arrow 2007.

90 Ramsey 1928.

91 Stern and Taylor 2007.

92 However, the case does not rest on economics alone. Arrow (2007) has shown that if the costs and benefits of mitigation suggested in the Stern Review are accepted, then the case for early action now is only rejected with a rate of pure time preference above 8.5 percent—a value that not even the strongest critics of Stern would advocate.

93 Wolf 2006b; Weitzman 2007.

94 Schelling 2007.

95 Dasgupta 2001.

96 HSBC 2007.

97 Pew Center on Global Climatic Change 2006.

98 Pew Center on Global Climatic Change 2006.

99 Leiserowitz 2007.

100 Leiserowitz 2006.

101 Leiserowitz 2006.

102 European Commission, Directorate General for Energy and Transport 2006.

103 HSBC 2007; The Economist 2007a.

104 Bernstein 1998.

105 Boykoff and Roberts 2007.

106 Boykoff and Roberts 2007; Boykoff and Boykoff 2004.

Chapter 2

1 de Montesquiou 2005.

2 Itano 2002.

3 Personal interview with Kaseyitu Agumas, 22 March 2007, Ethiopia.

4 Raworth 2007b.

5 Personal interview with Instar Husain, 2 February 2007, Bangladesh.

6 Sen 1999.

7 OFDA and CRED 2007.

8 ABI 2005a.

9 WMO 2006.

10 OFDA and CRED 2007.

11 Reliefweb 2007; BBC News 2007.

12 IFRC 2006.

13 OFDA and CRED 2007.

14 Skutsch et al. 2004.

15 IPCC 2007e.

16 Dercon 2005; Chambers 2006.

17 Calvo and Dercon 2005.

18 Our framework for looking at risk and vulnerability differs in emphasis from the conceptual framework used by the disaster-risk community. The standard approach is based on the following formulae: risk=hazard exposure*vulnerability (with hazard exposure a function of degree of hazard*elements exposed) (Maskrey et al. 2007).

19 ADB 2001.

20 GSS, NMIMR and ORC Macro 2004; CBS, MOH and ORC Macro 2004.

21 Roberts and Parks 2007.

22 USAID FEWS NET 2006.

23 OFDA and CRED 2007.

24 WEDO 2007.

25 Watt-Cloutier 2006.

26 Chafe 2007.

27 Rosenzweig and Binswanger 1993.

28 Dercon 1996.

29 Elbers and Gunning 2003.

30 OECD 2006b.

31 GAO 2007.

32 Devereux 2002.

33 Dercon, Hoddinott and Woldehanna 2005.

34 Dercon 2005.

35 Carter et al. 2007.

36 WFP 2005; IFRC 2005b.

37 Behrman and Deolalikar 1990; Dercon and Krishnan 2000; Rose 1999.

38 Baez and Santos 2007; de Janvry et al. 2006a.

39 de la Fuente and Fuentes 2007.

40 Devereux 2006b.

41 Hoddinott and Kinsley 2001.

42 Banerjee Bénabou and Mookherjee 2006.

43 Carter and Barrett 2006.

44 IPCC 2007d, 2007e.

45 The IPCC uses two-sided confidence intervals of 90 percent.

46 IPCC 2007e.

47 Warren et al. 2006.

48 World Bank 2006b.

49 World Bank 2003.

50 World Bank 2003.

51 Delgado et al. 1998.

52 Cline 2007.

53 Fischer et al. 2005; Agoumi 2003 cited in IPCC 2007b, Chapter 9: Africa.

54 Kurukulasuriya and Mendelsohn 2006.

55 UNEP and GRID – Arendal 2001.

56 Carvajal 2007.

57 UNEP 2007a.

58 Vaid et al. 2006.

59 World Bank 2006f.

60 Stern 2006.

61 Government of India 2007.

62 Government of the People's Republic of Bangladesh 2005b.

63 Kelkar and Bhadwal 2007.

64 PEACE 2007.

65 Jones and Thornton 2003.

66 IPCC 2001.

67 FAO 2004.

68 Water stress is defined as a situation where the per capita availability of renewable fresh water is between 1,000 cubic metres and 1,667 cubic metres. Water scarcity refers to a situation of living with an annual per capita availability of renewable fresh water of 1,000 cubic metres or less.

69 Bou-Zeid and El-Fadel 2002.

70 IPCC 2007b, Chapter 9: Africa.

71 Bou-Zeid and El-Fadel 2002.

72 UNEP 2007b.

73 Carvajal 2007.

74 Khoday 2007.

75 UNEP 2007b.

76 Regmi and Adhikari 2007.

77 Khoday 2007.

78 UNDP 2006b; Rosegrant, Cai and Cline 2002.

79 Vergara et al. 2007.

80 Maskrey et al. 2007.

81 Emanuel 2005.

82 Pierce et al. 2005.

83 Maskrey et al. 2007.

84 Arnell 2004.

85 Anthoff et al. 2006; Dasgupta et al. 2007.

86 Hemming 2007.

87 Hemming 2007; Brown 2007.

88 Brown 2007.

89 Agrawala et al. 2003.

90 World Bank 2006c.

91 IPCC 2007b, Chapter 16: Small Islands; Dasgupta et al. 2007.

92 UN-HABITAT 2006.

93 Millennium Ecosystem Assessment 2005.

94 World Watch Institute 2005.

95 Finlayson and Spiers 2000.

96 Hansen 2006.

97 IPCC 2007b, Chapter 4: Ecosystems, their Properties, Goods, and Services.

98 ACIA 2004.

99 Government of the United States 2006b.

100 The United Nations Convention on the Law of the Sea entered into force in 1994. It is a set of rules for the use of the world's

oceans, which cover 70 percent of the Earth's surface.

101 ACIA 2004; Perelet, Pegov and Yulkin 2007.
102 Hare 2005; Henderson 2007.
103 Henderson 2006b.
104 PEACE 2007.
105 Gardner et al. 2003.
106 Caldeira 2007.
107 Caldeira 2007.
108 Caldeira 2007.
109 Carvajal 2007.
110 McMichael et al. 2003.
111 WHO and UNICEF 2005; WHO 2006.
112 Tanser, Sharp and le Seur 2003.
113 van Lieshout et al. 2004.
114 Chretien et al. 2007.
115 Stern 2006.
116 PEACE 2007.
117 WMO 2006.
118 Epstein and Mills 2005.
119 Epstein and Rogers 2004.
120 New York Climate & Health Project 2004.
121 New York Climate & Health Project 2004.

Chapter 3

1 Government of the United Kingdom 2007a.
2 Government of France 2006.
3 Government of France 2006.
4 Government of Germany 2007.
5 G8 2007.
6 Hanemann and Farrell 2006.
7 These states include: Arizona, California, Connecticut, Florida, Hawaii, Illinois, Maine, Massachusetts, Minnesota, New Hampshire, New Jersey, New Mexico, New York, Oregon, Rhode Island, Vermont and Washington (Pew Center on Global Climate Change 2007c).
8 The Governors of Connecticut, Delaware, Maine, New Hampshire, New Jersey, New York, and Vermont established the RGGI in 2005. Maryland, Massachusetts and Rhode Island joined in 2007 (Pew Center on Global Climate Change 2007c).
9 Arroyo and Linguiti 2007.
10 Claussen 2007.
11 Brammer et al. 2006.
12 Pew Center on Global Climate Change 2007a.
13 USCAP 2007.
14 Arroyo and Linguiti 2007.
15 Arroyo and Linguiti 2007.
16 UNFCCC 2006.
17 EIA 2006; Arroyo and Linguiti 2007.
18 IPCC 2007c, Chapter 5: Transport and its infrastructure.
19 Cairns and Newson 2006.
20 Doniger, Herzog and Lashof 2006.
21 Sullivan 2007.
22 UNFCCC 2006.
23 Government of Australia 2007.
24 Henderson 2007.
25 Government of New South Wales 2007.
26 Acuiti Legal 2003.

27 Pederson 2007; Nippon Keidanren 2005.
28 Examples in this paragraph are taken from Pew Center on Global Climate Change 2007b.
29 Roosevelt 2006.
30 On the case for carbon taxation and the critique of cap-and-trade see Cooper 2000, 2005; Nordhaus 2005; Shapiro 2007.
31 Hanson and Hendricks 2006.
32 Nordhaus 2006.
33 HDR calculation based on data from Indicator Table 24; OECD emissions of CO_2 in 2004 were 13.3 Gt.
34 Stern 2006.
35 Shapiro 2007.
36 Shapiro 2007; EPA 2006.
37 IPCC 2007f. The global mitigation potential relative to the IPCC's SRES A1B non-mitigation scenario is estimated at 17–26 Gt CO_2e/yr with a carbon price of US$1/t CO_2e, or 25–38 percent.
38 Toder 2007.
39 Sierra Club 2006.
40 EEA 2004.
41 International Network for Sustainable Energy 2006.
42 Cairns and Newson 2006.
43 During Phase II the scheme will cover 27 countries.
44 There are three flexibility mechanisms introduced by the Kyoto Protocol: Emissions Trading, the Clean Development Mechanism (CDM) and Joint Implementation (JI). Unlike the CDM that links mitigation efforts in developing and developed countries (Annex I and non-Annex I parties), through the Joint Implementation, Annex I parties may fund emission reducing projects in other Annex I parties, typically countries in Eastern Europe.
45 UNFCCC 2007e.
46 Point Carbon 2007.
47 Carbon Trust 2006.
48 Grubb and Neuhoff 2006.
49 Carbon Trust 2006.
50 Government of the United Kingdom 2006b.
51 Sijm, Neuhoff and Chen 2006.
52 EU 2007c.
53 Hoffmann 2006.
54 Hoffmann 2006.
55 WWF 2007a, 2007b.
56 Reece et al. 2006; WWF 2006b, 2007a.
57 WWF 2007a, 2007b.
58 IEA 2006c.
59 IEA 2006c.
60 Government of the United States 2007a.
61 IEA 2006c.
62 NEA 2006.
63 The Economist 2007b.
64 Greenpeace and GWEC 2006.
65 NCEP 2004a.
66 Philibert 2006.
67 Arroyo and Linguiti 2007.

68 Greenpeace and GWEC 2006.
69 NCEP 2004a.
70 NCEP 2004a.
71 Ürge-Vorsatz et al. 2007a; IEA 2006b.
72 Ürge-Vorsatz, Mirasgedis and Koeppel 2007b.
73 Ürge-Vorsatz, Mirasgedis and Koeppel 2007b; EC 2005a.
74 IEA 2003.
75 IEA 2003; World Bank 2007d.
76 IEA 2003, page 128.
77 IEA 2006b.
78 Ürge-Vorsatz, Mirasgedis and Koeppel 2007b.
79 IEA 2003, 2006a.
80 Ürge-Vorsatz, Mirasgedis and Koeppel 2007b.
81 EC 2006a.
82 IPCC 2007c, Chapter 5: Transport and its infrastructure.
83 Merrill Lynch and WRI 2005.
84 Merrill Lynch and WRI 2005; NCEP 2004a.
85 Arroyo and Linguiti 2007.
86 NCEP 2004a.
87 EFTA 2007.
88 CEC 2007c.
89 CEC 2007c.
90 EFTA 2007.
91 Baumert, Herzog and Pershing 2005.
92 Government of the United States 2007c.
93 EC 2007a.
94 Steenblik 2007.
95 Runnalls 2007.
96 Runge and Senauer 2007.
97 Runge and Senauer 2007.
98 EC 2007a.
99 Summa 2007.
100 The payment is currently set at €45 per hectare with a minimum guaranteed area of 1.5 million hectares (CEC 2005b).
101 IEA 2006c; IPCC 2007c, Chapter 5: Transport and its infrastructure.
102 CEC 2006c; Jank et al. 2007.
103 Elobeid and Tokgoz 2006.
104 Tolgfors, Erlandsson and Carlgren 2007.
105 Schnepf 2006.
106 NASA 2005; Smithsonian National Air and Space Museum 1999.
107 Stern 2006.
108 Thermal efficiency describes the rate at which fuel is transformed into energy.
109 Watson et al. 2007.
110 IEA 2006b.
111 Government of the United States 2007b.
112 Government of the United States 2007b.
113 Watson et al. 2007; Rubin 2007.
114 Government of the United States 2005.
115 Government of the United States 2006a.
116 European Technology Platform on Zero Emission Fossil Fuel Power Plants (ZEP) 2007.
117 EC 2005b.
118 European Technology Platform on Zero Emission Fossil Fuel Power Plants (ZEP) 2007.

119 Government of the United Kingdom 2006c.
120 Rubin 2007a; Claussen 2007.
121 CEC 2007d.
122 Government of the United States 2007a.
123 Watson 2007.
124 OECD 2005c.
125 Watson 2007.
126 OECD 2005c.
127 Watson et al. 2007.
128 Data in this section is derived from tables in Annex A of IEA 2006c.
129 Winkler and Marquard 2007.
130 Watson et al. 2007.
131 Davidson et al. 2006.
132 Government of India 2006a, 2006b.
133 Watson et al. 2007.
134 Winkler and Marquand 2007.
135 IEA 2006c.
136 IEA 2006c.
137 Watson 2007.
138 Watson 2007.
139 Victor 2001.
140 UNFCCC 2007c.
141 World Bank 2007f.
142 World Bank 2007b.
143 FAO 2007b.
144 This value refers to the change in carbon stocks above and below ground biomass. To convert the reported values from carbon to carbon dioxide, a conversion factor of 3.664 has been applied (FAO 2007b).
145 PEACE 2007.
146 There are wide variations in estimates of CO_2 emissions linked to changes in forest areas. FAO Forest Resources Assessment data on carbon stocks in forests 1990–2005 suggests that approximately 1.1 Gt CO_2 are released a year from Brazilian forests—only from living biomass (above and below ground) (FAO 2007b).
147 Butler 2006.
148 The values used in this example are drawn from Chomitz et al. 2007.
149 Chomitz et al. 2007.
150 Pearce 2001.
151 Volpi 2007.
152 Chomitz et al. 2007.
153 Tauli-Corpuz and Tamang 2007.
154 INPE 2007.

Chapter 4

1 ABI 2007b.
2 IPCC 2007b, Chapter 12: Europe.
3 Linklater 2007.
4 CEC 2007b.
5 Huisman 2002; EEA 2007.
6 UKCIP 2007; The Economist 2007c; ABI 2007b.
7 Hulme and Sheard 1999b; British Oceanographic Data Centre 2007; Government of Japan 2002; EEA 2007.
8 EEA 2007; WWF 2002; Schröter, Zebisch and Grothmann 2005.
9 California Public Utilities Commission 2005; California Department of Water Resources 2006; Franco 2005; Government of California 2006; Cayan et al. 2005.
10 National Audit Office 2001.
11 CEC 2007b.
12 Branosky 2006; EPA 2006.
13 NFU 2005.
14 Practical Action 2006a.
15 Rahman et al. 2007; Raworth 2007b.
16 Chaudhry and Rysschaert 2007.
17 Cornejo 2007.
18 WRI, UNEP, and World Bank 2005; Narain 2006.
19 Practical Action 2006b.
20 Birch and Grahn 2007.
21 Glemarec 2007a.
22 Glemarec 2007b.
23 Washington et al. 2006.
24 Washington et al. 2006; Intsiful et al. 2007.
25 Meteo France 2007.
26 Regional Hunger and Vulnerability Programme 2007.
27 Intsiful et al. 2007.
28 IRI 2007.
29 G8 2005.
30 Intsiful et al. 2007.
31 Global Climate Observing System, UN Economic Commission for Africa and African Union Commission 2006.
32 OECD 2007a.
33 Nguyen 2007.
34 Jha 2007.
35 DFID 2006.
36 IFRC 2002.
37 Tanner et al. 2007.
38 The conversion factor is around 15 percent.
39 Ethiopia, for example, has a harvesting potential of 11,800m³ per capita compared to 1,600m³ per capita of renewable river and groundwater. Similarly for Kenya: 12,300m³ compared to 600m³ per capita, and Tanzania: 24,700m³ compared to 2,200m³ per capita for water harvesting potential, and renewable river and groundwater potential, respectively (UNEP 2005).
40 Narain 2006.
41 Devereux 2006b.
42 Grinspun 2005.
43 de Janvry and Sadoulet 2004.
44 de Janvry et al. 2006b; Barrientos and Holmes 2006.
45 Schubert 2005; Barrientos and Holmes 2006; Randel 2007. Calculations based on data in Indicator Tables 14 and 18.
46 ISDR 2007a.
47 de la Fuente 2007a.
48 ISDR 2007b.
49 IFRCa 2005; Catholic Relief Services 2004; Carvajal 2007; OFDA and CRED 2007.
50 Thompson and Gaviria 2004; IFRC 2005a. By comparison, there were 36 deaths in Florida.
51 IFRC 2006.
52 The UNFCCC deals with adaptation in several articles. Under Article 4.1(f): All Parties shall "take climate change considerations into account, to the extent feasible, in their relevant social, economic and environmental policies and actions, and employ appropriate methods, for example impact assessments, formulated and determined nationally, with a view to minimizing adverse effects on the economy, on public health and on the quality of the environment, of projects or measures undertaken by them to mitigate or adapt to climate change." Under Article 4.4: "The developed country Parties and other developed Parties included in Annex II shall also assist the developing country Parties that are particularly vulnerable to the adverse effects of climate change in meeting costs of adaptation to those adverse effects."
53 Heimann and Bernstein 2007.
54 Alaska Oil Spill Commission 1990.
55 Gurría and Manning 2007.
56 Gurría and Manning 2007.
57 As of April 30th 2007, see GEF 2007a. Corporate costs and administrative expenses and fees paid to the three implementing agencies—the World Bank, UNDP and UNEP—accounts for another US$2 million, or around 20 percent of total disbursements to date.
58 Proposals already in advanced stages, such as Bangladesh, Bhutan, Malawi, Mauritania and Niger are expected to receive an average of US$3–3.5 million each to start implementing the first priorities of their NAPA.
59 The fund also covers technology transfer.
60 GEF 2007a, 2007c.
61 GEF 2007b.
62 Müller and Hepburn 2006.
63 Abbott 2004.
64 Frankel-Reed 2006. The sample included projects where climate change risks and vulnerability were explicit considerations. Business-as-usual development activities (e.g. increased water provision, public health capacity) that may have reduced vulnerability to climate change but were not designed intentionally to support adaptation were not considered.
65 World Bank 2007g.
66 Agrawala 2005.
67 For a summary of the methodology and list of DAC sector and purpose codes used, see Agrawala 2005.
68 The World Bank estimates costs of adaptation as 5–20 percent of development investment sensitive to climate. For 2005 ODA, this amounts to between US$1.0 billion and US$8.1 billion, with US$4.5 billion as the mid-range value.

69 Gurría and Manning 2007.

70 OCHA Financial Tracking System [www.reliefweb.int/fts,] cited in Oxfam International 2005.

71 Müller and Hepburn 2006; Oxfam International 2007.

72 Oxfam International 2007.

73 This figure is based on the assumption that adaptation financing requirements in developing countries will represent around 0.1 percent of developed country GDP (the approximate level in 2005 based on World Bank methodology).

74 This figure would represent around 0.5 percent GDP for low income and lower-middle income countries.

75 SIPRI 2007.

76 Landau 2004.

77 Müller and Hepburn 2006.

78 Bouwer and Aerts 2006.

79 Oxfam International 2007.

80 OECD 2005b, 2006e.

81 Barrientos and Holmes 2006.

82 Government of the People's Republic of Bangladesh 2005a; UNDP 2005; Rahman et al. 2007; Mallick et al. 2005.

83 Government of the People's Republic of Bangladesh, mimeo.

84 ISDR 2007c.

85 ISDR and the World Bank GFDRR 2006, 2007.

86 Initial financial arrangements are given for 2006-2016 (ISDR and the World Bank GFDRR 2006). The mainstreaming track, projected to cost some US$350 million is to be met through a multi-donor trust fund, of which some US$42 million has been pledged as of August 2007. See: http://siteresources.worldbank.org/EXTDISMGMT/Resources/GfdrrDonorPledgesAugust7.pdf

Bibliography

Background Papers

Arroyo, Vicki and Peter Linguiti. 2007. "Current Directions in the Climate Change Debate in the United States."

Barker, Terry and Katie Jenkins. 2007. "The Costs of Avoiding Dangerous Climate Change: Estimates Derived from a Meta-Analysis of the Literature."

Boykoff, Maxwell T. and J. Timmons Roberts. 2007. "Media Coverage of Climate Change: Current Trends, Strengths, Weaknesses."

de la Fuente, Alejandro and Ricardo Fuentes. 2007. "The Impact of Natural Disasters on Children Morbidity in Rural Mexico."

Fuentes, Ricardo and Papa Seck. 2007. "The Short and Long-term Human Development Effects of Climate-Related Shocks: Some Empirical Evidence."

Helm, Dieter. 2007. "Climate Change: Sustainable Growth, Markets, and Institutions."

Henderson, Caspar. 2007. "Carbon Budget—the Agenda for Mitigation. Australia, Canada, the European Union and Japan."

IGAD (Intergovernmental Authority on Development) Climate Prediction and Applications Centre (ICPAC). 2007. "Climate Change and Human Development in Africa: Assessing the Risks and Vulnerability of Climate Change in Kenya, Malawi and Ethiopia."

O'Brien, Karen and Robin Leichenko. 2007. "Human Security, Vulnerability and Sustainable Adaptation."

Osbahr, Henny. 2007. "Building Resilience: Adaptation Mechanisms and Mainstreaming for the Poor."

Perelet, Renat. 2007. "Central Asia: Background Paper on Climate Change."

Perelet, Renat, Serguey Pegov and Mikhail Yulkin. 2007. "Climate Change. Russia Country Paper. Perelet, Renat, Serguey Pegov and Mikhail Yulkin. 2007. "Climate Change. Russia Country Paper."

Rahman, Atiq, Mozaharul Alam, Sarder Shafiqul Alam, Md. Rabi Uzzaman, Mariam Rashid and Golam Rabbani. 2007. "Risks, Vulnerability and Adaptation in Bangladesh."

Reid, Hannah and Saleemul Huq. 2007. "International and National Mechanisms and Politics of Adaptation: An Agenda for Reform."

Seck, Papa. 2007a. "Links between Natural Disasters, Humanitarian Assistance and Disaster Risk Reduction: A Critical Perspective."

Watson, Jim, Gordon MacKerron, David Ockwell and Tao Wang. 2007. "Technology and Carbon Mitigation in Developing Countries: Are Cleaner Coal Technologies a Viable Option?"

Thematic Papers

Brown, Oli. 2007. "Climate Change and Forced Migration: Observations, Projections and Implications."

Carvajal, Liliana. 2007. "Impacts of Climate Change on Human Development."

Conceição, P., Y. Zhang and R. Bandura. 2007. "Brief on Discounting in the Context of Climate Change Economics."

Conde, Cecilia, Sergio Saldaña and Víctor Magaña. 2007. "Thematic Regional Papers. Latin America."

de Buen, Odón. 2007. "Decarbonizing Growth in Mexico."

de la Fuente, Alejandro. 2007a. "Private and Public Responses to Climate Shocks."

———. **2007b.** "Climate Shocks and their Impact on Assets."

Dobie, Philip, Barry Shapiro, Patrick Webb and Mark Winslow. 2007. "How do Poor People Adapt to Weather Variability and Natural Disasters Today?"

Gaye, Amie. 2007. "Access to Energy and Human Development."

Intsiful, Joseph D, Richard Jones, Philip Beauvais and Vicky Pope. 2007. "Meteorological Capacity in Africa."

Kelkar, Ulka and Suruchi Bhadwal. 2007. "South Asian Regional Study on Climate Change Impacts and Adaptation: Implications for Human Development."

Khoday, Kishan. 2007. "Climate Change and the Right to Development. Himalayan Glacial Melting and the Future of Development on the Tibetan Plateau."

Krznaric, Roman. 2007. "For God's Sake, Do Something! How Religions Can Find Unexpected Unity Around Climate Change."

Kuonqui, Christopher. 2007. "Responding to Clear and Present Dangers: A New Manhattan Project for Climate Change?"

Leiserowitz, Anthony. 2007. "Public Perception, Opinion and Understanding of Climate Change—Current Patterns, Trends and Limitations."

Li, Junfeng. 2007. "Mitigation Country Study—China."

Mathur, Ritu and Preety Bhandari. 2007. "Living Within a Carbon Budget—the Agenda for Mitigation."

Matus Kramer, Arnoldo. 2007. "Adaptation to Climate Change in Poverty Reduction Strategies."

Menon, Roshni. 2007a. "Famine in Malawi: Causes and Consequences."

———. **2007b.** "Managing Disaster, Mitigating Vulnerability: Social Safety Nets in Ethiopia."

Newell, Peter. 2007. "The Kyoto Protocol and Beyond: The World After 2012."

Tolan, Sandy. 2007. "Coverage of Climate Change in Chinese Media."

Volpi, Giulio. 2007. "Climate Mitigation, Deforestation and Human Development in Brazil."

Winkler, Harald and Andrew Marquard. 2007. "Energy Development and Climate Change: Decarbonising Growth in South Africa."

Yue, Li, Lin Erda and Li Yan. 2007. "Impacts of, and Vulnerability and Adaptation to, Climate Change in Water Resources and Agricultural Sectors in China."

Issue Notes

Arredondo Brun, Juan Carlos. 2007. "Adapting to Impacts of Climate Change on Water Supply in Mexico City."

Bambaige, Albertina. 2007. "National Adaptation Strategies to Climate Change Impacts. A Case Study of Mozambique."

Bhadwal, Suruchi and Sreeja Nair. 2007. "India Case Study." Tata Energy Resources Institute (TERI), Mumbai.

Birch, Isobel and Richard Grahn. 2007. "Pastoralism—Managing Multiple Stressors and the Threat of Climate Variability and Change."

Chaudhry, Peter and Greet Ruysschaert. 2007. "Climate Change and Human Development in Viet Nam."

Canales Davila, Caridad and Alberto Carillo Pineda. 2007. "Spain Country Study."

Cornejo, Pilar. 2007. "Ecuador Case Study: Climate Change Impact on Fisheries."

Donner, Simon D. 2007. "Canada Country Study."

Lemos, Maria Carmen. 2007. "Drought, Governance and Adaptive Capacity in North East Brazil: a Case Study of Ceará."

Meinshausen, Malte. 2007. "Stylized Emission Path."

Nangoma, Everhart. 2007. "National Adaptation Strategy to Climate Change Impacts: A Case Study of Malawi".

Nguyen, Huu Ninh. 2007. "Flooding in Mekong River Delta, Viet Nam."

Orindi, Victor A., Anthony Nyong and Mario Herrero. 2007. "Pastoral Livelihood Adaptation to Drought and Institutional Interventions in Kenya."

Painter, James. 2007. "Deglaciation in the Andean Region."

Pederson, Peter D. 2007. "Japan—Country Study."

Regmi, Bimal R. and Adhikari, A. 2007. "Climate Change and Human Development—Risk and Vulnerability in a Warming World. Country Case Study Nepal."

Salem, Boshra. 2007. "Sustainable Management of the North African Marginal Drylands."

Schmid, Jürgen. 2007. "Mitigation Country Study for Germany."

Seck, Papa. 2007b. "The Rural Energy challenge in Senegal: A Mission Report."

Sullivan, Rory. 2007. "Australia Country Study."

Trigoso Rubio, Erika. 2007. "Climate Change Impacts and Adaptation in Peru: the Case of Puno and Piura."

References

ABI (Association of British Insurers). 2004. "A Changing Climate for Insurance. A Summary Report for Chief Executives and Policymakers." Association of British Insurers, London. [http://www.abi.org.uk/Display/File/Child/552/A_Changing_Climate_for_Insurance_2004.pdf]. July 2007.

————. 2005a. "Financial Risks of Climate Change." Summary Report. Association of British Insurers, London.

————. 2005b. "A Changing Climate for Risk Insurance." [http://www.abi.org.uk/Display/File/Child/552/A_Changing_Climate_for_Insurance_2004.pdf]. August 2007.

————. 2007a. "Adapting to Our Changing Climate: A Manifesto for Business, Government and the Public." Association of British Insurers, London.

————. 2007b. "Flooding and Insurance." Association of British Insurers. London. [http://www.abi.org.uk/Display/Display_Popup/default.asp?Menu_ID=1090&Menu_All=1,1088,1090&Child_ID=553]. July 2007.

Abbott, Alison. 2004. "Saving Venice." Nature. London. [http://www.nature.com/news/2004/040112/full/040112-8.html;jsessionid=26CC93DEBA2BEDF8762546E0413759D5]. January 2007.

ACIA (Arctic Climate Impacts Assessment). 2004. Impacts of a Warming Arctic—Arctic Climate Impacts Assessment. Cambridge University Press, Cambridge.

ActionAid. 2006. "Climate Change and Smallholder Farmers in Malawi. Understanding Poor People's Experiences in Climate Change Adaptation." ActionAid International, London and Johannesburg.

Acuiti Legal. 2003. "Overview of the NSW Greenhouse Gas Abatement Scheme." Research Paper No. 20. Independent Pricing and Regulatory Tribunal of New South Wales, Sydney.

Adan, Mohamud and Ruto Pkalya. 2005. "Closed to Progress: An Assessment of the Socio-Economic Impacts of Conflict on Pastoral and Semi Pastoral Economies in Kenya and Uganda." Practical Action–Eastern Africa, Nairobi.

ADB (Asian Development Bank). 2001. "Technical Assistance to the Republic of the Philippines for preparing the Metro-Manila Urban Services for the Poor Project." Manila.

Agoumi, A. 2003. "Vulnerability of North African Countries to Climatic Changes, Adaptation and Implementation Strategies for Climatic Change." International Institute for Sustainable Development (IISD), Winnipeg.

Agrawala, Shardul (ed). 2005. "Bridge Over Troubled Waters. Linking Climate Change and Development." OECD (Organisation for Economic Co-operation and Development), Paris.

Agrawala, Shardul, Tomoko Ota, Ahsan Uddin Ahmed, Joel Smith and Maarten van Aalst. 2003. "Development and Climate Change in Bangladesh: Focus on Coastal Flooding and the Sundarbans." OECD (Organisation for Economic Co-operation and Development), Paris.

Alaska Oil Spill Commission. 1990. "Spill, the Wreck of the Exxon Valdez: Implications for Safe Transportation of Oil." Final Report. Alaska Oil Spill Commission, Juneau, Alaska.

Anand, Sudhir and Amartya K. Sen. 1996. "Sustainable Human Development: Concepts and Priorities." Discussion Paper Series No.1. Office of Development Studies, United Nations Development Programme, New York.

Anderson, Kevin and Alice Bows. 2007. "A Response to the Draft Climate Change Bill's Carbon Reduction Targets." Tyndall Briefing Note 17. March 2007. Tyndall Centre for Climate Change Research, University of Manchester, Manchester.

Anthoff, David, Robert J. Nichols, Richard S.J. Tol and Athanasios T. Vafeidis. 2006. "Global and Regional Exposure to Large Rises in Sea-level: A Sensitivity Analysis." Working Paper No. 96. Tyndall Centre for Climate Change Research, University of East Anglia, Norwich.

Appiah, Kwame Anthony. 2006. Cosmopolitanism: Ethics in a World of Strangers. W.W. Norton, New York.

Arnell, N.W. 2004. "Climate Change and Global Water Resources: SRES Emissions and Socio-Economic Scenarios." Global Environmental Change 14: 31–52.

————. 2006. "Climate Change and Water Resources: A Global Perspective. Avoiding Dangerous Climate Change." Symposium on Stabilization of Greenhouse Gases, 1–3 February 2005, Met Office Hadley Centre for Climate Change, Exeter, UK. Department for Environment, Food and Rural Affairs, London.

Arrhenius, Svante. 1896. "On the Influence of Carbonic Acid in the Air upon the Temperature of the Ground." London, Edinburgh and Dublin Philosophical Magazine and Journal of Science. [Fifth series]. April 1896. 41: 237–275.

Arrow, Kenneth. 2007. "Global Climate Change: A Challenge to Policy." Economists' Voice 4(3), Article 2. [http://www.bepress.com/ev/vol4/iss3/art2]. September 2007.

Baez, Javier Eduardo and Indhira Vanessa Santos. 2007. "Children's Vulnerability to Weather Shocks: A Natural Disaster as a Natural Experiment." Social Science Research Network, New York.

BBC News. 2007. "Devastating Floods hit South Asia." 3 August 2007. [http://news.bbc.co.uk/2/hi/south_asia/6927389.stm]. August 2007.

Banerjee, Abhijit Vinayak, Roland Bénabou and Dilip Mookherjee, eds. 2006. Understanding Poverty. Oxford University Press, Oxford.

Barrientos, Armando and Rebecca Holmes. 2006. "Social Assistance in Developing Countries Database." Institute of Development Studies, University of Sussex, Brighton.

Baumert, Kevin, Timothy Herzog and Jonathan Pershing. 2005. Navigating the Numbers: Greenhouse Gas Data and International Climate Policy. World Resources Institute, Washington, DC.

Behrman, Jere R. and Anil Deolalikar. 1990. "The Intra-Household Demand for Nutrients in Rural South India: Individual Estimates, Fixed Effects and Permanent Income." Journal of Human Resources 24(4): 655–96.

Bernstein, Carl. 1998. "The Best Obtainable Version of the Truth." Speech to the Annual Convention of the Radio and Television News Directors Association, 26 September. San Antonio, Texas.

Bouwer, L.M. and J.C. Aerts. 2006. "Financing Climate Change Adaptation." *Disasters* 30(1): 49–63.

Bou-Zeid, E. and M. El-Fadel. 2002. "Climate Change and Water Resources in Lebanon and the Middle East." *Journal of Water Resources Planning and Management* 128(5): 343–355.

Boykoff, M. T. and J. M. Boykoff. 2004. "Bias as Balance: Global Warming and the U.S. Prestige Press." *Global Environmental Change* 14(2): 125–136.

Brammer, Marc, Dan Miner, Jeff Perlman, Richard Klein, Dick Koral and John Nettleton. 2006. "New York City Energy Policy for 2006 and Beyond." The American Lung Association, Bright Power Inc., Clean Air Cool Planet, The Long Island City Business Development Council, Natural Resources Defense Council, New York Climate Rescue, NYPIRG and Quixotic Systems, Inc., New York. [http://www.climaterescue.org/New%20York%20Energy%20Policy%20Proposal-2006%20Exec%20Sum.pdf]. August 2007.

Bramley, Matthew. 2005. "The Case for Deep Reductions: Canada's Role in Preventing Dangerous Climate Change." David Suzuki Foundation and the Pembina Institute. Vancouver. 24 November 2005. [http://www.pembina.org/climate-change/pubs/doc.php?id=536]. August 2007.

Branosky, Evan. 2006. "Agriculture and Climate Change: The Policy Context." World Resources Institute Policy Note, Climate: Agriculture No.1. World Resources Institute, Washington, DC.

Brieger. T., T. Fleck and D. Macdonald. 2001. "Political Action by the Canadian Insurance Industry on Climate Change." *Environmental Politics* 10: 111–126.

British Antarctic Survey. 2006. "Climate Change – Our View." [http://www.antarctica.ac.uk/bas_research/our_views/climate_change.php]. September 2007.

British Oceanographic Data Centre. 2007. "GLOSS Station Handbook: Station Information Sheet for Kuchiro." [http://www.bodc.ac.uk/data/information_and_inventories/gloss_handbook/stations/89/]. September 2007.

Broome, John. 2006a. "Should We Value Population." *The Journal of Political Philosophy* 13(4): 399–413.

————. **2006b.** "Valuing Policies in Response to Climate Change: Some Ethical Issues." A Contribution to the Work of the Stern Review on the Economics of Climate Change. Cambridge University Press, Cambridge.

Brundtland, Gro Harlem. 2007. "UN Special Envoy for Climate Change Gro Harlem Brundtland addresses the 15th Session of the UN Commission on Sustainable Development." Speech at the UN Commission on Sustainable Development. 9 May 2007. [http://www.regjeringen.no/en/dep/ud/selected-topics/un/Brundtland_speech_CSD.html?id=465906]. September 2007.

Burke, Tom. 2007. "Is Nuclear Inevitable? Policy and Politics in a Carbon Constrained World." The Professor David Hall Memorial Lecture, 17 May. The Law Society, London.

Butler, Rhett A. 2006. "A World Imperiled: Forces Behind Forest Loss." Mongabay.com / A Place Out of Time: Tropical Rainforests and the Perils They Face. [http://rainforests.mongabay.com/0801.htm]. January 2007.

Butler, Lucy and Karsten Neuhoff. 2005. "Comparison of Feed in Tariff, Quota and Auction Mechanisms to Support Wind Power Development." CMI Working Paper 70. Department of Applied Economics, University of Cambridge.

Cafiero, Carlo and Renos Vakis. 2006. "Risk and Vulnerability Considerations in Poverty Analysis: Recent Advances and Future Directions." Social Protection Discussion Paper No. 0610. World Bank, Washington, DC.

Cai, Ximing. 2006. "Water Stress, Water Transfer and Social Equity in Northern China: Implications for Policy Reforms." Issue note for the *Human Development Report 2006: Beyond Scarcity: Power, Poverty and the Global Water Crisis.* Palgrave Macmillan, New York.

Cairns, Sally and Carey Newson with Brenda Boardman and Jillian Anable. 2006. "Predict and Decide. Aviation, Climate Change and UK Policy." Final Report. Environmental Change Institute, University of Oxford.

Caldeira, Ken. 2007. "Climate Change and Acidification Are Affecting Our Oceans." Written testimony to *Wildlife and Oceans in a Changing Climate*, Subcommittee on Fisheries, Wildlife and Oceans, House Committee on Natural Resources, 17 April 2007, Washington, DC.

Caldeira Ken, A.K. Jain and M.I. Hoffert. 2003. "Climate Sensitivity Uncertainty and the Need for Energy without CO_2 Emission." Science 299 (5615): 2052–4.

Calvo, Cesar and Stefan Dercon. 2005. "Measuring Individual Vulnerability." Department of Economics Working Paper Series No. 229. University of Oxford.

California Department of Water Resources. 2006. "Progress on Incorporating Climate Change into Planning and Management of California's Water Resources." Technical Memorandum Report. San Francisco, July 2006.

California Public Utilities Commission. 2005. "Water Action Plan." San Francisco, 15 December 2005. [http://www.cpuc.ca.gov/Static/hottopics/3water/water_action_plan_final_12_27_05.pdf]. September 2007.

Carbon Trust. 2006. "Allocation and Competitiveness in the EU Emissions Trading Scheme. Options for Phase II and Beyond." Carbon Trust, London.

Carter, Michael and Christopher Barrett. 2006. "The Economics of Poverty Traps and Persistent Poverty: An Asset-Based Approach," *The Journal of Development Studies.* 42(2): 178–199.

Carter, Michael, R., Peter D. Little, Tewodaj Mogues and Workneh Negatu. 2005. "Shocks, Sensitivity and Resilience: Tracking the Economic Impacts of Environmental Disaster on Assets in Ethiopia and Honduras." Staff Paper No. 489. Department of Agricultural and Applied Economics, University of Wisconsin–Madison.

Carter, Michael, Peter Little, Tewodaj Mogues and Workneh Negatu. 2007. "Poverty Traps and Natural Disasters in Ethiopia and Honduras." *World Development* 35(5): 835–856.

CASS (Chinese Academy of Social Sciences). 2006. "Understanding China's Energy Policy: Economic Growth and Energy Use, Fuel Diversity, Energy/Carbon Intensity, and International Cooperation." Background Paper Prepared for Stern Review on the Economics of Climate Change. Research Centre for Sustainable Development, Beijing.

Catholic Relief Services. 2004. "CRS Allocates $200,000 for Relief Efforts in Haiti and the Dominican Republic." InterAction. 28 May. [http://www.interaction.org/newswire/detail.php?id=2938]. September 2007.

Cayan, Dan, Ed Maurer, Mike Dettinger, Mary Tyree, Katharine Hayhoe, Celine Bonfils, Phil Duffy and Ben Santer. 2005. "Climate Scenarios for California." Draft White Paper. California Climate Change Centre, Sacramento.

CBS (Central Bureau of Statistics, Kenya) MOH (Ministry of Health, Kenya) and ORC Macro. 2004. "Kenya Demographic and Health Survey 2003." Calverton, Maryland.

CDIAC (Carbon Dioxide Information Analysis Center). 2007. Correspondence on carbon dioxide emissions. US Department of Energy, Oak Ridge National Laboratory, Tennessee.

CEC (Commission of the European Communities). 2005a. "Winning the Battle Against Global Climate Change." Communication from the Commission to the Council, the European Parliament, the European Economic and Social Committee and the Committee of the Regions. COM. 2005. 35 final. Brussels.

————. **2005b.** "Biomass Action Plan." Communication from the Commission to the Council, the European Parliament, the European Economic and Social Committee and the Committee of the Regions, COM. 2005. 628 Final. Brussels.

———. **2006a.** "Building a Global Carbon Market—Report Pursuant to Article 30 of Directive 2003/87/EC." Communication from the Commission to the Council, the European Parliament, the European Economic and Social Committee and the Committee of the Regions, COM. 2006. 676 Final. Brussels.

———. **2006b.** *Green Paper: A European Strategy for Sustainable, Competitive and Secure Energy.* COM. 2006. 105 Final. Brussels.

———. **2006c.** "An EU Strategy for Biofuels." Communication from the Commission to the Council, the European Parliament, the European Economic and Social Committee and the Committee of the Regions. COM. 2006. 34 Final. Brussels.

———. **2007a.** "Renewable Energy Road Map. Renewable Energies in the 21st Century: Building a More Sustainable Future." COM. 2006. 848 Final. Brussels.

———. **2007b.** "Green Paper from the Commission to the Council, the European Parliament, the European Economic and Social Committee and the Committee of the Region. Adapting to Climate Change in Europe—Options for EU Action." COM. 2007. 354 Final. Brussels.

———. **2007c.** "On the Review of the Community Strategy to Reduce CO$_2$ Emissions and Improve Fuel Efficiency from Passenger Cars." Communication from the Commission to the European Parliament and Council, SEC 200760. Brussels.

———. **2007d.** "Sustainable power generation from fossil fuels: aiming for near zero emissions from coal after 2020." Communication from the Commission to the Council and the European Parliament. COM.2006. 843 Final. Brussels.

CEI (Committee of European Insurers). 2005. "Climate Change. Insurers Present Risk Management Recommendations for a Safer, Unpolluted World." Press Release. 9 November. Brussels. [http://www.cea.assur.org/cea/v1.1/actu/pdf/uk/communique239. pdf]. July 2007.

Chafe, Zoë. 2007. "Reducing Natural Disaster Risk in Cities." In Linda Stark, (ed.) *State of the World 2007: Our Urban Future.* 24th edition. A Worldwatch Institute Report on Progress Toward a Sustainable Society. Earthscan, London.

Chambers, Robert. 2006. "Editorial Introduction: Vulnerability, Coping and Policy." IDS Bulletin 37(4): 33–40.

Chen, Dorothée and Nicolas Meisel. 2006. "The Integration of Food Aid Programmes in Niger's Development Policies: the 2004–2005 food crisis." Working Paper 26. Agence Française de Développement, Paris.

Chhibber, Ajay and Rachid Laajaj. 2006. "Disasters, Climate Change, and Economic Development in sub-Saharan Africa: Lessons and Directions." Independent Evaluation Group, World Bank, Washington, DC.

Chomitz, Kenneth M. with Piet Buys, Giacomo de Luca, Timothy S. Thoas and Sheila Wertz-Kanounnikoff. 2007. *At Loggerheads? Agricultural Expansion, Poverty Reduction, and Environment in the Tropical Forests.* A World Bank Policy Research Report. World Bank, Washington, DC.

Chretien, Jean-Paul, Assaf Anyamba, Sheryl A. Bedno, Robert F. Breiman, Rosemary Sang, Kibet Sergon, Ann M. Powers, Clayton O. Onyango, Jennifer Small, Compton J. Tucker and Kenneth J. Linthicum. 2007. "Drought-Associated Chikungunya Emergence Along Coastal East Africa." *American Journal of Tropical Medicine and Hygiene* 76(3): 405–407.

Claussen, Eileen. 2007a. "Speech by Eileen Claussen, President, Pew Center on Global Climate Change." American College and University Presidents Climate Commitment Summit. 12 June 2007. Washington, DC. [http://www.pewclimate.org/press_room/ speech_transcripts/ec_acupcc]. August 2007.

———. **2007b.** "Can Technology Transform the Climate Debate?" Remarks by Eileen Claussen, President, Pew Center on Global Climate Change at the Exxonmobil Longer Range Research Meeting, 16 May 2007. Paulboro, New Jersey.

Climate Institute, The. 2006. "Common Belief. Australia's Faith Communities on Climate Change." The Climate Institute (Australia), Sydney.

Cline, William. 2007. *Global Warming and Agriculture: Impact Estimates by Country.* Center for Global Development, Peterson Institute for International Economics, Washington, DC.

CNA (Center for Naval Analyses) Corporation. 2007. *National Security and the Threat of Climate Change.* Center for Naval Analyses, Alexandria, Virginia. [http://securityandclimate.cna. org/report/National%20Security%20and%20the%20Threat%20of %20Climate%20Change.pdf]. August 2007.

Coal Industry Advisory Board, International Energy Agency. 2006. *Case Studies in Sustainable Development in the Coal Industry.* OECD/IEA, Paris.

Colchester, Marcus, Norman Jiwan, Andiko, Martua Sirait, Asep Yunan Firdaus, A. Surambo and Herbert Pane. 2006a. "Promised Land: Palm Oil and Land Acquisition in Indonesia." Forest Peoples Programme, Perkumpulan Sawit Watch, HuMA and the World Agroforestry Centre, Moreton-in-the-Marsh and West Java.

Colchester, Marcus with Nalua Silva Monterrey, Ramon Tomedes, Henry Zaalman, Georgette Kumanajare, Louis Biswana, Grace Watalmaleo, Michel Barend, Sylvia Oeloekanamoe, Steven Majarawai, Harold Galgren, Ellen-Rose Kambel, Caroline de Jong, Belmond Tchoumba, John Nelson, George Thierry Handja, Stephen Nounah, Emmanuel Minsolo, Beryl David, Percival Isaacs, Angelbert Johnny, Larry Johnson, Maxi Pugsley, Claudine Ramacindo, Gavin Winter and Yolanda Winter, Peter Poole, Tom Griffiths, Fergus MacKay and Maurizio Farhan Ferrari. 2006b. "Forest Peoples, Customary Use and State Forests: The Case for Reform." Draft paper to be presented to the 11th Biennial Congress of the International Association for the Study of Common Property, Bali, Indonesia, 19–23 June 2006. Forest Peoples Programme, Moreton-in-the-Marsh.

Commission for Africa. 2005. "Our Common Interest: Report of the Commission for Africa." London. [http://www. commissionforafrica.org/english/report/introduction. html#report]. September 2007.

CONAM (Consejo Nacional del Ambiente). 2004. "Estado del Ambiente de Cusco y el Cambio Climático a Nivel Nacional." Reporte Ambiental No. 4. [http://www.conam.gob.pe/Modulos/home/ reportes.asp]. September 2007.

Cooper, Richard N. 2000. "International Approaches to Global Climate Change." *The World Bank Research Observer* 15: 2 (August): 145–72.

———. **2005.** "Alternative to Kyoto: the Case for a Carbon Tax." [http://www.economics.harvard.edu/faculty/cooper/papers.html]. July 2007.

Coudrain, Anne, Bernard Francou and Zbifniew Kundzewicz. 2005. "Glacial shrinkage in the Andes and consequences for water resources – Editorial" *Hydrological Sciences–Journal des Sciences Hydrologiques* 50(6) December: 925–932.

Council of the European Union. 2007. "Presidency Conclusions 8/9 March 2007." 7224/1/07 REV 1. 2 May. Brussels.

Council on Foreign Relations. 2006. "National Security Consequences of US Oil Dependency." Independent Task Force Report No. 58. Council on Foreign Relations, New York.

Dasgupta, Partha. 2001. *Human Well-Being and the Natural Environment.* Oxford University Press, Oxford.

Dasgupta, Nandini with Mitra Associates. 2005 "Chars Baseline Survey 2005: Volume I. Household." Chars Livelihoods Programme. [http://www.livelihoods.org/lessons/project_summaries/comdev7_ projsum.html]. May 2007.

Dasgupta, Susmita, Benoit Laplante, Craig Meisner, David Wheeler and Jinping Yan. 2007. "The Impact of Sea Level Rise on Developing Countries: A Comparative Analysis." Policy Research Working Paper 4136. World Bank, Washington, DC.

Davidson, Ogunlade, Harald Winkler, Andrew Kenny, Gisela Prasad, Jabavu Nkomo, Debbie Sparks, Mark Howells and Thomas Alfstad with Stanford Mwakasonda, Bill Cowan and Eugene Visagie. 2006. *Energy Policies for Sustainable Development in South Africa: Options for the Future.* (Harald Winkler, ed.). Energy Research Centre, University of Cape Town.

de Janvry, Alain and Elisabeth Sadoulet. 2004. "Conditional Cash Transfer Programs: Are They Really Magic Bullets?" Department of Agricultural and Resource Economics, University of California, Berkeley.

de Janvry, Alain, Elisabeth Sadoulet, Pantelis Solomon and Renos Vakis. 2006a. "Uninsured Risk and Asset Protection: Can Conditional Transfer Programs Serve as Safety Nets?" Social Protection Discussion Paper No. 0604. World Bank, Washington, DC.

————. 2006b. "Can Conditional Cash Transfer Programs Serve as Safety Nets in Keeping Children at School and from Working when Exposed to Shocks?" *Journal of Development Economics* 79: 349–373.

————. 2006c. "Evaluating Brazil's Bolsa Escola Program: Impact on Schooling and Municipal Roles." University of California, Berkeley.

de Montesquiou, Alfred. 2005. "Haitian Town Struggles to Recover One Year after Devastating Floods." The Associated Press. 19 September.

Deaton, Angus. 2001. "Health, inequality and economic development." Based on a paper prepared for the Working Group 1 of the WHO Commission on Macroeconomics and Health. Princeton University.

DEFRA (Department for Environment, Food and Rural Affairs). 2007. "New Bill and Strategy Lay Foundations for Tackling Climate Change – Miliband." News Release. 13 March. London. [http://www.defra.gov.uk/news/2007/070313a.htm]. July 2007.

Delgado, Christopher L., Jane Hopkins, and Valerie A. Kelly with Peter Hazell, Anna A. McKenna, Peter Gruhn, Behjat Hojjati, Jayashree Sil, and Claude Courbois. 1998. "Agricultural Growth Linkages in sub-Saharan Africa." IFPRI Research Report No. 107. International Food Policy Research Institute, Washington, DC.

del Ninno, Carlo, and Lisa C. Smith. 2003. "Public Policy, Markets and Household Coping Strategies in Bangladesh: Avoiding a Food Security Crisis Following the 1998 Floods." *World Development* 31(7): 1221–1238.

den Elzen, M. G. J. and M. Meinshausen. 2005. "Meeting the EU 2°C Climate Target: Global and Regional Emission Implications." Report 728001031/2005. Netherlands Environmental Assessment Agency, Amsterdam.

Denning, Glenn and Jeffrey Sachs. 2007. "How the Rich World Can Help Africa Help Itself." *The Financial Times.* May 29. [http://www.ft.com/cms/s/2/81059fb4-0e02-11dc-8219-000b5df10621,dwp_uuid=8806bae8-0dc4-11dc-8219-000b5df10621.html]. August 2007.

Dercon, Stefan. 1996. "Risk, Crop Choice and Savings: Evidence from Tanzania." *Economic Development Cultural Change.* 44(3): 385–514.

————. 2004. "Growth and Shocks: Evidence from Rural Ethiopia." *Journal of Development Economics* 74: 309–329.

————. 2005. "Vulnerability: A Micro-perspective." Paper presented at the Annual Bank Conference on Development Economics (ABCDE) Conference. Amsterdam, May 2005. World Bank, Washington, DC.

Dercon, Stefan and Pramila Krishnan. 2000. "In Sickness and in Health: Risk Sharing within Households in Rural Ethiopia." *Journal of Political Economy* 108(4): 668–727.

Dercon, Stefan, John Hoddinott and Tassew Woldehanna. 2005. "Shocks and Consumption in 15 Ethiopian Villages, 1999–2004." International Food Policy Research Institute, Washington, DC.

Devereux, Stephen. 1999. "Making Less Last Longer. Informal Safety Nets in Malawi." IDS Discussion Paper No. 373. Institute of Development Studies, University of Sussex, Brighton.

————. 2002. "State of Disaster. Causes, Consequences and Policy Lessons from Malawi." ActionAid Malawi, Lilongwe.

————. 2006a. "Vulnerable Livelihoods in Somali Region, Ethiopia." Institute of Development Studies, University of Sussex, Brighton.

————. 2006b. "Cash Transfers and Social Protection." Paper prepared for the regional workshop on "Cash Transfer Activities in Southern Africa", 9–10 October 2006, Johannesburg, South Africa. Southern African Regional Poverty Network (SARPN), Regional Hunger and Vulnerability Programme (RHVP) and Oxfam GB. Johannesburg.

————. 2006c. "The Impacts of Droughts and Floods on Food Security and Policy Options to Alleviate Negative Effects." Paper submitted for plenary session on "Economics of Natural Disasters" International Association of Agricultural Economists (IAAE) conference. Gold Coast Convention and Exhibition Center, Queensland, Australia. 12–18 August. Institute of Development Studies, University of Sussex, Brighton.

Devereux, Stephen, Rachel Sabates-Wheeler, Mulugeta Tefera and Hailemichael Taye. 2006. "Ethiopia's Productive Safety Net Programme (PSNP): Trends in PSNP Transfers Within Targeted Households." Final Report for the Department for International Development. DFID, Ethiopia and the Institute of Development Studies (IDS), University of Sussex. Brighton and Addis Ababa.

Devereux, Stephen and Zoltan Tiba. 2007. "Malawi's First Famine, 2001–2002." In Stephen Devereux (ed.), *The New Famines. Why Famines Persist in an Era of Globalization.* Routledge, London.

DFID (Department for International Development). 2002. "Bangladesh. Chars Livelihood Programme." London.

————. 2004 "Adaptation to Climate Change: Can Insurance Reduce Vulnerability of the Poor?" Key Sheet No. 8, London.

————. 2006. "Natural Disaster and Disaster Risk Reduction Measures – A Desk Review of Costs and Benefits." Environmental Resources Management, DFID, London.

————. 2007. "A Record Maize Harvest in Malawi." Case Studies. [http://www.dfid.gov.uk/casestudies/fi les/africa%5Cmalawiharvest.asp]. July 2007.

Diamond, Jared. 2005. *Collapse: How Societies Choose to Fail or Succeed.* Viking, New York.

Doniger, David D., Antonia V. Herzog and Daniel A. Lashof. 2006. "Climate Change: An Ambitious, Centrist Approach to Global Warming Legislation." *Science* 314: 764.

EEA (European Environment Agency). 2004. "Energy Subsidies in the European Union: A Brief Overview." DEA Technical Report 1/2004. Brussels.

————. 2006. "Greenhouse Gas Emission Trends and Projections in Europe 2006." EEA Report No. 9/2006. Copenhagen.

————. 2007. "Climate Change and Water Adaptation Issues." EEA Technical Report No. 2/2007. Office for Official Publications of the European Communities, Luxembourg. [http://reports.eea.europa.eu/technical_report_2007_2/en/eea_technical_report_2_2007.pdf]. July 2007.

EFTA (European Federation for Transport and Environment). 2007. "Regulating Fuel Efficiency of New Cars." Background Briefing. January 2007. Brussels.

Elbers, Chris and Jan Willem Gunning. 2003. "Growth and Risk: Methodology and Micro-Evidence." Tin Bergen Institute Discussion Papers 03-068/2. University of Amsterdam.

Elobeid, Amani and Simla Tokgoz. 2006. "Removal of US Ethanol Domestic and Trade Distortions: Impact on US and Brazilian Ethanol Markets." Working Paper 06-WP 427. Center for Agricultural and Rural Development, Iowa University, Ames.

Emanuel, Kerry. 2005. "Increasing Destructiveness of Tropical Cyclones over the Past 30 Years." Nature 436: 686–688.

EIA (Energy Information Administration). 2006. "Emission of Greenhouse Gases in the United States 2005." Washington, DC.

EPA (Environment Protection Agency). 2006. "Clean Air Markets — Data and Publications." [www.epa.gov/airmarkets/auctions/index.html]. August 2007.

Epstein, Paul R. and Christine Rogers. 2004. *Inside the Greenhouse. The Impacts of CO₂ and Climate Change on Public Health in the Inner City.* Center for Health and the Global Environment, Boston, Massachusetts.

Epstein, Paul R. and Evan Mills (eds.). 2005. *Climate Change Futures: Health, Ecological and Economic Dimensions.* The Center for Health and the Global Environment, Harvard Medical School, Cambridge, Massachusetts.

EC (European Commission). 2005a. "Doing More With Less." Green Paper on Energy Efficiency. Brussels.

———. **2005b.** "Zero emissions technology platform: Commission Fosters CO₂-free Energy in the Future." IP/05/1512. Information and Communication Unit, Research DG, Brussels.

———. **2006a.** "Action Plan for Energy Efficiency: Realizing the Potential." Communication from the Commission. Brussels. [http://ec.europa.eu/energy/action_plan_energy_efficiency/doc/com_2006_0545_en.pdf]. September 2007.

———. **2006b.** "Clean Coal Technology." EUROPA, Brussels. [http://ec.europa.eu/energy/coal/clean_coal/index_en.htm]. September 2007.

———. **2006c.** EU Greenhouse Gas Emission Trends and Projections. [http://reports.eea.europa.eu/eea_report_2006_9/en/eea_report_9_2006.pdf]. September 2007.

———. **2007a.** "The Impact of a Minimum 10% Obligation for Biofuel Use in the EU-27 in 2020 on Agricultural Markets." Directorate-General for Agriculture and Rural Development, Brussels. [http://ec.europa.eu/agriculture/analysis/markets/biofuel/impact042007/text_en.pdf]. August 2007.

———. **2007b.** "Commission Proposes an Integrated Energy and Climate Change Package to Cut Emissions for the 21ˢᵗ Century." Press Release, 10 January. EUROPA, Brussels.

———. **2007b.** "Energy for a Changing World." EUROPA. On President José Manuel Barroso. [http://ec.europa.eu/commission_barroso/president/focus/energy_en.htm]. July 2007.

———. **2007c.** "Commission Reports on the Application of State Aid Rules to the Coal Industry in the EU." Press Release. EUROPA, Brussels.

EC (European Commission), Directorate General for Energy and Transport. 2006. "European Survey—Attitude on Issues Related to EU Energy Policy." Press Release. EUROPA. Brussels.

———. **2007.** "Energy for a Changing World. An Energy Policy for Europe—the Need for Action." Brussels.

EFTA (European Federation for Transport and Environment). 2007. "Regulating Fuel Efficiency of New Cars." Background Briefing January 2007. Brussels. [http://www.transportenvironment.org/docs/Publications/2007/2007-01_background_briefing_cars_co2_regulation.pdf]. August 2007.

European Technology Platform on Zero Emission Fossil Fuel Power Plants (ZEP). 2007. "European Technology Platform for Zero Emission Fossil Fuel Power Plants (ZEP): Strategic Overview." ZEP Secretariat, Brussels.

———. **2006a.** *Green Paper: A European Strategy for Sustainable, Competitive and Secure Energy.* European Union, Brussels.

———. **2006b.** "An EU Strategy for Bio-fuels' Communication from the Commission." COM. 2006. 34 Final. Brussels.

EU (European Union). 2007a. "EU almost On Track in Reaching its 2010 Renewable Electricity Target." Press Release. 10 January. MEMO/07/12. EUROPA. Brussels.

———. **2007b.** "Limiting Global Climate Change to 2 degrees Celsius." Press Release. 10 January. MEMO/07/16. EUROPA. Brussels.

———. **2007c.** "Emissions Trading: Commission Adopts Decision on Finland's National Allocation Plan for 2008–2012." Press Release. 4 June. [IP/07/749]. EUROPA. Brussels. 3 53

EWEA (European Wind Energy Association). 2006. "Large Scale Integration of Wind Energy in the European Power Supply: Analysis, Issues and Recommendations." EWEA Grid Report. Brussels.

FAO (Food and Agriculture Organization of the United Nations). 2004. 28ava Conferencia regional de la FAO para América Latina y el Caribe. Seguridad Alimentaria como estrategia de Desarrollo rural. Ciudad de Guatemala (Guatemala), 26 al 30 de abril de 2004.

———. **2007a.** *State of the Worlds Forests* 2007. Forestry Department, Rome.

———. **2007b.** "Forest Resources Assessment." Correspondence on carbon stocks in forests. Extract from database. August 2007. Forestry Department, Rome.

Finlayson, C.M. and A.G. Spiers. 2000. "Global Review of Wetland Resources." In *World Resources 2000–2001.* World Resources Institute, Washington, DC.

Fischer, G., M. Shah, N. Tubiello and H. van Velthuizen. 2005. "Socio-economic and Climate Change Impacts on Agriculture: An Integrated Assessment, 1990–2000." *Philosophical Transactions of the Royal Society* 360: 2067–2083.

Flannery, Tim. 2005. *The Weather Makers: The History and Future Impact of Climate Change.* Penguin, London.

Franco, Guido. 2005. "Climate Change Impacts and Adaptation in California." Support document to the 2005 Integrated Energy Policy Report. Staff Paper. California Energy Commission, Sacramento.

Frankel-Reed, Jenny. 2006. "Adaptation Through Development: A Review of Bilateral Development Agency Programmes, Methods and Projects." Global Environment Fund (GEF), New York.

Friends of the Earth Middle East. 2007. "Climate Change May Further Erode Political Stability in the Middle East." [http://www.foeme.org/press.php?ind=49]. June 2007.

GAO (US Government Accountability Office). 2007. "Climate Change: Financial Risks to Federal and Private Insurers in Coming Decades are Potentially Significant." March 2007. GAO-07-285. Report to the Committee on Homeland Security and Government Affairs, US Senate. Washington, DC.

G8 (Group of Eight). 2005. "Geneagles Plan of Action. Climate Change, Clean Energy and Sustainable Development." Gleneagles.

———. **2007.** "Growth and Responsibility in the World Economy." Summit Declaration Heiligendamm. [http://www.whitehouse.gov/g8/2007/g8agenda.pdf]. September 2007.

Gardner, T.A., Isabelle M. Côté, Jennifer A. Gill, Alastair Grant and Andrew R. Watkinson. 2003. "Long Term Region-wide Declines in Carribbean Corals." *Science* 301(5635): 958–960. 15 August.

GCOS (Global Climate Observing System), UN Economic Commission for Africa and African Union Commission. 2006. "Climate Information for Development Needs: An Action Plan for Africa. Report and Implementation Strategy." 18-21 April, Addis Ababa.

GEF (Global Environment Facility). 2007a. "Status Report on the Climate Change Funds as of April 30, 2007." Report of the Trustee. GEF Secretariat, Washington, DC.

———. **2007b.** "SPA (Strategic Priority on Adaptation) Status Report June 2007." GEF Secretariat, Washington, DC.

———. **2007c.** "Pledging Meeting for Climate Change Funds 15 June 2007." GEF Secretariat, Washington, DC.

Glemarec, Yannick. 2007a. "Embedding climate resilience thinking into national planning in Egypt." Internal Communication.

———. **2007b.** "The impacts of climate change: creating an uncertain future for fisheries in Namibia." Internal Communication.

Global Representation for the Wind Energy Sector and Greenpeace. 2006. *Global Wind Energy Outlook* 2006. Greenpeace and Global Wind Energy Council, London. [http://www.greenpeace.org/raw/content/international/press/reports/globalwindenergyoutlook.pdf]. August 2007.

Government of Australia. 2007. *National Greenhouse Gas Inventory 2005.* Canberra: Australian Greenhouse Office, Department of the Environment and Water Resources. [http://www.greenhouse.gov.au/inventory/2005/index.html]. March 2007.

Government of California. 2006. "Proposition 1E. Disaster Preparedness and Flood Prevention Bond Act of 2006." Legislative Analyst's Office, Sacramento, California. [http://www.lao.ca.gov/ballot/2006/1E_11_2006.htm]. September 2007.

Government of Canada. 2005. "Canada's Greenhouse Gas Inventory, 1990–2003." Greenhouse Gas Division, Environment Canada, Ottawa. [http://www.ec.gc.ca/pdb/ghg/inventory_report/2003_report/sum_e.cfm]. September 2007.

———. 2006. "Canada's Greenhouse Gas Emissions Reporting Program. Overview of the Reported 2005 Facility Level GHG Emissions." Environment Canada, Ottawa. [http://www.ec.gc.ca/pdb/ghg]. August 2007.

———. 2007. "Regulatory Framework for Air Emissions." Ministry of Environment, Ottawa.

Government of the Federal Democratic Republic of Ethiopia. 2006. "Productive Safety Net Programme: Programme Implementation Manual." Ministry of Agriculture and Rural development, Addis Ababa.

Government of France. 2006. "Report from the Working Group on Achieving a fourfold reduction in greenhouse gas emissions in France by 2050." Chaired by Christian de Boisseau. Ministère de l'économie des finances et de l'industrie and Ministère de l'écologie et du développement durable, Paris.

———. 2007. "Actions futures et facteur 4." Ministère de l'écologie, du développement et de l'aménagement, Paris. [http://www.ecologie.gouv.fr/-Actions-futures-et-facteur-4-.html]. August 2007.

Government of Germany. 2007. "Sigmar Gabriel: Klimaschutz nutzt auch Verbrauchern und Wirtschaft." Pressemitteilungen Nr. 224/07. 24 August. Bundesministerium für Umwelt, Naturschutz und Reaktorsicherheit, Berlin.

Government of India. 2006a. *Integrated Energy Policy. Report of the Expert Committee.* New Delhi: Planning Commission.

———. 2006b. *Towards Faster and More Inclusive Growth. An Approach to the 11th Five Year Plan (2007–2012).* Planning Commission, New Delhi.

Government of India. 2007. "2005–2006 National Family Health Survey (NFHS-3)." Ministry of Health and Family Welfare, International Institute for Population Sciences, Mumbai.

Government of Japan. 2002. "Japan's Third National Communication under the United Nations Framework Convention on Climate Change." [http://unfccc.int/resource/docs/natc/japnc3.pdf]. July 2007.

Government of New South Wales. 2007. "Greenhouse Gas Abatement Scheme (GGAS)." Sydney. [http://www.greenhousegas.nsw.gov.au/overview/scheme_overview/overview.asp]. September 2007.

Government of Norway. 2007. "The Prime Minister sets New Climate Goals." Office of the Prime Minister, Oslo.

Government of Pakistan. 2005. Annual Report 2005–06. Oil and Gas Regulatory Authority, Islamabad.

Government of the People's Republic of Bangladesh. 2005a. *Bangladesh. Unlocking the Potential. National Strategy for Accelerated Poverty Reduction.* Dhaka: General Economics Division.

———. 2005b. *National Adaptation Plan of Action. Final Report.* Dhaka: Ministry of Environment and Forests.

———. Mimeo. "Comprehensive Disaster Management Bangladesh Experience." Comprehensive Disaster Management Programme, Ministry of Food and Disaster Management, Dhaka.

Government of Sweden. 2006. "Making Sweden an OIL-FREE Society." Commission on Oil Independence, Stockholm.

———. 2007. "Regeringens proposition 2005/06: 172. Nationell klimatpolitik i global samverkan." Harpsund.

[http://www.regeringen.se/content/1/c6/06/07/78/a096b1c8.pdf]. September 2007.

Government of the United Kingdom. 2006a. *Climate Change. The UK Programme 2006.* Presented to Parliament by the Secretary of State for the Environment, Food and Rural Affairs. Her Majesty's Stationery Office, Norwich.

———. 2006b. "UK Energy and CO_2 Emissions Projections. Updated Projections to 2020." Department of Trade and Industry, London.

———. 2006c. *The Energy Challenge: Energy Review Report 2006.* London: Department of Trade and Industry.

———. 2007a. *Draft Climate Change Bill.* Presented to Parliament by the Secretary of State for Environment, Food and Rural Affairs. Her Majesty's Stationery Office, Norwich.

———. 2007b. *Draft Climate Change Bill. Partial Regulatory Impact Assessment.* London: Department for Environment, Food and Rural Affairs. [http://www.defra.gov.uk/corporate/consult/climatechange-bill/ria.pdf]. September 2007.

———. 2007c. *Energy Trends and Quarterly Energy Prices.* Department of Trade and Industry, London. [http://www.dti.gov.uk/energy/statistics/publications/dukes/page29812.html]. March 2007.

———. 2007d. "Funding UK Flood Management." Department for Environment, Food and Rural Affairs, London. [http://www.defra.gov.uk/environ/fcd/policy/funding.htm]. July 2007.

———. 2007e. *Meeting the Energy Challenge: A White Paper on Energy.* London: Department of Trade and Industry. [http://www.berr.gov.uk/files/file39387.pdf]. May 2007.

Government of the United States. 2005. "Regional Carbon Sequestration Partnerships: Phase I Accomplishments." Department of Energy, NETL (National Energy Technology Laboratory), Pittsburg, Pennsylvania.

———. 2006a. "*FutureGen*—A Sequestration and Hydrogen Initiative." Project Update: December 2006. Department of Energy, Office of Fossil Energy, Washington, DC. [http://www.fossil.energy.gov/programs/powersystems/futuregen/index.html]. August 2007.

———. 2006b. "Interior Secretary Kempthorne Announces Proposal to List Polar Bears as Threatened Under Endangered Species Act." Department of the Interior. Press Release. [http://www.doi.gov/news/06_News_Releases/061227.html]. December 2006.

———. 2007a. "Tracking New Coal-Fired Power Plants. Coal's Resurgence in Electric Power Generation." Department of Energy, NETL (National Energy Technology Laboratory), Pittsburg, Pennsylvania. [http://www.netl.doe.gov/coal/refshelf/ncp.pdf]. September 2007.

———. 2007b. "Carbon Sequestration Technology; Roadmap and Program Plan 2007. Ensuring the Future of Fossil Energy Systems through the Successful Deployment of Carbon Capture and Storage Technologies." Department of Energy, NETL (National Energy Technology Laboratory), Pittsburg, Pennsylvania.

———. 2007c. "President Bush Delivers State of the Union Address." United States Capitol, Washington, DC. [http://www.whitehouse.gov/news/releases/2007/01/20070123-2.html]. August 2007.

Greenpeace and GWEC (Global Wind Energy Council). 2006. *Global Wind Energy Outlook 2006.* GWEC and Greenpeace, Brussels and Amsterdam.

Grinspun, Alejandro. 2005. "Three models of social protection." One Pager No. 17. UNDP-International Poverty Agenda, Brasilia.

Grubb, Michael and Karsten Neuhoff. 2006. "Allocation and Competitiveness in the EU Emissions Trading Scheme: Policy Overview." *Climate Policy* 6: 7–30.

GSS (Ghana Statistical Service), NMIMR (Noguchi Memorial Institute for Medical Research), and ORC Macro. 2004. "Ghana Demographic and Health Survey 2003." Calverton, Maryland.

Gurría, Angel, and Richard Manning. 2007. "Statement by Angel Gurría, OECD Secretary-General, and Richard Manning, Chairman,

OECD Development Assistance Committee (DAC)." Meeting. Washington, 15 April 2007. OECD, Washington, DC.

Główny Urząd Statystyczny (GUS) [Central Statistical Office, Poland]. 2006. *Energy Consumption Efficiency, 1994–2004.* Warsaw.

Hanemann, Michael and A. Farrel. 2006. Managing Greenhouse Gas Emissions in California. The California Climate Change Center at University of California, Berkeley. [http://calclimate.berkeley. edu/managing GHGs in CA.html]. January 2006.

Hansen, James. 2006. "The Threat to the Planet." *New York Review of Books* 55 (12). [http://www.nybooks.com/articles/19131]. July 2007.

———. **2007a.** "Scientific Reticence and Sea Level Rise." *Environmental Research Letters* 2 024002 (6pp). [http://www.iop. org/EJ/article/1748-9326/2/2/024002/erl7_2_024002.html]. March 2007.

———. **2007b.** "Why We Can't Wait." *The Nation*. 7 May. New York.

———. **2007c.** "Dangerous Human-Made Interference with Climate." Testimony to Select Committee on Energy Independence and Global Warming, United States House of Representatives, 26 April, Washington, DC.

Hansen, James, Makiko Sato, Reto Ruedy, Ken Lo, David W. Lea and Martin Medina-Elizade. 2006. "Global Temperature Change." *Proceedings of the National Academy of Sciences* 103 (39): 14288–14293.

Hansen, J., Mki Sato, R. Ruedy, P. Kharecha, A. Lacis, R.L. Miller, L. Nazarenko, K. Lo, G.A. Schmidt, G. Russell, I. Aleinov, S. Bauer, E. Baum, B. Cairns, V. Canuto, M. Chandler, Y. Cheng, A. Cohen, A. Del Genio, G. Faluvegi, E. Fleming, A. Friend, T. Hall, C. Jackman, J. Jonas, M. Kelley, N.Y. Kiang, D. Koch, G. Labow, J. Lerner, S. Menon, T. Novakov, V. Oinas, Ja. Perlwitz, Ju. Perlwitz, D. Rind, A. Romanou, R. Schmunk, D. Shindell, P. Stone, S. Sun, D. Streets, N. Tausnev, D. Thresher, N. Unger, M. Yao, and S. Zhang. 2007. Dangerous Human-made Interference with Climate: A GISS modelE study. *Atmospheric Chemistry and Physics* 7: 2287–2312.

Hanson, Craig and James R. Hendricks Jr. 2006. "Taxing Carbon to Finance Tax Reform." Issue Brief. Duke Energy and World Resources Institute. Charlotte, North Carolina and Washington, DC.

Hare, William. 2005. "Relationship Between Increases in Global Mean Temperature and Impacts on Ecosystems, Food Production, Water and Socio-Economic Systems." In *Avoiding Dangerous Climate Change*. Conference Report for Symposium on Stabilization of Greenhouse Gases, 1–3 February, 2005. Hadley Centre, Exeter, Department for Environment, Food and Rural Affairs, London.

Heimann, Lieff Cabraser and L.L.P. Bernstein. 2007. "Tobacco and Smokers Litigation." [http://www.lieffcabraser.com/tobacco.htm]. April 2007.

Hemming, D. 2007. "Impacts of Mean Sea Level Rise Based on Current State-of-the-Art Modelling." Hadley Centre, Exeter University.

Henderson, Caspar. 2006a. "Ocean acidification: The Other CO_2 Problem." NewScientist.com news service. 5 August 2006. [http:// environment.newscientist.com/channel/earth/mg19125631.200- ocean-acidification-the-iotheri-cosub2sub-problem.html]. September 2007.

———. **2006b.** "Paradise Lost," *New Scientist* 191 (2563): 28–33. 5 August 2006.

High-Level Task Force on UK Energy Security, Climate Change and Development Assistance. 2007. *Energy, Politics, and Poverty. A Strategy for Energy Security, Climate Change and Development Assistance*. University of Oxford.

Hoddinott, John and Bill Kinsley. 2000. "Adult Health in the Time of Drought." Food Consumption and Nutrition Division (FCND) Discussion Paper No. 79. International Food Policy Research Institute, Washington, DC.

———. **2001.** "Child Growth in the Times of Drought." *Oxford Bulletin of Economics and Statistics* 63(4):0305–0949.

Hoffmann, Yvonne. 2006. "Auctioning of CO_2 Emission Allowances in the EU ETS." Report under the project "Review of EU Emissions Trading Scheme." European Commission Directorate General for Environment, Brussels.

Houghton, R.A. 2005. "Tropical Deforestation as a Source of Greenhouse Gas Emission." In *Tropical Deforestation and Climate Change* (P. Mutinho and S. Schwartzman eds). Belém: Instituto de Pesquisa Ambiental da Amazônia (IPAM). Environmental Defense, Washington, DC.

Hoyois, P., J-M. Scheuren, R. Below and D. Guha-Sapir. 2007. *Annual Disaster Statistical Review: Numbers and Trends 2006*. Centre for Research on the Epidemiology of Disasters (CRED). Brussels.

HSBC (Hong Kong Shanghai Bank of Commerce). 2007. "HSBC Climate Confidence Index 2007." HSBC Holdings plc.

Huisman, Pieter. 2002. "How the Netherlands Finance Public Water Management." European Water Management Online. Official Publication of the European Water Association. [http://www. ewaonline.de/journal/2002_03.pdf]. May 2007.

Hulme, Mike and Nicola Sheard. 1999a. "Climate Change Scenarios for Australia." Climatic Research Unit, Norwich. [http://www.cru. uea.ac.uk/~mikeh/research/australia.pdf]. August 2007.

———. **1999b.** "Climate Change Scenarios for Japan." Climate Research Unit. Norwich. [http://www.cru.uea.ac.uk/~mikeh/ research/wwf.japan.pdf]. September 2007.

Ikkatai, Seiji. 2007. "Current Status of Japanese Climate Change Policy and Issues on Emission Trading Scheme in Japan." The Research Center for Advanced Policy Studies Institute of Economic Research, Kyoto University, Kyoto.

IEA (International Energy Agency). 2003. "Cool Appliances: Policy Strategies for Energy-Efficient Homes." Energy Efficiency Policy Profiles. OECD (Organisation for Economic Co-operation and Development)/IEA, Paris.

———. **2006a.** "Energy Policies of IEA Countries. 2006 Review." OECD (Organisation for Economic Co-operation and Development)/IEA, Paris.

———. **2006b.** *Energy Technology Perspectives. Scenarios and Strategies to 2050*. OECD (Organisation for Economic Co-operation and Development)/IEA, Paris.

———. **2006c.** *World Energy Outlook*. OECD (Organisation for Economic Co-operation and Development)/IEA, Paris.

IFEES (Islamic Foundation for Ecology and Environmental Sciences). 2006. "EcoIslam." Newsletter. Issue No.02. [http:// ifees.org.uk/newsletter_2_small.pdf]. August 2007.

IFRC (International Federation of the Red Cross and Red Crescent Societies). 2002. *World Disasters Report 2002*. Geneva.

———. **2005a.** *World Disasters Report 2005: Focus on Information in Disasters*. Geneva.

———. **2005b.** Operations Update No 3. Kenya: Drought. 4 February. [www.reliefweb.int/library/documents/2005/IFRC/ifrc-drought- 04feb.pdf]. July 2007.

———. **2006.** *World Disasters Report 2006: Focus on Neglected Crises*. Geneva.

IMF (International Monetary Fund). 2006. *World Economic Outlook Report 2006: Financial Systems and Economic Cycles*. September. Washington, DC.

———. **2007.** *World Economic Outlook Database*. April 2007. Washington, DC.

International Network for Sustainable Energy – Europe. 2006. "Subsidies and Public Support for Energy." [http://www.inforse. org/europe/subsidies.htm]. August 2007.

INPE (Instituto Nacional de Pesquisas Espaciais). 2007. "Sistema de Detecção do Desmatamento em Tempo Real (DETER)." Database. São José dos Campos.

International Network for Sustainable Energy. 2006. "Subsidies and Public Support for Energy." [http://www.inforse.org/europe/subsidies.htm]. August 2007.

IRI (International Research Institute for Climate and Society). 2007. "Climate Risk Management in Africa: Learning from Practice." *Climate and Society* No 1. The Earth Institute, Columbia University, New York.

ISSC (International Scientific Steering Committee). 2005. *Report of the International Scientific Steering Committee. International Symposium on Stabilization on Greenhouse Gas Concentrations—Avoiding Dangerous Climate Change, 1–3 February, 2005 Met-Office Hadley Centre for Climate Change, Exeter, UK.* Department for Environment, Food and Rural Affairs, London.

IPCC (Intergovernmental Panel on Climate Change). 1999. "Summary for Policymakers. Aviation and the Global Atmosphere." A Special Report of IPCC Working Groups I and III in collaboration with the Scientific Assessment Panel to the Montreal Protocol on Substances that Deplete the Ozone Layer. (Joyce E. Penner, David H. Lister, David J. Griggs, David J. Dokken and Mack McFarland, eds.). Cambridge University Press, Cambridge and New York.

———. 2001. "Technical Summary." In *Climate Change 2001: Impacts, Adaptation, and Vulnerability. Contribution of Working Group II to the Third Assessment Report of the Intergovernmental Panel on Climate Change* (James J. McCarthy, Osvaldo F. Canziani, Neil A. Leary, David J. Dokken and Kasey S. White, eds.). Cambridge University Press, Cambridge and New York.

———. 2007a. *Climate Change 2007—The Physical Science Basis. Contribution of Working Group I to the Fourth Assessment Report of the Intergovernmental Panel on Climate Change.* (S. Solomon, D. Qin, M. Manning, Z. Chen, M. Marquis, K.B. Averyt, M. Tignor and H.L. Miller, eds.). Cambridge University Press, Cambridge and New York.

———. 2007b. *Climate Change 2007: Climate Change Impacts, Adaptation and Vulnerability. Working Group II Contribution to the Fourth Assessment Report of the Intergovernmental Panel on Climate Change.* (S. Solomon, D. Qin, M. Manning, Z. Chen, M. Marquis, K.B. Averyt, M. Tignor and H.L. Miller, eds.). Cambridge University Press, Cambridge and New York.

———. 2007c. *Climate Change 2007: Mitigation of Climate Change. Working Group III Contribution to the Fourth Assessment Report of the Intergovernmental Panel on Climate Change.* (S. Solomon, D. Qin, M. Manning, Z. Chen, M. Marquis, K.B. Averyt, M. Tignor and H.L. Miller, eds.). Cambridge University Press, Cambridge and New York.

———. 2007d. "Summary for Policymakers." In *Climate Change 2007— The Physical Science Basis. Contribution of Working Group I to the Fourth Assessment Report of the Intergovernmental Panel on Climate Change.* (S. Solomon, D. Qin, M. Manning, Z. Chen, M. Marquis, K.B. Averyt, M. Tignor and H.L. Miller,eds.). Cambridge University Press, Cambridge and New York.

———. 2007e. "Summary for Policymakers." In *Climate Change 2007: Climate Change Impacts, Adaptation and Vulnerability. Working Group II Contribution to the Fourth Assessment Report of the Intergovernmental Panel on Climate Change.* (S. Solomon, D. Qin, M. Manning, Z. Chen, M. Marquis, K.B. Averyt, M. Tignor and H.L. Miller, eds.). Cambridge University Press, Cambridge and New York.

———. 2007f. "Summary for Policymakers." *In Climate Change 2007: Mitigation of Climate Change. Working Group III Contribution to the Fourth Assessment Report of the Intergovernmental Panel on Climate Change.* (S. Solomon, D. Qin, M. Manning, Z. Chen, M. Marquis, K.B. Averyt, M. Tignor and H.L. Miller, eds.). Cambridge University Press, Cambridge and New York.

———. 2007g. "Technical Summary." In *Climate Change 2007: Climate Change Impacts, Adaptation and Vulnerability. Working Group II Contribution to the Fourth Assessment Report of the Intergovernmental Panel on Climate Change.* (S. Solomon, D. Qin,

M. Manning, Z. Chen, M. Marquis, K.B. Averyt, M. Tignor and H.L. Miller, eds.). Cambridge University Press, Cambridge and New York.

IRI (International Research Institute for Climate and Society). 2007. "Climate Risk Management in Africa: Learning from Practice." *Climate and Society* No 1. The Earth Institute, Columbia University, New York.

ISDR (International Strategy for Disaster Reduction). 2007a. "Drought Risk Reduction Framework and Practices: Contributing to the Implementation of the Hyogo Framework for Action." Geneva.

———. 2007b. "Building Disaster Resilient Communities. Good Practices and Lessons Learned." Geneva.

———. 2007c. "Words into Action: A Guide for Implementing the Hyogo Framework. Hyogo Framework for Action 2005–2015: Building the Resilience of Nations and Communities to Disasters." Geneva.

ISDR (International Strategy for Disaster Reduction) and World Bank GFDRR (Global Facility for Disaster Reduction and Recovery). 2006. "A Partnership for Mainstreaming Disaster Mitigation in Poverty Reduction Strategies." Geneva and Washington, DC.

———. 2007. "Committed to Reducing Vulnerabilities to Hazards by Mainstreaming Disaster Reduction and Recovery in Development. Progress Report 1. Geneva and Washington, DC.

Itano, Nicole. 2002. "Famine, AIDS Devastating Malawi Women." WOMENSENEWS. 26 February. [http://www.sahims.net/doclibrary/2004/02_February/11%20Wed/Regional%20abstract/Famine,%20AIDS%20Devastating%20Malawi%20Women.pdf]. August 2007.

Jank, Marcos J., Géraldine Kutas, Luiz Fernando do Amaral and André M. Nassar. 2007. "EU and US Policies on Biofuels: Potential Impacts on Developing Countries." The German Marshall Fund of the United States, Washington, DC.

Jacquet, Pierre and Laurence Tubiana (eds.) 2007. *Regards sur la terre: L'annuel du développement durable. 2007. Energie et changements climatiques.* Presses de Sciences Pos, Paris.

Jha, Saroj Kumar. 2007. "GFDRR. Track II. Multi-donor Trust Fund for Mainstreaming Disaster Reduction for Sustainable Poverty Reduction." ISDR and the Global Facility for Disaster Reduction and Recovery, The World Bank Group. Washington, DC. [http://www.unisdr.org/eng/partner-netw/wb-isdr/Twb-isdr-rackII-ApproachPaper-Results-CG-comments.doc]. August 2007.

Jones, P. and P.K. Thornton. 2003. "The Potential Impacts of Climate Change on Maize Production in Africa and Latin America in 2055." *Global Environmental Change* 13: 51–59.

Jones, Chris, Peter Cox and Chris Huntingford. 2005. "Impact of climate-carbon cycle feedbacks on emissions scenarios to achieve stabilization." In *Avoiding Dangerous Climate Change.* Conference Report for Symposium on Stabilization of Greenhouse Gases, 1–3 February, 2005 Met Office Hadley Centre for Climate Change, Exeter, UK. Department for Environment, Food and Rural Affairs, London.

Kennedy, John F. 1963. Address before the Irish Parliament, June 28, 1963. [http://www.jfklibrary.org/Asset+Tree/Asset+Viewers/Audio+Video+Asset+Viewer.htm?guid={D8A7601E-F3DA-451F-86B4-43B3EE316F64}&type=Audio]. August 2007.

Klein, R.J.T., S.E.H.Eriksen, L.O. Næss, A. Hammill, C. Robledo, K.L.O. Brien and T.M.Tanner. 2007. "Portfolio Screening to Support the Mainstreaming of Adaptation to Climate Change into Development Assistance." Working Paper 102. Tyndall Centre for Climate Change Research, University of East Anglia, Norwich.

Kurukulasuriya, Pradeep and Robert Mendelsohn. 2006. "A Ricardian Analysis of the Impact of Climate Change on African Cropland." CEEPA Discussion Paper No. 8. Centre for Environmental Economics and Policy in Africa (CEEPA), University of Pretoria.

Landau, J.P. 2004. "Rapport à Monsieur Jacques Chirac, Président de la Republique, Group de travail sur les nouvelles contributions financières internationales." [http://www.diplomatie.gouv.fr/en/IMG/pdf/LandauENG1.pdf]. August 2007.

Leiserowitz, Anthony. 2006. "Climate Change, Risk Perception and Policy Preferences." *Climate Change* 77 (Spring): 45–72.

Lindert, Kathy, Anja Linder, Jason Hobbs and Bénédicte de la Brière. 2007. "The Nuts and Bolts of Brazil's Bolsa Família Program: Implementing Conditional Cash Transfers in a Decentralized Context". Social Protection Discussion Paper 0709. World Bank, Washington, DC.

Linklater, Magnus. 2007. "A Brilliantly Swiss Scheme to Ignore Global Warming." *The Times*. London. 18 July. [http://www.timesonline.co.uk/tol/comment/columnists/magnus_linklater/article2093516.ece]. September 2007.

Lockwood, Mike and Claus Fröhlich. 2007. "Recent Oppositely Directed Trends in Solar Climate Forcings and the Global Mean Surface Air Temperature." *Proceedings of the Royal Society A* 463 (2086): 2447–2460. [http://www.journals.royalsoc.ac.uk/content/h844264320314105/]. August 2007.

Lopez, Humberto. 2006. "Did Growth Become Less Pro-Poor in the 1990s?" World Bank Policy Research Working Paper Series No. 3931. World Bank, Washington, DC. [http://econ.worldbank.org]. June 2006.

Mallick, Dwijendra Lal, Atiq Rahman, Mozaharul Alam, Abu Saleh Md Juel, Azra N. Ahmad and Sarder Shafiqul Alam. 2005. "Floods in Bangladesh: A Shift from Disaster Management Towards Disaster Preparedness." *IDS Bulletin* 36(4): 53–70.

Maskrey, A., Gabriella Buescher, Pascal Peduzzi and Carolin Schaerpf. 2007. Disaster Risk Reduction: 2007 Global Review. Consultation Edition. Prepared for the Global Platform for Disaster Risk Reduction First Session, Geneva, Switzerland, 5–7 June 2007. Geneva.

McMichael, A.J., D.H. Campbell-Lendrum, C.F. Corvalán, K.L. Ebi, A. Githeko, J.D. Scheraga and A. Woodward. 2003. "Chapter 1: Global Climate Change and Health: An Old Story Writ Large." In: *Climate Change and Human Health—Risks and Responses*. Geneva: World Health Organization.

Mechler, Reinhard, Joanne Linnerooth-Bayer and David Peppiatt. 2006. Disaster Insurance for the Poor? A Review of Micro-Insurance for Natural Disaster Risks in Developing Countries." Provention/IIASA Study. Provention Consortium, Geneva.

Meinshausen, Malte. 2005. "On the Risk of Overshooting 2°C." Paper presented at Scientific Symposium: *Avoiding Dangerous Climate Change*. Symposium on Stabilisation of Greenhouse Gases, 1–3 February, 2005. MetOffice Hadley Centre Exeter, UK. London: Department for Environment, Food and Rural Affairs.

Mendonca, Miguel. 2007. *Feed-in Tariffs – Accelerating the Development of Renewable Energy*. Earthscan, London.

Merrill Lynch and WRI (World Resources Institute). 2005. "Energy Security and Climate Change. Investing in the Clean Car Revolution." Washington, DC.

Meteo France. 2007. "L'établissement Météo-France." [http://www.meteofrance.com/FR/qui_sommes_nous/enbref/enbref.jsp]. September 2007.

Met Office. 2006. "Effects of Climate Change in Developing Countries. Met Office Hadley Centre for Climate Change." Exeter.

Millennium Ecosystem Assessment. 2005. *Ecosystems and Human Well-being—Synthesis*. Island Press, Washington, DC. [http://www.millenniumassessment.org/documents/document.356.aspx.pdf]. September 2007.

Mills, Evan. 2006. "The Role of NAIC in Responding to Climate Change." Testimony to the National Association of Insurance Commissioners. University of California, Berkeley.

Mills, E., R.J. Roth and E. Lecomte. 2005. "Availability and Affordability of Insurance Under Climate Change. A Growing Challenge for the U.S." Prepared for The National Association of Insurance Commissioners. University of California, Berkeley.

MIT (Massachusetts Institute of Technology). 2007. *The Future of Coal: Options for a Carbon Constrained World*. Boston.

Modi, Vijay, Susan McDade, Dominique Lallement and Jamal Saghir. 2005. "Energy Services for the Millennium Development Goals." Energy Sector Management Assistance Programme, UN Millennium Project, United Nations Development Programme and World Bank, New York.

Monbiot, George. 2006. *Heat*. Penguin Books, London.

Morris, S., O. Neidecker-Gonzales, C. Carletto, M. Munguia, J.M. Medina and Q. Wodon. 2001. "Hurricane Mitch and Livelihoods of the Rural Poor in Honduras," *World Development* 30(1): 39–60.

Mosley, P. 2000. "Insurance Against Poverty? Design and Impact of "New Generation" Agricultural Micro-Insurance Schemes." University of Sheffield.

Mousseau, Frederic and Anuradha Mittal. 2006. *Sahel: A Prisoner of Starvation? A case study of the 2005 food crisis in Niger*. The Oakland Institute, California.

Müller, Benito and Cameron Hepburn. 2006. "IATAL – an Outline Proposal for an International Air Travel Adaptation Levy." Oxford Institute for Energy Studies, Oxford.

Narain, Sunita. 2006. "Community-let Alternatives to Water Management: India Case Study. Paper commissioned for *Human Development Report 2006: Beyond Scarcity: Power, Poverty and the Global Water Crisis*. Palgrave Macmillan, New York.

National Audit Office. 2001. "Inland Flood Defence." Report by the Comptroller and Auditor General, London.

NASA (North American Space Agency). 2005. "NASA History — Human Space Flight." [http://spaceflight.nasa.gov/history/]. September 2007.

NCEP (National Commission on Energy Policy). 2004a. "Ending the Energy Stalemate. A Bipartisan Strategy to Meet America's Energy Challenges. Summary of Recommendations." National Commission on Energy Policy, Washington, DC.

———. **2004b.** "Taking Climate Change into Account in US Transportation." In Innovative Policy Solutions to Global Climate Change, Brief No.6, National Commission on Energy Policy, Washington, DC.

NEA (Nuclear Energy Authority). 2006. *Annual Report*. OECD (Organisation for Economic Co-operation and Development), Paris.

NERC (Natural Environment Research Council) British Antarctic Survey. 2000. "Future Changes in the Size of the Antarctic Ice Sheet." [http://www.antarctica.ac.uk/Key_Topics/IceSheet_SeaLevel/ice_sheet_change.html]. August 2007.

NETL (National Energy Technology Laboratory). 2007. "Tracking New Coal-fired Power Plants." United States Department of Energy, Pittsburgh, Pennsylvania. [http://www.netl.doe.gov/coal/refshelf/ncp.pdf]. August 2007.

New York Climate & Health Project. 2004. "Assessing Potential Public Health and Air Quality Impacts of Changing Climate and Land Use in Metropolitan New York." Columbia University, New York.

NFU (National Farmers Union). 2005. *Agriculture and Climate Change*. London. [http://www.nfuonline.com/documents/Policy%20Services/Environment/Climate%20Change/NFU%20Climate%20Change.pdf]. May 2007.

Nippon Keidanren. 2005. "Results of the Fiscal 2005 Follow-up to the Keidanren Voluntary Action Plan on the Environment (Summary)." Section on Global Warming Measures–Performance in Fiscal 2004. Tokyo. [http://www.keidanren.or.jp/english/policy/2005/086.pdf]. September 2007.

Nobre, Carlos. 2007. "Climate Policy: It's Good to be in the "RED."" News Release. 10 May. Carnegie Institution, Washington, DC.

Nordhaus, William D. 2005. "Life after Kyoto: Alternative Approaches to Global Warming Policies." National Bureau of Economic Research, Working Paper 11889. Cambridge, Massachusetts.

———. **2006.** "The Stern Review on the Economics of Climate Change." National Bureau of Economic Research, Working Paper 12741. Cambridge, Massachusetts. [http://papers.ssrn.com/sol3/papers.cfm?abstract_id=948654]. December 2006.

———. **2007.** "Critical Assumptions in the Stern Review on Climate Change." Science 317 (5835): 203–204. 13 July.

NREL (National Renewable Energy Laboratory) Energy Analysis Office. 2005a. Renewable Energy Cost Trends. Presentation. [http://www.nrel.gov/analysis/docs/cost_curves_2005.ppt]. November 2005.

———. **2005b.** Global Competitiveness in Fuel Economy and Greenhouse Gas Emission Standards for Vehicles. Presentation by Amanda Sauer, 10 February. World Resources Institute, Washington, DC. [http://www.nrel.gov/analysis/seminar/docs/2005/ea_seminar_feb_10.ppt]. September 2007.

OECD (Organisation for Economic Co-operation and Development). 2005b. "Harmonisation, Alignment, Results: Report on Progress, Challenges and Opportunities." Paris.

———. **2005c.** "Reducing Greenhouse Gas Emissions: the Potential of Coal." Paris. [http://www.iea.org/Textbase/work/2006/gb/publications/ciab_ghg.pdf]. September 2007.

———. **2006a.** *Declaration on Integrating Climate Change Adaptation into Development Cooperation.* Paris.

———. **2006b.** *Agricultural Policies in OECD Countries: At a Glance, 2006 Edition.* Paris.

———. **2006c.** "DAC Members' net ODA 1990–2005 and SAC Secretariat Simulation of net ODA in 2006 and 2010." [http://www.oecd.org/dac/stats]. March 2007.

———. **2006d.** "Japan Floods." OECD Studies in Risk Management. Paris.

———. **2006e.** "Survey on Harmonisation and Alignment of Donor Practices." Paris.

———. **2007a.** "Climate Change and Africa." Paper prepared by the AFP Support Unit and NEPAD Secretariat for the 8th Meeting of the Africa Partnership Forum. 22–23 May, Berlin.

———. **2007b.** "International Development Statistics (CRS)." Online Database on Aid and Other Resource Flows. Paris [http://www.oecd.org/dac/stats/idsonline]. July 2007.

OFDA (Office of US Foreign Disaster Assistance) and CRED (Collaborating Centre for Research on the Epidemiology of Disasters). 2007. *Emergency Events Database (EM-DAT).* Database. Brussels. [http://www.em-dat.net/who.htm]. September 2007.

Olshanskaya, Marina. 2007. "Russia and the Kyoto Protocol: Global and National Human Development Perspectives." UNDP Bratislava Regional Centre, Bratislava.

Oxfam International. 2005. "Predictable Funding for Humanitarian Emergencies: a Challenge to Donors." Oxfam Briefing Note. [http://www.oxfam.org/en/files/bn051024_CERF_predictablefunding/download]. October 2005.

———. **2007.** "Adapting to Climate Change. What's Needed in Poor Countries, and Who Should Pay." Oxfam Briefing Paper 104. Oxford.

Page, Edward A. 2006. *Climate Change, Justice and Future Generations.* Cheltenham: Edward Elgar. [http://www2.warwick.ac.uk/fac/soc/pais/staff/page/publications/]. July 2007.

PEACE (Pelangi Energi Abadi Citra Enviro). 2007. "Indonesia and Climate Change." Working Paper on Current Status and Policies. Department for International Development, World Bank, and State Ministry of Environment in Indonesia, Jakarta.

Pearce, David. 2001. "The Economic Value of Forest Ecosystems." CSERGE–Economics, University College London, London. [http://www.cserge.ucl.ac.uk/web-pa_1.HTM]. September 2007.

Pembina Institute. 2007a. "Canada's Implementation of the Kyoto Protocol." Gatineau. [http://www.pembina.org/climate-change/work-kyoto.php]. April 2007.

———. **2007b.** "Future Greenhouse Gas Emission Reductions."Gatineau. [http://www.pembina.org/climate-change/work-future.php]. April 2007.

People's Republic of China. 2007. *China's National Climate Change Programme.* People's Republic of China: National Development Reform Commission.

Perry, Michael, Adrianne Dulio, Samantha Artiga, Adele Shartzer and David Rousseau. 2006. "Voices of the Storm. Health Experiences of Low-Income Katrina Survivors." Henry J. Kaiser Foundation, California.

Pew Center on Global Climate Change. 2006. "Little Consensus on Global Warming. Partisanship Drives Opinion." Survey Report. Arlington, Virginia.

———. **2007a.** "Senate Greenhouse Gas Cap-And-Trade Proposals in the 110th Congress." Washington, DC. [http://www.earthscape.org/l2/ES17454/PEW_SenateGreenHouse.pdf]. September 2007.

———. **2007b.** "What's Being Done in the Business Community." [http://www.pewclimate.org/what_s_being_done/in_the_business_community/]. August 2007.

———. **2007c.** "A Look at Emission Targets." [http://www.pewclimate.org/what_s_being_done/targets]. September 2007.

Philibert, Cedric. 2006. "Barriers to Technology Diffusion. The Case of Solar Thermal Technologies." OECD/IEA (Organisation for Economic Co-operation and Development/International Energy Authority), Paris.

Philibert, Cédric and Jacek Podkanski. 2005. "International Energy Technology Collaboration and Climate Change Mitigation. Case Study 4: Clean Coal Technologies." OECD/IEA (Organisation for Economic Co-operation and Development/International Energy Authority), Paris.

Phiri, Frank. 2006. "Challenges 2005–2006: A Difficult Year Ahead for Famine-Hit Malawi." IPS Terraviva Online. [http://www.ipsterraviva.net/Africa/print.asp?idnews=484]. January 2006.

Pierce, David W., Tim P. Barnett, Krishna M. AchutaRao, Peter J. Gleckler, Jonathan M. Gregory and Warren M. Washington. 2005. "Anthropogenic Warming of the Oceans: Observations and Model Results." (Version 2). Scripps Institution of Oceanography, San Diego, California.

Point Carbon. 2007. "Carbon 2007—A New Climate for Carbon Trading." K. Roine and H. Hasselknippe (eds.). Report published at Point Carbon's 4th Annual Conference, Carbon Market Insights 2007. Copenhagen, 13–15 March.

Practical Action. 2006a. "Shouldering the burden. Adapting to climate change in Kenya." [http://practicalaction.org/?id=climatechange_panniers]. August 2007.

———. **2006b.** "Rainwater harvesting." [http://practicalaction.org/?id=rainwater_case_study]. October 2007.

Pritchard, H. D., and D. G. Vaughan. 2007. "Widespread Acceleration of Tidewater Glaciers on the Antarctic Peninsula." *Journal of Geophysical Research* 112 online (F03S29, doi:10.1029/2006JF000597). September 2007.

Ramsey, Frank. 1928. "A Mathematical Theory of Saving." *The Economic Journal* 38(152) December: 543–559.

Randel, Judith. 2007. "Social Protection in Zambia, Bangladesh, Nicaragua, Ethiopia, Viet Nam and Uganda." Development Initiatives, Somerton, Somerset.

Raworth, Kate. 2007a. "Adapting to Climate Change. What's Needed in Poor Countries and Who Should Pay." Oxfam Briefing Paper No.104. Oxfam International, Oxford.

———. **2007b.** "West Bengal River Basin Programme. Climate Change Research Visit Note." Oxfam–GB, Oxford.

Reece, Gemma, Dian Phylipsen, Max Rathmann, Max Horstink and Tana Angelini. 2006. "Use of JI/CDM Credits by Participants in Phase II of the EU Emissions Trading Scheme." Final report. Ecofys UK, London.

Regional Hunger and Vulnerability Programme. 2007. "Malawi: Summary of Information Systems." [http://www.wahenga.net/uploads/documents/nationalsp/Malawi_SP_Info_systems_Jan2007.pdf]. September 2007.

Reliefweb. 2007. Information on Complex Emergencies and Natural Disasters. [http://www.reliefweb.int/]. September 2007.

Republic of Malawi. 2006. *Malawi's National Adaptation Programmes of Action.* Ministry of Mines, Natural Resources and Environment, Lilongwe.

Republic of Niger. 2006. *National Adaptation Programme of Action.* Cabinet of Prime Minister, Niamey.

RGGI (Regional Greenhouse Gas Initiative). 2005. "Memorandum of Understanding." [http://www.rggi.org/docs/mou_12_20_05.pdf]. September 2007.

Roberts, Paul. 2005. *The End of Oil: On the Edge of a Perilous New World.* Houghton Mifflin, Boston.

Roberts, Timmons and Bradley C. Parks. 2007. *A Climate of Injustice: Global Inequality, North-South Politics and Climate Policy.* MIT Press, Cambridge, Massachusetts.

Roosevelt, Theodore IV. 2006. "Solutions Testimony at the US House of Representatives Committee on Government Reform regarding Climate Change: Understanding the Degree of the Problem—and the Nature of its Solutions." Pew Center on Global Climate Change, Washington DC. [http://www.pewclimate.org/what_s_being_done/in_the_congress/roosevelt_7_20_06.cfm]. August 2007.

Rose, Elaina. 1999. "Consumption Smoothing and Excess Female Mortality in Rural India." *Review of Economics and Statistics.* 81(1): 41–49.

Rosegrant, Mark W., Ximing Cai and Sarah A. Cline. 2002. "Global Water Outlook 2025: Dealing with Scarcity." International Food Policy Research Institute, Washington, DC.

Rosenzweig, Mark, R. and Hans P. Binswagner. 1993. "Wealth, Weather Risk and the Composition and Profitability of Agricultural Investments." *The Economic Journal* 103:56–78.

Rowland, Diane. 2007. "Health Care: Squeezing the Middle Class with More Costs and Less Coverage." Testimony before the US House of Representatives, Ways and Means Committee: *Economic Challenges Facing Middle Class Families.* January 2007. Washington, DC.

Royal Government of Cambodia. 2006. *National Adaptation Programme of Action to Climate Change (NAPA).* Ministry of Environment, Phnom Penh.

Rubin, Edward S. 2007. "Accelerating Deployment of CCS at US Coal-Based Power Plants." Presentation to the Sixth Annual Carbon Capture and Sequestration Conference. 8 May 2007. Department of Engineering and Public Policy, Carnegie Mellon University, Pittsburg, Pennyslvania.

Runge, C. Ford and Benjamin Senauer. 2007. "How Biofuels Could Starve the Poor." *Foreign Affairs* 86(3). [http://www.foreignaffairs.org/20070501faessay86305/c-ford-runge-benjamin-senauer/how-biofuels-could-starve-the-poor.html]. June 2007.

Runnalls, David. 2007. "Subsidizing Biofuels Backfires." IISD Commentary. International Institute for Sustainable Development, Winnipeg.

Scheer, Hermann. 2001. *A Solar Manifesto.* Second Edition. James and James (Science Publishers), London.

Schelling, Thomas. 2007. "Climate Change: The Uncertainties, the Certainties, and What They Imply About Action." *Economists' Voice* 4(3): Article 3. [http://www.bepress.com/ev/vol4/iss3/art3/]. September 2007.

Schellnhuber, John. 2006. "The Irregular Side of Climate Change". Presentation made at the Cambridge University Business and Environment Programme Climate Science Meeting. 15 December. London. Mimeo.

Schellnhuber, John and Janica Lane. 2006. In *Avoiding Dangerous Climate Change.* Conference Report for Symposium on Stabilization of Greenhouse Gases, 1–3 February, 2005 Met Office Hadley Centre for Climate Change, Exeter. Department for Environment, Food and Rural Affairs, London. [http://www.stabilisation2005.com/outcomes.html]. August 2007.

Schlesinger, Michael E., Jianjun Yin, Gary Yohe, Natalia G. Andronova, Sergey Malyshev and Bin Li. 2005. "Assessing the Risk of a Collapse of the Atlantic Thermohaline Circulation." In Avoiding Dangerous Climate Change. Conference Report for Symposium on Stabilization of Greenhouse Gases, 1–3 February, 2005. Met Office Hadley Centre for Climate Change, Exeter. Department for Environment, Food and Rural Affairs, London.

Schnepf, Randy. 2006. "European Union Biofuels Policy and Agriculture: An Overview." Congressional Research Service (CRS) Report for Congress, Washington, DC.

Schröter, D., M. Zebisch and T. Grothmann. 2005. "Climate Change in Germany - Vulnerability and Adaptation of Climate-Sensitive Sectors." Klimastatusbericht. [http://www.schroeter-patt.net/Schroeter-et-al-KSB06.pdf]. July 2007.

Schubert, Bernd. 2005. "The Pilot Social Cash Transfer Scheme. Kalomo District, Zambia." CPRC Working Paper 52. Chronic Poverty Research Centre, Institute for Development Policy and Management, University of Manchester.

Seager, Ashley and Mark Milner. 2007. "No Policies, no Cash: The Result: Missed Targets." The Guardian, London. 13 August. [http://www.guardian.co.uk/environment/2007/aug/13/renewableenergy.climatechange]. September 2007.

Sen, Amartya. 1999. *Development as Freedom.* Anchor Books, New York.

———. **2004.** "Why We Should Preserve the Spotted Owl." *London Review of Books* 26(3). [http://www.lrb.co.uk/v26/n03/sen_01_.html]. August 2007.

Shapiro, Robert J. 2007. "Addressing the Risks of Climate Change: The Environmental Effectiveness and Economic Efficiency of Emissions Caps and Tradeable Permits, Compared to Carbon Taxes." February. [http://www.theamericanconsumer.org/shapiro.pdf]. August 2007.

Sharp, Kay, Taylor Brown and Amdissa Teshome. 2006. "Targeting Ethiopia's Productive Safety Net Programme (PSNP)." Overseas Development Institute, London and the IDL Group Ltd., Bristol.

Shen, Dajun and Ruiju Liang. 2003. "State of China's Water." Research Report. Third World Centre for Water Management with the Nippon Foundation. [www.thirdworldcentre.org/epubli.html]. August 2007.

Sierra Club. 2006. "Dirty Coal Power—Clean Air." [http://www.sierraclub.org/cleanair/factsheets/power.asp]. August 2007.

Sijm, Jos, Karsten Neuhoff and Yihsu Chen. 2006. "CO_2 Cost Pass-through and Windfall Profits in the Power Sector." *Climate Policy* 6: 49–72.

Singer, Peter. 1993. *Practical Ethics.* 2nd Edition. Cambridge University Press, Cambridge.

———. **2002.** *One World: The Ethics of Globalization.* 2nd Edition. Yale University Press, New Haven, Connecticut.

SIPRI (Stockholm International Peace Research Institute). 2007. "World and regional military expenditure estimates 1988-2006." [http://www.sipri.org/contents/milap/milex/mex_wnr_table.html]. June 2007.

Skutsch, Margaret, Ulrike Roehr, Gotelind Alber, Joanne Rose and Roselyne van der Heul. 2004. "Mainstreaming Gender into the Climate Change Regime." *Gender and Climate Change.* [http://www.gencc.interconnection.org/Gender&CCCOP10.pdf]. August 2007.

Slater, Rachel, Steve Ashley, Mulugeta Tefera, Mengistu Buta and Delelegne Esubalew. 2006. Ethiopia Productive Safety Net Programme (PSNP). Policy, Programme and Institutional

Linkages. Final Report. Overseas Development Institute,
London; the IDL Group Ltd., Bristol; and Indak International
Pvt., Addis Ababa.

**Smale, Robin, Murray Hartley, Cameron Hepburn, John Ward and
Michael Grubb. 2006.** "The Impact of CO_2 Emissions Trading on
Firm Profits and Market Prices." *Climate Policy* 6: 29–46.

Smith, Adam. 1854. *The Theory of Moral Sentiments.* Paperback edition
2004. Kessinger Publishing, Oxford.

Smith, Joseph and David Shearman. 2006. *Climate Change Litigation.
Analysing the Law, Scientific Evidence and Impacts on the Environment,
Health and Property.* Presidian Legal Publications, Adelaide.

Smithsonian National Air and Space Museum. 1999. "Apollo to
the Moon." [http://www.nasm.si.edu/exhibitions/attm/attm.html].
September 2007.

**Solórzano, Raúl, Ronnie de Camino, Richard Woodward, Joseph
Tosi, Vicente Watson, Alexis Vásquez, Carlos Villalobos,
Jorge Jiménez, Roberth Repetto and Wilfrido Cruz. 1991.**
Accounts Overdue: Natural Resource Depreciation in Costa Rica.
World Resources Institute, Washington, DC.

Sperling, Daniel and James S. Cannon. 2007. *Driving Climate
Change. Cutting Carbon from Transportation.* Elsevier, New York.

State of California. 2005. "Executive Order S-3-05 by the Governor of the
State of California." Executive Department. Sacramento, California.

State of California. 2006. *Chapter 488, Assembly Bill No. 32.*
27 September.

Steenblik, Ronald. 2007. "Born Subsidized: Biofuel Production in
the USA." Global Subsidies Initiative. International Institute for
Sustainable Development, Winnipeg.

Stern, Nicholas. 2006. *The Economics of Climate Change. The Stern
Review.* Cambridge University Press, Cambridge and New York.

Stern, Nicholas and Chris Taylor. 2007. "Climate Change: Risk,
Ethics and the Stern Review," *Science* 317 (5835): 203–204.

Sumaila, Ussif R. and Carl Walters. 2005. "Intergenerational
Discounting: a New Intuitive Approach." *Ecological Economics* 52:
135–142.

Sumaila, Ussif R. and Kevin Stephanus. 2006. "Declines in
Namibia's Pilchard Catch : the Reasons and Consequences." In
Climate Change and the Economics of the World's Fisheries. (R.
Hannesson, Manuel Barange and Samuel Herrick Jr., eds.) Edward
Elgar Publishing, Cheltenham.

Summa, Hilkka. 2007. "Energy Crops and the Common Agricultural
Policy." Speech for the Third International European Conference
on GMO-free Regions, Biodiversity and Rural Development.
European Commission, Directorate-General Agriculture and Rural
Development. 19–20 April 2007. Brussels.

**Tanner T.M., A. Hassan, K.M.N. Islam, D. Conway, R. Mechler,
A.U. Ahmed and M. Alam. 2007.** "ORCHID: Piloting Climate
Risk Screening in DFID Bangladesh." Research Report. Institute of
Development Studies, University of Sussex, Brighton.

Tanser, F.C., B. Sharp and D. le Sueur. 2003. "Potential Effect of
Climate Change on Malaria Transmission in Africa."
Lancet Infectious Diseases 362: 1792–1798.

Tauli-Corpuz, Victor and Parshuram Tamang. 2007. "Oil Palm
and Other Commercial Tree Plantations, Monocropping: Impacts
on Indigenous People's Land Tenure and Resource Management
Systems and Livelihoods." Paper presented to the Sixth Session,
United Nations Permanent Forum on Indigenous Issues,
14–25 May 2007, New York.

TERI (The Energy and Resources Institute). 2006. "Modeling a Low
Carbon Pathway for India." Presentation at CoP 12/MoP2. November.

——. **2007.** "Adaptation to Climate Change in the Context of
Sustainable Development." Background Paper No.9. Mumbai.

The Economist. 2007a. "Losing Sleep over Climate Change." 16 July.
London.

——. **2007b.** "Cleaning up." 31 May. London.

——. **2007c.** "Double Deluge." 26 July. London.

The Japan Times. 2007. "Japan to Seek 50% Global Emission Cut at
G-8 Meet." 9 May. Tokyo. [http://search.japantimes.co.jp/print/
nn20070509al.html]. August 2007.

The Spectator. 2007. "The Leader : Climate of Opinion". 2007. I
The Spectator. 10 March 2007. Pg. 5. London. [http://www.
spectator.co.uk/archive/the-week/28377/climate-of-opinion.
thtml]. August 2007.

Thompson, Martha and Izaskun Gaviria. 2004. "Cuba, Weathering
the Storm. Lessons in Risk Reduction from Cuba." Oxfam
America, Boston.

Thorpe, Donald. 2007. "Broader, Deeper—and Less Risky?"
Environmental Finance. February print edition: 20–21.

Time Magazine. 1962. "The Thalidomide Disaster." Friday, 10 August.
[http://www.time.com/time/magazine/article/0,9171,873697,00.
html]. August 2007.

Toder, Eric. 2007. "Eliminating Tax Expenditures with Adverse
Environmental Effects." Tax Reform, Energy and the Environment
Policy Brief. The Brookings Institute and World Resources Institute,
Washington, DC.

Tolgfors, Sten, Eskil Erlandsson and Andreas Carlgren. 2007.
"The EU Should Scrap High Tariffs on Ethanol." Government Offices
of Sweden, Stockholm.

Turner, Margery Austin and Sheila R. Zedlewski. 2006. "After
Katrina. Rebuilding Opportunity and Equity into the New New
Orleans." The Urban Institute, Washington, DC.

UKCIP (United Kingdom Climate Information Programme). 2007.
"UKCIP Climate Digest: April." [http://www.ukcip.org.uk/news_
releases/38.pdf]. May 2007.

UN (United Nations). 2005a. "In Larger Freedom: Towards
Development, Security and Human Rights for All." Report of the
Secretary-General. A/59/2005. UN General Assembly, Fifty-ninth
session. Agenda items 45 and 55. New York.

——. **2005b.** *Report on the World Conference on Disaster Reduction.*
18–22 January, Kobe, Hyogo, Japan 2005. UN, New York.

——. **2007a.** "Press Conference by Security Council President."
4 April 2007. Department of Public Information, News and
Media Division, New York. [http://www.un.org/News/briefings/
docs/2007/070404_Parry.doc.htm]. October 2007.

——. **2007b.** *The Millennium Development Goals Report.* New York.

——. **2007c.** *Energy Statistics Year book 2004.* DESA (Department of
Economic and Social Affairs) Statistics Division, New York.

UNDP (United Nations Development Programme). 2005.
*Human Development Report 2005. International Cooperation at a
Crossroads: Aid, Trade and Security in an Unequal World.* Palgrave
Macmillan, New York.

——. **2006a.** "Human Security and Human Development: A
Deliberate Choice." National Human Development Report for
Kenya 2006. Nairobi.

——. **2006b.** *Human Development Report 2006. Beyond Scarcity:
Power, Poverty and the Global Water Crisis.* Palgrave Macmillan,
New York.

——. **2007.** "MDG Carbon Facility: Leveraging carbon finance for
Sustainable Development." New York. [http://www.undp.org/
mdgcarbonfacility/docs/brochure-eng-29may07.pdf].
September 2007.

**UNDP (United Nations Development Programme)-Dryland
Development Centre/Bureau for Conflict Prevention and
Recovery and UN (United Nations)-International Strategy
for Disaster Reduction. 2005.** "Drought Risk and Development
Policy." Discussion paper prepared for the UNDP-DDC/BCPR and
UN-ISDR Expert Workshop *Drought Risk and Development Policy*,
31 January–2 February, 2005, Nairobi.

UNDP and AusAID 2004. "The Regional Poverty Assessment Mekong
River Region." UNDP and AusAID. [http://siteresources.worldbank.
org/INTVIET NAM/Resources/Mekong_PPA_English.pdf].
September 2007.

UNDP (United Nations Development Programme)–Global Environment Facility (GEF). 2003. "The Adaptation Policy Framework. User's Guidebook." UNDP, New York.

UNDP (United Nations Development Programme) Ukraine. 2005. "The New Wave of Reform : On Track to Succeed. Analysis of policy developments in January – June 2005 and further recommendations." The Blue Ribbon Commission for Ukraine, Kiev. [http://www.un.org.ua/brc/brci/docs/BRC2Final190705Eng.pdf?id=1123140007&cm=doc&fn=brc2final190705eng.pdf&l=e]. September 2007.

——. **2006.** "The State and the Citizen: Delivering on Promises." Blue Ribbon Commission Report for Ukraine, Kiev. [http://www.un.org.ua/files/BRC3_Eng.pdf]. September 2007.

UNESCO (United Nations Educational, Scientific and Cultural Organization). 2006. *EFA Global Monitoring Report 2006: Education for All, Literacy for Life.* Paris.

UN–E (United Nations – Energy). 2005. "The Energy Challenge for Achieving the Millennium Development Goals." [http://es.un.org/un-energy]. August 2007.

UNEP (United Nations Environment Programme). 2005. "Potential for Rainwater Harvesting in Africa. A GIS Overview." Nairobi.

——. **2007a.** *Sudan. Post-Conflict Environmental Assessment.* Nairobi. [http://sudanreport.unep.ch/UNEP_Sudan.pdf]. September 2007.

——. **2007b.** "Global Outlook for Ice and Snow." DEWA (Division of Early Warning and Assessment), Nairobi.

UNEP (United Nations Environment Programme) and GRID (Global Resource Information Database)–Arendal. 2001. "Vital Climate Graphics." Arendal, Norway. [http://www.grida.no/climate/vital/36.htm]. May 2007.

UNFCCC (United Nations Framework Convention on Climate Change). 1998. "Kyoto Protocol to the United Nations Framework Convention on Climate Change." Climate Change Secretariat, Bonn. [http://unfccc.int/resource/docs/convkp/kpeng.pdf]. September 2007.

——. **2006.** "National Greenhouse Gas Inventory Data for the Period 1990 to 2004 and Status of Reporting." Document number FCCC/SPI/2006/26. Note by the Secretariat. Bonn.

——. **2007a.** Vulnerability and Adaptation to Climate Change in Small Island Developing States. Background paper for Expert Meeting on Adaptation for Small Island Developing States, 5–7 February Jamaica and 26–28 February Cook Islands. Climate Change Secretariat, Bonn.

——. **2007b.** "Registered Project Activities by Host Party". [http://cdm.unfccc.int/Statistics/Issuance/CERsIssuedByHostPartyPieChart.html]. July 2007.

——. **2007c.** "Report on the analysis of existing and potential investment and financial flows relevant to the development of an effective and appropriate international response to climate change." Dialogue on Long term Cooperative Action to Address Climate Change by Enhancing Implementation of the Convention. Dialogue Working Paper 8. Bonn.

——. **2007d.** "Clean Development Mechanism (CDM)." Webpage. [http://cdm.unfccc.int/index.html]. September 2007.

——. **2007e.** "CDM Statistics." [http://cdm.unfccc.int/Statistics/index.html]. September 2007.

UN-HABITAT (United Nations Human Settlements Programme). 2006. *The State of the World's Cities Report 2006/07.* Nairobi.

UNICEF (United Nations Children's Fund). 2006. "Schools Empty as Drought Effects Linger in Ethiopia." Press Report. New York. [http://www.unicef.org.uk/press/news_detail.asp?news_id=724]. January 2007.

Urban Institute. 2005. "Katrina: Demographics of a Disaster." The Urban Institute, Washington, DC.

USAID FEWS NET (United States Agency for International Development Famine Early Warning Systems Network). 2006. "Guatemala Food Security Update." [http://www.fews.net/centers/innerSections.aspx]. April 2006.

——. **2007.** "Hurricane Stan Affecting Household Stocks." [http://www.fews.net/centers/innerSections.aspx]. August 2007.

USCAP (United States Climate Action Partnership). 2007. "A Call for Action." [www.us-cap.org/uscap/callforaction.pdf]. September 2007.

Ürge-Vorsatz, Diana, Gergana Miladinova and László Paizs. 2006. "Energy in Transition: From the Iron Curtain to the European Union." Energy Policy 34(15): 2279–2297.

Ürge-Vorsatz, Diana, L.D. Danny Harvey, Sevastianos Mirasgedis and Mark Levine. 2007a. "Mitigating CO_2 Emissions from Energy Use in the World's Buildings." *Building Research and Information* 35(4) 370–398.

Ürge-Vorsatz, Diana, Sebastian Mirasgedis and Sojia Koeppel. 2007b. "Appraisal of Policy Instruments for Reducing Buildings' CO_2 Emissions." *Building Research and Information* 35(4): 458–477.

Vaid, B.H., C. Gnanaseelan, P.S. Polito and P.S. Salvekar. 2006. *Influence of El Nino on the Biennial and Annual Rossby Waves Propagation in the Indian Ocean with Special Emphasis on Indian Ocean Dipole.* Indian Institute of Tropical Meteorology, Pune.

Vakis, Renos. 2006. "Complementing Natural Disasters Management: The Role of Social Protection." Social Protection Discussion Paper No. 0543. World Bank, Washington, DC.

Van Lieshout, M., R.S. Kovats, M.T.J. Livermore and P. Martens. 2004. "Climate Change and Malaria: Analysis of the SRES Climate and Socio-Economic Scenarios." *Global Environmental Change* 14: 87–99.

Vergara, W., A. M. Deeb, A. M. Valencia, R. S. Bradley, B. Francou, A. Zarzar, A. Grünwaldt and S. M. Haeussling. 2007. Economic Impacts of Rapid Glacier Retreat in the Andes, Eos. *Transactions of the American Geophysical Union,* 88(25): 261.

Victor, David G. 2001. *The Collapse of the Kyoto Protocol and the Struggle to Slow Global Warming.* A Council on Foreign Relations Book. Princeton University Press, Princeton and Oxford.

Wagstaff, Adam and Mariam Claeson. 2004. *The Millennium Development Goals for Health. Rising to the Challenges.* World Bank, Washington, DC.

Warren, Rachel, Nigel Arnell, Robert Nicholls, Peter Levy and Jeff Price. 2006. "Understanding the Regional Impacts of Climate Change. Research Report Prepared for the Stern Review on the Economics of Climate Change." Research Working Paper No. 90. Tyndall Centre for Climate Change, Norwich.

Washington, Richard, Mike Harrison, Declan Conway, Emily Black, Andrew Challinor, David Grimes, Richard Jones, Andy Morse, Gillian Kay and Martin Todd. 2006. "African Climate Change. Taking the Shorter Route." *Bulletin of the American Meteorological Society* 87(10): 1355–1366.

Watson, Robert. 2007. "Financing the Transition to a Low Carbon Economy. Beyond Stern: Financing International Investment in Low Carbon." World Bank, Washington, DC.

Watt-Cloutier, Sheila. 2006. "The Canadian Environment Awards Citation of Lifetime Achievement. Remarks by Sheila Watt-Cloutier." Inuit Circumpolar Conference, Canada. 5 June. Vancouver. [http://www.inuitcircumpolar.com/index.php?auto_slide=&ID=357&Lang=En&Parent_ID=&current_slide_num=]. August 2007.

Watt-Cloutier, Sheila, Terry Fenge and Paul Crowley. 2004. "Responding to Global Climate Change: The Perspective of the Inuit Circumpolar Conference on the Arctic Climate Impact Assessment." Inuit Circumpolar Conference. Ontario.

WEDO (Women's Environment and Development Organization). 2007. "Changing the Climate: Why Women's Perspectives Matter." New York.

Weitzman, Martin L. 2007. "The Stern Review of the Economics of Climate Change." Book review for *Journal of Economic Literature (JEL).* Harvard University, Cambridge, Massachusetts. [http://www.

economics.harvard.edu/faculty/Weitzman/papers/JELSternReport.
pdf]. July 2007.

Wolf, Martin. 2006a. "Curbs on Emissions Will Take a Change of
Political Climate." Financial Times. 7 November 2006. London.
[http://www.ft.com/cms/s/cb25e5a4-6e7f-11db-b5c4-
0000779e2340.html]. August 2007.

Wolf, Martin. 2006b. "Figures Still Justify Swift Climate Action."
Financial Times. 14 November 2006. London. [http://www.
ft.com/cms/s/8dc6191a-740e-11db-8dd7-0000779e2340.html].
July 2007.

World Bank. 2003. *Reaching the Rural Poor: A Renewed Strategy for
Rural Development*. Washington, DC.

————. **2004a.** *Saving Fish and Fishers: Toward Sustainable and
Equitable Governance of the Global Fishing Sector*. Agriculture and
Rural Development Department, Washington, DC.

————. **2004b.** "Coral Reef Targeted Research and Capacity Building
for Management Project." Project Appraisal Document. World
Bank, Washington, DC. and East Asia Environment and Social
Development Unit (EASES), Environment Department, University of
Queensland, Brisbane.

————. **2005a.** "World Bank Group Progress on Renewable Energy and
Energy Efficiency: Fiscal Year 2005." The Energy and Mining Sector
Board, Washington, DC.

————. **2005b.** "Learning the Lessons from Disasters Recovery,
The Case of Mozambique." Disaster Risk Management Working
Paper Series No.12, Hazard Management Unit, Washington, DC.

————. **2006a.** *Hazards of Nature, Risks to Development: An IEG
(Independent Evaluation Group) Evaluation of World Bank Assistance
for Natural Disasters*. Washington, DC.

————. **2006b.** *Re-engaging in Agricultural Water Management.
Challenges and Options*. Washington, DC.

————. **2006c.** "Not If, But When: Adapting to Natural Hazards
in the Pacific Islands Region, A Policy Note." Washington, DC.

————. **2006d.** *"Clean Energy and Development: Towards an
Investment Framework."* Washington, DC.

————. **2006e.** *Global Monitoring Report 2006*. Washington, DC.

————. **2006f.** "Overcoming Drought: Adaptation Strategies for Andhra
Pradesh." Washington, DC.

————. **2006g.** *World Development Report 2006: Equity and
Development*. Washington, DC.

————. **2007a.** "An Investment Framework for Clean Energy and
Development. A Platform for Convergence of Public and Private
Investments." Washington, DC.

————. **2007b.** "Clean Energy for Development Investment Framework:
World Bank Group Action Plan." Development Committee (Joint
Ministerial Committee of the Boards of Governors of the Bank
and the Fund On the Transfer of Real Resources to Developing
Countries), Washington, DC.

————. **2007c.** *Global Monitoring Report 2007: Confronting the
Challenges of Gender Equality and Fragile States*. Washington, DC.

————. **2007d.** *World Development Indicators*. CD-ROM. Washington, DC.

————. **2007e.** *Global Economic Prospects 2007: Managing the Next
Wave of Globalization*. Washington, DC.

————. **2007f.** State and Trends of the Carbon Market 2007.
Washington, DC.

————. **2007g.** "Climate Change. Frequently Asked Questions."
[http://web.worldbank.org/WBSITE/EXTERNAL/
EXTSITETOOLS/ 0,,contentMDK:20205607~menuPK:43533
2~pagePK:98400~piPK:98424~theSitePK:95474,00.html].
August 2007.

World Commission on Environment and Development. 1987. *Our
Common Future*. Oxford University Press, Oxford.

WFP (World Food Programme). 2005a. "Emergency Assessment
Brief: Niger." August. Rome.

————. **2005b.** "Emergency Report No. 18, 29 April 2005." [http://
www.wfp.org/english/?ModuleID=78&Key=631#404]. July 2007.

————. **2007.** "Mozambique Emergency Situation Report." 30 March.
WFP, Rome.

WHO (World Health Organization). 2006. *The World Health Report
2006 – Working Together for Health*. Geneva.

**WHO (World Health Organization) and UNICEF (United Nations
Children's Fund). 2005.** *World Malaria Report 2005*. WHO
and UNICEF, Geneva and New York. [http://www.rbm.who.
int/wmr2005/index.html]. March 2007.

WMO (World Meteorological Organization). 2006. *Statement on the
Status of the Global Climate in 2005*. Geneva.

————. **2007.** "Observing Stations." Publication No. 9, Volume A, (9
July 2007). [http://www.wmo.int/pages/prog/www/ois/volume-
a/vola-home.htm]. September 2007.

WRI (World Resources Institute). 2007a. "Climate Analysis
Indicators Tool (CAIT)." [http://www.wri.org/climate/project_
description2.cfm?pid=93]. July 2007.

————. **2007b.** *Earth Trends, the Environmental Information Portal*. Online
database. Accessed July 2007.

**WRI (World Resources Institute), UNEP (United Nations
Environment Programme) and World Bank in collaboration
with United Nations Development Programme (UNDP).
2005.** *World Resources 2005: The Wealth of the Poor – Managing
Ecosystems to Fight Poverty*. World Resources Institute,
Washington, DC.

World Watch Institute. 2005. *Vital Signs*. [http://www.amazon.
com/Vital-Signs-2006-2007-Trends-Shaping/dp/0393328724].
August 2007.

WWF (World Wide Fund for Nature). 2002. "Managing Floods in
Europe: The Answers Already Exist." WWF Danube-Carpathian
Programme and WWF Loving Waters Programme-Europe. [http://
assets.panda.org/downloads/managingfloodingbriefingpaper.pdf].
August 2007.

————. **2006a.** "Including aviation into the EU Emissions Trading
Scheme—WWF Position Statement." London.

————. **2006b.** "Use of CDM/JI Project Credits by Participant in Phase
II of the EU Emissions Trading Scheme—A WWF Summary of the
Ecofys UK Report." London.

————. **2007a.** "Emission Impossible: access to JI/CDM credits
in Phase II of the EU Emissions Trading Scheme WWF–UK."
London. [http://www.panda.org/about_wwf/where_we_work/
europe/what_we_do/epo/intiatives/climate/eu_emissions_
trading/index.cfm]. August 2007.

————. **2007b.** "The EU Emissions Trading Scheme." London.

World Wind Energy Association. 2007. "New World Record in
Wind Power Capacity: 14,9 GW added in 2006 – Worldwide
Capacity at 73,9 GW." 29 January. [http://www.wwindea.
org/home/index.php?option=com_content&task=view&id=167
&Itermid=43]. August 2007.

**Wu, Zongxin, Pat de la Quil, Eric D. Larson, Chen Wenying
and Gao Pengfei. 2001.** "Future Implications of China's
Energy-Technology Choices." Prepared for the Working Group
on Energy Strategies and Technologies. China Council for
International Cooperation on Environment and Development
(CCICED), Beijing.

Zeitlin, June. 2007. "Statement by June Zeitlin, Women's Environment
and Development Organization in informal thematic debate on
Climate Change as a Global Challenge. United Nations General
Assembly." UNDP (United Nations Development Programme),
Washington, DC.

**Zero Emissions Fossil Fuel Power Plants Technology Platform.
2006.** "A Vision for Zero Emission Fossil Fuel Power Plants." EUR
22043. European Commission, Luxembourg.

Human development indicators

Readers guide and notes to tables

Human development indicator tables

The human development indicator tables provide a global assessment of country achievements in different areas of human development. The main tables are organized thematically, as described by their titles. The tables include data for 175 UN member states—those for which the human development index (HDI) could be calculated—along with Hong Kong Special Administrative Region of China, and the Occupied Palestinian Territories. Because of insufficient cross-nationally comparable data of good quality, the HDI has not been calculated for the remaining 17 UN member countries. Instead a set of basic human development indicators for these countries is presented in Table 1a.

In the tables, countries and areas are ranked by their HDI value. To locate a country in the tables, refer to the *Key to countries* on the back cover flap where countries with their HDI ranks are listed alphabetically. Most of the data in the tables are for 2005 and are those available to the Human Development Report Office (HDRO) as of 1 July 2007, unless otherwise specified.

Sources and definitions

HDRO is primarily a user, not a producer, of statistics. It relies on international data agencies with the mandate, resources and expertise to collect and compile international data on specific statistical indicators. Sources for all data used in compiling the indicator tables are given in short citations at the end of each table. These correspond to full references in *Statistical references*. When an agency provides data that it has collected from another source, both sources are credited in the table notes. But when an agency has built on the work of many other contributors, only that agency is given as the source. In order to ensure that all calculations can be easily replicated the source notes also show the original data components used in any calculations by HDRO. Indicators for which short, meaningful definitions can be given are included in *Definitions of statistical terms*. Other relevant information appears in the notes at the end of each table. For more detailed technical information about these indicators, please consult the relevant websites of the source agencies through the *Human Development Report* website at http://hdr.undp.org/statistics/.

Inconsistencies between national and international estimates

When compiling international data series, international data agencies often apply international standards and harmonization procedures to improve comparability across countries. When international data are based on national statistics, as they usually are, national data may need to be adjusted. When data for a country are missing, an international agency may produce an estimate if other relevant information can be used. And because of the difficulties in coordination between national and international data agencies, international data series may not incorporate the most recent national data. All these factors can lead to substantial differences between national and international estimates.

This Report has often brought such inconsistencies to light. When data inconsistencies have arisen, HDRO has helped to link national and international data authorities to address those inconsistencies. In many cases this has led to better statistics in the Report. HDRO continues to advocate improving international data and plays an active role in supporting efforts to enhance data quality. It works with national agencies and international bodies to improve data consistency through more systematic reporting and monitoring of data quality.

Comparability over time

Statistics presented in different editions of the Report may not be comparable, due to revisions to data or changes in methodology. For this reason HDRO strongly advises against trend analysis based on data from different editions. Similarly, HDI values and ranks are not comparable across editions of the Report. For HDI trend analysis based on consistent data and methodology, refer to Table 2 (Human development index trends).

Country classifications

Countries are classified in four ways: by human development level, by income, by major world aggregates and by region (see the *Classification of countries*). These designations do not necessarily express a judgement about the development stage of a particular country or area. The term *country* as used in the text and tables refers, as appropriate, to territories or areas.

Human development classifications. All countries included in the HDI are classified into one of three clusters of achievement in human development: high human development (with an HDI of 0.800 or above), medium human development (HDI of 0.500–0.799) and low human development (HDI of less than 0.500).

Income classifications. All countries are grouped by income using World Bank classifications: high income (gross national income per capita of US$10,726 or more in 2005), middle income (US$876–$10,725) and low income (US$875 or less).

Major world classifications. The three global groups are *developing countries, Central and Eastern Europe and the Commonwealth of Independent States (CIS)* and the *Organization for Economic Co-operation and Development (OECD)*. These groups are not mutually ex-

clusive. (Replacing the OECD group with the high-income OECD group and excluding the Republic of Korea would produce mutually exclusive groups). Unless otherwise specified, the classification *world* represents the universe of 194 countries and areas covered—192 UN member countries plus Hong Kong Special Administrative Region of China, and the Occupied Palestinian Territories.

Regional classifications. Developing countries are further classified into regions: *Arab States, East Asia and the Pacific, Latin America and the Caribbean (including Mexico), South Asia, Southern Europe and Sub-Saharan Africa.* These regional classifications are consistent with the Regional Bureaux of the United Nations Development Programme. An additional classification is *least developed countries,* as defined by the United Nations (UN-OHRLLS 2007).

Aggregates and growth rates

Aggregates. Aggregates for the classifications described above are presented at the end of tables when it is analytically meaningful to do so and sufficient data are available. Aggregates that are the total for the classification (such as for population) are indicated by a T. All other aggregates are weighted averages.

In general, an aggregate is shown for a country grouping only when data are available for at least half the countries and represent at least two-thirds of the available weight in that classification. HDRO does not supply missing data for the purpose of aggregation. Therefore, unless otherwise specified, aggregates for each classification represent only the countries: for which data are available; refer to the year or period specified; and refer only to data from the primary sources listed. Aggregates are not shown where appropriate weighting procedures are unavailable.

Aggregates for indices, growth rates and indicators covering more than one point in time are based only on countries for which data exist for all necessary points in time. When no aggregate is shown for one or more regions, aggregates are not always shown for the world classification, which refers only to the universe of 194 countries and areas.

Aggregates in this Report will not always conform to those in other publications because of differences in country classifications and methodology. Where indicated, aggregates are calculated by the statistical agency providing the data for the indicator.

Growth rates. Multiyear growth rates are expressed as average annual rates of change. In calculating growth rates, HDRO uses only the beginning and end points. Year-to-year growth rates are expressed as annual percentage changes.

Country notes

Unless otherwise noted, data for China do not include Hong Kong Special Administrative Region of China, Macao Special Administrative Region of China, or Taiwan Province of China. In most cases data for Eritrea before 1992 are included in the data for Ethiopia. Data for Germany refer to the unified Germany, unless otherwise noted. Data for Indonesia include Timor-Leste through 1999, unless otherwise noted. Data for Jordan refer to the East Bank only. Economic data for the United Republic of Tanzania cover the mainland only. Data for Sudan are often based on information collected from the northern part of the country. While Serbia and Montenegro became two independent States in June 2006, data for the union of the two States have been used where data do not yet exist separately for the independent States. Where this is the case, a note has been included to that effect. And data for Yemen refer to that country from 1990 onwards, while data for earlier years refer to aggregated data for the former People's Democratic Republic of Yemen and the former Yemen Arab Republic.

Changes to existing indicator tables and introduction of new tables

This year, a number of changes have been introduced into some existing indicator tables and three new tables have been included. This is with a view to making the indicator tables more policy-relevant and also to make a link to the theme of this year's Report. New indicators have also been introduced in response to some of the recommendations of the GDI-GEM review held in 2006. As a consequence, some tables do not correspond to the indicator table bearing that number in HDR 2006.

Changes to existing tables

The 'Energy and environment' table (formerly Table 21 in HDR 2006) has been extended and split into four tables: energy and the environment (Table 22), energy sources (Table 23), carbon dioxide (CO_2) emissions and stocks (Table 24) and status of major internationl environmental treaties (Table 25).

The following new indicators have been introduced in the 'Energy and the environment' table (Table 22);

- Percentage change in electricity consumption between 1990 and 2004
- Electrification rate
- Population without access to electricity
- Change in GDP per capita per unit of energy use between 1990 and 2004
- Forest as a percentage of total land.
- Total area of forest cover in 2005
- Absolute change in area of forest cover between 1990 and 2005
- Average annual percentage change in forest cover between 1990 and 2005.

These indicators can be used: to monitor progress in improving access to modern energy; in reducing energy intensity of GDP growth; and to assess rates of deforestation or afforestation in countries.

The 'Energy sources' table (Table 23) is an entirely new table describing the share of total primary energy supply from different sources: fossil fuels (coal, oil and natural gas), renewable energy (from hydro, solar, wind, geothermal as well as biomass and waste) and other sources (nuclear). The total primary energy supply is also given in this table.

The 'Carbon dioxide emissions and stocks' table (Table 24) brings together indicators on CO_2 emissions previously contained in the original energy and environment table and introduces a number of new indicators including:

- Total CO_2 emissions and the average annual percentage change between 1990 and 2004
- Countries' share of the world's total CO_2 emissions
- CO_2 emissions per capita (carbon footprints)
- CO_2 emissions per unit of energy use (carbon intensity of energy)
- CO_2 emissions per unit of GDP (carbon intensity of growth)
- CO_2 emissions from forest biomass and total carbon stocks in forests.

The 'Status of major environmental treaties' table (Table 25) extends the range of environmental treaties covered in the original table on energy and environment and presents them all in a single table.

The 'Victims of Crime' table (formerly Table 23 in HDR 2006) has been dropped for this Report in the absence of a new round of the International Crime Victims Survey on which the table was based since 2000–01. It has been replaced by a table on crime and justice (Table 27) which presents information on homicide rates, prison populations and the abolition or retention of capital punishment.

Tables introduced in response to some of the GDI-GEM review recommendations

Cross-nationally comparable gender disaggregated statistics are a major challenge to assessing progress towards the elimination of all forms of discrimination against women and men. In response to some of the recommendations from the GDI-GEM review, new gender disaggregated indicators of labour force participation in non-OECD countries have been introduced and an existing indicator table was also modified to provide more information.

Previously, unemployment information was presented for OECD countries only because of insufficient comparable data for other countries. In the new Table 21, in addition to data for men and women, such labour force statistics as total employment and unemploy-

ment, the distribution of employment by economic activity and participation in the informal sector are presented.

Table 32 'Gender work and time allocation' is a modification of Table 28 in HDR 2006, which provides information on how women and men share their time between market and nonmarket activities. Nonmarket activities have been broken down further to provide information on how much time women and men spend daily on cooking and cleaning, caring for children, on such other activities as personal care, and on free time for leisure and other social activities.

HDRO will continue to work with national, regional and international agencies towards improving availability and quality of gender-disaggregated data.

Currency conversion

Throughout the Report, for currency units that were originally reported in currencies other than US dollars (US$), the estimated equivalent value in US$ has been provided right next to them. The exchange rates used for these conversions are the 'average period' rates for the specific year, while for currencies with no specified year, the yearly rate for the most recently available 'average period' was used, as reported in the September 2007 International Monetary Fund's *International Financial Statistics* report.

Symbols

In the absence of the words *annual, annual rate* or *growth rate*, a dash between two years, such as in 1995–2000, indicates that the data were collected during one of the years shown. A slash between two years, such as in 1998/2001, indicates an average for the years shown unless otherwise specified. The following symbols are used:

..   Data not available

(.)   Greater (or less) than zero but small enough to be rounded off to zero at the displayed number of decimal points

<   Less than

—   Not applicable

T   Total.

Note to Table 1: about this year's human development index

The human development index (HDI) is a composite index that measures the average achievements in a country in three basic dimensions of human development: a long and healthy life; access to knowledge; and a decent standard of living. These basic dimensions are measured by life expectancy at birth, adult literacy and combined gross enrolment in primary, secondary and tertiary level education, and gross domestic product (GDP) per capita in Purchasing Power Parity US dollars (PPP US$), respectively. The index is constructed from indicators that are available globally using a methodology that is simple and transparent (see *Technical note 1*).

While the concept of human development is much broader than any single composite index can measure, the HDI offers a powerful alternative to GDP per capita as a summary measure of human well-being. It provides a useful entry point into the rich information contained in the subsequent indicator tables on different aspects of human development.

Data availability determines HDI country coverage

The HDI in this Report refers to 2005. It covers 175 UN member countries, along with Hong Kong Special Administrative Region of China, and the Occupied Palestinian Territories.

To enable cross-country comparisons, the HDI is, to the extent possible, calculated based on data from leading international data agencies available at the time the Report was prepared (see *Primary international data sources* below). But, for a number of countries, data are missing from these agencies for one or more of the four HDI components. For this reason, 17 UN member countries cannot be included in the HDI ranking this year. Instead a set of basic HDIs for these countries is presented in Table 1a.

In very rare cases, HDRO has made special efforts to obtain estimates from other international, regional or national sources when the primary international data agencies lack data for one or two HDI components of a country. In a very few cases HDRO has produced an estimate. These estimates from sources other than the primary international agencies are clearly documented in the footnotes to Table 1. They are of varying quality and reliability and are not presented in other indicator tables showing similar data.

Primary international data sources

Life expectancy at birth. The life expectancy at birth estimates are taken from *World Population Prospects 1950–2050: The 2006 Revision* (UN 2007e) the official source of UN population estimates and projections. They are prepared biennially by the United Nations Department of Economic and Social Affairs Population Division (UNPD) using data from national vital registration systems, population censuses and surveys.

In *The 2006 Revision* UNPD incorporated available national data through the end of 2006. For assessing the impact of HIV/AIDS, the latest HIV prevalence estimates prepared by the Joint United Nations Programme on HIV/AIDS (UNAIDS) are combined with a series of assumptions about the demographic trends and mortality of both infected and non-infected people in each of the 62 countries for which the impact of the disease is explicitly modelled.

The availability of new empirical evidence on the HIV/AIDS epidemic and demographic trends often requires adjustments to earlier estimates. Recent UNAIDS estimates indicate a decline in the rate of transition of new individuals into the high risk group. Based on these and other factors, *World Population Prospects 1950–2050: The 2006 Revision* made several methodological changes, which resulted in significant increases in estimates of life expectancy at birth for some of the countries. Firstly, *The 2006 Revision* incorporates a longer survival for infected persons receiving treatment. Secondly, the rate of mother to child transmission is also projected to decline at varying rates depending on the progress made by each country in increasing access to treatment. The life expectancy estimates published by UNPD are usually five-year averages although it does also produce annual life expectancy estimates interpolated from the five-year averages. The life expectancy estimates for 2005 shown in Table 1 and those underlying Table 2 are from these interpolated data. For details on *World Population Prospects 1950–2050: The 2006 Revision* see www.un.org/esa/population/unpop.htm.

Adult literacy rate. This Report uses data on adult literacy rates from the United Nations Educational, Scientific and Cultural Organization (UNESCO) Institute for Statistics (UIS) April 2007 Assessment (UNESCO Institute for Statistics 2007a), that combines direct national estimates with recent estimates based on its Global age-specific literacy projections model developed in 2007. The national estimates, made available through targeted efforts by UIS to collect recent literacy data from countries, are obtained from national censuses or surveys between 1995 and 2005. Where recent estimates are not available, older UIS estimates, produced in July 2002 and based mainly on national data collected before 1995, have been used instead.

Many high-income countries, having attained high levels of literacy, no longer collect basic literacy statistics and thus are not included in the UIS data. In calculating the HDI, a literacy rate of 99.0% is assumed for high-income countries that do not report adult literacy information.

In collecting literacy data, many countries estimate the number of literate people based on self-reported data. Some use educational attainment data as a proxy, but measures of school attendance or grade completion may differ. Because definitions and data collection methods vary across countries, literacy estimates should be used with caution.

The UIS, in collaboration with partner agencies, is actively pursuing an alternative methodology for measuring literacy, the Literacy Assessment and Monitoring Programme (LAMP). LAMP seeks to go beyond the current simple categories of literate and illiterate by providing information on a continuum of literacy skills. It is hoped that literacy rates from LAMP will eventually provide more reliable estimates.

Combined gross enrolment ratios in primary, secondary and tertiary education. Gross enrolment ratios are produced by the UIS (UNESCO Institute for Statistics 2007c) based on enrolment data collected from national governments (usually from administrative sources) and population data from the *World Population Prospects 1950–2040: The 2004 Revision.* The ratios are calculated by dividing the number of students enrolled in primary, secondary and tertiary levels of education by the total population in the theoretical age group corresponding to these levels. The theoretical age group for tertiary education is assumed to be the five-year age group immediately following on the end of upper secondary school in all countries.

Although intended as a proxy for educational attainment, combined gross enrolment ratios do not reflect the quality of educational outcomes. Even when used to capture access to educational opportunities, combined gross enrolment ratios can hide important differences among countries because of differences in the age range corresponding to a level of education and in the duration of education programmes. Grade repetition and dropout rates can also distort the data. Measures such as the mean years of schooling of a population or school life expectancy could more adequately capture educational attainment and should ideally supplant the gross enrolment ratio in the HDI. However, such data are not yet regularly available for a sufficient number of countries.

As currently defined, the combined gross enrolment ratio measures enrolment in the country of study and therefore excludes students studying abroad from the enrolment ratio of their home country. Current data for many smaller countries, for which pursuit of a tertiary education abroad is common, could substantially under estimate access to education or educational attainment of the population and thus lead to a lower HDI value.

GDP per capita (PPP US$). In comparing standards of living across countries, economic statistics must be converted into purchasing power parity (PPP) terms to eliminate differences in national price levels. The GDP per capita (PPP US$) data for the HDI are provided by the World Bank (World Bank 2007b) for 168 countries based on price data from the last International Comparison Program (ICP surveys and GDP in local currency from national accounts data. The last round of ICP surveys conducted between 1993 and 1996 covered 118 countries. PPPs for these countries are estimated directly by extrapolating from the latest benchmark results. For countries not included in the ICP surveys, estimates are derived through econometric regression. For countries not covered by the World Bank, PPP estimates provided by the Penn World Tables of the University of Pennsylvania (Heston, Summers and Aten 2006) are used.

Though much progress has been made in recent decades, the current PPP data set suffers from several deficiencies, including lack of universal coverage, of timeliness of the data and of uniformity in the quality of results from different regions and countries. Filling gaps in country coverage with econometric regression requires strong assumptions, while extrapolation over time implies that the results become weaker as the distance lengthens between the reference survey year and the current year. The importance of PPPs in economic analysis underlines the need for improvement in PPP data. A new Millennium Round of the ICP has been launched and promises much improved PPP data for economic policy analysis. First results are expected to be published in late 2007 or early 2008. For details on the ICP and the PPP methodology, see the ICP website at www.worldbank.org/data/icp.

Comparisons over time and across editions of the Report

The HDI is an important tool for monitoring long-term trends in human development. To facilitate trend analyses across countries, the HDI is calculated at five-year intervals for the period 1975–2005. These estimates, presented in Table 2, are based on a consistent method-

ology and on comparable trend data available when the Report is prepared.

As international data agencies continually improve their data series, including updating historical data periodically, the year to year changes in the HDI values and rankings across editions of the *Human Development Report* often reflect revisions to data—both specific to a country and relative to other countries—rather than real changes in a country. In addition, occasional changes in country coverage could also affect the HDI ranking of a country, even when consistent methodology is used to calculate the HDI. As a result, a country's HDI rank could drop considerably between two consecutive Reports. But when comparable, revised data are used to reconstruct the HDI for recent years, the HDI rank and value may actually show an improvement.

For these reasons HDI trend analysis should not be based on data from different editions of the Report. Table 2 provides up-to-date HDI trend data based on consistent data and methodology.

HDI for high human development countries

The HDI in this Report is constructed to compare country achievements across the most basic dimensions of human development. Thus, the indicators chosen are not necessarily those that best differentiate between rich countries. The indicators currently used in the index yield very small differences among the top HDI countries, and thus the top of the HDI ranking often reflects only very small differences in these underlying indicators. For these high-income countries, an alternative index—the human poverty index (shown in Table 4)—can better reflect the extent of human deprivation that still exists among the populations of these countries and can help direct the focus of public policies.

For further discussions on the use and limitations of the HDI and its component indicators, see http://hdr.undp.org/statistics.

Acronyms and abbreviations

CDIAC	Carbon Dioxide Information Analysis Center	ISCO	International Standard Classification of Occupations
CIS	Commonwealth of Independent States	ISIC	International Standard Industrial Classification
CO_2	Carbon dioxide		
CO_2e	Carbon dioxide equivalent	ITU	International Telecommunication Union
DAC	Development Assistance Committee (of OECD)	LIS	Luxembourg Income Studies
		MDG	Millennium Development Goals
DHS	Demographic and Health Survey	MICS	Multiple Indicator Cluster Survey
DOTS	Directly Observed Treatment Short courses (method of detection and treatment of tuberculosis)	Mt	Megatonne (one million tonnes)
		ODA	Official development assistance
EM-DAT	Emergency disasters database	OECD	Organization for Economic Co-operation and Development
FAO	Food and Agriculture Organization		
GDI	Gender-related development index	PPP	Purchasing power parity
GDP	Gross domestic product	R&D	Research and development
GEM	Gender empowerment measure	SAR	Special Administrative Region (of China)
GER	Gross enrolment ratio	SIPRI	Stockholm International Peace Research Institute
GNI	Gross national income		
Gt	Gigatonne (one billion tonnes)	SITC	Standard International Trade Classification
HDI	Human development index	TFYR	The former Yugoslav Republic (of Macedonia)
HDRO	Human Development Report Office	UN	United Nations
HIV/AIDS	Human Immunodeficiency Virus/Acquired Immune Deficiency Syndrome	UNAIDS	Joint United Nations Programme on HIV/AIDS
HPI-1	Human poverty index (for developing countries)	UNCTAD	United Nations Conference on Trade and Development
HPI-2	Human poverty index (for OECD countries, Central and Eastern Europe and the CIS)	UNODC	United Nations Office on Drugs and Crime
		UNESCO	United Nations Educational, Scientific and Cultural Organization
IALS	International Adult Literacy Survey		
ICPS	International Centre for Prison Studies	UNDP	United Nations Development Programme
ICSE	International Classification of Status in Employment	UNFPA	United Nations Population Fund
		UNHCR	Office of the United Nations High Commissioner for Refugees
IDMC	Internal Displacement Monitoring Centre		
IEA	International Energy Agency	UNICEF	United Nations Children's Fund
IISS	International Institute for Strategic Studies	UN-ORHLLS	United Nations Office of the High Representative for the Least Developed Countries, Landlocked Developing Countries and Small Island Developing States
ILO	International Labour Organization		
ILOLEX	ILO database on International Labour Standards		
IPU	Inter-Parliamentary Union	WHO	World Health Organization
ISCED	International Standard Classification of Education	WIPO	World Intellectual Property Organization

Human development index

HDI rank [a]	Human development index (HDI) value 2005	Life expectancy at birth (years) 2005	Adult literacy rate (% aged 15 and above) 1995-2005[b]	Combined gross enrolment ratio for primary, secondary and tertiary education (%) 2005	GDP per capita (PPP US$) 2005	Life expectancy index	Education index	GDP index	GDP per capita (PPP US$) rank minus HDI rank[c]
HIGH HUMAN DEVELOPMENT									
1 Iceland	0.968	81.5	.. [d]	95.4 [e]	36,510	0.941	0.978	0.985	4
2 Norway	0.968	79.8	.. [d]	99.2	41,420 [f]	0.913	0.991	1.000	1
3 Australia	0.962	80.9	.. [d]	113.0 [g]	31,794	0.931	0.993	0.962	13
4 Canada	0.961	80.3	.. [d]	99.2 [e,h]	33,375	0.921	0.991	0.970	6
5 Ireland	0.959	78.4	.. [d]	99.9	38,505	0.890	0.993	0.994	-1
6 Sweden	0.956	80.5	.. [d]	95.3	32,525	0.925	0.978	0.965	7
7 Switzerland	0.955	81.3	.. [d]	85.7	35,633	0.938	0.946	0.981	-1
8 Japan	0.953	82.3	.. [d]	85.9	31,267	0.954	0.946	0.959	9
9 Netherlands	0.953	79.2	.. [d]	98.4	32,684	0.904	0.988	0.966	3
10 France	0.952	80.2	.. [d]	96.5	30,386	0.919	0.982	0.954	8
11 Finland	0.952	78.9	.. [d]	101.0 [g]	32,153	0.898	0.993	0.964	3
12 United States	0.951	77.9	.. [d]	93.3	41,890 [f]	0.881	0.971	1.000	-10
13 Spain	0.949	80.5	.. [d]	98.0	27,169	0.925	0.987	0.935	11
14 Denmark	0.949	77.9	.. [d]	102.7 [g]	33,973	0.881	0.993	0.973	-6
15 Austria	0.948	79.4	.. [d]	91.9	33,700	0.907	0.966	0.971	-6
16 United Kingdom	0.946	79.0	.. [d]	93.0 [e]	33,238	0.900	0.970	0.969	-5
17 Belgium	0.946	78.8	.. [d]	95.1	32,119	0.897	0.977	0.963	-2
18 Luxembourg	0.944	78.4	.. [d]	84.7 [i]	60,228 [f]	0.891	0.942	1.000	-17
19 New Zealand	0.943	79.8	.. [d]	108.4 [g]	24,996	0.913	0.993	0.922	9
20 Italy	0.941	80.3	98.4	90.6	28,529	0.922	0.958	0.944	1
21 Hong Kong, China (SAR)	0.937	81.9	.. [j]	76.3	34,833	0.949	0.885	0.977	-14
22 Germany	0.935	79.1	.. [d]	88.0 [e]	29,461	0.902	0.953	0.949	-2
23 Israel	0.932	80.3	97.1 [k]	89.6	25,864	0.921	0.946	0.927	3
24 Greece	0.926	78.9	96.0	99.0	23,381	0.898	0.970	0.910	5
25 Singapore	0.922	79.4	92.5	87.3 [h,k]	29,663	0.907	0.908	0.950	-6
26 Korea (Republic of)	0.921	77.9	.. [d]	96.0	22,029	0.882	0.980	0.900	6
27 Slovenia	0.917	77.4	99.7 [d,l]	94.3	22,273	0.874	0.974	0.902	4
28 Cyprus	0.903	79.0	96.8	77.6 [e]	22,699 [h]	0.900	0.904	0.905	2
29 Portugal	0.897	77.7	93.8 [l]	89.8	20,410	0.879	0.925	0.888	6
30 Brunei Darussalam	0.894	76.7	92.7	77.7	28,161 [h,m]	0.862	0.877	0.941	-8
31 Barbados	0.892	76.6	.. [d,j]	88.9 [h]	17,297 [h,m]	0.861	0.956	0.860	8
32 Czech Republic	0.891	75.9	.. [d]	82.9	20,538	0.849	0.936	0.889	2
33 Kuwait	0.891	77.3	93.3	74.9	26,321 [n]	0.871	0.871	0.930	-8
34 Malta	0.878	79.1	87.9	80.9	19,189	0.901	0.856	0.877	2
35 Qatar	0.875	75.0	89.0	77.7	27,664 [h,m]	0.834	0.852	0.938	-12
36 Hungary	0.874	72.9	.. [d,j]	89.3	17,887	0.799	0.958	0.866	2
37 Poland	0.870	75.2	.. [d,j]	87.2	13,847	0.836	0.951	0.823	11
38 Argentina	0.869	74.8	97.2	89.7 [h]	14,280	0.831	0.947	0.828	9
39 United Arab Emirates	0.868	78.3	88.7 [l]	59.9 [e,h]	25,514 [n]	0.889	0.791	0.925	-12
40 Chile	0.867	78.3	95.7	82.9	12,027	0.889	0.914	0.799	15
41 Bahrain	0.866	75.2	86.5	86.1	21,482	0.837	0.864	0.896	-8
42 Slovakia	0.863	74.2	.. [d]	78.3	15,871	0.821	0.921	0.846	-1
43 Lithuania	0.862	72.5	99.6 [d]	91.4	14,494	0.792	0.965	0.831	3
44 Estonia	0.860	71.2	99.8 [d]	92.4	15,478	0.770	0.968	0.842	0
45 Latvia	0.855	72.0	99.7 [d]	90.2	13,646	0.784	0.961	0.821	4
46 Uruguay	0.852	75.9	96.8	88.9 [e,h]	9,962	0.848	0.942	0.768	16
47 Croatia	0.850	75.3	98.1	73.5 [h]	13,042	0.839	0.899	0.813	4
48 Costa Rica	0.846	78.5	94.9	73.0 [e]	10,180 [n]	0.891	0.876	0.772	13
49 Bahamas	0.845	72.3	.. [j]	70.8	18,380 [h]	0.789	0.875	0.870	-12
50 Seychelles	0.843	72.7 [h,k]	91.8	82.2 [e]	16,106	0.795	0.886	0.848	-10
51 Cuba	0.838	77.7	99.8 [d]	87.6	6,000 [o]	0.879	0.952	0.683	43
52 Mexico	0.829	75.6	91.6	75.6	10,751	0.843	0.863	0.781	7
53 Bulgaria	0.824	72.7	98.2	81.5	9,032	0.795	0.926	0.752	11

Human development indicators

TABLE 1

Human development index

HDI rank[a]		Human development index (HDI) value 2005	Life expectancy at birth (years) 2005	Adult literacy rate (% aged 15 and above) 1995-2005[b]	Combined gross enrolment ratio for primary, secondary and tertiary education (%) 2005	GDP per capita (PPP US$) 2005	Life expectancy index	Education index	GDP index	GDP per capita (PPP US$) rank minus HDI rank[c]
54	Saint Kitts and Nevis	0.821	70.0 [h,p]	97.8 [k]	73.1 [e]	13,307 [h]	0.750	0.896	0.816	-4
55	Tonga	0.819	72.8	98.9	80.1 [e]	8,177 [n]	0.797	0.926	0.735	15
56	Libyan Arab Jamahiriya	0.818	73.4	84.2 [l]	94.1 [e,h]	10,335 [h,m]	0.806	0.875	0.774	4
57	Antigua and Barbuda	0.815	73.9 [h,p]	85.8 [q]	.. [r]	12,500 [h]	0.815	0.824	0.806	-4
58	Oman	0.814	75.0	81.4	67.1	15,602 [h]	0.833	0.766	0.843	-15
59	Trinidad and Tobago	0.814	69.2	98.4 [l]	64.9 [e]	14,603	0.737	0.872	0.832	-14
60	Romania	0.813	71.9	97.3	76.8	9,060	0.782	0.905	0.752	3
61	Saudi Arabia	0.812	72.2	82.9	76.0	15,711 [n]	0.787	0.806	0.844	-19
62	Panama	0.812	75.1	91.9	79.5	7,605	0.836	0.878	0.723	15
63	Malaysia	0.811	73.7	88.7	74.3 [h]	10,882	0.811	0.839	0.783	-6
64	Belarus	0.804	68.7	99.6 [d]	88.7	7,918	0.728	0.956	0.730	8
65	Mauritius	0.804	72.4	84.3	75.3 [e]	12,715	0.790	0.813	0.809	-13
66	Bosnia and Herzegovina	0.803	74.5	96.7	69.0 [h,s]	7,032 [h,t]	0.825	0.874	0.710	17
67	Russian Federation	0.802	65.0	99.4 [d]	88.9 [e]	10,845	0.667	0.956	0.782	-9
68	Albania	0.801	76.2	98.7	68.6 [h]	5,316	0.853	0.887	0.663	30
69	Macedonia (TFYR)	0.801	73.8	96.1	70.1	7,200	0.814	0.875	0.714	11
70	Brazil	0.800	71.7	88.6	87.5 [h]	8,402	0.779	0.883	0.740	-3
MEDIUM HUMAN DEVELOPMENT										
71	Dominica	0.798	75.6 [h,q]	88.0 [q]	81.0 [e]	6,393 [h]	0.844	0.857	0.694	19
72	Saint Lucia	0.795	73.1	94.8 [q]	74.8	6,707 [h]	0.802	0.881	0.702	15
73	Kazakhstan	0.794	65.9	99.5 [d]	93.8	7,857	0.682	0.973	0.728	1
74	Venezuela (Bolivarian Republic of)	0.792	73.2	93.0	75.5 [e,h]	6,632	0.804	0.872	0.700	14
75	Colombia	0.791	72.3	92.8	75.1	7,304 [n]	0.788	0.869	0.716	4
76	Ukraine	0.788	67.7	99.4 [d]	86.5	6,848	0.711	0.948	0.705	9
77	Samoa	0.785	70.8	98.6 [l]	73.7 [e]	6,170	0.763	0.903	0.688	14
78	Thailand	0.781	69.6	92.6	71.2 [e]	8,677	0.743	0.855	0.745	-13
79	Dominican Republic	0.779	71.5	87.0	74.1 [e,h]	8,217 [n]	0.776	0.827	0.736	-10
80	Belize	0.778	75.9	75.1 [q]	81.8 [e]	7,109	0.849	0.773	0.712	1
81	China	0.777	72.5	90.9	69.1 [e]	6,757 [u]	0.792	0.837	0.703	5
82	Grenada	0.777	68.2	96.0 [q]	73.1 [e]	7,843 [h]	0.720	0.884	0.728	-7
83	Armenia	0.775	71.7	99.4 [d]	70.8	4,945	0.779	0.896	0.651	20
84	Turkey	0.775	71.4	87.4	68.7 [e]	8,407	0.773	0.812	0.740	-18
85	Suriname	0.774	69.6	89.6	77.1 [e]	7,722	0.743	0.854	0.725	-9
86	Jordan	0.773	71.9	91.1	78.1	5,530	0.782	0.868	0.670	11
87	Peru	0.773	70.7	87.9	85.8 [e]	6,039	0.761	0.872	0.684	6
88	Lebanon	0.772	71.5	.. [j]	84.6	5,584	0.775	0.871	0.671	8
89	Ecuador	0.772	74.7	91.0	.. [r]	4,341	0.828	0.858	0.629	21
90	Philippines	0.771	71.0	92.6	81.1	5,137	0.767	0.888	0.657	11
91	Tunisia	0.766	73.5	74.3	76.3	8,371	0.808	0.750	0.739	-23
92	Fiji	0.762	68.3	.. [j]	74.8 [e]	6,049	0.722	0.879	0.685	0
93	Saint Vincent and the Grenadines	0.761	71.1	88.1 [q]	68.9	6,568	0.768	0.817	0.698	-4
94	Iran (Islamic Republic of)	0.759	70.2	82.4	72.8 [e]	7,968	0.754	0.792	0.731	-23
95	Paraguay	0.755	71.3	93.5 [l]	69.1 [e,h]	4,642 [n]	0.771	0.853	0.641	10
96	Georgia	0.754	70.7	100.0 [d,v]	76.3	3,365	0.761	0.914	0.587	24
97	Guyana	0.750	65.2	.. [j]	85.0	4,508 [n]	0.670	0.943	0.636	12
98	Azerbaijan	0.746	67.1	98.8	67.1	5,016	0.702	0.882	0.653	4
99	Sri Lanka	0.743	71.6	90.7 [w]	62.7 [e,h]	4,595	0.776	0.814	0.639	7
100	Maldives	0.741	67.0	96.3	65.8 [e]	5,261 [h,m]	0.701	0.862	0.661	-1
101	Jamaica	0.736	72.2	79.9	77.9 [e]	4,291	0.787	0.792	0.627	11
102	Cape Verde	0.736	71.0	81.2 [l]	66.4	5,803 [n]	0.766	0.763	0.678	-7
103	El Salvador	0.735	71.3	80.6 [l]	70.4	5,255 [n]	0.772	0.772	0.661	-3
104	Algeria	0.733	71.7	69.9	73.7 [e]	7,062 [n]	0.778	0.711	0.711	-22
105	Viet Nam	0.733	73.7	90.3	63.9	3,071	0.812	0.815	0.572	18
106	Occupied Palestinian Territories	0.731	72.9	92.4	82.4 [e]	.. [x]	0.799	0.891	0.505	33

Human development indicators

TABLE 1

HDI rank[a]		Human development index (HDI) value 2005	Life expectancy at birth (years) 2005	Adult literacy rate (% aged 15 and above) 1995-2005[b]	Combined gross enrolment ratio for primary, secondary and tertiary education (%) 2005	GDP per capita (PPP US$) 2005	Life expectancy index	Education index	GDP index	GDP per capita (PPP US$) rank minus HDI rank[c]
107	Indonesia	0.728	69.7	90.4	68.2[e]	3,843	0.745	0.830	0.609	6
108	Syrian Arab Republic	0.724	73.6	80.8	64.8[e]	3,808	0.811	0.755	0.607	7
109	Turkmenistan	0.713	62.6	98.8	..[r]	3,838[h]	0.627	0.903	0.609	5
110	Nicaragua	0.710	71.9	76.7	70.6[e]	3,674[n]	0.782	0.747	0.601	6
111	Moldova	0.708	68.4	99.1[d,l]	69.7[e]	2,100	0.724	0.892	0.508	25
112	Egypt	0.708	70.7	71.4	76.9[e]	4,337	0.761	0.732	0.629	-1
113	Uzbekistan	0.702	66.8	..[d,j]	73.8[e,h]	2,063	0.696	0.906	0.505	25
114	Mongolia	0.700	65.9	97.8	77.4	2,107	0.682	0.910	0.509	21
115	Honduras	0.700	69.4	80.0	71.2[e]	3,430[n]	0.739	0.771	0.590	3
116	Kyrgyzstan	0.696	65.6	98.7	77.7	1,927	0.676	0.917	0.494	29
117	Bolivia	0.695	64.7	86.7	86.0[e,h]	2,819	0.662	0.865	0.557	7
118	Guatemala	0.689	69.7	69.1	67.3[e]	4,568[n]	0.746	0.685	0.638	-11
119	Gabon	0.677	56.2	84.0[l]	72.4[e,h]	6,954	0.521	0.801	0.708	-35
120	Vanuatu	0.674	69.3	74.0	63.4[e]	3,225[n]	0.738	0.705	0.580	2
121	South Africa	0.674	50.8	82.4	77.0[h]	11,110[n]	0.430	0.806	0.786	-65
122	Tajikistan	0.673	66.3	99.5[d]	70.8	1,356	0.689	0.896	0.435	32
123	Sao Tome and Principe	0.654	64.9	84.9	65.2	2,178	0.665	0.783	0.514	10
124	Botswana	0.654	48.1	81.2	69.5[e]	12,387	0.385	0.773	0.804	-70
125	Namibia	0.650	51.6	85.0	64.7[e]	7,586[n]	0.444	0.783	0.723	-47
126	Morocco	0.646	70.4	52.3	58.5[e]	4,555	0.757	0.544	0.637	-18
127	Equatorial Guinea	0.642	50.4	87.0	58.1[e,h]	7,874[h,n]	0.423	0.773	0.729	-54
128	India	0.619	63.7	61.0	63.8[e]	3,452[n]	0.645	0.620	0.591	-11
129	Solomon Islands	0.602	63.0	76.6[k]	47.6	2,031[n]	0.633	0.669	0.503	14
130	Lao People's Democratic Republic	0.601	63.2	68.7	61.5	2,039	0.637	0.663	0.503	11
131	Cambodia	0.598	58.0	73.6	60.0[e]	2,727[n]	0.550	0.691	0.552	-6
132	Myanmar	0.583	60.8	89.9	49.5[e]	1,027[h,y]	0.596	0.764	0.389	35
133	Bhutan	0.579	64.7	47.0[v]	..[r]	..[h,z]	0.662	0.485	0.589	-14
134	Comoros	0.561	64.1	..[j]	46.4[e]	1,993[n]	0.651	0.533	0.499	10
135	Ghana	0.553	59.1	57.9	50.7[e]	2,480[n]	0.568	0.555	0.536	-8
136	Pakistan	0.551	64.6	49.9	40.0[e]	2,370	0.659	0.466	0.528	-8
137	Mauritania	0.550	63.2	51.2	45.6	2,234[n]	0.637	0.493	0.519	-5
138	Lesotho	0.549	42.6	82.2	66.0[e]	3,335[n]	0.293	0.768	0.585	-17
139	Congo	0.548	54.0	84.7[l]	51.4[e]	1,262	0.484	0.736	0.423	16
140	Bangladesh	0.547	63.1	47.5	56.0[h]	2,053	0.635	0.503	0.504	0
141	Swaziland	0.547	40.9	79.6	59.8[e]	4,824	0.265	0.730	0.647	-37
142	Nepal	0.534	62.6	48.6	58.1[e]	1,550	0.626	0.518	0.458	8
143	Madagascar	0.533	58.4	70.7	59.7[e]	923	0.557	0.670	0.371	27
144	Cameroon	0.532	49.8	67.9	62.3[e]	2,299	0.414	0.660	0.523	-13
145	Papua New Guinea	0.530	56.9	57.3	40.7[e,h]	2,563[n]	0.532	0.518	0.541	-19
146	Haiti	0.529	59.5	..[j]	..[r]	1,663[n]	0.575	0.542	0.469	2
147	Sudan	0.526	57.4	60.9[aa]	37.3[e]	2,083[n]	0.540	0.531	0.507	-10
148	Kenya	0.521	52.1	73.6	60.6[e]	1,240	0.451	0.693	0.420	9
149	Djibouti	0.516	53.9	..[j]	25.3	2,178[n]	0.482	0.553	0.514	-15
150	Timor-Leste	0.514	59.7	50.1[ab]	72.0[e]	..[h,ac]	0.578	0.574	0.390	16
151	Zimbabwe	0.513	40.9	89.4[l]	52.4[e,h]	2,038	0.265	0.770	0.503	-9
152	Togo	0.512	57.8	53.2	55.0[e]	1,506[n]	0.547	0.538	0.453	-1
153	Yemen	0.508	61.5	54.1[l]	55.2	930	0.608	0.545	0.372	16
154	Uganda	0.505	49.7	66.8	63.0[e]	1,454[n]	0.412	0.655	0.447	-2
155	Gambia	0.502	58.8	..[j]	50.1[e,h]	1,921[n]	0.563	0.450	0.493	-9
LOW HUMAN DEVELOPMENT										
156	Senegal	0.499	62.3	39.3	39.6[e]	1,792	0.622	0.394	0.482	-9
157	Eritrea	0.483	56.6	..[j]	35.3[e]	1,109[n]	0.527	0.521	0.402	6
158	Nigeria	0.470	46.5	69.1[l]	56.2[e]	1,128	0.359	0.648	0.404	4
159	Tanzania (United Republic of)	0.467	51.0	69.4	50.4[e]	744	0.434	0.631	0.335	15

TABLE 1

Human development index

HDI rank [a]	Human development index (HDI) value 2005	Life expectancy at birth (years) 2005	Adult literacy rate (% aged 15 and above) 1995-2005 [b]	Combined gross enrolment ratio for primary, secondary and tertiary education (%) 2005	GDP per capita (PPP US$) 2005	Life expectancy index	Education index	GDP index	GDP per capita (PPP US$) rank minus HDI rank [c]
160 Guinea	0.456	54.8	29.5	45.1 [e]	2,316	0.497	0.347	0.524	-30
161 Rwanda	0.452	45.2	64.9	50.9 [e]	1,206 [n]	0.337	0.602	0.416	-1
162 Angola	0.446	41.7	67.4	25.6 [e,h]	2,335 [n]	0.279	0.535	0.526	-33
163 Benin	0.437	55.4	34.7	50.7 [e]	1,141	0.506	0.400	0.406	-2
164 Malawi	0.437	46.3	64.1	63.1 [e]	667	0.355	0.638	0.317	13
165 Zambia	0.434	40.5	68.0	60.5 [e]	1,023	0.259	0.655	0.388	3
166 Côte d'Ivoire	0.432	47.4	48.7	39.6 [e,h]	1,648	0.373	0.457	0.468	-17
167 Burundi	0.413	48.5	59.3	37.9 [e]	699 [n]	0.391	0.522	0.325	9
168 Congo (Democratic Republic of the)	0.411	45.8	67.2	33.7 [e,h]	714 [n]	0.346	0.560	0.328	7
169 Ethiopia	0.406	51.8	35.9	42.1 [e]	1,055 [n]	0.446	0.380	0.393	-5
170 Chad	0.388	50.4	25.7	37.5 [e]	1,427 [n]	0.423	0.296	0.444	-17
171 Central African Republic	0.384	43.7	48.6	29.8 [e,h]	1,224 [n]	0.311	0.423	0.418	-13
172 Mozambique	0.384	42.8	38.7	52.9	1,242 [n]	0.296	0.435	0.421	-16
173 Mali	0.380	53.1	24.0	36.7	1,033	0.469	0.282	0.390	-8
174 Niger	0.374	55.8	28.7	22.7	781 [n]	0.513	0.267	0.343	-1
175 Guinea-Bissau	0.374	45.8	.. [i]	36.7 [e,h]	827 [n]	0.347	0.421	0.353	-4
176 Burkina Faso	0.370	51.4	23.6	29.3	1,213 [n]	0.440	0.255	0.417	-17
177 Sierra Leone	0.336	41.8	34.8	44.6 [h]	806	0.280	0.381	0.348	-5
Developing countries	0.691	66.1	76.7	64.1	5,282	0.685	0.725	0.662	..
Least developed countries	0.488	54.5	53.9	48.0	1,499	0.492	0.519	0.452	..
Arab States	0.699	67.5	70.3	65.5	6,716	0.708	0.687	0.702	..
East Asia and the Pacific	0.771	71.7	90.7	69.4	6,604	0.779	0.836	0.699	..
Latin America and the Caribbean	0.803	72.8	90.3	81.2	8,417	0.797	0.873	0.740	..
South Asia	0.611	63.8	59.5	60.3	3,416	0.646	0.598	0.589	..
Sub-Saharan Africa	0.493	49.6	60.3	50.6	1,998	0.410	0.571	0.500	..
Central and Eastern Europe and the CIS	0.808	68.6	99.0	83.5	9,527	0.726	0.938	0.761	..
OECD	0.916	78.3	..	88.6	29,197	0.888	0.912	0.947	..
High-income OECD	0.947	79.4	..	93.5	33,831	0.906	0.961	0.972	..
High human development	0.897	76.2	..	88.4	23,986	0.854	0.922	0.915	..
Medium human development	0.698	67.5	78.0	65.3	4,876	0.709	0.738	0.649	..
Low human development	0.436	48.5	54.4	45.8	1,112	0.391	0.516	0.402	..
High income	0.936	79.2	..	92.3	33,082	0.903	0.937	0.968	..
Middle income	0.776	70.9	89.9	73.3	7,416	0.764	0.843	0.719	..
Low income	0.570	60.0	60.2	56.3	2,531	0.583	0.589	0.539	..
World	0.743	68.1	78.6	67.8	9,543	0.718	0.750	0.761	..

NOTES

a. The HDI rank is determined using HDI values to the sixth decimal point.

b. Data refer to national literacy estimates from censuses or surveys conducted between 1995 and 2005, unless otherwise specified. Due to differences in methodology and timeliness of underlying data, comparisons across countries and over time should be made with caution. For more details, see http://www.uis.unesco.org/.

c. A positive figure indicates that the HDI rank is higher than the GDP per capita (PPP US$) rank, a negative the opposite.

d. For purposes of calculating the HDI, a value of 99.0% was applied.

e. National or UNESCO Institute for Statistics estimate.

f. For purposes of calculating the HDI, a value of 40,000 (PPP US$) was applied.

g. For purposes of calculating the HDI, a value of 100% was applied.

h. Data refer to a year other than that specified.

i. Statec 2006. Data refer to nationals enrolled both in the country and abroad and thus differ from the standard definition.

j. In the absence of recent data, estimates from UNESCO Institute for Statistics 2003, based on outdated census or survey information, were used and should be interpreted with caution: Bahamas 95.8, Barbados 99.7, Comoros 56.8, Djibouti 70.3, Eritrea 60.5, Fiji 94.4, Gambia 42.5, Guinea-Bissau 44.8, Guyana 99.0, Haiti 54.8, Hong Kong, China (SAR) 94.6, Hungary 99.4, Lebanon 88.3, Poland 99.8 and Uzbekistan 99.4.

k. Data are from national sources.

l. UNESCO Institute for Statistics estimates based on its Global age-specific literacy projections model, April 2007.

m. Heston, Summers and Aten 2006. Data differ from the standard definition.

n. World Bank estimate based on regression.

o. Efforts to produce a more accurate estimate are ongoing (see Readers guide and notes to tables for details). A preliminary estimate of 6,000 (PPP US$) was used.

p. Data are from the Secretariat of the Organization of Eastern Caribbean States, based on national sources.

q. Data are from the Secretariat of the Caribbean Community, based on national sources.

r. Because the combined gross enrolment ratio was unavailable, the following HDRO estimates were used: Antigua and Barbuda 76, Bhutan 52, Ecuador 75, Haiti 53 and Turkmenistan 73.

s. UNDP 2007.

t. World Bank 2006.

u. World Bank estimate based on a bilateral comparison between China and the United States (Ruoen and Kai 1995).

v. UNICEF 2004.

w. Data refer to 18 of the 25 states of the country only.

x. In the absence of an estimate of GDP per capita (PPP US$), the HDRO estimate of 2,056 (PPP US$) was used, derived from the value of GDP in US$ and the weighted average ratio of PPP US$ to US$ in the Arab States.

y. Heston, Summers and Aten 2001. Data differ from the standard definition.

z. In the absence of an estimate of GDP per capita (PPP US$), the HDRO estimate of 3,413 (PPP US$) was used, derived from the value of GDP per capita in PPP US$ estimated by Heston, Summers and Aten 2006 adjusted to reflect the latest population estimates from UN 2007e.

aa. Data refer to North Sudan only.

ab. UNDP 2006.

ac. For the purposes of calculating the HDI, a national estimate of 1,033 (PPP US$) was used.

SOURCES

Column 1: calculated on the basis of data in columns 6–8; see Technical note 1 for details.

Column 2: UN 2007e, unless otherwise specified.

Column 3: UNESCO Institute for Statistics 2007a, unless otherwise specified.

Column 4: UNESCO Institute for Statistics 2007c, unless otherwise specified.

Column 5: World Bank 2007b, unless otherwise specified; aggregates calculated for the HDRO by the World Bank.

Column 6: calculated on the basis of data in column 2.

Column 7: calculated on the basis of data in columns 3 and 4.

Column 8: calculated on the basis of data in column 5.

Column 9: calculated on the basis of data in columns 1 and 5.

Monitoring human development: enlarging people's choices . . .

Basic indicators for other UN member states

	Human development index components						MDG	MDG		MDG	MDG
	Life expectancy at birth (years) 2005	Adult literacy rate (% aged 15 and above) 1995–2005 b	Combined gross enrolment ratio for primary, secondary and tertiary education (%) 2005	GDP per capita (PPP US$) 2005	Total population (thousands) 2005	Total fertility rate (births per woman) 2000–05	Under-five mortality rate (per 1,000 live births) 2005	Net primary enrolment rate (%) 2005	HIV prevalence a (% aged 15–49) 2005	Population under-nourished (% of total population) 2002/04 c	Population using an improved water source (%) 2004
Afghanistan	42.9	28.0	42.8 d	..	25,067	7.5	257	..	<0.1 [<0.2]	..	39
Andorra	..	..	62.6 d	..	73	..	3	80 d	..	..	100
Iraq	57.7	74.1	59.6 d	..	27,996	4.9	125	88 d	[<0.2]	..	81
Kiribati	..	...	75.1 d	4,597	92	..	65	97 d,e	..	7	65
Korea (Democratic People's Rep. of)	66.8	..	..	..	23,616	1.9	55	..	[<0.2]	33	100
Liberia	44.7	51.9 f	57.4 e	..	3,442	6.8	235	66 e	[2.0–5.0]	50	61
Liechtenstein	..	..	86.4 d,e	..	35	..	4	88 d,e	..	..	..
Marshall Islands	..	...	71.1 d	..	57	..	58	90 d,e	..	..	87
Micronesia (Federated States of)	68.0	..	..	7,242	110	4.2	42	..	..	..	94
Monaco	..	..	..	..	33	..	5	..	..	..	100
Montenegro	74.1	96.4 g,h	74.5 d,e,h	..	608	1.8	15 h	96 d,e,h	0.2 [0.1–0.3] h	9 h	93 h
Nauru	..	..	50.6 d,e	..	10	..	30	..	..	..	..
Palau	..	...	96.9 d,e	..	20	..	11	96 d,e	..	..	85
San Marino	..	..	..	..	30	..	3	..	..	..	..
Serbia	73.6	96.4 g,h	74.5 d,e,h	..	9,863	1.7	15 h	96 d,e,h	0.2 [0.1–0.3] h	9 h	93 h
Somalia	47.1	...	..	..	8,196	6.4	225	..	0.9 [0.5–1.6]	..	29
Tuvalu	..	...	69.2 d,e	..	10	..	38	..	..	..	100

NOTES

a. Data are point and range estimates based on new estimation models developed by UNAIDS. Range estimates are presented in square brackets.

b. Data refer to national literacy estimates from censuses or surveys conducted between 1995 and 2005, unless otherwise specified. Due to differences in methodology and timeliness of underlying data, comparisons across countries and over time should be made with caution. For more details, see http://www.uis.unesco.org/.

c. Data refer to the average for the years specified.

d. National or UNESCO Institute for Statistics estimate.

e. Data refer to a year other than that specified.

f. UNESCO Institute for Statistics estimates based on its Global age-specific literacy projections model, April 2007.

g. Data exclude Kosovo and Metohia.

h. Data refer to Serbia and Montenegro prior to its separation into two independent states in June 2006.

SOURCES

Column 1: UN 2007e, unless otherwise specified.
Column 2: UNESCO Institute for Statistics. 2007a, unless otherwise specified.
Column 3: UNESCO Institute for Statistics. 2007c, unless otherwise specified.
Column 4: World Bank 2007b.
Columns 5 and 6: UN 2007e, unless otherwise specified.
Column 7: UNICEF 2006.
Column 8: UNESCO Institute for Statistics 2007c.
Column 9: UNAIDS 2006.
Column 10: FAO 2007a.
Column 11: UN 2006a, based on a joint effort by UNICEF and WHO.

TABLE 2   Monitoring human development: enlarging people's choices...

Human development index trends

HDI rank	1975	1980	1985	1990	1995	2000	2005
HIGH HUMAN DEVELOPMENT							
1 Iceland	0.868	0.890	0.899	0.918	0.923	0.947	0.968
2 Norway	0.870	0.889	0.900	0.913	0.938	0.958	0.968
3 Australia	0.851	0.868	0.880	0.894	0.934	0.949	0.962
4 Canada	0.873	0.888	0.911	0.931	0.936	0.946	0.961
5 Ireland	0.823	0.835	0.851	0.875	0.898	0.931	0.959
6 Sweden	0.872	0.882	0.893	0.904	0.935	0.952	0.956
7 Switzerland	0.883	0.895	0.902	0.915	0.926	0.946	0.955
8 Japan	0.861	0.886	0.899	0.916	0.929	0.941	0.953
9 Netherlands	0.873	0.885	0.899	0.914	0.934	0.947	0.953
10 France	0.856	0.872	0.884	0.907	0.925	0.938	0.952
11 Finland	0.846	0.866	0.884	0.906	0.918	0.940	0.952
12 United States	0.870	0.890	0.904	0.919	0.931	0.942	0.951
13 Spain	0.846	0.863	0.877	0.896	0.914	0.932	0.949
14 Denmark	0.875	0.883	0.890	0.898	0.916	0.935	0.949
15 Austria	0.848	0.862	0.876	0.899	0.918	0.938	0.948
16 United Kingdom	0.853	0.860	0.870	0.890	0.929	0.931	0.946
17 Belgium	0.852	0.869	0.883	0.903	0.931	0.943	0.946
18 Luxembourg	0.836	0.850	0.863	0.890	0.913	0.929	0.944
19 New Zealand	0.854	0.860	0.871	0.880	0.908	0.927	0.943
20 Italy	0.845	0.861	0.869	0.892	0.910	0.926	0.941
21 Hong Kong, China (SAR)	0.763	0.803	0.830	0.865	0.886	0.919	0.937
22 Germany	..	0.863	0.871	0.890	0.913	0.928	0.935
23 Israel	0.805	0.830	0.850	0.869	0.891	0.918	0.932
24 Greece	0.841	0.856	0.869	0.877	0.882	0.897	0.926
25 Singapore	0.729	0.762	0.789	0.827	0.865	..	0.922
26 Korea (Republic of)	0.713	0.747	0.785	0.825	0.861	0.892	0.921
27 Slovenia	..	..	..	0.851	0.857	0.891	0.917
28 Cyprus	..	0.809	0.828	0.851	0.870	0.893	0.903
29 Portugal	0.793	0.807	0.829	0.855	0.885	0.904	0.897
30 Brunei Darussalam	..	..	..	..	..	..	0.894
31 Barbados	..	..	..	..	..	..	0.892
32 Czech Republic	..	..	..	0.845	0.854	0.866	0.891
33 Kuwait	0.771	0.789	0.794	..	0.826	0.855	0.891
34 Malta	0.738	0.772	0.799	0.833	0.857	0.877	0.878
35 Qatar	..	..	..	..	..	..	0.875
36 Hungary	0.786	0.801	0.813	0.813	0.817	0.845	0.874
37 Poland	..	..	..	0.806	0.822	0.852	0.870
38 Argentina	0.790	0.804	0.811	0.813	0.836	0.862	0.869
39 United Arab Emirates	0.734	0.769	0.790	0.816	0.825	0.837	0.868
40 Chile	0.708	0.743	0.761	0.788	0.819	0.845	0.867
41 Bahrain	..	0.747	0.783	0.808	0.834	0.846	0.866
42 Slovakia	..	..	..	..	..	..	0.863
43 Lithuania	..	..	..	0.827	0.791	0.831	0.862
44 Estonia	..	0.811	0.820	0.813	0.792	0.829	0.860
45 Latvia	..	0.797	0.810	0.804	0.771	0.817	0.855
46 Uruguay	0.762	0.782	0.787	0.806	0.821	0.842	0.852
47 Croatia	..	..	..	0.812	0.805	0.828	0.850
48 Costa Rica	0.746	0.772	0.774	0.794	0.814	0.830	0.846
49 Bahamas	..	0.809	0.822	0.831	0.820	0.825	0.845
50 Seychelles	..	..	..	..	..	..	0.843
51 Cuba	..	..	..	..	..	..	0.838
52 Mexico	0.694	0.739	0.758	0.768	0.786	0.814	0.829
53 Bulgaria	..	0.771	0.792	0.794	0.785	0.800	0.824

TABLE 2

HDI rank	1975	1980	1985	1990	1995	2000	2005
54 Saint Kitts and Nevis	..	..	..	..	..	..	0.821
55 Tonga	..	..	..	..	..	..	0.819
56 Libyan Arab Jamahiriya	..	..	..	..	..	..	0.818
57 Antigua and Barbuda	..	..	..	..	..	..	0.815
58 Oman	0.487	0.547	0.641	0.697	0.741	0.779	0.814
59 Trinidad and Tobago	0.756	0.784	0.782	0.784	0.785	0.796	0.814
60 Romania	..	0.786	0.792	0.777	0.772	0.780	0.813
61 Saudi Arabia	0.611	0.666	0.684	0.717	0.748	0.788	0.812
62 Panama	0.718	0.737	0.751	0.752	0.775	0.797	0.812
63 Malaysia	0.619	0.662	0.696	0.725	0.763	0.790	0.811
64 Belarus	..	..	..	0.790	0.755	0.778	0.804
65 Mauritius	..	0.662	0.692	0.728	0.751	0.781	0.804
66 Bosnia and Herzegovina	..	..	..	..	..	..	0.803
67 Russian Federation	..	..	..	0.815	0.771	0.782	0.802
68 Albania	..	0.675	0.694	0.704	0.705	0.746	0.801
69 Macedonia (TFYR)	..	..	..	..	..	..	0.801
70 Brazil	0.649	0.685	0.700	0.723	0.753	0.789	0.800
MEDIUM HUMAN DEVELOPMENT							
71 Dominica	..	..	..	..	..	..	0.798
72 Saint Lucia	..	..	..	..	..	..	0.795
73 Kazakhstan	..	..	..	0.771	0.724	0.738	0.794
74 Venezuela (Bolivarian Republic of)	0.723	0.737	0.743	0.762	0.770	0.776	0.792
75 Colombia	0.663	0.694	0.709	0.729	0.753	0.772	0.791
76 Ukraine	..	..	..	0.809	0.756	0.761	0.788
77 Samoa	..	..	0.709	0.721	0.740	0.765	0.785
78 Thailand	0.615	0.654	0.679	0.712	0.745	0.761	0.781
79 Dominican Republic	0.628	0.660	0.684	0.697	0.723	0.757	0.779
80 Belize	..	0.712	0.718	0.750	0.777	0.795	0.778
81 China	0.530	0.559	0.595	0.634	0.691	0.732	0.777
82 Grenada	..	..	..	..	..	..	0.777
83 Armenia	..	..	..	0.737	0.701	0.738	0.775
84 Turkey	0.594	0.615	0.651	0.683	0.717	0.753	0.775
85 Suriname	..	..	..	..	..	..	0.774
86 Jordan	..	0.647	0.669	0.684	0.710	0.751	0.773
87 Peru	0.647	0.676	0.699	0.710	0.737	0.763	0.773
88 Lebanon	..	..	..	0.692	0.730	0.748	0.772
89 Ecuador	0.636	0.678	0.699	0.714	0.734	..	0.772
90 Philippines	0.655	0.688	0.692	0.721	0.739	0.758	0.771
91 Tunisia	0.519	0.575	0.626	0.662	0.702	0.741	0.766
92 Fiji	0.665	0.688	0.702	..	0.743	0.747	0.762
93 Saint Vincent and the Grenadines	..	..	..	..	..	..	0.761
94 Iran (Islamic Republic of)	0.571	0.578	0.615	0.653	0.693	0.722	0.759
95 Paraguay	0.667	0.701	0.707	0.718	0.737	0.749	0.755
96 Georgia	..	..	..	..	..	..	0.754
97 Guyana	0.682	0.684	0.675	0.679	0.699	0.722	0.750
98 Azerbaijan	..	..	..	..	..	..	0.746
99 Sri Lanka	0.619	0.656	0.683	0.702	0.721	0.731	0.743
100 Maldives	..	..	..	..	..	..	0.741
101 Jamaica	0.686	0.689	0.690	0.713	0.728	0.744	0.736
102 Cape Verde	..	..	0.589	0.627	0.678	0.709	0.736
103 El Salvador	0.595	0.590	0.611	0.653	0.692	0.716	0.735
104 Algeria	0.511	0.562	0.613	0.652	0.672	0.702	0.733
105 Viet Nam	..	..	0.590	0.620	0.672	0.711	0.733
106 Occupied Palestinian Territories	..	..	..	..	..	..	0.731

Human development indicators

TABLE 2

Human development index trends

HDI rank	1975	1980	1985	1990	1995	2000	2005
107 Indonesia	0.471	0.533	0.585	0.626	0.670	0.692	0.728
108 Syrian Arab Republic	0.547	0.593	0.628	0.646	0.676	0.690	0.724
109 Turkmenistan	..	..	..	..	..	..	0.713
110 Nicaragua	0.583	0.593	0.601	0.610	0.637	0.671	0.710
111 Moldova	..	0.700	0.722	0.740	0.684	0.683	0.708
112 Egypt	0.434	0.482	0.532	0.575	0.613	0.659	0.708
113 Uzbekistan	..	..	..	0.704	0.683	0.691	0.702
114 Mongolia	..	..	0.637	0.654	0.638	0.667	0.700
115 Honduras	0.528	0.578	0.611	0.634	0.653	0.668	0.700
116 Kyrgyzstan	..	..	..	..	..	..	0.696
117 Bolivia	0.519	0.553	0.580	0.606	0.639	0.677	0.695
118 Guatemala	0.514	0.550	0.566	0.592	0.626	0.667	0.689
119 Gabon	..	..	..	..	..	..	0.677
120 Vanuatu	..	..	..	..	..	..	0.674
121 South Africa	0.650	0.670	0.699	0.731	0.745	0.707	0.674
122 Tajikistan	..	..	0.705	0.703	0.638	0.640	0.673
123 Sao Tome and Principe	..	..	..	..	..	..	0.654
124 Botswana	0.509	0.571	0.624	0.674	0.658	0.631	0.654
125 Namibia	..	..	..	..	0.698	0.657	0.650
126 Morocco	0.435	0.483	0.519	0.551	0.581	0.613	0.646
127 Equatorial Guinea	..	..	0.484	0.505	0.529	0.606	0.642
128 India	0.419	0.450	0.487	0.521	0.551	0.578	0.619
129 Solomon Islands	..	..	..	..	..	..	0.602
130 Lao People's Democratic Republic	..	..	0.448	0.478	0.524	0.563	0.601
131 Cambodia	..	..	..	..	0.540	0.547	0.598
132 Myanmar	..	..	..	..	..	..	0.583
133 Bhutan	..	..	..	..	..	..	0.579
134 Comoros	..	0.483	0.500	0.506	0.521	0.540	0.561
135 Ghana	0.442	0.471	0.486	0.517	0.542	0.568	0.553
136 Pakistan	0.367	0.394	0.427	0.467	0.497	0.516	0.551
137 Mauritania	0.383	0.410	0.435	0.455	0.487	0.509	0.550
138 Lesotho	0.499	0.541	0.571	0.605	0.616	0.581	0.549
139 Congo	0.478	0.520	0.567	0.559	0.546	0.518	0.548
140 Bangladesh	0.347	0.365	0.392	0.422	0.453	0.511	0.547
141 Swaziland	0.527	0.561	0.588	0.633	0.641	0.592	0.547
142 Nepal	0.301	0.338	0.380	0.427	0.469	0.502	0.534
143 Madagascar	0.407	0.444	0.440	0.450	0.463	0.493	0.533
144 Cameroon	0.422	0.468	0.523	0.529	0.513	0.525	0.532
145 Papua New Guinea	0.431	0.462	0.481	0.495	0.532	0.544	0.530
146 Haiti	..	0.442	0.462	0.472	0.487	..	0.529
147 Sudan	0.354	0.381	0.400	0.429	0.463	0.491	0.526
148 Kenya	0.466	0.514	0.534	0.556	0.544	0.529	0.521
149 Djibouti	..	..	..	0.476	0.485	0.490	0.516
150 Timor-Leste	..	..	..	..	..	..	0.514
151 Zimbabwe	0.550	0.579	0.645	0.654	0.613	0.541	0.513
152 Togo	0.423	0.473	0.469	0.496	0.514	0.521	0.512
153 Yemen	..	..	..	0.402	0.439	0.473	0.508
154 Uganda	..	..	0.420	0.434	0.433	0.480	0.505
155 Gambia	0.290	..	..	..	0.436	0.472	0.502
LOW HUMAN DEVELOPMENT							
156 Senegal	0.342	0.367	0.401	0.428	0.449	0.473	0.499
157 Eritrea	..	..	..	..	0.435	0.459	0.483
158 Nigeria	0.321	0.378	0.391	0.411	0.432	0.445	0.470
159 Tanzania (United Republic of)	..	..	..	0.421	0.419	0.433	0.467

TABLE 2

HDI rank	1975	1980	1985	1990	1995	2000	2005
160 Guinea	..	..	..	..	..	..	0.456
161 Rwanda	0.337	0.385	0.403	0.340	0.330	0.418	0.452
162 Angola	..	..	..	..	..	..	0.446
163 Benin	0.312	0.344	0.367	0.374	0.403	0.424	0.437
164 Malawi	0.330	0.355	0.370	0.388	0.444	0.431	0.437
165 Zambia	0.470	0.478	0.489	0.477	0.439	0.420	0.434
166 Côte d'Ivoire	0.419	0.448	0.453	0.450	0.436	0.432	0.432
167 Burundi	0.290	0.318	0.352	0.366	0.347	0.368	0.413
168 Congo (Democratic Republic of the)	0.414	0.423	0.430	0.423	0.391	0.375	0.411
169 Ethiopia	..	..	0.311	0.332	0.347	0.379	0.406
170 Chad	0.296	0.298	0.342	0.364	0.377	0.397	0.388
171 Central African Republic	0.350	0.371	0.394	0.398	0.390	0.394	0.384
172 Mozambique	..	0.304	0.291	0.317	0.335	0.375	0.384
173 Mali	0.245	0.268	0.272	0.296	0.321	0.352	0.380
174 Niger	0.246	0.264	0.261	0.279	0.296	0.321	0.374
175 Guinea-Bissau	0.267	0.271	0.300	0.322	0.350	0.365	0.374
176 Burkina Faso	0.257	0.280	0.305	0.321	0.337	0.353	0.370
177 Sierra Leone	..	..	..	..	..	..	0.336

NOTE

The human development index values in this table were calculated using a consistent methodology and data series. They are not strictly comparable with those in earlier Human Development Reports. For detailed discussion, see *Readers guide and notes on tables*.

SOURCES

Columns 1–6: calculated on the basis of data on life expectancy from UN 2007e; data on adult literacy rates from UNESCO Institute for Statistics 2003 and 2007a; data on combined gross enrolment ratios from UNESCO Institute for Statistics 1999 and 2007c and data on GDP per capita (2005 PPP US$) from World Bank 2007b.

Column 7: column 1 of indicator table 1.

Human development indicators

Human and income poverty: developing countries

HDI rank	Human poverty index (HPI-1) Rank	Human poverty index (HPI-1) Value (%)	Probability at birth of not surviving to age 40 [a,†] (% of cohort) 2000–05	Adult illiteracy rate [b,†] (% aged 15 and older) 1995–2005	Population not using an improved water source[†] (%) 2004	MDG Children under weight for age[†] (% under age 5) 1996–2005[d]	MDG Population below income poverty line (%) $1 a day 1990–2005[d]	MDG Population below income poverty line (%) $2 a day 1990–2005[d]	MDG Population below income poverty line (%) National poverty line 1990–2004[d]	HPI-1 rank minus income poverty rank[c]
HIGH HUMAN DEVELOPMENT										
21 Hong Kong, China (SAR)	..	..	1.5 [e]	..	..	..	..	..	..	..
25 Singapore	7	5.2	1.8	7.5	0	3	..	..	..	..
26 Korea (Republic of)	..	..	2.5	1.0	8	..	<2	<2	..	..
28 Cyprus	..	..	2.4	3.2	0	..	..	..	..	..
30 Brunei Darussalam	..	..	3.0	7.3	..	..	..	..	..	..
31 Barbados	1	3.0	3.7	.. [f]	0	6 [e,g]	..	..	..	..
33 Kuwait	..	..	2.7	6.7	..	10	..	..	..	..
35 Qatar	13	7.8	3.7	11.0	0	6 [e]	..	..	..	..
38 Argentina	4	4.1	4.9	2.8	4	4	6.6	17.4	..	-14
39 United Arab Emirates	17	8.4	2.1	11.3 [h]	0	14 [e]	..	..	..	..
40 Chile	3	3.7	3.5	4.3	5	1	<2	5.6	17.0	1
41 Bahrain	..	..	3.4	13.5	..	9 [e]	..	..	..	..
46 Uruguay	2	3.5	4.3	3.2	0	5 [e]	<2	5.7	..	0
48 Costa Rica	5	4.4	3.7	5.1	3	5	3.3	9.8	22.0	-10
49 Bahamas	..	..	10.6	..	3	..	..	..	..	..
50 Seychelles	..	..	..	8.2	12	6 [e,g]	..	..	..	..
51 Cuba	6	4.7	3.1	.. [i]	9	4	..	..	..	..
52 Mexico	10	6.8	5.8	8.4	3	8	3.0	11.6	17.6	-7
54 Saint Kitts and Nevis	..	..	..	2.2 [j]	0	..	..	..	..	..
55 Tonga	..	..	5.0	1.1	0	..	..	..	..	..
56 Libyan Arab Jamahiriya	..	..	4.6	15.8 [h]	..	5 [e]	..	..	..	..
57 Antigua and Barbuda	..	..	..	14.2 [k]	9	10 [e,g]	..	..	..	..
58 Oman	..	..	3.7	18.6	..	18	..	..	..	..
59 Trinidad and Tobago	12	7.3	9.1	1.6 [h]	9	6	12.4	39.0	21.0	-19
61 Saudi Arabia	..	..	5.7	17.1	..	14	..	..	..	..
62 Panama	15	8.0	6.5	8.1	10	8	7.4	18.0	37.3	-10
63 Malaysia	16	8.3	4.4	11.3	1	11	<2	9.3	15.5 [e]	9
65 Mauritius	27	11.4	5.1 [e]	15.7	0	15 [e]	..	..	..	..
70 Brazil	23	9.7	9.2	11.4	10	6	7.5	21.2	21.5	-6
MEDIUM HUMAN DEVELOPMENT										
71 Dominica	..	..	..	12.0 [k]	3	5 [e,g]	..	..	..	..
72 Saint Lucia	8	6.5	5.6	5.2 [k]	2	14 [e,g]	..	..	..	..
74 Venezuela (Bolivarian Republic of)	21	8.8	7.3	7.0	17	5	18.5	40.1	31.3 [e]	-24
75 Colombia	14	7.9	9.2	7.2	7	7	7.0	17.8	64.0	-10
77 Samoa	..	..	6.6	1.4 [h]	12	..	..	..	..	..
78 Thailand	24	10.0	12.1	7.4	1	18 [e]	<2	25.2	13.6	15
79 Dominican Republic	26	10.5	10.5	13.0	5	5	2.8	16.2	42.2	6
80 Belize	43	17.5	5.4	24.9 [k]	9	6 [e,g]	..	..	..	..
81 China	29	11.7	6.8 [e]	9.1	23	8	9.9	34.9	4.6	-3
82 Grenada	..	..	9.7	4.0 [k]	5	..	..	..	..	..
84 Turkey	22	9.2	6.5	12.6	4	4	3.4	18.7	27.0	-1
85 Suriname	25	10.2	9.8	10.4	8	13	..	..	..	..
86 Jordan	11	6.9	6.4	8.9	3	4	<2	7.0	14.2	5
87 Peru	28	11.6	9.7	12.1	17	8	10.5	30.6	53.1	-5
88 Lebanon	18	8.5	6.3	.. [f]	0	4	..	..	..	..
89 Ecuador	19	8.7	8.1	9.0	6	12	17.7	40.8	46.0	-25
90 Philippines	37	15.3	7.0	7.4	15	28	14.8	43.0	36.8	-6
91 Tunisia	45	17.9	4.6	25.7	7	4	<2	6.6	7.6	27
92 Fiji	50	21.2	6.9	.. [f]	53	8 [e,g]	..	..	..	..
93 Saint Vincent and the Grenadines	..	..	6.7	11.9 [k]	..	..	..	..	..	..
94 Iran (Islamic Republic of)	30	12.9	7.8	17.6	6	11	<2	7.3	..	19
95 Paraguay	20	8.8	9.7	6.5 [h]	14	5	13.6	29.8	21.8	-16
97 Guyana	33	14.0	16.6	.. [f]	17	14	..	..	..	..

TABLE 3

HDI rank	Human poverty index (HPI-1) Rank	Value (%)	Probability at birth of not surviving to age 40 [a,†] (% of cohort) 2000–05	Adult illiteracy rate [b,†] (% aged 15 and older) 1995–2005	Population not using an improved water source† (%) 2004	MDG Children under weight for age† (% under age 5) 1996-2005[d]	MDG Population below income poverty line (%) $1 a day 1990–2005[d]	$2 a day 1990–2005[d]	National poverty line 1990–2004[d]	HPI-1 rank minus income poverty rank[c]
99 Sri Lanka	44	17.8	7.2	9.3 [e]	21	29	5.6	41.6	25.0	11
100 Maldives	42	17.0	12.1	3.7	17	30	..	..	..	..
101 Jamaica	34	14.3	8.3	20.1	7	4	<2	14.4	18.7	21
102 Cape Verde	38	15.8	7.5	18.8 [h]	20	14 [e,g]	..	..	..	..
103 El Salvador	35	15.1	9.6	19.4 [h]	16	10	19.0	40.6	37.2	-15
104 Algeria	51	21.5	7.7	30.1	15	10	<2	15.1	22.6	31
105 Viet Nam	36	15.2	6.7	9.7	15	27	..	..	28.9	..
106 Occupied Palestinian Territories	9	6.6	5.2	7.6	8	5	..	..	..	..
107 Indonesia	47	18.2	8.7	9.6	23	28	7.5	52.4	27.1	10
108 Syrian Arab Republic	31	13.6	4.6	19.2	7	7	..	..	..	..
110 Nicaragua	46	17.9	9.5	23.3	21	10	45.1	79.9	47.9	-28
112 Egypt	48	20.0	7.5	28.6	2	6	3.1	43.9	16.7	18
114 Mongolia	40	16.3	11.6	2.2	38	7	10.8	44.6	36.1	0
115 Honduras	41	16.5	12.9	20.0	13	17	14.9	35.7	50.7	-5
117 Bolivia	32	13.6	15.5	13.3	15	8	23.2	42.2	62.7	-21
118 Guatemala	54	22.5	12.5	30.9	5	23	13.5	31.9	56.2	6
119 Gabon	49	20.4	27.1	16.0 [h]	12	12	..	..	..	..
120 Vanuatu	56	24.6	8.8	26.0	40	20 [e,g]	..	..	..	..
121 South Africa	55	23.5	31.7	17.6	12	12	10.7	34.1	..	10
123 Sao Tome and Principe	39	15.8	15.1	15.1	21	13	..	..	..	..
124 Botswana	63	31.4	44.0	18.8	5	13	28.0	55.5	..	-9
125 Namibia	58	26.5	35.9	15.0	13	24	34.9	55.8	..	-16
126 Morocco	68	33.4	8.2	47.7	19	10	<2	14.3	19.0	41
127 Equatorial Guinea	66	32.4	35.6	13.0	57	19	..	..	..	..
128 India	62	31.3	16.8	39.0 [e]	14	47	34.3	80.4	28.6	-13
129 Solomon Islands	53	22.4	16.1	23.4 [j]	30	21 [e,g]	..	..	..	..
130 Lao People's Democratic Republic	70	34.5	16.6	31.3	49	40	27.0	74.1	38.6	-2
131 Cambodia	85	38.6	24.1	26.4	59	45	34.1	77.7	35.0	6
132 Myanmar	52	21.5	21.0	10.1	22	32	..	..	..	..
133 Bhutan	86	38.9	16.8	53.0 [l]	38	19	..	..	..	..
134 Comoros	61	31.3	15.3 [e]	.. [f]	14	25	..	..	..	..
135 Ghana	65	32.3	23.8	42.1	25	22	44.8	78.5	39.5	-16
136 Pakistan	77	36.2	15.4	50.1	9	38	17.0	73.6	32.6	15
137 Mauritania	87	39.2	14.6	48.8	47	32	25.9	63.1	46.3	12
138 Lesotho	71	34.5	47.8	17.8	21	20	36.4	56.1	..	-10
139 Congo	57	26.2	30.1	15.3 [h]	42	15	..	..	..	..
140 Bangladesh	93	40.5	16.4	52.5	26	48	41.3	84.0	49.8	4
141 Swaziland	73	35.4	48.0	20.4	38	10	47.7	77.8	..	-13
142 Nepal	84	38.1	17.4	51.4	10	48	24.1	68.5	30.9	11
143 Madagascar	75	35.8	24.4	29.3	50	42	61.0	85.1	71.3	-20
144 Cameroon	64	31.8	35.7	32.1	34	18	17.1	50.6	40.2	4
145 Papua New Guinea	90	40.3	20.7	42.7 [e]	61	35 [e,g]	..	..	37.5	..
146 Haiti	74	35.4	21.4	.. [f]	46	17	53.9	78.0	65.0 [e]	-13
147 Sudan	69	34.4	26.1	39.1 [e]	30	41	..	..	..	..
148 Kenya	60	30.8	35.1	26.4	39	20	22.8	58.3	52.0	-4
149 Djibouti	59	28.5	28.6	.. [f]	27	27	..	..	..	..
150 Timor-Leste	95	41.8	21.2	49.9 [m]	42	46	..	..	..	..
151 Zimbabwe	91	40.3	57.4	10.6 [h]	19	17	56.1	83.0	34.9	-4
152 Togo	83	38.1	24.1	46.8	48	25	..	..	32.3 [e]	..
153 Yemen	82	38.0	18.6	45.9 [h]	33	46	15.7	45.2	41.8	21
154 Uganda	72	34.7	38.5	33.2	40	23	..	..	37.7	..
155 Gambia	94	40.9	20.9	.. [f]	18	17	59.3	82.9	57.6	-4

Human development indicators

TABLE 3

Human and income poverty: developing countries

HDI rank		Human poverty index (HPI-1) Rank	Human poverty index (HPI-1) Value (%)	Probability at birth of not surviving to age 40 [a,†] (% of cohort) 2000–05	Adult illiteracy rate [b,†] (% aged 15 and older) 1995–2005	Population not using an improved water source[†] (%) 2004	MDG Children under weight for age[†] (% under age 5) 1996-2005[d]	MDG Population below income poverty line (%) $1 a day 1990–2005[d]	MDG Population below income poverty line (%) $2 a day 1990–2005[d]	MDG Population below income poverty line (%) National poverty line 1990–2004[d]	HPI-1 rank minus income poverty rank[c]
LOW HUMAN DEVELOPMENT											
156	Senegal	97	42.9	17.1	60.7	24	17	17.0	56.2	33.4	28
157	Eritrea	76	36.0	24.1	..[f]	40	40	..	..	53.0	..
158	Nigeria	80	37.3	39.0	30.9[h]	52	29	70.8	92.4	34.1	-19
159	Tanzania (United Republic of)	67	32.5	36.2	30.6	38	22	57.8	89.9	35.7	-22
160	Guinea	103	52.3	28.6	70.5	50	26	..	..	40.0	..
161	Rwanda	78	36.5	44.6	35.1	26	23	60.3	87.8	60.3	-16
162	Angola	89	40.3	46.7	32.6	47	31	..	..	..	..
163	Benin	100	47.6	27.9	65.3	33	23	30.9	73.7	29.0	16
164	Malawi	79	36.7	44.4	35.9	27	22	20.8	62.9	65.3	11
165	Zambia	96	41.8	53.9	32.0	42	20	63.8	87.2	68.0	-7
166	Côte d'Ivoire	92	40.3	38.6	51.3	16	17	14.8	48.8	..	29
167	Burundi	81	37.6	38.2	40.7	21	45	54.6	87.6	36.4	-8
168	Congo (Democratic Republic of the)	88	39.3	41.1	32.8	54	31	..	..	..	..
169	Ethiopia	105	54.9	33.3	64.1	78	38	23.0	77.8	44.2	27
170	Chad	108	56.9	32.9	74.3	58	37	..	..	64.0	..
171	Central African Republic	98	43.6	46.2	51.4	25	24	66.6	84.0	..	-6
172	Mozambique	101	50.6	45.0	61.3	57	24	36.2	74.1	69.4	12
173	Mali	107	56.4	30.4	76.0	50	33	36.1	72.1	63.8	18
174	Niger	104	54.7	28.7	71.3	54	40	60.6	85.8	63.0[e]	1
175	Guinea-Bissau	99	44.8	40.5	..[f]	41	25	..	..	..	..
176	Burkina Faso	106	55.8	26.5	76.4	39	38	27.2	71.8	46.4	23
177	Sierra Leone	102	51.7	45.6	65.2	43	27	57.0[e]	74.5[e]	70.2	4

NOTES

† Denotes indicators used to calculate the human poverty index (HPI-1). For further details, see *Technical note 1*.

a. Data refer to the probability at birth of not surviving to age 40, multiplied by 100.

b. Data refer to national illiteracy estimates from censuses or surveys conducted between 1995 and 2005, unless otherwise specified. Due to differences in methodology and timeliness of underlying data, comparisons across countries and over time should be made with caution. For more details, see http://www.uis.unesco.org/.

c. Income poverty refers to the share of the population living on less than $1 a day. All

countries with an income poverty rate of less than 2% were given equal rank. The rankings are based on countries for which data are available for both indicators. A positive figure indicates that the country performs better in income poverty than in human poverty, a negative the opposite.

d. Data refer to the most recent year available during the period specified.

e. Data refer to a year or period other than that specified, differ from the standard definition or refer to only part of a country.

f. In the absence of recent data, estimates from UNESCO Institute for Statistics 2003 based on outdated census or survey information, were used and should be interpreted with caution: Barbados

0.3, Comoros 43.2, Djibouti 29.7, Eritrea 39.5, Fiji 5.6, Gambia 57.5, Guinea-Bissau 55.2, Guyana 1.0, Haiti 45.2, and Lebanon 11.7.

g. UNICEF 2005.

h. UNESCO Institute for Statistics estimates based on its Global Age-specific Literacy Projections model (2007).

i. An adult illiteracy rate of 0.2 was used to calculate the HPI-1 for Cuba.

j. Data are from national sources.

k. Data are from the Secretariat of the Caribbean Community, based on national sources.

l. UNICEF 2004.

m. UNDP 2006.

SOURCES

Column 1: determined on the basis of HPI-1 values in column 2.

Column 2: calculated on the basis of data in columns 3–6, see *Technical note 1* for details.

Column 3: UN 2007e.

Column 4: calculated on the basis of data on adult literacy rates from UNESCO Institute for Statistics 2007a.

Column 5: UN 2006a, based on a joint effort by UNICEF and WHO.

Column 6: UNICEF 2006.

Columns 7–9: World Bank 2007b.

Column 10: calculated on the basis of data in columns 1 and 7.

HPI-1 ranks for 108 developing countries and areas

1	Barbados	22	Turkey	45	Tunisia	68	Morocco
2	Uruguay	23	Brazil	46	Nicaragua	69	Sudan
3	Chile	24	Thailand	47	Indonesia	70	Lao People's Democratic Republic
4	Argentina	25	Suriname	48	Egypt	71	Lesotho
5	Costa Rica	26	Dominican Republic	49	Gabon	72	Uganda
6	Cuba	27	Mauritius	50	Fiji	73	Swaziland
7	Singapore	28	Peru	51	Algeria	74	Haiti
8	Saint Lucia	29	China	52	Myanmar	75	Madagascar
9	Occupied Palestinian Territories	30	Iran (Islamic Republic of)	53	Solomon Islands	76	Eritrea
10	Mexico	31	Syrian Arab Republic	54	Guatemala	77	Pakistan
11	Jordan	32	Bolivia	55	South Africa	78	Rwanda
12	Trinidad and Tobago	33	Guyana	56	Vanuatu	79	Malawi
13	Qatar	34	Jamaica	57	Congo	80	Nigeria
14	Colombia	35	El Salvador	58	Namibia	81	Burundi
15	Panama	36	Viet Nam	59	Djibouti	82	Yemen
16	Malaysia	37	Philippines	60	Kenya	83	Togo
17	United Arab Emirates	38	Cape Verde	61	Comoros	84	Nepal
18	Lebanon	39	Sao Tome and Principe	62	India	85	Cambodia
19	Ecuador	40	Mongolia	63	Botswana	86	Bhutan
20	Paraguay	41	Honduras	64	Cameroon	87	Mauritania
21	Venezuela (Bolivarian Republic of)	42	Maldives	65	Ghana	88	Congo (Democratic Republic of the)
		43	Belize	66	Equatorial Guinea	89	Angola
		44	Sri Lanka	67	Tanzania (United Republic of)	90	Papua New Guinea
						91	Zimbabwe
						92	Côte d'Ivoire
						93	Bangladesh
						94	Gambia
						95	Timor-Leste
						96	Zambia
						97	Senegal
						98	Central African Republic
						99	Guinea-Bissau
						100	Benin
						101	Mozambique
						102	Sierra Leone
						103	Guinea
						104	Niger
						105	Ethiopia
						106	Burkina Faso
						107	Mali
						108	Chad

Human development indicators

Human and income poverty: OECD countries, Central and Eastern Europe and the CIS

HDI rank	Human poverty index (HPI-2) [a] Rank	Human poverty index (HPI-2) [a] Value (%)	Probability at birth of not surviving to age 60 [b, †] (% of cohort) 2000–05	People lacking functional literacy skills [c, †] (% aged 16–65) 1994–2003 [e]	Long-term unemployment [†] (as % of labour force) 2006	Population below income poverty line (%) 50% of median income [†] 2000–04 [e]	Population below income poverty line (%) $11 a day 1994–95 [e]	Population below income poverty line (%) $4 a day 2000–04 [e]	HPI-2 rank minus income poverty rank [d]
HIGH HUMAN DEVELOPMENT									
1 Iceland	..	..	5.9	..	0.2	..	..	..	..
2 Norway	2	6.8	7.9 [f]	7.9	0.5	6.4	4.3	..	-2
3 Australia	13	12.1	7.3 [f]	17.0 [g]	0.9	12.2	17.6	..	-1
4 Canada	8	10.9	8.1	14.6	0.5	11.4	7.4	..	-4
5 Ireland	18	16.0	8.7	22.6 [g]	1.5	16.2	..	..	0
6 Sweden	1	6.3	6.7	7.5 [g]	1.1	6.5	6.3	..	-4
7 Switzerland	7	10.7	7.2	15.9	1.5	7.6	..	..	-1
8 Japan	12	11.7	6.9	.. [h]	1.3	11.8 [i]	..	..	-1
9 Netherlands	3	8.1	8.3	10.5 [g]	1.8	7.3 [j]	7.1	..	-3
10 France	11	11.2	8.9	.. [h]	4.1	7.3	9.9	..	5
11 Finland	4	8.1	9.4 [f]	10.4 [g]	1.8	5.4	4.8	..	3
12 United States	17	15.4	11.6	20.0	0.5	17.0	13.6	..	-2
13 Spain	15	12.5	7.7	.. [h]	2.2	14.2	..	..	-2
14 Denmark	5	8.2	10.3	9.6 [g]	0.8	5.6	..	..	3
15 Austria	10	11.1	8.8	.. [h]	1.3	7.7	..	..	1
16 United Kingdom	16	14.8	8.7	21.8 [g]	1.2	12.5 [j]	15.7	..	1
17 Belgium	14	12.4	9.3	18.4 [f,g]	4.6	8.0	..	..	4
18 Luxembourg	9	11.1	9.2	.. [h]	1.2 [k]	6.0	0.3	..	6
19 New Zealand	..	..	8.3	18.4 [g]	0.2	..	..	..	..
20 Italy	19	29.8	7.7	47.0	3.4	12.7	..	..	3
22 Germany	6	10.3	8.6	14.4 [g]	5.8	8.4	7.3	..	-5
23 Israel	..	..	7.2	..	..	15.6	..	..	..
24 Greece	..	..	8.2	..	4.9	14.3	..	..	..
27 Slovenia	..	..	10.8	..	..	8.2 [j]	..	..	..
29 Portugal	..	..	9.5	..	3.8	..	..	..	..
32 Czech Republic	..	..	11.6	..	3.9	4.9 [j]	..	1.0 [j]	..
34 Malta	..	..	7.6 [f]	..	..	..	..	..	..
36 Hungary	..	..	17.9	..	3.4	6.7 [j]	..	15.9	..
37 Poland	..	..	14.5	..	7.0	8.6 [j]	..	20.6	..
42 Slovakia	..	..	14.6	..	9.7	7.0 [j]	..	11.4 [j]	..
43 Lithuania	..	..	20.0	..	..	..	..	36.0	..
44 Estonia	..	..	21.4	..	..	12.4	..	33.2	..
45 Latvia	..	..	19.8	..	..	..	..	26.3	..
47 Croatia	..	..	12.7	..	..	..	..	10.0	..
53 Bulgaria	..	..	15.9	..	..	..	..	39.9	..
60 Romania	..	..	17.7	..	..	8.1 [j]	..	54.8	..
64 Belarus	..	..	24.8	..	..	..	..	15.9	..
66 Bosnia and Herzegovina	..	..	13.5	..	..	..	..	..	..
67 Russian Federation	..	..	32.4	..	..	18.8	..	45.3	..
68 Albania	..	..	11.3	..	..	..	..	48.0	..
69 Macedonia (TFYR)	..	..	13.5 [f]	..	..	..	..	22.0	..

TABLE 4

Human and income poverty: OECD countries, Central and Eastern Europe and the CIS

HDI rank	Human poverty index (HPI-2) [a] Rank	Human poverty index (HPI-2) [a] Value (%)	Probability at birth of not surviving to age 60 [b, †] (% of cohort) 2000–05	People lacking functional literacy skills [c, †] (% aged 16–65) 1994–2003 [e]	Long-term unemployment [†] (as % of labour force) 2006	Population below income poverty line (%) 50% of median income [†] 2000–04 [e]	Population below income poverty line (%) $11 a day 1994–95 [e]	Population below income poverty line (%) $4 a day 2000–04 [e]	HPI-2 rank minus income poverty rank [d]
MEDIUM HUMAN DEVELOPMENT									
73 Kazakhstan	..	..	31.1	..	..	..	..	56.7	..
76 Ukraine	..	..	26.5	..	..	..	..	44.7	..
83 Armenia	..	..	17.6	..	..	..	..	80.5	..
96 Georgia	..	..	19.1	..	..	..	..	61.9	..
98 Azerbaijan	..	..	24.5	..	..	..	..	85.9 [j]	..
109 Turkmenistan	..	..	31.3	..	..	..	..	79.4 [j]	..
111 Moldova	..	..	24.2	..	..	..	..	64.7	..
113 Uzbekistan	..	..	25.9	..	..	..	..	16.9	..
116 Kyrgyzstan	..	..	26.9	..	..	..	..	72.5	..
122 Tajikistan	..	..	25.9	..	..	..	..	84.7	..

NOTES

This table includes Israel and Malta, which are not OECD member countries, but excludes the Republic of Korea, Mexico and Turkey, which are. For the human poverty index (HPI-1) and related indicators for these countries, see Table 3.

† Denotes indicator used to calculate HPI-2; for details see *Technical note 1*.

a. HPI-2 is calculated for selected high-income OECD countries only.

b. Data refer to the probability at birth of not surviving to age 60, multiplied by 100.

c. Based on scoring at level 1 on the prose literacy scale of the IALS.

d. Income poverty refers to the share of the population living on less than 50% of the median adjusted disposable household income. A positive figure indicates that the country performs better in income poverty than in human poverty, a negative the opposite.

e. Data refer to the most recent year available during the period specified.

f. Data refer to a year or period other than that specified, differ from the standard definition or refer to only part of a country.

g. Based on OECD and Statistics Canada 2000.

h. For calculating HPI-2 an estimate of 16.4%, the unweighted average of countries with available data, was applied.

i. Smeeding 1997.

j. Data refer to a year between 1996 and 1999.

k. Data refer to 2005.

SOURCES

Column 1: determined on the basis of HPI-2 values in column 2.

Column 2: calculated on the basis of data in columns 3–6; see *Technical note 1* for details.

Column 3: calculated on the basis of survival data from UN 2007e.

Column 4: OECD and Statistics Canada 2005, unless otherwise specified.

Column 5: calculated on the basis of data on long-term unemployment and labour force from OECD 2007.

Column 6: LIS 2007.

Column 7: Smeeding, Rainwater and Burtless 2000.

Column 8: World Bank 2007a.

Column 9: calculated on the basis of data in columns 1 and 6.

HPI-2 ranks for 19 selected OECD countries

1 Sweden	9 Luxembourg	17 United States
2 Norway	10 Austria	18 Ireland
3 Netherlands	11 France	19 Italy
4 Finland	12 Japan	
5 Denmark	13 Australia	
6 Germany	14 Belgium	
7 Switzerland	15 Spain	
8 Canada	16 United Kingdom	

TABLE 5

... to lead a long and healthy life ...

Demographic trends

HDI rank	Total population (millions)			Annual population growth rate (%)		Urban population[a] (% of total)			Population under age 15 (% of total)		Population aged 65 and older (% of total)		Total fertility rate (births per woman)	
	1975	2005	2015[b]	1975–2005	2005–2015[b]	1975	2005	2015[b]	2005	2015[b]	2005	2015[b]	1970–1975[c]	2000–2005[c]
HIGH HUMAN DEVELOPMENT														
1 Iceland	0.2	0.3	0.3	1.0	0.8	86.7	92.8	93.6	22.1	20.0	11.7	14.2	2.8	2.0
2 Norway	4.0	4.6	4.9	0.5	0.6	68.2	77.4	78.6	19.6	17.7	14.7	17.0	2.2	1.8
3 Australia	13.6	20.3	22.4	1.3	1.0	85.9	88.2	89.9	19.5	17.9	13.1	16.1	2.5	1.8
4 Canada	23.1	32.3	35.2	1.1	0.9	75.6	80.1	81.4	17.6	15.6	13.1	16.1	2.0	1.5
5 Ireland	3.2	4.1	4.8	0.9	1.5	53.6	60.5	63.8	20.7	21.1	11.1	12.4	3.8	2.0
6 Sweden	8.2	9.0	9.4	0.3	0.4	82.7	84.2	85.1	17.4	16.7	17.2	20.2	1.9	1.7
7 Switzerland	6.3	7.4	7.7	0.5	0.4	55.7	75.2	78.7	16.7	14.5	15.4	18.7	1.8	1.4
8 Japan	111.5	127.9	126.6	0.5	-0.1	56.8	65.8	68.2	13.9	12.5	19.7	26.2	2.1	1.3
9 Netherlands	13.7	16.3	16.6	0.6	0.2	63.2	80.2	84.9	18.4	16.5	14.2	18.0	2.1	1.7
10 France	52.7	61.0	63.7	0.5	0.4	72.9	76.7	79.0	18.4	17.8	16.3	18.5	2.3	1.9
11 Finland	4.7	5.2	5.4	0.4	0.3	58.3	61.1	62.7	17.4	16.5	15.9	20.1	1.6	1.8
12 United States	220.2	299.8	329.0	1.0	0.9	73.7	80.8	83.7	20.8	19.8	12.3	14.1	2.0	2.0
13 Spain	35.7	43.4	46.0	0.7	0.6	69.6	76.7	78.3	14.4	15.4	16.8	18.3	2.9	1.3
14 Denmark	5.1	5.4	5.5	0.2	0.2	82.1	85.6	86.9	18.8	17.0	15.1	18.8	2.0	1.8
15 Austria	7.6	8.3	8.5	0.3	0.3	65.6	66.0	67.7	15.8	14.1	16.2	18.6	2.0	1.4
16 United Kingdom	56.2	60.2	62.8	0.2	0.4	82.7	89.7	90.6	18.0	17.2	16.1	18.1	2.0	1.7
17 Belgium	9.8	10.4	10.6	0.2	0.2	94.5	97.2	97.5	17.0	15.8	17.3	19.0	2.0	1.6
18 Luxembourg	0.4	0.5	0.5	0.8	1.1	77.3	82.8	82.1	18.5	17.0	14.2	14.6	1.7	1.7
19 New Zealand	3.1	4.1	4.5	0.9	0.8	82.8	86.2	87.4	21.5	19.4	12.2	14.7	2.8	2.0
20 Italy	55.4	58.6	59.0	0.2	0.1	65.6	67.6	69.5	14.0	13.5	19.7	22.1	2.3	1.3
21 Hong Kong, China (SAR)	4.4	7.1	7.7	1.6	0.9	89.7	100.0	100.0	15.1	12.3	12.0	14.5	2.9	0.9
22 Germany	78.7	82.7	81.8	0.2	-0.1	72.7	75.2	76.3	14.4	12.9	18.8	20.9	1.6	1.3
23 Israel	3.4	6.7	7.8	2.3	1.5	86.6	91.6	91.9	27.9	26.2	10.1	11.5	3.8	2.9
24 Greece	9.0	11.1	11.3	0.7	0.2	55.3	59.0	61.0	14.3	13.7	18.3	19.9	2.3	1.3
25 Singapore	2.3	4.3	4.8	2.2	1.1	100.0	100.0	100.0	19.5	12.8	8.5	13.5	2.6	1.4
26 Korea (Republic of)	35.3	47.9	49.1	1.0	0.3	48.0	80.8	83.1	18.6	13.7	9.4	13.3	4.3	1.2
27 Slovenia	1.7	2.0	2.0	0.5	(.)	42.4	51.0	53.3	14.1	13.4	15.6	18.2	2.2	1.2
28 Cyprus	0.6	0.8	0.9	1.1	1.0	47.3	69.3	71.5	19.9	17.3	12.1	14.2	2.5	1.6
29 Portugal	9.1	10.5	10.8	0.5	0.3	40.8	57.6	63.6	15.7	15.3	16.9	18.5	2.7	1.5
30 Brunei Darussalam	0.2	0.4	0.5	2.8	1.9	62.0	73.5	77.6	29.6	25.8	3.2	4.3	5.4	2.5
31 Barbados	0.2	0.3	0.3	0.6	0.3	40.8	52.7	58.8	18.9	16.1	9.2	11.6	2.7	1.5
32 Czech Republic	10.0	10.2	10.1	0.1	-0.1	63.7	73.5	74.0	14.8	13.8	14.2	18.2	2.2	1.2
33 Kuwait	1.0	2.7	3.4	3.3	2.2	89.4	98.3	98.5	23.8	22.5	1.8	3.1	6.9	2.3
34 Malta	0.3	0.4	0.4	0.9	0.4	89.7	95.3	97.2	17.4	14.6	13.2	17.7	2.1	1.5
35 Qatar	0.2	0.8	1.0	5.1	1.9	88.9	95.4	96.2	21.7	20.6	1.3	2.1	6.8	2.9
36 Hungary	10.5	10.1	9.8	-0.1	-0.3	62.2	66.3	70.3	15.8	14.2	15.2	17.3	2.1	1.3
37 Poland	34.0	38.2	37.6	0.4	-0.2	55.3	62.1	64.0	16.3	14.2	13.3	15.5	2.3	1.3
38 Argentina	26.0	38.7	42.7	1.3	1.0	81.0	90.1	91.6	26.4	23.9	10.2	11.1	3.1	2.4
39 United Arab Emirates	0.5	4.1	5.3	6.8	2.5	83.6	76.7	77.4	19.8	19.7	1.1	1.6	6.4	2.5
40 Chile	10.4	16.3	17.9	1.5	1.0	78.4	87.6	90.1	24.9	20.9	8.1	10.5	3.6	2.0
41 Bahrain	0.3	0.7	0.9	3.3	1.7	85.0	96.5	98.2	26.3	22.2	3.1	4.2	5.9	2.5
42 Slovakia	4.7	5.4	5.4	0.4	(.)	46.3	56.2	58.0	16.8	14.6	11.7	13.8	2.5	1.2
43 Lithuania	3.3	3.4	3.3	0.1	-0.5	55.7	66.6	66.8	16.8	14.0	15.3	16.8	2.3	1.3
44 Estonia	1.4	1.3	1.3	-0.2	-0.3	67.6	69.1	70.1	15.2	16.0	16.6	17.3	2.2	1.4
45 Latvia	2.5	2.3	2.2	-0.2	-0.5	64.2	67.8	68.9	14.4	14.2	16.6	17.7	2.0	1.2
46 Uruguay	2.8	3.3	3.4	0.5	0.3	83.4	92.0	93.1	23.8	21.4	13.5	14.4	3.0	2.2
47 Croatia	4.3	4.6	4.5	0.2	-0.2	45.1	56.5	59.5	15.5	13.9	17.2	18.7	2.0	1.3
48 Costa Rica	2.1	4.3	5.0	2.5	1.4	41.3	61.7	66.9	28.4	23.8	5.8	7.4	4.3	2.3
49 Bahamas	0.2	0.3	0.4	1.8	1.2	71.5	90.4	92.2	27.6	23.0	6.2	8.2	3.4	2.1
50 Seychelles	0.1	0.1	0.1	1.1	0.4	46.3	52.9	58.2	..	..	..	..	..	..
51 Cuba	9.4	11.3	11.3	0.6	(.)	64.2	75.5	74.7	19.2	15.7	11.2	14.3	3.6	1.6
52 Mexico	60.7	104.3	115.8	1.8	1.0	62.8	76.0	78.7	30.8	25.6	5.8	7.5	6.5	2.4
53 Bulgaria	8.7	7.7	7.2	-0.4	-0.8	57.6	70.0	72.8	13.8	13.5	17.2	19.2	2.2	1.3

TABLE 5

Demographic trends

HDI rank	Total population (millions)			Annual population growth rate (%)		Urban population[a] (% of total)			Population under age 15 (% of total)		Population aged 65 and older (% of total)		Total fertility rate (births per woman)	
	1975	2005	2015 [b]	1975–2005	2005–2015 [b]	1975	2005	2015 [b]	2005	2015 [b]	2005	2015 [b]	1970–1975 [c]	2000–2005 [c]
54 Saint Kitts and Nevis	(.)	(.)	0.1	0.3	1.2	35.0	32.2	33.5	..	..	..	..	..	..
55 Tonga	0.1	0.1	0.1	0.2	0.4	20.3	24.0	27.4	37.5	33.9	6.4	6.8	5.5	3.7
56 Libyan Arab Jamahiriya	2.5	5.9	7.1	2.9	1.9	57.3	84.8	87.4	30.3	29.4	3.8	4.9	7.6	3.0
57 Antigua and Barbuda	0.1	0.1	0.1	0.3	1.1	34.2	39.1	44.7	..	..	..	..	..	..
58 Oman	0.9	2.5	3.1	3.4	2.0	34.1	71.5	72.3	33.8	28.6	2.6	3.6	7.2	3.7
59 Trinidad and Tobago	1.0	1.3	1.4	0.9	0.4	11.4	12.2	15.8	22.2	20.8	6.5	8.2	3.5	1.6
60 Romania	21.2	21.6	20.6	0.1	-0.5	42.8	53.7	56.1	15.7	14.7	14.8	15.7	2.6	1.3
61 Saudi Arabia	7.3	23.6	29.3	3.9	2.1	58.3	81.0	83.2	34.5	30.7	2.8	3.3	7.3	3.8
62 Panama	1.7	3.2	3.8	2.1	1.6	49.0	70.8	77.9	30.4	27.2	6.0	7.5	4.9	2.7
63 Malaysia	12.3	25.7	30.0	2.5	1.6	37.7	67.3	75.4	31.4	27.3	4.4	5.8	5.2	2.9
64 Belarus	9.4	9.8	9.3	0.1	-0.6	50.6	72.2	76.7	15.7	14.4	14.4	13.7	2.3	1.2
65 Mauritius	0.9	1.2	1.3	1.1	0.7	43.4	42.4	44.1	24.4	20.9	6.6	8.3	3.2	1.9
66 Bosnia and Herzegovina	3.7	3.9	3.9	0.1	(.)	31.3	45.7	51.8	17.6	13.9	13.7	16.3	2.6	1.3
67 Russian Federation	134.2	144.0	136.5	0.2	-0.5	66.9	73.0	72.6	15.1	15.9	13.8	13.1	2.0	1.3
68 Albania	2.4	3.2	3.3	0.9	0.6	32.7	45.4	52.8	26.3	22.3	8.4	10.6	4.7	2.2
69 Macedonia (TFYR)	1.7	2.0	2.0	0.6	(.)	50.6	68.9	75.1	19.7	16.2	11.1	13.0	3.0	1.6
70 Brazil	108.1	186.8	210.0	1.8	1.2	61.7	84.2	88.2	27.8	25.4	6.1	7.7	4.7	2.3
MEDIUM HUMAN DEVELOPMENT														
71 Dominica	0.1	0.1	0.1	(.)	-0.1	55.3	72.9	76.4	..	..	..	..	..	..
72 Saint Lucia	0.1	0.2	0.2	1.3	1.1	25.2	27.6	29.0	27.9	25.4	7.2	7.3	5.7	2.2
73 Kazakhstan	14.1	15.2	16.3	0.2	0.7	52.6	57.3	60.3	24.2	24.9	8.0	7.5	3.5	2.0
74 Venezuela (Bolivarian Republic of)	12.7	26.7	31.3	2.5	1.6	75.8	93.4	95.9	31.3	27.9	5.0	6.6	4.9	2.7
75 Colombia	25.3	44.9	50.7	1.9	1.2	60.0	72.7	75.7	30.3	25.4	5.1	6.8	5.0	2.5
76 Ukraine	49.0	46.9	43.4	-0.1	-0.8	58.4	67.8	70.2	14.7	13.9	16.1	15.9	2.2	1.2
77 Samoa	0.2	0.2	0.2	0.7	0.8	21.0	22.4	24.9	40.8	33.8	4.6	4.8	5.7	4.4
78 Thailand	42.2	63.0	66.8	1.3	0.6	23.8	32.3	36.2	21.7	19.7	7.8	10.2	5.0	1.8
79 Dominican Republic	5.3	9.5	10.9	2.0	1.4	45.7	66.8	73.6	33.5	30.5	5.6	6.7	5.7	3.0
80 Belize	0.1	0.3	0.3	2.4	2.0	50.2	48.3	51.2	37.6	32.0	4.2	4.6	6.3	3.4
81 China	927.8 [d]	1,313.0 [d]	1,388.6 [d]	1.2 [d]	0.6 [d]	17.4	40.4	49.2	21.6	18.5	7.7	9.6	4.9	1.7
82 Grenada	0.1	0.1	0.1	0.4	0.1	32.6	30.6	32.2	34.2	26.7	6.8	6.0	4.6	2.4
83 Armenia	2.8	3.0	3.0	0.2	-0.1	63.6	64.1	64.1	20.8	17.5	12.1	11.0	3.0	1.3
84 Turkey	41.2	73.0	82.1	1.9	1.2	41.6	67.3	71.9	28.3	24.4	5.6	6.5	5.3	2.2
85 Suriname	0.4	0.5	0.5	0.7	0.5	49.5	73.9	77.4	29.8	26.2	6.3	7.3	5.3	2.6
86 Jordan	1.9	5.5	6.9	3.5	2.2	57.7	82.3	85.3	37.2	32.2	3.2	3.9	7.8	3.5
87 Peru	15.2	27.3	30.8	2.0	1.2	61.5	72.6	74.9	31.8	27.4	5.6	6.7	6.0	2.7
88 Lebanon	2.7	4.0	4.4	1.3	1.0	67.0	86.6	87.9	28.6	24.6	7.2	7.6	4.8	2.3
89 Ecuador	6.9	13.1	14.6	2.1	1.1	42.4	62.8	67.6	32.6	28.2	5.9	7.5	6.0	2.8
90 Philippines	42.0	84.6	101.1	2.3	1.8	35.6	62.7	69.6	36.2	32.5	3.8	4.7	6.0	3.5
91 Tunisia	5.7	10.1	11.2	1.9	1.0	49.9	65.3	69.1	26.0	22.5	6.3	6.7	6.2	2.0
92 Fiji	0.6	0.8	0.9	1.2	0.5	36.7	50.8	56.1	32.9	28.7	4.2	6.0	4.2	3.0
93 Saint Vincent and the Grenadines	0.1	0.1	0.1	0.7	0.4	27.0	45.9	50.0	29.3	26.8	6.5	7.0	5.5	2.3
94 Iran (Islamic Republic of)	33.3	69.4	79.4	2.4	1.3	45.7	66.9	71.9	28.8	25.6	4.5	4.9	6.4	2.1
95 Paraguay	2.8	5.9	7.0	2.5	1.7	39.0	58.5	64.4	35.8	31.4	4.8	5.8	5.4	3.5
96 Georgia	4.9	4.5	4.2	-0.3	-0.7	49.5	52.2	53.8	18.9	15.9	14.3	14.4	2.6	1.5
97 Guyana	0.7	0.7	0.7	(.)	-0.3	30.0	28.2	29.4	31.1	25.3	5.7	8.2	4.9	2.4
98 Azerbaijan	5.7	8.4	9.0	1.3	0.8	51.9	51.5	52.8	25.3	20.6	7.2	6.8	4.3	1.7
99 Sri Lanka	13.7	19.1	20.0	1.1	0.4	19.5	15.1	15.7	24.2	21.4	6.5	9.3	4.1	2.0
100 Maldives	0.1	0.3	0.4	2.6	1.8	17.3	29.6	34.8	34.0	29.0	3.8	3.9	7.0	2.8
101 Jamaica	2.0	2.7	2.8	1.0	0.5	44.1	53.1	56.7	31.7	27.9	7.5	7.9	5.0	2.6
102 Cape Verde	0.3	0.5	0.6	2.0	2.1	21.4	57.3	64.3	39.5	35.6	4.3	3.3	7.0	3.8
103 El Salvador	4.1	6.7	7.6	1.6	1.3	41.5	59.8	63.2	34.1	29.7	5.5	6.5	6.1	2.9
104 Algeria	16.0	32.9	38.1	2.4	1.5	40.3	63.3	69.3	29.6	26.7	4.5	5.0	7.4	2.5
105 Viet Nam	48.0	85.0	96.5	1.9	1.3	18.8	26.4	31.6	29.6	25.0	5.6	5.8	6.7	2.3
106 Occupied Palestinian Territories	1.3	3.8	5.1	3.7	3.0	59.6	71.6	72.9	45.9	41.9	3.1	3.0	7.7	5.6

TABLE 5

HDI rank	Total population (millions)			Annual population growth rate (%)		Urban population[a] (% of total)			Population under age 15 (% of total)		Population aged 65 and older (% of total)		Total fertility rate (births per woman)	
	1975	2005	2015[b]	1975–2005	2005–2015[b]	1975	2005	2015[b]	2005	2015[b]	2005	2015[b]	1970–1975[c]	2000–2005[c]
107 Indonesia	135.4	226.1	251.6	1.7	1.1	19.3	48.1	58.5	28.4	24.9	5.5	6.6	5.3	2.4
108 Syrian Arab Republic	7.5	18.9	23.5	3.1	2.2	45.1	50.6	53.4	36.6	33.0	3.2	3.6	7.5	3.5
109 Turkmenistan	2.5	4.8	5.5	2.2	1.3	47.6	46.2	50.8	31.8	27.0	4.7	4.4	6.2	2.8
110 Nicaragua	2.8	5.5	6.3	2.2	1.4	48.9	59.0	63.0	37.9	32.0	4.0	4.8	6.8	3.0
111 Moldova	3.8	3.9	3.6	(.)	-0.6	36.2	46.7	50.0	20.0	17.2	11.1	11.8	2.6	1.5
112 Egypt	39.2	72.8	86.2	2.1	1.7	43.5	42.8	45.4	33.3	30.7	4.8	5.6	5.9	3.2
113 Uzbekistan	14.0	26.6	30.6	2.1	1.4	39.1	36.7	38.0	33.2	28.3	4.7	4.4	6.3	2.7
114 Mongolia	1.4	2.6	2.9	1.9	1.0	48.7	56.7	58.8	28.9	24.3	3.9	4.3	7.3	2.1
115 Honduras	3.1	6.8	8.3	2.6	1.9	32.1	46.5	51.4	40.0	34.3	4.1	4.6	7.1	3.7
116 Kyrgyzstan	3.3	5.2	5.8	1.5	1.1	38.2	35.8	38.1	31.0	27.3	5.9	5.1	4.7	2.5
117 Bolivia	4.8	9.2	10.9	2.2	1.7	41.3	64.2	68.8	38.1	33.5	4.5	5.2	6.5	4.0
118 Guatemala	6.2	12.7	16.2	2.4	2.4	36.7	47.2	52.0	43.1	39.5	4.3	4.7	6.2	4.6
119 Gabon	0.6	1.3	1.5	2.6	1.5	43.0	83.6	87.7	35.9	31.8	4.7	4.8	5.0	3.4
120 Vanuatu	0.1	0.2	0.3	2.5	2.3	13.4	23.5	28.1	39.8	35.1	3.3	3.8	6.1	4.2
121 South Africa	25.7	47.9	50.3	2.1	0.5	48.1	59.3	64.1	32.1	30.2	4.2	5.5	5.5	2.8
122 Tajikistan	3.4	6.6	7.7	2.1	1.6	35.5	24.7	24.6	39.4	33.6	3.9	3.5	6.8	3.8
123 Sao Tome and Principe	0.1	0.2	0.2	2.1	1.6	31.6	58.0	65.8	41.6	38.1	4.4	3.5	6.5	4.3
124 Botswana	0.8	1.8	2.1	2.7	1.2	11.8	57.4	64.6	35.6	32.1	3.4	3.8	6.5	3.2
125 Namibia	0.9	2.0	2.3	2.7	1.2	23.7	35.1	41.1	39.1	33.2	3.5	4.0	6.6	3.6
126 Morocco	17.3	30.5	34.3	1.9	1.2	37.8	58.7	65.0	30.3	26.8	5.2	5.9	6.9	2.5
127 Equatorial Guinea	0.2	0.5	0.6	2.6	2.4	27.4	38.9	41.1	42.4	41.3	4.1	3.9	5.7	5.6
128 India	613.8	1,134.4	1,302.5	2.0	1.4	21.3	28.7	32.0	33.0	28.7	5.0	5.8	5.3	3.1
129 Solomon Islands	0.2	0.5	0.6	3.0	2.2	9.1	17.0	20.5	40.5	35.9	2.9	3.3	7.2	4.4
130 Lao People's Democratic Republic	2.9	5.7	6.7	2.2	1.7	11.1	20.6	24.9	39.8	32.8	3.5	3.4	6.4	3.6
131 Cambodia	7.1	14.0	16.6	2.3	1.8	10.3	19.7	26.1	37.6	32.1	3.1	4.0	5.5	3.6
132 Myanmar	29.8	48.0	52.0	1.6	0.8	23.9	30.6	37.4	27.3	23.1	5.6	6.3	5.9	2.2
133 Bhutan	0.4	0.6	0.7	1.9	1.5	4.6	11.1	14.8	33.0	24.9	4.6	5.4	6.7	2.9
134 Comoros	0.3	0.8	1.0	3.1	2.3	21.2	37.0	44.0	42.0	38.5	2.7	3.1	7.1	4.9
135 Ghana	10.3	22.5	27.3	2.6	1.9	30.1	47.8	55.1	39.0	35.1	3.6	4.3	6.7	4.4
136 Pakistan	68.3	158.1	190.7	2.8	1.9	26.3	34.9	39.6	37.2	32.1	3.9	4.3	6.6	4.0
137 Mauritania	1.3	3.0	3.8	2.7	2.4	20.6	40.4	43.1	40.3	36.9	3.6	3.6	6.6	4.8
138 Lesotho	1.1	2.0	2.1	1.8	0.6	10.8	18.7	22.0	40.4	37.4	4.7	4.7	5.8	3.8
139 Congo	1.5	3.6	4.5	2.8	2.1	43.3	60.2	64.2	41.9	39.8	3.2	3.3	6.3	4.8
140 Bangladesh	79.0	153.3	180.1	2.2	1.6	9.9	25.1	29.9	35.2	31.1	3.5	4.3	6.2	3.2
141 Swaziland	0.5	1.1	1.2	2.5	0.6	14.0	24.1	27.5	39.8	36.5	3.2	3.8	6.9	3.9
142 Nepal	13.5	27.1	32.8	2.3	1.9	4.8	15.8	20.9	39.0	34.1	3.7	4.2	5.8	3.7
143 Madagascar	7.9	18.6	24.1	2.9	2.6	16.3	26.8	30.1	43.8	40.4	3.1	3.3	6.7	5.3
144 Cameroon	7.8	17.8	21.5	2.7	1.9	27.3	54.6	62.7	41.8	38.4	3.5	3.6	6.3	4.9
145 Papua New Guinea	2.9	6.1	7.3	2.5	1.9	11.9	13.4	15.0	40.6	35.8	2.4	2.7	6.1	4.3
146 Haiti	5.1	9.3	10.8	2.0	1.5	21.7	38.8	45.5	38.0	34.1	4.1	4.6	5.6	4.0
147 Sudan	16.8	36.9	45.6	2.6	2.1	18.9	40.8	49.4	40.7	36.4	3.5	4.1	6.6	4.8
148 Kenya	13.5	35.6	46.2	3.2	2.6	12.9	20.7	24.1	42.6	42.5	2.7	2.6	8.0	5.0
149 Djibouti	0.2	0.8	1.0	4.3	1.7	67.1	86.1	89.6	38.5	33.5	3.0	3.7	7.2	4.5
150 Timor-Leste	0.7	1.1	1.5	1.5	3.4	14.6	26.5	31.2	45.0	44.0	2.7	3.0	6.2	7.0
151 Zimbabwe	6.2	13.1	14.5	2.5	1.0	19.9	35.9	40.9	39.5	35.2	3.5	3.7	7.4	3.6
152 Togo	2.4	6.2	8.0	3.1	2.5	22.8	40.1	47.4	43.3	40.0	3.1	3.3	7.1	5.4
153 Yemen	7.1	21.1	28.3	3.6	2.9	14.8	27.3	31.9	45.9	42.4	2.3	2.5	8.7	6.0
154 Uganda	10.9	28.9	40.0	3.3	3.2	7.0	12.6	14.5	49.4	48.0	2.5	2.3	7.1	6.7
155 Gambia	0.6	1.6	2.1	3.5	2.5	24.4	53.9	61.8	41.2	38.3	3.7	4.5	6.6	5.2
LOW HUMAN DEVELOPMENT														
156 Senegal	5.1	11.8	14.9	2.8	2.3	33.7	41.6	44.7	42.2	39.0	4.2	4.4	7.0	5.2
157 Eritrea	2.1	4.5	6.2	2.5	3.1	13.5	19.4	24.3	43.0	42.6	2.3	2.5	6.5	5.5
158 Nigeria	61.2	141.4	175.7	2.8	2.2	23.4	48.2	55.9	44.3	41.3	2.9	3.0	6.9	5.8
159 Tanzania (United Republic of)	16.0	38.5	49.0	2.9	2.4	11.1	24.2	28.9	44.4	42.8	3.0	3.2	6.8	5.7

TABLE 5

Demographic trends

HDI rank	Total population (millions) 1975	2005	2015[b]	Annual population growth rate (%) 1975–2005	2005–2015[b]	Urban population[a] (% of total) 1975	2005	2015[b]	Population under age 15 (% of total) 2005	2015[b]	Population aged 65 and older (% of total) 2005	2015[b]	Total fertility rate (births per woman) 1970–1975[c]	2000–2005[c]
160 Guinea	4.0	9.0	11.4	2.7	2.4	19.5	33.0	38.1	43.4	41.5	3.1	3.4	7.0	5.8
161 Rwanda	4.4	9.2	12.1	2.5	2.7	4.0	19.3	28.7	43.5	43.7	2.5	2.2	8.3	6.0
162 Angola	6.8	16.1	21.2	2.9	2.8	19.1	53.3	59.7	46.4	45.3	2.4	2.4	7.2	6.8
163 Benin	3.2	8.5	11.3	3.2	2.9	21.9	40.1	44.6	44.2	41.9	2.7	2.9	7.1	5.9
164 Malawi	5.3	13.2	17.0	3.1	2.5	7.7	17.2	22.1	47.1	44.6	3.0	3.1	7.4	6.0
165 Zambia	5.0	11.5	13.8	2.7	1.9	34.9	35.0	37.0	45.7	43.4	2.9	3.0	7.4	5.6
166 Côte d'Ivoire	6.6	18.6	22.3	3.5	1.8	32.2	45.0	49.8	41.7	37.9	3.2	3.5	7.4	5.1
167 Burundi	3.7	7.9	11.2	2.5	3.6	3.2	10.0	13.5	45.1	45.9	2.6	2.4	6.8	6.8
168 Congo (Democratic Republic of the)	24.0	58.7	80.6	3.0	3.2	29.5	32.1	38.6	47.2	47.8	2.6	2.5	6.5	6.7
169 Ethiopia	34.2	79.0	101.0	2.8	2.5	9.5	16.0	19.1	44.5	41.0	2.9	3.1	6.8	5.8
170 Chad	4.2	10.1	13.4	3.0	2.8	15.6	25.3	30.5	46.2	45.2	3.0	2.8	6.6	6.5
171 Central African Republic	2.1	4.2	5.0	2.4	1.8	32.0	38.0	40.4	42.7	39.9	3.9	3.7	5.7	5.0
172 Mozambique	10.6	20.5	24.7	2.2	1.8	8.7	34.5	42.4	44.2	43.2	3.2	3.4	6.6	5.5
173 Mali	5.4	11.6	15.7	2.5	3.0	16.2	30.5	36.5	47.7	46.4	3.6	3.0	7.6	6.7
174 Niger	4.9	13.3	18.8	3.3	3.5	11.4	16.8	19.3	48.0	47.3	3.1	3.4	8.1	7.4
175 Guinea-Bissau	0.7	1.6	2.2	3.0	3.0	16.0	29.6	31.1	47.4	47.9	3.0	2.7	7.1	7.1
176 Burkina Faso	6.1	13.9	18.5	2.8	2.8	6.4	18.3	22.8	46.2	44.2	3.1	2.6	7.8	6.4
177 Sierra Leone	2.9	5.6	6.9	2.1	2.2	21.2	40.7	48.2	42.8	42.8	3.3	3.3	6.5	6.5
Developing countries	2,972.0 T	5,215.0 T	5,956.6 T	1.9	1.3	26.5	42.7	47.9	30.9	28.0	5.5	6.4	5.4	2.9
Least developed countries	357.6 T	765.7 T	965.2 T	2.5	2.3	14.8	26.7	31.6	41.5	39.3	3.3	3.5	6.6	4.9
Arab States	144.4 T	313.9 T	380.4 T	2.6	1.9	41.8	55.1	58.8	35.2	32.1	3.9	4.4	6.7	3.6
East Asia and the Pacific	1,312.3 T	1,960.6 T	2,111.2 T	1.3	0.7	20.5	42.8	51.1	23.8	20.6	7.1	8.8	5.0	1.9
Latin America and the Caribbean	323.9 T	556.6 T	626.5 T	1.8	1.2	61.1	77.3	80.6	29.8	26.3	6.3	7.7	5.0	2.5
South Asia	835.4 T	1,587.4 T	1,842.2 T	2.1	1.5	21.2	30.2	33.8	33.6	29.5	4.7	5.4	5.5	3.2
Sub-Saharan Africa	314.1 T	722.7 T	913.2 T	2.8	2.3	21.2	34.9	39.6	43.6	41.7	3.1	3.2	6.8	5.5
Central and Eastern Europe and the CIS	366.6 T	405.2 T	398.6 T	0.3	-0.2	57.7	63.2	63.9	18.1	17.4	12.8	12.9	2.5	1.5
OECD	928.0 T	1,172.6 T	1,237.3 T	0.8	0.5	66.9	75.6	78.2	19.4	17.8	13.8	16.1	2.6	1.7
High-income OECD	766.8 T	931.5 T	976.6 T	0.6	0.5	69.3	77.0	79.4	17.6	16.5	15.3	18.0	2.2	1.7
High human development	1,280.6 T	1,658.7 T	1,751.1 T	0.9	0.5	66.4	76.8	79.4	20.2	18.8	12.7	14.5	2.7	1.8
Medium human development	2,514.9 T	4,239.6 T	4,759.8 T	1.7	1.2	23.8	39.3	44.9	29.3	26.0	5.8	6.8	5.3	2.6
Low human development	218.5 T	508.7 T	653.0 T	2.8	2.5	18.6	33.2	38.6	44.9	43.0	2.9	3.0	6.9	6.0
High income	793.3 T	991.5 T	1,047.2 T	0.7	0.5	69.4	77.6	80.0	18.1	17.0	14.8	17.3	2.3	1.7
Middle income	2,054.2 T	3,084.7 T	3,339.7 T	1.4	0.8	34.7	53.9	60.3	25.1	22.5	7.3	8.6	4.6	2.1
Low income	1,218.0 T	2,425.5 T	2,894.7 T	2.3	1.8	20.5	30.0	34.2	36.6	33.3	4.2	4.7	5.9	3.8
World	4,076.1 T[e]	6,514.8 T[e]	7,295.1 T[e]	1.6	1.1	37.2	48.6	52.8	28.3	26.0	7.3	8.3	4.5	2.6

NOTES

a. Because data are based on national definitions of what constitutes a city or metropolitan area, cross-country comparisons should be made with caution.
b. Data refer to medium-variant projections.
c. Data refer to estimates for the period specified.
d. Population estimates include Taiwan Province of China.
e. Data are aggregates provided by original data source. The total population of the 177 countries included in the main indicator tables was estimated to be 4,013.6 million in 1975, 6,406.9 million in 2005 and projected to be 7,164.3 million in 2015.

SOURCES

Columns 1–3 and 9–14: UN 2007e.
Columns 4 and 5: calculated on the basis of columns 1 and 2.
Columns 6–8: UN 2006b.

Commitment to health: resources, access and services

	Health expenditure			MDG One-year-olds fully immunized		Children with diarrhoea receiving oral rehydration and continued feeding	MDG Contraceptive prevalence rate[a]	MDG Births attended by skilled health personnel	Physicians
	Public (% of GDP)	Private (% of GDP)	Per capita (PPP US$)	Against tuberculosis (%)	Against measles (%)	(% under age 5)	(% of married women aged 15–49)	(%)	(per 100,000 people)
HDI rank	2004	2004	2004	2005	2005	1998–2005[b]	1997–2005[b]	1997–2005[b]	2000–04[b]
HIGH HUMAN DEVELOPMENT									
1 Iceland	8.3	1.6	3,294	..	90	..	..	..	362
2 Norway	8.1	1.6	4,080	..	90	..	..	100[c,d]	313
3 Australia	6.5	3.1	3,123	..	94	..	..	100	247
4 Canada	6.8	3.0	3,173	..	94	..	75[d]	98	214
5 Ireland	5.7	1.5	2,618	93	84	..	..	100	279
6 Sweden	7.7	1.4	2,828	16	94	..	78[c,d]	100[c,d]	328
7 Switzerland	6.7	4.8	4,011	..	82	..	82[d]	..	361
8 Japan	6.3	1.5	2,293	..	99	..	56	100[d]	198
9 Netherlands	5.7	3.5	3,092	94	96	..	79[d]	100	315
10 France	8.2	2.3	3,040	84	87	..	75[d]	99[d]	337
11 Finland	5.7	1.7	2,203	98	97	..	..	100	316
12 United States	6.9	8.5	6,096	..	93	..	76[d]	99	256
13 Spain	5.7	2.4	2,099	..	97	..	81[d]	..	330[e]
14 Denmark	7.1	1.5	2,780	..	95	..	..	100[c,d]	293
15 Austria	7.8	2.5	3,418	..	75	..	51[d]	100[d]	338
16 United Kingdom	7.0	1.1	2,560	..	82	..	84	99	230
17 Belgium	6.9	2.8	3,133	..	88	..	78[d]	100[c,d]	449
18 Luxembourg	7.2	0.8	5,178	..	95	..	..	100	266
19 New Zealand	6.5	1.9	2,081	..	82	..	75[d]	100[d]	237
20 Italy	6.5	2.2	2,414	..	87	..	60[d]	..	420
21 Hong Kong, China (SAR)	..	..	..	..	..	..	..	..	..
22 Germany	8.2	2.4	3,171	..	93	..	75[d]	100[c,d]	337
23 Israel	6.1	2.6	1,972	61	95	..	..	99[c,d]	382
24 Greece	4.2	3.7	2,179	88	88	..	..	..	438
25 Singapore	1.3	2.4	1,118	98	96	..	62	100	140
26 Korea (Republic of)	2.9	2.7	1,135	97	99	..	81	100	157
27 Slovenia	6.6	2.1	1,815	98[c]	94	..	74[d]	100	225
28 Cyprus	2.6	3.2	1,128	..	86	..	..	100[c,d]	234
29 Portugal	7.0	2.8	1,897	89	93	..	..	100	342
30 Brunei Darussalam	2.6	0.6	621	96	97	..	..	99	101
31 Barbados	4.5	2.6	1,151	..	93	..	55	100	121[e]
32 Czech Republic	6.5	0.8	1,412	99	97	..	72	100	351
33 Kuwait	2.2	0.6	538	..	99	..	50[d]	98[d]	153
34 Malta	7.0	2.2	1,733	..	86	..	..	98[d]	318
35 Qatar	1.8	0.6	688	99	99	..	43	99	222
36 Hungary	5.7	2.2	1,308	99	99	..	77[d]	100	333
37 Poland	4.3	1.9	814	94	98	..	49[d]	100	247
38 Argentina	4.3	5.3	1,274	99	99	..	..	99	301[e]
39 United Arab Emirates	2.0	0.9	503	98	92	..	28[d]	99[d]	202
40 Chile	2.9	3.2	720	95	90	..	56[d]	100	109
41 Bahrain	2.7	1.3	871	70[c]	99	..	62[d]	98[d]	109
42 Slovakia	5.3	1.9	1,061	98	98	..	74[d]	99	318
43 Lithuania	4.9	1.6	843	99	97	..	47[d]	100	397
44 Estonia	4.0	1.3	752	99	96	..	70[d]	100	448
45 Latvia	4.0	3.1	852	99	95	..	48[d]	100	301
46 Uruguay	3.6	4.6	784	99	95	..	84	100	365
47 Croatia	6.1[d]	1.5[d]	917	98	96	..	..	100	244
48 Costa Rica	5.1	1.5	592	88	89	..	80	99	132
49 Bahamas	3.4	3.4	1,349	..	85	..	..	99	105[e]
50 Seychelles	4.6	1.5	634	99	99	..	..	..	151
51 Cuba	5.5	0.8	229	99	98	..	73	100	591
52 Mexico	3.0	3.5	655	99	96	..	74	83	198
53 Bulgaria	4.6	3.4	671	98	96	..	42	99	356

Human development indicators

TABLE 6

Commitment to health: resources, access and services

	Health expenditure			One-year-olds fully immunized MDG		Children with diarrhoea receiving oral rehydration and continued feeding	MDG Contraceptive prevalence rate [a]	MDG Births attended by skilled health personnel	Physicians
	Public (% of GDP)	**Private** (% of GDP)	**Per capita** (PPP US$)	**Against tuberculosis** (%)	**Against measles** (%)	(% under age 5)	(% of married women aged 15–49)	(%)	(per 100,000 people)
HDI rank	2004	2004	2004	2005	2005	1998–2005 [b]	1997–2005 [b]	1997–2005 [b]	2000–04 [b]
54 Saint Kitts and Nevis	3.3	1.9	710	99	99	..	41	100	119 [e]
55 Tonga	5.0	1.3	316	99	99	..	33	95	34
56 Libyan Arab Jamahiriya	2.8	1.0	328	99	97	..	45 [d]	94 [d]	129 [e]
57 Antigua and Barbuda	3.4	1.4	516	..	99	..	53	100	17 [e]
58 Oman	2.4	0.6	419	98	98	..	32	95	132
59 Trinidad and Tobago	1.4	2.1	523	98	93	31	38	96	79 [e]
60 Romania	3.4	1.7	433	98	97	..	70	99	190
61 Saudi Arabia	2.5	0.8	601	96	96	..	32 [d]	91 [d]	137
62 Panama	5.2	2.5	632	99	99	..	..	93	150
63 Malaysia	2.2	1.6	402	99	90	..	55 [d]	97	70
64 Belarus	4.6	1.6	427	99	99	..	50 [d]	100	455
65 Mauritius	2.4	1.9	516	99	98	..	76	98	106
66 Bosnia and Herzegovina	4.1	4.2	603	95	90	23	48	100	134
67 Russian Federation	3.7	2.3	583	97	99	..	..	99	425
68 Albania	3.0	3.7	339	98	97	51	75	98	131
69 Macedonia (TFYR)	5.7	2.3	471	99	96	..	..	99	219
70 Brazil	4.8	4.0	1,520	99	99	28 [d]	77 [d]	97	115
MEDIUM HUMAN DEVELOPMENT									
71 Dominica	4.2	1.7	309	98	98	..	50	100	50 [e]
72 Saint Lucia	3.3	1.8	302	99	94	..	47	99	517 [e]
73 Kazakhstan	2.3	1.5	264	69	99	22	66	99	354
74 Venezuela (Bolivarian Republic of)	2.0	2.7	285	95	76	51	77	95	194
75 Colombia	6.7	1.1	570	87	89	39	78	96	135
76 Ukraine	3.7	2.8	427	96	96	..	68	100	295
77 Samoa	4.1	1.2	218	86	57	..	30 [d]	100	70 [e]
78 Thailand	2.3	1.2	293	99	96	..	79	99	37
79 Dominican Republic	1.9	4.1	377	99	99	42	70	99	188
80 Belize	2.7	2.4	339	96	95	..	56	83	105
81 China	1.8 [d]	2.9 [d]	277	86	86	..	87	97	106
82 Grenada	5.0	1.9	480	..	99	..	54	100	50 [e]
83 Armenia	1.4	4.0	226	94	94	48	53	98	359
84 Turkey	5.2 [d]	2.1 [d]	557	89	91	19	71	83	135
85 Suriname	3.6	4.2	376	..	91	43	42	85	45
86 Jordan	4.7 [d]	5.1 [d]	502	89	95	44	56	100	203
87 Peru	1.9	2.2	235	93	80	57	71	73	117 [e]
88 Lebanon	3.2	8.4	817	..	96	..	58	89 [d]	325
89 Ecuador	2.2	3.3	261	99	93	..	73	75	148
90 Philippines	1.4	2.0	203	91	80	76	49	60	58
91 Tunisia	2.8 [f]	2.8 [f]	502	97 [c]	96	..	66	90	134
92 Fiji	2.9	1.7	284	90	70	..	44	99	34 [e]
93 Saint Vincent and the Grenadines	3.9	2.2	418	95	97	..	58	100	87 [e]
94 Iran (Islamic Republic of)	3.2	3.4	604	99	94	..	74	90	87
95 Paraguay	2.6	5.1	327	78	90	..	73	77	111
96 Georgia	1.5	3.8	171	95	92	..	47	92	409
97 Guyana	4.4	0.9	329	96	92	40	37	86	48
98 Azerbaijan	0.9	2.7	138	98	98	40	55	88	355
99 Sri Lanka	2.0	2.3	163	99	99	..	70	96	55
100 Maldives	6.3	1.4	494	99	97	..	39	70	92
101 Jamaica	2.8	2.4	223	95	84	21	69	97	85
102 Cape Verde	3.9	1.3	225	78	65	..	53	89	49
103 El Salvador	3.5	4.4	375	84	99	..	67	92	124
104 Algeria	2.6	1.0	167	98	83	..	57	96	113
105 Viet Nam	1.5	4.0	184	95	95	39	77	85	53
106 Occupied Palestinian Territories	7.8 [f]	5.2 [f]	..	99	99	..	51	97	..

Human development indicators

TABLE 6

HDI rank	Health expenditure			One-year-olds fully immunized MDG		Children with diarrhoea receiving oral rehydration and continued feeding (% under age 5)	MDG Contraceptive prevalence rate[a] (% of married women aged 15–49)	MDG Births attended by skilled health personnel (%)	Physicians (per 100,000 people)
	Public (% of GDP)	Private (% of GDP)	Per capita (PPP US$)	Against tuberculosis (%)	Against measles (%)				
	2004	2004	2004	2005	2005	1998–2005[b]	1997–2005[b]	1997–2005[b]	2000–04[b]
107 Indonesia	1.0	1.8	118	82	72	56	57	72	13
108 Syrian Arab Republic	2.2	2.5	109	99	98	..	48	77[d]	140
109 Turkmenistan	3.3	1.5	245	99	99	..	62	97	418
110 Nicaragua	3.9	4.3	231	88[c]	96	49	69	67	37
111 Moldova	4.2	3.2	138	97	97	52	68	100	264
112 Egypt	2.2	3.7	258	98	98	29	59	74	54
113 Uzbekistan	2.4	2.7	160	93	99	33	68	96	274
114 Mongolia	4.0	2.0	141	99	99	66	69	97	263
115 Honduras	4.0	3.2	197	91	92	..	62	56	57
116 Kyrgyzstan	2.3	3.3	102	96	99	16[d]	60	98	251
117 Bolivia	4.1	2.7	186	93	64	54	58	67	122
118 Guatemala	2.3	3.4	256	96	77	22	43	41	90[e]
119 Gabon	3.1	1.4	264	89	55	44	33	86	29
120 Vanuatu	3.1	1.0	123	65	70	..	28	88	11[e]
121 South Africa	3.5	5.1	748	97	82	37	60	92	77
122 Tajikistan	1.0	3.4	54	98	84	29	34	71	203
123 Sao Tome and Principe	9.9	1.6	141	98	88	44	29	76	49
124 Botswana	4.0	2.4	504	99	90	7	48	94	40
125 Namibia	4.7	2.1	407	95	73	39	44	76	30
126 Morocco	1.7	3.4	234	95	97	46	63	63	51
127 Equatorial Guinea	1.2	0.4	223	73	51	36	..	65	30
128 India	0.9	4.1	91	75	58	22	47	43	60
129 Solomon Islands	5.6	0.3	114	84	72	..	11[d]	85	13[e]
130 Lao People's Democratic Republic	0.8	3.1	74	65	41	37	32	19	..
131 Cambodia	1.7	5.0	140	87	79	59	24	32	16
132 Myanmar	0.3	1.9	38	76	72	48	34	57	36
133 Bhutan	3.0	1.6	93	99	93	..	31	37	5
134 Comoros	1.6	1.2	25	90	80	31	26	62	15
135 Ghana	2.8	3.9	95	99	83	40	25	47	15
136 Pakistan	0.4	1.8	48	82	78	33[d]	28	31	74
137 Mauritania	2.0	0.9	43	87	61	28	8	57	11
138 Lesotho	5.5	1.0	139	96	85	53	37	55	5
139 Congo	1.2	1.3	30	85[c]	56	..	44	86	20
140 Bangladesh	0.9	2.2	64	99	81	52	58	13	26
141 Swaziland	4.0	2.3	367	84	60	24	48	74	16
142 Nepal	1.5	4.1	71	87	74	43	38	11	21
143 Madagascar	1.8	1.2	29	72	59	47	27	51	29
144 Cameroon	1.5	3.7	83	77	68	43	26	62	19
145 Papua New Guinea	3.0	0.6	147	73	60	..	26[d]	41	5
146 Haiti	2.9	4.7	82	71	54	41	28	24	25[e]
147 Sudan	1.5	2.6	54	57	60	38	7	87	22
148 Kenya	1.8	2.3	86	85	69	33	39	42	14
149 Djibouti	4.4	1.9	87	52	65	..	9	61	18
150 Timor-Leste	8.8	2.4	143	70	48	..	10	18	10
151 Zimbabwe	3.5	4.0	139	98	85	80	54	73	16
152 Togo	1.1	4.4	63	96	70	25	26	61	4
153 Yemen	1.9	3.1	82	66	76	23[d]	23	27	33
154 Uganda	2.5	5.1	135	92	86	29	20	39	8
155 Gambia	1.8	5.0	88	89	84	38	18	55	11
LOW HUMAN DEVELOPMENT									
156 Senegal	2.4	3.5	72	92	74	33	12	58	6
157 Eritrea	1.8	2.7	27	91	84	54	8	28	5
158 Nigeria	1.4	3.2	53	48	35	28	13	35	28
159 Tanzania (United Republic of)	1.7	2.3	29	91	91	53	26	43	2

Human development indicators

Commitment to health: resources, access and services

	Health expenditure			One-year-olds fully immunized	MDG	Children with diarrhoea receiving oral rehydration and continued feeding	MDG Contraceptive prevalence rate[a]	MDG Births attended by skilled health personnel	Physicians
	Public (% of GDP)	**Private** (% of GDP)	**Per capita** (PPP US$)	**Against tuberculosis** (%)	**Against measles** (%)	(% under age 5)	(% of married women aged 15–49)	(%)	(per 100,000 people)
HDI rank	2004	2004	2004	2005	2005	1998–2005[b]	1997–2005[b]	1997–2005[b]	2000–04[b]
160 Guinea	0.7	4.6	96	90	59	44	7	56	11
161 Rwanda	4.3	3.2	126	91	89	16	17	39	5
162 Angola	1.5	0.4	38	61	45	32	6	45	8
163 Benin	2.5	2.4	40	99	85	42	19	66	4
164 Malawi	9.6	3.3	58	97 c	82	51	33	56	2
165 Zambia	3.4	2.9	63	94	84	48	34	43	12
166 Côte d'Ivoire	0.9	2.9	64	51 c	51	34	15	68	12
167 Burundi	0.8	2.4	16	84	75	16	16	25	3
168 Congo (Democratic Republic of the)	1.1	2.9	15	84	70	17	31	61	11
169 Ethiopia	2.7	2.6	21	67	59	38	15	6	3
170 Chad	1.5	2.7	42 ·	40	23	27	3	14	4
171 Central African Republic	1.5	2.6	54	70	35	47	28	44	8
172 Mozambique	2.7	1.3	42	87	77	47	17	48	3
173 Mali	3.2	3.4	54	82	86	45	8	41	8
174 Niger	2.2	2.0	26	93	83	43	14	16	2
175 Guinea-Bissau	1.3	3.5	28	80	80	23	8	35	12
176 Burkina Faso	3.3	2.8	77	99	84	47	14	38	5
177 Sierra Leone	1.9	1.4	34	83 c	67	39	4	42	3
Developing countries	..	..	..	83	74	..	..	60	..
Least developed countries	..	..	..	82	72	..	..	35	..
Arab States	..	..	..	86	86	..	..	74	..
East Asia and the Pacific	..	..	..	87	84	..	..	87	..
Latin America and the Caribbean	..	..	..	96	92	..	..	87	..
South Asia	..	..	..	79	65	..	..	39	..
Sub-Saharan Africa	..	..	..	76	65	..	..	43	..
Central and Eastern Europe and the CIS	..	..	..	95	97	..	..	97	..
OECD	..	..	..	92	93	..	..	95	..
High-income OECD	..	..	..	86	92	..	..	99	..
High human development	..	..	..	96	95	..	..	97	..
Medium human development	..	..	..	84	75	..	..	63	..
Low human development	..	..	..	71	61	..	..	38	..
High income	..	..	..	87	93	..	..	99	..
Middle income	..	..	..	90	87	..	..	88	..
Low income	..	..	..	77	65	..	..	41	..
World	..	..	..	83 g	77 g	..	..	63 g	..

NOTES

a. Data usually refer to women aged 15-49 who are married or in union; the actual age range covered may vary across countries.

b. Data refer to the most recent year available during the period specified.

c. UNICEF 2005.

d. Data refer to a year or period other than that specified, differ from the standard definition or refer to only part of a country.

e. Data refer to a year between 1997 and 1999.

f. Data refer to 2003.

g. Data are aggregates provided by original data source.

SOURCES

Columns 1 and 2: World Bank 2007b.

Column 3: WHO 2007a.

Columns 4–8: UNICEF 2006.

Column 9: calculated on the basis of data on physicians per 1000 population from WHO 2007a.

Human development indicators

TABLE 7 . . . to lead a long and healthy life . . .

Water, sanitation and nutritional status

	MDG Population using improved sanitation (%)		MDG Population using an improved water source (%)		MDG Population undernourished (% of total population)		MDG Children under weight for age (% of children under age 5)	Children under height for age (% of children under age 5)	Infants with low birthweight (%)
HDI rank	1990	2004	1990	2004	1990/92[a]	2002/04[a]	1996–2005[b]	1996–2005[b]	1998–2005[b]
HIGH HUMAN DEVELOPMENT									
1 Iceland	100	100	100	100	<2.5	<2.5	..	..	4
2 Norway	..	..	100	100	<2.5	<2.5	..	..	5
3 Australia	100	100	100	100	<2.5	<2.5	..	..	7
4 Canada	100	100	100	100	<2.5	<2.5	..	..	6
5 Ireland	..	..	..	..	<2.5	<2.5	..	..	6
6 Sweden	100	100	100	100	<2.5	<2.5	..	..	4
7 Switzerland	100	100	100	100	<2.5	<2.5	..	..	6
8 Japan	100	100	100	100	<2.5	<2.5	..	..	8
9 Netherlands	100	100	100	100	<2.5	<2.5	..	..	..
10 France	..	..	100	100	<2.5	<2.5	..	..	7
11 Finland	100	100	100	100	<2.5	<2.5	..	..	4
12 United States	100	100	100	100	<2.5	<2.5	2	3	8
13 Spain	100	100	100	100	<2.5	<2.5	..	..	6[c]
14 Denmark	..	..	100	100	<2.5	<2.5	..	..	5
15 Austria	100	100	100	100	<2.5	<2.5	..	..	7
16 United Kingdom	..	..	100	100	<2.5	<2.5	..	..	8
17 Belgium	..	..	..	..	<2.5	<2.5	..	..	8[c]
18 Luxembourg	..	..	100	100	<2.5	<2.5	..	..	8
19 New Zealand	..	..	97	..	<2.5	<2.5	..	..	6
20 Italy	..	..	..	..	<2.5	<2.5	..	..	6
21 Hong Kong, China (SAR)	..	..	..	..	..	..	..	..	..
22 Germany	100	100	100	100	<2.5	<2.5	..	..	7
23 Israel	..	..	100	100	<2.5	<2.5	..	..	8
24 Greece	..	..	..	..	<2.5	<2.5	..	..	8
25 Singapore	100	100	100	100	..	..	3	4	8
26 Korea (Republic of)	..	..	..	92	<2.5	<2.5	..	..	4
27 Slovenia	..	..	..	..	3[d]	3	..	..	6
28 Cyprus	100	100	100	100	<2.5	<2.5	..	..	..
29 Portugal	..	..	..	..	<2.5	<2.5	..	..	8
30 Brunei Darussalam	..	..	..	..	4	4	..	..	10
31 Barbados	100	100	100	100	<2.5	<2.5	6[c,e]	..	11
32 Czech Republic	99	98	100	100	..	<2.5	1[c,e]	3	7
33 Kuwait	..	..	..	..	24	5	10	7	7
34 Malta	..	..	100	100	<2.5	<2.5	..	..	6
35 Qatar	100	100	100	100	..	..	6[c]	..	10
36 Hungary	..	95	99	99	..	<2.5	2[c,e]	..	9
37 Poland	..	..	..	..	..	<2.5	..	..	6
38 Argentina	81	91	94	96	<2.5	3	4	8	8
39 United Arab Emirates	97	98	100	100	4	<2.5	14[c]	..	15[c]
40 Chile	84	91	90	95	8	4	1	3	6
41 Bahrain	..	..	..	..	..	..	9[c]	..	8
42 Slovakia	99	99	100	100	4[d]	7	..	..	7
43 Lithuania	..	..	..	..	4[d]	<2.5	..	..	4
44 Estonia	97	97	100	100	9[d]	<2.5	..	..	4
45 Latvia	..	78	99	99	3[d]	3	..	..	5
46 Uruguay	100	100	100	100	7	<2.5	5[c]	14	8
47 Croatia	100	100	100	100	16[d]	7	1	..	6
48 Costa Rica	..	92	..	97	6	5	5	..	7
49 Bahamas	100	100	..	97	9	8	..	..	7
50 Seychelles	..	..	88	88	14	9	6[c,e]	..	..
51 Cuba	98	98	..	91	7	<2.5	4	10	5
52 Mexico	58	79	82	97	5	5	8	16	8
53 Bulgaria	99	99	99	99	8[d]	8	..	9	10

TABLE 7

Water, sanitation and nutritional status

	MDG Population using improved sanitation (%)		MDG Population using an improved water source (%)		MDG Population undernourished (% of total population)		MDG Children under weight for age (% of children under age 5)	Children under height for age (% of children under age 5)	Infants with low birthweight (%)
HDI rank	1990	2004	1990	2004	1990/92[a]	2002/04[a]	1996–2005[b]	1996–2005[b]	1998–2005[b]
54 Saint Kitts and Nevis	95	95	100	100	13	10	..	..	9
55 Tonga	96	96	100	100	..	..	..	..	0
56 Libyan Arab Jamahiriya	97	97	71	..	<2.5	<2.5	5[c]	..	7[c]
57 Antigua and Barbuda	..	95	..	91	..	..	10[c,e]	..	8
58 Oman	83	..	80	..	..	..	18	16	8
59 Trinidad and Tobago	100	100	92	91	13	10	6	5	23
60 Romania	..	..	..	57	..	<2.5	3	13	8
61 Saudi Arabia	..	..	90	..	4	4	14	..	11[c]
62 Panama	71	73	90	90	21	23	8	22	10
63 Malaysia	..	94	98	99	3	3	11	20	9
64 Belarus	..	84	100	100	..	4	..	..	5
65 Mauritius	..	94	100	100	6	5	15[c]	..	14
66 Bosnia and Herzegovina	..	95	97	97	9[d]	9	4	12	4
67 Russian Federation	87	87	94	97	4[d]	3	3[c]	..	6
68 Albania	..	91	96	96	5[d]	6	14	39	5
69 Macedonia (TFYR)	..	..	..	..	15[d]	5	6	1	6
70 Brazil	71	75	83	90	12	7	6	..	8
MEDIUM HUMAN DEVELOPMENT									
71 Dominica	..	84	..	97	4	8	5[c,e]	..	11
72 Saint Lucia	..	89	98	98	8	5	14[c,e]	..	10
73 Kazakhstan	72	72	87	86	..	6	4	14	8
74 Venezuela (Bolivarian Republic of)	..	68	..	83	11	18	5	17	9
75 Colombia	82	86	92	93	17	13	7	16	9
76 Ukraine	..	96	..	96	..	<2.5	1	6	5
77 Samoa	98	100	91	88	11	4	..	9	4[c]
78 Thailand	80	99	95	99	30	22	18[c]	16	9
79 Dominican Republic	52	78	84	95	27	29	5	12	11
80 Belize	..	47	..	91	7	4	6[c,e]	..	6
81 China	23	44	70	77	16[f]	12[f]	8	19	4
82 Grenada	97	96	..	95	9	7	..	..	8
83 Armenia	..	83	..	92	52[d]	24	4	18	7
84 Turkey	85	88	85	96	<2.5	3	4	19	16
85 Suriname	..	94	..	92	13	8	13	15	13
86 Jordan	93	93	97	97	4	6	4	12	12
87 Peru	52	63	74	83	42	12	8	31	11
88 Lebanon	..	98	100	100	<2.5	3	4	6	6
89 Ecuador	63	89	73	94	8	6	12	29	16
90 Philippines	57	72	87	85	26	18	28	34	20
91 Tunisia	75	85	81	93	<2.5	<2.5	4	16	7
92 Fiji	68	72	..	47	10	5	8[c,e]	..	10
93 Saint Vincent and the Grenadines	..	..	..	..	22	10	..	..	10
94 Iran (Islamic Republic of)	83	..	92	94	4	4	11	20	7[c]
95 Paraguay	58	80	62	86	18	15	5	..	9
96 Georgia	97	94	80	82	44[d]	9	3	15	7
97 Guyana	..	70	..	83	21	8	14	14	13
98 Azerbaijan	..	..	68	77	34[d]	7	7	24	12
99 Sri Lanka	69	91	68	79	28	22	29	18	22
100 Maldives	..	59	96	83	17	10	30	32	22
101 Jamaica	75	80	92	93	14	9	4	5	10
102 Cape Verde	..	43	..	80	..	..	14[c,e]	..	13
103 El Salvador	51	62	67	84	12	11	10	25	7
104 Algeria	88	92	94	85	5	4	10	22	7
105 Viet Nam	36	61	65	85	31	16	27	43	9
106 Occupied Palestinian Territories	..	73	..	92	..	16	5	..	9

Human development indicators

TABLE 7

HDI rank		MDG Population using improved sanitation (%)		MDG Population using an improved water source (%)		MDG Population undernourished (% of total population)		MDG Children under weight for age (% of children under age 5)	Children under height for age (% of children under age 5)	Infants with low birthweight (%)
		1990	2004	1990	2004	1990/92 [a]	2002/04 [a]	1996–2005 [b]	1996–2005 [b]	1998–2005 [b]
107	Indonesia	46	55	72	77	9	6	28	29	9
108	Syrian Arab Republic	73	90	80	93	5	4	7	24	6
109	Turkmenistan	..	62	..	72	12 [d]	7	12	28	6
110	Nicaragua	45	47	70	79	30	27	10	25	12
111	Moldova	..	68	..	92	5 [d]	11	4	11	5
112	Egypt	54	70	94	98	4	4	6	24	12
113	Uzbekistan	51	67	94	82	8 [d]	25	8	26	7
114	Mongolia	..	59	63	62	34	27	7	24	7
115	Honduras	50	69	84	87	23	23	17	30	14
116	Kyrgyzstan	60	59	78	77	21 [d]	4	11	33	7 [c]
117	Bolivia	33	46	72	85	28	23	8	33	7
118	Guatemala	58	86	79	95	16	22	23	54	12
119	Gabon	..	36	..	88	10	5	12	26	14
120	Vanuatu	..	50	60	60	12	11	20 [c,e]	..	6
121	South Africa	69	65	83	88	<2.5	<2.5	12	31	15
122	Tajikistan	..	51	..	59	22 [d]	56	..	42	15
123	Sao Tome and Principe	..	25	..	79	18	10	13	35	20
124	Botswana	38	42	93	95	23	32	13	29	10
125	Namibia	24	25	57	87	34	24	24	30	14
126	Morocco	56	73	75	81	6	6	10	23	15
127	Equatorial Guinea	..	53	..	43	..	..	19	43	13
128	India	14	33	70	86	25	20	47	51	30
129	Solomon Islands	..	31	..	70	33	21	21 [c,e]	..	13 [c]
130	Lao People's Democratic Republic	..	30	..	51	29	19	40	48	14
131	Cambodia	..	17	..	41	43	33	45	49	11
132	Myanmar	24	77	57	78	10	5	32	41	15
133	Bhutan	..	70	..	62	..	..	19	48	15
134	Comoros	32	33	93	86	47	60	25	47	25
135	Ghana	15	18	55	75	37	11	22	36	16
136	Pakistan	37	59	83	91	24	24	38	42	19 [c]
137	Mauritania	31	34	38	53	15	10	32	40	..
138	Lesotho	37	37	..	79	17	13	20	53	13
139	Congo	..	27	..	58	54	33	15	31	..
140	Bangladesh	20	39	72	74	35	30	48	51	36
141	Swaziland	..	48	..	62	14	22	10	37	9
142	Nepal	11	35	70	90	20	17	48	57	21
143	Madagascar	14	34	40	50	35	38	42	53	17
144	Cameroon	48	51	50	66	33	26	18	35	13
145	Papua New Guinea	44	44	39	39	..	..	35 [c,e]	44	11 [c]
146	Haiti	24	30	47	54	65	46	17	28	21
147	Sudan	33	34	64	70	31	26	41	48	31
148	Kenya	40	43	45	61	39	31	20	36	10
149	Djibouti	79	82	72	73	53	24	27	29	16
150	Timor-Leste	..	36	..	58	11	9	46	56	12
151	Zimbabwe	50	53	78	81	45	47	17	34	11
152	Togo	37	35	50	52	33	24	25	30	18
153	Yemen	32	43	71	67	34	38	46	60	32 [c]
154	Uganda	42	43	44	60	24	19	23	45	12
155	Gambia	..	53	..	82	22	29	17	24	17
LOW HUMAN DEVELOPMENT										
156	Senegal	33	57	65	76	23	20	17	20	18
157	Eritrea	7	9	43	60	70 [d]	75	40	44	14
158	Nigeria	39	44	49	48	13	9	29	43	14
159	Tanzania (United Republic of)	47	47	46	62	37	44	22	44	10

TABLE 7

Water, sanitation and nutritional status

	MDG Population using improved sanitation (%)		MDG Population using an improved water source (%)		MDG Population undernourished (% of total population)		MDG Children under weight for age (% of children under age 5)	Children under height for age (% of children under age 5)	Infants with low birthweight (%)
HDI rank	1990	2004	1990	2004	1990/92 [a]	2002/04 [a]	1996–2005 [b]	1996–2005 [b]	1998–2005 [b]
160 Guinea	14	18	44	50	39	24	26	39	16
161 Rwanda	37	42	59	74	43	33	23	48	9
162 Angola	29	31	36	53	58	35	31	51	12
163 Benin	12	33	63	67	20	12	23	39	16
164 Malawi	47	61	40	73	50	35	22	53	16
165 Zambia	44	55	50	58	48	46	20	53	12
166 Côte d'Ivoire	21	37	69	84	18	13	17	32	17
167 Burundi	44	36	69	79	48	66	45	63	16
168 Congo (Democratic Republic of the)	16	30	43	46	31	74	31	44	12
169 Ethiopia	3	13	23	22	69 [d]	46	38	51	15
170 Chad	7	9	19	42	58	35	37	45	22
171 Central African Republic	23	27	52	75	50	44	24	45	14
172 Mozambique	20	32	36	43	66	44	24	47	15
173 Mali	36	46	34	50	29	29	33	43	23
174 Niger	7	13	39	46	41	32	40	54	13
175 Guinea-Bissau	..	35	..	59	24	39	25	36	22
176 Burkina Faso	7	13	38	61	21	15	38	43	19
177 Sierra Leone	..	39	..	57	46	51	27	38	23
Developing countries	33	49	71	79	21	17	..	..	..
Least developed countries	22	37	51	59	38	35	..	..	..
Arab States	61	71	84	86	..	..	..	..	..
East Asia and the Pacific	30	50	72	79	17	12	..	..	..
Latin America and the Caribbean	67	77	83	91	14	10	..	..	..
South Asia	18	37	72	85	25	21	..	..	..
Sub-Saharan Africa	32	37	48	55	36	32	..	..	..
Central and Eastern Europe and the CIS	..	..	93	94	..	..	..	..	..
OECD	94	96	97	99	..	..	..	..	..
High-income OECD	100	100	100	100	..	..	..	..	..
High human development	90	92	96	98	..	..	..	..	..
Medium human development	30	48	73	82	20	16	..	..	..
Low human development	26	34	43	49	36	34	..	..	..
High income	..	..	100	100	..	..	..	..	..
Middle income	46	61	78	84	14	11	..	..	..
Low income	21	38	64	76	28	24	..	..	..
World	49 [g]	59 [g]	78 [g]	83 [g]	20	17	..	..	..

NOTES

a. Data refer to the average for the years specified.
b. Data refer to the most recent year available during the period specified.
c. Data refer to a year or period other than that specified, differ from the standard definition or refer only to part of a country.
d. Data refer to the period 1993/95.
e. UNICEF 2005.
f. Data for China include Hong Kong SAR, Macao SAR and Taiwan Province.
g. Data are aggregates provided by original data source.

SOURCES

Columns 1–4: UN 2006a, based on a joint effort by UNICEF and WHO.
Columns 5 and 6: FAO 2007a.
Columns 7 and 9: UNICEF 2006.
Column 8: WHO 2007a.

Inequalities in maternal and child health

		Births attended by skilled health personnel (%)		One-year-olds fully immunized [a] (%)		Children under height for age (% under age 5)		Infant mortality rate [b] (per 1,000 live births)		Under-five mortality rate [b] (per 1,000 live births)	
HDI rank	Survey year	Poorest 20%	Richest 20%	Poorest 20%	Richest 20%	Poorest 20%	Richest 20%	Poorest 20%	Richest 20%	Poorest 20%	Richest 20%
HIGH HUMAN DEVELOPMENT											
70 Brazil	1996	72	99	57	74	23	2	83	29	99	33
MEDIUM HUMAN DEVELOPMENT											
73 Kazakhstan	1999	99	99	69	62 [c]	15	8	68	42	82	45
75 Colombia	2005	72	99	47	72	20	3	32	14	39	16
78 Thailand [d]	2005–06	93	100	92 [e]	86 [e]	16	7	..	..	..	..
79 Dominican Republic	1996	89	98	34	47	14	2	67	23	90	27
83 Armenia	2005	96	100	59 [e]	51 [c,e]	15	8	41	14	52	23
84 Turkey	1998	53	98	28	70	29	4	68	30	85	33
86 Jordan	1997	91	99	21	17	14	5	35	23	42	25
87 Peru	2004–05	34	100	65 [e]	73 [e]	46	4	46	6	63	11
90 Philippines	2003	25	92	56	83	..	..	42	19	66	21
95 Paraguay	1990	41	98	20	53	23	3	43	16	57	20
105 Viet Nam	2002	58	100	44	92	..	..	39	14	53	16
107 Indonesia	1997	21	89	43	72	..	..	78	23	109	29
109 Turkmenistan	2000	97	98	85	78	25	17	89	58	106	70
110 Nicaragua	2001	78	99	64	71	35	5	50	16	64	19
111 Moldova	2005	99	100	86 [c,f]	86 [f]	14	6	20	16	29	17
112 Egypt	2005	51	96	85 [e]	91 [e]	24	14	59	23	75	25
113 Uzbekistan	1996	92	100	81	78	40	31	54	46	70	50
116 Kyrgyzstan	1997	96	100	69	73	34	14	83	46	96	49
117 Bolivia	2003	27	98	48 [e]	57 [e]	42	5	72 [g]	27 [g]	105 [g]	32 [g]
118 Guatemala	1998–99	9	92	66	56	65	8	58	39	78	39
119 Gabon	2000	67	97	6	24	33	12	57	36	93	55
121 South Africa	1998	68	98	51	70	..	..	62	17	87	22
122 Tajikistan [d,h]	2006	69	91	..	..	32	21	..	..	..	..
125 Namibia	2000	55	97	60	68	27	15	36	23	55	31
126 Morocco	2003–04	30	95	81 [e]	97 [e]	29	10	62	24	78	26
128 India	1998–99	16	84	21	64	58	27	97	38	141	46
131 Cambodia	2005	21	90	56 [e]	76 [e]	47	19	101	34	127	43
134 Comoros	1996	26	85	40	82	45	23	87	65	129	87 [i]
135 Ghana [d,h]	2006	..	..	62 [e]	86 [e]	31	7	75	64	118	100
136 Pakistan	1990	5	55	23	55	61	33	89	63	125	74
137 Mauritania	2000–01	15	93	16	45	39	23	61	62	98	79
138 Lesotho	2004	34	83	66 [e]	69 [e]	47	25	88	70	114	82
139 Congo	2005	70	98	29 [e]	73 [e]	32	20	91	56	135	85
140 Bangladesh	2004	3	40	57 [e]	87 [e]	54	25	90	65	121	72
142 Nepal	2001	4	45	54	82	62	36	86	53	130	68
143 Madagascar	2003–04	30	94	32	80	51	38	87	33	142	49
144 Cameroon	2004	29	94	36	60	41	12	101	51	189	88
146 Haiti	2005–06	6	68	34	56	34	5	78	45	125	55
148 Kenya	2003	17	75	40 [f]	65 [f]	38	19	96	62	149	91
151 Zimbabwe	1999	57	94	64	64	33	19	59	44	100	62
152 Togo	1998	25	91	22	52	29	11	84	66	168	97
153 Yemen	1997	7	50	8	56	58	35	109	60	163	73
154 Uganda	2000–01	20	77	27	43	43	25	106	60	192	106

TABLE 8

Inequalities in maternal and child health

HDI rank	Survey year	Births attended by skilled health personnel (%)		One-year-olds fully immunized [a] (%)		Children under height for age (% under age 5)		Infant mortality rate [b] (per 1,000 live births)		Under-five mortality rate [b] (per 1,000 live births)	
		Poorest 20%	Richest 20%	Poorest 20%	Richest 20%	Poorest 20%	Richest 20%	Poorest 20%	Richest 20%	Poorest 20%	Richest 20%
LOW HUMAN DEVELOPMENT											
156 Senegal	2005	20	89	59	65	26	6	89	41	183	64
157 Eritrea	2002	7	81	74	91	45	18	48	38	100	65
158 Nigeria	2003	12	84	3	40	49	18	133	52	257	79
159 Tanzania (United Republic of)	1999	29	83	53	78	50	23	115	92	160	135
160 Guinea	2005	15	87	29	45	41	22	127	68	217	113
161 Rwanda	2005	27	66	74	74	55	30	114	73	211	122
163 Benin	2001	50	99	49	73	35	18	112	50	198	93
164 Malawi [d,h]	2000	43	83	65	81	26	23	132	86	231	149
165 Zambia	2001–02	20	91	64	80	54	32	115	57	192	92
166 Côte d'Ivoire	2005	27	88	..	..	..	..	93	79	150	100
169 Ethiopia	2005	1	27	14	36	48	35	80	60	130	92
170 Chad	2004	4	55	1	24	51	32	109	101	176	187
171 Central African Republic	1994–95	14	82	18	64	42	25	132	54	193	98
172 Mozambique	2003	25	89	45	90	49	20	143	71	196	108
173 Mali	2001	8	82	20	56	45	20	137	90	248	148
174 Niger	2006	21	71	20	48	54	37	91	67	206	157
176 Burkina Faso	2003	39	91	34	61	46	21	97	78	206	144
177 Sierra Leone [d,h]	2005	27	83	..	..	44	26	159	108	268	179

NOTES

This table presents data for developing countries based on data from DHS conducted since 1990. Quintiles are defined by socioeconomic status in terms of assets or wealth, rather than in terms of income or consumption. For details, see Macro International 2007b.

a. Includes tuberculosis (BCG), measles or measles, mumps and rubella (MMR) and diphtheria, pertussis and tetanus (DPT) vaccinations.

b. Based on births in the 10 years preceding the survey.

c. Figure is based on less than 50 unweighted cases.

d. Data are obtained from UNICEF 2007b.

e. Includes BCG, measles or MMR, DPT or Pentavalente, and polio vaccinations.

f. Data are from preliminary MICS reports.

g. Includes BCG, measles or MMR, DPT, polio and other vaccinations.

h. Data pertain to 5-year period preceding the survey.

i. Large sampling error due to small number of cases.

SOURCES

All columns: Macro International 2007a and 2007b, unless otherwise specified.

TABLE 9

... to lead a long and healthy life ...

Leading global health crises and risks

HDI rank		HIV prevalence[a] (% aged 15–49) 2005	MDG Condom use at last high-risk sex[b] (% aged 15–24) Women 1999–2005[g]	Men 1999–2005[g]	MDG Antimalarial measures Use of insecticide-treated bednets (% of children under five) 1999–2005[g]	MDG Fevers treated with antimalarial drugs 1999–2005[g]	MDG Tuberculosis cases Prevalence[c] (per 100,000 people) 2005	MDG Detected under DOTS[d] (%) 2005	MDG Cured under DOTS[e] (%) 2004	Prevalence of smoking (% of adults)[f] Women 2002–04[g]	Men 2002–04[g]
HIGH HUMAN DEVELOPMENT											
1	Iceland	0.2 [0.1–0.3]	..	..	..	..	2	53	50	20	25
2	Norway	0.1 [0.1–0.2]	..	..	..	..	4	44	89	25	27
3	Australia	0.1 [<0.2]	..	..	..	..	6	42	85	16	19
4	Canada	0.3 [0.2–0.5]	..	..	..	..	4	64	62	17	22
5	Ireland	0.2 [0.1–0.4]	..	..	..	..	10	0	..	26	28
6	Sweden	0.2 [0.1–0.3]	..	..	..	..	5	56	64	18	17
7	Switzerland	0.4 [0.3–0.8]	..	..	..	..	6	0	..	23	27
8	Japan	<0.1 [<0.2]	..	..	..	..	38	57	57	15	47
9	Netherlands	0.2 [0.1–0.4]	..	..	..	..	5	47	83	28	36
10	France	0.4 [0.3–0.8]	..	..	..	..	10	0[h]	..	21	30
11	Finland	0.1 [<0.2]	..	..	..	..	5	0[h]	..	19	26
12	United States	0.6 [0.4–1.0]	..	..	..	..	3	85	61	19	24
13	Spain	0.6 [0.4–1.0]	..	..	..	..	22	0	..	25[h]	39[h]
14	Denmark	0.2 [0.1–0.4]	..	..	..	..	6	71	88	25	31
15	Austria	0.3 [0.2–0.5]	..	..	..	..	9	56	69	..	..
16	United Kingdom	0.2 [0.1–0.4]	..	..	..	..	11	0	..	25	27
17	Belgium	0.3 [0.2–0.5]	..	..	..	..	10	64	72	25	30
18	Luxembourg	0.2 [0.1–0.4]	..	..	..	..	9	59	..	26	39
19	New Zealand	0.1 [<0.2]	..	..	..	..	9	51	66	22	24
20	Italy	0.5 [0.3–0.9]	..	..	..	..	5	72	95[h]	17	31
21	Hong Kong, China (SAR)	..	..	..	..	..	77[i]	55[h,i]	78[h,i]	4[h]	22[h]
22	Germany	0.1 [0.1–0.2]	..	..	..	..	6	52	68	28	37
23	Israel	[<0.2]	..	..	..	..	6	42	80	18	32
24	Greece	0.2 [0.1–0.3]	..	..	..	..	15	0	..	29[h]	47[h]
25	Singapore	0.3 [0.2–0.7]	..	..	..	..	28	100	81	4[h]	24[h]
26	Korea (Republic of)	<0.1 [<0.2]	..	..	..	..	135	18	80	..	..
27	Slovenia	<0.1 [<0.2]	..	..	..	..	15	84	90	20[h]	28[h]
28	Cyprus	[<0.2]	..	..	..	..	5	57	20	..	..
29	Portugal	0.4 [0.3–0.9]	..	..	..	..	25	85	84	..	..
30	Brunei Darussalam	<0.1 [<0.2]	..	..	..	..	63	112	71	..	..
31	Barbados	1.5 [0.8–2.5]	..	..	..	..	12	135[h]	100[h]	..	..
32	Czech Republic	0.1 [<0.2]	..	..	..	..	11	65	73	20	31
33	Kuwait	[<0.2]	..	..	..	..	28	66	63	..	..
34	Malta	0.1 [0.1–0.2]	..	..	..	..	4	50	100	18	30
35	Qatar	[<0.2]	..	..	..	..	65	47	78	..	..
36	Hungary	0.1 [<0.2]	..	..	..	..	25	43	54	28	41
37	Poland	0.1 [0.1–0.2]	..	..	..	..	29	62	79	25	40
38	Argentina	0.6 [0.3–1.9]	..	..	..	..	51	67	58	25	32
39	United Arab Emirates	[<0.2]	..	..	..	..	24	19	70	1	17
40	Chile	0.3 [0.2–1.2]	..	..	..	..	16	112	83	37	48
41	Bahrain	[<0.2]	..	..	..	..	43	77	82	3[h]	15[h]
42	Slovakia	<0.1 [<0.2]	..	..	..	..	20	39	88	..	..
43	Lithuania	0.2 [0.1–0.6]	..	..	..	..	63	100	72	13	44
44	Estonia	1.3 [0.6–4.3]	..	..	..	..	46	64	71	18	45
45	Latvia	0.8 [0.5–1.3]	..	..	..	..	66	83	73	19	51
46	Uruguay	0.5 [0.2–6.1]	..	..	..	..	33	83	86[h]	24	35
47	Croatia	<0.1 [<0.2]	..	..	..	..	65	0[h]	..	27[h]	34[h]
48	Costa Rica	0.3 [0.1–3.6]	..	..	..	..	17	118	94[h]	10[h]	29[h]
49	Bahamas	3.3 [1.3–4.5]	..	..	..	..	49	67[h]	62[h]	..	..
50	Seychelles	..	..	..	..	..	56	65	92	..	..
51	Cuba	0.1 [<0.2]	..	..	..	..	11	98	93	..	..
52	Mexico	0.3 [0.2–0.7]	..	..	..	..	27	110	82	5	13
53	Bulgaria	<0.1 [<0.2]	..	..	..	..	41	90	80	23[h]	44[h]

TABLE 9

Leading global health crises and risks

		HIV prevalence [a] (% aged 15–49) 2005	MDG Condom use at last high-risk sex [b] (% aged 15–24) Women 1999–2005 [g]	Men 1999–2005 [g]	MDG Antimalarial measures Use of insecticide-treated bednets 1999–2005 [g]	MDG Fevers treated with antimalarial drugs (% of children under five) 1999–2005 [g]	MDG Tuberculosis cases Prevalence [c] (per 100,000 people) 2005	Detected under DOTS [d] (%) 2005	Cured under DOTS [e] (%) 2004	Prevalence of smoking (% of adults) [f] Women 2002–04 [g]	Men 2002–04 [g]
54	Saint Kitts and Nevis	..	..	..	..	..	17	0	50 [h]	..	..
55	Tonga	..	..	..	..	..	32	96	83 [h]	11 [h]	53 [h]
56	Libyan Arab Jamahiriya	[<0.2]	..	..	..	..	18	178	64	..	..
57	Antigua and Barbuda	..	..	..	..	..	9	246	100	..	..
58	Oman	[<0.2]	..	..	..	..	11	108	90	..	..
59	Trinidad and Tobago	2.6 [1.4–4.2]	..	..	..	..	13	..	..	..	..
60	Romania	<0.1 [<0.2]	..	..	..	..	146	82	82	10 [h]	32 [h]
61	Saudi Arabia	[<0.2]	..	..	..	..	58	38	82	8 [h]	19 [h]
62	Panama	0.9 [0.5–3.7]	..	..	..	..	46	131	78	..	..
63	Malaysia	0.5 [0.2–1.5]	..	..	..	..	131	73	56	2	43
64	Belarus	0.3 [0.2–0.8]	..	..	..	..	70	46	74	7	53
65	Mauritius	0.6 [0.3–1.8]	..	..	..	..	132	32	89	1	32
66	Bosnia and Herzegovina	<0.1 [<0.2]	..	..	..	..	57	71	98	30	49
67	Russian Federation	1.1 [0.7–1.8]	..	..	..	..	150	30	59	16 [h]	60 [h]
68	Albania	[<0.2]	..	..	..	..	28	25	78	18 [h]	60 [h]
69	Macedonia (TFYR)	<0.1 [<0.2]	..	..	..	..	33	66	84	..	..
70	Brazil	0.5 [0.3–1.6]	..	..	..	..	76	53	81	14	22
MEDIUM HUMAN DEVELOPMENT											
71	Dominica	..	..	..	..	..	24	35 [h]	100 [h]	..	..
72	Saint Lucia	..	..	..	..	..	22	92	64	..	..
73	Kazakhstan	0.1 [0.1–3.2]	32	65	..	..	155	72	72	9 [h]	65 [h]
74	Venezuela (Bolivarian Republic of)	0.7 [0.3–8.9]	..	..	..	..	52	73	81	..	..
75	Colombia	0.6 [0.3–2.5]	30	..	1 [j]	..	66	26	85	..	..
76	Ukraine	1.4 [0.8–4.3]	..	..	..	..	120	..	..	11 [h]	53 [h]
77	Samoa	..	..	..	..	..	27	66	100	..	..
78	Thailand	1.4 [0.7–2.1]	..	..	..	..	204	73	74	3 [h]	49 [h]
79	Dominican Republic	1.1 [0.9–1.3]	29	52	..	..	116	76	80	11	16
80	Belize	2.5 [1.4–4.0]	..	..	..	..	55	102	60	..	..
81	China	0.1 [<0.2]	..	..	..	..	208	80	94	4 [k]	67 [k]
82	Grenada	..	..	..	..	..	8	..	..	..	..
83	Armenia	0.1 [0.1–0.6]	..	44	..	..	79	60	71	2 [h]	62 [h]
84	Turkey	[<0.2]	..	..	..	..	44	3	91	18	49
85	Suriname	1.9 [1.1–3.1]	..	..	3	..	99	..	..	..	..
86	Jordan	[<0.2]	..	..	..	..	6	63	85	8	51
87	Peru	0.6 [0.3–1.7]	19	..	..	..	206	86	90	..	..
88	Lebanon	0.1 [0.1–0.5]	..	..	..	..	12	74	90	31	42
89	Ecuador	0.3 [0.1–3.5]	..	..	..	..	202	28	85	..	..
90	Philippines	<0.1 [<0.2]	..	..	..	..	450	75	87	8	41
91	Tunisia	0.1 [0.1–0.3]	..	..	..	..	28	82	90	2	50
92	Fiji	0.1 [0.1–0.4]	..	..	..	..	30	72	86 [h]	4	26
93	Saint Vincent and the Grenadines	..	..	..	..	..	42	39	86	..	..
94	Iran (Islamic Republic of)	0.2 [0.1–0.4]	..	..	..	..	30	64	84	2 [h]	22 [h]
95	Paraguay	0.4 [0.2–4.6]	..	..	..	..	100	33	83	7	23
96	Georgia	0.2 [0.1–2.7]	..	..	..	..	86	91	68	6 [h]	53 [h]
97	Guyana	2.4 [1.0–4.9]	..	..	6	3	194	40	72	..	..
98	Azerbaijan	0.1 [0.1–0.4]	..	..	1	1	85	55	60	1 [h]	..
99	Sri Lanka	<0.1 [<0.2]	..	..	..	..	80	86	85	2	23
100	Maldives	[<0.2]	..	..	..	..	53	94	95	16 [h]	37 [h]
101	Jamaica	1.5 [0.8–2.4]	..	..	..	..	10	61	46	..	..
102	Cape Verde	..	..	..	..	..	327	34	71	..	..
103	El Salvador	0.9 [0.5–3.8]	..	..	..	..	68	67	90	15 [h]	42 [h]
104	Algeria	0.1 [<0.2]	..	..	..	..	55	106	91	(.)	32
105	Viet Nam	0.5 [0.3–0.9]	..	68	16	7	235	84	93	2	35
106	Occupied Palestinian Territories	..	..	..	..	..	36	1 [h,i]	80 [h,i]	..	..

TABLE 9

HDI rank	HIV prevalence [a] (% aged 15–49) 2005	MDG Condom use at last high-risk sex [b] (% aged 15–24)		MDG Antimalarial measures		MDG Tuberculosis cases			Prevalence of smoking [f] (% of adults)	
		Women 1999–2005 [g]	Men 1999–2005 [g]	Use of insecticide-treated bednets (% of children under five) 1999–2005 [g]	Fevers treated with antimalarial drugs 1999–2005 [g]	Prevalence [c] (per 100,000 people) 2005	Detected under DOTS [d] (%) 2005	Cured under DOTS [e] (%) 2004	Women 2002–04 [g]	Men 2002–04 [g]
107 Indonesia	0.1 [0.1–0.2]	..	..	26	1	262	66	90	3 [h]	58 [h]
108 Syrian Arab Republic	[<0.2]	..	..	..	..	46	42	86	..	..
109 Turkmenistan	<0.1 [<0.2]	..	..	..	..	90	43	86	..	..
110 Nicaragua	0.2 [0.1–0.6]	17	..	..	2	74	88	87	5 [h]	..
111 Moldova	1.1 [0.6–2.6]	44	63	..	..	149	65	62	2	34
112 Egypt	<0.1 [<0.2]	..	..	..	..	32	63	70	18 [h]	40 [h]
113 Uzbekistan	0.2 [0.1–0.7]	..	50	..	..	139	39	78	1	24
114 Mongolia	<0.1 [<0.2]	..	..	..	..	206	82	88	26 [h]	68 [h]
115 Honduras	1.5 [0.8–2.4]	..	..	..	..	99	82	85	..	..
116 Kyrgyzstan	0.1 [0.1–1.7]	..	..	..	..	133	67	85	5 [h]	51 [h]
117 Bolivia	0.1 [0.1–0.3]	20	37	..	..	280	72	80	..	..
118 Guatemala	0.9 [0.5–2.7]	..	..	1	..	110	55	85	2 [h]	21 [h]
119 Gabon	7.9 [5.1–11.5]	33	48	..	..	385	57	40	..	..
120 Vanuatu	..	..	..	..	..	84	61	90	..	..
121 South Africa	18.8 [16.8–20.7]	20 [j]	..	..	..	511	103	70	8	23
122 Tajikistan	0.1 [0.1–1.7]	..	..	2	69	297	22	84	..	..
123 Sao Tome and Principe	..	..	..	..	61	258	..	..	..	..
124 Botswana	24.1 [23.0–32.0]	75	88	..	..	556	69	65	..	..
125 Namibia	19.6 [8.6–31.7]	48	69	3	14	577	90	68	10	23
126 Morocco	0.1 [0.1–0.4]	..	..	..	..	73	101	87	(.)	29
127 Equatorial Guinea	3.2 [2.6–3.8]	..	..	1	49	355	81 [h]	51 [h]	..	..
128 India	0.9 [0.5–1.5]	51	59	..	12	299	61	86	17	47
129 Solomon Islands	..	..	..	..	..	201	55	87	..	..
130 Lao People's Democratic Republic	0.1 [0.1–0.4]	..	..	18	9	306	68	86	13	59
131 Cambodia	1.6 [0.9–2.6]	..	..	..	..	703	66	91	..	..
132 Myanmar	1.3 [0.7–2.0]	..	..	..	..	170	95	84	12	36
133 Bhutan	<0.1 [<0.2]	..	..	..	..	174	31	83	..	..
134 Comoros	<0.1 [<0.2]	..	..	9	63	89	49	94	..	..
135 Ghana	2.3 [1.9–2.6]	33	52	4	63	380	37	72	1	7
136 Pakistan	0.1 [0.1–0.2]	..	..	..	..	297	37	82	..	..
137 Mauritania	0.7 [0.4–2.8]	..	..	2	33	590	28	22	..	..
138 Lesotho	23.2 [21.9–24.7]	50	48	..	..	588	85	69	..	..
139 Congo	5.3 [3.3–7.5]	20	38	..	..	449	57	63	..	..
140 Bangladesh	<0.1 [<0.2]	..	..	..	..	406	59	90	27	55
141 Swaziland	33.4 [21.2–45.3]	..	..	0	26	1,211	42	50	3	11
142 Nepal	0.5 [0.3–1.3]	..	..	..	..	244	67	87	24	49
143 Madagascar	0.5 [0.2–1.2]	5	12	..	34	396	67	71	..	..
144 Cameroon	5.4 [4.9–5.9]	46	57	1	53	206	106	71	..	..
145 Papua New Guinea	1.8 [0.9–4.4]	..	..	..	..	475	21	65	..	..
146 Haiti	3.8 [2.2–5.4]	19	30	..	12	405	57	80	6 [k]	15 [k]
147 Sudan	1.6 [0.8–2.7]	..	..	0	50	400	35	77	..	..
148 Kenya	6.1 [5.2–7.0]	25	47	5	27	936	43	80	1	21
149 Djibouti	3.1 [0.8–6.9]	..	..	..	..	1,161	42	80	..	..
150 Timor-Leste	[<0.2]	..	..	8 [j]	19	713	44	80	..	..
151 Zimbabwe	20.1 [13.3–27.6]	42	69	..	..	631	41	54	2	20
152 Togo	3.2 [1.9–4.7]	22 [j]	54 [j]	54	60	753	18	67	..	..
153 Yemen	[<0.2]	..	..	..	..	136	41	82	..	..
154 Uganda	6.7 [5.7–7.6]	53	55	0	..	559	45	70	3 [h]	25 [h]
155 Gambia	2.4 [1.2–4.1]	..	..	15	55	352	69	86	..	..
LOW HUMAN DEVELOPMENT										
156 Senegal	0.9 [0.4–1.5]	36	52	14	29	466	51	74	..	..
157 Eritrea	2.4 [1.3–3.9]	..	..	4	4	515	13	85	..	..
158 Nigeria	3.9 [2.3–5.6]	24	46	1	34	536	22	73	1	..
159 Tanzania (United Republic of)	6.5 [5.8–7.2]	42	47	16	58	496	45	81	..	..

Human development indicators

TABLE 9

Leading global health crises and risks

HDI rank		HIV prevalence [a] (% aged 15–49) 2005	MDG Condom use at last high-risk sex [b] (% aged 15–24) Women 1999–2005 [g]	Men 1999–2005 [g]	MDG Antimalarial measures Use of insecticide-treated bednets (% of children under five) 1999–2005 [g]	MDG Fevers treated with antimalarial drugs (% of children under five) 1999–2005 [g]	MDG Tuberculosis cases Prevalence [c] (per 100,000 people) 2005	MDG Detected under DOTS [d] (%) 2005	MDG Cured under DOTS [e] (%) 2004	Prevalence of smoking (% of adults) [f] Women 2002–04 [g]	Men 2002–04 [g]
160	Guinea	1.5 [1.2–1.8]	17	32	4	56	431	56	72	..	..
161	Rwanda	3.1 [2.9–3.2]	26	40	5	13	673	29	77	..	..
162	Angola	3.7 [2.3–5.3]	..	..	2	63	333	85	68	..	..
163	Benin	1.8 [1.2–2.5]	19	34	7	60	144	83	83	..	..
164	Malawi	14.1 [6.9–21.4]	35	47	15	28	518	39	71	5	21
165	Zambia	17.0 [15.9–18.1]	35	40	7	52	618	52	83	1	16
166	Côte d'Ivoire	7.1 [4.3–9.7]	25 [j]	56 [j]	4	58	659	38	71	..	..
167	Burundi	3.3 [2.7–3.8]	..	..	1	31	602	30	78	..	..
168	Congo (Democratic Republic of the)	3.2 [1.8–4.9]	..	..	1	45	541	72	85	..	..
169	Ethiopia	[0.9–3.5]	17	30	1	3	546	33	79	(.)	6
170	Chad	3.5 [1.7–6.0]	17	25	1 [j]	44	495	22	69	..	..
171	Central African Republic	10.7 [4.5–17.2]	..	..	2	69	483	40	91	..	..
172	Mozambique	16.1 [12.5–20.0]	29	33	..	15	597	49	77	..	..
173	Mali	1.7 [1.3–2.1]	14	30	8	38	578	21	71	..	..
174	Niger	1.1 [0.5–1.9]	7 [j]	30 [j]	6	48	294	50	61	..	..
175	Guinea-Bissau	3.8 [2.1–6.0]	..	..	7	58	293	79	75	..	..
176	Burkina Faso	2.0 [1.5–2.5]	54	67	2	50	461	18	67	..	..
177	Sierra Leone	1.6 [0.9–2.4]	..	..	2	61	905	37	82	..	..

NOTES

a. Data are point and range estimates based on new estimation models developed by UNAIDS. Range estimates are presented in square brackets.

b. Because of data limitations, comparisons across countries should be made with caution. Data for some countries may refer only to part of the country or differ from the standard definition.

c. Data refer to all forms of tuberculosis.

d. Calculated by dividing the new smear-positive cases of tuberculosis detected under DOTS, the internationally recommended tuberculosis control strategy, by the estimated annual incidence of new smear-positive cases. Values can exceed 100% because of intense case detection in an area

with a backlog of chronic cases, overreporting (for example, double counting), overdiagnosis or underestimation of incidence (WHO 2007b).

e. Data are the share of new smear-positive cases registered for treatment under the DOTS case detection and treatment strategy that were successfully treated.

f. The age range varies among countries, but in most is 18 and older or 15 and older.

g. Data refer to the most recent year available during the period specified.

h. Data refer to a period other than that specified.

i. UN 2006a.

j. UNICEF 2005.

k. Data refer to 2005.

SOURCES

Column 1: UNAIDS 2006.

Columns 2–5: UNICEF 2006.

Columns 6–8: WHO 2007a.

Columns 9 and 10: World Bank 2007b, based on data from the Tobacco Atlas, 2nd edition (2006).

Human development indicators

TABLE 10

... to lead a long and healthy life ...

Survival: progress and setbacks

HDI rank	Life expectancy at birth (years)		MDG Infant mortality rate (per 1,000 live births)		MDG Under-five mortality rate (per 1,000 live births)		Probability at birth of surviving to age 65 [a] (% of cohort)		MDG Maternal mortality ratio (per 100,000 live births)	
							Female	Male	Reported [b]	Adjusted [c]
	1970–75 [d]	2000–05 [d]	1970	2005	1970	2005	2000–05 [d]	2000–05 [d]	1990–2005 [e]	2005
HIGH HUMAN DEVELOPMENT										
1 Iceland	74.3	81.0	13	2	14	3	92.4	88.7	..	4
2 Norway	74.4	79.3	13	3	15	4	91.7	85.1	6	7
3 Australia	71.7	80.4	17	5	20	6	92.2	86.2	..	4
4 Canada	73.2	79.8	19	5	23	6	91.0	84.9	..	7
5 Ireland	71.3	77.8	20	5	27	6	90.0	83.2	6	1
6 Sweden	74.7	80.1	11	3	15	4	92.3	87.0	5	3
7 Switzerland	73.8	80.7	15	4	18	5	92.6	86.1	5	5
8 Japan	73.3	81.9	14	3	21	4	93.8	86.1	8	6
9 Netherlands	74.0	78.7	13	4	15	5	90.4	84.4	7	6
10 France	72.4	79.6	18	4	24	5	92.2	82.1	10	8
11 Finland	70.7	78.4	13	3	16	4	91.8	81.0	6	7
12 United States	71.5	77.4	20	6	26	7	87.0	79.4	8	11
13 Spain	72.9	80.0	27	4	34	5	93.5	83.9	6	4
14 Denmark	73.6	77.3	14	4	19	5	87.4	81.3	10	3
15 Austria	70.6	78.9	26	4	33	5	91.9	82.4	..	4
16 United Kingdom	72.0	78.5	18	5	23	6	89.6	83.7	7	8
17 Belgium	71.6	78.2	21	4	29	5	91.0	81.9	..	8
18 Luxembourg	70.6	78.2	19	4	26	5	90.8	82.4	0	12
19 New Zealand	71.7	79.2	17	5	20	6	90.0	84.9	15	9
20 Italy	72.1	79.9	30	4	33	4	92.5	84.6	7	3
21 Hong Kong, China (SAR)	72.0	81.5	..	..	..	..	93.6	86.3	..	..
22 Germany	71.0	78.7	22	4	26	5	91.0	82.9	8	4
23 Israel	71.6	79.7	24	5	27	6	92.3	85.8	5	4
24 Greece	72.3	78.3	38	4	54	5	91.3	83.7	1	3
25 Singapore	69.5	78.8	22	3	27	3	90.8	84.4	6	14
26 Korea (Republic of)	62.6	77.0	43	5	54	5	90.8	78.6	20	14
27 Slovenia	69.8	76.8	25	3	29	4	90.1	77.6	17	6
28 Cyprus	71.4	79.0	29	4	33	5	92.3	86.1	0	10
29 Portugal	68.0	77.2	53	4	62	5	90.9	81.0	8	11
30 Brunei Darussalam	68.3	76.3	58	8	78	9	87.7	84.5	0	41
31 Barbados	69.4	76.0	40	11	54	12	88.3	79.0	0	16
32 Czech Republic	70.1	75.4	21	3	24	4	89.0	75.3	4	4
33 Kuwait	67.7	76.9	49	9	59	11	88.9	83.8	5	4
34 Malta	70.6	78.6	25	5	32	6	90.4	86.0	..	8
35 Qatar	62.1	74.3	45	18	65	21	80.1	78.7	10	12
36 Hungary	69.3	72.4	36	7	39	8	84.4	64.4	7	6
37 Poland	70.5	74.6	32	6	36	7	88.0	69.7	4	8
38 Argentina	67.1	74.3	59	15	71	18	85.6	72.5	40	77
39 United Arab Emirates	62.2	77.8	63	8	84	9	90.2	85.3	3	37
40 Chile	63.4	77.9	78	8	98	10	88.6	79.1	17	16
41 Bahrain	63.3	74.8	55	9	82	11	85.9	80.2	46	32
42 Slovakia	70.0	73.8	25	7	29	8	87.3	68.9	4	6
43 Lithuania	71.3	72.1	23	7	28	9	85.6	60.0	3	11
44 Estonia	70.5	70.9	21	6	26	7	84.3	57.2	8	25
45 Latvia	70.1	71.3	21	9	26	11	84.8	60.0	14	10
46 Uruguay	68.7	75.3	48	14	57	15	87.1	74.4	26	20
47 Croatia	69.6	74.9	34	6	42	7	88.5	73.4	8	7
48 Costa Rica	67.8	78.1	62	11	83	12	88.6	81.0	36	30
49 Bahamas	66.5	71.1	38	13	49	15	75.9	65.2	..	16
50 Seychelles	..	..	46	12	59	13	..	..	57	..
51 Cuba	70.7	77.2	34	6	43	7	86.8	80.6	37	45
52 Mexico	62.4	74.9	79	22	110	27	84.5	76.2	63	60
53 Bulgaria	71.0	72.4	28	12	32	15	85.3	68.3	6	11

HDI rank	Life expectancy at birth (years)		MDG Infant mortality rate (per 1,000 live births)		MDG Under-five mortality rate (per 1,000 live births)		Probability at birth of surviving to age 65 [a] (% of cohort)		MDG Maternal mortality ratio (per 100,000 live births)	
	1970–75 [d]	2000–05 [d]	1970	2005	1970	2005	Female 2000–05 [d]	Male 2000–05 [d]	Reported [b] 1990–2005 [e]	Adjusted [c] 2005
54 Saint Kitts and Nevis	..	..	..	18	..	20	..	..	250	..
55 Tonga	65.6	72.3	40	20	50	24	78.2	73.8	..	..
56 Libyan Arab Jamahiriya	52.8	72.7	105	18	160	19	82.1	72.2	77	97
57 Antigua and Barbuda	..	..	..	11	..	12	..	..	65	..
58 Oman	52.1	74.2	126	10	200	12	84.9	79.5	23	64
59 Trinidad and Tobago	65.9	69.0	49	17	57	19	72.1	63.8	45	45
60 Romania	69.2	71.3	46	16	57	19	83.7	66.3	17	24
61 Saudi Arabia	53.9	71.6	118	21	185	26	82.0	73.7	..	18
62 Panama	66.2	74.7	46	19	68	24	85.9	77.4	40	83
63 Malaysia	63.0	73.0	46	10	70	12	83.1	72.9	30	62
64 Belarus	71.5	68.4	31	10	37	12	81.3	50.7	17	18
65 Mauritius	62.9	72.0	64	13	86	15	80.9	66.4	22	15
66 Bosnia and Herzegovina	67.5	74.1	60	13	82	15	85.3	74.4	8	3
67 Russian Federation	69.0	64.8	29	14	36	18	76.0	42.1	32	28
68 Albania	67.7	75.7	78	16	109	18	89.5	79.7	17	92
69 Macedonia (TFYR)	67.5	73.4	85	15	119	17	84.3	75.3	21	10
70 Brazil	59.5	71.0	95	31	135	33	78.5	64.2	72	110
MEDIUM HUMAN DEVELOPMENT										
71 Dominica	..	..	..	13	..	15	..	..	67	..
72 Saint Lucia	65.3	72.5	..	12	..	14	78.2	72.3	35	..
73 Kazakhstan	63.1	64.9	..	63	..	73	73.7	45.8	42	140
74 Venezuela (Bolivarian Republic of)	65.7	72.8	48	18	62	21	82.6	71.9	58	57
75 Colombia	61.6	71.7	68	17	105	21	81.8	69.0	84	120
76 Ukraine	70.1	67.6	22	13	27	17	79.5	50.4	13	18
77 Samoa	56.1	70.0	73	24	101	29	78.6	65.1	..	..
78 Thailand	60.4	68.6	74	18	102	21	75.5	57.8	24	110
79 Dominican Republic	59.6	70.8	91	26	127	31	76.7	65.7	180	150
80 Belize	67.6	75.6	..	15	..	17	86.8	77.3	140	52
81 China	63.2 [f]	72.0 [f]	85	23	120	27	80.9 [f]	73.8 [f]	51	45
82 Grenada	64.6	67.7	..	17	..	21	73.8	67.0	1	..
83 Armenia	70.8	71.4	..	26	..	29	81.9	66.9	22	39
84 Turkey	57.0	70.8	150	26	201	29	82.3	71.9	130 [g]	44
85 Suriname	64.0	69.1	..	30	..	39	76.9	63.3	150	72
86 Jordan	56.5	71.3	77	22	107	26	78.2	70.9	41	62
87 Peru	55.4	69.9	119	23	174	27	77.5	68.0	190	240
88 Lebanon	65.4	71.0	45	27	54	30	80.6	72.1	100 [g]	150
89 Ecuador	58.8	74.2	87	22	140	25	84.0	74.0	80	110
90 Philippines	58.1	70.3	56	25	90	33	79.3	70.7	170	230
91 Tunisia	55.6	73.0	135	20	201	24	85.3	76.5	69	100
92 Fiji	60.6	67.8	50	16	65	18	72.9	62.0	38	210
93 Saint Vincent and the Grenadines	61.6	70.6	..	17	..	20	79.9	71.3	93	..
94 Iran (Islamic Republic of)	55.2	69.5	122	31	191	36	78.3	71.1	37	140
95 Paraguay	65.8	70.8	58	20	78	23	77.7	70.8	180	150
96 Georgia	68.2	70.5	..	41	..	45	83.0	66.1	52	66
97 Guyana	60.0	63.6	..	47	..	63	66.8	55.0	120	470
98 Azerbaijan	65.6	66.8	..	74	..	89	76.0	61.2	19	82
99 Sri Lanka	65.0	70.8	65	12	100	14	81.3	62.8	43	58
100 Maldives	51.4	65.6	157	33	255	42	67.7	66.2	140	120
101 Jamaica	69.0	72.0	49	17	64	20	78.3	69.1	110	26
102 Cape Verde	57.5	70.2	..	26	..	35	80.3	68.3	76	210
103 El Salvador	58.2	70.7	111	23	162	27	78.5	68.3	170	170
104 Algeria	54.5	71.0	143	34	220	39	78.9	75.9	120	180
105 Viet Nam	50.3	73.0	55	16	87	19	82.7	76.0	170	150
106 Occupied Palestinian Territories	56.5	72.4	..	21	..	23	81.8	75.5	..	..

TABLE 10

HDI rank	Life expectancy at birth (years)		MDG Infant mortality rate (per 1,000 live births)		MDG Under-five mortality rate (per 1,000 live births)		Probability at birth of surviving to age 65 [a] (% of cohort)		MDG Maternal mortality ratio (per 100,000 live births)	
	1970–75 [d]	2000–05 [d]	1970	2005	1970	2005	Female 2000–05 [d]	Male 2000–05 [d]	Reported [b] 1990–2005 [e]	Adjusted [c] 2005
107 Indonesia	49.2	68.6	104	28	172	36	75.8	68.1	310	420
108 Syrian Arab Republic	57.3	73.1	90	14	123	15	83.6	76.4	65	130
109 Turkmenistan	59.1	62.4	..	81	..	104	70.8	52.1	14	130
110 Nicaragua	55.2	70.8	113	30	165	37	77.3	67.0	83	170
111 Moldova	64.8	67.9	53	14	70	16	75.5	56.7	22	22
112 Egypt	51.1	69.8	157	28	235	33	80.2	70.4	84	130
113 Uzbekistan	63.6	66.5	83	57	101	68	73.3	60.0	30	24
114 Mongolia	53.8	65.0	..	39	..	49	68.0	55.3	93	46
115 Honduras	53.9	68.6	116	31	170	40	76.6	62.1	110	280
116 Kyrgyzstan	61.2	65.3	104	58	130	67	74.4	56.3	49	150
117 Bolivia	46.7	63.9	147	52	243	65	69.0	61.0	30	290
118 Guatemala	53.7	69.0	115	32	168	43	77.6	65.4	150	290
119 Gabon	48.7	56.8	..	60	..	91	53.8	48.9	520	520
120 Vanuatu	54.0	68.4	107	31	155	38	75.6	68.2	68	..
121 South Africa	53.7	53.4	..	55	..	68	46.0	33.9	150	400
122 Tajikistan	60.9	65.9	108	59	140	71	72.0	61.9	37	170
123 Sao Tome and Principe	56.5	64.3	..	75	..	118	72.7	65.2	100	..
124 Botswana	56.0	46.6	99	87	142	120	31.9	24.4	330	380
125 Namibia	53.9	51.5	85	46	135	62	41.9	34.3	270	210
126 Morocco	52.9	69.6	119	36	184	40	79.4	71.2	230	240
127 Equatorial Guinea	40.5	49.3	..	123	..	205	44.7	39.7	..	680
128 India	50.7	62.9	127	56	202	74	66.1	57.4	540	450
129 Solomon Islands	55.5	62.3	70	24	97	29	63.6	59.6	550 [g]	220
130 Lao People's Democratic Republic	46.5	61.9	145	62	218	79	63.7	57.9	410	660
131 Cambodia	40.3	56.8	..	98	..	143	57.8	43.7	440	590
132 Myanmar	53.1	59.9	122	75	179	105	64.1	50.7	230	380
133 Bhutan	41.8	63.5	156	65	267	75	67.6	61.3	260	440
134 Comoros	48.9	63.0	159	53	215	71	66.9	58.3	380	400
135 Ghana	49.9	58.5	111	68	186	112	56.5	54.3	210 [g]	560
136 Pakistan	51.9	63.6	120	79	181	99	66.6	63.2	530	320
137 Mauritania	48.4	62.2	151	78	250	125	69.4	60.4	750	820
138 Lesotho	49.8	44.6	140	102	186	132	30.7	21.9	760	960
139 Congo	54.9	53.0	100	81	160	108	45.9	39.7	..	740
140 Bangladesh	45.3	62.0	145	54	239	73	63.2	59.0	320	570
141 Swaziland	49.6	43.9	132	110	196	160	31.1	22.9	230	390
142 Nepal	44.0	61.3	165	56	250	74	61.3	58.4	540	830
143 Madagascar	44.9	57.3	109	74	180	119	58.1	52.1	470	510
144 Cameroon	47.0	49.9	127	87	215	149	42.5	39.9	670	1,000
145 Papua New Guinea	44.7	56.7	110	55	158	74	54.3	40.3	370 [g]	470
146 Haiti	48.0	58.1	148	84	221	120	57.5	50.8	520	670
147 Sudan	45.1	56.4	104	62	172	90	55.3	49.7	550 [g]	450
148 Kenya	53.6	51.0	96	79	156	120	42.5	37.0	410	560
149 Djibouti	44.4	53.4	..	88	..	133	50.4	43.7	74	650
150 Timor-Leste	40.0	58.3	..	52	..	61	57.3	52.9	..	380
151 Zimbabwe	55.6	40.0	86	81	138	132	18.0	15.0	1,100	880
152 Togo	49.8	57.6	128	78	216	139	61.2	52.8	480	510
153 Yemen	39.8	60.3	202	76	303	102	61.7	55.0	370	430
154 Uganda	51.0	47.8	100	79	170	136	36.6	33.6	510	550
155 Gambia	38.3	58.0	180	97	311	137	61.4	54.8	730	690
LOW HUMAN DEVELOPMENT										
156 Senegal	45.8	61.6	164	77	279	136	69.7	60.7	430	980
157 Eritrea	44.1	55.2	143	50	237	78	50.2	36.4	1,000	450
158 Nigeria	42.8	46.6	140	100	265	194	40.6	37.0	..	1,100
159 Tanzania (United Republic of)	47.6	49.7	129	76	218	122	41.0	36.0	580	950

Human development indicators

TABLE 10

Survival: progress and setbacks

HDI rank	Life expectancy at birth (years)		MDG Infant mortality rate (per 1,000 live births)		MDG Under-five mortality rate (per 1,000 live births)		Probability at birth of surviving to age 65 [a] (% of cohort)		MDG Maternal mortality ratio (per 100,000 live births)	
	1970–75 [d]	2000–05 [d]	1970	2005	1970	2005	Female 2000–05 [d]	Male 2000–05 [d]	Reported [b] 1990–2005 [e]	Adjusted [c] 2005
160 Guinea	38.8	53.7	197	98	345	150	55.7	48.9	530	910
161 Rwanda	44.6	43.4	124	118	209	203	34.5	28.3	1,100	1,300
162 Angola	37.9	41.0	180	154	300	260	33.9	27.5	..	1,400
163 Benin	47.0	54.4	149	89	252	150	55.7	48.6	500	840
164 Malawi	41.8	45.0	204	79	341	125	33.7	27.4	980	1,100
165 Zambia	50.1	39.2	109	102	181	182	21.9	18.6	730	830
166 Côte d'Ivoire	49.8	46.8	158	118	239	195	40.7	34.9	600	810
167 Burundi	44.1	47.4	138	114	233	190	41.1	35.9	..	1,100
168 Congo (Democratic Republic of the)	46.0	45.0	148	129	245	205	38.8	33.3	1,300	1,100
169 Ethiopia	43.5	50.7	160	109	239	164	46.9	41.4	870	720
170 Chad	45.6	50.5	154	124	261	208	50.5	43.7	1,100	1,500
171 Central African Republic	43.5	43.3	145	115	238	193	32.1	25.7	1,100	980
172 Mozambique	40.3	44.0	168	100	278	145	35.3	29.2	410	520
173 Mali	40.0	51.8	225	120	400	218	54.1	44.3	580	970
174 Niger	40.5	54.5	197	150	330	256	54.4	56.8	590	1,800
175 Guinea-Bissau	36.5	45.5	..	124	..	200	40.9	34.2	910	1,100
176 Burkina Faso	43.6	50.7	166	96	295	191	54.5	44.0	480	700
177 Sierra Leone	35.4	41.0	206	165	363	282	37.6	30.4	1,800	2,100
Developing countries	55.8	65.5	109 [h]	57 [h]	167 [h]	83 [h]	70.3	62.6	..	..
Least developed countries	44.6 [h]	52.7 [h]	152 [h]	97 [h]	245 [h]	153 [h]	49.9 [h]	44.3 [h]	..	..
Arab States	51.9	66.7	129	46	196	58	73.5	66.4	..	..
East Asia and the Pacific	60.6	71.1	84	25	123	31	79.6	71.8	..	..
Latin America and the Caribbean	61.2	72.2	86	26	123	31	80.8	69.3	..	..
South Asia	50.3	62.9	130	60	206	80	66.0	58.4	..	..
Sub-Saharan Africa	46.0	49.1	144	102	244	172	43.3	37.8	..	..
Central and Eastern Europe and the CIS	68.7	68.2	39	22	48	27	79.5	54.9	..	..
OECD	70.3	77.8	41	9	54	11	89.2	80.5	..	..
High-income OECD	71.7	78.9	22	5	28	6	90.3	82.4	..	..
High human development	69.4	75.7	43	13	59	15	86.6	74.8	..	..
Medium human development	56.6	66.9	106	45	162	59	72.6	64.5	..	..
Low human development	43.7	47.9	155	108	264	184	42.6	37.4	..	..
High income	71.5	78.7	24	6	32	7	90.2	82.2	..	..
Middle income	61.8	70.3	87	28	127	35	78.9	68.4	..	..
Low income	49.1	59.2	..130	75	209	113	60.0	53.2	..	..
World	58.3 [h]	66.0 [h]	96 [h]	52 [h]	148 [h]	76 [h]	72.0 [h]	63.1 [h]	..	..

NOTES

a. Data refer to the probability at birth of surviving to age 65, multiplied by 100.
b. Data reported by national authorities.
c. Data adjusted based on reviews by UNICEF, WHO and UNFPA to account for well-documented problems of underreporting and misclassifications.
d. Data are estimates for the period specified.
e. Data refer to the most recent year available during the period specified.
f. For statistical purposes, the data for China do not include Hong Kong and Macao, SARs of China.
g. Data refer to years or periods other than those specified in the column heading, differ from the standard definition or refer to only part of a country.
h. Data are aggregates provided by original data source.

SOURCES

Columns 1, 2, 7 and 8: UN 2007e.
Columns 3–6 and 9: UNICEF 2006.
Columns 10: UNICEF 2007a.

TABLE 11

... to acquire knowledge ...

Commitment to education: public spending

	Public expenditure on education				Current public expenditure on education by level [a] (% of total current public expenditure on education)					
	As a % of GDP		As a % of total government expenditure		Pre-primary and primary		Secondary and post-secondary non-tertiary		Tertiary	
HDI rank	1991	2002–05 [b]	1991	2002–05 [b]	1991	2002–05 [b]	1991	2002–05 [b]	1991	2002–05 [b]
HIGH HUMAN DEVELOPMENT										
1 Iceland	..	8.1	..	16.6	..	40	..	35	..	19
2 Norway	7.1	7.7	14.6	16.6	38	28	27	35	16	33
3 Australia	4.9	4.7	14.8	13.3 [c]	..	34	..	41	..	25
4 Canada	6.5	5.2	14.2	12.5 [c]	.. [d]	..	68	..	31	34 [e]
5 Ireland	5.0	4.8	9.7	14.0	37	33	40	43	21	24
6 Sweden	7.1	7.4	13.8	12.9	48	34	20	38	13	28
7 Switzerland	5.3	6.0	18.8	13.0	50	33	26	37	19	28
8 Japan	..	3.6	..	9.8	..	38 [c,e]	..	40 [c,e]	..	14 [c,e]
9 Netherlands	5.6	5.4	14.3	11.2	23	33	37	40	32	27
10 France	5.5	5.9	..	10.9	26	31	40	48	14	21
11 Finland	6.5	6.5	11.9	12.8	30	26	41	41	28	33
12 United States	5.1	5.9	12.3	15.3	..	..	..	..	..	..
13 Spain	4.1	4.3	..	11.0	29	39	45	41	16	20
14 Denmark	6.9	8.5	11.8	15.3	..	31	..	35	..	30
15 Austria	5.3	5.5	7.6	10.8	24	26	46	48	20	26
16 United Kingdom	4.8	5.4	..	12.1	30	..	44	..	20	..
17 Belgium	5.0	6.1	..	12.2	24	33	42	43	16	22
18 Luxembourg	3.0	3.6 [c,e]	10.8	8.5 [c,e]	..	..	..	..	..	..
19 New Zealand	6.1	6.5	..	20.9	31	29	25	46	37	23
20 Italy	3.0	4.7	..	9.6	35	35	62	48	..	17
21 Hong Kong, China (SAR)	2.8	4.2	17.4	23.0	..	26	..	36	..	32
22 Germany	..	4.6	..	9.8	..	22	..	51	..	24
23 Israel	6.5	6.9	11.4	13.7	41	47	31	30	26	17
24 Greece	2.3	4.3	..	8.5	34	30 [e]	45	37	20	30
25 Singapore	3.1	3.7 [c]	18.2	..	..	23 [c]	..	43 [c]	..	23 [c]
26 Korea (Republic of)	3.8	4.6	25.6	16.5	45	35	39	43	7	13
27 Slovenia	4.8	6.0	16.1	12.6	43	28 [e]	37	48 [e]	17	24
28 Cyprus	3.7	6.3	11.6	14.4	39	35	50	50	4	14
29 Portugal	4.6	5.7	..	11.5	43	39	35	41	15	16
30 Brunei Darussalam	3.5	..	..	9.1 [c,e]	22	..	30	..	2	..
31 Barbados	7.8	6.9	22.2	16.4	..	35 [e]	..	33	..	33
32 Czech Republic	..	4.4	..	10.0	..	24	..	53	..	20
33 Kuwait	4.8	5.1	3.4	12.7	..	31	..	38	..	30
34 Malta	4.4	4.5	8.5	10.1	23	32	40	48	19	20
35 Qatar	3.5	1.6 [e]	..	..	..	..	..	..	..	..
36 Hungary	6.1	5.5	7.8	11.1	55	34	25	46	15	17
37 Poland	5.2	5.4	14.6	12.7	..	42	..	37	..	21
38 Argentina	3.3	3.8	..	13.1	..	45	..	38	..	17
39 United Arab Emirates	2.0	1.3	15.0	27.4 [e]	..	..	..	..	..	..
40 Chile	2.4	3.5	10.0	18.5	..	47	..	39	..	15
41 Bahrain	3.9	..	12.8	..	..	..	..	..	..	..
42 Slovakia	5.6	4.3	..	10.8	..	23	..	51	..	22
43 Lithuania	5.5	5.2	20.6	15.6	..	28	..	52	..	20
44 Estonia	..	5.3	..	14.9	..	31	..	50	..	18
45 Latvia	4.1	5.3	16.9	15.4	..	..	..	..	..	..
46 Uruguay	2.5	2.6	16.6	7.9	36	42 [c,e]	29	38 [c,e]	24	20 [c,e]
47 Croatia	5.5	4.7	..	10.0	..	29 [e]	..	49 [e]	..	19
48 Costa Rica	3.4	4.9	21.8	18.5	38	66	22	34	36	—
49 Bahamas	3.7	3.6 [c,e]	16.3	19.7 [c,e]	..	..	..	..	..	..
50 Seychelles	6.5	5.4 [e]	11.6	..	..	40 [e]	..	42 [e]	..	18 [e]
51 Cuba	9.7	9.8	10.8	16.6	27	41	37	38	15	22
52 Mexico	3.8	5.4	15.3	25.6	39	50	28	30	17	17
53 Bulgaria	5.4	4.2	..	..	70	36	..	45	14	19

Human development indicators

TABLE 11

Commitment to education: public spending

		Public expenditure on education			Current public expenditure on education by level [a] (% of total current public expenditure on education)					
		As a % of GDP		As a % of total government expenditure		Pre-primary and primary		Secondary and post-secondary non-tertiary		Tertiary	
HDI rank		1991	2002–05 [b]	1991	2002–05 [b]	1991	2002–05 [b]	1991	2002–05 [b]	1991	2002–05 [b]
54	Saint Kitts and Nevis	2.7	9.3	11.6	12.7	43	42	56	58	—	—
55	Tonga	..	4.8	..	13.5	..	59	..	34	..	—
56	Libyan Arab Jamahiriya	..	2.7 c	..	..	..	12 c,e	..	19 c,e	..	69 c
57	Antigua and Barbuda	..	3.8	..	..	..	32	..	46	..	7
58	Oman	3.0	3.6	15.8	24.2	52	50	40	41	7	8
59	Trinidad and Tobago	4.1	4.2 e	12.4	13.4 c	..	42 c	..	39 c	..	11 c
60	Romania	3.5	3.4	..	..	..	25 e	..	42 e	..	18
61	Saudi Arabia	5.8	6.8	17.8	27.6	..	..	..	..	..	..
62	Panama	4.6	3.8 e	18.9	8.9 e	36	..	22	..	20	26 c
63	Malaysia	5.1	6.2	18.0	25.2	34	30	35	35	20	35
64	Belarus	5.7	6.0	..	11.3	..	27 e	..	48 e	..	25
65	Mauritius	3.8	4.5	11.8	14.3	38	32	36	43	17	12
66	Bosnia and Herzegovina	..	..	..	..	..	..	..	..	..	..
67	Russian Federation	3.6	3.6 e	..	12.9 e	..	..	..	..	..	..
68	Albania	..	2.9 e	..	8.4 e	..	..	..	..	..	..
69	Macedonia (TFYR)	..	3.5	..	15.6	..	..	..	..	..	..
70	Brazil	..	4.4	..	10.9	..	41	..	40	..	19
MEDIUM HUMAN DEVELOPMENT											
71	Dominica	..	5.0 c,e	..	..	..	..	..	..	..	..
72	Saint Lucia	..	5.8	..	16.9	..	40	..	41	..	0
73	Kazakhstan	3.9	2.3	19.1	12.1 c	..	..	..	..	..	..
74	Venezuela (Bolivarian Republic of)	4.6	..	17.0	..	..	..	..	..	..	..
75	Colombia	2.4	4.8	14.3	11.1	..	51	..	36	..	13
76	Ukraine	6.2	6.4	18.9	18.9	..	..	..	..	..	..
77	Samoa	..	4.5 e	..	13.7 e	..	34 c,e	..	29 c,e	..	37 c
78	Thailand	3.1	4.2	20.0	25.0	56	44 c,e	22	19 c,e	15	20 c,e
79	Dominican Republic	..	1.8	..	9.7	..	66 e	..	29 e	..	..
80	Belize	4.6	5.4	18.5	18.1	..	48	..	48	..	1
81	China	2.2	1.9 c	12.7	13.0 c	..	36 c,e	..	38 c,e	..	21 c,e
82	Grenada	4.9	5.2	11.9	12.9	..	41 e	..	39 e	..	11 e
83	Armenia	..	3.2 c	..	..	..	16 c,e	..	53 c,e	..	30 c
84	Turkey	2.4	3.7	..	..	59	40 c,e	29	32 c,e	..	28 c,e
85	Suriname	5.9	..	..	..	59	..	15	..	9	..
86	Jordan	8.0	4.9 c	19.1	20.6 c	..	..	..	..	..	..
87	Peru	2.8	2.4	..	13.7	..	51	..	36 e	..	11
88	Lebanon	..	2.6	..	11.0	..	33 e	..	30 e	..	31
89	Ecuador	2.5	1.0 c,e	17.5	8.0 c	..	..	..	..	..	..
90	Philippines	3.0	2.7	10.5	16.4	..	55	..	27	..	14
91	Tunisia	6.0	7.3	14.3	20.8	..	35 e	..	43 e	..	22
92	Fiji	5.1	6.4	..	20.0	..	40	..	34	..	16
93	Saint Vincent and the Grenadines	5.9	8.2	13.8	16.1	64	50	32	36	..	5
94	Iran (Islamic Republic of)	4.1	4.7	22.4	22.8	..	24	..	37	..	14
95	Paraguay	1.9	4.3	10.3	10.8	..	54	..	28	..	18
96	Georgia	..	2.9	..	13.1	..	..	..	..	..	..
97	Guyana	2.2	8.5	6.5	14.5	..	44	..	13	..	4
98	Azerbaijan	7.7	2.5	24.7	19.6	..	25 e	..	56 e	..	6
99	Sri Lanka	3.2	..	8.4	..	..	..	..	..	..	..
100	Maldives	7.0	7.1	16.0 e	15.0	..	54 e	..	..	..	..
101	Jamaica	4.5	5.3	12.8	8.8	37	37 e	33	44 e	21	20 e
102	Cape Verde	3.6	6.6	19.9	25.4	..	54	..	36	..	10
103	El Salvador	1.8	2.8	15.2	20.0	..	60 e	..	29 e	..	11 e
104	Algeria	5.1	..	22.0	..	95	..	.. f	..	.. f	..
105	Viet Nam	1.8	..	9.7	..	..	..	..	..	..	..
106	Occupied Palestinian Territories	..	..	..	..	..	..	..	..	..	..

TABLE 11

	Public expenditure on education				Current public expenditure on education by level[a] (% of total current public expenditure on education)					
	As a % of GDP		As a % of total government expenditure		Pre-primary and primary		Secondary and post-secondary non-tertiary		Tertiary	
HDI rank	1991	2002–05[b]	1991	2002–05[b]	1991	2002–05[b]	1991	2002–05[b]	1991	2002–05[b]
107 Indonesia	1.0	0.9	..	9.0[e]	..	39[e]	..	42[e]	..	19[e]
108 Syrian Arab Republic	3.9	..	14.2	..	..	..	..	..	..	..
109 Turkmenistan	3.9	..	19.7	..	..	..	..	..	..	..
110 Nicaragua	3.4	3.1[e]	12.1	15.0	..	..	..	..	..	..
111 Moldova	5.3	4.3	21.6	21.1	..	36[e]	..	55[e]	..	9
112 Egypt	3.9	..	..	..	..	..	..	..	..	..
113 Uzbekistan	9.4	..	17.8	..	..	..	..	..	..	..
114 Mongolia	11.5	5.3	22.7	..	..	43	..	37	..	19
115 Honduras	3.8	..	..	..	..	..	..	..	..	..
116 Kyrgyzstan	6.0	4.4[e]	22.7	18.6[c]	..	23[e]	..	46[e]	..	19
117 Bolivia	2.4	6.4	..	18.1	..	49	..	25	..	23
118 Guatemala	1.3	..	13.0	..	..	..	..	..	..	..
119 Gabon	..	3.9[c,e]	..	..	..	..	..	..	..	..
120 Vanuatu	4.6	9.6	18.8	26.7[c]	..	44[c]	..	41[c]	..	9[c]
121 South Africa	5.9	5.4	..	17.9	76	43	..	33	22	16
122 Tajikistan	9.1	3.5	24.4	18.0	..	31[e]	..	54[e]	..	5
123 Sao Tome and Principe	..	..	..	..	..	..	..	..	..	..
124 Botswana	6.2	10.7	17.0	21.5	..	25	..	41	..	32
125 Namibia	7.9	6.9	..	21.0[c]	..	60[c,e]	..	29[c,e]	..	11[c,e]
126 Morocco	5.0	6.7	26.3	27.2	35	45	49	38	16	16
127 Equatorial Guinea	..	0.6[e]	..	4.0[e]	..	35[c,e]	..	..	..	34[c]
128 India	3.7	3.8	12.2	10.7	..	31[c,e]	..	..	..	18[c,e]
129 Solomon Islands	3.8	3.3[c,e]	7.9	..	57	..	30	..	14	..
130 Lao People's Democratic Republic	..	2.3	..	11.7	..	49	..	35	..	15
131 Cambodia	..	1.9	..	14.6[c]	..	74[c]	..	21[c]	..	5[c]
132 Myanmar	..	1.3[c]	..	18.1[c,e]	..	..	..	..	..	..
133 Bhutan	..	5.6[c]	..	12.9[c]	..	27[c,e]	..	54[c,e]	..	20[c,e]
134 Comoros	..	3.9	..	24.1	..	..	..	..	..	..
135 Ghana	..	5.4	..	..	..	39	..	42	..	18
136 Pakistan	2.6	2.3	7.4	10.9	..	..	..	..	..	..
137 Mauritania	4.6	2.3	13.9	8.3	..	62[e]	..	33[e]	..	5[e]
138 Lesotho	6.2	13.4	12.2	29.8	..	39[e]	..	21[e]	..	42[e]
139 Congo	7.4	2.2	..	8.1	..	30	..	44	..	26
140 Bangladesh	1.5	2.5	10.3	14.2	..	38[e]	..	48	..	14
141 Swaziland	5.7	6.2	19.5	..	..	38[e]	..	30[e]	..	27
142 Nepal	2.0	3.4	8.5	14.9	..	53[e]	..	28	..	12
143 Madagascar	2.5	3.2	..	25.3	..	47	..	23	..	12
144 Cameroon	3.2	1.8[e]	19.6	8.6[e]	..	68[e]	..	8[e]	..	24[e]
145 Papua New Guinea	..	..	..	..	..	..	..	..	..	..
146 Haiti	1.4	..	20.0	..	53	..	19	..	9	..
147 Sudan	6.0	..	2.8	..	..	..	..	..	..	..
148 Kenya	6.7	6.7	17.0	29.2	..	64	..	25	..	11
149 Djibouti	3.5	7.9	11.1	27.3	53	44	21	42	14	15
150 Timor-Leste	..	..	..	..	..	..	..	..	..	..
151 Zimbabwe	7.7	4.6[c,e]	..	..	54	..	29	..	..	..
152 Togo	..	2.6	..	13.6	..	45[c,e]	..	31[c]	..	19[c]
153 Yemen	..	9.6[c,e]	..	32.8[c]	..	..	..	..	..	..
154 Uganda	1.5	5.2[e]	11.5	18.3[e]	..	62[e]	..	24[e]	..	12[e]
155 Gambia	3.8	2.0[e]	14.6	8.9	42	..	21	..	18	..
LOW HUMAN DEVELOPMENT										
156 Senegal	3.9	5.4	26.9	18.9	..	48[e]	..	28[e]	..	24[e]
157 Eritrea	..	5.4	..	..	..	25	..	13	..	48
158 Nigeria	0.9	..	..	..	..	..	..	..	..	..
159 Tanzania (United Republic of)	2.8	2.2[c,e]	11.4	..	..	..	..	..	..	..

Human development indicators

TABLE 11

Commitment to education: public spending

| | Public expenditure on education | | | | Current public expenditure on education by level [a] (% of total current public expenditure on education) | | | | | |
| | As a % of GDP | | As a % of total government expenditure | | Pre-primary and primary | | Secondary and post-secondary non-tertiary | | Tertiary | |
HDI rank	1991	2002–05 [b]	1991	2002–05 [b]	1991	2002–05 [b]	1991	2002–05 [b]	1991	2002–05 [b]
160 Guinea	2.0	2.0	25.7	25.6 c,e	..	..	..	..	..	..
161 Rwanda	..	3.8	..	12.2	..	55	..	11	..	34
162 Angola	..	2.6 c,e	..	6.4 c,e	..	..	..	..	..	..
163 Benin	..	3.5 e	..	14.1 e	..	50	..	28	..	22
164 Malawi	3.2	5.8	11.1	24.6 c	..	63	..	..	..	..
165 Zambia	2.8	2.0	7.1	14.8	..	59	..	15	..	26
166 Côte d'Ivoire	..	4.6 c,e	..	21.5 c	..	43 c	..	36 c	..	20 c
167 Burundi	3.5	5.1	17.7	17.7	43	52	28	33	27	15
168 Congo (Democratic Republic of the)	..	..	..	..	..	..	..	..	..	..
169 Ethiopia	2.4	6.1 g	9.4	17.5 g	54	51 g	28	..	..	17 g
170 Chad	1.6	2.1	..	10.1	47	48	21	29	8	23
171 Central African Republic	2.2	..	..	..	55	..	17	..	24	..
172 Mozambique	..	3.7	..	19.5	..	70	..	17	..	13
173 Mali	..	4.3	..	14.8	..	50 c,e	..	34 c,e	..	16 c,e
174 Niger	3.3	2.3	18.6	..	..	..	..	..	..	..
175 Guinea-Bissau	..	5.2 c	..	11.9 c	..	..	..	..	..	..
176 Burkina Faso	2.6	4.7	..	16.6	..	71	..	18	..	9
177 Sierra Leone	..	4.6 e	..	..	..	52 e	..	27 e	..	20 e

NOTES

a. Expenditures by level may not sum to 100 as a result of rounding or the omission of expenditures not allocated by level.

b. Data refer to the most recent year available during the period specified.

c. Data refer to an earlier year than that specified (in the period 1999 to 2001).

d. Expenditure included in secondary category.

e. National or UNESCO Institute for Statistics estimate.

f. Expenditure included in pre-primary and primary category.

g. Data refer to 2006.

SOURCES

Columns 1–4, 7, 9 and 10: UNESCO Institute for Statistics 2007b.

Columns 5 and 6: calculated on the basis of data on public expenditure on pre-primary and primary levels of education from UNESCO Institute for Statistics 2007b.

Column 8: calculated on the basis of data on public expenditure on secondary and post-secondary non-tertiary levels of education from UNESCO Institute for Statistics 2007b.

TABLE

12

. . . to acquire knowledge . . .

Literacy and enrolment

HDI rank	Adult literacy rate (% aged 15 and older)		MDG Youth literacy rate (% aged 15–24)		MDG Net primary enrolment rate (%)		Net secondary enrolment rate [a] (%)		MDG Children reaching grade 5 (% of grade 1 students)		Tertiary students in science, engineering, manufacturing and construction (% of tertiary students)
	1985–1994[b]	1995–2005[c]	1985–1994[b]	1995–2005[c]	1991	2005	1991	2005	1991	2004	1999–2005[d]
HIGH HUMAN DEVELOPMENT											
1 Iceland	..	..	..	..	100[e]	99[e]	..	88[e]	..	100[f]	16
2 Norway	..	..	..	..	100	98	88	97	100	100	16
3 Australia	..	..	..	..	99	97	79[e]	86[e]	99	..	22
4 Canada	..	..	..	..	98	99[e,f]	89	..	97	..	20[g]
5 Ireland	..	..	..	..	90	96	80	88	100	100[e]	23[g]
6 Sweden	..	..	..	..	100	96	85	99	100	..	26
7 Switzerland	..	..	..	..	84	93	80	84	..	...	24
8 Japan	..	..	..	..	100	100	97	100[e]	100	..	19
9 Netherlands	..	..	..	..	95	99	84	87	..	99	15
10 France	..	..	..	..	100	99	..	99	96	98[f]	..
11 Finland	..	..	..	..	98[e]	98	93	95	100	99	38
12 United States	..	..	..	..	97	92	85	89	..	..	16[g]
13 Spain	96.5	..	99.6	..	100	99	..	98	..	100[e]	30
14 Denmark	..	..	..	..	98	95	87	..	94	93	18
15 Austria	..	..	..	..	88[e]	97[e]	..	..	..	..	24
16 United Kingdom	..	..	..	..	98[e]	99	81	95	..	..	22
17 Belgium	..	..	..	..	96	99	87	97	91	..	17
18 Luxembourg	..	..	..	..	..	95	..	82	..	92[e,f]	..
19 New Zealand	..	..	..	..	98	99	85	91	..	..	17
20 Italy	..	98.4	..	99.8	100[e]	99	..	92	..	100	24
21 Hong Kong, China (SAR)	..	..	..	..	..	93[e]	..	80[e]	100	100	31[e]
22 Germany	..	..	..	..	84[e]	96[e]	..	..	..	..	..
23 Israel	..	..	..	..	92[e]	97	..	89	..	100	28
24 Greece	92.6	96.0	99.0	98.9	95	99	83	91	100	99	32
25 Singapore	89.1	92.5	99.0	99.5	..	..	..	..	..	..	..
26 Korea (Republic of)	..	..	..	..	100	99	86	90	99	98	40
27 Slovenia	99.5	99.7[h]	99.8	99.8[h]	96[e]	98	..	94	..	..	21
28 Cyprus	94.4	96.8	99.6	99.8	87	99[e]	69	94[e]	100	99	18
29 Portugal	87.9	93.8[h]	99.2	99.6[h]	98	98	..	83	..	..	29
30 Brunei Darussalam	87.8	92.7	98.1	98.9	92	93	71	87	..	100	10
31 Barbados	..	..	..	..	80[e]	98	..	96	..	98	..
32 Czech Republic	..	..	..	..	87[e]	92[e]	..	..	..	98	29
33 Kuwait	74.5	93.3	87.5	99.7	49[e]	87	..	78[e]	..	..	..
34 Malta	..	87.9	..	96.0	97	86	78	84	99	99[f]	14
35 Qatar	75.6	89.0	89.5	95.9	89	96	70	90	64	..	19
36 Hungary	..	..	..	..	91	89	75	90	98	..	18
37 Poland	..	..	..	..	97	96	76	93	98	99	20
38 Argentina	96.1	97.2	98.3	98.9	..	99[f]	..	79[f]	..	97[f]	19
39 United Arab Emirates	79.5[h]	88.7[h]	93.6[h]	97.0[h]	99	71	60	57	80	97	..
40 Chile	94.3	95.7	98.4	99.0	89	90[e]	55	..	92	100	28
41 Bahrain	84.0	86.5	96.9	97.0	99	97	85	90	89	99	17
42 Slovakia	..	..	..	..	..	92[e]	..	..	..	..	26
43 Lithuania	98.4	99.6	99.7	99.7	..	89	..	91	..	..	25
44 Estonia	99.7	99.8	99.9	99.8	99[e]	95	..	91	..	99	23
45 Latvia	99.5	99.7	99.8	99.8	92[e]	88[e]	..	..	..	..	15
46 Uruguay	95.4	96.8	98.6	98.6	91	93[e,f]	..	..	97	91[f]	..
47 Croatia	96.7	98.1	99.6	99.6	79	87[f]	63[e]	85	..	..	24
48 Costa Rica	..	94.9	..	97.6	87	..	38	..	84	87	23
49 Bahamas	..	..	..	..	90[e]	91	..	84	84	99[e]	..
50 Seychelles	87.8	91.8	98.8	99.1	..	99[e,f]	..	97[e]	93	99[f]	..
51 Cuba	..	99.8	..	100.0	93	97	70	87	92	97	..
52 Mexico	87.6	91.6	95.4	97.6	98	98	44	65	80	94	31
53 Bulgaria	..	98.2	..	98.2	86	93	63	88	91	..	27

Human development indicators

TABLE 12 Literacy and enrolment

HDI rank	Adult literacy rate (% aged 15 and older)		MDG Youth literacy rate (% aged 15–24)		MDG Net primary enrolment rate (%)		Net secondary enrolment rate [a] (%)		MDG Children reaching grade 5 (% of grade 1 students)		Tertiary students in science, engineering, manufacturing and construction (% of tertiary students)
	1985–1994 [b]	1995–2005 [c]	1985–1994 [b]	1995–2005 [c]	1991	2005	1991	2005	1991	2004	1999–2005 [d]
54 Saint Kitts and Nevis	..	..	..	..	..	93 [e]	..	86 [e]	..	87 [f]	..
55 Tonga	..	98.9	..	99.3	..	95 [e]	..	68 [e,f]	..	89 [e]	..
56 Libyan Arab Jamahiriya	74.7 [h]	84.2 [h]	94.9 [h]	98.0 [h]	96 [e]	..	..	..	..	..	31
57 Antigua and Barbuda	..	..	..	..	..	..	..	..	..	..	..
58 Oman	..	81.4	..	97.3	69	76	..	75	97	98	20 [e,g]
59 Trinidad and Tobago	97.1 [h]	98.4 [h]	99.3 [h]	99.5 [h]	91	90 [e]	..	69 [e]	..	91 [e]	36
60 Romania	96.7	97.3	99.1	97.8	81 [e]	93	..	80	..	..	25 [g]
61 Saudi Arabia	70.8	82.9	87.9	95.8	59	78	31	66	83	96	17
62 Panama	88.8	91.9	95.1	96.1	..	98	..	64	..	85	20 [g]
63 Malaysia	82.9	88.7	95.6	97.2	..	95 [f]	..	76 [f]	97	98 [f]	40
64 Belarus	97.9	99.6	99.8	99.8	86 [e]	89	..	89	..	..	27
65 Mauritius	79.9	84.3	91.2	94.5	91	95	..	82 [e]	97	97	26
66 Bosnia and Herzegovina	..	96.7	..	99.8	..	..	..	..	..	..	..
67 Russian Federation	98.0	99.4	99.7	99.7	99 [e]	92 [e]	..	..	..	..	..
68 Albania	..	98.7	..	99.4	95 [e]	94 [f]	..	74 [e,f]	..	..	12
69 Macedonia (TFYR)	94.1	96.1	98.9	98.7	94	92	..	82	..	..	26
70 Brazil	..	88.6	..	96.8	85	95 [f]	17	78 [f]	73	..	16
MEDIUM HUMAN DEVELOPMENT											
71 Dominica	..	..	..	..	..	84 [e]	..	..	75	93	..
72 Saint Lucia	..	..	..	..	95 [e]	97	..	68 [e]	96	96	..
73 Kazakhstan	97.5	99.5	99.7	99.8	89 [e]	91	..	92	..	..	..
74 Venezuela (Bolivarian Republic of)	89.8	93.0	95.4	97.2	87	91	18	63	86	91	..
75 Colombia	81.4	92.8	90.5	98.0	69	87	34	55 [e]	76	81	33
76 Ukraine	..	99.4	..	99.8	80 [e]	83	..	79	..	..	27
77 Samoa	98.1 [h]	98.6 [h]	99.1 [h]	99.3 [h]	..	90 [e,f]	..	66 [e,f]	..	94 [f]	14
78 Thailand	..	92.6	..	98.0	76 [e]	88 [i]	..	64 [i]	..	..	..
79 Dominican Republic	..	87.0	..	94.2	57 [e]	88	..	53	..	86	..
80 Belize	70.3	..	76.4	..	94 [e]	94	31	71 [e]	67	91 [f]	9 [g]
81 China	77.8	90.9	94.3	98.9	97	..	..	..	86	..	..
82 Grenada	..	..	..	..	..	84 [e]	..	79 [e]	..	79 [f]	..
83 Armenia	98.8	99.4	99.9	99.8	..	79	..	84	..	..	7 [g]
84 Turkey	79.2	87.4	92.5	95.6	89	89	42	67 [e]	98	97	21 [g]
85 Suriname	..	89.6	..	94.9	81 [e]	94	..	75 [e]	..	..	19
86 Jordan	..	91.1	..	99.0	94	89	..	79	..	96	22
87 Peru	87.2	87.9	95.4	97.1	..	96	..	70	..	90	..
88 Lebanon	..	..	..	..	73 [e]	92	..	..	..	93	24
89 Ecuador	88.3	91.0	96.2	96.4	98 [e]	98 [e,f]	..	52 [f]	..	76 [e,f]	..
90 Philippines	93.6	92.6	96.6	95.1	96 [e]	94	..	61	..	75	27 [g]
91 Tunisia	..	74.3	..	94.3	94	97	..	65 [e]	86	97	31 [g]
92 Fiji	..	..	..	..	..	96 [e]	..	83 [e]	87	99 [f]	..
93 Saint Vincent and the Grenadines	..	..	..	..	..	90	..	64 [e]	..	88 [e,f]	..
94 Iran (Islamic Republic of)	65.5	82.4	87.0	97.4	92 [e]	95	..	77	90	88 [f]	40
95 Paraguay	90.3	93.5 [h]	95.6	95.9 [h]	94	88 [f]	26	..	74	81 [f]	..
96 Georgia	..	..	..	..	97 [e]	93 [f]	..	81 [f]	..	..	23
97 Guyana	..	..	..	..	89	..	67	..	..	64 [e,f]	14
98 Azerbaijan	..	98.8	..	99.9	89	85	..	78	..	..	..
99 Sri Lanka	..	90.7 [j]	..	95.6 [j]	.	97 [e,f]	..	..	92	..	..
100 Maldives	96.0	96.3	98.2	98.2	..	79	..	63 [e]	..	92	..
101 Jamaica	..	79.9 [k]	..	.. [k]	96	90 [e]	64	78 [e]	..	90 [f]	..
102 Cape Verde	62.8	81.2 [h]	88.2	96.3 [h]	91 [e]	90	..	58	..	93	..
103 El Salvador	74.1	80.6 [h]	84.9	88.5 [h]	..	93	..	53 [e]	58	69 [e]	23
104 Algeria	49.6	69.9	74.3	90.1	89	97	53	66 [e,f]	95	96	18 [g]
105 Viet Nam	87.6	90.3	93.7	93.9	90 [e]	88	..	69 [e]	..	87 [e,f]	20
106 Occupied Palestinian Territories	..	92.4	..	99.0	..	80	..	95	..	..	18

Human development indicators

TABLE 12

HDI rank	Adult literacy rate (% aged 15 and older)		MDG Youth literacy rate (% aged 15–24)		MDG Net primary enrolment rate (%)		Net secondary enrolment rate[a] (%)		MDG Children reaching grade 5 (% of grade 1 students)		Tertiary students in science, engineering, manufacturing and construction (% of tertiary students)
	1985–1994[b]	1995–2005[c]	1985–1994[b]	1995–2005[c]	1991	2005	1991	2005	1991	2004	1999–2005[d]
107 Indonesia	81.5	90.4	96.2	98.7	97	96[e]	39	58[e]	84	89[e]	..
108 Syrian Arab Republic	..	80.8	..	92.5	91	95[f]	43	62	96	92[f]	..
109 Turkmenistan	..	98.8	..	99.8	..	..	..	..	..	..	..
110 Nicaragua	..	76.7	..	86.2	73	87	..	43	44	54	..
111 Moldova	96.4	99.1[h]	99.7	99.7[h]	89[e]	86[e]	..	76[e]	..	..	..
112 Egypt	44.4	71.4	63.3	84.9	84[e]	94[e]	..	82[e]	..	94[e]	..
113 Uzbekistan	..	..	..	..	78[e]	..	..	..	..	..	..
114 Mongolia	..	97.8	..	97.7	90[e]	84	..	84	..	..	23
115 Honduras	..	80.0	..	88.9	89[e]	91[e]	21	..	..	70[e]	23
116 Kyrgyzstan	..	98.7	..	99.7	92[e]	87	..	80	..	..	17
117 Bolivia	80.0	86.7	93.9	97.3	..	95[e,f]	..	73[e,f]	..	85[e,f]	..
118 Guatemala	64.2	69.1	76.0	82.2	..	94	..	34[e,f]	..	68	19[g]
119 Gabon	72.2	84.0[h]	93.2	96.2[h]	85[e]	77[e,f]	..	..	..	69[e,f]	..
120 Vanuatu	..	74.0			..	94[e]	17	39[e,f]	..	78[e]	..
121 South Africa	..	82.4	..	93.9	90	87[f]	45	62[e]	..	82[f]	20
122 Tajikistan	97.7	99.5	99.7	99.8	77[e]	97	..	80	..	..	18
123 Sao Tome and Principe	73.2	84.9	93.8	95.4	..	97	..	32	..	76	..
124 Botswana	68.6	81.2	89.3	94.0	83	85[e]	35	60[e]	84	90[e,f]	17[g]
125 Namibia	75.8	85.0	88.1	92.3	..	72	..	39	62	86	12
126 Morocco	41.6	52.3	58.4	70.5	56	86	..	35[e]	75	79	21
127 Equatorial Guinea	..	87.0	..	94.9	91[e]	81[f]	..	24[e]	..	33[e,f]	..
128 India	48.2	61.0[l]	61.9	76.4[l]	..	89[e]	..	..	..	73	22[g]
129 Solomon Islands	..	..	..	..	..	63[e,f]	..	26[e]	88	..	..
130 Lao People's Democratic Republic	..	68.7	..	78.5	63[e]	84	..	38	..	63	6[g]
131 Cambodia	..	73.6	..	83.4	69[e]	99	..	24[e]	..	63	19
132 Myanmar	..	89.9	..	94.5	98[e]	90	..	37	..	70	42
133 Bhutan	..	..	..	..	..	..	..	..	..	91[f]	..
134 Comoros	..	..	..	..	57[e]	55[e,f]	..	..	..	80[e]	11
135 Ghana	..	57.9	..	70.7	54[e]	65	..	37[e]	80	63[f]	26
136 Pakistan	..	49.9	..	65.1	33[e]	68	..	21[e]	..	70	24[g]
137 Mauritania	..	51.2	..	61.3	35[e]	72	..	15	75	53	6[g]
138 Lesotho	..	82.2	..	..	71	87	15	25	66	73	24
139 Congo	73.8[h]	84.7[h]	93.7[h]	97.4[h]	79[e]	44	..	..	60	66[f]	11[g]
140 Bangladesh	35.3	47.5	44.7	63.6	..	94[e,f]	..	44[f]	..	65[f]	20[g]
141 Swaziland	67.2	79.6	83.7	88.4	75[e]	80[e]	30	33[e]	77	77[f]	9
142 Nepal	33.0	48.6	49.6	70.1	..	79[e,f]	..	..	51	61[e]	..
143 Madagascar	..	70.7	..	70.2	64[e]	92	..	..	21	43	20
144 Cameroon	..	67.9	..	..	74[e]	..	..	..	..	64[e,f]	23[e]
145 Papua New Guinea	..	57.3	..	66.7	..	..	..	..	69	68[e,f]	..
146 Haiti	..	..	..	..	22	..	..	..	..	..	..
147 Sudan	..	60.9[m]	..	77.2[m]	40[e]	43[e,f]	..	..	94	79	..
148 Kenya	..	73.6	..	80.3	..	79	..	42[e]	77	83[e]	29
149 Djibouti	..	..	..	..	29	33	..	23[e]	87	77[f]	9[g]
150 Timor-Leste	..	..	..	..	..	98[e]	..	..	..	..	..
151 Zimbabwe	83.5	89.4[h]	95.4	97.7[h]	..	82[f]	..	34	76	70[e,f]	..
152 Togo	..	53.2	..	74.4	64	78	15	22[e]	48	75	8
153 Yemen	37.1	54.1[h]	60.2	75.2[h]	51[e]	75[e,f]	..	..	..	73[e,f]	..
154 Uganda	56.1	66.8	69.8	76.6	..	..	..	15[e]	36	49[e]	10
155 Gambia	..	..	..	..	48[e]	77[e,f]	..	45[e]	..	..	21
LOW HUMAN DEVELOPMENT											
156 Senegal	26.9	39.3	37.9	49.1	43[e]	69	..	17[e,f]	85	73	..
157 Eritrea	..	..	..	..	16[e]	47	..	25	..	79	37
158 Nigeria	55.4	69.1[h]	71.2	84.2[h]	58[e]	68[e]	..	27	89	73[e,f]	..
159 Tanzania (United Republic of)	59.1	69.4	81.8	78.4	49	91	..	..	81[e]	84	24[e,g]

Human development indicators

TABLE 12 Literacy and enrolment

HDI rank	Adult literacy rate (% aged 15 and older)		MDG Youth literacy rate (% aged 15–24)		MDG Net primary enrolment rate (%)		Net secondary enrolment rate [a] (%)		MDG Children reaching grade 5 (% of grade 1 students)		Tertiary students in science, engineering, manufacturing and construction (% of tertiary students)
	1985–1994 [b]	1995–2005 [c]	1985–1994 [b]	1995–2005 [c]	1991	2005	1991	2005	1991	2004	1999–2005 [d]
160 Guinea	..	29.5	..	46.6	27 [e]	66	..	24 [e]	59	76	34
161 Rwanda	57.9	64.9	74.9	77.6	66	74 [e]	7	..	60	46 [f]	..
162 Angola	..	67.4	..	72.2	50 [e]	..	..	..	..	..	18
163 Benin	27.2	34.7	39.9	45.3	41 [e]	78	..	17 [e]	55	52	..
164 Malawi	48.5	64.1	59.0	76.0	48	95	..	24	64	42	..
165 Zambia	65.0	68.0	66.4	69.5	..	89	..	26 [e]	..	94 [f]	..
166 Côte d'Ivoire	34.1	48.7	48.5	60.7	45	56 [e,f]	..	20 [e]	73	88 [e,f]	..
167 Burundi	37.4	59.3	53.6	73.3	53 [e]	60	..	..	62	67	10 [g]
168 Congo (Democratic Republic of the)	..	67.2	..	70.4	54	..	..	..	55	..	..
169 Ethiopia	27.0	35.9	33.6	49.9	22 [e]	61	..	28 [e]	18	..	17
170 Chad	12.2	25.7	17.0	37.6	35 [e]	61 [e,f]	..	11 [e]	51 [e]	33	..
171 Central African Republic	33.6	48.6	48.2	58.5	52	..	..	..	23	..	..
172 Mozambique	..	38.7	..	47.0	43	77	..	7	34	62	24
173 Mali	..	24.0	..	..	21 [e]	51	5 [e]	..	70 [e]	87	..
174 Niger	..	28.7	..	36.5	22	40	5	8	62	65	..
175 Guinea-Bissau	..	..	..	..	38 [e]	45 [e,f]	..	9 [e]	..	..	..
176 Burkina Faso	13.6	23.6	20.2	33.0	29	45	..	11	70	76	..
177 Sierra Leone	..	34.8	..	47.9	43 [e]	..	..	..	..	..	8
Developing countries	68.2 [n]	77.1 [n]	80.2 [n]	85.6 [n]	80	85	..	53 [n]	..	..	..
Least developed countries	47.4 [n]	53.4 [n]	56.3 [n]	65.5 [n]	47	77	..	27 [n]	..	..	..
Arab States	58.2 [n]	70.3 [n]	74.8 [n]	85.2 [n]	71	83	..	59 [n]	..	..	..
East Asia and the Pacific	..	90.7	..	97.8	..	93	..	69 [n]	..	..	..
Latin America and the Caribbean	87.6 [n]	89.9 [n]	93.7 [n]	96.6 [n]	86	95	..	68 [n]	..	..	..
South Asia	47.6 [n]	59.7 [n]	60.7 [n]	74.7 [n]	..	87	..	..	..	..	..
Sub-Saharan Africa	54.2 [n]	59.3 [n]	64.4 [n]	71.2 [n]	52	72	..	26 [n]	..	..	..
Central and Eastern Europe and the CIS	97.5	99.1	..	99.6	90	91	..	84 [n]	..	..	..
OECD	..	..	..	..	97	96	..	87 [n]	..	..	..
High-income OECD	98.9 [n]	99.1 [n]	99.4 [n]	..	97	96	..	92 [n]	..	..	..
High human development	..	94.1	..	98.1	93	95	..	..	..	..	..
Medium human development	..	78.3	..	87.3	..	87	..	..	..	..	..
Low human development	43.5	54.1	55.9	66.4	45	69	..	..	..	..	..
High income	98.4 [n]	98.6 [n]	99.0 [n]	..	96	95	..	91 [n]	..	..	..
Middle income	82.3 [n]	90.1 [n]	93.1 [n]	96.8 [n]	92	93	..	70 [n]	..	..	..
Low income	51.5 [n]	60.8 [n]	63.0 [n]	73.4 [n]	..	81	..	40 [n]	..	..	..
World	76.4 [n]	82.4 [n]	83.5 [n]	86.5 [n]	83	87	..	59 [n]	..	..	..

NOTES

a. Enrolment rates for the most recent years are based on the new International Standard Classification of Education, adopted in 1997 (UNESCO 1997), and so may not be strictly comparable with those for 1991.

b. Data refer to national literacy estimates from censuses or surveys conducted between 1985 and 1994, unless otherwise specified. Due to differences in methodology and timeliness of underlying data, comparisons across countries and over time should be made with caution. For more details, see http://www.uis.unesco.org/.

c. Data refer to national literacy estimates from censuses or surveys conducted between 1995 and 2005, unless otherwise specified. Due to differences in methodology and timeliness of underlying data, comparisons across countries and over time should be made with caution. For more details, see http://www.uis.unesco.org/.

d. Data refer to the most recent year available during the period specified.

e. National or UNESCO Institute for Statistics estimate.

f. Data refer to an earlier year than that specified.

g. Figure should be treated with caution because the reported number of enrolled students in the "Not known or unspecified" category represents more than 10% of total enrolment.

h. UNESCO Institute for Statistics estimates based on its Global Age-specific Literacy Projections model, April 2007.

i. Data refer to 2006.

j. Data refer to 18 of the 25 states of the country only.

k. Data are based on a literacy assessment.

l. Data exclude three sub-divisions of Senapati district of Manipur: Mao Maram, Paomata and Purul.

m. Data refer to North Sudan only.

n. Data refer to aggregates calculated by UNESCO Institute for Statistics.

SOURCES

Columns 1–4: UNESCO Institute for Statistics 2007a.
Columns 5–11: UNESCO Institute for Statistics 2007c.

TABLE
13
... to acquire knowledge ...

Technology: diffusion and creation

		MDG Telephone mainlines [a] (per 1,000 people)		MDG Cellular subscribers [a] (per 1,000 people)		MDG Internet users (per 1,000 people)		Patents granted to residents (per million people)	Receipts of royalties and licence fees (US$ per person)	Research and development (R&D) expenditures (% of GDP)	Researchers in R&D (per million people)
HDI rank		1990	2005	1990	2005	1990	2005	2000–05 [b]	2005	2000–05 [b]	1990–2005 [b]
HIGH HUMAN DEVELOPMENT											
1	Iceland	512	653	39	1,024	0	869	0	0.0	3.0	6,807
2	Norway	503	460	46	1,028	7	735	103	78.4	1.7	4,587
3	Australia	456	564	11	906	6	698	31	25.0	1.7	3,759
4	Canada	550	566	21	514	4	520	35	107.6	1.9	3,597
5	Ireland	280	489	7	1,012	0	276	80	142.2	1.2	2,674
6	Sweden	683	717 [c]	54	935	6	764	166	367.7	3.7	5,416
7	Switzerland	587	689	19	921	6	498	77	..	2.6	3,601
8	Japan	441	460	7	742	(.)	668	857	138.0	3.1	5,287
9	Netherlands	464	466	5	970	3	739	110	236.8	1.8	2,482
10	France	495	586	5	789	1	430	155	97.1	2.2	3,213
11	Finland	535	404	52	997	4	534	214	230.0	3.5	7,832
12	United States	545	606 [c]	21	680	8	630 [c]	244	191.5	2.7	4,605
13	Spain	325	422	1	952	(.)	348	53	12.9	1.1	2,195
14	Denmark	566	619	29	1,010	1	527	19	..	2.6	5,016
15	Austria	418	450	10	991	1	486	92	21.3	2.3	2,968
16	United Kingdom	441	528	19	1,088	1	473	62	220.8	1.9	2,706
17	Belgium	393	461 [c]	4	903	(.)	458	51	106.5	1.9	3,065
18	Luxembourg	481	535	2	1,576	0	690	31	627.9	1.8	4,301
19	New Zealand	426	422	16	861	0	672	10	24.8	1.2	3,945
20	Italy	394	427	5	1,232	(.)	478	71	19.3	1.1	1,213
21	Hong Kong, China (SAR)	434	546	23	1,252	0	508	5	31.2 [c]	0.6	1,564
22	Germany	401	667	3	960	1	455	158	82.6	2.5	3,261
23	Israel	349	424	3	1,120	1	470 [c]	48	91.2	4.5	..
24	Greece	389	568	0	904	0	180	29	5.4	0.6	1,413
25	Singapore	346	425	17	1,010	0	571 [c]	96	125.8	2.3	4,999
26	Korea (Republic of)	310	492	2	794	(.)	684	1,113	38.2	2.6	3,187
27	Slovenia	211	408	0	879	0	545	113	8.2	1.6	2,543
28	Cyprus	424	554	5	949	0	430	7	18.1	0.4	630
29	Portugal	240	401	1	1,085	0	279	14	5.7	0.8	1,949
30	Brunei Darussalam	136	224	7	623	0	277 [c]	..	..	0.0	274
31	Barbados	281	500	0	765	0	594	..	5.8	..	..
32	Czech Republic	157	314	0	1,151	0	269	34	6.2	1.3	1,594
33	Kuwait	156	201	10	939	0	276	..	0.0	0.2	..
34	Malta	356	501	0	803	0	315	0	7.5	0.3	681
35	Qatar	197	253	8	882	0	269	..	..	..	..
36	Hungary	96	333	(.)	924	0	297	13	82.7	0.9	1,472
37	Poland	86	309	0	764	0	262	28	1.6	0.6	1,581
38	Argentina	93	227	(.)	570	0	177	4	1.4	0.4	720
39	United Arab Emirates	224	273	19	1,000	0	308	..	..	..	..
40	Chile	66	211	1	649	0	172	1	3.3	0.6	444
41	Bahrain	191	270	10	1,030	0	213	..	..	..	..
42	Slovakia	135	222	0	843	0	464	9	9.2 [d]	0.5	1,984
43	Lithuania	211	235	0	1,275	0	358	21	0.6	0.8	2,136
44	Estonia	204	328	0	1,074	0	513	56	4.0	0.9	2,523
45	Latvia	232	318	0	814	0	448	36	4.3	0.4	1,434
46	Uruguay	134	290	0	333	0	193	1	(.)	0.3	366
47	Croatia	172	425	(.)	672	0	327	4	16.1	1.1	1,296
48	Costa Rica	92	321	0	254	0	254	..	0.0	0.4	..
49	Bahamas	274	439 [c]	8	584 [c]	0	319	..	..	..	..
50	Seychelles	124	253	0	675	0	249	..	..	0.1	19
51	Cuba	32	75	0	12	0	17	3	..	0.6	..
52	Mexico	64	189	1	460	0	181	1	0.7	0.4	268
53	Bulgaria	250	321	0	807	0	206	10	0.7	0.5	1,263

TABLE 13

Technology: diffusion and creation

	MDG Telephone mainlines [a] (per 1,000 people)		MDG Cellular subscribers [a] (per 1,000 people)		MDG Internet users (per 1,000 people)		Patents granted to residents (per million people)	Receipts of royalties and licence fees (US$ per person)	Research and development (R&D) expenditures (% of GDP)	Researchers in R&D (per million people)
HDI rank	1990	2005	1990	2005	1990	2005	2000–05 [b]	2005	2000–05 [b]	1990–2005 [b]
54 Saint Kitts and Nevis	231	532 [c]	0	213 [c]	0	..	..	0.0	..	..
55 Tonga	46	..	0	161 [c]	0	29	..	..	..	45,454
56 Libyan Arab Jamahiriya	51	133 [d]	0	41 [c]	0	36 [c]	..	0.0 [c]	..	361
57 Antigua and Barbuda	252	467 [c]	0	663 [c]	0	350	..	0.0	..	..
58 Oman	57	103	1	519	0	111	..	..	..	..
59 Trinidad and Tobago	136	248	0	613	0	123 [c]	..	..	0.1	..
60 Romania	102	203	0	617	0	208 [c]	24	2.2	0.4	976
61 Saudi Arabia	75	164	1	575	0	70 [c]	(.)	0.0	..	..
62 Panama	90	136	0	418	0	64	..	0.0	0.3	97
63 Malaysia	89	172	5	771	0	435	..	1.1	0.7	299
64 Belarus	154	336	0	419	0	347	76	0.3	0.6	..
65 Mauritius	53	289	2	574	0	146 [c]	..	(.)	0.4	360
66 Bosnia and Herzegovina	..	248	0	408	0	206	3	..	..	..
67 Russian Federation	140	280	0	838	0	152	135	1.8	1.2	3,319
68 Albania	12	88 [c]	0	405 [c]	0	60	..	0.2	..	..
69 Macedonia (TFYR)	150	262	0	620	0	79	11	1.5	0.3	504
70 Brazil	63	230 [c]	(.)	462	0	195	1	0.5	1.0	344
MEDIUM HUMAN DEVELOPMENT										
71 Dominica	161	293 [c]	0	585 [c]	0	361	..	0.0	..	..
72 Saint Lucia	127	..	0	573 [c]	0	339 [c]	0	..	0.4 [e]	..
73 Kazakhstan	82	167 [c]	0	327	0	27 [c]	83	(.)	0.2	629
74 Venezuela (Bolivarian Republic of)	75	136	(.)	470	0	125	1	0.0	0.3	..
75 Colombia	69	168	0	479	0	104	(.)	0.2	0.2	109
76 Ukraine	135	256 [c]	0	366	0	97	52	0.5	1.2	..
77 Samoa	25	73 [d]	0	130	0	32	0	..	..	..
78 Thailand	24	110	1	430 [c]	0	110	1	0.3	0.3	287
79 Dominican Republic	48	101	(.)	407	0	169	..	0.0	..	..
80 Belize	92	114	0	319	0	130	..	..	..	..
81 China	6	269	(.)	302	0	85	16	0.1	1.4	708
82 Grenada	162	309 [c]	2	410 [c]	0	182	..	0.0	..	..
83 Armenia	158	192 [c]	0	106	0	53	39	..	0.3	..
84 Turkey	122	263	1	605	0	222	1	0.0 [c]	0.7	341
85 Suriname	91	180	0	518	0	71	..	..	..	..
86 Jordan	78	119 [c]	(.)	304 [c]	0	118 [c]	..	..	..	1,927
87 Peru	26	80	(.)	200	0	164	(.)	0.1	0.1	226
88 Lebanon	144	277	0	277	0	196	..	0.0 [c]	..	..
89 Ecuador	48	129	0	472	0	47	0	0.0 [c]	0.1	50
90 Philippines	10	41	0	419	0	54 [c]	(.)	0.1	0.1	48
91 Tunisia	37	125	(.)	566	0	95	..	1.4	0.6	1,013
92 Fiji	59	122 [d]	0	229	0	77	..	..	..	..
93 Saint Vincent and the Grenadines	120	189	0	593	0	84	0	..	0.2	..
94 Iran (Islamic Republic of)	40	278	0	106	0	103	8	..	0.7	1,279
95 Paraguay	27	54	0	320	0	34	..	33.2	0.1	79
96 Georgia	99	151 [c]	0	326	0	39 [c]	42	2.1	0.3	..
97 Guyana	22	147	0	375	0	213	..	47.9	..	..
98 Azerbaijan	87	130	0	267	0	81	..	(.)	0.3	..
99 Sri Lanka	7	63	(.)	171	0	14 [c]	3	..	0.1	128
100 Maldives	29	98	0	466	0	59 [c]	..	8.6	..	..
101 Jamaica	44	129	0	1,017	0	404 [c]	1	4.7	0.1	..
102 Cape Verde	23	141	0	161	0	49	..	0.2 [d]	..	127
103 El Salvador	24	141	0	350	0	93	..	0.4	0.1 [e]	47
104 Algeria	32	78	(.)	416	0	58	1	..	..	..
105 Viet Nam	1	191	0	115	0	129	(.)	..	0.2	115
106 Occupied Palestinian Territories	..	96	0	302	0	67	..	..	..	..

TABLE 13

HDI rank	MDG Telephone mainlines[a] (per 1,000 people)		MDG Cellular subscribers[a] (per 1,000 people)		MDG Internet users (per 1,000 people)		Patents granted to residents (per million people)	Receipts of royalties and licence fees (US$ per person)	Research and development (R&D) expenditures (% of GDP)	Researchers in R&D (per million people)
	1990	2005	1990	2005	1990	2005	2000–05[b]	2005	2000–05[b]	1990–2005[b]
107 Indonesia	6	58	(.)	213	0	73	..	1.2	0.1	207
108 Syrian Arab Republic	39	152	0	155	0	58	2	..	..	29
109 Turkmenistan	60	80 [d]	0	11 [c]	0	8 [c]	..	..	..	..
110 Nicaragua	12	43	0	217	0	27	1	0.0	0.0	73
111 Moldova	106	221	0	259	0	96 [c]	67	0.4	0.8 [e]	..
112 Egypt	29	140	(.)	184	0	68	1	1.9	0.2	493
113 Uzbekistan	68	67 [d]	0	28	0	34 [c]	10	..	..	1,754
114 Mongolia	32	61	0	218	0	105	44	..	0.3	..
115 Honduras	18	69	0	178	0	36	1	0.0	0.0	..
116 Kyrgyzstan	71	85	0	105	0	54	17	0.4	0.2	..
117 Bolivia	27	70	0	264	0	52	..	0.2	0.3	120
118 Guatemala	21	99	(.)	358	0	79	(.)	(.) [c]	..	..
119 Gabon	22	28	0	470	0	48	..	..	..	..
120 Vanuatu	17	33 [c]	(.)	60	0	38	..	..	..	..
121 South Africa	94	101	(.)	724	0	109	..	0.9	0.8	307
122 Tajikistan	45	39 [d]	0	41	0	1 [c]	2	0.2	..	660
123 Sao Tome and Principe	19	46 [c]	0	77	0	131 [c]	..	..	..	..
124 Botswana	18	75	0	466	0	34	..	0.3	..	..
125 Namibia	38	64 [c]	0	244	0	37 [c]	..	0.0 [d]	..	..
126 Morocco	17	44	(.)	411	0	152	1	0.4	0.6	..
127 Equatorial Guinea	4	20	0	192	0	14	..	..	..	..
128 India	6	45	0	82	0	55	1	(.) [d]	0.8	119
129 Solomon Islands	15	16	0	13	0	8	..	..	..	..
130 Lao People's Democratic Republic	2	13	0	108	0	4	..	..	..	..
131 Cambodia	(.)	3 [d]	0	75	0	3 [c]	..	(.)	..	..
132 Myanmar	2	9	0	4	0	2	..	0.0 [d]	0.1	17
133 Bhutan	3	51	0	59	0	39	..	..	..	..
134 Comoros	8	28	0	27	0	33	..	..	..	..
135 Ghana	3	15	0	129	0	18	..	0.0	..	..
136 Pakistan	8	34	(.)	82	0	67	0	0.1	0.2	75
137 Mauritania	3	13	0	243	0	7	..	..	..	..
138 Lesotho	8	27	0	137	0	24 [c]	..	9.1	0.0	..
139 Congo	6	4 [c]	0	123	0	13	..	..	..	30
140 Bangladesh	2	8	0	63	0	3	..	(.)	0.6	51
141 Swaziland	18	31	0	177	0	32 [c]	..	(.)	..	..
142 Nepal	3	17	0	9	0	4	..	..	0.7	59
143 Madagascar	3	4	0	27	0	5	(.)	(.)	0.1	15
144 Cameroon	3	6 [c]	0	138	0	15	..	(.) [d]	..	..
145 Papua New Guinea	7	11 [c]	0	4	0	23	..	..	..	..
146 Haiti	7	17 [c]	0	48 [c]	0	70	..	0.0	..	..
147 Sudan	2	18	0	50	0	77	..	0.0	0.3	..
148 Kenya	7	8	0	135	0	32	..	0.5	..	..
149 Djibouti	10	14	0	56	0	13	..	..	..	..
150 Timor-Leste	..	..	..	..	..	..	..	..	..	..
151 Zimbabwe	12	25	0	54	0	77	0	..	..	..
152 Togo	3	10	0	72	0	49	..	0.0 [c]	..	102
153 Yemen	10	39 [c]	0	95	0	9 [c]	..	..	..	..
154 Uganda	2	3	0	53	0	17	..	0.3	0.8	..
155 Gambia	7	29	0	163	0	33 [c]	..	..	..	..
LOW HUMAN DEVELOPMENT										
156 Senegal	6	23	0	148	0	46	..	0.0 [c]	..	..
157 Eritrea	..	9	0	9	0	16	..	..	..	..
158 Nigeria	3	9	0	141	0	38	..	..	..	..
159 Tanzania (United Republic of)	3	4 [c]	0	52 [c]	0	9 [c]	..	0.0	..	..

TABLE 13

Technology: diffusion and creation

HDI rank	MDG Telephone mainlines [a] (per 1,000 people)		MDG Cellular subscribers [a] (per 1,000 people)		MDG Internet users (per 1,000 people)		Patents granted to residents (per million people)	Receipts of royalties and licence fees (US$ per person)	Research and development (R&D) expenditures (% of GDP)	Researchers in R&D (per million people)
	1990	2005	1990	2005	1990	2005	2000–05 [b]	2005	2000–05 [b]	1990–2005 [b]
160 Guinea	2	3 [c]	0	20	0	5	..	0.0 [c]	..	..
161 Rwanda	1	3 [c]	0	32	0	6	..	0.0	..	..
162 Angola	7	6	0	69	0	11	..	3.1	..	..
163 Benin	3	9	0	89	0	50	..	0.0 [c]	..	..
164 Malawi	3	8	0	33	0	4	0	..	..	..
165 Zambia	8	8	0	81	0	20 [c]	..	..	0.0 [e]	51
166 Côte d'Ivoire	6	14 [c]	0	121	0	11	..	(.) [c]	..	..
167 Burundi	1	4 [c]	0	20	0	5	..	0.0	..	..
168 Congo (Democratic Republic of the)	1	(.)	0	48	0	2	..	..	..	..
169 Ethiopia	2	9	0	6	0	2	..	(.)	..	..
170 Chad	1	1 [c]	0	22	0	4	..	..	..	..
171 Central African Republic	2	2	0	25	0	3	..	..	..	47
172 Mozambique	4	4 [c]	0	62	0	7 [c]	..	0.1	0.6	..
173 Mali	1	6	0	64	0	4	..	(.) [c]	..	..
174 Niger	1	2	0	21	0	2	..	..	..	..
175 Guinea-Bissau	6	7 [d]	0	42	0	20	..	..	..	..
176 Burkina Faso	2	7	0	43	0	5	..	..	0.2 [e]	17
177 Sierra Leone	3	..	0	22 [d]	0	2 [c]	..	0.2 [c]	..	..
Developing countries	21	132	(.)	229	(.)	86	..	..	1.0	..
Least developed countries	3	9	0	48	0	12	..	0.2	..	..
Arab States	34	106	(.)	284	0	88	..	0.9	..	..
East Asia and the Pacific	18	223	(.)	301	(.)	106	..	1.7	1.6	722
Latin America and the Caribbean	61	..	(.)	439	0	156	..	1.1	0.6	256
South Asia	7	51	(.)	81	0	52	..	(.)	0.7	119
Sub-Saharan Africa	10	17	(.)	130	0	26	..	0.3	..	..
Central and Eastern Europe and the CIS	125	277	(.)	629	0	185	73	4.1	1.0	2,423
OECD	390	441	10	785	3	445	239	104.2	2.4	3,096
High-income OECD	462	..	12	828	3	524	299	130.4	2.4	3,807
High human development	308	394	7	743	2	365	189	75.8	2.4	3,035
Medium human development	16	135	(.)	209	0	73	..	0.3	0.8	..
Low human development	3	7	0	74	0	17	..	0.2	..	..
High income	450	500	12	831	3	525	286	125.3	2.4	3,781
Middle income	40	211	(.)	379	0	115	..	1.0	0.8	725
Low income	6	37	(.)	77	0	45	..	(.)	0.7	..
World	98	180	2	341	1	136	..	21.6	2.3	..

NOTES

a. Telephone mainlines and cellular subscribers combined form an indicator for MDG 8; see *Index to Millennium Development Goal Indicators in the indicator tables.*
b. Data refer to the most recent year available during the period specified.
c. Data refer to 2004.
d. Data refer to 2003.
e. Data refer to year other than specified.

SOURCES

Columns 1–6, 9 and 10: World Bank 2007b; aggregates calculated for HDRO by the World Bank.
Column 7: calculated on the basis of data on patents from WIPO 2007 and data on population from UN 2007e.
Column 8: calculated on the basis of data on royalties and license fees from World Bank 2007b and data on pupulation from UN 2007e; aggregates calculated for HDRO by the World Bank.

TABLE 14

... to have access to the resources needed for a decent standard of living ...

Economic performance

	GDP		GDP per capita						Average annual change in consumer price index (%)	
HDI rank	US$ billions 2005	PPP US$ billions 2005	US$ 2005	2005 PPP US$[a] 2005	Annual growth rate (%) 1975–2005	Annual growth rate (%) 1990–2005	Highest value during 1975–2005 2005 PPP US$[a]	Year of highest value	1990–2005	2004–05
HIGH HUMAN DEVELOPMENT										
1 Iceland	15.8	10.8	53,290	36,510	1.8	2.2	36,510	2005	3.3	4.2
2 Norway	295.5	191.5	63,918	41,420	2.6	2.7	41,420	2005	2.2	1.5
3 Australia	732.5	646.3	36,032	31,794	2.0	2.5	31,794	2005	2.5	2.7
4 Canada	1,113.8	1,078.0	34,484	33,375	1.6	2.2	33,375	2005	1.9	2.2
5 Ireland	201.8	160.1	48,524	38,505	4.5	6.2	38,505	2005	2.9	2.4
6 Sweden	357.7	293.5	39,637	32,525	1.6	2.1	32,525	2005	1.6	0.5
7 Switzerland	367.0	265.0	49,351	35,633	1.0	0.6	35,633	2005	1.2	1.2
8 Japan	4,534.0	3,995.1	35,484	31,267	2.2	0.8	31,267	2005	0.2	-0.3
9 Netherlands	624.2	533.4	38,248	32,684	1.8	1.9	32,684	2005	2.5	1.7
10 France	2,126.6	1,849.7	34,936	30,386	1.8	1.6	30,386	2005	1.6	1.7
11 Finland	193.2	168.7	36,820	32,153	2.0	2.5	32,153	2005	1.6	0.9
12 United States	12,416.5	12,416.5	41,890	41,890	2.0	2.1	41,890	2005	2.6	3.4
13 Spain	1,124.6	1,179.1	25,914	27,169	2.3	2.5	27,169	2005	3.4	3.4
14 Denmark	258.7	184.0	47,769	33,973	1.7	1.9	33,973	2005	2.1	1.8
15 Austria	306.1	277.5	37,175	33,700	2.1	1.9	33,700	2005	2.0	2.3
16 United Kingdom	2,198.8	2,001.8	36,509	33,238	2.2	2.5	33,238	2005	2.7	2.8
17 Belgium	370.8	336.6	35,389	32,119	1.9	1.7	32,119	2005	1.9	2.8
18 Luxembourg	36.5	27.5	79,851	60,228	3.8	3.3	60,228	2005	2.0	2.5
19 New Zealand	109.3	102.5	26,664	24,996	1.1	2.1	24,996	2005	1.9	3.0
20 Italy	1,762.5	1,672.0	30,073	28,529	2.0	1.3	28,944	2002	3.1	2.0
21 Hong Kong, China (SAR)	177.7	241.9	25,592	34,833	4.2	2.4	34,833	2005	2.5	0.9
22 Germany	2,794.9	2,429.6	33,890	29,461	2.0	1.4	29,461	2005	1.7	2.0
23 Israel	123.4	179.1	17,828	25,864	1.8	1.5	25,864	2005	6.6	1.3
24 Greece	225.2	259.6	20,282	23,381	1.3	2.5	23,381	2005	6.5	3.6
25 Singapore	116.8	128.8	26,893	29,663	4.7	3.6	29,663	2005	1.2	0.5
26 Korea (Republic of)	787.6	1,063.9	16,309	22,029	6.0	4.5	22,029	2005	4.3	2.7
27 Slovenia	34.4	44.6	17,173	22,273	3.2[b]	3.2	22,273[b]	2005	9.2	2.5
28 Cyprus	15.4[c]	16.3[c]	20,841[c]	22,699[c]	4.0[b]	2.3	22,699[b]	2004	3.3	2.6
29 Portugal	183.3	215.3	17,376	20,410	2.7	2.1	20,679	2002	3.8	2.3
30 Brunei Darussalam	6.4	..	17,121	..	-1.9[b]	-0.8[b]	..	..	1.3	1.2
31 Barbados	3.1	..	11,465	..	1.3[b]	1.5[b]	..	..	2.2	6.1
32 Czech Republic	124.4	210.2	12,152	20,538	1.9[b]	1.9	20,538[b]	2005	5.2	1.8
33 Kuwait	80.8	66.7[d]	31,861	26,321[d]	-0.5[b]	0.6[b]	34,680[b]	1979	1.8	4.1
34 Malta	5.6	7.7	13,803	19,189	4.1	2.7	19,862	2002	2.8	3.0
35 Qatar	42.5	..	52,240	..	..	..	..	..	2.7	8.8
36 Hungary	109.2	180.4	10,830	17,887	1.3	3.1	17,887	2005	15.0	3.6
37 Poland	303.2	528.5	7,945	13,847	4.3[b]	4.3	13,847[b]	2005	16.0	2.1
38 Argentina	183.2	553.3	4,728	14,280	0.3	1.1	14,489	1998	7.1	9.6
39 United Arab Emirates	129.7	115.7[d]	28,612	25,514[d]	-2.6	-0.9	50,405	1981	..	..
40 Chile	115.2	196.0	7,073	12,027	3.9	3.8	12,027	2005	6.3	3.1
41 Bahrain	12.9	15.6	17,773	21,482	1.5[b]	2.3	21,482[b]	2005	0.5	2.6
42 Slovakia	46.4	85.5	8,616	15,871	1.0[b]	2.8	15,871[b]	2005	7.8	2.7
43 Lithuania	25.6	49.5	7,505	14,494	1.9[b]	1.9	14,494[b]	2005	14.6	2.7
44 Estonia	13.1	20.8	9,733	15,478	1.1[b]	4.2	15,478[b]	2005	12.0	4.1
45 Latvia	15.8	31.4	6,879	13,646	0.6	3.6	13,646	2005	15.5	6.8
46 Uruguay	16.8	34.5	4,848	9,962	1.1	0.8	10,459	1998	22.3	4.7
47 Croatia	38.5	57.9	8,666	13,042	2.6[b]	2.6	13,042[b]	2005	40.6	3.3
48 Costa Rica	20.0	44.1[d]	4,627	10,180[d]	1.5	2.3	10,180	2005	13.5	13.8
49 Bahamas	5.5[e]	5.3[f]	17,497[e]	18,380[f]	1.3[b]	0.4[b]	19,162[b]	2000	2.0	1.6
50 Seychelles	0.7	1.4	8,209	16,106	2.6	1.5	18,872	2000	2.5	0.9
51 Cuba	..	..	..	..	..	3.5[b]	..	..	..	..
52 Mexico	768.4	1,108.3	7,454	10,751	1.0	1.5	10,751	2005	14.8	4.0
53 Bulgaria	26.6	69.9	3,443	9,032	0.7[b]	1.5	9,032[b]	2005	67.6	5.0

Human development indicators

TABLE 14

Economic performance

		GDP		GDP per capita							
		US$ billions 2005	PPP US$ billions 2005	US$ 2005	2005 PPP US$[a] 2005	Annual growth rate (%) 1975–2005	Annual growth rate (%) 1990–2005	Highest value during 1975–2005 2005 PPP US$[a]	Year of highest value	Average annual change in consumer price index (%) 1990–2005	Average annual change in consumer price index (%) 2004–05
HDI rank											
54	Saint Kitts and Nevis	0.5	0.6[c]	9,438	13,307[c]	4.9[b]	2.9	13,307[b]	2004	3.0	1.8
55	Tonga	0.2	0.8[d]	2,090	8,177[d]	1.8[b]	1.9	8,177[b]	2005	5.2	8.3
56	Libyan Arab Jamahiriya	38.8	..	6,621	..	2.5[b]	..	..	..	1.9	..
57	Antigua and Barbuda	0.9	1.0[c]	10,578	12,500[c]	3.7[b]	1.5	12,500[b]	2004	..	..
58	Oman	24.3[c]	38.4[c]	9,584[c]	15,602[c]	2.4[b]	1.8	15,602[b]	2004	0.1	1.2
59	Trinidad and Tobago	14.4	19.1	11,000	14,603	0.6	4.3	14,603	2005	5.1	6.9
60	Romania	98.6	196.0	4,556	9,060	-0.3[b]	1.6	9,060[b]	2005	66.5	9.0
61	Saudi Arabia	309.8	363.2[d]	13,399	15,711[d]	-2.0	0.1	27,686	1977	0.4	0.7
62	Panama	15.5	24.6	4,786	7,605	1.0	2.2	7,605	2005	1.0	3.3
63	Malaysia	130.3	275.8	5,142	10,882	3.9	3.3	10,882	2005	2.9	3.0
64	Belarus	29.6	77.4	3,024	7,918	2.2[b]	2.2	7,918[b]	2005	144.6	10.3
65	Mauritius	6.3	15.8	5,059	12,715	4.4[b]	3.8	12,715[b]	2005	5.8	4.9
66	Bosnia and Herzegovina	9.9	..	2,546	..	..	12.7[b]	..	..	..	..
67	Russian Federation	763.7	1,552.0	5,336	10,845	-0.7[b]	-0.1	11,947[b]	1989	53.5	12.7
68	Albania	8.4	16.6	2,678	5,316	0.9[b]	5.2	5,316[b]	2005	15.6	2.4
69	Macedonia (TFYR)	5.8	14.6	2,835	7,200	-0.1[b]	-0.1	7,850[b]	1990	5.7	(.)
70	Brazil	796.1	1,566.3	4,271	8,402	0.7	1.1	8,402	2005	86.0	6.9
MEDIUM HUMAN DEVELOPMENT											
71	Dominica	0.3	0.4[c]	3,938	6,393[c]	3.1[b]	1.3	6,393[b]	2004	1.6	2.2
72	Saint Lucia	0.8	1.1[c]	5,007	6,707[c]	3.6[b]	0.9	6,707[b]	2004	2.7	3.9
73	Kazakhstan	57.1	119.0	3,772	7,857	2.0[b]	2.0	7,857[b]	2005	29.7	7.6
74	Venezuela (Bolivarian Republic of)	140.2	176.3[d]	5,275	6,632	-1.0	-1.0	8,756	1977	37.6	16.0
75	Colombia	122.3	333.1[d]	2,682	7,304[d]	1.4	0.6	7,304	2005	15.2	5.0
76	Ukraine	82.9	322.4	1,761	6,848	-3.8[b]	-2.4	10,587[b]	1989	63.9	13.5
77	Samoa	0.4	1.1	2,184	6,170	1.4[b]	2.5	6,170[b]	2005	4.0	1.8
78	Thailand	176.6	557.4	2,750	8,677	4.9	2.7	8,677	2005	3.7	4.5
79	Dominican Republic	29.5	73.1[d]	3,317	8,217[d]	2.1	3.9	8,217	2005	10.5	4.2
80	Belize	1.1	2.1	3,786	7,109	3.1	2.3	7,120	2004	1.8	3.6
81	China	2,234.3	8,814.9[g]	1,713	6,757[g]	8.4	8.8	6,757	2005	5.1	1.8
82	Grenada	0.5	0.8[c]	4,451	7,843[c]	3.4[b]	2.5	8,264[b]	2003	2.0	..
83	Armenia	4.9	14.9	1,625	4,945	4.4[b]	4.4	4,945[b]	2005	27.3	0.6
84	Turkey	362.5	605.9	5,030	8,407	1.8	1.7	8,407	2005	64.2	8.2
85	Suriname	1.3	3.5	2,986	7,722	-0.5	1.1	8,634	1978	60.7	..
86	Jordan	12.7	30.3	2,323	5,530	0.5	1.6	5,613	1986	2.8	3.5
87	Peru	79.4	168.9	2,838	6,039	-0.3	2.2	6,097	1981	15.0	1.6
88	Lebanon	21.9	20.0	6,135	5,584	3.2[b]	2.8	5,586[b]	2004	..	..
89	Ecuador	36.5	57.4	2,758	4,341	0.3	0.8	4,341	2005	34.1	2.4
90	Philippines	99.0	426.7	1,192	5,137	0.4	1.6	5,137	2005	6.6	7.6
91	Tunisia	28.7	84.0	2,860	8,371	2.3	3.3	8,371	2005	3.6	2.0
92	Fiji	2.7	5.1	3,219	6,049	0.9[b]	1.4[b]	6,056[b]	2004	3.1	2.4
93	Saint Vincent and the Grenadines	0.4	0.8	3,612	6,568	3.2	1.6	6,568	2005	1.8	3.7
94	Iran (Islamic Republic of)	189.8	543.8	2,781	7,968	-0.2	2.3	9,311	1976	21.3	13.4
95	Paraguay	7.3	27.4[d]	1,242	4,642[d]	0.5	-0.6	5,430	1981	11.1	6.8
96	Georgia	6.4	15.1	1,429	3,365	-3.9	0.2	6,884	1985	12.8	8.2
97	Guyana	0.8	3.4[d]	1,048	4,508[d]	0.9	3.2	4,618	2004	5.5	6.3
98	Azerbaijan	12.6	42.1	1,498	5,016	(.)[b]	(.)	5,310[b]	1990	66.4	9.5
99	Sri Lanka	23.5	90.2	1,196	4,595	3.2	3.7	4,595	2005	9.5	11.6
100	Maldives	0.8	..	2,326	..	..	3.8[b]	..	..	4.3	3.3
101	Jamaica	9.6	11.4	3,607	4,291	1.0	0.7	4,291	2005	16.6	15.3
102	Cape Verde	1.0	2.9[d]	1,940	5,803[d]	2.9[b]	3.4	5,803[b]	2005	3.9	0.4
103	El Salvador	17.0	36.2[d]	2,467	5,255[d]	0.3	1.6	5,745	1978	5.9	4.7
104	Algeria	102.3	232.0[d]	3,112	7,062[d]	0.1	1.1	7,062	2005	10.7	1.6
105	Viet Nam	52.4	255.3	631	3,071	5.2[b]	5.9	3,071[b]	2005	3.3	8.3
106	Occupied Palestinian Territories	4.0	..	1,107	..	..	-2.9[b]	..	..	..	..

TABLE 14

HDI rank	GDP US$ billions 2005	GDP PPP US$ billions 2005	GDP per capita US$ 2005	GDP per capita 2005 PPP US$[a] 2005	Annual growth rate (%) 1975–2005	Annual growth rate (%) 1990–2005	Highest value during 1975–2005 2005 PPP US$[a]	Year of highest value	Average annual change in consumer price index (%) 1990–2005	Average annual change in consumer price index (%) 2004–05
107 Indonesia	287.2	847.6	1,302	3,843	3.9	2.1	3,843	2005	13.3	10.5
108 Syrian Arab Republic	26.3	72.5	1,382	3,808	0.9	1.4	3,808	2005	4.9	..
109 Turkmenistan	8.1	15.4 [h]	1,669	3,838 [h]	..	-6.8 [b]	6,752 [b]	1988	..	..
110 Nicaragua	4.9	18.9 [d]	954	3,674 [d]	-2.1	1.8	7,187	1977	18.9	9.4
111 Moldova	2.9	8.8	694	2,100	-4.4 [b]	-3.5	4,168 [b]	1989	16.5	13.1
112 Egypt	89.4	321.1	1,207	4,337	2.8	2.4	4,337	2005	6.6	4.9
113 Uzbekistan	14.0	54.0	533	2,063	-0.4 [b]	0.3	2,080 [b]	1989	..	..
114 Mongolia	1.9	5.4	736	2,107	1.2 [b]	2.2	2,107 [b]	2005	19.2	8.9
115 Honduras	8.3	24.7 [d]	1,151	3,430 [d]	0.2	0.5	3,430	2005	15.0	8.8
116 Kyrgyzstan	2.4	9.9	475	1,927	-2.3 [b]	-1.3	2,806 [b]	1990	13.2	4.4
117 Bolivia	9.3	25.9	1,017	2,819	-0.2	1.3	3,025	1977	6.3	5.4
118 Guatemala	31.7	57.6 [d]	2,517	4,568 [d]	0.4	1.3	4,568	2005	8.6	8.4
119 Gabon	8.1	9.6	5,821	6,954	-1.4	-0.4	13,812	1976	3.0	(.)
120 Vanuatu	..	..	..	3,225	0.1 [b]	..	3,833 [b]	1984	..	..
121 South Africa	239.5	520.9 [d]	5,109	11,110 [d]	-0.3	0.6	11,617	1981	7.4	3.4
122 Tajikistan	2.3	8.8	355	1,356	-6.3 [b]	-4.0	3,150 [b]	1988	..	..
123 Sao Tome and Principe	0.1	0.3	451	2,178	0.3 [b]	0.5	2,178 [b]	2005	..	..
124 Botswana	10.3	21.9	5,846	12,387	5.9	4.8	12,387	2005	7.9	8.6
125 Namibia	6.1	15.4 [d]	3,016	7,586 [d]	0.1 [b]	1.4	7,586 [b]	2005	..	2.3
126 Morocco	51.6	137.4	1,711	4,555	1.4	1.5	4,555	2005	2.8	1.0
127 Equatorial Guinea	3.2	3.8 [c,d]	6,416	7,874 [c,d]	11.7 [b]	16.6	7,874 [b]	2004	7.6	..
128 India	805.7	3,779.0 [d]	736	3,452 [d]	3.4	4.2	3,452	2005	7.2	4.2
129 Solomon Islands	0.3	1.0 [d]	624	2,031 [d]	1.1	-2.4	2,804	1996	9.6	7.2
130 Lao People's Democratic Republic	2.9	12.1	485	2,039	3.4 [b]	3.8	2,039 [b]	2005	28.0	7.2
131 Cambodia	6.2	38.4 [d]	440	2,727 [d]	..	5.5 [b]	2,727 [b]	2005	3.9	5.7
132 Myanmar	..	..	..	..	2.6 [b]	6.6 [b]	..	..	25.2	9.4
133 Bhutan	0.8	..	1,325	..	5.4 [b]	5.6 [b]	..	..	7.0	5.3
134 Comoros	0.4	1.2 [d]	645	1,993 [d]	-0.6 [b]	-0.4	2,272 [b]	1984	..	..
135 Ghana	10.7	54.8 [d]	485	2,480 [d]	0.7	2.0	2,480	2005	25.6	15.1
136 Pakistan	110.7	369.2	711	2,370	2.5	1.3	2,370	2005	7.5	9.1
137 Mauritania	1.9	6.9 [d]	603	2,234 [d]	-0.1	0.3	2,338	1976	5.8	12.1
138 Lesotho	1.5	6.0 [d]	808	3,335 [d]	2.7	2.3	3,335	2005	8.5	3.4
139 Congo	5.1	5.0	1,273	1,262	-0.1	-1.0	1,758	1984	6.4	5.3
140 Bangladesh	60.0	291.2	423	2,053	2.0	2.9	2,053	2005	5.1	7.0
141 Swaziland	2.7	5.5	2,414	4,824	1.6	0.2	4,824	2005	8.7	4.8
142 Nepal	7.4	42.1	272	1,550	2.0	2.0	1,550	2005	6.8	6.8
143 Madagascar	5.0	17.2	271	923	-1.6	-0.7	1,450	1975	14.7	18.5
144 Cameroon	16.9	37.5	1,034	2,299	-0.4	0.6	3,175	1986	4.7	2.0
145 Papua New Guinea	4.9	15.1 [d]	840	2,563 [d]	0.5	0.2	2,986	1994	10.1	1.7
146 Haiti	4.3	14.2 [d]	500	1,663 [d]	-2.2	-2.0	3,151	1980	19.6	15.7
147 Sudan	27.5	75.5 [d]	760	2,083 [d]	1.3	3.5	2,083	2005	41.8	8.5
148 Kenya	18.7	42.5	547	1,240	0.1	-0.1	1,263	1990	11.6	10.3
149 Djibouti	0.7	1.7 [d]	894	2,178 [d]	-2.7 [b]	-2.7	3,200 [b]	1990	..	..
150 Timor-Leste	0.3	..	358	..	..	..	..	..	..	..
151 Zimbabwe	3.4	26.5	259	2,038	-0.5	-2.1	3,228	1998	36.1	..
152 Togo	2.2	9.3 [d]	358	1,506 [d]	-1.1	(.)	2,133	1980	5.7	6.8
153 Yemen	15.1	19.5	718	930	1.5 [b]	1.5	943 [b]	2002	20.8	..
154 Uganda	8.7	41.9 [d]	303	1,454 [d]	2.4 [b]	3.2	1,454 [b]	2005	7.1	8.2
155 Gambia	0.5	2.9 [d]	304	1,921 [d]	-0.1	0.1	1,932	1984	5.0	3.2
LOW HUMAN DEVELOPMENT										
156 Senegal	8.2	20.9	707	1,792	(.)	1.2	1,792	2005	3.7	1.7
157 Eritrea	1.0	4.9 [d]	220	1,109 [d]	..	0.3 [b]	1,435 [b]	1997	..	..
158 Nigeria	99.0	148.3	752	1,128	-0.1	0.8	1,177	1977	23.5	13.5
159 Tanzania (United Republic of)	12.1	28.5	316	744	1.4 [b]	1.7	744 [b]	2005	13.8	8.6

Human development indicators

TABLE

14

Economic performance

	GDP		GDP per capita						Average annual change in consumer price index (%)	
HDI rank	US$ billions 2005	PPP US$ billions 2005	US$ 2005	2005 PPP US$[a] 2005	Annual growth rate (%) 1975–2005	1990–2005	Highest value during 1975–2005 2005 PPP US$ [a]	Year of highest value	1990–2005	2004–05
160 Guinea	3.3	21.8	350	2,316	1.0 [b]	1.2	2,316 [b]	2005	..	..
161 Rwanda	2.2	10.9 [d]	238	1,206 [d]	-0.3	0.1	1,358	1983	11.2	9.1
162 Angola	32.8	37.2 [d]	2,058	2,335 [d]	-0.6 [b]	1.5	2,335 [b]	2005	393.3	23.0
163 Benin	4.3	9.6	508	1,141	0.4	1.4	1,141	2005	5.6	5.4
164 Malawi	2.1	8.6	161	667	-0.2	1.0	719	1979	28.4	15.4
165 Zambia	7.3	11.9	623	1,023	-1.8	-0.3	1,559	1976	40.0	18.3
166 Côte d'Ivoire	16.3	29.9	900	1,648	-2.1	-0.5	3,195	1978	5.4	3.9
167 Burundi	0.8	5.3 [d]	106	699 [d]	-1.0	-2.8	1,047	1991	13.8	13.0
168 Congo (Democratic Republic of the)	7.1	41.1 [d]	123	714 [d]	-4.9	-5.2	2,488	1975	424.3	21.3
169 Ethiopia	11.2	75.1 [d]	157	1,055 [d]	-0.2 [b]	1.5	1,055 [b]	2005	4.2	11.6
170 Chad	5.5	13.9 [d]	561	1,427 [d]	0.5	1.7	1,427	2005	5.3	7.9
171 Central African Republic	1.4	4.9 [d]	339	1,224 [d]	-1.5	-0.6	1,935	1977	3.9	2.9
172 Mozambique	6.6	24.6 [d]	335	1,242 [d]	2.3 [b]	4.3	1,242 [b]	2005	22.1	7.2
173 Mali	5.3	14.0	392	1,033	0.2	2.2	1,033	2005	3.8	6.4
174 Niger	3.4	10.9 [d]	244	781 [d]	-1.7	-0.5	1,293	1979	4.4	7.8
175 Guinea-Bissau	0.3	1.3 [d]	190	827 [d]	-0.6	-2.6	1,264	1997	20.2	3.3
176 Burkina Faso	5.2	16.0 [d]	391	1,213 [d]	0.9	1.3	1,213	2005	4.1	6.4
177 Sierra Leone	1.2	4.5	216	806	-2.1	-1.4	1,111	1982	19.7	12.1
Developing countries	9,812.5 T	26,732.3 T	1,939	5,282	2.5	3.1	..	..	..	..
Least developed countries	306.2 T	1,081.8 T	424	1,499	0.9	1.8	..	..	..	..
Arab States	1,043.4 T	1,915.2 T	3,659	6,716	0.7	2.3	..	..	..	..
East Asia and the Pacific	4,122.5 T	12,846.6 T	2,119	6,604	6.1	5.8	..	..	..	..
Latin America and the Caribbean	2,469.5 T	4,639.2 T	4,480	8,417	0.7	1.2	..	..	..	..
South Asia	1,206.1 T	5,152.2 T	800	3,416	2.6	3.4	..	..	..	..
Sub-Saharan Africa	589.9 T	1,395.6 T	845	1,998	-0.5	0.5	..	..	..	..
Central and Eastern Europe and the CIS	1,873.0 T	3,827.2 T	4,662	9,527	1.4	1.4	..	..	..	..
OECD	34,851.2 T	34,076.8 T	29,860	29,197	2.0	1.8	..	..	..	..
High-income OECD	32,404.5 T	30,711.7 T	35,696	33,831	2.1	1.8	..	..	..	..
High human development	37,978.4 T	39,633.4 T	22,984	23,986	1.9	1.8	..	..	..	..
Medium human development	5,881.2 T	20,312.6 T	1,412	4,876	3.2	4.0	..	..	..	..
Low human development	236.4 T	544.2 T	483	1,112	-0.7	0.6	..	..	..	..
High income	34,338.1 T	32,680.7 T	34,759	33,082	2.1	1.8	..	..	..	..
Middle income	8,552.0 T	22,586.3 T	2,808	7,416	2.1	3.0	..	..	..	..
Low income	1,416.2 T	5,879.1 T	610	2,531	2.2	2.9	..	..	..	..
World	44,155.7 T	60,597.3 T	6,954	9,543	1.4	1.5	..	..	..	..

NOTES

a. GDP values expressed in 2005 constant prices.
b. Data refer to a period shorter than that specified.
c. Data refer to 2004.
d. World Bank estimates based on regression.
e. Data refer to 2003.
f. Data refer to 2002.
g. Estimate based on a bilateral comparison between China and the United States (Ruoen and Kai 1995).
h. Data refer to 2000.
i. Data refer to 2001.

SOURCES

Columns 1–4: World Bank 2007b; aggregates calculated for HDRO by the World Bank.
Columns 5 and 6: World Bank 2007b; aggregates calculated for HDRO by the World Bank using the least squares method.
Columns 7 and 8: calculated based on GDP per capita (PPP US$) time series from World Bank 2007b.
Columns 9 and 10: calculated based on data on the consumer price index from World Bank 2007b.

TABLE 15

... to have access to the resources needed for a decent standard of living ...

Inequality in income or expenditure

HDI rank	Survey year	MDG Share of income or expenditure (%)				Inequality measures		Gini index [b]
		Poorest 10%	Poorest 20%	Richest 20%	Richest 10%	Richest 10% to poorest 10% [a]	Richest 20% to poorest 20% [a]	
HIGH HUMAN DEVELOPMENT								
1 Iceland	..	..	..	..	..	..	..	..
2 Norway	2000 [c]	3.9	9.6	37.2	23.4	6.1	3.9	25.8
3 Australia	1994 [c]	2.0	5.9	41.3	25.4	12.5	7.0	35.2
4 Canada	2000 [c]	2.6	7.2	39.9	24.8	9.4	5.5	32.6
5 Ireland	2000 [c]	2.9	7.4	42.0	27.2	9.4	5.6	34.3
6 Sweden	2000 [c]	3.6	9.1	36.6	22.2	6.2	4.0	25.0
7 Switzerland	2000 [c]	2.9	7.6	41.3	25.9	9.0	5.5	33.7
8 Japan	1993 [c]	4.8	10.6	35.7	21.7	4.5	3.4	24.9
9 Netherlands	1999 [c]	2.5	7.6	38.7	22.9	9.2	5.1	30.9
10 France	1995 [c]	2.8	7.2	40.2	25.1	9.1	5.6	32.7
11 Finland	2000 [c]	4.0	9.6	36.7	22.6	5.6	3.8	26.9
12 United States	2000 [c]	1.9	5.4	45.8	29.9	15.9	8.4	40.8
13 Spain	2000 [c]	2.6	7.0	42.0	26.6	10.3	6.0	34.7
14 Denmark	1997 [c]	2.6	8.3	35.8	21.3	8.1	4.3	24.7
15 Austria	2000 [c]	3.3	8.6	37.8	23.0	6.9	4.4	29.1
16 United Kingdom	1999 [c]	2.1	6.1	44.0	28.5	13.8	7.2	36.0
17 Belgium	2000 [c]	3.4	8.5	41.4	28.1	8.2	4.9	33.0
18 Luxembourg	..	..	..	..	..	..	..	..
19 New Zealand	1997 [c]	2.2	6.4	43.8	27.8	12.5	6.8	36.2
20 Italy	2000 [c]	2.3	6.5	42.0	26.8	11.6	6.5	36.0
21 Hong Kong, China (SAR)	1996 [c]	2.0	5.3	50.7	34.9	17.8	9.7	43.4
22 Germany	2000 [c]	3.2	8.5	36.9	22.1	6.9	4.3	28.3
23 Israel	2001 [c]	2.1	5.7	44.9	28.8	13.4	7.9	39.2
24 Greece	2000 [c]	2.5	6.7	41.5	26.0	10.2	6.2	34.3
25 Singapore	1998 [c]	1.9	5.0	49.0	32.8	17.7	9.7	42.5
26 Korea (Republic of)	1998 [c]	2.9	7.9	37.5	22.5	7.8	4.7	31.6
27 Slovenia	1998 [d]	3.6	9.1	35.7	21.4	5.9	3.9	28.4
28 Cyprus	..	..	..	..	..	..	..	..
29 Portugal	1997 [c]	2.0	5.8	45.9	29.8	15.0	8.0	38.5
30 Brunei Darussalam	..	..	..	..	..	..	..	..
31 Barbados	..	..	..	..	..	..	..	..
32 Czech Republic	1996 [c]	4.3	10.3	35.9	22.4	5.2	3.5	25.4
33 Kuwait	..	..	..	..	..	..	..	..
34 Malta	..	..	..	..	..	..	..	..
35 Qatar	..	..	..	..	..	..	..	..
36 Hungary	2002 [d]	4.0	9.5	36.5	22.2	5.5	3.8	26.9
37 Poland	2002 [d]	3.1	7.5	42.2	27.0	8.8	5.6	34.5
38 Argentina [e]	2004 [c]	0.9	3.1	55.4	38.2	40.9	17.8	51.3
39 United Arab Emirates	..	..	..	..	..	..	..	..
40 Chile	2003 [c]	1.4	3.8	60.0	45.0	33.0	15.7	54.9
41 Bahrain	..	..	..	..	..	..	..	..
42 Slovakia	1996 [c]	3.1	8.8	34.8	20.9	6.7	4.0	25.8
43 Lithuania	2003 [d]	2.7	6.8	43.2	27.7	10.4	6.3	36.0
44 Estonia	2003 [d]	2.5	6.7	42.8	27.6	10.8	6.4	35.8
45 Latvia	2003 [d]	2.5	6.6	44.7	29.1	11.6	6.8	37.7
46 Uruguay [e]	2003 [c]	1.9	5.0	50.5	34.0	17.9	10.2	44.9
47 Croatia	2001 [d]	3.4	8.3	39.6	24.5	7.3	4.8	29.0
48 Costa Rica	2003 [c]	1.0	3.5	54.1	37.4	37.8	15.6	49.8
49 Bahamas	..	..	..	..	..	..	..	..
50 Seychelles	..	..	..	..	..	..	..	..
51 Cuba	..	..	..	..	..	..	..	..
52 Mexico	2004 [d]	1.6	4.3	55.1	39.4	24.6	12.8	46.1
53 Bulgaria	2003 [d]	3.4	8.7	38.3	23.9	7.0	4.4	29.2

TABLE 15

Inequality in income or expenditure

		MDG Share of income or expenditure (%)				Inequality measures		
HDI rank	Survey year	Poorest 10%	Poorest 20%	Richest 20%	Richest 10%	Richest 10% to poorest 10% [a]	Richest 20% to poorest 20% [a]	Gini index [b]
54 Saint Kitts and Nevis	..	..	..	..	..	..	..	..
55 Tonga	..	..	..	..	..	..	..	..
56 Libyan Arab Jamahiriya	..	..	..	..	..	..	..	..
57 Antigua and Barbuda	..	..	..	..	..	..	..	..
58 Oman	..	..	..	..	..	..	..	..
59 Trinidad and Tobago	1992 [c]	2.2	5.9	44.9	28.8	12.9	7.6	38.9
60 Romania	2003 [d]	3.3	8.1	39.2	24.4	7.5	4.9	31.0
61 Saudi Arabia	..	..	..	..	..	..	..	..
62 Panama	2003 [c]	0.7	2.5	59.9	43.0	57.5	23.9	56.1
63 Malaysia	1997 [c]	1.7	4.4	54.3	38.4	22.1	12.4	49.2
64 Belarus	2002 [d]	3.4	8.5	38.3	23.5	6.9	4.5	29.7
65 Mauritius	..	..	..	..	..	..	..	..
66 Bosnia and Herzegovina	2001 [d]	3.9	9.5	35.8	21.4	5.4	3.8	26.2
67 Russian Federation	2002 [d]	2.4	6.1	46.6	30.6	12.7	7.6	39.9
68 Albania	2004 [d]	3.4	8.2	39.5	24.4	7.2	4.8	31.1
69 Macedonia (TFYR)	2003 [d]	2.4	6.1	45.5	29.6	12.5	7.5	39.0
70 Brazil	2004 [c]	0.9	2.8	61.1	44.8	51.3	21.8	57.0
MEDIUM HUMAN DEVELOPMENT								
71 Dominica	..	..	..	..	..	..	..	..
72 Saint Lucia	..	..	..	..	..	..	..	..
73 Kazakhstan	2003 [d]	3.0	7.4	41.5	25.9	8.5	5.6	33.9
74 Venezuela (Bolivarian Republic of)	2003	0.7	3.3	52.1	35.2	48.3	16.0	48.2
75 Colombia	2003 [c]	0.7	2.5	62.7	46.9	63.8	25.3	58.6
76 Ukraine	2003 [d]	3.9	9.2	37.5	23.0	5.9	4.1	28.1
77 Samoa	..	..	..	..	..	..	..	..
78 Thailand	2002 [d]	2.7	6.3	49.0	33.4	12.6	7.7	42.0
79 Dominican Republic	2004 [c]	1.4	4.0	56.7	41.1	28.5	14.3	51.6
80 Belize	..	..	..	..	..	..	..	..
81 China	2004 [c]	1.6	4.3	51.9	34.9	21.6	12.2	46.9
82 Grenada	..	..	..	..	..	..	..	..
83 Armenia	2003 [d]	3.6	8.5	42.8	29.0	8.0	5.0	33.8
84 Turkey	2003 [d]	2.0	5.3	49.7	34.1	16.8	9.3	43.6
85 Suriname	..	..	..	..	..	..	..	..
86 Jordan	2002-03 [d]	2.7	6.7	46.3	30.6	11.3	6.9	38.8
87 Peru	2003 [c]	1.3	3.7	56.7	40.9	30.4	15.2	52.0
88 Lebanon	..	..	..	..	..	..	..	..
89 Ecuador	1998 [d]	0.9	3.3	58.0	41.6	44.9	17.3	53.6
90 Philippines	2003 [d]	2.2	5.4	50.6	34.2	15.5	9.3	44.5
91 Tunisia	2000 [d]	2.3	6.0	47.3	31.5	13.4	7.9	39.8
92 Fiji	..	..	..	..	..	..	..	..
93 Saint Vincent and the Grenadines	..	..	..	..	..	..	..	..
94 Iran (Islamic Republic of)	1998 [d]	2.0	5.1	49.9	33.7	17.2	9.7	43.0
95 Paraguay	2003 [c]	0.7	2.4	61.9	46.1	65.4	25.7	58.4
96 Georgia	2003 [d]	2.0	5.6	46.4	30.3	15.4	8.3	40.4
97 Guyana	..	..	..	..	..	..	..	..
98 Azerbaijan	2001 [d]	3.1	7.4	44.5	29.5	9.7	6.0	36.5
99 Sri Lanka	2002 [d]	3.0	7.0	48.0	32.7	11.1	6.9	40.2
100 Maldives	..	..	..	..	..	..	..	..
101 Jamaica	2004 [d]	2.1	5.3	51.6	35.8	17.3	9.8	45.5
102 Cape Verde	..	..	..	..	..	..	..	..
103 El Salvador	2002 [c]	0.7	2.7	55.9	38.8	57.5	20.9	52.4
104 Algeria	1995 [d]	2.8	7.0	42.6	26.8	9.6	6.1	35.3
105 Viet Nam	2004 [d]	4.2	9.0	44.3	28.8	6.9	4.9	34.4
106 Occupied Palestinian Territories	..	..	..	..	..	..	..	..

Human development indicators

TABLE 15

HDI rank	Survey year	MDG Share of income or expenditure (%)				Inequality measures		
		Poorest 10%	Poorest 20%	Richest 20%	Richest 10%	Richest 10% to poorest 10% [a]	Richest 20% to poorest 20% [a]	Gini index [b]
107 Indonesia	2002 [d]	3.6	8.4	43.3	28.5	7.8	5.2	34.3
108 Syrian Arab Republic	..	..	..	..	..	..	..	..
109 Turkmenistan	1998 [d]	2.6	6.1	47.5	31.7	12.3	7.7	40.8
110 Nicaragua	2001 [d]	2.2	5.6	49.3	33.8	15.5	8.8	43.1
111 Moldova	2003 [d]	3.2	7.8	41.4	26.4	8.2	5.3	33.2
112 Egypt	1999-00 [d]	3.7	8.6	43.6	29.5	8.0	5.1	34.4
113 Uzbekistan	2003 [d]	2.8	7.2	44.7	29.6	10.6	6.2	36.8
114 Mongolia	2002 [d]	3.0	7.5	40.5	24.6	8.2	5.4	32.8
115 Honduras	2003 [c]	1.2	3.4	58.3	42.2	34.2	17.2	53.8
116 Kyrgyzstan	2003 [d]	3.8	8.9	39.4	24.3	6.4	4.4	30.3
117 Bolivia	2002 [c]	0.3	1.5	63.0	47.2	168.1	42.3	60.1
118 Guatemala	2002 [c]	0.9	2.9	59.5	43.4	48.2	20.3	55.1
119 Gabon	..	..	..	..	..	..	..	..
120 Vanuatu	..	..	..	..	..	..	..	..
121 South Africa	2000 [d]	1.4	3.5	62.2	44.7	33.1	17.9	57.8
122 Tajikistan	2003 [d]	3.3	7.9	40.8	25.6	7.8	5.2	32.6
123 Sao Tome and Principe	..	..	..	..	..	..	..	..
124 Botswana	1993 [d]	1.2	3.2	65.1	51.0	43.0	20.4	60.5
125 Namibia	1993 [c]	0.5	1.4	78.7	64.5	128.8	56.1	74.3
126 Morocco	1998-99 [d]	2.6	6.5	46.6	30.9	11.7	7.2	39.5
127 Equatorial Guinea	..	..	..	..	..	..	..	..
128 India	2004-05 [d]	3.6	8.1	45.3	31.1	8.6	5.6	36.8
129 Solomon Islands	..	..	..	..	..	..	..	..
130 Lao People's Democratic Republic	2002 [d]	3.4	8.1	43.3	28.5	8.3	5.4	34.6
131 Cambodia	2004 [d]	2.9	6.8	49.6	34.8	12.2	7.3	41.7
132 Myanmar	..	..	..	..	..	..	..	..
133 Bhutan	..	..	..	..	..	..	..	..
134 Comoros	..	..	..	..	..	..	..	..
135 Ghana	1998-99 [d]	2.1	5.6	46.6	30.0	14.1	8.4	40.8
136 Pakistan	2002 [d]	4.0	9.3	40.3	26.3	6.5	4.3	30.6
137 Mauritania	2000 [d]	2.5	6.2	45.7	29.5	12.0	7.4	39.0
138 Lesotho	1995 [d]	0.5	1.5	66.5	48.3	105.0	44.2	63.2
139 Congo	..	..	..	..	..	..	..	..
140 Bangladesh	2000 [d]	3.7	8.6	42.7	27.9	7.5	4.9	33.4
141 Swaziland	2000-01 [c]	1.6	4.3	56.3	40.7	25.1	13.0	50.4
142 Nepal	2003-04 [d]	2.6	6.0	54.6	40.6	15.8	9.1	47.2
143 Madagascar	2001 [d]	1.9	4.9	53.5	36.6	19.2	11.0	47.5
144 Cameroon	2001 [d]	2.3	5.6	50.9	35.4	15.7	9.1	44.6
145 Papua New Guinea	1996 [d]	1.7	4.5	56.5	40.5	23.8	12.6	50.9
146 Haiti	2001 [c]	0.7	2.4	63.4	47.7	71.7	26.6	59.2
147 Sudan	..	..	..	..	..	..	..	..
148 Kenya	1997 [d]	2.5	6.0	49.1	33.9	13.6	8.2	42.5
149 Djibouti	..	..	..	..	..	..	..	..
150 Timor-Leste	..	..	..	..	..	..	..	..
151 Zimbabwe	1995-96 [d]	1.8	4.6	55.7	40.3	22.0	12.0	50.1
152 Togo	..	..	..	..	..	..	..	..
153 Yemen	1998 [d]	3.0	7.4	41.2	25.9	8.6	5.6	33.4
154 Uganda	2002 [d]	2.3	5.7	52.5	37.7	16.6	9.2	45.7
155 Gambia	1998 [d]	1.8	4.8	53.4	37.0	20.2	11.2	50.2
LOW HUMAN DEVELOPMENT								
156 Senegal	2001 [d]	2.7	6.6	48.4	33.4	12.3	7.4	41.3
157 Eritrea	..	..	..	..	..	..	..	..
158 Nigeria	2003 [d]	1.9	5.0	49.2	33.2	17.8	9.7	43.7
159 Tanzania (United Republic of)	2000-01 [d]	2.9	7.3	42.4	26.9	9.2	5.8	34.6

Human development indicators

TABLE

15 Inequality in income or expenditure

HDI rank	Survey year	MDG Share of income or expenditure (%)				Inequality measures		
		Poorest 10%	Poorest 20%	Richest 20%	Richest 10%	Richest 10% to poorest 10% [a]	Richest 20% to poorest 20% [a]	Gini index [b]
160 Guinea	2003 [d]	2.9	7.0	46.1	30.7	10.5	6.6	38.6
161 Rwanda	2000 [d]	2.1	5.3	53.0	38.2	18.6	9.9	46.8
162 Angola	..	..	..	..	..	..	..	..
163 Benin	2003 [d]	3.1	7.4	44.5	29.0	9.4	6.0	36.5
164 Malawi	2004-05 [d]	2.9	7.0	46.6	31.8	10.9	6.7	39.0
165 Zambia	2004 [d]	1.2	3.6	55.1	38.8	32.3	15.3	50.8
166 Côte d'Ivoire	2002 [d]	2.0	5.2	50.7	34.0	16.6	9.7	44.6
167 Burundi	1998 [d]	1.7	5.1	48.0	32.8	19.3	9.5	42.4
168 Congo (Democratic Republic of the)	..	..	..	..	..	..	..	..
169 Ethiopia	1999-00 [d]	3.9	9.1	39.4	25.5	6.6	4.3	30.0
170 Chad	..	..	..	..	..	..	..	..
171 Central African Republic	1993 [d]	0.7	2.0	65.0	47.7	69.2	32.7	61.3
172 Mozambique	2002-03 [d]	2.1	5.4	53.6	39.4	18.8	9.9	47.3
173 Mali	2001 [d]	2.4	6.1	46.6	30.2	12.5	7.6	40.1
174 Niger	1995 [d]	0.8	2.6	53.3	35.4	46.0	20.7	50.5
175 Guinea-Bissau	1993 [d]	2.1	5.2	53.4	39.3	19.0	10.3	47.0
176 Burkina Faso	2003 [d]	2.8	6.9	47.2	32.2	11.6	6.9	39.5
177 Sierra Leone	1989 [d]	0.5	1.1	63.4	43.6	87.2	57.6	62.9

NOTES

Because the underlying household surveys differ in method and in the type of data collected, the distribution data are not strictly comparable across countries.

a. Data show the ratio of the income or expenditure share of the richest group to that of the poorest. Because of rounding, results may differ from ratios calculated using the income or expenditure shares in columns 2-5.

b. A value of 0 represents absolute equality, and a value of 100 absolute inequality.

c. Data refer to income shares by percentiles of population, ranked by per capita income.

d. Data refer to expenditure shares by percentiles of population, ranked by per capita expenditure.

e. Data refer to urban areas only.

SOURCES

Columns 1–5 and 8: World Bank 2007b.
Columns 6 and 7: calculated based on data on income or expenditure from World Bank 2007b.

TABLE 16

. . . to have access to the resources needed for a decent standard of living . . .

Structure of trade

HDI rank	Imports of goods and services (% of GDP)		Exports of goods and services (% of GDP)		Primary exports [a] (% of merchandise exports)		Manufactured exports (% of merchandise exports)		High-technology exports (% of manufactured exports)		Terms of trade (2000=100) [b]
	1990	2005	1990	2005	1990	2005	1990	2005	1990	2005	2004–05 [c]
HIGH HUMAN DEVELOPMENT											
1 Iceland	32	45	34	32	91	80	8	19	10.0	27.1	..
2 Norway	34	28	40	45	67	80	32	17	12.4	17.3	122
3 Australia	16	21 d	16	18 d	73	67	27	25	11.9	12.7	131
4 Canada	26	34 d	26	39 d	36	37	59	58	13.7	14.4	111
5 Ireland	52	68 d	57	83 d	26	10	70	86	..	..	99
6 Sweden	30	41	30	49	16	15	83	79	13.3	16.7	90
7 Switzerland	34	39 d	36	46 d	6	6	94	93	12.1	21.7	..
8 Japan	10	11 d	10	13 d	3	4	96	92	23.8	22.5	83
9 Netherlands	52	63	56	71	37	31	59	68	16.4	30.1	100
10 France	23	27	21	26	23	18	77	80	16.1	20.0	111
11 Finland	24	35	22	39	17	15	83	84	7.6	25.2	86
12 United States	11	15 d	10	10 d	21	15	75	82	33.7	31.8	97
13 Spain	19	31	16	25	24	22	75	77	6.4	7.1	102
14 Denmark	33	44	37	49	35	31	60	65	15.2	21.6	104
15 Austria	37	48	38	53	12	16	88	80	7.8	12.8	102
16 United Kingdom	27	30	24	26	19	18	79	77	23.6	28.0	105
17 Belgium	68	85	69	87	19 e	19	77 e	79	..	8.7	99
18 Luxembourg	88	136	102	158	..	14	..	82	..	11.8	..
19 New Zealand	27	30 d	27	29 d	72	66	26	31	9.5	14.2	112
20 Italy	19	26	19	26	11	12	88	85	7.6	7.8	101
21 Hong Kong, China (SAR)	122	185	131	198	7	3	92	96	12.1 f	33.9	98
22 Germany	25	35	25	40	10	10	89	83	11.1	16.9	101
23 Israel	45	51	35	46	13	4	87	83	10.4	13.9	95
24 Greece	28	28	18	21	46	41	54	56	2.2	10.2	95
25 Singapore	..	213	..	243	27	15	72	81	39.7	56.6	87
26 Korea (Republic of)	29	40	28	42	6	9	94	91	17.8	32.3	77
27 Slovenia	79	65	91	65	14 f	12	86 f	88	3.2 f	4.6	..
28 Cyprus	57	..	52	..	42	36	58	63	8.2	46.3	..
29 Portugal	38	37	31	29	19	16	80	75	4.4	8.7 d	102 d
30 Brunei Darussalam	..	..	..	..	97	88 d	3	12 d	..	4.9 d	..
31 Barbados	52	69	49	58	55	56	43	43	20.2 f	14.8 d	..
32 Czech Republic	43	70	45	72	..	10	..	88	..	12.9 d	..
33 Kuwait	58	30	45	68	94	93 d	6	7 d	3.5	1.0 d	..
34 Malta	99	82	85	71	7	4	93	95	43.6	53.5	85
35 Qatar	..	33	..	68	82	84	18	7	0.4 f	1.2	..
36 Hungary	29	69	31	66	35	11	63	84	4.0 f	24.5	97
37 Poland	22	37	29	37	36	20	58	78	3.7 f	3.8	107
38 Argentina	5	19	10	25	71	68	29	31	7.1 f	6.6	107
39 United Arab Emirates	41	76	66	94	88 f	76 d	12 f	24 d	(.) f	10.2 d	..
40 Chile	31	34	34	42	87	84	11	14	4.6	4.8 d	115
41 Bahrain	95	64 d	116	82 d	54	93	45	7	..	2.0	..
42 Slovakia	36	83	27	79	..	16	..	84	..	7.3	..
43 Lithuania	61	65	52	58	38 f	44	59 f	56	0.4 f	6.1	..
44 Estonia	54 f	90	60 f	84	..	22	..	69	..	17.6	..
45 Latvia	49	62	48	48	..	40	..	57	..	5.3	..
46 Uruguay	18	28	24	30	61	68	39	32	..	2.4 d	108
47 Croatia	86 f	56	78 f	47	32 f	32	68 f	68	5.3 f	11.5	..
48 Costa Rica	36	54	30	48	66	34	27	66	..	38.0	102
49 Bahamas	..	..	..	..	81 f	58 d	19 f	42 d	..	4.9 d	..
50 Seychelles	67	121	62	110	74	93	26	6	59.4 f	18.2	99 d
51 Cuba	..	..	..	..	..	81 d	..	19 d	..	29.1 d	..
52 Mexico	20	32	19	30	56	23	43	77	8.3	19.6	98
53 Bulgaria	37	77	33	61	..	37	..	59	..	4.7	..

TABLE 16

Structure of trade

HDI rank	Imports of goods and services (% of GDP)		Exports of goods and services (% of GDP)		Primary exports [a] (% of merchandise exports)		Manufactured exports (% of merchandise exports)		High-technology exports (% of manufactured exports)		Terms of trade (2000=100) [b]
	1990	2005	1990	2005	1990	2005	1990	2005	1990	2005	2004–05 [c]
54 Saint Kitts and Nevis	83	61 [d]	52	49 [d]	..	4	..	96	..	0.7 [d]	..
55 Tonga	65	44 [d]	34	10 [d]	74 [g]	93 [d]	24	5 [d]	..	0.3 [d]	..
56 Libyan Arab Jamahiriya	31	36 [d]	40	48 [d]	96 [f,g]	..	4 [f]	..	..	..	186 [d]
57 Antigua and Barbuda	87	69 [d]	89	62 [d]	..	71	..	29	..	16.1 [d]	..
58 Oman	28	43 [d]	47	57 [d]	94	89	5	6	2.1	2.2	..
59 Trinidad and Tobago	29	46 [d]	45	58 [d]	73	74	27	26	0.8 [f]	1.3	..
60 Romania	26	43	17	33	26	20	73	80	2.5	3.4	..
61 Saudi Arabia	32	26	41	61	92	90	8	9	0.7 [f]	1.3	..
62 Panama	79	72	87	69	78	91	21	9	..	0.9	94
63 Malaysia	72	100	75	123	46	24	54	75	38.2	54.7	99
64 Belarus	44	60	46	61	..	46	..	52	..	2.6	..
65 Mauritius	71	61	64	57	34	29	66	70	0.5	21.3	85
66 Bosnia and Herzegovina	..	81 ·	..	36	..	..	..	..	..	..	..
67 Russian Federation	18	22	18	35	..	60	..	19	..	8.1	..
68 Albania	23	46	15	22	..	20	..	80	..	1.0	..
69 Macedonia (TFYR)	36	62	26	45	..	28	..	72	..	1.1	..
70 Brazil	7	12	8	17	47	46	52	54	7.1	12.8	101
MEDIUM HUMAN DEVELOPMENT											
71 Dominica	81	69	55	45	65	40	35	60	..	7.2	..
72 Saint Lucia	84	70 [d]	73	60 [d]	68	63	32	36	4.5 [f]	20.1 [d]	..
73 Kazakhstan	75 [f]	45	74 [f]	54	..	84 [d]	..	16 [d]	..	2.3 [d]	..
74 Venezuela (Bolivarian Republic of)	20	21	39	41	90	91	10	9	3.9	2.7 [d]	108
75 Colombia	15	21	21	21	74	64	25	36	5.2 [f]	4.9	93
76 Ukraine	29	53	28	54	..	30	..	69	..	3.7	..
77 Samoa	..	51 [d]	..	27 [d]	90	23 [d]	10	77 [d]	..	0.1 [d]	..
78 Thailand	42	75	34	74	36	22	63	77	20.7	26.6	93
79 Dominican Republic	44	38	34	34	22 [f]	60 [d]	78 [f]	34 [d]	..	1.3 [d]	95
80 Belize	60	63	62	55	88 [g]	86 [d]	15	13 [d]	10.4 [f]	2.8 [d]	..
81 China	16	32	19	37	27	8	72	92	6.1 [f]	30.6	92
82 Grenada	63	76 [d]	42	43 [d]	66	64 [d]	34	36 [d]	..	4.7 [d]	..
83 Armenia	46	40	35	27	..	29	..	71	..	0.7	..
84 Turkey	18	34	13	27	32	17	68	82	1.2	1.5	101
85 Suriname	44	60	42	41	26	27 [d]	74	80 [d]	..	0.2 [d]	..
86 Jordan	93	93	62	52	44	28	56	72	6.8	5.2	88
87 Peru	14	19	16	25	82	83	18	17	1.6 [f]	2.6	109
88 Lebanon	100	44	18	19	..	29 [d]	..	70 [d]	..	2.4 [d]	..
89 Ecuador	32	32	33	31	98	91	2	9	0.3	7.6	108
90 Philippines	33	52	28	47	31	11	38	89	32.5 [f]	71.0	89
91 Tunisia	51	51	44	48	31	22 [d]	69	78 [d]	2.1	4.9 [d]	99
92 Fiji	67	..	62	74 [d]	64	74	35	25	12.1	3.2	..
93 Saint Vincent and the Grenadines	77	65	66	44	..	75	..	25	..	7.7 [d]	..
94 Iran (Islamic Republic of)	23	30	15	39	..	88	..	9	..	2.6 [d]	..
95 Paraguay	39	54	33	47	90 [g]	87 [d]	10	13 [d]	0.2	6.6 [d]	112 [d]
96 Georgia	46	54	40	42	..	60	..	40	..	22.6	..
97 Guyana	80	124	63	88	..	78	..	20	..	1.1	..
98 Azerbaijan	39	54	44	57	..	87	..	13	..	0.8	..
99 Sri Lanka	38	46	29	34	42	28	54	70	0.6	1.5 [d]	101 [d]
100 Maldives	..	110	..	62	..	92	..	8	..	2.1	..
101 Jamaica	52	61	48	41	30	34 [d]	70	66 [d]	9.5 [f]	0.4 [d]	..
102 Cape Verde	44	66 [d]	13	32 [d]	..	65 [d]	..	90 [d]	..	(.) [d]	91
103 El Salvador	31	45	19	27	62	40 [d]	38	60 [d]	..	4.1 [d]	91
104 Algeria	25	23	23	48	97	98 [d]	3	2 [d]	1.3 [f]	1.0 [d]	126
105 Viet Nam	45	75	36	70	..	46 [d]	..	53 [d]	..	5.6 [d]	..
106 Occupied Palestinian Territories	..	68	..	14	..	..	..	..	..	..	..

TABLE 16

HDI rank	Imports of goods and services (% of GDP)		Exports of goods and services (% of GDP)		Primary exports [a] (% of merchandise exports)		Manufactured exports (% of merchandise exports)		High-technology exports (% of manufactured exports)		Terms of trade (2000=100) [b]
	1990	2005	1990	2005	1990	2005	1990	2005	1990	2005	2004–05 [c]
107 Indonesia	24	29	25	34	65	53	35	47	1.2	16.3	104
108 Syrian Arab Republic	28	40	28	37	64	87 d	36	11 d	..	1.0 d	..
109 Turkmenistan	..	48	..	65	..	92 d	..	7 d	..	4.9 d	..
110 Nicaragua	46	58	25	28	92	89	8	11	..	5.2	91
111 Moldova	51	91	48	53	..	61	..	39	..	2.7	..
112 Egypt	33	33	20	30	57	64 d	42	31 d	..	0.6 d	107
113 Uzbekistan	48	30	29	40	..	..	..	..	..	..	..
114 Mongolia	49	84	22	76	..	79	..	21	..	0.1	..
115 Honduras	40	61	37	41	91	64	9	36	..	2.2 d	90
116 Kyrgyzstan	50	58	29	39	..	35	..	27	..	2.2	..
117 Bolivia	24	33	23	36	95	89	5	11	6.8 f	9.2 d	108
118 Guatemala	25	30	21	16	76	43	24	57	..	3.2	93
119 Gabon	31	39	46	59	..	93 d	..	7 d	..	14.5 d	125
120 Vanuatu	77	..	49	..	87 g	92 d	13	8 d	19.8	1.2 d	..
121 South Africa	19	29	24	27	29 f,h	43 h	29 f,h	57 h	6.8 f	6.6	109
122 Tajikistan	35	73	28	54	..	87 d	..	13 d	..	41.8 d	..
123 Sao Tome and Principe	72	99	14	40	..	..	..	..	..	..	137
124 Botswana	50	35	55	51	.. i	13 d,i	.. i	86 d,i	..	0.2 d	92
125 Namibia	67	45	52	46	.. i	58 d,i	.. i	41 d,i	..	2.9 d	97
126 Morocco	32	43	26	36	48	35	52	65	..	10.1	100
127 Equatorial Guinea	70	..	32	..	..	..	..	..	..	..	124
128 India	9	24	7	21	28	29	70	70	2.4	4.9 d	76
129 Solomon Islands	73	46 d	47	48 d	109 f,g	..	..	..	..	..	..
130 Lao People's Democratic Republic	25	31	12	27	..	..	..	..	..	..	..
131 Cambodia	13	74	6	65	..	3 d	..	97 d	..	0.2 d	..
132 Myanmar	5	..	3	..	89 f	..	11 f	..	3.0 f	..	102
133 Bhutan	31	55	27	27	58 f	..	42 f	..	..	..	..
134 Comoros	37	35	14	12	..	89 d	..	8 d	..	0.5 d	58
135 Ghana	26	62	17	36	92 f	88 d	8 f	12 d	2.1 f	9.3 d	123
136 Pakistan	23	20	16	15	21	18	79	82	0.4	1.6	75
137 Mauritania	61	95	46	36	..	..	..	..	..	..	95
138 Lesotho	122	88	17	48	.. i	.. i	.. i	.. i	..	..	91
139 Congo	46	55	54	82	..	..	..	..	..	..	121
140 Bangladesh	14	23	6	17	22 g	10 d	77	90 d	0.1	(.) d	88
141 Swaziland	87	95	75	88	.. i	23 d,i	.. i	76 d,i	..	0.5 d	94
142 Nepal	21	33	11	16	17 g	26 d	83	74 d	..	0.1 d	..
143 Madagascar	28	40	17	26	85	76 d	14	22 d	7.5	0.8 d	82
144 Cameroon	17	25	20	23	91	85	9	3	3.1	2.0	112
145 Papua New Guinea	49	54 d	41	45 d	89	94 d	10	6 d	..	39.4 d	..
146 Haiti	20	45 d	18	16 d	15	..	85	..	13.8	..	87
147 Sudan	..	28	..	18	98 f,g	99	2 f	(.)	..	(.) d	121
148 Kenya	31	35	26	27	70	79 d	30	21 d	3.9	3.1 d	..
149 Djibouti	78	54	54	37	44	..	8	..	..	..	..
150 Timor-Leste	..	..	..	..	..	..	..	..	..	..	..
151 Zimbabwe	23	53	23	43	68	72 d	31	28 d	1.5	0.9 d	104
152 Togo	45	47	33	34	89	42	9	58	0.6 f	0.1	30
153 Yemen	20	38	14	46	85 f	96	15 f	4	..	5.3	..
154 Uganda	19	27	7	13	..	83	..	17	..	14.0	88
155 Gambia	72	65	60	45	..	84 g	..	17	..	5.9	115
LOW HUMAN DEVELOPMENT											
156 Senegal	30	42	25	27	77	55	23	43	..	11.7	96
157 Eritrea	45 f	56	11 f	9	..	..	..	..	..	..	93
158 Nigeria	29	35	43	53	99 f	98 d	1 f	2 d	..	1.7 d	122
159 Tanzania (United Republic of)	37	26	13	17	..	85	..	14	..	0.8	100

Human development indicators

TABLE **16** Structure of trade

HDI rank	Imports of goods and services (% of GDP)		Exports of goods and services (% of GDP)		Primary exports [a] (% of merchandise exports)		Manufactured exports (% of merchandise exports)		High-technology exports (% of manufactured exports)		Terms of trade (2000=100) [b]
	1990	2005	1990	2005	1990	2005	1990	2005	1990	2005	2004–05 [c]
160 Guinea	31	30	31	26	..	75 [d]	..	25 [d]	..	(.) [d]	106
161 Rwanda	14	31	6	11	..	90 [d]	..	10 [d]	..	25.4 [d]	89
162 Angola	21	48	39	74	100	..	(.)	..	..	..	121
163 Benin	26	26	14	13	87 [f]	87	13 [f]	13	..	0.3	93
164 Malawi	33	53	24	27	93	84	7	16	3.8	7.5	82
165 Zambia	37	25	36	16	..	91	..	9	..	1.1	119
166 Côte d'Ivoire	27	42	32	50	..	78 [d]	..	20 [d]	..	8.4 [d]	121
167 Burundi	28	36	8	8	..	94	..	6	..	5.9 [d]	84
168 Congo (Democratic Republic of the)	29	39	30	32	..	..	..	..	..	..	94
169 Ethiopia	9	39	6	16	..	89 [d]	..	11 [d]	..	0.2 [d]	91
170 Chad	28	39	13	59	..	..	..	..	..	..	101
171 Central African Republic	28	17 [d]	15	12 [d]	56 [f]	59	44 [f]	36	..	(.)	99
172 Mozambique	36	42	8	33	..	89	..	7	..	7.5	94
173 Mali	34	37	17	26	98 [g]	44 [d]	2	55 [d]	..	6.6 [d]	113 [d]
174 Niger	22	24	15	15	..	91 [d]	..	8 [d]	..	3.2 [d]	131
175 Guinea-Bissau	37	55	10	38	..	..	..	..	..	..	94
176 Burkina Faso	24	22	11	9	..	92 [d]	..	8 [d]	..	9.8 [d]	97
177 Sierra Leone	24	43	22	24	..	93 [d]	..	7 [d]	..	31.1 [d]	78
Developing countries	24	40	25	44	40	28	59	71	10.4 [f]	28.3	..
Least developed countries	22	34	13	24	..	..	31 [f]	..	..	..	..
Arab States	38	38	38	54	87 [f]	..	14 [f]	..	1.2 [f]	2.0 [d]	..
East Asia and the Pacific	32	59	34	66	25	13	73	86	15.3 [f]	36.4	..
Latin America and the Caribbean	15	23	17	26	63	46	36	54	6.6	14.5	..
South Asia	13	25	10	23	28	47	71	51	2.0 [f]	3.8 [d]	..
Sub-Saharan Africa	26	35	27	33	..	66 [d]	..	34 [d]	..	4.0 [d]	..
Central and Eastern Europe and the CIS	28	43	29	45	..	36	..	54	..	8.3	..
OECD	18	23 [d]	17	22 [d]	21	18	77	79	18.1	18.2	..
High-income OECD	18	22 [d]	17	21 [d]	19	17	79	79	18.5	18.8	..
High human development	19	25 [d]	19	25 [d]	24	20	74	76	18.1	20.3	..
Medium human development	21	34	20	35	42	30	55	69	7.2 [f]	24.3	..
Low human development	28	36	28	38	98 [f]	93 [d]	1 [f]	7 [d]	..	3.1 [d]	..
High income	19	24	18	24 [d]	21	18	77	78	18.3	20.9	..
Middle income	21	33	22	36	48	33	50	65	..	21.5	..
Low income	16	29	13	25	50 [f]	49 [d]	49 [f]	50 [d]	..	3.8 [d]	..
World	19	26	19	26 [d]	26	21	72	75	17.5	21.0	..

NOTES

a. Primary exports include exports of agricultural raw materials, food, fuels, ores and metals as defined in the Standard International Trade Classification.

b. The ratio of the export price index to the import price index measured relative to the base year 2000. A value of more than 100 means that the price of exports has risen relative to the price of imports.

c. Data refer to the most recent year available during the period specified, unless otherwise noted.

d. Data refer to an earlier year than that specified; from 2000 onwards.

e. Data before 1999 include Luxembourg.

f. Data refer to the closest available year between 1988 and 1992.

g. One or more of the components of primary exports are missing.

h. Data refer to the South African Customs Union, which includes Botswana, Lesotho, Namibia, South Africa, and Swaziland.

i Included in data for South Africa.

SOURCES

Columns 1–4 and 7–10: World Bank 2007b, based on data from UNCTAD; aggregates calculated for HDRO by the World Bank.

Columns 5 and 6: calculated on the basis of export data on agricultural raw materials, food, fuels, ores and metals and total merchandise from World Bank 2007b, based on data from UNCTAD; aggregates calculated for HDRO by the World Bank.

Column 11: World Bank 2007b.

Human development indicators

TABLE 17

... to have access to the resources needed for a decent standard of living ...

OECD-DAC country expenditures on aid

HDI rank	MDG Net official development assistance (ODA) disbursed Total [a] (US$ millions) 2005	As % of GNI 1990 [d]	As % of GNI 2005	ODA per capita of donor country (2005 US$) 1990	ODA per capita of donor country (2005 US$) 2005	MDG ODA to least developed countries [b] (% of total) 1990	MDG ODA to least developed countries [b] (% of total) 2005	MDG ODA to basic social services [c] (% of total allocable by sector) 1996/97 [e]	MDG ODA to basic social services [c] (% of total allocable by sector) 2004/05 [e]	MDG Untied bilateral ODA (% of total) 1990	MDG Untied bilateral ODA (% of total) 2005
HIGH HUMAN DEVELOPMENT											
2 Norway	2,786	1.17	0.94	453	600	44	37	12.9	14.3	61	100
3 Australia	1,680	0.34	0.25	76	83	18	25	12.0	10.7	33	72
4 Canada	3,756	0.44	0.34	115	116	30	28	5.7	30.4	47	66
5 Ireland	719	0.16	0.42	27	180	37	51	0.5	32.0	..	100
6 Sweden	3,362	0.91	0.94	256	371	39	33	10.3	15.2	87	98
7 Switzerland	1,767	0.32	0.44	148	237	43	23	8.6	7.2	78	97
8 Japan	13,147	0.31	0.28	91	103	19	18	2.5	4.6	89	90
9 Netherlands	5,115	0.92	0.82	247	313	33	32	13.1	22.0	56	96
10 France	10,026	0.60	0.47	166	165	32	24	..	6.3	64	95
11 Finland	902	0.65	0.46	174	171	38	27	6.5	13.4	31	95
12 United States	27,622	0.21	0.22	63	93	19	21	20.0	18.4	..	..
13 Spain	3,018	0.20	0.27	35	70	20	27	10.4	18.3	..	87
14 Denmark	2,109	0.94	0.81	315	388	39	39	9.6	17.6	..	87
15 Austria	1,573	0.11	0.52	29	191	63	16	4.5	13.9	32	89
16 United Kingdom	10,767	0.27	0.47	72	179	32	25	22.9	30.2	..	100
17 Belgium	1,963	0.46	0.53	123	188	41	31	11.3	16.5	..	96
18 Luxembourg	256	0.21	0.82	101	570	39	41	34.4	29.5	..	99
19 New Zealand	274	0.23	0.27	44	67	19	25	..	29.9	100	92
20 Italy	5,091	0.31	0.29	77	87	41	28	7.3	9.4	22	92
22 Germany	10,082	0.42	0.36	125	122	28	19	9.7	12.1	62	93
24 Greece	384	..	0.17	..	35	..	21	16.9	18.8	..	74
29 Portugal	377	0.24	0.21	25	36	70	56	8.5	2.7	..	61
DAC	106,777 T	0.33	0.33	93	122	28	24	7.3	15.3	68 [e]	92 [e]

NOTES

This table presents data for members of the Development Assistance Committee (DAC) of the Organisation for Economic Co-operation and Development (OECD).

a. Some non-DAC countries and areas also provide ODA. According to OECD-DAC 2007a., net ODA disbursed in 2005 by Taiwan Province of China, Czech Republic, Hungary, Iceland, Israel, Republic of Korea, Kuwait, Poland, Saudi Arabia, Slovakia, Turkey, United Arab Emirates and other small donors, including Estonia, Latvia, Lithuania and Slovenia totalled US$3,231 million. China also provides aid but does not disclose the amount.

b. Includes imputed multilateral flows that make allowance for contributions through multilateral organizations. These are calculated using the geographic distribution of disbursements for the year specified.

c. Data exclude technical cooperation and administrative costs.

d. Data include forgiveness of non-ODA claims, except for Total DAC.

e. Aggregates are considered incomplete as missing data comprises a significant portion of total disbursed ODA.

SOURCES

All columns: OECD-DAC 2007b; aggregates calculated for HDRO by OECD.

Human development indicators

TABLE 18

... to have access to the resources needed for a decent standard of living ...

Flows of aid, private capital and debt

HDI rank	Official development assistance (ODA) received [a] (net disbursements) Total (US$ millions) 2005	Per capita (US$) 2005	As % of GDP 1990	As % of GDP 2005	Net foreign direct investment inflows [b] (% of GDP) 1990	(% of GDP) 2005	Other private flows [b, c] (% of GDP) 1990	(% of GDP) 2005	MDG Total debt service As % of GDP 1990	As % of GDP 2005	As % of exports of goods, services and net income from abroad 1990	2005
HIGH HUMAN DEVELOPMENT												
1 Iceland	..	..	..	..	0.3	15.6	..	..	..	..	..	..
2 Norway	..	..	..	..	0.9	1.1	..	..	..	..	..	..
3 Australia	..	..	..	..	2.5	-4.7	..	..	..	..	..	..
4 Canada	..	..	..	..	1.3	3.1	..	..	..	..	..	..
5 Ireland	..	..	..	..	1.3	-14.7	..	..	..	..	..	..
6 Sweden	..	..	..	..	0.8	3.0	..	..	..	..	..	..
7 Switzerland	..	..	..	..	2.4	4.2	..	..	..	..	..	..
8 Japan	..	..	..	..	0.1	0.1	..	..	..	..	..	..
9 Netherlands	..	..	..	..	3.5	6.5	..	..	..	..	..	..
10 France	..	..	..	..	1.1	3.3	..	..	..	..	..	..
11 Finland	..	..	..	..	0.6	2.1	..	..	..	..	..	..
12 United States	..	..	..	..	0.8	0.9	..	..	..	..	..	..
13 Spain	..	..	..	..	2.7	2.0	..	..	..	..	..	..
14 Denmark	..	..	..	..	0.8	2.0	..	..	..	..	..	..
15 Austria	..	..	..	..	0.4	3.0	..	..	..	..	..	..
16 United Kingdom	..	..	..	..	3.4	7.2	..	..	..	..	..	..
17 Belgium	..	..	..	..	4.0	8.6	..	..	..	..	..	..
18 Luxembourg	..	..	..	..	..	301.3	..	..	..	..	..	..
19 New Zealand	..	..	..	..	4.0	1.8	..	..	..	..	..	..
20 Italy	..	..	..	..	0.6	1.1	..	..	..	..	..	..
21 Hong Kong, China (SAR)	..	..	(.)	..	..	20.2	..	..	..	..	..	..
22 Germany	..	..	..	..	0.2	1.1	..	..	..	..	..	..
23 Israel	..	..	2.6	..	0.3	4.5	..	..	..	..	..	..
24 Greece	..	..	..	..	1.2	0.3	..	..	..	..	..	..
25 Singapore	..	..	(.)	..	15.1	17.2	..	..	..	..	..	..
26 Korea (Republic of)	..	..	(.)	..	0.3	0.6	..	..	..	..	..	..
27 Slovenia	..	..	..	..	..	1.6	..	..	..	..	..	..
28 Cyprus	..	..	0.7	..	2.3	7.3 [d]	..	..	..	..	..	..
29 Portugal	..	..	..	..	3.5	1.7	..	..	..	..	..	..
30 Brunei Darussalam	..	..	0.1	..	..	..	..	..	..	..	..	..
31 Barbados	-2.1	-7.7	0.2	-0.1	0.7	2.0	-0.8	-0.3	8.2	3.1	15.1	4.7
32 Czech Republic	..	..	..	..	0.0	4.1 [d]	1.9	-3.8	3.0	4.8	..	..
33 Kuwait	..	..	(.)	..	0.0	0.3	..	..	..	..	..	..
34 Malta	..	..	0.2	..	..	..	..	..	..	..	..	..
35 Qatar	..	..	(.)	..	..	..	..	..	..	..	..	..
36 Hungary	..	..	..	..	1.9	5.9	-1.4	4.7	12.8	21.5	34.3	31.0
37 Poland	..	..	..	..	0.2	3.2	(.)	5.1	1.6	11.2	4.9	28.8
38 Argentina	99.7	2.6	0.1	0.1	1.3	2.6	-1.5	0.5	4.4	5.8	37.0	20.7
39 United Arab Emirates	..	..	(.)	..	..	..	..	..	..	..	..	..
40 Chile	151.7	9.3	0.3	0.1	2.1	5.8	4.9	4.2	8.8	6.7	25.9	15.4
41 Bahrain	..	..	3.2	..	..	..	..	..	..	..	..	..
42 Slovakia	..	..	..	..	0.6	4.1	0.0	-5.0	..	12.6	..	13.8 [e]
43 Lithuania	..	..	..	..	..	4.0	0.0	0.4	..	10.1	..	16.5
44 Estonia	..	..	..	..	..	22.9	0.0	-7.1	..	12.1	..	13.7
45 Latvia	..	..	..	..	..	4.6	0.0	15.8	..	19.6	..	37.4
46 Uruguay	14.6	4.2	0.6	0.1	0.4	4.2	-2.1	2.1	10.6	13.3	40.8	38.9
47 Croatia	125.4	28.2	..	0.3	..	4.6	..	4.6	..	12.8	..	23.9
48 Costa Rica	29.5	6.8	3.1	0.1	2.2	4.3	-1.9	1.3	6.8	3.0	23.9	5.9
49 Bahamas	..	..	0.1	..	-0.6	3.5 [e]	..	..	..	..	..	..
50 Seychelles	18.8	222.6	9.6	2.7	5.5	11.9	-1.7	2.6	5.8	7.9	8.9	7.4
51 Cuba	87.8	7.8	..	..	..	..	..	..	..	..	..	..
52 Mexico	189.4	1.8	0.1	(.)	1.0	2.4	2.7	0.5	4.3	5.7	20.7	17.2
53 Bulgaria	..	..	..	..	(.)	9.8	0.0	4.7	..	21.7	..	31.5

TABLE 18

		Official development assistance (ODA) received [a] (net disbursements)				Net foreign direct investment inflows [b] (% of GDP)		Other private flows [b, c] (% of GDP)		MDG Total debt service			
		Total (US$ millions)	Per capita (US$)	As % of GDP						As % of GDP		As % of exports of goods, services and net income from abroad	
HDI rank		2005	2005	1990	2005	1990	2005	1990	2005	1990	2005	1990	2005
54	Saint Kitts and Nevis	3.5	73.3	5.1	0.8	30.6	10.4	-0.3	-3.2	1.9	10.6	2.9	22.8
55	Tonga	31.8	310.3	26.2	14.8	0.2	2.1	-0.1	0.0	1.7	1.9	2.9	..
56	Libyan Arab Jamahiriya	24.4	..	(.)	0.1	..	..	..	..	..	..	..	..
57	Antigua and Barbuda	7.2	89.3	1.2	0.8	..	..	..	..	..	..	..	..
58	Oman	30.7	12.0	0.5	..	1.2	0.8 [d]	0.0	-0.1 [d]	..	4.1 [d]	..	7.5
59	Trinidad and Tobago	-2.1	-1.6	0.4	(.)	2.2	7.7	-3.5	-1.0	8.9	2.6	19.3	5.4 [d]
60	Romania	..	..	..	..	(.)	6.7	(.)	7.7	(.)	7.0	0.3	18.3
61	Saudi Arabia	26.3	1.1	(.)	(.)	..	..	..	..	..	..	..	..
62	Panama	19.5	6.0	1.9	0.1	2.6	6.6	-0.1	2.5	6.5	13.5	6.2	17.5
63	Malaysia	31.6	1.2	1.1	(.)	5.3	3.0	-4.2	-1.6	9.8	7.2	12.6	5.6
64	Belarus	53.8	..	..	0.2	..	1.0	0.0	0.1	..	2.3	..	3.7
65	Mauritius	31.9	25.6	3.7	0.5	1.7	0.6	1.9	(.)	6.5	4.5	8.8	7.2
66	Bosnia and Herzegovina	546.1	139.8	..	5.5	..	3.0	..	2.8	..	2.7	..	4.9
67	Russian Federation	..	..	..	..	..	2.0	0.0	5.6	..	5.5	..	14.6
68	Albania	318.7	101.8	0.5	3.8	..	3.1	0.0	0.4	..	1.0	..	2.5
69	Macedonia (TFYR)	230.3	113.2	..	4.0	..	1.7	0.0	2.8	..	4.1	..	8.6
70	Brazil	191.9	1.0	(.)	(.)	0.2	1.9	-0.1	1.0	1.8	7.9	22.2	44.8
MEDIUM HUMAN DEVELOPMENT													
71	Dominica	15.2	210.7	11.8	5.3	7.7	9.2	-0.3	-0.2	3.5	6.0	5.6	13.2
72	Saint Lucia	11.1	66.8	3.1	1.3	11.3	13.1	-0.1	-0.6	1.6	4.0	2.1	7.1
73	Kazakhstan	229.2	15.1	..	0.4	..	3.5	0.0	11.9	..	23.1	..	42.1
74	Venezuela (Bolivarian Republic of)	48.7	1.8	0.2	(.)	1.0	2.1	-1.2	3.5	10.6	4.0	23.3	9.1
75	Colombia	511.1	11.2	0.2	0.4	1.2	8.5	-0.4	-0.2	9.7	8.3	40.9	35.3
76	Ukraine	409.6	..	..	0.5	..	9.4	0.0	4.8	..	7.1	..	13.0
77	Samoa	44.0	237.6	42.4	10.9	5.9	-0.9	0.0	0.0	4.9	5.5	5.8	17.3
78	Thailand	-171.1	-2.7	0.9	-0.1	2.9	2.6	2.3	3.0	6.2	11.0	16.9	14.6
79	Dominican Republic	77.0	8.7	1.4	0.3	1.9	3.5	(.)	0.6	3.3	3.0	10.4	6.9
80	Belize	12.9	44.2	7.3	1.2	4.2	11.4	0.5	2.5	4.4	20.7	6.8	34.5
81	China	1,756.9	1.3	0.6	0.1	1.0	3.5	1.3	1.1	2.0	1.2	11.7	3.1
82	Grenada	44.9	421.3	6.2	9.5	5.8	5.6	0.1	-0.4	1.5	2.6	3.1	7.1
83	Armenia	193.3	64.1	..	3.9	81.4	5.3	0.0	1.7	..	2.8	..	7.9
84	Turkey	464.0	6.4	0.8	0.1	0.5	2.7	0.8	6.5	4.9	11.6	29.4	39.1
85	Suriname	44.0	97.9	15.3	3.3	..	..	..	..	..	..	..	..
86	Jordan	622.0	114.9	22.0	4.9	0.9	12.1	5.3	1.6	15.6	4.8	20.4	6.5
87	Peru	397.8	14.2	1.5	0.5	0.2	3.2	0.1	3.1	1.8	7.0	10.8	26.0
88	Lebanon	243.0	67.9	8.9	1.1	0.2	11.7	0.2	11.3	3.5	16.1	..	17.7
89	Ecuador	209.5	15.8	1.5	0.6	1.2	4.5	0.6	1.6	10.5	11.4	32.5	30.6
90	Philippines	561.8	6.8	2.9	0.6	1.2	1.1	0.2	2.6	8.1	10.0	27.0	16.7
91	Tunisia	376.5	37.6	3.2	1.3	0.6	2.5	-1.6	-0.4	11.6	7.2	24.5	13.0
92	Fiji	64.0	75.5	3.7	2.3	6.9	-0.1	-1.2	-0.1	7.9	0.6	12.0	..
93	Saint Vincent and the Grenadines	4.9	41.1	7.8	1.1	3.9	12.9	0.0	5.3	2.2	5.5	2.9	11.2
94	Iran (Islamic Republic of)	104.0	1.5	0.1	0.1	-0.3	(.)	(.)	0.3	0.6	1.3	3.2	..
95	Paraguay	51.1	8.3	1.1	0.7	1.5	0.9	-0.2	(.)	6.2	6.7	12.4	11.4
96	Georgia	309.8	69.2	..	4.8	..	7.0	0.0	0.8	..	2.9	..	7.4
97	Guyana	136.8	182.1	42.4	17.4	2.0	9.8	-4.1	-0.1	74.5	4.2	..	3.7
98	Azerbaijan	223.4	26.6	..	1.8	(.)	13.4	0.0	0.1	..	1.9	..	2.6
99	Sri Lanka	1,189.3	60.7	9.1	5.1	0.5	1.2	0.1	-1.3	4.8	1.9	13.8	4.5
100	Maldives	66.8	203.0	9.7	8.7	2.6	1.2	0.5	0.6	4.1	4.4	4.8	6.9
101	Jamaica	35.7	13.5	5.9	0.4	3.0	7.1	-1.0	9.8	14.4	10.1	26.9	16.3
102	Cape Verde	160.6	316.9	31.1	16.3	0.1	5.5	(.)	0.4	1.7	3.4	4.8	6.4
103	El Salvador	199.4	29.0	7.2	1.2	(.)	3.0	0.1	2.7	4.3	3.8	15.3	8.6
104	Algeria	370.6	11.3	0.2	0.4	(.)	1.1	-0.7	-0.8	14.2	5.8	63.4	..
105	Viet Nam	1,904.9	23.0	2.8	3.6	2.8	3.7	(.)	1.3	2.7	1.8	..	2.6
106	Occupied Palestinian Territories	1,101.6	303.8	..	27.4	..	..	..	..	..	..	..	..

Human development indicators

TABLE 18

Flows of aid, private capital and debt

	Official development assistance (ODA) received [a] (net disbursements)				Net foreign direct investment inflows [b] (% of GDP)		Other private flows [b, c] (% of GDP)		MDG Total debt service			
	Total (US$ millions)	Per capita (US$)	As % of GDP						As % of GDP		As % of exports of goods, services and net income from abroad	
HDI rank	2005	2005	1990	2005	1990	2005	1990	2005	1990	2005	1990	2005
107 Indonesia	2,523.5	11.4	1.5	0.9	1.0	1.8	1.6	0.5	8.7	6.3	33.3	22.0 d
108 Syrian Arab Republic	77.9	4.1	5.5	0.3	0.6	1.6	-0.1	(.)	9.7	0.8	21.8	1.9
109 Turkmenistan	28.3	5.8	..	0.4	..	0.8	0.0	-1.0	..	3.8	..	..
110 Nicaragua	740.1	134.9	32.6	15.1	0.1	4.9	2.0	0.3	1.6	3.5	3.9	6.9
111 Moldova	191.8	45.6	..	6.6	..	6.8	0.0	2.9	..	8.6	..	10.2
112 Egypt	925.9	12.5	12.6	1.0	1.7	6.0	-0.2	5.8	7.1	2.8	20.4	6.8
113 Uzbekistan	172.3	6.5	..	1.2	..	0.3	0.0	-1.7	..	5.6	..	..
114 Mongolia	211.9	82.9	0.6	11.3	..	9.7	0.0	(.)	..	2.4	..	2.9 d
115 Honduras	680.8	94.5	14.7	8.2	1.4	5.6	1.0	0.7	12.8	4.6	35.3	7.2
116 Kyrgyzstan	268.5	52.1	..	11.0	..	1.7	0.0	(.)	..	5.2	..	10.0
117 Bolivia	582.9	63.5	11.2	6.2	0.6	-3.0	-0.5	3.4	7.9	5.7	38.6	14.8
118 Guatemala	253.6	20.1	2.6	0.8	0.6	0.7	-0.1	(.)	3.0	1.5	13.6	5.8
119 Gabon	53.9	38.9	2.2	0.7	1.2	3.7	0.5	0.1	3.0	1.4	6.4	5.3 d
120 Vanuatu	39.5	186.8	32.9	11.6	8.7	3.9	-0.1	0.0	1.6	0.7	2.1	1.3
121 South Africa	700.0	15.5	..	0.3	-0.1	2.6	0.3	3.4	..	2.0	..	6.9
122 Tajikistan	241.4	37.1	..	10.4	..	2.4	0.0	-0.1	..	3.4	..	4.5
123 Sao Tome and Principe	31.9	203.8	94.0	45.2	..	9.9	-0.2	0.0	4.9	13.8	34.4	..
124 Botswana	70.9	40.2	3.8	0.7	2.5	2.7	-0.5	0.6	2.8	0.5	4.3	0.9
125 Namibia	123.4	60.7	5.1	2.0	..	..	..	..	..	..	..	..
126 Morocco	651.8	21.6	4.1	1.3	0.6	3.0	1.2	0.3	6.9	5.3	21.5	11.3
127 Equatorial Guinea	39.0	77.5	45.6	1.2	8.4	57.6	0.0	0.0	3.9	0.1	12.1	..
128 India	1,724.1	1.6	0.4	0.2	0.1	0.8	0.5	1.5	2.6	3.0	31.9	19.1 e
129 Solomon Islands	198.2	415.0	21.6	66.5	4.9	-0.3	-1.5	-2.1	5.5	4.7	11.8	..
130 Lao People's Democratic Republic	295.7	49.9	17.2	10.3	0.7	1.0	0.0	7.9	1.0	6.0	8.7	..
131 Cambodia	537.8	38.2	3.7	8.7	..	6.1	0.0	0.0	2.7	0.5	..	0.7
132 Myanmar	144.7	2.9	..	..	..	..	..	..	..	..	18.4	3.8 d
133 Bhutan	90.0	98.1	15.4	10.7	0.5	0.1	-0.9	0.0	1.7	0.8	..	..
134 Comoros	25.2	42.0	17.9	6.5	0.2	0.3	0.0	0.0	0.4	1.0	2.3	..
135 Ghana	1,119.9	50.6	9.5	10.4	0.3	1.0	-0.4	0.1	6.2	2.7	38.1	7.1
136 Pakistan	1,666.5	10.7	2.8	1.5	0.6	2.0	-0.2	1.3	4.8	2.2	21.3	10.2
137 Mauritania	190.4	62.0	23.2	10.3	0.7	6.2	-0.1	0.8	14.3	3.6	29.8	..
138 Lesotho	68.8	38.3	22.6	4.7	2.8	6.3	(.)	-0.5	3.8	3.7	4.2	5.0
139 Congo	1,448.9	362.3	7.8	28.5	-0.5	14.2	-3.6	0.0	19.0	2.3	35.3	2.4
140 Bangladesh	1,320.5	9.3	6.9	2.2	(.)	1.3	0.2	(.)	2.5	1.3	25.8	5.3
141 Swaziland	46.0	40.7	6.1	1.7	3.4	-0.6	-0.5	0.4	5.3	1.6	5.7	1.9
142 Nepal	427.9	15.8	11.7	5.8	0.2	(.)	-0.4	(.)	1.9	1.6	15.7	4.6
143 Madagascar	929.2	49.9	12.9	18.4	0.7	0.6	-0.5	(.)	7.2	1.5	45.5	17.0
144 Cameroon	413.8	25.4	4.0	2.5	-1.0	0.1	-0.1	-0.3	4.6	4.7	20.3	15.4 e
145 Papua New Guinea	266.1	45.2	12.8	5.4	4.8	0.7	1.5	-3.3	17.2	7.9	37.2	10.7
146 Haiti	515.0	60.4	5.8	12.1	0.3	0.2	0.0	0.0	1.3	1.4	11.1	3.7
147 Sudan	1,828.6	50.5	6.2	6.6	-0.2	8.4	0.0	0.2	0.4	1.4	8.7	6.5
148 Kenya	768.3	22.4	13.8	4.1	0.7	0.1	0.8	(.)	9.2	1.3	35.4	4.4
149 Djibouti	78.6	99.1	42.8	11.1	..	3.2	-0.1	0.0	3.3	2.6	..	..
150 Timor-Leste	184.7	189.4	..	52.9	..	..	..	..	..	..	..	..
151 Zimbabwe	367.7	28.3	3.8	10.9	-0.1	3.0	1.1	-0.5	5.4	6.7	23.1	..
152 Togo	86.7	14.1	15.9	3.9	1.1	0.1	0.3	0.0	5.3	0.8	11.9	2.2 d
153 Yemen	335.9	16.0	8.3	2.2	-2.7	-1.8	3.3	0.2	3.5	1.4	5.6	2.6
154 Uganda	1,198.0	41.6	15.4	13.7	-0.1	2.9	0.4	0.1	3.4	2.0	81.4	9.2
155 Gambia	58.2	38.3	30.7	12.6	4.5	11.3	-2.4	0.0	11.9	6.3	22.2	12.0
LOW HUMAN DEVELOPMENT												
156 Senegal	689.3	59.1	14.2	8.4	1.0	0.7	-0.2	0.2	5.7	2.3	19.9	11.8 d
157 Eritrea	355.2	80.7	..	36.6	..	1.2	..	0.0	..	2.1	..	..
158 Nigeria	6,437.3	48.9	0.9	6.5	2.1	2.0	-0.4	-0.2	11.7	9.0	22.6	15.8
159 Tanzania (United Republic of)	1,505.1	39.3	27.3	12.4	(.)	3.9	0.1	(.)	4.2	1.1	32.9	4.3

TABLE 18

HDI rank	Official development assistance (ODA) received[a] (net disbursements) Total (US$ millions) 2005	Per capita (US$) 2005	As % of GDP 1990	As % of GDP 2005	Net foreign direct investment inflows[b] (% of GDP) 1990	Net foreign direct investment inflows[b] (% of GDP) 2005	Other private flows[b, c] (% of GDP) 1990	Other private flows[b, c] (% of GDP) 2005	MDG Total debt service As % of GDP 1990	As % of GDP 2005	As % of exports of goods, services and net income from abroad 1990	As % of exports of goods, services and net income from abroad 2005
160 Guinea	182.1	19.4	10.3	5.5	0.6	3.1	-0.7	0.0	6.0	4.9	20.0	19.9 [d]
161 Rwanda	576.0	63.7	11.1	26.7	0.3	0.4	-0.1	0.0	0.8	1.1	14.2	8.1
162 Angola	441.8	27.7	2.6	1.3	-3.3	-4.0	5.6	4.7	3.2	6.8	8.1	9.2
163 Benin	349.1	41.4	14.5	8.1	3.4	0.5	(.)	-0.1	2.1	1.6	8.2	7.2 [d]
164 Malawi	575.3	44.7	26.6	27.8	1.2	0.1	0.1	-0.1	7.1	4.6	29.3	..
165 Zambia	945.0	81.0	14.4	13.0	6.2	3.6	-0.3	1.8	6.1	3.3	14.7	..
166 Côte d'Ivoire	119.1	6.6	6.4	0.7	0.4	1.6	0.1	-0.8	11.7	2.8	35.4	5.5
167 Burundi	365.0	48.4	23.2	45.6	0.1	0.1	-0.5	-0.6	3.7	4.9	43.4	41.4
168 Congo (Democratic Republic of the)	1,827.6	31.8	9.6	25.7	0.2	5.7	-0.1	(.)	3.7	3.0	..	..
169 Ethiopia	1,937.3	27.2	8.4	17.3	0.1	2.4	-0.5	1.0	2.0	0.8	39.0	4.1
170 Chad	379.8	39.0	17.9	6.9	0.5	12.9	(.)	(.)	0.7	1.1	4.4	..
171 Central African Republic	95.3	23.6	16.7	7.0	(.)	0.4	(.)	0.0	2.0	0.4	13.2	..
172 Mozambique	1,285.9	65.0	40.5	19.4	0.4	1.6	1.0	-0.3	3.2	1.4	26.2	4.2
173 Mali	691.5	51.1	19.8	13.0	0.2	3.0	(.)	0.2	2.8	1.7	12.3	7.2 [d]
174 Niger	515.4	36.9	15.6	15.1	1.6	0.4	0.4	-0.2	4.0	1.1	17.4	7.1 [d]
175 Guinea-Bissau	79.1	49.9	51.8	26.3	0.8	3.3	(.)	0.0	3.5	10.8	31.1	40.2 [d]
176 Burkina Faso	659.6	49.9	10.5	12.8	(.)	0.4	(.)	(.)	1.1	0.9	6.8	..
177 Sierra Leone	343.4	62.1	9.1	28.8	5.0	4.9	0.6	0.0	3.3	2.1	10.1	9.2
Developing countries	86,043.0 T	16.5	1.4	0.9	0.9	2.7	0.5	1.5	4.4	4.6	..	13.0
Least developed countries	25,979.5 T	33.9	11.8	9.3	0.3	2.6	0.5	0.8	3.0	2.3	16.9	7.0
Arab States	29,612.0 T	94.3	2.9	3.0	..	..	..	1.8	..	..	..	..
East Asia and the Pacific	9,541.6 T	4.9	0.8	0.2	..	..	..	..	..	..	..	..
Latin America and the Caribbean	6,249.5 T	11.3	0.5	0.3	0.8	2.9	0.5	1.2	4.0	6.6	23.7	22.9
South Asia	9,937.5 T	6.3	1.2	0.8	(.)	0.8	0.3	1.2	2.3	2.6	..	15.4
Sub-Saharan Africa	30,167.7 T	41.7	5.7	5.1	0.4	2.4	0.3	1.7	..	..	..	..
Central and Eastern Europe and the CIS	5,299.4 T	13.1	(.)	0.3	..	..	(.)	4.4	..	..	..	..
OECD	759.4 T [f]	..	..	(.)	1.0	1.6	..	..	..	..	..	..
High-income OECD	0.0 T	0.0	..	0.0	1.0	1.6	..	..	..	..	..	..
High human development	2,633.0 T	1.6	..	(.)	1.0	1.7	..	..	..	..	..	..
Medium human development	40,160.4 T	9.4	1.8	0.7	0.7	2.8	0.6	1.9	4.8	3.7	22.2	10.3
Low human development	21,150.9 T	42.0	9.7	9.0	0.7	1.5	0.4	0.6	6.4	5.6	22.0	12.2
High income	.. T	..	..	..	1.0	1.6	..	..	..	..	..	..
Middle income	42,242.2 T	13.7	0.7	1.3	0.9	3.1	0.4	2.2	4.5	5.5	20.3	14.3
Low income	44,123.0 T	18.2	4.1	3.2	0.4	1.4	0.3	1.0	3.7	3.1	27.1	13.7
World	106,372.9 T [g]	16.3	0.3	0.2	1.0	1.9	..	2.0	..	5.1	..	..

NOTES

This table presents data for countries included in Parts I and II of the DAC list of aid recipients (OECD-DAC 2007a). The denominator conventionally used when comparing official development assistance and total debt service to the size of the economy is GNI, not GDP (see *Definitions of statistical terms*). GDP is used here, however, to allow comparability throughout the table. With few exceptions the denominators produce similar results.

a. ODA receipts are total net ODA flows from DAC countries as well as Taiwan Province of China, Czech Republic, Hungary, Iceland, Israel, Republic of Korea, Kuwait, Poland, Saudi Arabia, Slovakia, Turkey, United Arab Emirates and other small donors, including Estonia, Latvia, Lithuania and Slovenia, and concessional lending from multilateral organizations. A negative value indicates that repayments of ODA loans exceed the amount of ODA received.

b. A negative value indicates that the capital flowing out of the country exceeds that flowing in.

c. Other private flows combine non-debt-creating portfolio equity investment flows, portfolio debt flows and bank and trade-related lending.

d. Data refer to 2004.

e. Data refer to 2003.

f. Mexico and Turkey were the only OECD member states to receive ODA from these sources in 2005.

g. World total includes US$ 14,614 million not allocated either to individual countries or to specific regions.

SOURCES

Column 1: OECD-DAC 2007b.

Column 2: Calculated on the basis of data on ODA and population from OECD-DAC 2007b.

Columns 3 and 4: Calculated on the basis of data on ODA from OECD-DAC 2007b and GDP from World Bank 2007b.

Columns 5 and 6: Calculated on the basis of data on foreign direct investment and GDP from World Bank 2007b and GDP from World Bank 2007b.

Columns 7 and 8: Calculated on the basis of data on portfolio investment, bank- and trade-related lending and GDP data from World Bank 2007b.

Columns 9 and 10: Calculated on the basis of data on debt service and GDP data from World Bank 2007b.

Columns 11 and 12: World Bank 2007b.

TABLE 19

... to have access to the resources needed for a decent standard of living ...

Priorities in public spending

HDI rank	Public expenditure on health (% of GDP)	Public expenditure on education (% of GDP)		Military expenditure[a] (% of GDP)		Total debt service[b] (% of GDP)	
	2004	1991	2002–05[c]	1990	2005	1990	2005
HIGH HUMAN DEVELOPMENT							
1 Iceland	8.3	..	8.1	0.0	0.0	..	..
2 Norway	8.1	7.1	7.7	2.9	1.7	..	..
3 Australia	6.5	4.9	4.7	2.0	1.8	..	..
4 Canada	6.8	6.5	5.2	2.0	1.1	..	..
5 Ireland	5.7	5.0	4.8	1.3	0.6	..	..
6 Sweden	7.7	7.1	7.4	2.6	1.5	..	..
7 Switzerland	6.7	5.3	6.0	1.8	1.0	..	..
8 Japan	6.3	..	3.6	0.9	1.0	..	..
9 Netherlands	5.7	5.6	5.4	2.5	1.5	..	..
10 France	8.2	5.5	5.9	3.4	2.5	..	..
11 Finland	5.7	6.5	6.5	1.6	1.4	..	..
12 United States	6.9	5.1	5.9	5.3	4.1	..	..
13 Spain	5.7	4.1	4.3	1.8	1.1	..	..
14 Denmark	7.1	6.9	8.5	2.0	1.8	..	..
15 Austria	7.8	5.3	5.5	1.2	0.9	..	..
16 United Kingdom	7.0	4.8	5.4	3.9	2.7	..	..
17 Belgium	6.9	5.0	6.1	2.4	1.1	..	..
18 Luxembourg	7.2	3.0	3.6 [d,e]	0.9	0.8	..	..
19 New Zealand	6.5	6.1	6.5	1.9	1.0	..	..
20 Italy	6.5	3.0	4.7	2.1	1.9	..	..
21 Hong Kong, China (SAR)	..	2.8	4.2	..	..	..	..
22 Germany	8.2	..	4.6	2.8 [f]	1.4	..	..
23 Israel	6.1	6.5	6.9	12.3	9.7	..	..
24 Greece	4.2	2.3	4.3	4.5	4.1	..	..
25 Singapore	1.3	3.1	3.7 [e]	4.9	4.7	..	..
26 Korea (Republic of)	2.9	3.8	4.6	3.7	2.6	..	..
27 Slovenia	6.6	4.8	6.0	2.2 [g]	1.5	..	..
28 Cyprus	2.6	3.7	6.3	5.0	1.4	..	..
29 Portugal	7.0	4.6	5.7	2.7	2.3	..	..
30 Brunei Darussalam	2.6	3.5	..	6.4	3.9	..	..
31 Barbados	4.5	7.8	6.9	0.8	0.8 [e]	8.2	3.1
32 Czech Republic	6.5	..	4.4	..	1.8	3.0	4.8
33 Kuwait	2.2	4.8	5.1	48.5	4.8	..	..
34 Malta	7.0	4.4	4.5	0.9	0.7	..	..
35 Qatar	1.8	3.5	1.6 [d]	..	..	..	..
36 Hungary	5.7	6.1	5.5	2.8	1.5	12.8	21.5
37 Poland	4.3	5.2	5.4	2.8	1.9	1.6	11.2
38 Argentina	4.3	3.3	3.8	1.2	1.0	4.4	5.8
39 United Arab Emirates	2.0	2.0	1.3 [d]	6.2	2.0	..	..
40 Chile	2.9	2.4	3.5	4.3	3.8	8.8	6.7
41 Bahrain	2.7	3.9	..	5.1	3.6	..	..
42 Slovakia	5.3	5.6	4.3	..	1.7	..	12.6
43 Lithuania	4.9	5.5	5.2	..	1.2	..	10.1
44 Estonia	4.0	..	5.3	0.5 [g]	1.5	..	12.1
45 Latvia	4.0	4.1	5.3	..	1.7	..	19.6
46 Uruguay	3.6	2.5	2.6	3.1	1.3	10.6	13.3
47 Croatia	6.2 [h,i]	5.5	4.7	7.6 [g]	1.6	..	12.8
48 Costa Rica	5.1	3.4	4.9	0.0	0.0	6.8	3.0
49 Bahamas	3.4	3.7	3.6 [d,e]	0.8	0.7	..	..
50 Seychelles	4.6	6.5	5.4 [d]	4.0	1.8	5.8	7.9
51 Cuba	5.5	9.7	9.8	..	..	..	..
52 Mexico	3.0	3.8	5.4	0.4	0.4	4.3	5.7
53 Bulgaria	4.6	5.4	4.2	3.5	2.4	..	21.7

TABLE 19

HDI rank	Public expenditure on health (% of GDP)	Public expenditure on education (% of GDP)		Military expenditure [a] (% of GDP)		Total debt service [b] (% of GDP)	
	2004	1991	2002–05 [c]	1990	2005	1990	2005
54 Saint Kitts and Nevis	3.3	2.7	9.3	..	..	1.9	10.6
55 Tonga	5.0	..	4.8	..	1.0 [e]	1.7	1.9
56 Libyan Arab Jamahiriya	2.8	..	2.7 [e]	..	2.0	..	..
57 Antigua and Barbuda	3.4	..	3.8	..	..	..	..
58 Oman	2.4	3.0	3.6	16.5	11.9	..	4.1
59 Trinidad and Tobago	1.4	4.1	4.2 [d]	..	..	8.9	2.6
60 Romania	3.4	3.5	3.4	4.6	2.0	(.)	7.0
61 Saudi Arabia	2.5	5.8	6.8	14.0	8.2	..	..
62 Panama	5.2	4.6	3.8 [d]	1.3	1.0 [e]	6.5	13.5
63 Malaysia	2.2	5.1	6.2	2.6	2.4	9.8	7.2
64 Belarus	4.6	5.7	6.0	1.5 [g]	1.2	..	2.3
65 Mauritius	2.4	3.8	4.5	0.3	0.2	6.5	4.5
66 Bosnia and Herzegovina	4.1	..	..	..	1.9	..	2.7
67 Russian Federation	3.7	3.6	3.6 [d]	12.3	4.1	..	5.5
68 Albania	3.0	..	2.9 [d]	5.9	1.4	..	1.0
69 Macedonia (TFYR)	5.7	..	3.5	..	2.2	..	4.1
70 Brazil	4.8	..	4.4	2.4	1.6	1.8	7.9
MEDIUM HUMAN DEVELOPMENT							
71 Dominica	4.2	..	5.0 [d,e]	..	..	3.5	6.0
72 Saint Lucia	3.3	..	5.8	..	..	1.6	4.0
73 Kazakhstan	2.3	3.9	2.3	..	1.1	..	23.1
74 Venezuela (Bolivarian Republic of)	2.0	4.6	..	1.8 [g]	1.2	10.6	4.0
75 Colombia	6.7	2.4	4.8	1.8	3.7	9.7	8.3
76 Ukraine	3.7	6.2	6.4	..	2.4	..	7.1
77 Samoa	4.1	..	4.5 [d]	..	..	4.9	5.5
78 Thailand	2.3	3.1	4.2	2.6	1.1	6.2	11.0
79 Dominican Republic	1.9	..	1.8	0.6	0.5	3.3	3.0
80 Belize	2.7	4.6	5.4	1.2	..	4.4	20.7
81 China	1.8 [i]	2.2	1.9 [e]	2.7	2.0	2.0	1.2
82 Grenada	5.0	4.9	5.2	..	..	1.5	2.6
83 Armenia	1.4	..	3.2 [e]	2.2 [g]	2.7	..	2.8
84 Turkey	5.6 [h,i]	2.4	3.7	3.5	2.8	4.9	11.6
85 Suriname	3.6	5.9	..	..	..	..	..
86 Jordan	4.7 [i]	8.0	4.9 [e]	6.9	5.3	15.6	4.8
87 Peru	1.9	2.8	2.4	0.1	1.4	1.8	7.0
88 Lebanon	3.2	..	2.6	7.6	4.5	3.5	16.1
89 Ecuador	2.2	2.5	1.0 [d,e]	1.9	2.6	10.5	11.4
90 Philippines	1.4	3.0	2.7	1.4	0.9	8.1	10.0
91 Tunisia	2.8 [e]	6.0	7.3	2.0	1.6	11.6	7.2
92 Fiji	2.9	5.1	6.4	2.3	1.2 [e]	7.9	0.6
93 Saint Vincent and the Grenadines	3.9	5.9	8.2	..	..	2.2	5.5
94 Iran (Islamic Republic of)	3.2	4.1	4.7	2.9	5.8	0.6	1.3
95 Paraguay	2.6	1.9	4.3	1.0	0.7	6.2	6.7
96 Georgia	1.5	..	2.9	..	3.5	..	2.9
97 Guyana	4.4	2.2	8.5	0.9	..	74.5	4.2
98 Azerbaijan	0.9	7.7	2.5	2.5 [g]	2.5	..	1.9
99 Sri Lanka	2.0	3.2	..	2.1	2.6	4.8	1.9
100 Maldives	6.3	7.0	7.1	..	..	4.1	4.4
101 Jamaica	2.8	4.5	5.3	0.6	0.6	14.4	10.1
102 Cape Verde	3.9	3.6	6.6	..	0.7 [e]	1.7	3.4
103 El Salvador	3.5	1.8	2.8	2.0	0.6	4.3	3.8
104 Algeria	2.6	5.1	..	1.5	2.9	14.2	5.8
105 Viet Nam	1.5	1.8	..	..	..	2.7	1.8
106 Occupied Palestinian Territories	7.8 [e]	..	..	..	..	..	..

TABLE 19

Priorities in public spending

		Public expenditure on health (% of GDP)	Public expenditure on education (% of GDP)		Military expenditure [a] (% of GDP)		Total debt service [b] (% of GDP)	
HDI rank		2004	1991	2002–05 [c]	1990	2005	1990	2005
107	Indonesia	1.0	1.0	0.9	1.8	1.2	8.7	6.3
108	Syrian Arab Republic	2.2	3.9	..	6.0	5.1	9.7	0.8
109	Turkmenistan	3.3	3.9	..	..	2.9 [e]	..	3.8
110	Nicaragua	3.9	3.4	3.1 [d]	4.0 [g]	0.7	1.6	3.5
111	Moldova	4.2	5.3	4.3	..	0.3	..	8.6
112	Egypt	2.2	3.9	..	4.7	2.8	7.1	2.8
113	Uzbekistan	2.4	9.4	..	..	0.5 [e]	..	5.6
114	Mongolia	4.0	11.5	5.3	4.3	1.6	..	2.4
115	Honduras	4.0	3.8	..	..	0.6	12.8	4.6
116	Kyrgyzstan	2.3	6.0	4.4 [d]	1.6 [g]	3.1	..	5.2
117	Bolivia	4.1	2.4	6.4	2.3	1.6	7.9	5.7
118	Guatemala	2.3	1.3	..	1.5	0.3	3.0	1.5
119	Gabon	3.1	..	3.9 [d,e]	..	1.5	3.0	1.4
120	Vanuatu	3.1	4.6	9.6	..	..	1.6	0.7
121	South Africa	3.5	5.9	5.4	3.8	1.5	..	2.0
122	Tajikistan	1.0	9.1	3.5	0.3 [g]	2.2 [e]	..	3.4
123	Sao Tome and Principe	9.9	..	..	..	..	4.9	13.8
124	Botswana	4.0	6.2	10.7	4.1	3.0	2.8	0.5
125	Namibia	4.7	7.9	6.9	5.6 [g]	3.2	..	..
126	Morocco	1.7	5.0	6.7	5.0	4.5	6.9	5.3
127	Equatorial Guinea	1.2	..	0.6 [d]	..	..	3.9	0.1
128	India	0.9	3.7	3.8	3.2	2.8	2.6	3.0
129	Solomon Islands	5.6	3.8	3.3 [d,e]	..	..	5.5	4.7
130	Lao People's Democratic Republic	0.8	..	2.3	..	2.1 [e]	1.0	6.0
131	Cambodia	1.7	..	1.9	3.1	1.8	2.7	0.5
132	Myanmar	0.3	..	1.3 [e]	..	..	..	..
133	Bhutan	3.0	..	5.6 [e]	..	..	1.7	0.8
134	Comoros	1.6	..	3.9	..	..	0.4	1.0
135	Ghana	2.8	..	5.4	0.4	0.7	6.2	2.7
136	Pakistan	0.4	2.6	2.3	5.8	3.5	4.8	2.2
137	Mauritania	2.0	4.6	2.3	3.8	3.6	14.3	3.6
138	Lesotho	5.5	6.2	13.4	4.5	2.3	3.8	3.7
139	Congo	1.2	7.4	2.2	..	1.4	19.0	2.3
140	Bangladesh	0.9	1.5	2.5	1.0	1.0	2.5	1.3
141	Swaziland	4.0	5.7	6.2	1.8	1.8 [e]	5.3	1.6
142	Nepal	1.5	2.0	3.4	0.9	2.1	1.9	1.6
143	Madagascar	1.8	2.5	3.2	1.2	1.1	7.2	1.5
144	Cameroon	1.5	3.2	1.8 [d]	1.5	1.3	4.6	4.7
145	Papua New Guinea	3.0	..	..	2.1	0.6	17.2	7.9
146	Haiti	2.9	1.4	..	..	..	1.3	1.4
147	Sudan	1.5	6.0	..	3.5	2.3 [e]	0.4	1.4
148	Kenya	1.8	6.7	6.7	2.9	1.7	9.2	1.3
149	Djibouti	4.4	3.5	7.9	5.9	4.2 [e]	3.3	2.6
150	Timor-Leste	8.8	..	..	..	..	..	..
151	Zimbabwe	3.5	7.7	4.6 [d,e]	4.4	2.3	5.4	6.7
152	Togo	1.1	..	2.6	3.1	1.5	5.3	0.8
153	Yemen	1.9	..	9.6 [d,e]	7.9	7.0	3.5	1.4
154	Uganda	2.5	1.5	5.2 [d]	3.1	2.3	3.4	2.0
155	Gambia	1.8	3.8	2.0 [d]	1.2	0.5 [e]	11.9	6.3
LOW HUMAN DEVELOPMENT								
156	Senegal	2.4	3.9	5.4	2.0	1.5	5.7	2.3
157	Eritrea	1.8	..	5.4	..	24.1 [e]	..	2.1
158	Nigeria	1.4	0.9	..	0.9	0.7	11.7	9.0
159	Tanzania (United Republic of)	1.7	2.8	2.2 [d,e]	2.0	1.1	4.2	1.1

TABLE 19

HDI rank	Public expenditure on health (% of GDP)	Public expenditure on education (% of GDP)		Military expenditure [a] (% of GDP)		Total debt service [b] (% of GDP)	
	2004	1991	2002–05 [c]	1990	2005	1990	2005
160 Guinea	0.7	2.0	2.0	2.4 [g]	2.0 [e]	6.0	4.9
161 Rwanda	4.3	..	3.8	3.7	2.9	0.8	1.1
162 Angola	1.5	..	2.6 [d,e]	2.7	5.7	3.2	6.8
163 Benin	2.5	..	3.5 [d]	..	..	2.1	1.6
164 Malawi	9.6	3.2	5.8	1.3	0.7 [e]	7.1	4.6
165 Zambia	3.4	2.8	2.0	3.7	2.3 [e]	6.1	3.3
166 Côte d'Ivoire	0.9	..	4.6 [d,e]	1.3	1.5 [e]	11.7	2.8
167 Burundi	0.8	3.5	5.1	3.4	6.2	3.7	4.9
168 Congo (Democratic Republic of the)	1.1	..	..	..	2.4	3.7	3.0
169 Ethiopia	2.7	2.4	6.1 [i]	8.5	2.6	2.0	0.8
170 Chad	1.5	1.6	2.1	..	1.0	0.7	1.1
171 Central African Republic	1.5	2.2	..	1.6 [g]	1.1	2.0	0.4
172 Mozambique	2.7	..	3.7	5.9	0.9	3.2	1.4
173 Mali	3.2	..	4.3	2.1	2.3	2.8	1.7
174 Niger	2.2	3.3	2.3	..	1.2 [e]	4.0	1.1
175 Guinea-Bissau	1.3	..	5.2 [e]	..	4.0	3.5	10.8
176 Burkina Faso	3.3	2.6	4.7	2.7	1.3	1.1	0.9
177 Sierra Leone	1.9	..	3.8 [d]	1.4	1.0	3.3	2.1

NOTES

a. Because of limitations in the data, comparisons across countries should be made with caution. For detailed notes on the data see SIPRI 2007c.
b. For aggregates, see Table 18.
c. Data refer to the most recent year available during the period specified.
d. National or UNESCO Institute for Statistics estimate.
e. Data refer to an earlier year than that specified; from 1999 onwards.

f. Data refer to the Federal Republic of Germany before reunification.
g. Data refer to the closest available year between 1991 and 1992.
h. Data refer to 2005.
i. Data differ from the standard definition or refer to only part of a country.
j. Data refer to 2006.

SOURCES

Column 1: World Bank 2007b.
Columns 2 and 3: UNESCO Institute for Statistics 2007b.
Column 4: SIPRI 2007b.
Column 5: SIPRI 2007c.
Columns 6 and 7: calculated on the basis of data on debt service and GDP data from World Bank 2007b.

TABLE 20

... to have access to the resources needed for a decent standard of living ...

Unemployment in OECD countries

HDI rank	Unemployed people (thousands) 2006	Unemployment rate			MDG Youth unemployment rate		Long-term unemployment (% of total unemployment)	
		Total (% of labour force) 2006	Average annual (% of labour force) 1996/2006	Female (% of male rate) 2006	Total (% of labour force aged 15–24) [a] 2006	Female (% of male rate) 2006	Women 2006	Men 2006
HIGH HUMAN DEVELOPMENT								
1 Iceland	5.2	3.0	2.9	110	8.4	81	5.3	9.2
2 Norway	83.8	3.5	3.9	94	8.6	101	11.1	16.8
3 Australia	527.0	4.9	6.6	104	10.4	90	15.2	20.1
4 Canada	1,106.0	6.3	7.7	94	11.6	80	8.3	9.1
5 Ireland	91.4	4.4	6.0	89	8.4	89	24.5	40.8
6 Sweden	331.9	7.0	6.9	103	21.3	102	12.2	16.1
7 Switzerland	168.7	4.0	3.7	138	7.7	94	42.6	35.0
8 Japan	2,730.0	4.1	4.5	91	8.0	81	20.8	40.9
9 Netherlands	365.0	3.9	3.9	126	7.6	117	43.6	46.8
10 France	2,729.0	9.4	9.9	121	23.9	115	43.3	44.8
11 Finland	204.0	7.7	10.1	109	18.8	95	21.8	28.0
12 United States	7,002.0	4.6	5.0	100	10.5	86	9.2	10.7
13 Spain	1,837.1	8.5	12.2	184	17.9	144	32.2	25.9
14 Denmark	114.2	3.9	5.0	136	7.6	100	20.2	20.7
15 Austria	195.5	4.8	4.3	118	9.1	105	25.1	29.5
16 United Kingdom	1,602.0	5.3	5.6	86	13.9	75	14.9	27.5
17 Belgium	381.8	8.2	8.3	126	18.9	106	56.5	54.7
18 Luxembourg	9.1 [b]	4.8	3.3	180	13.7 [b]	138 [b]	20.5 [b]	33.8 [b]
19 New Zealand	82.6	3.8	5.4	117	9.6	108	5.5	8.8
20 Italy	1,673.6	6.8	9.4	165	21.6	132	54.8	50.8
22 Germany	4,250.0	8.4	8.5	119	13.5	89	56.5	57.8
24 Greece	427.4	8.9	10.3	243	24.5	196	60.1	48.1
26 Korea (Republic of)	824.0	3.5	4.0	76	10.0	77	0.9	1.2
29 Portugal	427.8	7.7	5.9	138	16.2	126	53.3	50.3
32 Czech Republic	371.1	7.2	7.2	153	17.5	112	56.3	53.9
36 Hungary	316.8	7.5	7.1	108	19.1	107	45.1	47.1
37 Poland	2,344.3	13.8	15.7	116	29.8	112	52.0	49.0
42 Slovakia	353.1	13.4	15.8	120	26.6	103	72.3	73.9
52 Mexico	1,367.3	3.2	3.3	118	6.2	138	2.3	2.7
MEDIUM HUMAN DEVELOPMENT								
84 Turkey	2,445.0	9.9	8.6	106	18.7	109	44.2	32.6
OECD	34,366.6 T	6.0	6.7	112	12.5	98	32.0	32.4

NOTES

a. The age range may be 16–24 for some countries.
b. Data refer to 2005.

SOURCES

Columns 1—3, 5, 7 and 8: OECD 2007.
Columns 4 and 6: calculated on the basis of data on male and female unemployment rates from OECD 2007.

Human development indicators

TABLE 21

... to have access to the resources needed for a decent standard of living ...

Unemployment and informal sector work in non-OECD countries

	Unemployed people (thousands) 1996–2005[d]	Unemployment rate [a] Total (% of labour force) 1996–2005[d]	Female (% of male rate) 1996–2005[d]	Employment by economic activity [b] Total (thousands) 1996–2005[d]	Agriculture (%) 1996–2005[d]	Industry (%) 1996–2005[d]	Services (%) 1996–2005[d]	Employment in informal sector as a % of non-agricultural employment [c] Survey year	Both sexes (%)	Female (%)	Male (%)
HIGH HUMAN DEVELOPMENT											
21 Hong Kong, China (SAR)	201	5.6	68	3,386	(.)	15	85	..	..	..	..
23 Israel	246	9.0	112	2,494	2	22	76	..	..	..	..
25 Singapore	116	5.3	98	2,267	0	30	70	..	..	..	..
27 Slovenia	58	5.8	111	946	9	37	53	..	..	..	..
28 Cyprus	19	5.3	148	338	5	24	71	..	..	..	..
30 Brunei Darussalam	7 [e]	..	..	146	1	21	77	..	..	..	..
31 Barbados	14	9.8	118	132	3	17	70	..	..	..	..
33 Kuwait	15 [f]	1.1 [f]	173 [f]	..	..	..	..	..	..	..	..
34 Malta	12	7.5	142	149	2	29	68	..	..	..	..
35 Qatar	13	3.9	548	438	3	41	56	..	..	..	..
38 Argentina	1,141	10.6	135	9,639	1	24	75	2003 [g]	40 [g]	31 [g]	46 [g]
39 United Arab Emirates	41	2.3	118	1,779	8	33	59	..	..	..	..
40 Chile	440	6.9	139	5,905	13	23	64	1996 [h]	36 [h]	44 [h]	31 [h]
41 Bahrain	16	..	..	..	..	..	..	..	..	..	..
43 Lithuania	133	8.3	101	1,474	14	29	57	..	..	..	..
44 Estonia	52	7.9	81	607	5	34	61	..	..	..	..
45 Latvia	99	8.7	93	1,036 [g]	12 [g]	26 [g]	62 [g]	..	..	..	..
46 Uruguay	155	12.2	161	1,115 [g]	5 [g]	22 [g]	74 [g]	2000	30	25	34
47 Croatia	229	12.7	120	1,573	17	29	54	..	..	..	..
48 Costa Rica	126	6.6	192	1,777	15	22	63	2000	20	17	22
49 Bahamas	18	10.2	122	161	4	18	78	..	..	..	..
50 Seychelles	4	..	..	..	..	..	..	..	..	..	..
51 Cuba	88	1.9	129	4,642	21	19	59	..	..	..	..
53 Bulgaria	334	10.1	95	2,980	9	34	57	..	..	..	..
57 Antigua and Barbuda	..	..	..	28 [g]	4 [g]	19 [g]	74 [g]	..	..	..	..
58 Oman	53	..	..	282 [g]	6 [g]	11 [g]	82 [g]	..	..	..	..
59 Trinidad and Tobago	50	8.0	190	525	7	28	64	..	..	..	..
60 Romania	705	7.2	83	9,147	32	30	38	..	..	..	..
61 Saudi Arabia	327	5.2	274	5,913	5	21	74	..	..	..	..
62 Panama	137	10.3	173	1,188	16	17	67	2004	33	29	35
63 Malaysia	370	3.6	100	9,987	15	30	53	..	..	..	..
64 Belarus	68 [f]	1.5 [f]	325 [f]	4,701 [g]	21 [g]	35 [g]	40 [g]	..	..	..	..
65 Mauritius	52	9.6	284	490	10	32	57	2004	8	6	9
67 Russian Federation	5,775	7.8	105	68,169	10	30	60	2004	12	11	12
68 Albania	157	14.4	141	931	58	14	28 [i]	..	..	..	..
69 Macedonia (TFYR)	324	37.3	105	545	20	32	48	..	..	..	..
70 Brazil	8,264	8.9	172	84,596	21	21	58	2003	37	31	42
MEDIUM HUMAN DEVELOPMENT											
71 Dominica	3	11.0	80	26	24	18	54	..	..	..	..
72 Saint Lucia	13	16.4	164	59	11	18	53	..	..	..	..
73 Kazakhstan	659	8.4	140	7,182	34	17	49	..	..	..	..
74 Venezuela (Bolivarian Republic of)	1,823	15.8	127	9,994	11	20	69	2004	46	45	47
75 Colombia	2,406	11.8	174	18,217	22	19	59 [i]	2004 [g]	58 [g]	59 [g]	55 [g]
76 Ukraine	1,601	7.2	91	20,680	19	24	56 [i]	2004	4	4	4
78 Thailand	496	1.4	80	36,302	43	20	37	2002	72	..	..
79 Dominican Republic	716	17.9	254	3,315	16	21	63	1997 [h]	48 [h]	50 [h]	47 [h]
80 Belize	12	11.0	230	78	28	17	55	..	..	..	..
81 China	8,390	4.2	..	737,400	44	18	16	..	..	..	..
82 Grenada	..	..	..	35	14	24	59	..	..	..	..
83 Armenia	424	36.4	91	1,108	46	17	38	..	..	..	..
85 Suriname	12	14.0	200	73	6	15	75	..	..	..	..
86 Jordan	..	..	..	43	4	22	74	..	..	..	..
87 Peru	437	11.4	143	3,400	1	24	76	2004 [g]	56 [g]	55 [g]	57 [g]

TABLE 21

Unemployment and informal sector work in non-OECD countries

	Unemployed people (thousands) 1996–2005[d]	Unemployment rate [a] Total (% of labour force) 1996–2005[d]	Female (% of male rate) 1996–2005[d]	Employment by economic activity [b] Total (thousands) 1996–2005[d]	Agriculture (%) 1996–2005[d]	Industry (%) 1996–2005[d]	Services (%) 1996–2005[d]	Employment in informal sector as a % of non-agricultural employment [c] Survey year	Both sexes (%)	Female (%)	Male (%)
HDI rank											
88 Lebanon	116	..	..	..	..	..	..	..	..	..	..
89 Ecuador	334	7.9	186	3,892	8	21	70	2004[g]	40[g]	44[g]	37[g]
90 Philippines	2,619	7.4	99	32,875	37	15	48	1995[h]	72[h]	73[h]	71[h]
91 Tunisia	486	14.2	132	..	..	..	..	1994–95	50[h]	39[h]	53[h]
93 Saint Vincent and the Grenadines	..	..	..	35	15	20	56	..	..	..	..
94 Iran (Islamic Republic of)	2,556	11.5	170	19,760	25	30	45	..	..	..	..
95 Paraguay	206	8.1	151	2,247	32	16	53	1995[h]	66[h]	..	..
96 Georgia	279	13.8	85	1,745	54	9	36	..	..	..	..
97 Guyana	..	..	..	240	28	23	48	..	..	..	..
98 Azerbaijan	369	8.5	125	3,850[g]	39[g]	12[g]	49[g]	..	..	..	..
99 Sri Lanka	623	7.7	216	6,943	34	23	39	..	..	..	..
100 Maldives	2	..	..	86	14	19	50	..	..	..	..
101 Jamaica	130	10.9	207	1,063	18	18	64	..	..	..	..
103 El Salvador	184	6.8	44	2,526	19	24	57	1997[h]	57[h]	69[h]	46[h]
104 Algeria	1,475	15.3	103	7,798	21	26	53	1997[h]	43[h]	41[h]	43[h]
105 Viet Nam	926	2.1	131	42,316	58	17	25	..	..	..	..
106 Occupied Palestinian Territories	212	26.7	71	578	16	25	58	..	..	..	..
107 Indonesia	10,854	9.1	155	94,948	44	18	38	1998[h]	78[h]	77[h]	78[h]
108 Syrian Arab Republic	638	11.7	290	4,822	30	27	43	2003	22	7	24
110 Nicaragua	135	12.2	165	1,953	31	18	40	2000[g]	55[g]	59[g]	52[g]
111 Moldova	104	7.3	69	1,319	41	16	43	2004	8	5	11
112 Egypt	2,241	11.0	311	18,119	30	20	50	2003[g]	45[g]	59[g]	42[g]
113 Uzbekistan	..	..	..	8,885	39	19	35	..	..	..	..
114 Mongolia	33[f]	3.3[f]	120[f]	951	40	16	44	..	..	..	..
115 Honduras	108	4.1	197	2,544	39	21	40	1997[h]	58[h]	66[h]	74[h]
116 Kyrgyzstan	186	8.5	116	1,807	53	10	37	2003	43	39	45
117 Bolivia	222	5.5	161	2,091[g]	5[g]	28[g]	67[g]	1997[h]	64[h]	74[h]	55[h]
118 Guatemala	172	3.4	196	4,769	39	20	38	..	..	..	..
121 South Africa	4,385	26.6	100	11,622	10	25	65	2004	16	16	15
122 Tajikistan	51[f]	2.7[f]	121[f]	..	..	..	..	..	..	..	..
124 Botswana	144	23.8	123	567	23	22	50	..	..	..	..
125 Namibia	221	33.8	138	432	31	12	56	..	..	..	..
126 Morocco	1,226	11.0	106	9,603	44	20	36[i]	1995[h]	45[h]	47[h]	44[h]
128 India	16,634	4.3	100	308,760[g]	67[g]	13[g]	20[g,i]	2000[g]	56[g]	57[g]	55[g]
130 Lao People's Democratic Republic	38	..	..	2,165[g]	85[g]	4[g]	11[g]	..	..	..	..
131 Cambodia	503	1.8	147	6,243	70	11	19	..	..	..	..
132 Myanmar	190[f]	..	..	18,359	63	12	25[i]	..	..	..	..
135 Ghana	..	..	..	8,300	55	14	31	..	..	..	..
136 Pakistan	3,566	7.7	194	38,882	42	21	37	2003–04	70	66	70
138 Lesotho	216	39.3	153	353	57	15	23	..	..	..	..
140 Bangladesh	2,002	4.3	117	44,322	52	14	35	..	..	..	..
142 Nepal	178	1.8	85	7,459[g]	79[g]	6[g]	21[g]	..	..	..	..
143 Madagascar	383	4.5	160	8,099	78	7	15	..	..	..	..
144 Cameroon	468	7.5	82	5,806[g]	61[g]	9[g]	23[g]	..	..	..	..
145 Papua New Guinea	69	2.8	30	2,345	72	4	23	..	..	..	..
146 Haiti	..	..	..	..	51	11	39	..	..	..	..
148 Kenya	1,276	..	..	1,674	19	20	62	1999[h]	72[h]	83[h]	59[h]
149 Djibouti	..	..	..	77[g]	2[g]	8[g]	80[g]	..	..	..	..
151 Zimbabwe	298	6.0	63	..	..	..	..	..	..	..	..
153 Yemen	469	11.5	66	3,622	54	11	35	..	..	..	..
154 Uganda	346	3.2	156	9,257	69	8	22	..	..	..	..

Human development indicators

TABLE 21

HDI rank	Unemployed people (thousands) 1996–2005[d]	Unemployment rate [a] Total (% of labour force) 1996–2005[d]	Female (% of male rate) 1996–2005[d]	Employment by economic activity [b] Total (thousands) 1996–2005[d]	Agriculture (%) 1996–2005[d]	Industry (%) 1996–2005[d]	Services (%) 1996–2005[d]	Employment in informal sector as a % of non-agricultural employment [c] Survey year	Both sexes (%)	Female (%)	Male (%)
LOW HUMAN DEVELOPMENT											
157 Eritrea	..	..	..	82[g]	4[g]	19[g]	77[g]	..	..	..	..
158 Nigeria	..	..	..	5,229[g]	3[g]	22[g]	75[g]	..	..	..	..
159 Tanzania (United Republic of)	913	5.1	132	16,915	82	3	15	2001	43	41	46
160 Guinea	..	..	..	..	..	..	..	1991[h]	72[h]	87[h]	66[h]
161 Rwanda	16	0.6	38	3,143[g]	90[g]	3[g]	7[g]	..	..	..	..
162 Angola	19[e]	..	..	..	..	..	..	..	..	..	..
163 Benin	..	..	..	..	..	..	..	1992[h]	93[h]	97[h]	87[h]
165 Zambia	508	12.0	92	3,530	70	7	23	..	..	..	..
167 Burundi	1[e]	14.0[e]	88[e]	..	..	..	..	..	..	..	..
169 Ethiopia	1,654	5.0	312	20,843[g]	93[g]	3[g]	5[g]	2004	41	48	36
170 Chad	..	..	..	..	..	..	..	1993[h]	74[h]	95[h]	60[h]
171 Central African Republic	..	..	..	..	..	..	..	2003[g]	21[g]	21[g]	21[g]
172 Mozambique	192	..	..	..	..	..	..	1999[h]	74[h]	..	..
173 Mali	227	8.8	153	..	..	..	..	2004	71	80	63
176 Burkina Faso	7[e]	..	..	..	..	..	..	2000[h]	77[h]	..	..

NOTES

Data are not strictly comparable across countries as they were compiled using different sources. As a result data may differ from the standard definitions of unemployment and the informal sector.

a. Data refer to the ILO definition of unemployment unless otherwise specified.

b. Employment by economic activity may not sum to 100 as a result of rounding or the omission of employment in economic activity that is not adequately defined.

c. Informal sector may not be of the same year as data for employment and unemployment. As a result, they may not be strictly comparable.

d. Data refer to the most recent year during the period specified.

e. Data refer to work applicants.

f. Data refer to the registered unemployed.

g. Data refer to a year or period other than that specified, differ from the standard definition or refer to only part of a country.

h. Data are from Charmes and Rani 2007.

i. Services include persons engaged in extra-territorial organizations and bodies and/or persons not classifiable by economic activity.

SOURCES

Columns 1–3: ILO 2007b.
Columns 4–7: ILO 2005.
Columns 8–11: ILO Bureau of Statistics 2007, unless otherwise specified.

TABLE 22

... while preserving it for future generations ...

Energy and the environment

HDI rank	Electricity consumption per capita (kilowatt-hours) 2004	Electricity consumption per capita (% change) 1990–2004	Electrification rate (%) 2000–05 [a]	Population without electricity (millions) 2005	GDP per unit of energy use (2000 PPP US$ per kg of oil equivalent) 2004	GDP per unit of energy use (% change) 1990–2004	Forest area % of total land area (%) 2005	Forest area Total (thousand sq km) 2005	Forest area Total change (thousand sq km) 1990–2005	Forest area Average annual change (%) 1990–2005
HIGH HUMAN DEVELOPMENT										
1 Iceland	29,430	66.4	100	..	2.5	-12.1	0.5	0.5	0.2	5.6
2 Norway	26,657	6.5	100	..	5.9	15.9	30.7	93.9	2.6	0.2
3 Australia	11,849	30.4	100	..	4.8	21.3	21.3	1,636.8	-42.3	-0.2
4 Canada	18,408	5.9	100	..	3.4	12.5	33.6	3,101.3	..	..
5 Ireland	6,751	62.7	100	..	9.5	81.9	9.7	6.7	2.3	3.4
6 Sweden	16,670	-1.9	100	..	4.5	13.0	66.9	275.3	1.6	(.)
7 Switzerland	8,669 [b]	10.3 [b]	100	..	8.3	0.9	30.9	12.2	0.7	0.4
8 Japan	8,459	21.8	100	..	6.4	-1.4	68.2	248.7	-0.8	(.)
9 Netherlands	7,196	32.7	100	..	5.8	11.7	10.8	3.7	0.2	0.4
10 France	8,231 [c]	24.6 [c]	100	..	5.9	8.0	28.3	155.5	10.2	0.5
11 Finland	17,374	33.2	100	..	3.8	-1.1	73.9	225.0	3.1	0.1
12 United States	14,240	11.9	100	..	4.6	25.3	33.1	3,030.9	44.4	0.1
13 Spain	6,412	63.3	100	..	6.9	-4.9	35.9	179.2	44.4	2.2
14 Denmark	6,967	7.4	100	..	7.9	14.7	11.8	5.0	0.6	0.8
15 Austria	8,256	27.7	100	..	7.3	2.9	46.7	38.6	0.9	0.2
16 United Kingdom	6,756	15.9	100	..	7.3	22.2	11.8	28.5	2.3	0.6
17 Belgium	8,986	33.4	100	..	5.2	10.3	22.0	6.7	-0.1	-0.1
18 Luxembourg	16,630	21.1	100	..	6.1	77.5	33.5	0.9	(.)	0.1
19 New Zealand	10,238	6.7	100	..	5.1	25.0	31.0	83.1	5.9	0.5
20 Italy	6,029 [d]	36.1 [d]	100	..	8.2	-2.5	33.9	99.8	16.0	1.3
21 Hong Kong, China (SAR)	6,401	34.4	..	..	11.5	6.4	..	..	..	..
22 Germany	7,442	10.4	100	..	6.2	31.6	31.7	110.8	3.4	0.2
23 Israel	6,924	62.8	97	0.2	7.3	4.7	8.3	1.7	0.2	0.7
24 Greece	5,630	60.1	100	..	7.4	11.1	29.1	37.5	4.5	0.9
25 Singapore	8,685	67.7	100	0.0	4.4	30.6	3.4	(.)	0.0	0.0
26 Korea (Republic of)	7,710	178.3	100	..	4.2	-6.3	63.5	62.7	-1.1	-0.1
27 Slovenia	7,262	..	..	..	5.4	10.6	62.8	12.6	0.8	0.4
28 Cyprus	5,718	97.2	..	..	5.9	8.5	18.9	1.7	0.1	0.5
29 Portugal	4,925	69.9	100	..	7.1	-9.8	41.3	37.8	6.8	1.5
30 Brunei Darussalam	8,842	80.9	99	0.0	..	..	52.8	2.8	-0.4	-0.7
31 Barbados	3,304	85.0	..	..	..	..	4.0	(.)	..	..
32 Czech Republic	6,720	..	..	..	4.0	30.8	34.3	26.5	0.2	(.)
33 Kuwait	15,423	75.0	100	0.0	1.9	63.1	0.3	0.1	(.)	6.7
34 Malta	5,542	53.4	..	..	7.5	47.9	1.1	..	..	..
35 Qatar	19,840	101.8	71	0.2	..	..	(.)	..	..	..
36 Hungary	4,070	6.7	..	..	5.9	40.6	21.5	19.8	1.8	0.6
37 Poland	3,793	6.9	..	..	5.1	74.8	30.0	91.9	3.1	0.2
38 Argentina	2,714	70.6	95	1.8	7.4	15.8	12.1	330.2	-22.4	-0.4
39 United Arab Emirates	12,000	41.5	92	0.4	2.2	15.7	3.7	3.1	0.7	1.8
40 Chile	3,347	138.7	99	0.2	6.1	11.9	21.5	161.2	8.6	0.4
41 Bahrain	11,932	52.3	99	0.0	1.8	21.5	0.6	..	..	..
42 Slovakia	5,335	..	..	..	3.9	45.3	40.1	19.3	0.1	(.)
43 Lithuania	3,505	..	..	..	4.5	60.5	33.5	21.0	1.5	0.5
44 Estonia	6,168	..	..	..	3.5	113.2	53.9	22.8	1.2	0.4
45 Latvia	2,923	..	..	..	5.6	122.6	47.4	29.4	1.7	0.4
46 Uruguay	2,408	52.4	95	0.2	10.4	5.3	8.6	15.1	6.0	4.4
47 Croatia	3,818	..	..	..	5.6	12.0	38.2	21.4	0.2	0.1
48 Costa Rica	1,876	54.4	99	0.1	10.0	2.9	46.8	23.9	-1.7	-0.4
49 Bahamas	6,964 [e]	87.0	..	..	..	..	51.5	5.2	..	..
50 Seychelles	2,716 [e]	88.2	..	..	..	..	88.9	0.4	0.0	0.0
51 Cuba	1,380	0.6	96	0.5	..	..	24.7	27.1	6.6	2.1
52 Mexico	2,130	46.5	..	..	5.5	8.5	33.7	642.4	-47.8	-0.5
53 Bulgaria	4,582	-10.3	..	..	3.0	44.7	32.8	36.3	3.0	0.6

Human development indicators

TABLE 22

HDI rank	Electricity consumption per capita (kilowatt-hours) 2004	Electricity consumption per capita (% change) 1990–2004	Electrification rate (%) 2000–05 [a]	Population without electricity (millions) 2005	GDP per unit of energy use (2000 PPP US$ per kg of oil equivalent) 2004	GDP per unit of energy use (% change) 1990–2004	Forest area % of total land area (%) 2005	Forest area Total (thousand sq km) 2005	Forest area Total change (thousand sq km) 1990–2005	Forest area Average annual change (%) 1990–2005
54 Saint Kitts and Nevis	3,333 [e]	115.3	..	..	..	..	14.7	0.1	0.0	0.0
55 Tonga	327 [e]	30.8	..	..	..	..	5.0	(.)	0.0	0.0
56 Libyan Arab Jamahiriya	3,147	-22.2	97	0.2	..	..	0.1	2.2	0.0	0.0
57 Antigua and Barbuda	1,346 [e]	-10.7	..	..	..	..	21.4	0.1	..	..
58 Oman	5,079	83.2	96	0.1	3.0	-29.9	(.)	(.)	0.0	0.0
59 Trinidad and Tobago	4,921	67.1	99	0.0	1.3	-5.3	44.1	2.3	-0.1	-0.3
60 Romania	2,548	-19.9	..	..	4.5	80.9	27.7	63.7	(.)	0.0
61 Saudi Arabia	6,902	57.9	97	0.8	2.0	-28.2	1.3	27.3	0.0	0.0
62 Panama	1,807	51.0	85	0.5	8.4	13.5	57.7	42.9	-0.8	-0.1
63 Malaysia	3,196	129.6	98	0.6	4.1	-5.1	63.6	208.9	-14.9	-0.4
64 Belarus	3,508	..	..	..	2.4	89.6	38.0	78.9	5.2	0.5
65 Mauritius	1,775	147.2	94	0.1	..	..	18.2	0.4	(.)	-0.3
66 Bosnia and Herzegovina	2,690	..	..	..	5.3	..	43.1	21.9	-0.3	-0.1
67 Russian Federation	6,425	..	..	..	2.0	28.3	47.9	8,087.9	-1.6	0.0
68 Albania	1,847	82.3	..	..	5.9	55.2	29.0	7.9	0.1	(.)
69 Macedonia (TFYR)	3,863	..	..	..	4.6	13.7	35.8	9.1	0.0	0.0
70 Brazil	2,340	39.5	97	6.5	6.8	-6.7	57.2	4,777.0	-423.3	-0.5
MEDIUM HUMAN DEVELOPMENT										
71 Dominica	1,129	170.7	..	..	..	..	61.3	0.5	(.)	-0.5
72 Saint Lucia	1,879	136.6	..	..	..	..	27.9	0.2	0.0	0.0
73 Kazakhstan	4,320	..	..	..	1.9	86.7	1.2	33.4	-0.9	-0.2
74 Venezuela (Bolivarian Republic of)	3,770	23.6	99	0.4	2.6	0.5	54.1	477.1	-43.1	-0.6
75 Colombia	1,074 [e]	3.1	86	6.3	10.9	29.6	58.5	607.3	-7.1	-0.1
76 Ukraine	3,727	..	..	..	2.0	11.7	16.5	95.8	3.0	0.2
77 Samoa	619 [e]	103.0	..	..	..	..	60.4	1.7	0.4	2.1
78 Thailand	2,020 [e]	141.1	99	0.6	4.9	-14.0	28.4	145.2	-14.5	-0.6
79 Dominican Republic	1,536	197.7	93	0.7	7.6	7.0	28.4	13.8	..	..
80 Belize	686 [e]	13.8	..	..	..	..	72.5	16.5	..	..
81 China	1,684	212.4	99	8.5	4.4	108.6	21.2	1,972.9	401.5	1.7
82 Grenada	1,963	225.0	..	..	..	..	12.2	(.)	..	..
83 Armenia	1,744	..	..	..	5.6	122.8	10.0	2.8	-0.6	-1.2
84 Turkey	2,122	109.5	..	..	6.2	6.4	13.2	101.8	5.0	0.3
85 Suriname	3,437	-9.9	..	..	..	..	94.7	147.8	0.0	0.0
86 Jordan	1,738	53.4	100	0.0	3.6	4.3	0.9	0.8	0.0	0.0
87 Peru	927	44.6	72	7.7	10.9	30.0	53.7	687.4	-14.1	-0.1
88 Lebanon	2,691	374.6	100	0.0	3.5	29.9	13.3	1.4 [f]	0.2	0.8
89 Ecuador	1,092	77.3	90	1.3	4.8	-17.7	39.2	108.5	-29.6	-1.4
90 Philippines	677	68.8	81	16.2	7.9	-12.7	24.0	71.6	-34.1	-2.2
91 Tunisia	1,313	93.7	99	0.1	8.2	22.2	6.8	10.6	4.1	4.3
92 Fiji	926 [e]	44.9	..	..	..	..	54.7	10.0	0.2	0.1
93 Saint Vincent and the Grenadines	1,030	114.1	..	..	..	..	27.4	0.1	(.)	1.5
94 Iran (Islamic Republic of)	2,460	126.7	97	1.8	3.1	-13.6	6.8	110.8	0.0	0.0
95 Paraguay	1,146	99.3	86	0.9	6.4	-2.0	46.5	184.8	-26.8	-0.8
96 Georgia	1,577	..	..	..	4.1	236.3	39.7	27.6	..	..
97 Guyana	1,090	155.3	..	..	..	..	76.7	151.0 [f]	..	..
98 Azerbaijan	2,796	..	..	..	2.5	..	11.3	9.4	..	..
99 Sri Lanka	420	127.0	66	6.7	8.3	13.8	29.9	19.3	-4.2	-1.2
100 Maldives	539	385.6	..	..	..	..	3.0	(.)	0.0	0.0
101 Jamaica	2,697	160.8	87	0.3	2.5	-18.2	31.3	3.4	-0.1	-0.1
102 Cape Verde	529	330.1	..	..	..	..	20.7	0.8	0.3	3.0
103 El Salvador	732	62.7	80	1.4	7.0	-3.1	14.4	3.0	-0.8	-1.4
104 Algeria	889	40.7	98	0.6	6.0	4.5	1.0	22.8	4.9	1.8
105 Viet Nam	560	324.2	84	13.2	4.2	26.5	39.7	129.3	35.7	2.5
106 Occupied Palestinian Territories	513	..	..	..	..	..	1.5	0.1 [f]	0.0	0.0

TABLE 22

Energy and the environment

							Forest area			
	Electricity consumption per capita		Electrification rate	Population without electricity	GDP per unit of energy use		% of total land area	Total	Total change	Average annual change
	(kilowatt-hours)	(% change)	(%)	(millions)	(2000 PPP US$ per kg of oil equivalent)	(% change)	(%)	(thousand sq km)	(thousand sq km)	(%)
HDI rank	2004	1990–2004	2000–05 a	2005	2004	1990–2004	2005	2005	1990–2005	1990–2005
107 Indonesia	476 e	75.0	54	101.2	4.1	-0.1	48.8	885.0	-280.7	-1.6
108 Syrian Arab Republic	1,784	88.4	90	1.9	3.4	19.9	2.5	4.6	0.9	1.6
109 Turkmenistan	2,060	..	..	..	1.3 g	-21.3	8.8	41.3	0.0	0.0
110 Nicaragua	525	37.1	69	1.7	5.2	-2.3	42.7	51.9	-13.5	-1.4
111 Moldova	1,554	..	..	..	2.0	40.8	10.0	3.3	0.1	0.2
112 Egypt	1,465 e	93.0	98	1.5	4.9	-2.2	0.1	0.7	0.2	3.5
113 Uzbekistan	1,944	..	..	..	0.8	11.1	8.0	33.0	2.5	0.5
114 Mongolia	1,260	-25.2	65	1.0	..	..	6.5	102.5	-12.4	-0.7
115 Honduras	730	79.4	62	2.7	4.8	-3.9	41.5	46.5	-27.4	-2.5
116 Kyrgyzstan	2,320	..	..	..	3.3	92.3	4.5	8.7	0.3	0.3
117 Bolivia	493	42.1	64	3.3	4.5	-10.6	54.2	587.4	-40.6	-0.4
118 Guatemala	532	100.0	79	2.7	6.4	-3.6	36.3	39.4	-8.1	-1.1
119 Gabon	1,128	5.4	48	0.7	4.9	3.1	84.5	217.8	-1.5	(.)
120 Vanuatu	206 e	18.4	..	..	..	..	36.1	4.4	0.0	0.0
121 South Africa	4,818 h	20.8 h	70	14.0	3.7	-4.5	7.6	92.0	0.0	0.0
122 Tajikistan	2,638	..	..	..	2.1	139.6	2.9	4.1	(.)	(.)
123 Sao Tome and Principe	99 e	-23.8	..	..	..	..	28.4	0.3	0.0	0.0
124 Botswana	.. i	.. i	39	1.1	8.6	40.0	21.1	119.4	-17.8	-0.9
125 Namibia	.. i	.. i	34	1.4	10.2	-16.5	9.3	76.6	-11.0	-0.8
126 Morocco	652	84.7	85	4.5	10.3	-13.9	9.8	43.6	0.8	0.1
127 Equatorial Guinea	52 e	0	..	..	..	..	58.2	16.3	-2.3	-0.8
128 India	618	77.6	56	487.2	5.5	37.1	22.8	677.0	37.6	0.4
129 Solomon Islands	107 e	13.8	..	..	..	..	77.6	21.7	-6.0	-1.4
130 Lao People's Democratic Republic	126 e	80.0	..	..	..	..	69.9	161.4	-11.7	-0.5
131 Cambodia	10 e	-44.4	20	10.9	..	..	59.2	104.5	-25.0	-1.3
132 Myanmar	129	111.5	11	45.1	..	..	49.0	322.2	-70.0	-1.2
133 Bhutan	229 e	126.7	..	..	..	..	68.0	32.0	1.6	0.4
134 Comoros	31 e	3.3	..	..	..	..	2.9	0.1	-0.1	-3.9
135 Ghana	289	-22.3	49	11.3	5.4	18.3	24.2	55.2	-19.3	-1.7
136 Pakistan	564	61.6	54	71.1	4.2	7.7	2.5	19.0	-6.3	-1.6
137 Mauritania	112 e	60.0	..	..	..	..	0.3	2.7	-1.5	-2.4
138 Lesotho	.. i	.. i	11	1.9	..	..	0.3	0.1	(.)	4.0
139 Congo	229	-2.1	20	3.2	3.3	45.4	65.8	224.7	-2.6	-0.1
140 Bangladesh	154	111.0	32	96.2	10.5	7.2	6.7	8.7	-0.1	-0.1
141 Swaziland	.. i	.. i	..	..	..	..	31.5	5.4	0.7	1.0
142 Nepal	86	104.8	33	18.1	4.0	18.4	25.4	36.4	-11.8	-1.6
143 Madagascar	56	5.7	15	15.2	..	..	22.1	128.4	-8.5	-0.4
144 Cameroon	256	8.9	47	8.7	4.5	-4.4	45.6	212.5	-33.0	-0.9
145 Papua New Guinea	620 e	28.1	..	..	..	..	65.0	294.4	-20.9	-0.4
146 Haiti	61	-17.6	36	5.5	6.2	-39.9	3.8	1.1	-0.1	-0.6
147 Sudan	116	123.1	30	25.4	3.7	33.2	28.4	675.5	-88.4	-0.8
148 Kenya	169	26.1	14	29.4	2.1	-3.8	6.2	35.2	-1.9	-0.3
149 Djibouti	260 e	-46.8	..	..	..	..	0.2	0.1	..	..
150 Timor-Leste	294 e	..	..	..	..	..	53.7	8.0	-1.7	-1.2
151 Zimbabwe	924	-10.1	34	8.7	2.6	-13.4	45.3	175.4	-46.9	-1.4
152 Togo	102	1.0	17	5.1	3.1	-26.9	7.1	3.9	-3.0	-2.9
153 Yemen	208	34.2	36	13.2	2.8	-6.0	1.0	5.5	0.0	0.0
154 Uganda	63 e	61.5	9	24.6	..	..	18.4	36.3	-13.0	-1.8
155 Gambia	98 e	30.7	..	..	..	..	41.7	4.7	0.3	0.4
LOW HUMAN DEVELOPMENT										
156 Senegal	206	70.2	33	7.8	6.5	28.2	45.0	86.7	-6.8	-0.5
157 Eritrea	67	..	20	3.5	..	..	15.4	15.5	-0.7	-0.3
158 Nigeria	157	-1.9	46	71.1	1.4	22.7	12.2	110.9	-61.5	-2.4
159 Tanzania (United Republic of)	69	4.5	11	34.2	1.3	-12.5	39.9	352.6	-61.8	-1.0

TABLE 22

	Electricity consumption per capita		Electrification rate	Population without electricity	GDP per unit of energy use		Forest area			
	(kilowatt-hours)	(% change)	(%)	(millions)	(2000 PPP US$ per kg of oil equivalent)	(% change)	% of total land area (%)	Total (thousand sq km)	Total change (thousand sq km)	Average annual change (%)
HDI rank	2004	1990–2004	2000–05 ᵃ	2005	2004	1990–2004	2005	2005	1990–2005	1990–2005
160 Guinea	87 ᵉ	3.6	..	..	..	..	27.4	67.2	-6.8	-0.6
161 Rwanda	31 ᵉ	24.0	..	..	..	..	19.5	4.8	1.6	3.4
162 Angola	220	161.9	15	13.5	3.3	-12.4	47.4	591.0	-18.7	-0.2
163 Benin	81	72.3	22	6.5	3.3	25.8	21.3	23.5	-9.7	-1.9
164 Malawi	100 ᵉ	14.9	7	11.8	..	..	36.2	34.0	-4.9	-0.8
165 Zambia	721	-7.8	19	9.5	1.5	0.4	57.1	424.5	-66.7	-0.9
166 Côte d'Ivoire	224	7.7	50	9.1	3.7	-29.1	32.7	104.1	1.8	0.1
167 Burundi	22 ᵉ	-4.3	..	..	..	..	5.9	1.5	-1.4	-3.2
168 Congo (Democratic Republic of the)	92	-42.1	6	53.8	2.2	-55.8	58.9	1,336.1	-69.2	-0.3
169 Ethiopia	36	..	15	60.8	2.8	5.8	11.9	130.0	-21.1	-0.9
170 Chad	11 ᵉ	-31.3	..	..	..	..	9.5	119.2	-11.9	-0.6
171 Central African Republic	28 ᵉ	-12.5	..	..	..	..	36.5	227.6	-4.5	-0.1
172 Mozambique	545	856.1	6	18.6	2.6	105.8	24.6	192.6	-7.5	-0.2
173 Mali	41 ᵉ	36.7	..	..	..	..	10.3	125.7	-15.0	-0.7
174 Niger	40 ᵉ	-13.0	..	..	..	..	1.0	12.7	-6.8	-2.3
175 Guinea-Bissau	44 ᵉ	4.8	..	..	..	..	73.7	20.7	-1.4	-0.4
176 Burkina Faso	31 ᵉ	55.0	7	12.4	..	..	29.0	67.9	-3.6	-0.3
177 Sierra Leone	24	-54.7	..	..	..	..	38.5	27.5	-2.9	-0.6
Developing countries	1,221	..	68 ʲ	1,569.0 ʲ	4.6	..	27.9	21,147.8	-1,381.7	-0.4
Least developed countries	119	..	..	..	..	..	27.5	5,541.6	-583.6	-0.6
Arab States	1,841	..	..	..	3.4	..	7.2	877.7	-88.0	-0.6
East Asia and the Pacific	1,599	..	..	..	..	..	28.6	4,579.3	-75.5	0.1
Latin America and the Caribbean	2,043	..	90 ʲ	45.0 ʲ	6.2	..	45.9	9,159.0	-686.3	-0.5
South Asia	628	..	..	..	5.1	..	14.2	911.8	12.5	0.1
Sub-Saharan Africa	478	..	26 ʲ	547.0 ʲ	..	..	26.8	5,516.4	-549.6	-0.6
Central and Eastern Europe and the CIS	4,539	..	..	..	2.6	..	38.3	8,856.5	22.7	(.)
OECD	8,795	..	100	..	5.3	..	30.9	10,382.4	67.9	0.1
High-income OECD	10,360	..	100	..	5.3	..	31.2	9,480.8	105.6	0.1
High human development	7,518	..	99	..	5.0	..	36.2	24,327.1	-366.8	-0.1
Medium human development	1,146	..	72	..	4.5	..	23.3	10,799.6	-462.4	-0.2
Low human development	134	..	25	..	..	..	29.8	4,076.5	-379.5	-0.5
High income	10,210	..	100	..	5.2	..	29.2	9,548.4	107.1	0.1
Middle income	2,039	..	90	..	4.2	..	33.8	23,132.3	-683.1	-0.2
Low income	449	..	45	..	..	..	23.9	6,745.6	-676.2	-0.6
World	2,701 ʲ	..	76 ʲ	1,577.0 ʲ	4.8 ʲ	..	30.3 ʲ	39,520.3 ʲ	-1,252.7 ʲ	-0.2

NOTES

a. Data refer to the most recent year available during the period specified.

b. Includes Liechtenstein.

c. Includes Monaco.

d. Includes San Marino.

e. Data are estimates produced by the UN Statistics Division.

f. Estimate produced by the Food and Agriculture Organization based on information provided by the country.

g. Data refer to a year or period other than that specified.

h. Data refer to the South African Customs Union, which includes Botswana, Lesotho, Namibia and Swaziland.

i. Included in data for South Africa.

j. Data are aggregates provided by original data source.

SOURCES

Column 1: UN2007d.

Column 2: calculated based on data from UN 2007b.

Column 3-4: IEA 2002 and IEA 2006.

Column 5: World Bank 2007b, based on data from IEA.

Columns 6: calculated based on data from World Bank 2007b.

Column 7-8: FAO 2006.

Columns 9–10: calculated based on data from FAO 2006.

Human development indicators

	Total primary energy supply[a] (Mt of oil equivalent)		Share of TPES[a]											
			Fossil fuels						Renewable energy[b]				Other	
			Coal[c] (%)		Oil[d] (%)		Natural Gas (%)		Hydro, solar, wind and geothermal (%)		Biomass and waste[e] (%)		Nuclear (%)	
HDI rank	1990	2005	1990	2005	1990	2005	1990	2005	1990	2005	1990	2005	1990	2005
HIGH HUMAN DEVELOPMENT														
1 Iceland	2.2	3.6	3.0	2.7	32.6	24.6	0.0	0.0	64.5	72.6	0.0	0.1	0.0	0.0
2 Norway	21.5	32.1	4.0	2.4	39.8	44.1	9.2	16.1	48.5	36.6	4.8	4.1	0.0	0.0
3 Australia	87.5	122.0	40.0	44.5	37.1	31.1	16.9	18.9	1.5	1.2	4.5	4.3	0.0	0.0
4 Canada	209.4	272.0	11.6	10.3	36.9	35.8	26.1	29.6	12.2	11.5	3.9	4.6	9.3	8.8
5 Ireland	10.4	15.3	33.3	17.6	47.0	56.0	18.1	22.7	0.6	1.0	1.0	1.6	0.0	0.0
6 Sweden	47.6	52.2	6.2	5.0	30.8	28.5	1.2	1.6	13.1	12.7	11.6	17.2	37.4	36.2
7 Switzerland	25.0	27.2	1.4	0.6	53.8	47.1	6.5	10.2	10.5	10.5	3.7	7.1	24.7	22.5
8 Japan	444.5	530.5	17.4	21.1	57.4	47.4	9.9	13.3	2.3	2.0	1.1	1.2	11.9	15.0
9 Netherlands	66.8	81.8	13.4	10.0	36.5	40.2	46.1	43.1	(.)	0.3	1.4	3.2	1.4	1.3
10 France	227.8	276.0	8.9	5.2	38.3	33.1	11.4	14.9	2.1	1.7	5.1	4.3	35.9	42.6
11 Finland	29.2	35.0	18.2	14.1	35.1	30.6	7.5	10.3	3.2	3.9	15.6	19.6	17.2	17.3
12 United States	1,927.5	2,340.3	23.8	23.7	40.0	40.7	22.8	21.8	2.0	1.5	3.2	3.2	8.3	9.0
13 Spain	91.1	145.2	21.2	14.1	51.0	49.1	5.5	20.5	2.4	2.5	4.5	3.5	15.5	10.3
14 Denmark	17.9	19.6	34.0	18.9	45.7	41.8	10.2	22.4	0.3	3.0	6.4	13.2	0.0	0.0
15 Austria	25.1	34.4	16.3	11.8	42.4	42.2	20.7	24.0	10.9	9.7	9.8	11.6	0.0	0.0
16 United Kingdom	212.2	233.9	29.7	16.1	38.9	36.2	22.2	36.3	0.2	0.3	0.3	1.7	8.1	9.1
17 Belgium	49.2	56.7	21.7	9.0	38.1	40.2	16.6	24.9	0.1	0.2	1.5	2.8	22.6	21.9
18 Luxembourg	3.6	4.8	31.7	1.7	45.9	66.2	12.0	24.7	0.2	0.3	0.7	1.2	0.0	0.0
19 New Zealand	13.8	16.9	8.2	11.8	28.8	40.3	28.3	18.9	30.7	23.8	4.0	5.1	0.0	0.0
20 Italy	148.0	185.2	9.9	8.9	57.3	44.2	26.4	38.1	3.8	4.3	0.6	2.3	0.0	0.0
21 Hong Kong, China (SAR)	10.7	18.1	51.5	36.8	49.4	47.7	0.0	12.1	0.0	0.0	0.5	0.3	0.0	0.0
22 Germany	356.2	344.7	36.1	23.7	35.5	35.8	15.4	23.4	0.4	1.3	1.3	3.5	11.2	12.3
23 Israel	12.1	19.5	19.8	39.2	77.3	51.2	0.2	6.6	3.0	3.7	(.)	(.)	0.0	0.0
24 Greece	22.2	31.0	36.4	28.9	57.7	57.1	0.6	7.6	1.0	2.1	4.0	3.3	0.0	0.0
25 Singapore	13.4	30.1	0.2	(.)	99.8	80.3	0.0	19.7	0.0	0.0	0.0	0.0	0.0	0.0
26 Korea (Republic of)	93.4	213.8	27.4	23.1	53.6	45.0	2.9	12.8	0.6	0.2	0.8	1.0	14.8	17.9
27 Slovenia	5.6	7.3	25.4	20.2	31.7	35.8	13.6	12.7	4.5	4.1	4.8	6.7	21.5	21.0
28 Cyprus	1.6	2.6	3.7	1.5	95.9	96.3	0.0	0.0	0.0	1.6	0.4	0.6	0.0	0.0
29 Portugal	17.7	27.2	15.5	12.3	66.0	58.5	0.0	13.8	4.5	2.4	14.0	10.8	0.0	0.0
30 Brunei Darussalam	1.8	2.6	0.0	0.0	6.8	29.7	92.2	69.6	0.0	0.0	1.0	0.7	0.0	0.0
31 Barbados	..	..	..	..	..	..	..	..	..	..	..	..	..	..
32 Czech Republic	49.0	45.2	64.2	44.7	18.3	22.1	10.7	17.0	0.2	0.5	0.0	3.9	6.7	14.3
33 Kuwait	8.5	28.1	0.0	0.0	40.1	66.5	59.8	33.5	0.0	0.0	0.1	0.0	0.0	0.0
34 Malta	0.8	0.9	23.8	0.0	76.2	100.0	0.0	0.0	0.0	0.0	0.0	0.0	0.0	0.0
35 Qatar	6.3	15.8	0.0	0.0	12.1	15.7	87.8	84.3	0.0	0.0	0.1	(.)	0.0	0.0
36 Hungary	28.6	27.8	21.4	11.1	29.8	26.0	31.2	43.6	0.4	0.4	1.3	4.0	12.5	13.0
37 Poland	99.9	93.0	75.5	58.7	13.3	23.8	9.0	13.2	0.1	0.2	2.2	5.1	0.0	0.0
38 Argentina	46.1	63.7	2.1	1.4	45.7	36.7	40.8	50.4	3.4	4.6	3.7	3.5	4.1	2.8
39 United Arab Emirates	22.5	46.9	0.0	0.0	39.9	27.9	60.1	72.1	0.0	0.0	0.0	(.)	0.0	0.0
40 Chile	14.1	29.6	18.4	13.9	45.8	39.2	10.6	23.8	6.2	7.0	19.0	15.5	0.0	0.0
41 Bahrain	4.8	8.1	0.0	0.0	26.5	23.2	73.5	76.8	0.0	0.0	0.0	0.0	0.0	0.0
42 Slovakia	21.3	18.8	36.7	22.5	21.1	18.4	23.9	31.2	0.8	2.2	0.8	2.4	14.7	24.8
43 Lithuania	16.2	8.6	4.9	2.3	42.2	29.1	28.9	28.8	0.7	2.4	1.8	8.3	27.8	31.9
44 Estonia	9.6	5.1	59.9	59.3	31.7	15.5	12.8	15.7	0.0	0.1	2.0	12.1	0.0	0.0
45 Latvia	7.8	4.7	6.3	1.3	45.3	29.7	30.6	28.8	5.4	6.1	8.5	30.2	0.0	0.0
46 Uruguay	2.3	2.9	(.)	0.1	58.6	59.4	0.0	3.1	26.8	19.9	24.2	15.4	0.0	0.0
47 Croatia	9.1	8.9	9.0	7.5	53.4	50.7	24.2	26.7	3.6	6.1	3.4	4.0	0.0	0.0
48 Costa Rica	2.0	3.8	0.1	0.5	48.3	51.4	0.0	0.0	14.4	41.1	36.6	7.0	0.0	0.0
49 Bahamas	..	..	..	..	..	..	..	..	..	..	..	..	..	..
50 Seychelles	..	..	..	..	..	..	..	..	..	..	..	..	..	..
51 Cuba	16.8	10.2	0.8	0.2	64.1	73.4	0.2	6.0	(.)	0.1	34.9	20.3	0.0	0.0
52 Mexico	124.3	176.5	2.8	4.9	67.0	58.8	18.6	25.0	5.2	4.9	5.9	4.7	0.6	1.6
53 Bulgaria	28.8	20.1	32.1	34.6	33.7	24.6	18.7	14.0	0.6	2.0	0.6	3.7	13.3	24.3

TABLE 23

		Total primary energy supply[a] (Mt of oil equivalent)		\multicolumn{6}{c}{Share of TPES[a]}						Other	
				\multicolumn{2}{c}{Fossil fuels}					\multicolumn{2}{c}{Renewable energy[b]}				
				Coal[c] (%)		Oil[d] (%)		Natural Gas (%)		Hydro, solar, wind and geothermal (%)		Biomass and waste[e] (%)		Nuclear (%)	
HDI rank		1990	2005	1990	2005	1990	2005	1990	2005	1990	2005	1990	2005	1990	2005
54	Saint Kitts and Nevis	..	..	..	..	..	..	..	..	..	..	..	..	..	..
55	Tonga	..	..	..	..	..	..	..	..	..	..	..	..	..	..
56	Libyan Arab Jamahiriya	11.5	19.0	0.0	0.0	63.8	72.2	35.1	27.0	0.0	0.0	1.1	0.8	0.0	0.0
57	Antigua and Barbuda	..	..	..	..	..	..	..	..	..	..	..	..	..	..
58	Oman	4.6	14.0	0.0	0.0	46.6	33.3	53.4	66.7	0.0	0.0	0.0	0.0	0.0	0.0
59	Trinidad and Tobago	6.0	12.7	0.0	0.0	21.4	13.6	77.8	86.2	0.0	0.0	0.8	0.2	0.0	0.0
60	Romania	62.4	38.3	20.7	22.7	29.2	24.6	46.2	36.4	1.6	4.7	1.0	8.5	0.0	3.8
61	Saudi Arabia	61.3	140.3	0.0	0.0	64.7	63.6	35.3	36.4	0.0	0.0	(.)	(.)	0.0	0.0
62	Panama	1.5	2.6	1.3	0.0	57.1	71.7	0.0	0.0	12.8	12.3	28.3	16.1	0.0	0.0
63	Malaysia	23.3	61.3	4.4	9.6	55.8	43.3	29.2	41.8	1.5	0.8	9.1	4.5	0.0	0.0
64	Belarus	42.2	26.6	5.6	2.4	62.2	27.9	29.7	63.7	(.)	(.)	0.5	4.8	0.0	0.0
65	Mauritius	..	..	..	..	..	..	..	..	..	..	..	..	..	..
66	Bosnia and Herzegovina	7.0	5.0	59.4	55.3	29.0	26.6	5.5	7.4	3.7	9.5	2.3	3.7	0.0	0.0
67	Russian Federation	878.3	646.7	20.7	16.0	31.0	20.6	41.8	54.1	1.6	2.4	1.4	1.1	3.6	6.1
68	Albania	2.7	2.4	23.7	1.0	45.2	68.1	7.6	0.6	9.2	19.3	13.6	9.6	0.0	0.0
69	Macedonia (TFYR)	2.7	2.7	57.6	48.7	40.6	33.2	0.0	2.3	1.6	5.1	0.0	5.6	0.0	0.0
70	Brazil	134.0	209.5	7.2	6.5	43.9	42.2	2.4	8.0	13.3	13.9	31.1	26.5	0.4	1.2
MEDIUM HUMAN DEVELOPMENT															
71	Dominica	..	..	..	..	..	..	..	..	..	..	..	..	..	..
72	Saint Lucia	..	..	..	..	..	..	..	..	..	..	..	..	..	..
73	Kazakhstan	73.7	52.4	54.2	52.6	28.2	14.5	14.5	33.5	0.9	1.3	0.2	0.1	0.0	0.0
74	Venezuela (Bolivarian Republic of)	43.9	60.9	1.1	0.1	43.2	50.4	47.2	38.1	7.2	10.6	1.2	0.9	0.0	0.0
75	Colombia	24.7	28.6	12.4	9.4	42.0	43.3	13.6	21.4	9.6	12.0	22.3	14.4	0.0	0.0
76	Ukraine	251.7	143.2	32.0	26.0	24.1	10.3	36.5	47.1	0.4	0.7	0.1	0.2	7.9	16.1
77	Samoa	..	..	..	..	..	..	..	..	..	..	..	..	..	..
78	Thailand	43.9	100.0	8.7	11.2	45.2	45.5	11.6	25.9	1.0	0.5	33.4	16.5	0.0	0.0
79	Dominican Republic	4.1	7.4	0.3	4.0	74.8	75.1	0.0	0.1	0.7	2.2	24.2	18.6	0.0	0.0
80	Belize	..	..	..	..	..	..	..	..	..	..	..	..	..	..
81	China	863.2	1,717.2	61.2	63.3	12.8	18.5	1.5	2.3	1.3	2.0	23.2	13.0	0.0	0.8
82	Grenada	..	..	..	..	..	..	..	..	..	..	..	..	..	..
83	Armenia	7.9	2.6	3.1	0.0	48.9	16.6	45.2	52.3	1.7	6.0	(.)	(.)	0.0	27.7
84	Turkey	53.0	85.2	31.9	26.4	44.6	35.1	5.4	26.7	4.6	5.6	13.6	6.3	0.0	0.0
85	Suriname	..	..	..	..	..	..	..	..	..	..	..	..	..	..
86	Jordan	3.5	7.1	0.0	0.0	95.3	78.5	2.9	19.5	1.7	1.0	0.1	(.)	0.0	0.0
87	Peru	10.0	13.8	1.5	6.7	58.5	53.5	4.1	10.6	9.0	12.8	26.9	16.4	0.0	0.0
88	Lebanon	2.3	5.6	0.0	2.4	93.7	92.9	0.0	0.0	1.9	1.8	4.4	2.3	0.0	0.0
89	Ecuador	6.1	10.4	0.0	0.0	75.9	83.5	3.7	4.4	7.0	5.7	13.5	5.1	0.0	0.0
90	Philippines	26.2	44.7	5.0	13.6	45.9	35.4	0.0	5.9	20.0	20.7	29.2	24.4	0.0	0.0
91	Tunisia	5.5	8.5	1.4	0.0	57.5	50.0	22.3	36.6	0.1	0.2	18.7	13.3	0.0	0.0
92	Fiji	..	..	..	..	..	..	..	..	..	..	..	..	..	..
93	Saint Vincent and the Grenadines	..	..	..	..	..	..	..	..	..	..	..	..	..	..
94	Iran (Islamic Republic of)	68.8	162.5	0.9	0.7	71.9	47.5	25.4	50.5	0.8	0.9	1.0	0.5	0.0	0.0
95	Paraguay	3.1	4.0	..	..	..	..	..	..	..	..	..	..	..	..
96	Georgia	12.3	3.2	4.8	0.5	47.1	25.3	36.9	33.5	5.3	17.0	3.7	20.1	0.0	0.0
97	Guyana	..	..	..	..	..	..	..	..	..	..	..	..	..	..
98	Azerbaijan	26.0	13.8	0.3	0.0	45.2	38.6	54.7	58.7	0.2	1.9	(.)	(.)	0.0	0.0
99	Sri Lanka	5.5	9.4	0.1	0.7	24.0	43.2	0.0	0.0	4.9	3.2	71.0	52.9	0.0	0.0
100	Maldives	..	..	..	..	..	..	..	..	..	..	..	..	..	..
101	Jamaica	2.9	3.8	1.1	1.0	82.4	86.5	0.0	0.0	0.3	0.3	16.2	12.2	0.0	0.0
102	Cape Verde	..	..	..	..	..	..	..	..	..	..	..	..	..	..
103	El Salvador	2.5	4.6	0.0	(.)	32.0	44.4	0.0	0.0	19.8	22.6	48.1	32.4	0.0	0.0
104	Algeria	23.9	34.8	2.6	2.0	40.6	31.7	56.7	66.0	(.)	0.1	0.1	0.2	0.0	0.0
105	Viet Nam	24.3	51.3	9.1	15.8	11.3	24.3	(.)	9.6	1.9	3.6	77.7	46.7	0.0	0.0
106	Occupied Palestinian Territories	..	..	..	..	..	..	..	..	..	..	..	..	..	..

Human development indicators

TABLE 23

Energy sources

		Total primary energy supply[a] (Mt of oil equivalent)		Share of TPES[a]											
				Fossil fuels						Renewable energy[b]				Other	
				Coal[c] (%)		Oil[d] (%)		Natural Gas (%)		Hydro, solar, wind and geothermal (%)		Biomass and waste[e] (%)		Nuclear (%)	
HDI rank		1990	2005	1990	2005	1990	2005	1990	2005	1990	2005	1990	2005	1990	2005
107	Indonesia	103.2	179.5	3.8	14.2	33.2	36.6	17.9	17.1	1.5	3.7	43.6	28.5	0.0	0.0
108	Syrian Arab Republic	11.7	17.9	0.0	(.)	86.3	65.3	11.7	33.0	2.0	1.7	(.)	(.)	0.0	0.0
109	Turkmenistan	19.6	16.3	1.5	0.0	38.0	26.5	62.4	75.0	0.3	(.)	0.0	0.0	0.0	0.0
110	Nicaragua	2.1	3.3	0.0	0.0	29.2	41.4	0.0	0.0	17.3	8.1	53.2	50.5	0.0	0.0
111	Moldova	10.0	3.6	20.0	2.1	49.3	19.0	32.8	69.0	0.2	0.2	0.4	2.1	0.0	0.0
112	Egypt	31.9	61.3	2.4	1.5	70.5	49.2	21.1	45.3	2.7	1.9	3.3	2.3	0.0	0.0
113	Uzbekistan	46.4	47.0	7.3	2.2	21.8	12.1	70.0	84.6	1.2	1.1	(.)	(.)	0.0	0.0
114	Mongolia	3.4	2.6	73.6	75.0	24.5	22.7	0.0	0.0	0.0	0.0	1.3	1.7	0.0	0.0
115	Honduras	2.4	3.9	(.)	2.9	31.1	51.0	0.0	0.0	8.1	4.0	62.0	42.0	0.0	0.0
116	Kyrgyzstan	7.6	2.8	33.2	19.7	40.5	22.5	19.9	22.1	11.3	43.8	0.1	0.1	0.0	0.0
117	Bolivia	2.8	5.3	0.0	0.0	46.5	56.2	22.6	25.8	3.7	4.0	27.2	14.0	0.0	0.0
118	Guatemala	4.5	8.0	0.0	3.1	28.8	40.5	0.0	0.0	3.4	3.5	67.9	53.2	0.0	0.0
119	Gabon	1.2	1.7	0.0	0.0	28.2	31.0	7.2	6.1	4.9	4.1	59.7	58.8	0.0	0.0
120	Vanuatu	..	..	..	..	..	..	..	..	..	..	..	..	..	..
121	South Africa	91.2	127.6	72.9	72.0	11.6	12.2	1.6	2.8	0.1	0.2	11.4	10.5	2.4	2.3
122	Tajikistan	5.6	3.5	11.2	1.3	36.8	42.6	24.8	14.0	25.4	41.5	0.0	0.0	0.0	0.0
123	Sao Tome and Principe	..	..	..	..	..	..	..	..	..	..	..	..	..	..
124	Botswana	1.3	1.9	39.4	31.5	26.9	36.5	0.0	0.0	(.)	(.)	33.1	24.1	0.0	0.0
125	Namibia	..	1.4	..	0.2	..	66.8	..	0.0	..	10.3	..	13.5	..	0.0
126	Morocco	6.7	13.8	16.8	32.3	76.1	60.2	0.6	2.8	1.6	1.0	4.7	3.3	0.0	0.0
127	Equatorial Guinea	..	..	..	..	..	..	..	..	..	..	..	..	..	..
128	India	319.9	537.3	33.2	38.7	19.6	23.9	3.1	5.4	1.9	1.7	41.7	29.4	0.5	0.8
129	Solomon Islands	..	..	..	..	..	..	..	..	..	..	..	..	..	..
130	Lao People's Democratic Republic	..	..	..	..	..	..	..	..	..	..	..	..	..	..
131	Cambodia	..	4.8	..	0.0	..	26.6	..	0.0	..	0.1	..	73.2	..	0.0
132	Myanmar	10.7	14.7	0.6	0.6	6.9	13.7	7.1	14.4	1.0	1.8	84.4	69.6	0.0	0.0
133	Bhutan	..	..	..	..	..	..	..	..	..	..	..	..	..	..
134	Comoros	..	..	..	..	..	..	..	..	..	..	..	..	..	..
135	Ghana	5.3	8.9	0.0	0.0	18.9	28.7	0.0	0.0	9.2	5.1	73.1	66.0	0.0	0.0
136	Pakistan	43.4	76.3	4.8	5.3	25.2	21.9	23.2	33.0	3.4	3.5	43.2	35.5	0.2	0.8
137	Mauritania	..	..	..	..	..	..	..	..	..	..	..	..	..	..
138	Lesotho	..	..	..	..	..	..	..	..	..	..	..	..	..	..
139	Congo	1.1	1.2	0.0	0.0	26.5	38.2	0.0	0.0	4.0	2.5	69.4	56.3	0.0	0.0
140	Bangladesh	12.8	24.2	2.2	1.4	14.7	19.1	29.0	44.7	0.6	0.5	53.5	34.3	0.0	0.0
141	Swaziland	..	..	..	..	..	..	..	..	..	..	..	..	..	..
142	Nepal	5.8	9.2	0.8	2.0	4.5	9.2	0.0	0.0	1.3	2.3	93.4	86.6	0.0	0.0
143	Madagascar	..	..	..	..	..	..	..	..	..	..	..	..	..	..
144	Cameroon	5.0	7.0	0.0	0.0	19.5	16.6	0.0	0.0	4.5	4.8	75.9	78.6	0.0	0.0
145	Papua New Guinea	..	..	..	..	..	..	..	..	..	..	..	..	..	..
146	Haiti	1.6	2.5	0.5	0.0	20.5	23.2	0.0	0.0	2.5	0.9	76.5	75.8	0.0	0.0
147	Sudan	10.6	18.4	0.0	0.0	17.5	19.9	0.0	0.0	0.8	0.6	81.7	79.5	0.0	0.0
148	Kenya	12.5	17.2	0.7	0.4	16.8	19.1	0.0	0.0	4.0	5.9	78.4	74.6	0.0	0.0
149	Djibouti	..	..	..	..	..	..	..	..	..	..	..	..	..	..
150	Timor-Leste	..	..	..	..	..	..	..	..	..	..	..	..	..	..
151	Zimbabwe	9.4	9.7	36.6	23.1	8.7	7.1	0.0	0.0	4.0	5.2	50.4	61.9	0.0	0.0
152	Togo	1.4	2.0	0.0	0.0	15.6	18.2	0.0	0.0	0.6	0.3	82.6	79.4	0.0	0.0
153	Yemen	2.6	6.7	0.0	0.0	97.0	98.8	0.0	0.0	0.0	0.0	3.0	1.2	0.0	0.0
154	Uganda	..	..	..	..	..	..	..	..	..	..	..	..	..	..
155	Gambia	..	..	..	..	..	..	..	..	..	..	..	..	..	..
LOW HUMAN DEVELOPMENT															
156	Senegal	2.2	3.0	0.0	3.1	39.2	55.3	0.2	0.4	0.0	2.0	60.6	39.2	0.0	0.0
157	Eritrea	..	0.8	..	0.0	..	35.2	..	0.0	..	(.)	..	64.8	..	0.0
158	Nigeria	70.9	103.8	0.1	(.)	15.0	13.9	4.6	7.5	0.5	0.7	79.8	78.0	0.0	0.0
159	Tanzania (United Republic of)	9.8	20.4	(.)	0.2	7.6	6.3	0.0	0.6	1.4	0.7	91.0	92.1	0.0	0.0

Human development indicators

TABLE 23

HDI rank	Total primary energy supply[a] (Mt of oil equivalent)		Share of TPES[a]											
			Fossil fuels						Renewable energy[b]				Other	
			Coal[c] (%)		Oil[d] (%)		Natural Gas (%)		Hydro, solar, wind and geothermal (%)		Biomass and waste[e] (%)		Nuclear (%)	
	1990	2005	1990	2005	1990	2005	1990	2005	1990	2005	1990	2005	1990	2005
160 Guinea	..	..	..	..	..	..	..	..	..	..	..	..	..	..
161 Rwanda	..	..	..	..	..	..	..	..	..	..	..	..	..	..
162 Angola	6.3	9.9	0.0	0.0	23.2	28.5	7.0	6.2	1.0	1.5	68.8	63.8	0.0	0.0
163 Benin	1.7	2.6	0.0	0.0	5.8	33.3	0.0	0.0	0.0	(.)	93.2	64.7	0.0	0.0
164 Malawi	..	..	..	..	..	..	..	..	..	..	..	..	..	..
165 Zambia	5.5	7.1	4.0	1.3	12.6	9.6	0.0	0.0	12.5	10.7	73.4	78.7	0.0	0.0
166 Côte d'Ivoire	4.4	7.8	0.0	0.0	24.8	23.9	0.0	17.8	2.6	1.6	72.1	58.3	0.0	0.0
167 Burundi	..	..	..	..	..	..	..	..	..	..	..	..	..	..
168 Congo (Democratic Republic of the)	11.9	17.0	1.8	1.5	10.1	3.2	0.0	0.0	4.1	3.7	84.0	92.5	0.0	0.0
169 Ethiopia	15.2	21.6	0.0	0.0	6.6	8.2	0.0	0.0	0.6	1.1	92.8	90.6	0.0	0.0
170 Chad	..	..	..	..	..	..	..	..	..	..	..	..	..	..
171 Central African Republic	..	..	..	..	..	..	..	..	..	..	..	..	..	..
172 Mozambique	7.2	10.2	0.5	0.0	4.6	5.2	0.0	0.2	0.3	11.2	94.4	85.4	0.0	0.0
173 Mali	..	..	..	..	..	..	..	..	..	..	..	..	..	..
174 Niger	..	..	..	..	..	..	..	..	..	..	..	..	..	..
175 Guinea-Bissau	..	..	..	..	..	..	..	..	..	..	..	..	..	..
176 Burkina Faso	..	..	..	..	..	..	..	..	..	..	..	..	..	..
177 Sierra Leone	..	..	..	..	..	..	..	..	..	..	..	..	..	..
Developing countries	.. T	.. T	30.3	32.5	30.5	31.0	9.4	14.1	2.7	2.9	26.3	18.0	0.8	1.4
Least developed countries	.. T	.. T	..	..	..	17.4	..	..	..	..	..	..	..	..
Arab States	237.4 T	477.1 T	1.1	1.3	59.5	54.2	33.9	40.2	0.7	0.4	4.8	3.8	0.0	0.0
East Asia and the Pacific	.. T	.. T	..	..	..	25.1	..	..	..	..	..	..	..	..
Latin America and the Caribbean	.. T	.. T	4.5	4.8	51.9	48.7	16.8	21.7	7.9	9.0	17.7	14.3	0.7	1.1
South Asia	456.2 T	818.9 T	23.9	26.1	27.7	28.3	9.0	17.9	1.9	1.7	37.1	25.3	0.4	0.6
Sub-Saharan Africa	.. T	.. T	..	..	..	13.8	..	..	..	..	..	..	..	..
Central and Eastern Europe and the CIS	1,751.5 T	1,266.3 T	27.6	22.6	29.8	20.5	36.1	46.0	1.4	2.2	1.2	2.1	4.0	7.0
OECD	4,525.5 T	5,547.6 T	23.5	20.4	42.0	40.5	18.6	21.8	2.9	2.7	3.1	3.5	9.9	11.0
High-income OECD	4,149.4 T	5,101.1 T	22.2	19.9	42.3	40.6	19.0	21.7	2.9	2.6	3.0	3.4	10.6	11.6
High human development	5,950.8 T	6,981.2 T	21.7	18.3	40.9	39.3	22.8	26.0	2.8	2.9	3.4	3.9	8.3	9.5
Medium human development	.. T	3,816.7 T	36.8	40.6	24.7	25.1	12.9	13.8	2.0	2.5	22.7	16.8	1.0	1.2
Low human development	.. T	.. T	..	..	..	13.1	..	..	..	..	..	..	..	..
High income	4,300.4 T	5,423.2 T	21.7	19.0	42.9	41.5	19.5	22.7	2.8	2.5	2.9	3.2	10.2	11.0
Middle income	3,556.4 T	4,594.4 T	31.6	34.3	31.0	28.3	21.7	21.7	2.3	3.1	11.4	10.1	2.1	2.4
Low income	.. T	.. T	..	23.3	..	20.6	..	11.6	..	2.3	..	41.8	..	0.5
World	8,757.7 T[f]	11,433.9 T[f]	25.3	25.3[g]	36.8[g]	35.0[g]	19.1[g]	20.7[g]	2.5[g]	2.6[g]	10.3[g]	10.0[g]	6.0[g]	6.3[g]

NOTES

a. Total primary energy supply (TPES) is made up of 'indigenous production + imports - exports - international marine bunkers ± stock changes'. TPES is a measure of commercial energy consumption. In some instances, the sum of the shares by energy source may not sum up to 100% because pumped storage generation has not been deducted from hydroelectricity generation.

b. In 2005, 12.6% of the world's energy needs were supplied by renewable sources. Hydro-electric power constitutes 17% of this total, solar/wind/ other 1%, geothermal 3% and biomass and waste 79%. Shares for individual countries are different.

c. Coal and coal products.

d. Crude, natural gas liquids (NGLs), feedstocks and petroleum products.

e. Biomass, also referred to as traditional fuel, is comprised of animal and plant materials (wood, vegetal waste, ethanol, animal materials/wastes and sulphite lyes). Waste is comprised of municipal waste (wastes produced by the residential, commercial and public service sectors that are collected by local authorities for disposal in a central location for the production of heat and/or power) and industrial waste.

f. Data is a world aggregate from IEA 2007.

g. Data calculated based on world aggregates from IEA 2007.

SOURCES

Columns 1-2: IEA 2007.

Columns 3-14: calculated based on data on primary energy supply from IEA 2007.

Human development indicators

TABLE 24

... while preserving it for future generations ...

Carbon dioxide emissions and stocks

		Carbon dioxide emissions[a]												
		Total (Mt CO$_2$)		Annual change (%)	Share of world total[b] (%)		Per capita (t CO$_2$)		Carbon intensity of energy CO$_2$ emissions per unit of energy use (kt of CO$_2$ per kt of oil equivalent)		Carbon intensity of growth CO$_2$ emissions per unit of GDP (kt of CO$_2$ per million 2000 PPP US$)		Carbon dioxide emissions from forest biomass[c] (Mt CO$_2$ / year)	Carbon stocks in forest biomass[d] (Mt Carbon)
HDI rank		1990	2004	1990–2004	1990	2004	1990	2004	1990	2004	1990	2004	1990–2005	2005
HIGH HUMAN DEVELOPMENT														
1	Iceland	2.0	2.2	0.7	(.)	(.)	7.9	7.6	0.93	0.64	0.32	0.24	-0.1	1.5
2	Norway	33.2	87.5	11.7	0.1	0.3	7.8	19.1	1.54	3.17	0.31	0.53	-15.6	344.0
3	Australia	278.5	326.6	1.2	1.2	1.1	16.3	16.2	3.18	2.82	0.81	0.58	..	8,339.0
4	Canada	415.8	639.0	3.8	1.8	2.2	15.0	20.0	1.99	2.38	0.66	0.69	..	..
5	Ireland	30.6	42.3	2.7	0.1	0.1	8.8	10.5	2.94	2.78	0.55	0.31	-1.0	19.8
6	Sweden	49.5	53.0	0.5	0.2	0.2	5.8	5.9	1.04	0.98	0.26	0.21	-30.2	1,170.0
7	Switzerland	42.7	40.4	-0.4	0.2	0.1	6.2	5.4	1.71	1.49	0.21	0.17	-6.1	154.0
8	Japan	1,070.7	1,257.2	1.2	4.7	4.3	8.7	9.9	2.40	2.36	0.37	0.36	-118.5	1,892.0
9	Netherlands	141.0	142.0	(.)	0.6	0.5	9.4	8.7	2.11	1.73	0.41	0.30	-1.2	25.0
10	France	363.8	373.5	0.2	1.6	1.3	6.4	6.0	1.60	1.36	0.29	0.23	-44.2	1,165.0
11	Finland	51.2	65.8	2.0	0.2	0.2	10.3	12.6	1.76	1.73	0.46	0.45	-22.5	815.7
12	United States	4,818.3	6,045.8	1.8	21.2	20.9	19.3	20.6	2.50	2.60	0.68	0.56	-499.5	18,964.0
13	Spain	212.1	330.3	4.0	0.9	1.1	5.5	7.6	2.33	2.32	0.31	0.33	-28.3	392.0
14	Denmark	49.8	52.9	0.5	0.2	0.2	9.7	9.8	2.78	2.64	0.42	0.33	-1.0	26.0
15	Austria	57.6	69.8	1.5	0.3	0.2	7.4	8.6	2.30	2.10	0.32	0.29	..	..
16	United Kingdom	579.4	586.9	0.1	2.6	2.0	10.0	9.8	2.73	2.51	0.47	0.34	-4.2	112.0
17	Belgium	100.6	100.7	(.)	0.4	0.3	10.1	9.7	2.05	1.74	0.45	0.34	-3.7	65.3
18	Luxembourg	9.9	11.3	1.0	(.)	(.)	25.9	25.0	2.77	2.37	0.78	0.48	-0.5	9.0
19	New Zealand	22.6	31.6	2.8	0.1	0.1	6.7	7.7	1.65	1.79	0.39	0.35	..	..
20	Italy	389.7	449.7	1.1	1.7	1.6	6.9	7.8	2.63	2.44	0.32	0.30	-51.9	636.0
21	Hong Kong, China (SAR)	26.2	37.4	3.1	0.1	0.1	4.6	5.5	2.46	2.18	0.23	0.19	..	..
22	Germany	980.4 [h]	808.3	-1.3	4.3 [h]	2.8	12.3 [h]	9.8	2.75 [h]	2.32	0.58 [h]	0.38	-74.9	1,303.0
23	Israel	33.1	71.2	8.2	0.1	0.2	6.9	10.4	2.74	3.43	0.39	0.47	..	..
24	Greece	72.4	96.6	2.4	0.3	0.3	7.1	8.8	3.26	3.17	0.49	0.43	-1.7	58.7
25	Singapore	45.1	52.2	1.1	0.2	0.2	14.9	12.3	3.37	2.04	0.99	0.48	..	..
26	Korea (Republic of)	241.2	465.4	6.6	1.1	1.6	5.6	9.7	2.60	2.18	0.57	0.51	-32.2	258.0
27	Slovenia	12.3 [i]	16.2	2.6 [j]	0.1 [i]	0.1	6.2 [i]	8.1	2.46	2.26	0.51 [i]	0.43	-8.5	147.1
28	Cyprus	4.6	6.7	3.2	(.)	(.)	6.8	9.2	3.02	2.58	0.52	0.45	-0.1	2.8
29	Portugal	42.3	58.9	2.8	0.2	0.2	4.3	5.6	2.39	2.22	0.30	0.31	-8.9	113.8
30	Brunei Darussalam	5.8	8.8	3.7	(.)	(.)	23.0	24.0	3.20	3.27	..	..	1.2	39.3
31	Barbados	1.1	1.3	1.3	(.)	(.)	4.1	4.7	..	..	..	..	..	..
32	Czech Republic	138.4 [i]	116.9	-1.3 [j]	0.6 [i]	0.4	13.4 [i]	11.4	3.20	2.57	1.03 [i]	0.66	-12.6	326.3
33	Kuwait	43.4	99.3	9.2	0.2	0.3	20.3	37.1	5.13	3.95	..	1.81	..	..
34	Malta	2.2	2.5	0.7	(.)	(.)	6.3	6.1	2.88	2.70	0.53	0.36	0.0	0.1
35	Qatar	12.2	52.9	23.9	0.1	0.2	24.9	79.3	1.76	2.93	..	..	..	..
36	Hungary	60.1	57.1	-0.4	0.3	0.2	5.8	5.6	2.10	2.17	0.50	0.37	-6.2	173.0
37	Poland	347.6	307.1	-0.8	1.5	1.1	9.1	8.0	3.48	3.35	1.24	0.68	-44.1	895.6
38	Argentina	109.7	141.7	2.1	0.5	0.5	3.4	3.7	2.38	2.22	0.38	0.31	121.6	2,411.0
39	United Arab Emirates	54.7	149.1	12.3	0.2	0.5	27.2	34.1	2.43	3.40	1.19	1.57	-0.7	16.6
40	Chile	35.6	62.4	5.4	0.2	0.2	2.7	3.9	2.53	2.23	0.47	0.38	-105.9	1,945.9
41	Bahrain	11.7	16.9	3.2	0.1	0.1	24.2	23.9	2.43	2.26	1.92	1.30	..	..
42	Slovakia	44.3 [i]	36.3	-1.5 [j]	0.2 [i]	0.1	8.4 [i]	6.7	2.45	1.98	0.96 [i]	0.51	-9.8	202.9
43	Lithuania	21.4 [i]	13.3	-3.1 [j]	0.1 [i]	(.)	5.7 [i]	3.8	1.92	1.45	0.67 [i]	0.32	-6.3	128.9
44	Estonia	24.9 [i]	18.9	-2.0 [j]	0.1 [i]	0.1	16.1 [i]	14.0	3.96	3.66	2.46 [i]	1.12	..	167.2
45	Latvia	12.7 [i]	7.1	-3.7 [j]	0.1 [i]	(.)	4.8 [i]	3.0	2.15	1.54	0.85 [i]	0.28	-13.9	230.9
46	Uruguay	3.9	5.5	2.9	(.)	(.)	1.2	1.6	1.74	1.91	0.18	0.19	..	..
47	Croatia	17.4 [i]	23.5	2.9 [j]	0.1 [i]	0.1	3.9 [i]	5.3	2.59	2.66	0.52 [i]	0.48	-10.8	192.4
48	Costa Rica	2.9	6.4	8.5	(.)	(.)	1.0	1.5	1.44	1.73	0.15	0.17	3.4	192.8
49	Bahamas	1.9	2.0	0.2	(.)	(.)	7.6	6.7	..	..	0.46	..	..	..
50	Seychelles	0.1	0.5	27.2	(.)	(.)	1.6	6.7	..	..	0.13	0.44	0.0	3.7
51	Cuba	32.0	25.8	-1.4	0.1	0.1	3.0	2.3	1.91	2.41	..	..	-34.7	347.0
52	Mexico	413.3	437.8	0.4	1.8	1.5	5.0	4.2	3.32	2.65	0.65	0.46	..	..
53	Bulgaria	75.3	42.5	-3.1	0.3	0.1	8.4	5.5	2.61	2.25	1.29	0.72	-18.3	263.0

TABLE 24

Carbon dioxide emissions[a]

HDI rank	Total (Mt CO₂) 1990	Total (Mt CO₂) 2004	Annual change (%) 1990–2004	Share of world total[b] (%) 1990	Share of world total[b] (%) 2004	Per capita (t CO₂) 1990	Per capita (t CO₂) 2004	Carbon intensity of energy (kt of CO₂ per kt of oil equivalent) 1990	Carbon intensity of energy 2004	Carbon intensity of growth (kt of CO₂ per million 2000 PPP US$) 1990	Carbon intensity of growth 2004	Carbon dioxide emissions from forest biomass[c] (Mt CO₂/year) 1990–2005	Carbon stocks in forest biomass[d] (Mt Carbon) 2005
54 Saint Kitts and Nevis	0.1	0.1	6.3	(.)	(.)	1.5	3.2	..	..	0.20	0.22	..	..
55 Tonga	0.1	0.1	3.7	(.)	(.)	0.8	1.1	..	..	0.15	0.16	..	..
56 Libyan Arab Jamahiriya	37.8	59.9	4.2	0.2	0.2	9.1	9.3	3.27	3.29	..	..	0.0	6.4
57 Antigua and Barbuda	0.3	0.4	2.7	(.)	(.)	4.8	6.0	..	..	0.54	0.46	..	..
58 Oman	10.3	30.9	14.3	(.)	0.1	6.3	13.6	2.25	2.61	0.52	0.88	..	..
59 Trinidad and Tobago	16.9	32.5	6.6	0.1	0.1	13.9	24.9	2.80	2.88	1.98	2.05	0.2	23.6
60 Romania	155.1	90.4	-3.0	0.7	0.3	6.7	4.2	2.48	2.34	0.99	0.54	(.)	566.5
61 Saudi Arabia	254.8	308.2	1.5	1.1	1.1	15.9	13.6	3.78	2.19	1.18	1.02	0.0	17.5
62 Panama	3.1	5.7	5.8	(.)	(.)	1.3	1.8	2.10	2.22	0.29	0.28	9.8	620.0
63 Malaysia	55.3	177.5	15.8	0.2	0.6	3.0	7.5	2.44	3.13	0.56	0.76	3.4	3,510.0
64 Belarus	94.6 i	64.9	-2.6 i	0.4 i	0.2	9.2 i	6.6	2.43	2.42	1.96 i	1.03	-20.0	539.0
65 Mauritius	1.5	3.2	8.5	(.)	(.)	1.4	2.6	..	..	0.21	0.24	(.)	3.9
66 Bosnia and Herzegovina	4.7 i	15.6	19.2 i	(.) i	0.1	1.1 i	4.0	1.06	3.31	..	..	-10.9	175.5
67 Russian Federation	1,984.1 i	1,524.1	-1.9 i	8.8 i	5.3	13.4 i	10.6	2.56	2.38	1.61 i	1.17	71.8	32,210.0
68 Albania	7.3	3.7	-3.5	(.)	(.)	2.2	1.2	2.73	1.55	0.73	0.26	-0.7	52.0
69 Macedonia (TFYR)	10.6 i	10.4	-0.2 i	(.) i	(.)	5.2 i	5.1	3.63	3.86	0.91 i	0.83	0.0	20.3
70 Brazil	209.5	331.6	4.2	0.9	1.1	1.4	1.8	1.56	1.62	0.22	0.24	1,111.4	49,335.0
MEDIUM HUMAN DEVELOPMENT													
71 Dominica	0.1	0.1	5.8	(.)	(.)	0.8	1.5	..	..	0.17	0.26	..	..
72 Saint Lucia	0.2	0.4	9.1	(.)	(.)	1.2	2.2	..	..	0.24	0.38	..	..
73 Kazakhstan	259.2 i	200.2	-1.9 i	1.1 i	0.7	15.7 i	13.3	3.25	3.65	3.30 i	2.07	0.2	136.7
74 Venezuela (Bolivarian Republic of)	117.4	172.5	3.4	0.5	0.6	6.0	6.6	2.67	3.07	1.03	1.20	..	..
75 Colombia	58.0	53.6	-0.5	0.3	0.2	1.6	1.2	2.32	1.94	0.30	0.19	23.8	8,062.2
76 Ukraine	600.0 i	329.8	-3.8 i	2.6 i	1.1	11.5 i	7.0	2.86	2.35	1.59 i	1.18	-60.5	744.5
77 Samoa	0.1	0.2	1.5	(.)	(.)	0.8	0.8	..	..	0.19	0.16	..	..
78 Thailand	95.7	267.9	12.8	0.4	0.9	1.7	4.2	2.18	2.76	0.38	0.56	17.8	716.0
79 Dominican Republic	9.6	19.6	7.5	(.)	0.1	1.3	2.2	2.31	2.56	0.31	0.33	0.0	82.0
80 Belize	0.3	0.8	11.0	(.)	(.)	1.6	2.9	..	..	0.39	0.44	0.0	59.0
81 China	2,398.9	5,007.1	7.8	10.6	17.3	2.1	3.8	2.77	3.11	1.30	0.70	-334.9	6,096.0
82 Grenada	0.1	0.2	5.6	(.)	(.)	1.3	2.7	..	..	0.23	0.29	..	..
83 Armenia	3.7 i	3.6	-0.1 i	(.) i	(.)	1.0 i	1.2	0.86	1.71	0.65 i	0.31	0.4	18.1
84 Turkey	146.2	226.0	3.9	0.6	0.8	2.6	3.2	2.76	2.76	0.48	0.45	-18.0	816.8
85 Suriname	1.8	2.3	1.9	(.)	(.)	4.5	5.2	..	..	0.81	0.78	0.0	5,692.0
86 Jordan	10.2	16.5	4.4	(.)	0.1	3.1	2.9	2.91	2.52	0.84	0.66	0.0	2.3
87 Peru	21.0	31.5	3.5	0.1	0.1	1.0	1.1	2.11	2.38	0.25	0.22	..	..
88 Lebanon	9.1	16.3	5.6	(.)	0.1	3.3	4.2	3.94	3.01	1.24	0.92	..	1.8
89 Ecuador	16.7	29.3	5.4	0.1	0.1	1.6	2.2	2.73	2.90	0.50	0.60	..	..
90 Philippines	43.9	80.5	5.9	0.2	0.3	0.7	1.0	1.68	1.82	0.19	0.22	111.2	970.7
91 Tunisia	13.3	22.9	5.2	0.1	0.1	1.6	2.3	2.40	2.63	0.35	0.32	-0.9	9.8
92 Fiji	0.8	1.1	2.3	(.)	(.)	1.1	1.2	..	..	0.22 i	0.24	..	..
93 Saint Vincent and the Grenadines	0.1	0.2	10.4	(.)	(.)	0.8	1.7	..	..	0.16	0.29	..	..
94 Iran (Islamic Republic of)	218.3	433.3	7.0	1.0	1.5	4.0	6.4	3.17	2.97	0.85	0.93	-1.7	334.0
95 Paraguay	2.3	4.2	6.1	(.)	(.)	0.5	0.7	0.73	1.04	0.12	0.18	..	..
96 Georgia	15.1 i	3.9	-6.2 i	0.1 i	(.)	2.8 i	0.8	1.73	1.38	1.39 i	0.32	-4.6	210.0
97 Guyana	1.1	1.4	2.0	(.)	(.)	1.5	1.9	..	..	0.63	0.47	..	1,722.0
98 Azerbaijan	49.8 i	31.3	-3.1 i	0.2 i	0.1	6.9 i	3.8	2.99	2.42	1.92 i	1.06	0.0	57.9
99 Sri Lanka	3.8	11.5	14.8	(.)	(.)	0.2	0.6	0.68	1.22	0.09	0.15	3.2	40.0
100 Maldives	0.2	0.7	26.5	(.)	(.)	0.7	2.5	..	..	..	..	..	..
101 Jamaica	8.0	10.6	2.4	(.)	(.)	3.3	4.0	2.70	2.60	1.04	1.06	0.2	34.0
102 Cape Verde	0.1	0.3	15.2	(.)	(.)	0.3	0.7	..	..	0.08	0.11	-0.6	7.9
103 El Salvador	2.6	6.2	9.7	(.)	(.)	0.5	0.9	1.03	1.37	0.14	0.20	..	..
104 Algeria	77.0	193.9	10.8	0.3	0.7	3.0	5.5	3.23	5.89	0.56	0.99	-6.0	114.0
105 Viet Nam	21.4	98.6	25.8	0.1	0.3	0.3	1.2	0.88	1.96	0.28	0.47	-72.5	1,174.0
106 Occupied Palestinian Territories	..	0.6	..	..	(.)	..	0.2	..	..	..	..	..	..

Human development indicators

TABLE 24

Carbon dioxide emissions and stocks

		Carbon dioxide emissions[a]								Carbon intensity of energy CO2 emissions per unit of energy use (kt of CO2 per kt of oil equivalent)		Carbon intensity of growth CO2 emissions per unit of GDP (kt of CO2 per million 2000 PPP US$)		Carbon dioxide emissions from forest biomass[c] (Mt CO2 / year)	Carbon stocks in forest biomass[d] (Mt Carbon)
		Total (Mt CO2)		Annual change (%)	Share of world total[b] (%)		Per capita (t CO2)								
HDI rank		1990	2004	1990–2004	1990	2004	1990	2004	1990	2004	1990	2004	1990–2005	2005
107	Indonesia	213.8	378.0	5.5	0.9	1.3	1.2	1.7	2.19	2.17	0.54	0.53	2,271.5	5,897.0
108	Syrian Arab Republic	35.9	68.4	6.5	0.2	0.2	3.0	3.8	3.08	3.71	1.11	1.11	..	..
109	Turkmenistan	28.0[i]	41.7	4.1[j]	0.1[i]	0.1	7.0[i]	8.8	2.48	2.68	1.54[i]	..	-0.2	17.4
110	Nicaragua	2.6	4.0	3.7	(.)	(.)	0.7	0.7	1.25	1.22	0.24	0.24	45.4	716.0
111	Moldova	20.9[i]	7.7	-5.3[j]	0.1[i]	(.)	4.8[i]	1.8	3.03	2.27	2.23[i]	1.05	-0.7	13.2
112	Egypt	75.4	158.1	7.8	0.3	0.5	1.5	2.3	2.37	2.78	0.48	0.58	-0.6	7.1
113	Uzbekistan	118.1[i]	137.8	1.4[j]	0.5[i]	0.5	5.5[i]	5.3	2.62	2.55	3.55[i]	3.07	-1.7	12.4
114	Mongolia	10.0	8.5	-1.0	(.)	(.)	4.7	3.1	..	..	2.71	1.90	16.9	573.9
115	Honduras	2.6	7.6	13.8	(.)	(.)	0.5	1.1	1.07	1.97	0.19	0.36	..	..
116	Kyrgyzstan	11.0[i]	5.7	-4.0[j]	(.)[i]	(.)	2.4[i]	1.1	2.18	2.06	1.26[i]	0.65	-0.8	12.6
117	Bolivia	5.5	7.0	1.9	(.)	(.)	0.9	0.8	1.98	1.40	0.40	0.31	89.4	5,296.0
118	Guatemala	5.1	12.2	10.0	(.)	(.)	0.6	1.0	1.14	1.61	0.17	0.25	25.0	498.0
119	Gabon	6.0	1.4	-5.5	(.)	(.)	6.4	1.0	4.82	0.81	0.96	0.16	5.9	3,643.0
120	Vanuatu	0.1	0.1	2.4	(.)	(.)	0.5	0.4	..	..	0.16	0.15	..	..
121	South Africa	331.8	436.8	2.3	1.5	1.5	9.1	9.8	3.64	3.33	1.03	0.99	0.0	823.9
122	Tajikistan	20.6[i]	5.0	-6.3[j]	0.1[i]	(.)	3.7[i]	0.8	2.26	1.50	2.38[i]	0.68	0.1	2.8
123	Sao Tome and Principe	0.1	0.1	2.8	(.)	(.)	0.6	0.5	..	..	0.32	0.31	0.0	4.6
124	Botswana	2.2	4.3	7.0	(.)	(.)	1.7	2.4	1.71	2.30	0.27	0.23	5.1	141.5
125	Namibia	(.)	2.5	..	(.)	(.)	0.0	1.2	0.02	1.85	(.)	0.19	8.1	230.9
126	Morocco	23.5	41.1	5.4	0.1	0.1	1.0	1.4	3.49	3.59	0.29	0.34	-9.5	240.0
127	Equatorial Guinea	0.1	5.4	..	(.)	(.)	0.3	10.5	..	..	0.28	1.57	3.9	115.0
128	India	681.7	1,342.1	6.9	3.0	4.6	0.8	1.2	1.89	2.34	0.48	0.44	-40.8	2,343.0
129	Solomon Islands	0.2	0.2	0.6	(.)	(.)	0.5	0.3	..	..	0.23	0.21	..	..
130	Lao People's Democratic Republic	0.2	1.3	32.4	(.)	(.)	0.1	0.2	..	..	0.05	0.13	26.4	1,487.0
131	Cambodia	0.5	0.5	1.3	(.)	(.)	(.)	(.)	..	..	..	0.02	80.6	1,266.0
132	Myanmar	4.3	9.8	9.2	(.)	(.)	0.1	0.2	0.40	0.69	..	..	156.6	3,168.0
133	Bhutan	0.1	0.4	15.9	(.)	(.)	0.1	0.2	..	..	..	..	-7.3	345.0
134	Comoros	0.1	0.1	2.4	(.)	(.)	0.1	0.1	..	..	0.08	0.09	0.2	0.8
135	Ghana	3.8	7.2	6.5	(.)	(.)	0.3	0.3	0.71	0.86	0.15	0.16	40.9	496.4
136	Pakistan	68.0	125.6	6.0	0.3	0.4	0.6	0.8	1.57	1.69	0.39	0.41	22.2	259.0
137	Mauritania	2.6	2.6	-0.2	(.)	(.)	1.3	0.8	..	..	0.70	0.44	0.9	6.6
138	Lesotho	..	..	..	..	..	..	..	..	..	..	..	..	..
139	Congo	1.2	3.5	14.4	(.)	(.)	0.5	1.0	1.11	3.33	0.38	0.86	14.2	5,181.0
140	Bangladesh	15.4	37.1	10.1	0.1	0.1	0.1	0.3	1.20	1.63	0.12	0.15	1.2	31.0
141	Swaziland	0.4	1.0	8.9	(.)	(.)	0.5	0.8	..	..	0.13	0.20	0.2	23.4
142	Nepal	0.6	3.0	27.3	(.)	(.)	(.)	0.1	0.11	0.34	0.03	0.08	-26.9	485.0
143	Madagascar	0.9	2.7	13.6	(.)	(.)	0.1	0.1	..	..	0.08	0.19	50.8	3,130.0
144	Cameroon	1.6	3.8	9.9	(.)	(.)	0.1	0.3	0.32	0.55	0.07	0.12	72.1	1,902.0
145	Papua New Guinea	2.4	2.4	0.1	(.)	(.)	0.7	0.4	..	..	0.31	0.19	..	..
146	Haiti	1.0	1.8	5.5	(.)	(.)	0.1	0.2	0.63	0.80	0.07	0.14	0.2	8.3
147	Sudan	5.4	10.4	6.6	(.)	(.)	0.2	0.3	0.51	0.59	0.19	0.17	48.9	1,530.7
148	Kenya	5.8	10.6	5.8	(.)	(.)	0.3	0.3	0.47	0.63	0.22	0.30	5.5	334.7
149	Djibouti	0.4	0.4	0.3	(.)	(.)	1.0	0.5	..	..	0.22	0.25	0.0	0.4
150	Timor-Leste	..	0.2	..	..	(.)	..	0.2	..	..	..	..	..	..
151	Zimbabwe	16.6	10.6	-2.6	0.1	(.)	1.6	0.8	1.77	1.13	0.58	0.42	34.2	535.0
152	Togo	0.8	2.3	14.8	(.)	(.)	0.2	0.4	0.52	0.86	0.13	0.29	..	..
153	Yemen	10.1[i]	21.1	8.3[j]	(.)[i]	0.1	0.9[i,k]	1.0	3.25	3.31	1.15[i]	1.25	0.0	5.1
154	Uganda	0.8	1.8	8.9	(.)	(.)	(.)	0.1	..	..	0.06	0.05	12.1	138.2
155	Gambia	0.2	0.3	3.6	(.)	(.)	0.2	0.2	..	..	0.12	0.12	-0.5	33.2
LOW HUMAN DEVELOPMENT														
156	Senegal	3.1	5.0	4.2	(.)	(.)	0.4	0.4	1.40	1.81	0.28	0.28	6.8	371.0
157	Eritrea	..	0.8	..	..	(.)	..	0.2	..	..	..	0.17	..	..
158	Nigeria	45.3	114.0	10.8	0.2	0.4	0.5	0.9	0.64	1.15	0.59	0.92	181.6	1,401.5
159	Tanzania (United Republic of)	2.3	4.3	6.2	(.)	(.)	0.1	0.1	0.24	0.23	0.17	0.18	167.3	2,254.0

Human development indicators

TABLE 24

Carbon dioxide emissions[a]

HDI rank	Total (Mt CO$_2$) 1990	Total 2004	Annual change (%) 1990–2004	Share of world total[b] (%) 1990	Share 2004	Per capita (t CO$_2$) 1990	Per capita 2004	Carbon intensity of energy CO$_2$ emissions per unit of energy use (kt of CO$_2$ per kt of oil equivalent) 1990	2004	Carbon intensity of growth CO$_2$ emissions per unit of GDP (kt of CO$_2$ per million 2000 PPP US$) 1990	2004	Carbon dioxide emissions from forest biomass[c] (Mt CO$_2$ / year) 1990–2005	Carbon stocks in forest biomass[d] (Mt Carbon) 2005
160 Guinea	1.0	1.3	2.3	(.)	(.)	0.2	0.1	..	..	0.09	0.07	15.9	636.0
161 Rwanda	0.5	0.6	0.6	(.)	(.)	0.1	0.1	..	..	0.07	0.06	-2.1	44.1
162 Angola	4.6	7.9	5.0	(.)	(.)	0.5	0.7	0.74	0.83	0.25	0.29	37.6	4,829.3
163 Benin	0.7	2.4	16.7	(.)	(.)	0.1	0.3	0.43	0.96	0.16	0.29	..	..
164 Malawi	0.6	1.0	5.3	(.)	(.)	0.1	0.1	..	..	0.13	0.14	5.6	161.0
165 Zambia	2.4	2.3	-0.5	(.)	(.)	0.3	0.2	0.45	0.33	0.31	0.23	44.4	1,156.1
166 Côte d'Ivoire	5.4	5.2	-0.3	(.)	(.)	0.5	0.3	1.22	0.74	0.26	0.20	-9.0	1,864.0
167 Burundi	0.2	0.2	0.9	(.)	(.)	(.)	(.)	..	..	0.04	0.05	..	..
168 Congo (Democratic Republic of the)	4.0	2.1	-3.4	(.)	(.)	0.1	(.)	0.33	0.13	0.07	0.06	293.1	23,173.0
169 Ethiopia	3.0	8.0	12.1	(.)	(.)	0.1	0.1	0.20	0.38	0.07	0.13	13.4	252.0
170 Chad	0.1	0.1	-0.9	(.)	(.)	(.)	0.0	..	..	0.03	0.01	5.6	236.0
171 Central African Republic	0.2	0.3	2.0	(.)	(.)	0.1	0.1	..	..	0.05	0.06	13.7	2,801.0
172 Mozambique	1.0	2.2	8.4	(.)	(.)	0.1	0.1	0.14	0.25	0.12	0.11	5.7	606.3
173 Mali	0.4	0.6	2.4	(.)	(.)	(.)	(.)	..	..	0.07	0.05	7.1	241.9
174 Niger	1.0	1.2	1.1	(.)	(.)	0.1	0.1	..	..	0.16	0.13	1.7	12.5
175 Guinea-Bissau	0.2	0.3	2.1	(.)	(.)	0.2	0.2	..	..	0.21	0.24	0.5	61.0
176 Burkina Faso	1.0	1.1	0.7	(.)	(.)	0.1	0.1	..	..	0.13	0.08	19.1	298.0
177 Sierra Leone	0.3	1.0	14.1	(.)	(.)	0.1	0.2	..	..	0.10	0.27	..	..
Developing countries	6,831.1 T	12,303.3 T	5.7	30.1	42.5	1.7	2.4	2.34	2.59	0.64	0.56	5,091.5	190,359.7
Least developed countries	74.1 T	146.3 T	7.0	0.3	0.5	0.2	0.2	..	..	0.14	0.17	1,097.8	50,811.2
Arab States	733.6 T	1,348.4 T	6.0	3.2	4.7	3.4	4.5	3.02	2.94	0.75	0.86	44.4	2,393.3
East Asia and the Pacific	3,413.5 T	6,682.0 T	6.8	15.0	23.1	2.1	3.5	..	..	0.90	0.63	2,293.8	27,222.9
Latin America and the Caribbean	1,087.7 T	1,422.6 T	2.2	4.8	4.9	2.5	2.6	2.25	2.19	0.40	0.36	1,667.0	97,557.2
South Asia	990.7 T	1,954.6 T	7.0	4.4	6.7	0.8	1.3	1.94	2.34	0.49	0.46	-49.3	3,843.5
Sub-Saharan Africa	454.8 T	663.1 T	3.3	2.0	2.3	1.0	1.0	..	..	0.55	0.57	1,153.6	58,523.2
Central and Eastern Europe and the CIS	4,182.0 T	3,168.0 T	-2.0	18.4	10.9	10.3	7.9	2.71	2.51	1.49	0.97	-165.9	37,592.0
OECD	11,205.2 T	13,318.6 T	1.3	49.4	46.0	10.8	11.5	2.47	2.42	0.54	0.45	-999.7	59,956.6
High-income OECD	10,055.4 T	12,137.5 T	1.5	44.3	41.9	12.0	13.2	2.42	2.39	0.52	0.45	-979.6	45,488.9
High human development	14,495.5 T	16,615.8 T	1.0	63.9	57.3	9.8	10.1	2.45	2.40	0.60	0.48	89.8	152,467.3
Medium human development	5,944.4 T	10,215.2 T	5.1	26.2	35.2	1.8	2.5	2.39	2.76	0.83	0.61	3,026.5	86,534.2
Low human development	77.6 T	161.7 T	7.7	0.3	0.6	0.3	0.3	..	..	0.24	0.36	858.0	41,254.0
High income	10,572.1 T	12,975.1 T	1.6	46.6	44.8	12.1	13.3	2.44	2.40	0.53	0.46	-937.4	54,215.3
Middle income	8,971.5 T	12,162.9 T	2.5	39.5	42.0	3.4	4.0	2.57	2.76	0.95	0.65	3,693.1	170,735.6
Low income	1,323.4 T	2,083.9 T	4.1	5.8	7.2	0.8	0.9	..	..	0.47	0.43	1,275.1	56,686.1
World	22,702.5 T[b]	28,982.7 T[b]	2.0	100.0	100.0	4.3	4.5	2.64	2.63	0.68	0.55	4,038.1	282,650.1

NOTES

a. Refers to carbon dioxide emissions stemming from consumption of solid, liquid and gaseous fossil fuels as well as from gas flaring and the production of cement. Original values were reported in terms of metric carbon tonnes, in order to convert these values to metric tonnes of carbon dioxide a conversion factor of 3.664 (relative molecular weights 44/12) has been applied.

b. The world total includes carbon dioxide emissions not included in national totals, such as those from bunker fuels, oxidation of non-fuel hydrocarbon products (e.g., asphalt) and emissions by countries not shown in the main indicator tables. These emissions amount to approximately 5% of the world total. Thus the shares listed for individual countries in this table do not sum to 100%.

c. Refers to net emissions or sequestration due to changes in carbon stock of forest biomass. A positive number suggests carbon emissions while a negative number suggests carbon sequestration. It is assumed that all negative carbon stock changes are released as emissions.

d. Refers only to living biomass - above and below ground. Carbon in deadwood, soil and litter is not included.

e. Includes Monaco.

f. Includes American Samoa, Guam, Puerto Rico, Turks and Caicos and the US Virgin Islands.

g. Includes San Marino.

h. Data refers to the sum of the emissions from the former Federal Republic of Germany and the former German Democratic Republic in 1990.

i. In cases where data for 1990 are not available, data for the closest year between 1991 and 1992 have been used.

j. Refers to the 1992-2004 period.

SOURCES

Columns 1, 2 and 4–7: calculated based on data from CDIAC 2007.

Column 3: calculated on the basis of data in columns 1 and 2.

Columns 8–11: calculated based on data from CDIAC 2007 and World Bank 2007b.

Column 12: calculated based on data from FAO 2007b; aggregates calculated for HDRO by FAO.

Column 13: FAO 2007b; aggregates calculated for HDRO by FAO.

Human development indicators

Status of major international environmental treaties

HDI rank	Cartagena Protocol on Biosafety 2000	Framework Convention on Climate Change 1992	Kyoto Protocol to the Framework Convention on Climate Change 1997	Convention on Biological Diversity 1992	Vienna Convention for the Protection of the Ozone Layer 1988	Montreal Protocol on Substances that deplete the Ozone Layer 1989	Stockholm Convention on Persistent Organic Pollutants 2001	Convention of the Law of the Sea 1982	Convention to Combat Desertification 1994
HIGH HUMAN DEVELOPMENT									
1 Iceland	**2001**	1993	2002	1994	1989	1989	2002	1985	1997
2 Norway	2001	1993	2002	1993	1986	1988	2002	1996	1996
3 Australia	..	1992	**1998**	1993	1987	1989	2004	1994	2000
4 Canada	**2001**	1992	2002	1992	1986	1988	2001	2003	1995
5 Ireland	2003	1994	2002	1996	1988	1988	**2001**	1996	1997
6 Sweden	2002	1993	2002	1993	1986	1988	2002	1996	1995
7 Switzerland	2002	1993	2003	1994	1987	1988	2003	**1984**	1996
8 Japan	2003	1993	2002	1993	1988	1988	2002	1996	1998
9 Netherlands	2002	1993	2002	1994	1988	1988	2002	1996	1995
10 France	2003	1994	2002	1994	1987	1988	2004	1996	1997
11 Finland	2004	1994	2002	1994	1986	1988	2002	1996	1995
12 United States	..	1992	**1998**	**1993**	1986	1988	2001	..	2000
13 Spain	2002	1993	2002	1993	1988	1988	2004	1997	1996
14 Denmark	2002	1993	2002	1993	1988	1988	2003	2004	1995
15 Austria	2002	1994	2002	1994	1987	1989	2002	1995	1997
16 United Kingdom	2003	1993	2002	1994	1987	1988	2005	1997	1996
17 Belgium	2004	1996	2002	1996	1988	1988	2006	1998	1997
18 Luxembourg	2002	1994	2002	1994	1988	1988	2003	2000	1997
19 New Zealand	2005	1993	2002	1993	1987	1988	2004	1996	2000
20 Italy	2004	1994	2002	1994	1988	1988	**2001**	1995	1997
21 Hong Kong, China (SAR)	..	..	..	..	..	..	..	..	..
22 Germany	2003	1993	2002	1993	1988	1988	2002	1994	1996
23 Israel	..	1996	2004	1995	1992	1992	**2001**	..	1996
24 Greece	2004	1994	2002	1994	1988	1988	2006	1995	1997
25 Singapore	..	1997	2006	1995	1989	1989	2005	1994	1999
26 Korea (Republic of)	**2000**	1993	2002	1994	1992	1992	..	1996	1999
27 Slovenia	2002	1995	2002	1996	1992	1992	2004	1995	2001
28 Cyprus	2003	1997	1999	1996	1992	1992	2005	1988	2000
29 Portugal	2004	1993	2002	1993	1988	1988	2004	1997	1996
30 Brunei Darussalam	..	..	..	..	1990	1993	**2002**	1996	2002
31 Barbados	2002	1994	2000	1993	1992	1992	2004	1993	1997
32 Czech Republic	2001	1993	2001	1993	1993	1993	2002	1996	2000
33 Kuwait	..	1994	2005	2002	1992	1992	2006	1986	1997
34 Malta	2007	1994	2001	2000	1988	1988	**2001**	1993	1998
35 Qatar	2007	1996	2005	1996	1996	1996	2004	2002	1999
36 Hungary	2004	1994	2002	1994	1988	1989	**2001**	2002	1999
37 Poland	2003	1994	2002	1996	1990	1990	**2001**	1998	2001
38 Argentina	**2000**	1994	2001	1994	1990	1990	2005	1995	1997
39 United Arab Emirates	..	1995	2005	2000	1989	1989	2002	**1982**	1998
40 Chile	**2000**	1994	2002	1994	1990	1990	2005	1997	1997
41 Bahrain	..	1994	2006	1996	1990	1990	2006	1985	1997
42 Slovakia	2003	1994	2002	1994	1993	1993	2002	1996	2002
43 Lithuania	2003	1995	2003	1996	1995	1995	2006	2003	2003
44 Estonia	2004	1994	2002	1994	1996	1996	..	2005	..
45 Latvia	2004	1995	2002	1995	1995	1995	2004	2004	2002
46 Uruguay	**2001**	1994	2001	1993	1989	1991	2004	1992	1999
47 Croatia	2002	1996	**1999**	1996	1992	1992	2007	1995	2000
48 Costa Rica	2007	1994	2002	1994	1991	1991	2007	1992	1998
49 Bahamas	2004	1994	1999	1993	1993	1993	2005	1983	2000
50 Seychelles	2004	1992	2002	1992	1993	1993	**2002**	1991	1997
51 Cuba	2002	1994	2002	1994	1992	1992	**2001**	1984	1997
52 Mexico	2002	1993	2000	1993	1987	1988	2003	1983	1995
53 Bulgaria	2000	1995	2002	1996	1990	1990	2004	1996	2001

Human development indicators

TABLE 25

HDI rank	Cartagena Protocol on Biosafety 2000	Framework Convention on Climate Change 1992	Kyoto Protocol to the Framework Convention on Climate Change 1997	Convention on Biological Diversity 1992	Vienna Convention for the Protection of the Ozone Layer 1988	Montreal Protocol on Substances that deplete the Ozone Layer 1989	Stockholm Convention on Persistent Organic Pollutants 2001	Convention of the Law of the Sea 1982	Convention to Combat Desertification 1994
54 Saint Kitts and Nevis	2001	1993	..	1993	1992	1992	2004	1993	1997
55 Tonga	2003	1998	..	1998	1998	1998	**2002**	1995	1998
56 Libyan Arab Jamahiriya	2005	1999	2006	2001	1990	1990	2005	**1984**	1996
57 Antigua and Barbuda	2003	1993	1998	1993	1992	1992	2003	1989	1997
58 Oman	2003	1995	2005	1995	1999	1999	2005	1989	1996
59 Trinidad and Tobago	2000	1994	1999	1996	1989	1989	2002	1986	2000
60 Romania	2003	1994	2001	1994	1993	1993	2004	1996	1998
61 Saudi Arabia	..	1994	2005	2001	1993	1993	**2002**	1996	1997
62 Panama	2002	1995	1999	1995	1989	1989	2003	1996	1996
63 Malaysia	2003	1994	2002	1994	1989	1989	**2002**	1996	1997
64 Belarus	2002	2000	2005	1993	1986	1988	2004	2006	2001
65 Mauritius	2002	1992	2001	1992	1992	1992	2004	1994	1996
66 Bosnia and Herzegovina	..	2000	2007	2002	1993	1993	**2001**	1994	2002
67 Russian Federation	..	1994	2004	1995	1986	1988	**2002**	1997	2003
68 Albania	2005	1994	2005	1994	1999	1999	2004	2003	2000
69 Macedonia (TFYR)	2005	1998	2004	1997	1994	1994	2004	1994	2002
70 Brazil	2003	1994	2002	1994	1990	1990	2004	1988	1997
MEDIUM HUMAN DEVELOPMENT									
71 Dominica	2004	1993	2005	1994	1993	1993	2003	1991	1997
72 Saint Lucia	2005	1993	2003	1993	1993	1993	2002	1985	1997
73 Kazakhstan	..	1995	**1999**	1994	1998	1998	**2001**	..	1997
74 Venezuela (Bolivarian Republic of)	2002	1994	2005	1994	1988	1989	2005	..	1998
75 Colombia	2003	1995	2001	1994	1990	1993	**2001**	1982	1999
76 Ukraine	2002	1997	2004	1995	1986	1988	**2001**	1999	2002
77 Samoa	2002	1994	2000	1994	1992	1992	2002	1995	1998
78 Thailand	2005	1994	2002	2003	1989	1989	2005	**1982**	2001
79 Dominican Republic	2006	1998	2002	1996	1993	1993	2007	**1982**	1997
80 Belize	2004	1994	2003	1993	1997	1998	**2002**	1983	1998
81 China	2005	1993	2002	1993	1989	1991	2004	1996	1997
82 Grenada	2004	1994	2002	1994	1993	1993	..	1991	1997
83 Armenia	2004	1993	2003	1993	1999	1999	2003	2002	1997
84 Turkey	2003	2004	..	1997	1991	1991	**2001**	..	1998
85 Suriname	..	1997	2006	1996	1997	1997	**2002**	1998	2000
86 Jordan	2003	1993	2003	1993	1989	1989	2004	1995	1996
87 Peru	2004	1993	2002	1993	1989	1993	2005	..	1995
88 Lebanon	..	1994	2006	1994	1993	1993	2003	1995	1996
89 Ecuador	2003	1993	2000	1993	1990	1990	2004	..	1995
90 Philippines	2006	1994	2003	1993	1991	1991	2004	1984	2000
91 Tunisia	2003	1993	2003	1993	1989	1989	2004	1985	1995
92 Fiji	2001	1993	1998	1993	1989	1989	2001	1982	1998
93 Saint Vincent and the Grenadines	2003	1996	2004	1996	1996	1996	2005	1993	1998
94 Iran (Islamic Republic of)	2003	1996	2005	1996	1990	1990	2006	**1982**	1997
95 Paraguay	2004	1994	1999	1994	1992	1992	2004	1986	1997
96 Georgia	..	1994	1999	1994	1996	1996	2006	1996	1999
97 Guyana	..	1994	2003	1994	1993	1993	..	1993	1997
98 Azerbaijan	2005	1995	2000	2000	1996	1996	2004	..	1998
99 Sri Lanka	2004	1993	2002	1994	1989	1989	2005	1994	1998
100 Maldives	2002	1992	1998	1992	1988	1989	2006	2000	2002
101 Jamaica	**2001**	1995	1999	1995	1993	1993	2007	1983	1997
102 Cape Verde	2005	1995	2006	1995	2001	2001	2006	1987	1995
103 El Salvador	2003	1995	1998	1994	1992	1992	**2001**	**1984**	1997
104 Algeria	2004	1993	2005	1995	1992	1992	2006	1996	1996
105 Viet Nam	2004	1994	2002	1994	1994	1994	2002	1994	1998
106 Occupied Palestinian Territories	..	..	..	..	..	..	..	..	..

Human development indicators

TABLE 25

Status of major international environmental treaties

HDI rank	Cartagena Protocol on Biosafety 2000	Framework Convention on Climate Change 1992	Kyoto Protocol to the Framework Convention on Climate Change 1997	Convention on Biological Diversity 1992	Vienna Convention for the Protection of the Ozone Layer 1988	Montreal Protocol on Substances that deplete the Ozone Layer 1989	Stockholm Convention on Persistent Organic Pollutants 2001	Convention of the Law of the Sea 1982	Convention to Combat Desertification 1994
107 Indonesia	2004	1994	2004	1994	1992	1992	**2001**	1986	1998
108 Syrian Arab Republic	2004	1996	2006	1996	1989	1989	2005	..	1997
109 Turkmenistan	..	1995	1999	1996	1993	1993	..	..	1996
110 Nicaragua	2002	1995	1999	1995	1993	1993	2005	2000	1998
111 Moldova	2003	1995	2003	1995	1996	1996	2004	2007	1999
112 Egypt	2003	1994	2005	1994	1988	1988	2003	1983	1995
113 Uzbekistan	..	1993	1999	1995	1993	1993	..	..	1995
114 Mongolia	2003	1993	1999	1993	1996	1996	2004	1996	1996
115 Honduras	**2000**	1995	2000	1995	1993	1993	2005	1993	1997
116 Kyrgyzstan	2005	2000	2003	1996	2000	2000	2006	..	1997
117 Bolivia	2002	1994	1999	1994	1994	1994	2003	1995	1996
118 Guatemala	2004	1995	1999	1995	1987	1989	**2002**	1997	1998
119 Gabon	2007	1998	2006	1997	1994	1994	2007	1998	1996
120 Vanuatu	..	1993	2001	1993	1994	1994	2005	1999	1999
121 South Africa	2003	1997	2002	1995	1990	1990	2002	1997	1997
122 Tajikistan	2004	1998	..	1997	1996	1998	2007	..	1997
123 Sao Tome and Principe	..	1999	..	1999	2001	2001	2006	1987	1998
124 Botswana	2002	1994	2003	1995	1991	1991	2002	1990	1996
125 Namibia	2005	1995	2003	1997	1993	1993	2005	1983	1997
126 Morocco	**2000**	1995	2002	1995	1995	1995	2004	2007	1996
127 Equatorial Guinea	..	2000	2000	1994	1988	2006	..	1997	1997
128 India	2003	1993	2002	1994	1991	1992	2006	1995	1996
129 Solomon Islands	2004	1994	2003	1995	1993	1993	2004	1997	1999
130 Lao People's Democratic Republic	2004	1995	2003	1996	1998	1998	2006	1998	1996
131 Cambodia	2003	1995	2002	1995	2001	2001	2006	**1983**	1997
132 Myanmar	**2001**	1994	2003	1994	1993	1993	2004	1996	1997
133 Bhutan	2002	1995	2002	1995	2004	2004	..	**1982**	2003
134 Comoros	..	1994	..	1994	1994	1994	2007	1994	1998
135 Ghana	2003	1995	2003	1994	1989	1989	2003	1983	1996
136 Pakistan	**2001**	1994	2005	1994	1992	1992	**2001**	1997	1997
137 Mauritania	2005	1994	2005	1996	1994	1994	2005	1996	1996
138 Lesotho	2001	1995	2000	1995	1994	1994	2002	2007	1995
139 Congo	2006	1996	2007	1996	1994	1994	2007	**1982**	1999
140 Bangladesh	2004	1994	2001	1994	1990	1990	2007	2001	1996
141 Swaziland	2006	1996	2006	1994	1992	1992	2006	**1984**	1996
142 Nepal	**2001**	1994	2005	1993	1994	1994	2007	1998	1996
143 Madagascar	2003	1999	2003	1996	1996	1996	2005	2001	1997
144 Cameroon	2003	1994	2002	1994	1989	1989	**2001**	1985	1997
145 Papua New Guinea	2005	1993	2002	1993	1992	1992	2003	1997	2000
146 Haiti	**2000**	1996	2005	1996	2000	2000	**2001**	1996	1996
147 Sudan	2005	1993	2004	1995	1993	1993	2006	1985	1995
148 Kenya	2002	1994	2005	1994	1988	1988	2004	1989	1997
149 Djibouti	2002	1995	2002	1994	1999	1999	2004	1991	1997
150 Timor-Leste	..	2006	..	2006	..	..	..	..	2003
151 Zimbabwe	2005	1992	..	1994	1992	1992	**2001**	1993	1997
152 Togo	2004	1995	2004	1995	1991	1991	2004	1985	1995
153 Yemen	2005	1996	2004	1996	1996	1996	2004	1987	1997
154 Uganda	2001	1993	2002	1993	1988	1988	2004	1990	1997
155 Gambia	2004	1994	2001	1994	1990	1990	2006	1984	1996
LOW HUMAN DEVELOPMENT									
156 Senegal	2003	1994	2001	1994	1993	1993	2003	1984	1995
157 Eritrea	2005	1995	2005	1996	2005	2005	2005	..	1996
158 Nigeria	2003	1994	2004	1994	1988	1988	2004	1986	1997
159 Tanzania (United Republic of)	2003	1996	2002	1996	1993	1993	2004	1985	1997

Human development indicators

TABLE 25

HDI rank	Cartagena Protocol on Biosafety 2000	Framework Convention on Climate Change 1992	Kyoto Protocol to the Framework Convention on Climate Change 1997	Convention on Biological Diversity 1992	Vienna Convention for the Protection of the Ozone Layer 1988	Montreal Protocol on Substances that deplete the Ozone Layer 1989	Stockholm Convention on Persistent Organic Pollutants 2001	Convention of the Law of the Sea 1982	Convention to Combat Desertification 1994
160 Guinea	**2000**	1993	2000	1993	1992	1992	**2001**	1985	1997
161 Rwanda	2004	1998	2004	1996	2001	2001	2002	**1982**	1998
162 Angola	..	2000	2007	1998	2000	2000	2006	1990	1997
163 Benin	2005	1994	2002	1994	1993	1993	2004	1997	1996
164 Malawi	**2000**	1994	2001	1994	1991	1991	**2002**	**1984**	1996
165 Zambia	2004	1993	2006	1993	1990	1990	2006	1983	1996
166 Côte d'Ivoire	..	1994	2007	1994	1993	1993	2004	1984	1997
167 Burundi	..	1997	2001	1997	1997	1997	2005	**1982**	1997
168 Congo (Democratic Republic of the)	2005	1995	2005	1994	1994	1994	2005	1989	1997
169 Ethiopia	2003	1994	2005	1994	1994	1994	2003	**1982**	1997
170 Chad	2006	1994	..	1994	1989	1994	2004	**1982**	1996
171 Central African Republic	**2000**	1995	..	1995	1993	1993	**2002**	**1984**	1996
172 Mozambique	2002	1995	2005	1995	1994	1994	2005	1997	1997
173 Mali	2002	1994	2002	1995	1994	1994	2003	1985	1995
174 Niger	2004	1995	2004	1995	1992	1992	2006	**1982**	1996
175 Guinea-Bissau	..	1995	2005	1995	2002	2002	**2002**	1986	1995
176 Burkina Faso	2003	1993	2005	1993	1989	1989	2004	2005	1996
177 Sierra Leone	..	1995	2006	1994	2001	2001	2003	1994	1997
Others [a]									
Afghanistan	..	2002	..	2002	2004	2004	..	**1983**	1995
Andorra	..	..	..	..	..	..	..	..	2002
Cook Islands	**2001**	1993	2001	1993	2003	2003	2004	1995	1998
Iraq	..	..	..	..	..	..	..	1985	..
Kiribati	2004	1995	2000	1994	1993	1993	2004	2003	1998
Korea (Democratic People's Rep. of)	2003	1994	2005	1994	1995	1995	2002	**1982**	2003
Liberia	2002	2002	2002	2000	1996	1996	2002	**1982**	1998
Liechtenstein	..	1994	2004	1997	1989	1989	2004	**1984**	1999
Marshall Islands	2003	1992	2003	1992	1993	1993	2003	1991	1998
Micronesia (Federated States of)	..	1993	1999	1994	1994	1995	2005	1991	1996
Monaco	**2000**	1992	2006	1992	1993	1993	2004	1996	1999
Montenegro	2006	2006	2007	2006	2006	2006	**2006**	2006	2007
Nauru	2001	1993	2001	1993	2001	2001	2002	1996	1998
Niue	2002	1996	1999	1996	2003	2003	2005	2006	1998
Palau	2003	1999	1999	1999	2001	2001	**2002**	1996	1999
San Marino	..	1994	..	1994	..	..	..	..	1999
Serbia [b]	2006	2001	..	2002	2001	2001	**2002**	2001	..
Somalia	..	..	..	..	2001	2001	..	1989	2002
Tuvalu	..	1993	1998	2002	1993	1993	2004	2002	1998
Total states parties [c]	**140**	**190**	**173**	**189**	**190**	**190**	**145**	**154**	**191**
Treaties signed, not yet ratified	**18**	**0**	**4**	**1**	**0**	**0**	**35**	**23**	**0**

NOTES

Data are as of 1 July 2007. Data refer to year of ratification, accession approval or succession unless otherwise specified. All these stages have the same legal effects. **Bold** signifies signature not yet followed by ratification.

a. Countries or areas, in addition to the countries or areas included in the main indicator tables, that have signed at least one of the nine environmental treaties listed in this table.

b. Following separation of Serbia and Montenegro into two independent states in June 2006, all treaty actions (ratification, signature etc.) continue in force for the Republic of Serbia.

c. Refers to ratification, acceptance, approval, accession or succession.

SOURCE
All columns: UN 2007a

TABLE

26

... protecting personal security ...

Refugees and armaments

| | | Refugees | | Conventional arms transfers [b] (1990 prices) | | | | Total armed forces | |
| | | | | Imports | | Exports | | | |
HDI rank	Internally displaced people [a] (thousands) 2006 [e]	By country of asylum (thousands) 2006 [e]	By country of origin [c] (thousands) 2006 [e]	Imports (US$ millions) 1996	2006	US$ millions 2006	Share [d] (%) 2002–2006	Thousands 2007	Index (1985=100) 2007
HIGH HUMAN DEVELOPMENT									
1 Iceland	..	(.)	(.)	..	..	..	..	0	..
2 Norway	..	43	..	183	501	2	(.)	23	62
3 Australia	..	69	(.)	582	768	4	(.)	52	74
4 Canada	..	152	(.)	389	100	227	1	63	76
5 Ireland	..	8	..	0	11	..	..	10	73
6 Sweden	..	80	(.)	104	122	472	2	28	43
7 Switzerland	..	49	(.)	187	72	144	1	4	..
8 Japan	..	2	(.)	813	400	0	(.)	240	99
9 Netherlands	..	101	(.)	181	171	1,481	3	53	50
10 France	..	146	(.)	28	121	1,557	8	255	55
11 Finland	..	12	(.)	605	84	31	(.)	29	79
12 United States	..	844	1	540	417	7,888	30	1,506	70
13 Spain	..	5	2	435	378	803	1	147	46
14 Denmark	..	37	(.)	70	133	3	(.)	22	74
15 Austria	..	25	(.)	10	0	61	(.)	40	73
16 United Kingdom	..	302	0	735	462	1,071	4	191	57
17 Belgium	..	17	(.)	4	4	50	(.)	40	44
18 Luxembourg	..	2	..	4	0	..	..	1	129
19 New Zealand	..	5	(.)	7	8	0	(.)	9	73
20 Italy	..	27	(.)	293	697	860	2	191	50
21 Hong Kong, China (SAR)	..	2	(.)	..	..	..	..	..	..
22 Germany	..	605	(.)	213	529	3,850	9	246	51
23 Israel	150–420 [f]	1	1	88	994	224	2	168	118
24 Greece	..	2	(.)	377	1,452	23	(.)	147	73
25 Singapore	..	..	(.)	153	54	0	(.)	73	133
26 Korea (Republic of)	..	(.)	1	1,759	1,292	89	(.)	687	115
27 Slovenia	..	(.)	2	14	2	..	..	7	..
28 Cyprus	210 [g]	1	(.)	169	26	0	(.)	10	100
29 Portugal	..	(.)	(.)	7	431	..	..	44	60
30 Brunei Darussalam	..	..	..	17	3	..	..	7	171
31 Barbados	..	..	(.)	..	..	..	..	1	61
32 Czech Republic	..	2	2	24	65	56	(.)	25	12
33 Kuwait	..	(.)	1	1,161	107	0	(.)	16	133
34 Malta	..	2	(.)	1	0	0	(.)	2	250
35 Qatar	..	(.)	(.)	201	0	0	(.)	12	200
36 Hungary	..	8	3	138	337	0	(.)	32	30
37 Poland	..	7	14	99	224	169	(.)	142	45
38 Argentina	..	3	1	57	53	0	(.)	72	67
39 United Arab Emirates	..	(.)	(.)	474	2,439	7	(.)	51	119
40 Chile	..	1	1	180	1,125	0	(.)	76	75
41 Bahrain	..	..	(.)	181	60	0	(.)	11	393
42 Slovakia	..	(.)	1	30	0	0	(.)	15	..
43 Lithuania	..	1	1	15	33	0	(.)	12	..
44 Estonia	..	(.)	1	1	8	0	(.)	4	..
45 Latvia	..	(.)	1	0	4	..	..	5	..
46 Uruguay	..	(.)	(.)	4	7	0	(.)	25	78
47 Croatia	4–7	2	94	14	0	0	(.)	21	..
48 Costa Rica	..	12	(.)	..	..	..	..	0	..
49 Bahamas	..	..	(.)	0	0	..	..	1	172
50 Seychelles	..	..	(.)	..	..	..	..	(.)	17
51 Cuba	..	1	34	..	..	..	..	49	30
52 Mexico	10–12 [g]	3	3	79	68	..	..	238	184
53 Bulgaria	..	5	3	123	20	0	(.)	51	34

Human development indicators

TABLE 26

HDI rank	Internally displaced people [a] (thousands) 2006 [e]	Refugees By country of asylum (thousands) 2006 [e]	Refugees By country of origin [c] (thousands) 2006 [e]	Conventional arms transfers [b] (1990 prices) Imports (US$ millions) 1996	Imports (US$ millions) 2006	Exports US$ millions 2006	Exports Share [d] (%) 2002–2006	Total armed forces Thousands 2007	Total armed forces Index (1985=100) 2007
54 Saint Kitts and Nevis	..	..	..	..	..	..	..	..	..
55 Tonga	..	..	(.)	0	0	..	..	..	..
56 Libyan Arab Jamahiriya	..	3	2	0	5	24	(.)	76	..
57 Antigua and Barbuda	..	..	(.)	..	..	..	..	(.)	170
58 Oman	..	(.)	(.)	284	406	0	(.)	42	144
59 Trinidad and Tobago	..	..	(.)	0	0	..	..	3	143
60 Romania	..	2	7	41	131	0	(.)	70	37
61 Saudi Arabia	..	241	1	1,725	148	0	(.)	225	360
62 Panama	..	2	(.)	0	0	..	..	0	0
63 Malaysia	..	37	1	38	654	0	(.)	109	99
64 Belarus	..	1	9	0	254	0	(.)	73	..
65 Mauritius	..	..	(.)	30	0	..	..	0	0
66 Bosnia and Herzegovina	180	10	200	52	0	0	(.)	12	..
67 Russian Federation	82–190	1	159	0	4	6,733	29	1,027	19
68 Albania	..	(.)	14	0	0	..	..	11	27
69 Macedonia (TFYR)	1	1	8	0	0	..	..	11	..
70 Brazil	..	3	1	531	323	1	(.)	288	104
MEDIUM HUMAN DEVELOPMENT									
71 Dominica	..	..	(.)	..	..	..	..	..	..
72 Saint Lucia	..	..	(.)	..	..	..	..	..	..
73 Kazakhstan	..	4	7	170	53	0	(.)	66	..
74 Venezuela (Bolivarian Republic of)	..	1	4	35	498	6	(.)	82	167
75 Colombia	1853–3833 [h]	(.)	73	57	33	..	..	209	316
76 Ukraine	..	2	64	..	..	133	1	188	..
77 Samoa	..	..	..	..	..	..	..	..	..
78 Thailand	..	133	3	611	47	0	(.)	307	130
79 Dominican Republic	..	..	(.)	4	0	..	..	25	113
80 Belize	..	(.)	(.)	0	0	..	..	1	167
81 China	..	301	141	1,274	3,261	564	2	2,255	58
82 Grenada	..	..	(.)	..	..	..	..	..	..
83 Armenia	8 [g]	114	15	104	0	..	..	44	..
84 Turkey	954–1201	3	227	1,510	454	45	(.)	515	82
85 Suriname	..	..	(.)	0	0	..	..	2	100
86 Jordan	..	500	2	76	117	13	(.)	101	144
87 Peru	60 [g]	1	7	138	365	0	(.)	80	63
88 Lebanon	216–800	20	12	20	0	0	(.)	72	414
89 Ecuador	..	12	1	29	0	..	..	57	134
90 Philippines	120	(.)	1	32	43	..	..	106	92
91 Tunisia	..	(.)	3	56	16	..	..	35	100
92 Fiji	..	..	2	0	0	..	..	4	148
93 Saint Vincent and the Grenadines	..	..	(.)	..	..	..	..	..	..
94 Iran (Islamic Republic of)	..	968	102	630	891	9	(.)	545	89
95 Paraguay	..	(.)	(.)	2	0	..	..	10	69
96 Georgia	222–241	1	6	0	0	0	(.)	11	..
97 Guyana	..	..	1	0	0	..	..	1	15
98 Azerbaijan	579–687 [i]	3	126	0	0	..	..	67	..
99 Sri Lanka	600 [g]	(.)	117	152	20	..	..	151	699
100 Maldives	..	..	(.)	0	0	..	..	..	..
101 Jamaica	..	..	1	0	25	..	..	3	143
102 Cape Verde	..	..	(.)	0	0	..	..	1	13
103 El Salvador	..	(.)	6	3	0	..	..	16	38
104 Algeria	1,000 [g]	94 [j]	8	87	173	..	..	138	81
105 Viet Nam	..	2	374	207	179	..	..	455	44
106 Occupied Palestinian Territories	25–57 [g,k]	..	334	9	0	..	..	..	..

Human development indicators

TABLE 26

Refugees and armaments

HDI rank	Internally displaced people [a] (thousands) 2006 [e]	Refugees By country of asylum (thousands) 2006 [e]	Refugees By country of origin [c] (thousands) 2006 [e]	Conventional arms transfers [b] (1990 prices) Imports (US$ millions) 1996	Imports (US$ millions) 2006	Exports US$ millions 2006	Exports Share [d] (%) 2002–2006	Total armed forces Thousands 2007	Total armed forces Index (1985=100) 2007
107 Indonesia	150–250	(.)	35	435	54	8	(.)	302	109
108 Syrian Arab Republic	305 [g]	702	12	21	9	3	(.)	308	77
109 Turkmenistan	0	1	1	0	0	..	..	26	..
110 Nicaragua	..	(.)	2	..	..	0	(.)	14	22
111 Moldova	..	(.)	12	0	0	0	(.)	7	..
112 Egypt	..	88	8	986	526	0	(.)	469	105
113 Uzbekistan	3 [g]	1	9	0	0	0	1	55	..
114 Mongolia	..	(.)	1	..	..	..	..	9	27
115 Honduras	..	(.)	1	..	..	..	..	12	72
116 Kyrgyzstan	..	(.)	2	0	1	0	(.)	13	..
117 Bolivia	..	1	(.)	0	26	..	..	46	167
118 Guatemala	242 [g]	(.)	7	0	0	..	..	16	50
119 Gabon	..	8	(.)	0	63	..	..	5	208
120 Vanuatu	..	..	..	..	..	..	..	..	..
121 South Africa	..	35	1	38	862	115	(.)	62	58
122 Tajikistan	..	1	1	0	13	..	..	8	..
123 Sao Tome and Principe	..	..	(.)	..	..	..	..	..	..
124 Botswana	..	3	(.)	29	0	..	..	9	225
125 Namibia	..	5	1	0	0	..	..	9	..
126 Morocco	..	1	5	86	49	..	..	201	135
127 Equatorial Guinea	..	..	(.)	0	0	..	..	1	45
128 India	600	158	18	996	1,672	11	(.)	1,316	104
129 Solomon Islands	..	..	(.)	..	..	..	..	..	..
130 Lao People's Democratic Republic	..	..	26	0	0	..	..	29	54
131 Cambodia	..	(.)	18	33	0	0	(.)	124	354
132 Myanmar	500 [l]	..	203	120	7	..	..	375	202
133 Bhutan	..	..	108	0	0	..	..	..	..
134 Comoros	..	..	(.)	..	..	..	..	..	..
135 Ghana	..	45	10	7	0	..	..	14	93
136 Pakistan	.. [m]	1,044 [n]	26	529	309	0	(.)	619	..
137 Mauritania	..	1	33	2	0	..	..	16	188
138 Lesotho	..	..	(.)	0	0	..	..	2	100
139 Congo	8 [g]	56	21	0	0	..	..	10	115
140 Bangladesh	500	26	8	5	208	..	..	127	139
141 Swaziland	..	1	(.)	0	0	..	..	..	..
142 Nepal	100–200	128	3	0	0	..	..	69	276
143 Madagascar	..	..	(.)	19	0	..	..	14	66
144 Cameroon	..	35	10	4	0	..	..	14	192
145 Papua New Guinea	..	10	(.)	0	0	..	..	3	94
146 Haiti	..	..	21	..	..	..	..	..	..
147 Sudan	5,355	202	686	29	48	..	..	105	186
148 Kenya	431	273	5	0	0	..	..	24	175
149 Djibouti	..	9	(.)	0	0	..	..	11	367
150 Timor-Leste	100	..	(.)	..	..	..	..	1	..
151 Zimbabwe	570 [g,o]	4	13	0	20	..	..	29	71
152 Togo	2	6	27	0	0	..	..	9	250
153 Yemen	..	96	1	0	0	..	..	67	105
154 Uganda	1200–1700	272	22	0	0	..	..	45	225
155 Gambia	..	14	1	0	0	..	..	1	200
LOW HUMAN DEVELOPMENT									
156 Senegal	64 [g]	21	15	0	0	..	..	14	139
157 Eritrea	40–45	5	187	15	70	0	(.)	202	..
158 Nigeria	..	9	13	16	72	..	..	85	90
159 Tanzania (United Republic of)	..	485	2	0	0	..	..	27	67

TABLE 26

	Internally displaced people[a] (thousands) 2006[e]	Refugees By country of asylum (thousands) 2006[e]	Refugees By country of origin[c] (thousands) 2006[e]	Conventional arms transfers[b] (1990 prices) Imports (US$ millions) 1996	Imports (US$ millions) 2006	Exports US$ millions 2006	Exports Share[d] (%) 2002–2006	Total armed forces Thousands 2007	Total armed forces Index (1985=100) 2007
HDI rank									
160 Guinea	19[g]	31	7	0	0	..	..	12	121
161 Rwanda	..	49	93	1	0	..	..	33	635
162 Angola	62[g]	13	207	9	0	0	(.)	107	216
163 Benin	..	11	(.)	0	0	..	..	5	111
164 Malawi	..	4	(.)	..	..	0	(.)	5	94
165 Zambia	..	120	(.)	5	15	..	..	15	93
166 Côte d'Ivoire	750	39	26	0	0	..	..	17	129
167 Burundi	100	13	397	0	0	..	..	35	673
168 Congo (Democratic Republic of the)	1,100	208	402	46	13	..	..	51	106
169 Ethiopia	100–280	97	83	0	0	..	..	153	71
170 Chad	113	287	36	0	2	..	..	17	139
171 Central African Republic	212	12	72	0	9	..	..	3	130
172 Mozambique	..	3	(.)	0	0	..	..	11	70
173 Mali	..	11	1	0	0	..	..	7	143
174 Niger	..	(.)	1	0	0	..	..	5	227
175 Guinea-Bissau	..	8	1	..	..	..	..	9	105
176 Burkina Faso	..	1	(.)	0	0	..	..	11	275
177 Sierra Leone	..	27	43	0	0	..	..	11	355
Developing countries	..	7,084	..	..	..	..	..	13,950 T	90
Least developed countries	..	2,177	..	..	..	..	..	1,781 T	152
Arab States	..	2,001	..	..	..	..	..	2,167 T	80
East Asia and the Pacific	..	..	..	..	..	..	..	5,952 T	80
Latin America and the Caribbean	..	..	..	..	..	..	..	1,327 T	99
South Asia	..	2,326	..	..	..	..	..	2,877 T	113
Sub-Saharan Africa	..	2,227	..	..	..	..	..	1,102 T	130
Central and Eastern Europe and the CIS	..	168	..	..	..	..	..	2,050 T	..
OECD	..	2,556	..	..	..	..	..	4,995 T	69
High-income OECD	..	2,533	..	..	..	..	..	4,028	69
High human development	..	2,885	..	..	..	25,830	..	7,101	52
Medium human development	..	5,389	..	..	..	..	..	10,143	91
Low human development	..	1,453	..	..	..	..	..	835	146
High income	..	..	..	..	..	..	..	4,611	74
Middle income	..	3,267	..	..	..	..	..	9,440	..
Low income	..	3,741	..	..	..	..	..	5,413	110
World	23,700 T[p]	9,894 T[p]	9,894 T[p]	22,115 T[p]	26,130 T[p]	26,742 T[p]	..	19,801 T	73

NOTES

a. Estimates maintained by the IDMC based on various sources. Estimates are associated with high levels of uncertainty.

b. Data are as of 10 May 2007. Figures are trend indicator values, which are an indicator only of the volume of international arms transfers, not of the actual financial value of such transfers. Published reports of arms transfers provide partial information, as not all transfers are fully reported. The estimates presented are conservative and may understate actual transfers of conventional weapons.

c. The country of origin for many refugees is unavailable or unreported. These data may therefore be underestimates.

d. Calculated using the 2002-06 totals for all countries and non-state actors with exports of major conventional weapons as defined in SIPRI 2007a.

e. Data refer to the end of 2006 unless otherwise specified.

f. Higher figure includes estimate of Bedoin internally displaced people.

g. Data refer to a year or period other than that specified.

h. Lower estimate is cumulative since 1994. Higher figure is cumulative since 1985.

i. Figures do not include an estimated 30,000 ethnic Armenians displaced to Nagorno Karabakh.

j. According to the Government of Algeria, there are an estimated 165,000 Saharawi refugees in Tindouf camps.

k. Lower estimate includes only internally displaced people evicted mainly by dwelling demolitions since 2000. Higher figure is cumulative since 1967.

l. Estimate excludes certain parts of the country or some groups of internally displaced people.

m. Conflict-induced displacement has taken place in Balochistan and Waziristan, but no estimates are available due to lack of access.

n. Figures are only for Afghans living in camps and assisted by UNHCR.

o. Not including people previously displaced by land acquisitions or political violence. Also not including people recently displaced due to losing their businesses or other forms livelihood.

p. Data are aggregates provided by original data source.

SOURCES

Column 1: IDMC 2007.
Columns 2 and 3: UNHCR 2007.
Columns 4 – 6: SIPRI 2007a.
Column 7: calculated on the basis of data on arms transfers from SIPRI 2007a.
Column 8: IISS 2007.
Column 9: calculated on the basis of data on armed forces from IISS 2007.

Human development indicators

TABLE 27 ... protecting personal security ...

27 Crime and justice

| HDI rank | Intentional homicides[a] (per 100,000 people) 2000–04 [c] | Prison population | | | Year in which countries have partially or completely abolished the death penalty [b] |
		Total 2007 [d]	(per 100,000 people) 2007 [d]	Female (% of total) 2007 [e]	
HIGH HUMAN DEVELOPMENT					
1 Iceland	1.0	119	40	6	1928
2 Norway	0.8	3,048	66	5	1979
3 Australia	1.3	25,353	126	7	1985
4 Canada	1.9	34,096 [f]	107 [f]	5	1998
5 Ireland	0.9	3,080	72	4	1990
6 Sweden	2.4	7,450	82	5	1972
7 Switzerland	2.9	6,111	83	5	1992
8 Japan	0.5	79,055	62	6	.. [g]
9 Netherlands	1.0	21,013	128	9	1982
10 France	1.6	52,009 [f]	85 [f]	4	1981
11 Finland	2.8	3,954	75	6	1972
12 United States	5.6	2,186,230	738	9	.. [g]
13 Spain	1.2	64,215	145	8	1995
14 Denmark	0.8	4,198	77	5	1978
15 Austria	0.8	8,766	105	5	1968
16 United Kingdom	2.1	88,458 [f]	124 [f]	6 [f]	1998
17 Belgium	1.5	9,597	91	4	1996
18 Luxembourg	0.9	768	167	5	1979
19 New Zealand	1.3	7,620	186	6	1989
20 Italy	1.2	61,721 [f]	104 [f]	5	1994
21 Hong Kong, China (SAR)	0.6	11,580	168	20	..
22 Germany	1.0	78,581	95	5	1987
23 Israel	2.6	13,909	209	2	1954 [h]
24 Greece	0.8	9,984	90	6	2004
25 Singapore	0.5	15,038 [f]	350 [f]	11	.. [g]
26 Korea (Republic of)	2.2	45,882	97	5	.. [g]
27 Slovenia	1.5	1,301	65	4	1989
28 Cyprus	1.7	580 [f]	76 [f]	3	2002
29 Portugal	1.8	12,870	121	7	1976
30 Brunei Darussalam	1.4	529	140	8	1957 [i]
31 Barbados	7.5	997	367	5	.. [g]
32 Czech Republic	2.2	18,950	185	5	1990
33 Kuwait	1.0	3,500	130	15	.. [g]
34 Malta	1.8	352	86	4	2000
35 Qatar	0.8	465	55	1	.. [g]
36 Hungary	2.1	15,720	156	6	1990
37 Poland	1.6	87,901	230	3	1997
38 Argentina	9.5	54,472	140	5	1984 [h]
39 United Arab Emirates	0.6	8,927	288	11	.. [g]
40 Chile	1.7	39,916	240	7	2001 [h]
41 Bahrain	1.0	701	95	..	.. [g]
42 Slovakia	2.3	8,493	158	5	1990
43 Lithuania	9.4	8,124	240	3	1998
44 Estonia	6.8	4,463	333	4	1998
45 Latvia	8.6	6,676	292	6	1999 [h]
46 Uruguay	5.6	6,947	193	6	1907
47 Croatia	1.8	3,594	81	5	1990
48 Costa Rica	6.2	7,782	181	7	1877
49 Bahamas	15.9 [f]	1,500	462	2	.. [g]
50 Seychelles	7.4	193	239	8	1993
51 Cuba	..	55,000	487	..	.. [g]
52 Mexico	13.0	214,450	196	5	2005
53 Bulgaria	3.1	11,436	148	3	1998

TABLE 27

HDI rank	Intentional homicides[a] (per 100,000 people) 2000–04[c]	Prison population Total 2007[d]	Prison population (per 100,000 people) 2007[d]	Female (% of total) 2007[e]	Year in which countries have partially or completely abolished the death penalty[b]
54 Saint Kitts and Nevis	4.8[f]	214	547	1	..[g]
55 Tonga	2.0[f]	128	114	6	1982[i]
56 Libyan Arab Jamahiriya	..	11,790	207	3	..[g]
57 Antigua and Barbuda	..	176	225	3	..[g]
58 Oman	0.6	2,020	81	5	..[g]
59 Trinidad and Tobago	..	3,851	296	3	..[g]
60 Romania	2.4	35,429	164	5	1989
61 Saudi Arabia	0.9	28,612	132	6	..[g]
62 Panama	9.6	11,649	364	7	1922
63 Malaysia	2.4	35,644	141	7	..[g]
64 Belarus	8.3	41,583	426	8	..[g]
65 Mauritius	2.5	2,464	205	6	1995
66 Bosnia and Herzegovina	..	1,526	59	3	2001
67 Russian Federation	19.9	869,814	611	7	1999[i]
68 Albania	5.7	3,491	111	3	2007
69 Macedonia (TFYR)	2.3	2,026	99	2	1991
70 Brazil	..	361,402	191	6	1979[h]
MEDIUM HUMAN DEVELOPMENT					
71 Dominica	2.8	289	419	(.)	..[g]
72 Saint Lucia	..	503	303	2	..[g]
73 Kazakhstan	16.8[f]	49,292	340	7	..[g]
74 Venezuela (Bolivarian Republic of)	33.2	19,853	74	6	1863
75 Colombia	62.7	62,216	134	6	1910
76 Ukraine	7.4	165,716	356	6	1999
77 Samoa	..	223	123	9	2004
78 Thailand	8.5	164,443	256	17	..[g]
79 Dominican Republic	..	12,725	143	3	1966
80 Belize	..	1,359	487	2	..[g]
81 China	2.1[f]	1,548,498[f]	118[f]	5	..[g]
82 Grenada	..	237	265	1	1978[i]
83 Armenia	2.5	2,879	89	3	2003
84 Turkey	3.8	65,458	91	3	2004
85 Suriname	10.3	1,600	356	6	1982[i]
86 Jordan	0.9[f]	5,589	104	2	..[g]
87 Peru	5.5	35,642	126	7	1979[h]
88 Lebanon	5.7[f]	5,971	168	4	..[g]
89 Ecuador	18.3	12,251	93	11	1906
90 Philippines	4.3	89,639	108	8	2006
91 Tunisia	1.2	26,000	263	..	1991[i]
92 Fiji	1.7[f]	1,113	131	2	1979[h]
93 Saint Vincent and the Grenadines	..	367	312	3	..[g]
94 Iran (Islamic Republic of)	2.9	147,926	214	4	..[g]
95 Paraguay	12.6	5,063	86	5	1992
96 Georgia	6.2	11,731	276	2	1997
97 Guyana	13.8[f]	1,524	199	4	..[g]
98 Azerbaijan	2.4	18,259	219	2	1998
99 Sri Lanka	6.7	23,613	114	4	1976[i]
100 Maldives	1.3	1,125[f]	343[f]	22	1952[i]
101 Jamaica	34.4	4,913	182	5	..[g]
102 Cape Verde	..	755	178	5	1981
103 El Salvador	31.5	12,176	174	6	1983[h]
104 Algeria	1.4	42,000	127	1	1993[i]
105 Viet Nam	..	88,414	105	12	..[g]
106 Occupied Palestinian Territories	4.0	..	..	..	..[g]

Human development indicators

TABLE 27

Crime and justice

HDI rank	Intentional homicides[a] (per 100,000 people) 2000–04 [c]	Prison population Total 2007 [d]	Prison population (per 100,000 people) 2007 [d]	Female (% of total) 2007 [e]	Year in which countries have partially or completely abolished the death penalty [b]
107 Indonesia	1.1	99,946	45	5	.. [g]
108 Syrian Arab Republic	1.1	10,599	58	7	.. [g]
109 Turkmenistan	..	22,000	489	..	1999
110 Nicaragua	12.8 [f]	5,610	98	7	1979
111 Moldova	6.7	8,876 [f]	247 [f]	5	1995
112 Egypt	0.4 [f]	61,845	87	4	.. [g]
113 Uzbekistan	..	48,000	184	..	.. [g]
114 Mongolia	12.8	6,998	269	4	.. [g]
115 Honduras	..	11,589	161	3	1956
116 Kyrgyzstan	8.0	15,744	292	5	1998 [i]
117 Bolivia	2.8	7,710	83	7	1997 [h]
118 Guatemala	25.5	7,227	57	5	.. [g]
119 Gabon	..	2,750 [j]	212 [j]	..	..
120 Vanuatu	0.7 [f]	138	65	4	1980 [i]
121 South Africa	47.5	157,402	335	2	1997
122 Tajikistan	7.6 [f]	10,804	164	4	.. [g]
123 Sao Tome and Principe	6.2 [f]	155	82	2	1990
124 Botswana	0.5 [f]	6,259	348	5	.. [g]
125 Namibia	6.3	4,814	267	2	1990
126 Morocco	0.5	54,542	175	2	1993 [i]
127 Equatorial Guinea	..	..	..	..	.. [g]
128 India	3.7 [f]	332,112	30	4	.. [g]
129 Solomon Islands	..	297	62	1	1966 [h]
130 Lao People's Democratic Republic	..	4,020	69	11	.. [g]
131 Cambodia	..	8,160	58	6	1989
132 Myanmar	0.2	60,000	120	18	..
133 Bhutan	..	..	..	..	2004
134 Comoros	..	200	30	..	.. [g]
135 Ghana	..	12,736	55	2	1957 [i]
136 Pakistan	0.0	89,370	57	2	.. [g]
137 Mauritania	..	815	26	3 [k]	1987 [i]
138 Lesotho	50.7 [f]	2,924	156	3	.. [g]
139 Congo	..	918	38	..	1982 [i]
140 Bangladesh	..	71,200	50	3	.. [g]
141 Swaziland	13.6	2,734	249	3	1968 [i]
142 Nepal	3.4	7,135	26	8	1997
143 Madagascar	0.5 [f]	20,294	107	3	1958 [i]
144 Cameroon	..	20,000	125	..	.. [g]
145 Papua New Guinea	9.1	4,056	69	5	1950 [i]
146 Haiti	..	3,670	43	7	1987
147 Sudan	0.3 [f]	12,000	36	2	.. [g]
148 Kenya	..	47,036	130	4	1987 [i]
149 Djibouti	..	384	61	..	1995
150 Timor-Leste	..	320	41	(.)	1999
151 Zimbabwe	8.4	18,033	139	3	.. [g]
152 Togo	..	3,200	65	2	1960 [i]
153 Yemen	4.0	14,000 [f]	83 [f]	.. [l]	.. [g]
154 Uganda	7.4	26,126	95	3	.. [g]
155 Gambia	..	450	32	1	1981 [i]
LOW HUMAN DEVELOPMENT					
156 Senegal	..	5,360	54	4	2004
157 Eritrea	..	..	..	..	.. [g]
158 Nigeria	1.5 [f]	40,444	30	2	.. [g]
159 Tanzania (United Republic of)	7.5 [f]	43,911	113	3	.. [g]

TABLE 27

HDI rank	Intentional homicides[a] (per 100,000 people) 2000–04 [c]	Prison population		Female (% of total) 2007 [e]	Year in which countries have partially or completely abolished the death penalty [b]
		Total 2007 [d]	(per 100,000 people) 2007 [d]		
160 Guinea	..	3,070	37	2	..[g]
161 Rwanda	8.0[f]	67,000[f]	691[f,j]	3	..[g]
162 Angola	..	6,008	44	3	1992
163 Benin	..	5,834	75	4	1987[i]
164 Malawi	..	9,656	74	1	1992[i]
165 Zambia	8.1	14,347	120	3	..[g]
166 Côte d'Ivoire	4.1	9,274[f]	49[f]	2	2000
167 Burundi	..	7,969	106	3	..[g]
168 Congo (Democratic Republic of the)	..	30,000	57	3	..[g]
169 Ethiopia	..	65,000	92	..	..[g]
170 Chad	..	3,416	35	2	..[g]
171 Central African Republic	..	4,168	110	..	1981[i]
172 Mozambique	..	10,000	51	6	1990
173 Mali	..	4,407	33	2	1980[i]
174 Niger	..	5,709	46	3	1976[i]
175 Guinea-Bissau	..	..	..	..	1993
176 Burkina Faso	..	2,800	23	1	1988[i]
177 Sierra Leone	..	1,740	32	..	..[g]

NOTES

a. Because of differences in the legal definition of offences, data are not strictly comparable across countries.

b. Data are as of 4 April 2007 and refer to the year of abolition for all crimes (unless otherwise specified).

c. Data were collected during one of the years specified.

d. Data are as of January 2007.

e. Data are as of May 2007 unless otherwise specified.

f. Data refer to years or periods other than those specified in the column heading, differ from the standard definition or refer to only part of a country.

g. Country retaining the death penalty.

h. Death penalty abolished for ordinary crimes only.

i. Death penalty abolished in practice if not in law. No execution since the year reported.

j. Data are downloaded directly from http://www.kcl.ac.uk/depsta/rel/icps/worldbrief/highest_to_lowest_rates.php.

k. In 2005, six of the 435 prisoners in Nouakchott main prison were women.

l. In 2005 Parliamentary Committee on Human Rights reported that 2.7% of prisoners in Sana'a central prisons were women.

SOURCES

Column 1: UNODC 2007.
Columns 2–4: ICPS 2007.
Column 5: Amnesty International 2007.

TABLE 28

28 ... and achieving equality for all women and men

Gender-related development index

HDI rank	Gender-related development index (GDI)		Life expectancy at birth (years) 2005		Adult literacy rate[a] (% aged 15 and older) 1995–2005		Combined gross enrolment ratio for primary, secondary and tertiary education[b] (%) 2005		Estimated earned income[c] (PPP US$) 2005		HDI rank minus GDI rank[d]
	Rank	Value	Female	Male	Female	Male	Female	Male	Female	Male	
HIGH HUMAN DEVELOPMENT											
1 Iceland	1	0.962	83.1	79.9	..[e]	..[e]	101[f]	90[f]	28,637[f]	40,000[f]	0
2 Norway	3	0.957	82.2	77.3	..[e]	..[e]	103[f]	95[f]	30,749[f]	40,000[f]	-1
3 Australia	2	0.960	83.3	78.5	..[e]	..[e]	114[f]	112[f]	26,311	37,414	1
4 Canada	4	0.956	82.6	77.9	..[e]	..[e]	101[f,g]	98[f,g]	25,448[f,h]	40,000[f,h]	0
5 Ireland	15	0.940	80.9	76.0	..[e]	..[e]	102[f]	98[f]	21,076[f]	40,000[f]	-10
6 Sweden	5	0.955	82.7	78.3	..[e]	..[e]	100[f]	91[f]	29,044	36,059	1
7 Switzerland	9	0.946	83.7	78.5	..[e]	..[e]	83	88	25,056[f]	40,000[f]	-2
8 Japan	13	0.942	85.7	78.7	..[e]	..[e]	85	87	17,802[f]	40,000[f]	-5
9 Netherlands	6	0.951	81.4	76.9	..[e]	..[e]	98	99	25,625	39,845	3
10 France	7	0.950	83.7	76.6	..[e]	..[e]	99	94	23,945	37,169	3
11 Finland	8	0.947	82.0	75.6	..[e]	..[e]	105[f]	98[f]	26,795	37,739	3
12 United States	16	0.937	80.4	75.2	..[e]	..[e]	98	89	25,005[f,h]	40,000[f,h]	-4
13 Spain	12	0.944	83.8	77.2	..[e]	..[e]	101[f]	95[f]	18,335[h]	36,324[h]	1
14 Denmark	11	0.944	80.1	75.5	..[e]	..[e]	107[f]	99[f]	28,766	39,288	3
15 Austria	19	0.934	82.2	76.5	..[e]	..[e]	93	91	18,397[f]	40,000[f]	-4
16 United Kingdom	10	0.944	81.2	76.7	..[e]	..[e]	96	90	26,242[f]	40,000[f]	6
17 Belgium	14	0.940	81.8	75.8	..[e]	..[e]	97	94	22,182[f]	40,000[f]	3
18 Luxembourg	23	0.924	81.4	75.4	..[e]	..[e]	85[i]	84[i]	20,446[f]	40,000[f]	-5
19 New Zealand	18	0.935	81.8	77.7	..[e]	..[e]	115[f]	102[f]	20,666	29,479	1
20 Italy	17	0.936	83.2	77.2	98.0	98.8	93	88	18,501[h]	39,163[h]	3
21 Hong Kong, China (SAR)	22	0.926	84.9	79.1	97.3[j]	97.3[j]	73	79	22,433[f]	40,000[f]	-1
22 Germany	20	0.931	81.8	76.2	..[e]	..[e]	87	88	21,823	37,461	2
23 Israel	21	0.927	82.3	78.1	97.7[j]	97.7[j]	92	87	20,497[h]	31,345[h]	2
24 Greece	24	0.922	80.9	76.7	94.2	97.8	101[f]	97[f]	16,738	30,184	0
25 Singapore	..	..	81.4	77.5	88.6	96.6	..	..	20,044	39,150	..
26 Korea (Republic of)	26	0.910	81.5	74.3	..[e]	..[e]	89[f]	102[f]	12,531	31,476	-1
27 Slovenia	25	0.914	81.1	73.6	99.6[f,k]	99.7[f,k]	99	90	17,022[h]	27,779[h]	1
28 Cyprus	27	0.899	81.5	76.6	95.1	98.6	78	77	16,805[i]	27,808[i]	0
29 Portugal	28	0.895	80.9	74.5	92.0[k]	95.8[k]	93	87	15,294	25,881	0
30 Brunei Darussalam	31	0.886	79.3	74.6	90.2	95.2	79	76	15,658[h,m]	37,506[h,m]	-2
31 Barbados	30	0.887	79.3	73.6	99.7[f,j]	99.7[f,j]	94[g]	84[g]	12,868[h,m]	20,309[h,m]	0
32 Czech Republic	29	0.887	79.1	72.7	..[e]	..[e]	84	82	13,992	27,440	2
33 Kuwait	32	0.884	79.6	75.7	91.0	94.4	79	71	12,623[h]	36,403[h]	0
34 Malta	33	0.873	81.1	76.8	89.2	86.4	81	81	12,834	25,623	0
35 Qatar	37	0.863	75.8	74.6	88.6	89.1	85	71	9,211[h,m]	37,774[h,m]	-3
36 Hungary	34	0.872	77.0	68.8	..[e]	..[e]	93	86	14,058	22,098	1
37 Poland	35	0.867	79.4	71.0	..[e]	..[e]	91	84	10,414[h]	17,493[h]	1
38 Argentina	36	0.865	78.6	71.1	97.2	97.2	94[g]	86[g]	10,063[h]	18,686[h]	1
39 United Arab Emirates	43	0.855	81.0	76.8	87.8[k]	89.0[k]	68[g]	54[g]	8,329[h]	33,555[h]	-5
40 Chile	40	0.859	81.3	75.3	95.6	95.8	82	84	6,871[h]	17,293[h]	-1
41 Bahrain	42	0.857	77.0	73.9	83.6	88.6	90	82	10,496	29,796	-2
42 Slovakia	39	0.860	78.2	70.3	..[e]	..[e]	80	77	11,777[h]	20,218[h]	2
43 Lithuania	38	0.861	78.0	66.9	99.6[f]	99.6[f]	97	87	12,000	17,349	4
44 Estonia	41	0.858	76.8	65.5	99.8[f]	99.8[f]	99	86	12,112[h]	19,430[h]	2
45 Latvia	44	0.853	77.3	66.5	99.7[f]	99.8[f]	97	83	10,951	16,842	0
46 Uruguay	45	0.849	79.4	72.2	97.3	96.2	95[g]	83[g]	7,203[h]	12,890[h]	0
47 Croatia	46	0.848	78.8	71.8	97.1[f]	99.3[f]	75[g]	72[g]	10,587	15,687	0
48 Costa Rica	47	0.842	80.9	76.2	95.1	94.7	74	72	6,983	13,271	0
49 Bahamas	48	0.841	75.0	69.6	95.0[j]	95.0[j]	71	71	14,656[h,l]	20,803[h,l]	0
50 Seychelles	..	..	..	..	92.3	91.4	84	81	..[h]	..[h]	..
51 Cuba	49	0.839	79.8	75.8	99.8[f]	99.8[f]	92	83	4,268[h,m]	9,489[h,m]	0
52 Mexico	51	0.820	78.0	73.1	90.2	93.2	76	75	6,039	15,680	-1
53 Bulgaria	50	0.823	76.4	69.2	97.7	98.7	81	82	7,176	11,010	1

TABLE 28

HDI rank	Gender-related development index (GDI)		Life expectancy at birth (years) 2005		Adult literacy rate [a] (% aged 15 and older) 1995–2005		Combined gross enrolment ratio for primary, secondary and tertiary education [b] (%) 2005		Estimated earned income [c] (PPP US$) 2005		HDI rank minus GDI rank [d]
	Rank	Value	Female	Male	Female	Male	Female	Male	Female	Male	
54 Saint Kitts and Nevis	..	..	..	..	..	..	74	72	..[h,l]	..[h,l]	..
55 Tonga	53	0.814	73.8	71.8	99.0	98.8	81	79	5,243[h]	10,981[h]	-1
56 Libyan Arab Jamahiriya	62	0.797	76.3	71.1	74.8[k]	92.8[k]	97[g]	91[g]	4,054[h,m]	13,460[h,m]	-9
57 Antigua and Barbuda	..	..	..	..	..	..	..	..	..[h,l]	..[h,l]	..
58 Oman	67	0.788	76.7	73.6	73.5	86.9	67	67	4,516[h,l]	23,880[h,l]	-13
59 Trinidad and Tobago	56	0.808	71.2	67.2	97.8[k]	98.9[k]	66	64	9,307[h]	20,053[h]	-1
60 Romania	54	0.812	75.6	68.4	96.3	98.4	79	75	7,443	10,761	2
61 Saudi Arabia	70	0.783	74.6	70.3	76.3	87.5	76	76	4,031[h]	25,678[h]	-13
62 Panama	55	0.810	77.8	72.7	91.2	92.5	83	76	5,537	9,636	3
63 Malaysia	58	0.802	76.1	71.4	85.4	92.0	77[g]	72[g]	5,751	15,861	1
64 Belarus	57	0.803	74.9	62.7	99.4[f]	99.8[f]	91	87	6,236	9,835	3
65 Mauritius	63	0.796	75.8	69.1	80.5	88.2	75	76	7,407[h]	18,098[h]	-2
66 Bosnia and Herzegovina	..	..	77.1	71.8	94.4[f]	99.0[f]	..	..	2,864[h,m]	4,341[h,m]	..
67 Russian Federation	59	0.801	72.1	58.6	99.2[f]	99.7[f]	93	85	8,476[h]	13,581[h]	3
68 Albania	61	0.797	79.5	73.1	98.3[f]	99.2[f]	68[g]	69[g]	3,728[h]	6,930[h]	2
69 Macedonia (TFYR)	64	0.795	76.3	71.4	94.1	98.2	71	69	4,676[h]	9,734[h]	0
70 Brazil	60	0.798	75.5	68.1	88.8	88.4	89[g]	86[g]	6,204	10,664	5
MEDIUM HUMAN DEVELOPMENT											
71 Dominica	..	..	..	..	..	..	84	78	..[h,l]	..[h,l]	..
72 Saint Lucia	..	..	75.0	71.3	..	..	78	72	4,501[h,l]	8,805[h,l]	..
73 Kazakhstan	65	0.792	71.5	60.5	99.3[f]	99.8[f]	97	91	6,141	9,723	1
74 Venezuela (Bolivarian Republic of)	68	0.787	76.3	70.4	92.7	93.3	76[g]	73[g]	4,560[h]	8,683[h]	-1
75 Colombia	66	0.789	76.0	68.7	92.9	92.8	77	74	5,680	8,966	2
76 Ukraine	69	0.785	73.6	62.0	99.2[f]	99.7[f]	87	86	4,970	9,067	0
77 Samoa	72	0.776	74.2	67.8	98.3[k]	98.9[k]	76	72	3,338[h]	8,797[h]	-2
78 Thailand	71	0.779	74.5	65.0	90.5	94.9	72	71	6,695	10,732	0
79 Dominican Republic	74	0.773	74.8	68.6	87.2	86.8	78[g]	70[g]	4,907[h]	11,465[h]	-2
80 Belize	52	0.814	79.1	73.1	94.6[j]	94.6[j]	81	83	4,022[h]	10,117[h]	21
81 China	73	0.776	74.3[n]	71.0[n]	86.5	95.1	69	70	5,220[h]	8,213[h]	1
82 Grenada	..	..	69.8	66.5	..	..	74	72	..[h,l]	..[h,l]	..
83 Armenia	75	0.772	74.9	68.2	99.2[f]	99.7[f]	74	68	3,893[h]	6,150[h]	0
84 Turkey	79	0.763	73.9	69.0	79.6	95.3	64	73	4,385	12,368	-3
85 Suriname	78	0.767	73.0	66.4	87.2	92.0	82	72	4,426[h]	11,029[h]	-1
86 Jordan	80	0.760	73.8	70.3	87.0	95.2	79	77	2,566[h]	8,270[h]	-2
87 Peru	76	0.769	73.3	68.2	82.5	93.7	87	85	4,269[h]	7,791[h]	3
88 Lebanon	81	0.759	73.7	69.4	93.6[j]	93.6[j]	86	83	2,701[h]	8,585[h]	-1
89 Ecuador	..	..	77.7	71.8	89.7	92.3	..	..	3,102[h]	5,572[h]	..
90 Philippines	77	0.768	73.3	68.9	93.6	91.6	83	79	3,883	6,375	4
91 Tunisia	83	0.750	75.6	71.5	65.3	83.4	79	74	3,748[h]	12,924[h]	-1
92 Fiji	82	0.757	70.6	66.1	95.9[j]	95.9[j]	76	74	3,928[h]	8,103[h]	1
93 Saint Vincent and the Grenadines	..	..	73.2	69.0	..	..	70	68	4,449[h]	8,722[h]	..
94 Iran (Islamic Republic of)	84	0.750	71.8	68.7	76.8	88.0	73	73	4,475[h]	11,363[h]	0
95 Paraguay	86	0.744	73.4	69.2	92.7[k]	94.3[k]	70[g]	69[g]	2,358	6,892	-1
96 Georgia	..	..	74.5	66.7	..	..	77	75	1,731	5,188	..
97 Guyana	88	0.742	68.1	62.4	99.2[f,j]	99.2[f,j]	87	84	2,665[h]	6,467[h]	-2
98 Azerbaijan	87	0.743	70.8	63.5	98.2[f]	99.5[f]	66	68	3,960[h]	6,137[h]	0
99 Sri Lanka	89	0.735	75.6	67.9	89.1[o]	92.3[o]	64[g]	63[g]	2,647	6,479	-1
100 Maldives	85	0.744	67.6	66.6	96.4	96.2	66	65	3,992[h,m]	7,946[h,m]	4
101 Jamaica	90	0.732	74.9	69.6	85.9[o]	74.1[o]	82	74	3,107[h]	5,503[h]	0
102 Cape Verde	93	0.723	73.8	67.5	75.5[k]	87.8[k]	66	67	3,087[h]	8,756[h]	-2
103 El Salvador	92	0.726	74.3	68.2	79.2[k]	82.1[k]	70	70	3,043	7,543	0
104 Algeria	95	0.720	73.0	70.4	60.1	79.6	74	73	3,546[h]	10,515[h]	-2
105 Viet Nam	91	0.732	75.7	71.9	86.9	93.9	62	66	2,540[h]	3,604[h]	3
106 Occupied Palestinian Territories	..	..	74.4	71.3	88.0	96.7	84	81	..	..	..

Human development indicators

TABLE 28

Gender-related development index

HDI rank	Gender-related development index (GDI) Rank	Value	Life expectancy at birth (years) 2005 Female	Male	Adult literacy rate[a] (% aged 15 and older) 1995–2005 Female	Male	Combined gross enrolment ratio for primary, secondary and tertiary education[b] (%) 2005 Female	Male	Estimated earned income[c] (PPP US$) 2005 Female	Male	HDI rank minus GDI rank[d]
107 Indonesia	94	0.721	71.6	67.8	86.8	94.0	67	70	2,410[h]	5,280[h]	1
108 Syrian Arab Republic	96	0.710	75.5	71.8	73.6	87.8	63	67	1,907[h]	5,684[h]	0
109 Turkmenistan	..	..	67.0	58.5	98.3[f]	99.3[f]	..	..	6,108[h,m]	9,596[h,m]	..
110 Nicaragua	99	0.696	75.0	69.0	76.6	76.8	72	70	1,773[h]	5,577[h]	-2
111 Moldova	97	0.704	72.0	64.7	98.6[f,k]	99.6[f,k]	73	67	1,634[h]	2,608[h]	1
112 Egypt	..	..	73.0	68.5	59.4	83.0	..	..	1,635	7,024	..
113 Uzbekistan	98	0.699	70.0	63.6	99.6[f,j]	99.6[f,j]	72[g]	75[g]	1,547[h]	2,585[h]	1
114 Mongolia	100	0.695	69.2	62.8	97.5	98.0	83	72	1,413[h]	2,799[h]	0
115 Honduras	101	0.694	73.1	65.8	80.2	79.8	74	68	2,160[h]	4,680[h]	0
116 Kyrgyzstan	102	0.692	69.6	61.7	98.1[f]	99.3[f]	80	76	1,414[h]	2,455[h]	0
117 Bolivia	103	0.691	66.9	62.6	80.7	93.1	84[g]	90[g]	2,059[h]	3,584[h]	0
118 Guatemala	104	0.675	73.2	66.2	63.3	75.4	64	70	2,267[h]	6,990[h]	0
119 Gabon	105	0.670	56.9	55.6	79.7[k]	88.5[k]	68[g]	72[g]	5,049[h]	8,876[h]	0
120 Vanuatu	..	..	71.3	67.5	..	..	61	66	2,601[h]	3,830[h]	..
121 South Africa	107	0.667	52.0	49.5	80.9	84.1	77[g]	77[g]	6,927[h]	15,446[h]	-1
122 Tajikistan	106	0.669	69.0	63.8	99.2[f]	99.7[f]	64	77	992[h]	1,725[h]	1
123 Sao Tome and Principe	110	0.637	66.7	63.0	77.9	92.2	65	65	1,022[h]	3,357[h]	-2
124 Botswana	109	0.639	48.4	47.6	81.8	80.4	70	69	5,913	19,094	0
125 Namibia	108	0.645	52.2	50.9	83.5	86.8	66	63	5,527[h]	9,679[h]	2
126 Morocco	112	0.621	72.7	68.3	39.6	65.7	55	62	1,846[h]	7,297[h]	-1
127 Equatorial Guinea	111	0.631	51.6	49.1	80.5	93.4	52[g]	64[g]	4,635[h,l]	10,814[h,l]	1
128 India	113	0.600	65.3	62.3	47.8[o]	73.4[o]	60	68	1,620[h]	5,194[h]	0
129 Solomon Islands	..	..	63.8	62.2	..	..	46	50	1,345[h]	2,672[h]	..
130 Lao People's Democratic Republic	115	0.593	64.5	61.9	60.9	77.0	56	67	1,385[h]	2,692[h]	-1
131 Cambodia	114	0.594	60.6	55.2	64.1	84.7	56	64	2,332[h]	3,149[h]	1
132 Myanmar	..	..	64.2	57.6	86.4	93.9	51	48	..	..	..
133 Bhutan	..	..	66.5	63.1	..	..	..	..	2,141[h,m]	4,463[h,m]	..
134 Comoros	116	0.554	66.3	62.0	63.9[j]	63.9[j]	42	50	1,337[h]	2,643[h]	0
135 Ghana	117	0.549	59.5	58.7	49.8	66.4	48	53	2,056[h]	2,893[h]	0
136 Pakistan	125	0.525	64.8	64.3	35.4	64.1	34	45	1,059[h]	3,607[h]	-7
137 Mauritania	118	0.543	65.0	61.5	43.4	59.5	45	47	1,489[h]	2,996[h]	1
138 Lesotho	119	0.541	42.9	42.1	90.3	73.7	67	65	2,340[h]	4,480[h]	1
139 Congo	120	0.540	55.2	52.8	79.0[k]	90.5[k]	48	54	841[h]	1,691[h]	1
140 Bangladesh	121	0.539	64.0	62.3	40.8	53.9	56[g]	56[g]	1,282[h]	2,792[h]	1
141 Swaziland	123	0.529	41.4	40.4	78.3	80.9	58	62	2,187	7,659	0
142 Nepal	128	0.520	62.9	62.1	34.9	62.7	54	62	1,038[h]	2,072[h]	-4
143 Madagascar	122	0.530	60.1	56.7	65.3	76.5	58	61	758[h]	1,090[h]	3
144 Cameroon	126	0.524	50.2	49.4	59.8	77.0	57	68	1,519[h]	3,086[h]	0
145 Papua New Guinea	124	0.529	60.1	54.3	50.9	63.4	38[g]	43[g]	2,140[h]	2,960[h]	3
146 Haiti	..	..	61.3	57.7	56.5[j]	56.5[j]	..	..	1,146[h]	2,195[h]	..
147 Sudan	131	0.502	58.9	56.0	51.8[o]	71.1[o]	35	39	832[h]	3,317[h]	-3
148 Kenya	127	0.521	53.1	51.1	70.2	77.7	59	62	1,126	1,354	2
149 Djibouti	129	0.507	55.2	52.6	79.9[j]	79.9[j]	22	29	1,422[h]	2,935[h]	1
150 Timor-Leste	..	..	60.5	58.9	..	..	71	73	..[h]	..[h]	..
151 Zimbabwe	130	0.505	40.2	41.4	86.2[k]	92.7[k]	51[g]	54[g]	1,499[h]	2,585[h]	1
152 Togo	134	0.494	59.6	56.0	38.5	68.7	46	64	907[h]	2,119[h]	-2
153 Yemen	136	0.472	63.1	60.0	34.7[k]	73.1[k]	43	67	424[h]	1,422[h]	-3
154 Uganda	132	0.501	50.2	49.1	57.7	76.8	62	64	1,199[h]	1,708[h]	2
155 Gambia	133	0.496	59.9	57.7	49.9[j]	49.9[j]	49[g]	51[g]	1,327[h]	2,525[h]	2
LOW HUMAN DEVELOPMENT											
156 Senegal	135	0.492	64.4	60.4	29.2	51.1	37	42	1,256[h]	2,346[h]	1
157 Eritrea	137	0.469	59.0	54.0	71.5[j]	71.5[j]	29	41	689	1,544	0
158 Nigeria	139	0.456	47.1	46.0	60.1[k]	78.2[k]	51	61	652[h]	1,592[h]	-1
159 Tanzania (United Republic of)	138	0.464	52.0	50.0	62.2	77.5	49	52	627[h]	863[h]	1

Human development indicators

TABLE 28

HDI rank	Gender-related development index (GDI)		Life expectancy at birth (years) 2005		Adult literacy rate [a] (% aged 15 and older) 1995–2005		Combined gross enrolment ratio for primary, secondary and tertiary education [b] (%) 2005		Estimated earned income [c] (PPP US$) 2005		HDI rank minus GDI rank [d]
	Rank	Value	Female	Male	Female	Male	Female	Male	Female	Male	
160 Guinea	141	0.446	56.4	53.2	18.1	42.6	38	52	1,876 [h]	2,734 [h]	-1
161 Rwanda	140	0.450	46.7	43.6	59.8	71.4	51	51	1,031 [h]	1,392 [h]	1
162 Angola	142	0.439	43.3	40.1	54.2	82.9	24 [g]	28 [g]	1,787 [h]	2,898 [h]	0
163 Benin	145	0.422	56.5	54.1	23.3	47.9	42	59	732 [h]	1,543 [h]	-2
164 Malawi	143	0.432	46.7	46.0	54.0	74.9	62	64	565 [h]	771 [h]	1
165 Zambia	144	0.425	40.6	40.3	59.8	76.3	58	63	725 [h]	1,319 [h]	1
166 Côte d'Ivoire	146	0.413	48.3	46.5	38.6	60.8	32 [g]	47 [g]	795 [h]	2,472 [h]	0
167 Burundi	147	0.409	49.8	47.1	52.2	67.3	34	42	611 [h]	791 [h]	0
168 Congo (Democratic Republic of the)	148	0.398	47.1	44.4	54.1	80.9	28 [g]	39 [g]	488 [h]	944 [h]	0
169 Ethiopia	149	0.393	53.1	50.5	22.8	50.0	36	48	796 [h]	1,316 [h]	0
170 Chad	152	0.370	51.8	49.0	12.8	40.8	28	47	1,126 [h]	1,735 [h]	-2
171 Central African Republic	153	0.368	45.0	42.3	33.5	64.8	23 [g]	36 [g]	933 [h]	1,530 [h]	-2
172 Mozambique	150	0.373	43.6	42.0	25.0	54.8	48	58	1,115 [h]	1,378 [h]	2
173 Mali	151	0.371	55.3	50.8	15.9	32.7	31	42	833 [h]	1,234 [h]	2
174 Niger	155	0.355	54.9	56.7	15.1	42.9	19	26	561 [h]	991 [h]	-1
175 Guinea-Bissau	156	0.355	47.5	44.2	60.0	60.0 [j]	29 [g]	45 [g]	558 [h]	1,103 [h]	-1
176 Burkina Faso	154	0.364	52.9	49.8	16.6	31.4	25	33	966 [h]	1,458 [h]	2
177 Sierra Leone	157	0.320	43.4	40.2	24.2	46.7	38 [g]	52 [g]	507 [h]	1,114 [h]	0

NOTES

a. Data refer to national literacy estimates from censuses or surveys conducted between 1995 and 2005, unless otherwise specified. Due to differences in methodology and timeliness of underlying data, comparisons across countries and over time should be made with caution. For more details, see http://www.uis.unesco.org/.

b. Data for some countries may refer to national or UNESCO Institute for Statistics estimates. For details, see http://www.uis.unesco.org/.

c. Because of the lack of gender-disaggregated income data, female and male earned income are crudely estimated on the basis of data on the ratio of the female nonagricultural wage to the male nonagricultural wage, the female and male shares of the economically active population, the total female and male population and GDP per capita in PPP US$ (see Technical note 1). The wage ratios used in this calculation are based on data for the most recent year available between 1996 and 2005.

d. The HDI ranks used in this calculation are recalculated for the 157 countries with a GDI value. A positive figure indicates that the GDI rank is higher than the HDI rank, a negative the opposite.

e. For the purposes of calculating the GDI, a value of 99.0 % was applied.

f. For the purpose of calculating the GDI, the female and male values appearing in this table were scaled downward to reflect the maximum values for adult literacy (99%), gross enrolment ratios (100%), and GDP per capita ($40,000). For more details, see Technical note 1.

g. Data refer to an earlier year than that specified.

h. No wage data are available. For the purposes of calculating the estimated female and male earned income, a value of 0.75 was used for the ratio of the female nonagricultural wage to the male nonagricultural wage.

i. Statec. 2006.

j. In the absence of recent data, estimates from UNESCO Institute for Statistics 2003, based on outdated census or survey information were used, and should be interpreted with caution.

k. UNESCO Institute for Statistics estimates based on its Global age-specific literacy projections model.

l. Data from earlier years were adjusted to reflect their values in 2005 prices.

m. Heston, Alan, Robert Summers and Bettina Aten. 2006. Data may differ from the standard definition.

n. For statistical purposes, the data for China do not include Hong Kong and Macao, SARs of China.

o. Data refer to years or periods other than those specified in the column heading, differ from the standard definition or refer to only part of a country.

SOURCES

Column 1: determined on the basis of the GDI values in column 2.

Column 2: calculated on the basis of data in columns 3–10; see Technical note 1 for details.

Columns 3 and 4: UN 2007e.

Columns 5 and 6: UNESCO Institute for Statistics 2007a.

Columns 7 and 8: UNESCO Institute for Statistics 2007c.

Columns 9 and 10: calculated on the basis of data on GDP per capita (PPP US$) and population data from World Bank 2007b unless otherwise specified; data on wages from ILO 2007b; data on the economically active population from ILO 2005.

Column 11: calculated on the basis of recalculated HDI ranks and GDI ranks in column 1.

GDI ranks for 157 countries and areas

1 Iceland	28 Portugal	55 Panama	81 Lebanon	108 Namibia	134 Togo
2 Australia	29 Czech Republic	56 Trinidad and Tobago	82 Fiji	109 Botswana	135 Senegal
3 Norway	30 Barbados	57 Belarus	83 Tunisia	110 Sao Tome and Principe	136 Yemen
4 Canada	31 Brunei Darussalam	58 Malaysia	84 Iran (Islamic Republic of)	111 Equatorial Guinea	137 Eritrea
5 Sweden	32 Kuwait	59 Russian Federation	85 Maldives	112 Morocco	138 Tanzania (United Republic of)
6 Netherlands	33 Malta	60 Brazil	86 Paraguay	113 India	139 Nigeria
7 France	34 Hungary	61 Albania	87 Azerbaijan	114 Cambodia	140 Rwanda
8 Finland	35 Poland	62 Libyan Arab Jamahiriya	88 Guyana	115 Lao People's Democratic Republic	141 Guinea
9 Switzerland	36 Argentina	63 Mauritius	89 Sri Lanka	116 Comoros	142 Angola
10 United Kingdom	37 Qatar	64 Macedonia (TFYR)	90 Jamaica	117 Ghana	143 Malawi
11 Denmark	38 Lithuania	65 Kazakhstan	91 Viet Nam	118 Mauritania	144 Zambia
12 Spain	39 Slovakia	66 Colombia	92 El Salvador	119 Lesotho	145 Benin
13 Japan	40 Chile	67 Oman	93 Cape Verde	120 Congo	146 Côte d'Ivoire
14 Belgium	41 Estonia	68 Venezuela (Bolivarian Republic of)	94 Indonesia	121 Bangladesh	147 Burundi
15 Ireland	42 Bahrain	69 Ukraine	95 Algeria	122 Madagascar	148 Congo (Democratic Republic of the)
16 United States	43 United Arab Emirates	70 Saudi Arabia	96 Syrian Arab Republic	123 Swaziland	149 Ethiopia
17 Italy	44 Latvia	71 Thailand	97 Moldova	124 Papua New Guinea	150 Mozambique
18 New Zealand	45 Uruguay	72 Samoa	98 Uzbekistan	125 Pakistan	151 Mali
19 Austria	46 Croatia	73 China	99 Nicaragua	126 Cameroon	152 Chad
20 Germany	47 Costa Rica	74 Dominican Republic	100 Mongolia	127 Kenya	153 Central African Republic
21 Israel	48 Bahamas	75 Armenia	101 Honduras	128 Nepal	154 Burkina Faso
22 Hong Kong, China (SAR)	49 Cuba	76 Peru	102 Kyrgyzstan	129 Djibouti	155 Niger
23 Luxembourg	50 Bulgaria	77 Philippines	103 Bolivia	130 Zimbabwe	156 Guinea-Bissau
24 Greece	51 Mexico	78 Suriname	104 Guatemala	131 Sudan	157 Sierra Leone
25 Slovenia	52 Belize	79 Turkey	105 Gabon	132 Uganda	
26 Korea (Republic of)	53 Tonga	80 Jordan	106 Tajikistan	133 Gambia	
27 Cyprus	54 Romania		107 South Africa		

Human development indicators

TABLE **29**

... and achieving equality for all women and men

Gender empowerment measure

HDI rank	Gender empowerment measure (GEM)		MDG Seats in parliament held by women [a] (% of total)	Female legislators, senior officials and managers [b] (% of total)	Female professional and technical workers [b] (% of total)	Ratio of estimated female to male earned income [c]
	Rank	Value				
HIGH HUMAN DEVELOPMENT						
1 Iceland	5	0.862	31.7	27	56	0.72
2 Norway	1	0.910	37.9	30	50	0.77
3 Australia	8	0.847	28.3	37	56	0.70
4 Canada	10	0.820	24.3	36	56	0.64
5 Ireland	19	0.699	14.2	31	52	0.53
6 Sweden	2	0.906	47.3	30	51	0.81
7 Switzerland	27	0.660	24.8	8	22	0.63
8 Japan	54	0.557	11.1	10 [d]	46 [d]	0.45
9 Netherlands	6	0.859	36.0	26	50	0.64
10 France	18	0.718	13.9	37	47	0.64
11 Finland	3	0.887	42.0	30	55	0.71
12 United States	15	0.762	16.3	42	56	0.63
13 Spain	12	0.794	30.5	32	48	0.50
14 Denmark	4	0.875	36.9	25	53	0.73
15 Austria	13	0.788	31.0	27	49	0.46
16 United Kingdom	14	0.783	19.3	34	47	0.66
17 Belgium	7	0.850	35.7	32	49	0.55
18 Luxembourg	..	..	23.3	..	..	0.51
19 New Zealand	11	0.811	32.2	36	53	0.70
20 Italy	21	0.693	16.1	32	46	0.47
21 Hong Kong, China (SAR)	..	..	..	27	40	0.56
22 Germany	9	0.831	30.6	37	50	0.58
23 Israel	28	0.660	14.2	26	54	0.65
24 Greece	37	0.622	13.0	26	49	0.55
25 Singapore	16	0.761	24.5	26	44	0.51
26 Korea (Republic of)	64	0.510	13.4	8	39	0.40
27 Slovenia	41	0.611	10.8	33	57	0.61
28 Cyprus	48	0.580	14.3	15	45	0.60
29 Portugal	22	0.692	21.3	34	50	0.59
30 Brunei Darussalam	..	..	..[e]	26	44	0.42
31 Barbados	30	0.649	17.6	43	52	0.63
32 Czech Republic	34	0.627	15.3	30	52	0.51
33 Kuwait	..	..	3.1 [f]	..	..	0.35
34 Malta	63	0.514	9.2	20	38	0.50
35 Qatar	84	0.374	0.0	8	24	0.24
36 Hungary	50	0.569	10.4	35	62	0.64
37 Poland	39	0.614	19.1	33	61	0.60
38 Argentina	17	0.728	36.8	33	53	0.54
39 United Arab Emirates	29	0.652	22.5	8	25	0.25
40 Chile	60	0.519	12.7	25 [d]	52 [d]	0.40
41 Bahrain	..	..	13.8	..	..	0.35
42 Slovakia	33	0.630	19.3	31	58	0.58
43 Lithuania	25	0.669	24.8	43	67	0.69
44 Estonia	31	0.637	21.8	37	70	0.62
45 Latvia	38	0.619	19.0	42	65	0.65
46 Uruguay	59	0.525	10.8	40	54	0.56
47 Croatia	40	0.612	21.7	24	50	0.67
48 Costa Rica	24	0.680	38.6	25	40	0.53
49 Bahamas	20	0.696	22.2	46	60	0.70
50 Seychelles	..	..	23.5	..	..	..
51 Cuba	26	0.661	36.0	34 [d]	62 [d]	0.45
52 Mexico	46	0.589	21.5	29	42	0.39
53 Bulgaria	42	0.606	22.1	34	60	0.65

TABLE 29

HDI rank	Gender empowerment measure (GEM)		MDG Seats in parliament held by women [a] (% of total)	Female legislators, senior officials and managers [b] (% of total)	Female professional and technical workers [b] (% of total)	Ratio of estimated female to male earned income [c]
	Rank	Value				
54 Saint Kitts and Nevis	..	..	0.0	..	..	..
55 Tonga	..	..	3.3	..	..	0.48
56 Libyan Arab Jamahiriya	..	..	7.7	..	..	0.30
57 Antigua and Barbuda	..	..	13.9	45	55	..
58 Oman	80	0.391	7.8	9	33	0.19
59 Trinidad and Tobago	23	0.685	25.4	43	53	0.46
60 Romania	68	0.497	10.7	29	57	0.69
61 Saudi Arabia	92	0.254	0.0	31	6	0.16
62 Panama	49	0.574	16.7	43	51	0.57
63 Malaysia	65	0.504	13.1	23	40	0.36
64 Belarus	..	..	29.8	..	..	0.63
65 Mauritius	51	0.562	17.1	25	43	0.41
66 Bosnia and Herzegovina	..	..	14.0	..	..	..
67 Russian Federation	71	0.489	8.0	39	65	0.62
68 Albania	..	..	7.1	..	..	0.54
69 Macedonia (TFYR)	35	0.625	28.3	29	52	0.48
70 Brazil	70	0.490	9.3	34	52	0.58
MEDIUM HUMAN DEVELOPMENT						
71 Dominica	..	..	12.9	48	55	..
72 Saint Lucia	66	0.502	10.3 [g]	55	53	0.51
73 Kazakhstan	74	0.469	8.6	38	67	0.63
74 Venezuela (Bolivarian Republic of)	56	0.542	18.6	27 [d]	61 [d]	0.53
75 Colombia	69	0.496	9.7	38 [d]	50 [d]	0.63
76 Ukraine	75	0.462	8.7	38	64	0.55
77 Samoa	..	..	6.1	..	..	0.38
78 Thailand	73	0.472	8.7	29	54	0.62
79 Dominican Republic	53	0.559	17.1	32	51	0.43
80 Belize	62	0.517	11.9	41	50	0.40
81 China	57	0.534	20.3	17	52	0.64
82 Grenada	..	..	28.6	..	..	..
83 Armenia	..	..	9.2	..	..	0.63
84 Turkey	90	0.298	4.4	7	32	0.35
85 Suriname	..	..	25.5	..	..	0.40
86 Jordan	..	..	7.9	..	..	0.31
87 Peru	32	0.636	29.2	34	46	0.55
88 Lebanon	..	..	4.7	..	..	0.31
89 Ecuador	43	0.600	25.0	35	48	0.56
90 Philippines	45	0.590	22.1	58	61	0.61
91 Tunisia	..	..	19.3	..	..	0.29
92 Fiji	..	..	.. [h]	..	..	0.48
93 Saint Vincent and the Grenadines	..	..	18.2	..	..	0.51
94 Iran (Islamic Republic of)	87	0.347	4.1	16	34	0.39
95 Paraguay	78	0.428	9.6	23	54 [d]	0.34
96 Georgia	79	0.414	9.4	26	62	0.33
97 Guyana	..	..	29.0	..	..	0.41
98 Azerbaijan	..	..	11.3	..	..	0.65
99 Sri Lanka	85	0.369	4.9	21	46	0.41
100 Maldives	76	0.437	12.0	15	40	0.50
101 Jamaica	..	..	13.6	..	..	0.56
102 Cape Verde	..	..	15.3	..	..	0.35
103 El Salvador	58	0.529	16.7	33	45	0.40
104 Algeria	..	..	6.2	..	32	0.34
105 Viet Nam	52	0.561	25.8	22	51	0.70
106 Occupied Palestinian Territories	..	..	..	11	35	..

TABLE **29** Gender empowerment measure

HDI rank	Gender empowerment measure (GEM)		MDG Seats in parliament held by women [a] (% of total)	Female legislators, senior officials and managers [b] (% of total)	Female professional and technical workers [b] (% of total)	Ratio of estimated female to male earned income [c]
	Rank	**Value**				
107 Indonesia	..	..	11.3	..	..	0.46
108 Syrian Arab Republic	..	..	12.0	..	40 [d]	0.34
109 Turkmenistan	..	..	16.0	..	..	0.64
110 Nicaragua	..	..	18.5	..	..	0.32
111 Moldova	55	0.547	21.8	39	66	0.63
112 Egypt	91	0.263	3.8	9	30	0.23
113 Uzbekistan	..	..	16.4	..	..	0.60
114 Mongolia	77	0.429	6.6	50	54	0.50
115 Honduras	47	0.589	23.4	41 [d]	52 [d]	0.46
116 Kyrgyzstan	89	0.302	0.0	25	57	0.58
117 Bolivia	67	0.500	14.6	36	40	0.57
118 Guatemala	..	..	8.2	..	..	0.32
119 Gabon	..	..	13.7	..	..	0.57
120 Vanuatu	..	..	3.8	..	..	0.68
121 South Africa	..	..	32.8 [i]	..	..	0.45
122 Tajikistan	..	..	19.6	..	..	0.57
123 Sao Tome and Principe	..	..	7.3	..	..	0.30
124 Botswana	61	0.518	11.1	33	51	0.31
125 Namibia	36	0.623	26.9	30	55	0.57
126 Morocco	88	0.325	6.4	12	35	0.25
127 Equatorial Guinea	..	..	18.0	..	..	0.43
128 India	..	..	9.0	..	..	0.31
129 Solomon Islands	..	..	0.0	..	..	0.50
130 Lao People's Democratic Republic	..	..	25.2	..	..	0.51
131 Cambodia	83	0.377	11.4	14	33	0.74
132 Myanmar	..	..	.. [j]	..	..	..
133 Bhutan	..	..	2.7	..	..	..
134 Comoros	..	..	3.0	..	..	0.51
135 Ghana	..	..	10.9	..	..	0.71
136 Pakistan	82	0.377	20.4	2	26	0.29
137 Mauritania	..	..	17.6	..	..	0.50
138 Lesotho	..	..	25.0	..	..	0.52
139 Congo	..	..	10.1	..	..	0.50
140 Bangladesh	81	0.379	15.1 [k]	23	12	0.46
141 Swaziland	..	..	16.8	..	..	0.29
142 Nepal	86	0.351	17.3 [l]	8	19	0.50
143 Madagascar	..	..	8.4	..	..	0.70
144 Cameroon	..	..	8.9	..	..	0.49
145 Papua New Guinea	..	..	0.9	..	..	0.72
146 Haiti	..	..	6.3	..	..	0.52
147 Sudan	..	..	16.4	..	..	0.25
148 Kenya	..	..	7.3	..	..	0.83
149 Djibouti	..	..	10.8	..	..	0.48
150 Timor-Leste	..	..	25.3 [m]	..	..	..
151 Zimbabwe	..	..	22.2	..	..	0.58
152 Togo	..	..	8.6	..	..	0.43
153 Yemen	93	0.129	0.7	4	15	0.30
154 Uganda	..	..	29.8	..	..	0.70
155 Gambia	..	..	9.4	..	..	0.53
LOW HUMAN DEVELOPMENT						
156 Senegal	..	..	19.2	..	..	0.54
157 Eritrea	..	..	22.0	..	..	0.45
158 Nigeria	..	..	..	..	..	0.41
159 Tanzania (United Republic of)	44	0.597	30.4	49	32	0.73

Human development indicators

TABLE 29

HDI rank	Gender empowerment measure (GEM)		MDG Seats in parliament held by women [a] (% of total)	Female legislators, senior officials and managers [b] (% of total)	Female professional and technical workers [b] (% of total)	Ratio of estimated female to male earned income [c]
	Rank	Value				
160 Guinea	..	..	19.3	..	..	0.69
161 Rwanda	..	..	45.3	..	..	0.74
162 Angola	..	..	15.0	..	..	0.62
163 Benin	..	..	8.4	..	..	0.47
164 Malawi	..	..	13.6	..	..	0.73
165 Zambia	..	..	14.6	..	..	0.55
166 Côte d'Ivoire	..	..	8.5	..	..	0.32
167 Burundi	..	..	31.7	..	..	0.77
168 Congo (Democratic Republic of the)	..	..	7.7	..	..	0.52
169 Ethiopia	72	0.477	21.4	20	30	0.60
170 Chad	..	..	6.5	..	..	0.65
171 Central African Republic	..	..	10.5	..	..	0.61
172 Mozambique	..	..	34.8	..	..	0.81
173 Mali	..	..	10.2	..	..	0.68
174 Niger	..	..	12.4	..	..	0.57
175 Guinea-Bissau	..	..	14.0	..	..	0.51
176 Burkina Faso	..	..	11.7	..	..	0.66
177 Sierra Leone	..	..	14.5	..	..	0.45

NOTES

a. Data are as of 31 May 2007, unless otherwise specified. Where there are lower and upper houses, data refer to the weighted average of women's shares of seats in both houses.

b. Data refer to the most recent year available between 1994 and 2005. Estimates for countries that have implemented the International Standard Classification of Occupations (ISCO-88) are not strictly comparable with those for countries using the previous classification (ISCO-1968).

c. Calculated on the basis of data in columns 9 and 10 in Table 27. Estimates are based on data for the most recent year available between 1996 and 2005. Following the methodology implemented in the calculation of the GDI, the income component of the GEM has been scaled downward for countries whose income exceeds the maximum goalpost GDP per capita value of 40,000 (PPP US$). For more details, see Technical note 1.

d. Data follow the ISCO-1968 classification.

e. Brunei Darussalam does not currently have a parliament.

f. No woman candidate was elected in the 2006 elections. One woman was appointed to the 16-member cabinet sworn in July 2006. A new cabinet sworn in March 2007 included two women. As cabinet ministers also sit in parliament, there are two women out of a total of 65 members.

g. No woman candidate was elected in the 2006 elections. However one woman was appointed Speaker of the House and therefore became a member of the House.

h. Parliament has been dissolved or suspended for an indefinite period.

i. The figures on the distribution of seats do not include the 36 special rotating delegates appointed on an ad hoc basis. All percentages given are therefore calculated on the basis of the 54 permanent seats.

j. The parliament elected in 1990 has never been convened nor authorized to sit, and many of its members were detained or forced into exile.

k. In 2004, the number of seats in parliament was raised from 300 to 345, with the additional 45 seats reserved for women. These reserved seats were filled in September and October 2005, being allocated to political parties in proportion to their share of the national vote received in the 2001 election.

l. A transitional assembly was established in January 2007. Elections for the constituent assembly will be held in 2007.

m. The purpose of the elections held on 30 August 2001 was to elect the members of the constituent assembly of Timor-Leste. This body became the national parliament on 20 May 2002, the date on which the country became independent, without any new elections.

SOURCES

Column 1: determined on the basis of GEM values in column 2.

Column 2: calculated on the basis of data in columns 3–6; see Technical note 1 for details.

Column 3: calculated on the basis of data on parliamentary seats from IPU 2007c.

Columns 4 and 5: calculated on the basis of occupational data from ILO 2007b.

Column 6: calculated on the basis of data in columns 9 and 10 of Table 28.

GEM ranks for 93 countries

1 Norway	18 France	33 Slovakia	49 Panama	64 Korea (Republic of)	81 Bangladesh
2 Sweden	19 Ireland	34 Czech Republic	50 Hungary	65 Malaysia	82 Pakistan
3 Finland	20 Bahamas	35 Macedonia (TFYR)	51 Mauritius	66 Saint Lucia	83 Cambodia
4 Denmark	21 Italy	36 Namibia	52 Viet Nam	67 Bolivia	84 Qatar
5 Iceland	22 Portugal	37 Greece	53 Dominican Republic	68 Romania	85 Sri Lanka
6 Netherlands	23 Trinidad and	38 Latvia	54 Japan	69 Colombia	86 Nepal
7 Belgium	Tobago	39 Poland	55 Moldova	70 Brazil	87 Iran (Islamic
8 Australia	24 Costa Rica	40 Croatia	56 Venezuela	71 Russian Federation	Republic of)
9 Germany	25 Lithuania	41 Slovenia	(Bolivarian	72 Ethiopia	88 Morocco
10 Canada	26 Cuba	42 Bulgaria	Republic of)	73 Thailand	89 Kyrgyzstan
11 New Zealand	27 Switzerland	43 Ecuador	57 China	74 Kazakhstan	90 Turkey
12 Spain	28 Israel	44 Tanzania (United	58 El Salvador	75 Ukraine	91 Egypt
13 Austria	29 United Arab	Republic of)	59 Uruguay	76 Maldives	92 Saudi Arabia
14 United Kingdom	Emirates	45 Philippines	60 Chile	77 Mongolia	93 Yemen
15 United States	30 Barbados	46 Mexico	61 Botswana	78 Paraguay	
16 Singapore	31 Estonia	47 Honduras	62 Belize	79 Georgia	
17 Argentina	32 Peru	48 Cyprus	63 Malta	80 Oman	

Human development indicators

TABLE 30

. . . and achieving equality for all women and men

Gender inequality in education

	Adult literacy [a]		MDG Youth literacy [a]		Net primary enrolment [b, c]		MDG Gross primary enrolment [b, d]		MDG Gross secondary enrolment [b, d]		MDG Gross tertiary enrolment [b, d]	
	Female rate (% aged 15 and older)	Ratio of female rate to male rate	Female rate (% aged 15–24)	Ratio of female rate to male rate	Female rate (%)	Ratio of female rate to male rate	Female ratio (%)	Ratio of female rate to male rate	Female ratio (%)	Ratio of female rate to male rate	Female ratio (%)	Ratio of female rate to male rate
HDI rank	1995–2005	1995–2005	1995–2005	1995–2005	2005	2005	2005	2005	2005	2005	2005	2005
HIGH HUMAN DEVELOPMENT												
1 Iceland	..	..	..	..	97 [e]	0.97 [e]	98 [e]	0.97 [e]	109 [e]	1.03 [e]	93 [e]	1.85 [e]
2 Norway	..	..	..	..	98	1.00	98	1.00	114	1.01	97	1.54
3 Australia	..	..	..	..	97	1.00	104	0.99	144	0.95	80	1.25
4 Canada	..	..	..	..	..	..	99 [e,f]	1.00 [e,f]	116 [e,f]	0.98 [e,f]	72 [e,f]	1.36 [e,f]
5 Ireland	..	..	..	..	96	1.00	106	0.99	118	1.09	67	1.27
6 Sweden	..	..	..	..	96	1.00	97	1.00	103	1.00	100	1.55
7 Switzerland	..	..	..	..	93	0.99	101	0.99	91	0.93	43	0.84
8 Japan	..	..	..	..	100	1.00	100	1.00	102	1.00	52	0.89
9 Netherlands	..	..	..	..	98	0.99	106	0.98	117	0.98	63	1.08
10 France	..	..	..	..	99 [f]	1.00 [f]	110	0.99	116	1.00	64	1.29
11 Finland	..	..	..	..	98	1.00	99	0.99	113	1.05	101	1.21
12 United States	..	..	..	..	93	1.01	99	0.99	95	1.02	97	1.40
13 Spain	..	..	..	..	99	0.99	105	0.98	127	1.05	74	1.22
14 Denmark	..	..	..	..	96	1.01	99	1.00	126	1.03	94	1.39
15 Austria	..	..	..	..	98 [e]	1.02 [e]	106	1.00	100	0.95	55	1.20
16 United Kingdom	..	..	..	..	99	1.00	107	1.00	107	1.03	70	1.39
17 Belgium	..	..	..	..	99	1.00	103	0.99	108	0.97	70	1.24
18 Luxembourg	..	..	..	..	95	1.01	100	1.00	97	1.06	13 [e,f]	1.18 [e,f]
19 New Zealand	..	..	..	..	99	1.00	102	1.00	127	1.07	99	1.50
20 Italy	98.0	0.99	99.8	1.00	98	0.99	102	0.99	99	0.99	76	1.36
21 Hong Kong, China (SAR)	..	..	..	..	90 [e]	0.94 [e]	101	0.94	85	0.96	31	0.95
22 Germany	..	..	..	..	96 [e]	1.01 [e]	101	1.00	99	0.98	..	..
23 Israel	..	..	..	..	98	1.01	110	1.01	92	0.99	66	1.34
24 Greece	94.2	0.96	99.0	1.00	99	1.00	101	1.00	101	0.98	95	1.14
25 Singapore	88.6	0.92	99.6	1.00	..	..	..	..	..	..	..	..
26 Korea (Republic of)	..	..	..	..	99	1.00	104	0.99	93	1.00	69	0.62
27 Slovenia	99.6 [g]	1.00 [g]	99.9 [g]	1.00 [g]	98	0.99	100	0.99	99	1.00	96	1.43
28 Cyprus	95.1	0.96	99.8	1.00	99 [e]	1.00 [e]	101 [e]	1.00 [e]	97 [e]	1.02 [e]	35 [e]	1.13 [e]
29 Portugal	92.0 [g]	0.96 [g]	99.6 [g]	1.00 [g]	98	1.00	112	0.96	104	1.10	64	1.30
30 Brunei Darussalam	90.2	0.95	98.9	1.00	94	1.01	107	1.00	98	1.04	20	2.02
31 Barbados	..	..	..	..	98	1.00	108	1.00	113	1.00	54 [f]	2.47 [f]
32 Czech Republic	..	..	..	..	93 [e]	1.02 [e]	100	0.98	97	1.02	52	1.16
33 Kuwait	91.0	0.96	99.8	1.00	86	0.99	97	0.98	98	1.06	29	2.66
34 Malta	89.2	1.03	97.8	1.04	84	0.95	95	0.94	101	1.03	37	1.36
35 Qatar	88.6	0.99	97.5	1.03	96	1.00	106	0.99	99	0.98	33	3.45
36 Hungary	..	..	..	..	88	0.98	97	0.98	96	0.99	78	1.46
37 Poland	..	..	..	..	97	1.00	98	0.99	99	0.99	74	1.41
38 Argentina	97.2	1.00	99.1	1.00	98 [f]	0.99 [f]	112 [f]	0.99 [f]	89 [f]	1.07 [f]	76 [f]	1.41 [f]
39 United Arab Emirates	87.8 [g]	0.99 [g]	95.5 [g]	0.98 [g]	70	0.97	82	0.97	66	1.05	39 [e,f]	3.24 [e,f]
40 Chile	95.6	1.00	99.2	1.00	89 [e]	0.98 [e]	101	0.96	91	1.01	47	0.96
41 Bahrain	83.6	0.94	97.3	1.00	97	1.00	104	0.99	102	1.06	50	2.23
42 Slovakia	..	..	..	..	92 [e]	1.01 [e]	98	0.99	95	1.01	46	1.29
43 Lithuania	99.6	1.00	99.7	1.00	89	1.00	95	1.00	96	0.99	93	1.57
44 Estonia	99.8	1.00	99.8	1.00	95	0.99	99	0.97	101	1.01	82	1.66
45 Latvia	99.7	1.00	99.8	1.00	89 [e]	1.03 [e]	90	0.96	98	1.01	96	1.79
46 Uruguay	97.3	1.01	99.0	1.01	93 [e,f]	1.01 [e,f]	108 [f]	0.98 [f]	113 [f]	1.16 [f]	55 [e,f]	2.03 [e,f]
47 Croatia	97.1	0.98	99.7	1.00	87 [f]	0.99 [f]	94 [f]	0.99 [f]	89 [f]	1.02 [f]	42 [f]	1.19 [f]
48 Costa Rica	95.1	1.00	98.0	1.01	..	..	109	0.99	82	1.06	28 [e]	1.26 [e]
49 Bahamas	..	..	..	..	92	1.03	101	1.00	91	1.00	..	..
50 Seychelles	92.3	1.01	99.4	1.01	100 [e,f]	1.01 [e,f]	116 [e]	1.01 [e]	105 [e]	0.99 [e]	..	..
51 Cuba	99.8	1.00	100.0	1.00	96	0.98	99	0.95	94	1.00	78 [e]	1.72 [e]
52 Mexico	90.2	0.97	97.6	1.00	98	1.00	108	0.98	83	1.07	24	0.99
53 Bulgaria	97.7	0.99	98.1	1.00	93	0.99	101	0.99	101	0.95	47	1.14

TABLE 30

HDI rank	Adult literacy [a] Female rate (% aged 15 and older) 1995–2005	Adult literacy [a] Ratio of female rate to male rate 1995–2005	MDG Youth literacy [a] Female rate (% aged 15–24) 1995–2005	MDG Youth literacy [a] Ratio of female rate to male rate 1995–2005	Net primary enrolment [b, c] Female rate (%) 2005	Net primary enrolment [b, c] Ratio of female rate to male rate 2005	MDG Gross primary enrolment [b, d] Female ratio (%) 2005	MDG Gross primary enrolment [b, d] Ratio of female rate to male rate 2005	MDG Gross secondary enrolment [b, d] Female ratio (%) 2005	MDG Gross secondary enrolment [b, d] Ratio of female rate to male rate 2005	MDG Gross tertiary enrolment [b, d] Female ratio (%) 2005	MDG Gross tertiary enrolment [b, d] Ratio of female rate to male rate 2005
54 Saint Kitts and Nevis	..	..	..	..	96 [e]	1.06 [e]	102 [e]	1.06 [e]	93 [e]	0.98 [e]	..	..
55 Tonga	99.0	1.00	99.4	1.00	93 [e]	0.96 [e]	112 [e]	0.95 [e]	102 [e,f]	1.08 [e,f]	8 [e,f]	1.67 [e,f]
56 Libyan Arab Jamahiriya	74.8 [g]	0.81 [g]	96.5 [g]	0.97 [g]	..	..	106	0.98	107 [e]	1.19 [e]	59 [e,f]	1.09 [e,f]
57 Antigua and Barbuda	..	..	..	..	..	..	..	..	..	..	..	..
58 Oman	73.5	0.85	96.7	0.99	76	1.01	85	1.00	85	0.96	19	1.09
59 Trinidad and Tobago	97.8 [g]	0.99 [g]	99.5 [g]	1.00 [g]	90 [e]	1.00 [e]	99 [e]	0.97 [e]	82 [e]	1.04 [e]	14 [e]	1.27 [e]
60 Romania	96.3	0.98	97.8	1.00	92	0.99	106	0.99	86	1.01	50	1.26
61 Saudi Arabia	76.3	0.87	94.7	0.98	79	1.03	91	1.00	86	0.96	34	1.47
62 Panama	91.2	0.99	95.6	0.99	98	0.99	109	0.97	73	1.07	55	1.63
63 Malaysia	85.4	0.93	97.3	1.00	95 [f]	1.00 [f]	96 [f]	1.00 [f]	81 [f]	1.14 [f]	36 [f]	1.31 [f]
64 Belarus	99.4	1.00	99.8	1.00	88 [e]	0.97 [e]	100	0.97	96	1.01	72	1.37
65 Mauritius	80.5	0.91	95.4	1.02	96	1.02	102	1.00	88 [e]	0.99 [e]	19	1.26
66 Bosnia and Herzegovina	94.4	0.95	99.8	1.00	..	..	..	..	..	..	..	..
67 Russian Federation	99.2	1.00	99.8	1.00	93 [e]	1.01 [e]	128	1.00	91	0.99	82 [e]	1.36 [e]
68 Albania	98.3	0.99	99.5	1.00	94 [f]	1.00 [f]	105 [f]	0.99 [f]	77 [f]	0.96 [f]	23 [f]	1.57 [f]
69 Macedonia (TFYR)	94.1	0.96	98.5	0.99	92	1.00	98	1.00	83	0.98	35	1.38
70 Brazil	88.8	1.00	97.9	1.02	95 [f]	1.00 [f]	135 [f]	0.93 [f]	111 [f]	1.10 [f]	27 [f]	1.32 [f]
MEDIUM HUMAN DEVELOPMENT												
71 Dominica	..	..	..	..	85 [e]	1.02 [e]	92 [e]	0.99 [e]	106 [e]	0.97 [e]	..	..
72 Saint Lucia	..	..	..	..	96	0.98	107	0.97	85	1.21	20	2.80
73 Kazakhstan	99.3	1.00	99.9	1.00	90	0.98	108	0.99	97	0.97	62	1.42
74 Venezuela (Bolivarian Republic of)	92.7	0.99	98.1	1.02	92	1.01	104	0.98	79	1.13	41 [e,f]	1.08 [e,f]
75 Colombia	92.9	1.00	98.4	1.01	87	1.00	111	0.98	82	1.11	31	1.09
76 Ukraine	99.2	0.99	99.8	1.00	83 [e]	1.00 [e]	107	1.00	85	0.92	75	1.20
77 Samoa	98.3 [g]	0.99 [g]	99.4 [g]	1.00 [g]	91 [e,f]	1.00 [e,f]	100 [e]	1.00 [e]	85 [e]	1.12 [e]	7 [e,f]	0.93 [e,f]
78 Thailand	90.5	0.95	97.8	1.00	86 [h]	0.96 [h]	94 [h]	0.96 [h]	72 [h]	1.05 [h]	44 [h]	1.06 [h]
79 Dominican Republic	87.2	1.00	95.4	1.03	88	1.01	110	0.95	78	1.21	41 [e,f]	1.64 [e,f]
80 Belize	..	..	..	..	96	1.03	125	0.96	85 [e]	1.02 [e]	4 [f]	2.43 [f]
81 China	86.5	0.91	98.5	0.99	..	..	112 [e]	0.99 [e]	74 [e]	1.00 [e]	20	0.95
82 Grenada	..	..	..	..	83 [e]	0.99 [e]	91 [e]	0.96 [e]	102 [e]	1.03 [e]	..	..
83 Armenia	99.2	0.99	99.9	1.00	81	1.05	96	1.04	89	1.03	31	1.22
84 Turkey	79.6	0.84	93.3	0.95	87	0.95	91 [e]	0.95 [e]	68 [e]	0.82 [e]	26	0.74
85 Suriname	87.2	0.95	94.1	0.98	96	1.04	120	1.00	100	1.33	15 [f]	1.62 [f]
86 Jordan	87.0	0.91	99.0	1.00	90	1.02	96	1.01	88	1.02	40	1.06
87 Peru	82.5	0.88	96.3	0.98	97	1.00	112	1.00	92	1.01	34 [e]	1.03 [e]
88 Lebanon	..	..	..	..	92	0.99	105	0.97	93	1.10	54	1.15
89 Ecuador	89.7	0.97	96.5	1.00	98 [e,f]	1.01 [e,f]	117 [e]	1.00 [e]	61 [e]	1.00 [e]	..	..
90 Philippines	93.6	1.02	96.6	1.03	95	1.02	112	0.99	90	1.12	31	1.23
91 Tunisia	65.3	0.78	92.2	0.96	97	1.01	108	0.97	88	1.09	35	1.40
92 Fiji	..	..	..	..	96 [e]	0.99 [e]	105 [e]	0.98 [e]	91 [e]	1.07 [e]	17 [e]	1.20 [e]
93 Saint Vincent and the Grenadines	..	..	..	..	88	0.95	105	0.90	83	1.24	..	..
94 Iran (Islamic Republic of)	76.8	0.87	96.7	0.99	100	1.10	122	1.22	78	0.94	25	1.09
95 Paraguay	92.7 [g]	0.98 [g]	96.1 [g]	1.00 [g]	88 [f]	1.00 [f]	103 [f]	0.97 [f]	64 [f]	1.02 [f]	28 [e,f]	1.34 [e,f]
96 Georgia	..	..	..	..	92 [f]	0.99 [f]	94	1.01	83	1.01	47	1.04
97 Guyana	..	..	..	..	..	..	131	0.98	103	1.02	13	2.13
98 Azerbaijan	98.2	0.99	99.9	1.00	84	0.98	95	0.98	81	0.96	14	0.90
99 Sri Lanka	89.1	0.97	96.1	1.01	98 [e,f]	1.00 [e,f]	101 [e,f]	0.99 [e,f]	83 [e,f]	1.00 [e,f]	..	..
100 Maldives	96.4	1.00	98.3	1.00	79	1.00	93	0.98	78 [e,f]	1.14 [e,f]	(.) [e,f]	2.37 [e,f]
101 Jamaica	85.9	1.16	..	..	90 [e]	1.00 [e]	94	1.00	89	1.03	26 [e,f]	2.29 [e,f]
102 Cape Verde	75.5 [g]	0.86 [g]	96.7 [g]	1.01 [g]	89	0.98	105	0.95	70	1.07	7	1.04
103 El Salvador	79.2 [g]	0.96 [g]	90.3 [g]	1.04 [g]	93	1.00	111	0.96	64	1.03	21	1.23
104 Algeria	60.1	0.76	86.1	0.92	95	0.98	107	0.93	86 [e]	1.07 [e]	24	1.37
105 Viet Nam	86.9	0.93	93.6	0.99	..	..	91	0.94	75	0.97	13	0.71
106 Occupied Palestinian Territories	88.0	0.91	98.8	1.00	80	0.99	88	0.99	102	1.07	39 [e]	1.04 [e]

Human development indicators

TABLE **30**

Gender inequality in education

HDI rank	Adult literacy[a] Female rate (% aged 15 and older) 1995–2005	Adult literacy[a] Ratio of female rate to male rate 1995–2005	MDG Youth literacy[a] Female rate (% aged 15–24) 1995–2005	MDG Youth literacy[a] Ratio of female rate to male rate 1995–2005	Net primary enrolment[b, c] Female rate (%) 2005	Net primary enrolment[b, c] Ratio of female rate to male rate 2005	MDG Gross primary enrolment[b, d] Female ratio (%) 2005	MDG Gross primary enrolment[b, d] Ratio of female rate to male rate 2005	MDG Gross secondary enrolment[b, d] Female ratio (%) 2005	MDG Gross secondary enrolment[b, d] Ratio of female rate to male rate 2005	MDG Gross tertiary enrolment[b, d] Female ratio (%) 2005	MDG Gross tertiary enrolment[b, d] Ratio of female rate to male rate 2005
107 Indonesia	86.8	0.92	98.5	1.00	94[e]	0.96[e]	115[e]	0.96[e]	63[e]	0.99[e]	15[e]	0.79[e]
108 Syrian Arab Republic	73.6	0.84	90.2	0.95	..	..	121	0.95	65	0.94	..	..
109 Turkmenistan	98.3	0.99	99.8	1.00	..	..	..	..	..	..	..	..
110 Nicaragua	76.6	1.00	88.8	1.06	86	0.98	110	0.97	71	1.15	19[e,f]	1.11[e,f]
111 Moldova	98.6[g]	0.99[g]	99.7[g]	1.00[g]	86[e]	0.99[e]	92[e]	0.99[e]	83[e]	1.03[e]	41[e]	1.48[e]
112 Egypt	59.4	0.71	78.9	0.88	91[e]	0.95[e]	97	0.94	82	0.92	..	..
113 Uzbekistan	..	..	..	..	..	..	99[e,f]	0.99[e,f]	93[e,f]	0.97[e,f]	14[e,f]	0.80[e,f]
114 Mongolia	97.5	1.00	98.4	1.01	85	1.03	94	1.02	98	1.13	54	1.62
115 Honduras	80.2	1.01	90.9	1.05	92[e]	1.02[e]	113[e]	1.00[e]	73[e]	1.24[e]	20[e,f]	1.46[e,f]
116 Kyrgyzstan	98.1	0.99	99.7	1.00	86	0.99	97	0.99	87	1.01	46	1.25 -
117 Bolivia	80.7	0.87	96.1	0.98	96[e,f]	1.01[e,f]	113[e,f]	1.00[e,f]	87[f]	0.97[f]	..	..
118 Guatemala	63.3	0.84	78.4	0.91	92	0.95	109	0.92	49	0.91	8[e,f]	0.72[e,f]
119 Gabon	79.7[g]	0.90[g]	95.1[g]	0.98[g]	..	..	129[e,f]	0.99[e,f]	42[e,f]	0.86[e,f]	..	..
120 Vanuatu	..	..	..	..	93[e]	0.98[e]	116[e]	0.97[e]	38[f]	0.86[f]	4[e,f]	0.58[e,f]
121 South Africa	80.9	0.96	94.3	1.01	87[f]	1.00[f]	102[f]	0.96[f]	97[f]	1.07[f]	17	1.22
122 Tajikistan	99.2	1.00	99.8	1.00	96	0.96	99	0.96	74	0.83	9	0.35
123 Sao Tome and Principe	77.9	0.85	94.9	0.99	96	0.99	132	0.98	46	1.08	..	..
124 Botswana	81.8	1.02	95.6	1.04	84[e]	1.00[e]	105	0.98	75[e]	1.05[e]	5	1.00
125 Namibia	83.5	0.96	93.5	1.03	74	1.07	100	1.01	60	1.15	7[f]	1.15[f]
126 Morocco	39.6	0.60	60.5	0.75	83	0.94	99	0.89	46[e]	0.85[e]	10	0.85
127 Equatorial Guinea	80.5	0.86	94.9	1.00	..	..	111	0.95	22[e,f]	0.57[e,f]	2[f]	0.43[f]
128 India	47.8	0.65	67.7	0.80	85[e]	0.93[e]	116[e]	0.94[e]	50	0.80	9	0.70
129 Solomon Islands	..	..	..	..	..	..	94	0.95	27	0.83	..	..
130 Lao People's Democratic Republic	60.9	0.79	74.7	0.90	81	0.95	108	0.88	40	0.76	7	0.72
131 Cambodia	64.1	0.76	78.9	0.90	98	0.98	129	0.92	24[e,f]	0.69[e,f]	2	0.46
132 Myanmar	86.4	0.92	93.4	0.98	91	1.02	101	1.02	40	0.99	..	..
133 Bhutan	..	..	..	..	..	..	..	..	..	..	..	..
134 Comoros	..	..	..	..	..	..	80[e]	0.88[e]	30[e]	0.76[e]	2[e,f]	0.77[e,f]
135 Ghana	49.8	0.75	65.5	0.86	65	0.99	87	0.96	40[e]	0.85[e]	4	0.56
136 Pakistan	35.4	0.55	53.1	0.69	59	0.76	75	0.76	23	0.74	4	0.88
137 Mauritania	43.4	0.73	55.5	0.82	72	1.00	94	1.01	19	0.85	2	0.33
138 Lesotho	90.3	1.23	..	..	89	1.06	131	1.00	43	1.26	4	1.27
139 Congo	79.0[g]	0.87[g]	96.5[g]	0.98[g]	48	1.20	84	0.92	35[e,f]	0.84[e,f]	1[e,f]	0.19[e,f]
140 Bangladesh	40.8	0.76	60.3	0.90	96[e,f]	1.03[e,f]	111[f]	1.03[f]	48[f]	1.03[f]	4	0.53
141 Swaziland	78.3	0.97	89.8	1.03	80[e]	1.01[e]	104[e]	0.93[e]	44[e]	0.96[e]	5	1.06
142 Nepal	34.9	0.56	60.1	0.75	74[e,f]	0.87[e,f]	108	0.91	42[e]	0.86[e]	3[f]	0.40[f]
143 Madagascar	65.3	0.85	68.2	0.94	92	1.00	136	0.96	..	..	2	0.89
144 Cameroon	59.8	0.78	..	..	..	..	107[e]	0.85[e]	39[e]	0.80[e]	5[e]	0.66[e]
145 Papua New Guinea	50.9	0.80	64.1	0.93	..	..	70[e,f]	0.88[e,f]	23[e,f]	0.79[e,f]	..	..
146 Haiti	..	..	..	..	..	..	..	..	..	..	..	..
147 Sudan	51.8	0.73	71.4	0.84	..	..	56	0.87	33	0.94	..	..
148 Kenya	70.2	0.90	80.7	1.01	79	1.01	110	0.96	48[e]	0.95[e]	2[f]	0.60[f]
149 Djibouti	..	..	..	..	30	0.81	36	0.82	19	0.66	2	0.73
150 Timor-Leste	..	..	..	..	..	..	145	0.92	52	1.00	12[e,f]	1.48[e,f]
151 Zimbabwe	86.2[g]	0.93[g]	97.9[g]	1.00[g]	82[f]	1.01[f]	95[f]	0.98[f]	35[f]	0.91[f]	3[e,f]	0.63[e,f]
152 Togo	38.5	0.56	63.6	0.76	72	0.86	92	0.85	27[e]	0.51[e]	1[e,f]	0.20[e,f]
153 Yemen	34.7[g]	0.47[g]	58.9[g]	0.65[g]	63[e,f]	0.73[e,f]	75	0.74	31	0.49	5	0.37
154 Uganda	57.7	0.75	71.2	0.86	..	..	119	1.00	17[e]	0.81[e]	3[f]	0.62[f]
155 Gambia	..	..	..	..	77[e,f]	0.99[e,f]	84[f]	1.06[f]	42[f]	0.82[f]	(.)[f]	0.23[f]
LOW HUMAN DEVELOPMENT												
156 Senegal	29.2	0.57	41.0	0.70	67	0.97	77	0.97	18	0.75	..	..
157 Eritrea	..	..	..	..	43	0.86	57	0.81	23	0.59	(.)[f]	0.15[f]
158 Nigeria	60.1[g]	0.77[g]	81.3[g]	0.94[g]	64[e]	0.88[e]	95	0.86	31	0.84	7[f]	0.55[f]
159 Tanzania (United Republic of)	62.2	0.80	76.2	0.94	91	0.98	104	0.96	..	..	1[e]	0.48[e]

TABLE 30

HDI rank	Adult literacy [a] Female rate (% aged 15 and older) 1995–2005	Adult literacy [a] Ratio of female rate to male rate 1995–2005	MDG Youth literacy [a] Female rate (% aged 15–24) 1995–2005	MDG Youth literacy [a] Ratio of female rate to male rate 1995–2005	Net primary enrolment [b, c] Female rate (%) 2005	Net primary enrolment [b, c] Ratio of female rate to male rate 2005	MDG Gross primary enrolment [b, d] Female ratio (%) 2005	MDG Gross primary enrolment [b, d] Ratio of female rate to male rate 2005	MDG Gross secondary enrolment [b, d] Female ratio (%) 2005	MDG Gross secondary enrolment [b, d] Ratio of female rate to male rate 2005	MDG Gross tertiary enrolment [b, d] Female ratio (%) 2005	MDG Gross tertiary enrolment [b, d] Ratio of female rate to male rate 2005
160 Guinea	18.1	0.43	33.7	0.57	61	0.87	74	0.84	21 [e]	0.53 [e]	1	0.24
161 Rwanda	59.8	0.84	76.9	0.98	75 [e]	1.04 [e]	121 [e]	1.02 [e]	13 [e]	0.89 [e]	2 [e]	0.62 [e]
162 Angola	54.2	0.65	63.2	0.75	..	..	..	..	15 [f]	0.78 [f]	1 [e,f]	0.66 [e,f]
163 Benin	23.3	0.49	33.2	0.56	70	0.81	85	0.80	23 [e]	0.57 [e]	1 [e,f]	0.25 [e,f]
164 Malawi	54.0	0.72	70.7	0.86	97	1.05	124	1.02	25	0.81	(.) [f]	0.54 [f]
165 Zambia	59.8	0.78	66.2	0.91	89	1.00	108	0.95	25 [e]	0.82 [e]	..	..
166 Côte d'Ivoire	38.6	0.63	52.1	0.74	50 [e,f]	0.80 [e,f]	63 [e,f]	0.79 [e,f]	18 [e,f]	0.55 [e,f]	..	..
167 Burundi	52.2	0.78	70.4	0.92	58	0.91	78	0.86	11 [e]	0.74 [e]	1 [e]	0.38 [e]
168 Congo (Democratic Republic of the)	54.1	0.67	63.1	0.81	..	..	54 [e,f]	0.78 [e,f]	16 [e,f]	0.58 [e,f]	..	..
169 Ethiopia	22.8	0.46	38.5	0.62	59	0.92	86	0.86	24	0.65	1	0.32
170 Chad	12.8	0.31	23.2	0.42	..	..	62	0.67	8 [e]	0.33 [e]	(.) [e]	0.14 [e]
171 Central African Republic	33.5	0.52	46.9	0.67	..	..	44 [e]	0.66 [e]	..	..	..	..
172 Mozambique	25.0	0.46	36.6	0.61	74	0.91	94	0.85	11	0.69	1	0.49
173 Mali	15.9	0.49	16.9	0.52	45	0.81	59	0.80	18 [e]	0.62 [e]	2 [e]	0.47 [e]
174 Niger	15.1	0.35	23.2	0.44	33	0.73	39	0.73	7	0.68	1	0.45
175 Guinea-Bissau	..	..	..	..	37 [e,f]	0.71 [e,f]	56 [e,f]	0.67 [e,f]	13 [e,f]	0.54 [e,f]	(.) [e,f]	0.18 [e,f]
176 Burkina Faso	16.6	0.53	26.5	0.66	40	0.79	51	0.80	12	0.70	1	0.45
177 Sierra Leone	24.2	0.52	37.4	0.63	..	..	65 [f]	0.71 [f]	22 [e,f]	0.71 [e,f]	1 [e,f]	0.40 [e,f]
Developing countries	69.9	0.91	81.4	0.91	83 [i]	0.95 [i]	104 [i]	0.94 [i]	58 [i]	0.93 [i]	16 [i]	0.91 [i]
Least developed countries	44.3	0.80	58.0	0.80	70 [i]	0.92 [i]	90 [i]	0.89 [i]	28 [i]	0.81 [i]	3 [i]	0.63 [i]
Arab States	59.4	0.88	79.5	0.88	77 [i]	0.92 [i]	88 [i]	0.90 [i]	65 [i]	0.92 [i]	21 [i]	1.01 [i]
East Asia and the Pacific	86.7	0.99	97.5	0.99	93 [i]	0.99 [i]	110 [i]	0.98 [i]	72 [i]	1.00 [i]	21 [i]	0.93 [i]
Latin America and the Caribbean	89.7	1.01	97.0	1.01	95 [i]	1.00 [i]	115 [i]	0.96 [i]	91 [i]	1.08 [i]	32 [i]	1.17 [i]
South Asia	47.4	0.81	66.6	0.81	82 [i]	0.92 [i]	109 [i]	0.93 [i]	48 [i]	0.83 [i]	9 [i]	0.74 [i]
Sub-Saharan Africa	51.2	0.84	65.1	0.84	68 [i]	0.93 [i]	92 [i]	0.89 [i]	28 [i]	0.79 [i]	4 [i]	0.62 [i]
Central and Eastern Europe and the CIS	98.7	1.00	99.6	1.00	91 [i]	1.00 [i]	107 [i]	0.99 [i]	90 [i]	0.98 [i]	63 [i]	1.30 [i]
OECD	..	..	..	..	96 [i]	1.00 [i]	101 [i]	0.99 [i]	98 [i]	1.00 [i]	65 [i]	1.17 [i]
High-income OECD	..	..	..	..	96 [i]	1.01 [i]	102 [i]	0.99 [i]	103 [i]	1.00 [i]	76 [i]	1.20 [i]
High human development	93.6	1.01	98.4	1.01	..	..	..	..	..	..	..	..
Medium human development	71.2	0.92	83.2	0.92	..	..	..	..	..	..	..	..
Low human development	43.8	0.80	58.9	0.80	..	..	..	..	..	..	..	..
High income	..	..	..	..	95 [i]	1.01 [i]	101 [i]	0.99 [i]	102 [i]	1.00 [i]	73 [i]	1.21 [i]
Middle income	86.5	0.99	96.2	0.99	92 [i]	0.99 [i]	110 [i]	0.97 [i]	78 [i]	1.01 [i]	28 [i]	1.09 [i]
Low income	48.8	0.82	65.8	0.82	76 [i]	0.92 [i]	99 [i]	0.91 [i]	41 [i]	0.82 [i]	7 [i]	0.68 [i]
World	72.7	0.92	82.5	0.92	85 [i]	0.96 [i]	104 [i]	0.95 [i]	64 [i]	0.94 [i]	25 [i]	1.05 [i]

NOTES

a. Data refer to national literacy estimates from censuses or surveys conducted between 1995 and 2005, unless otherwise specified. Due to differences in methodology and timeliness of underlying data, comparisons across countries and over time should be made with caution. For more details, see http://www.uis.unesco.org/.

b. Data for some countries may refer to national or UNESCO Institute for Statistics estimates. For more details, see http://www.uis.unesco.org/.

c. The net enrolment rate is the number of pupils of the theoretical school-age group for a given level of education level who are enrolled in that level, expressed as a percentage of the total population in that age group.

d. The gross enrolment ratio is the total number of pupils or students enrolled in a given level of education, regardless of age, expressed as a percentage of the population in the theoretical age group for the same level of education. For the tertiary level, the population used is the five-year age group following on from the secondary school leaving age. Gross enrolment ratios in excess of 100 indicate that there are pupils or students outside the theoretical age group who are enrolled in that level of education.

e. National or UNESCO Institute for Statistics estimate.

f. Data refer to an earlier year than that specified.

g. UNESCO Institute for Statistics estimate based on its Global Age-specific Literacy Projections model, April 2007.

h. Data refer to the 2006 school year.

i. Data refer to aggregates calculated by UNESCO Institute for Statistics.

SOURCES

Columns 1–4: UNESCO Institute for Statistics 2007a.
Columns 5–12: UNESCO Institute for Statistics 2007c.

TABLE 31

. . . and achieving equality for all women and men

Gender inequality in economic activity

	Female economic activity (aged 15 and older)			Employment by economic activity [a] (%)						Contributing family workers (%)	
				Agriculture		Industry		Services			
	Rate (%)	Index (1990=100)	As % of male rate	Women	Men	Women	Men	Women	Men	Women	Men
HDI rank	2005	2005	2005	1995–2005 [b]	1995–2005 [b]	1995–2005 [b]	1995–2005 [b]	1995–2005 [b]	1995–2005 [b]	1995–2005 [b]	1995–2005 [b]
HIGH HUMAN DEVELOPMENT											
1 Iceland	70.5	104	86	4	11	11	34	85	55	50	50
2 Norway	63.3	112	87	2	5	8	32	90	63	50	50
3 Australia	56.4	109	80	3	5	9	31	88	65	60	40
4 Canada	60.5	105	84	2	4	11	32	88	64	61	39
5 Ireland	53.2	150	74	1	9	12	39	86	51	53	47
6 Sweden	58.7	93	87	1	3	9	34	90	63	50	50
7 Switzerland	60.4	116	80	3	5	12	32	85	63	62	38
8 Japan	48.3	96	66	5	4	18	35	77	59	80	20
9 Netherlands	56.2	129	77	2	4	8	30	86	62	79	21
10 France	48.2	105	79	3	5	12	35	84	60	..	..
11 Finland	56.9	98	86	3	7	12	38	84	56	40	60
12 United States	59.6	105	82	1	2	10	30	90	68	62	38
13 Spain	44.9	132	66	4	6	12	41	84	52	64	36
14 Denmark	59.3	96	84	2	4	12	34	86	62	84	16
15 Austria	49.5	115	76	6	6	13	40	81	55	68	32
16 United Kingdom	55.2	104	80	1	2	9	33	90	65	60	40
17 Belgium	43.7	120	73	1	3	11	35	82	62	85	15
18 Luxembourg	44.6	124	69	3 [c]	3 [c]	8 [c]	42 [c]	89 [c]	55 [c]	..	..
19 New Zealand	60.4	113	82	5	9	11	32	84	59	66	34
20 Italy	37.4	104	62	3	5	18	39	79	56	54	46
21 Hong Kong, China (SAR)	53.7	114	76	(.)	(.)	7	22	93	77	..	..
22 Germany	50.8	114	77	2	3	16	41	82	56	76	24
23 Israel	50.1	122	85	1	3	11	32	88	64	72	28
24 Greece	43.5	121	67	14	12	10	30	76	58	68	32
25 Singapore	50.6	101	66	(.)	(.)	21	36	79	63	..	..
26 Korea (Republic of)	50.2	107	68	9	7	17	34	74	59	..	..
27 Slovenia	53.6	99	80	9	9	25	47	65	43	58	42
28 Cyprus	53.7	113	76	4	6	11	34	85	59	75	25
29 Portugal	55.7	113	79	13	12	21	42	66	46	65	35
30 Brunei Darussalam	44.1	98	55	(.)	2	11	29	88	69	..	..
31 Barbados	64.9	110	83	3	4	8	26	78	62	..	..
32 Czech Republic	51.9	85	77	3	5	27	49	71	46	74	26
33 Kuwait	49.0	141	58	..	..	..	..	..	..	..	..
34 Malta	34.0	159	49	1	2	18	34	81	63	..	..
35 Qatar	36.3	123	41	(.)	3	3	48	97	49	..	..
36 Hungary	42.1	91	73	3	7	21	42	76	51	69	31
37 Poland	47.7	83	78	17	18	17	39	66	43	60	40
38 Argentina	53.3	139	70	1	2	11	33	88	66	..	..
39 United Arab Emirates	38.2	152	42	(.)	9	14	36	86	55	..	..
40 Chile	36.6	114	52	6	17	12	29	83	54	..	..
41 Bahrain	29.3	103	33	..	..	..	..	..	..	..	..
42 Slovakia	51.8	87	76	3	6	25	50	72	44	74	26
43 Lithuania	51.7	87	82	11	17	21	37	68	46	62	38
44 Estonia	52.3	81	80	4	7	24	44	72	49	50	50
45 Latvia	49.0	78	77	8	15	16	35	75	49	43	57
46 Uruguay	56.4	123	72	2	7	13	29	86	64	..	..
47 Croatia	44.7	96	74	19	16	18	37	63	47	73	27
48 Costa Rica	44.9	137	56	5	21	13	26	82	52	..	..
49 Bahamas	64.4	105	91	(.)	6	5	30	94	64	..	..
50 Seychelles	..	..	..	..	..	..	..	..	..	..	..
51 Cuba	43.9	113	59	10	28	14	23	76	50	..	..
52 Mexico	40.2	116	50	5	21	19	30	76	49	..	..
53 Bulgaria	41.2	69	78	7	11	29	39	64	50	65	35

TABLE 31

	Female economic activity (aged 15 and older)			Employment by economic activity [a] (%)						Contributing family workers (%)	
				Agriculture		Industry		Services			
	Rate (%)	Index (1990=100)	As % of male rate	Women	Men	Women	Men	Women	Men	Women	Men
HDI rank	2005	2005	2005	1995–2005[b]	1995–2005[b]	1995–2005[b]	1995–2005[b]	1995–2005[b]	1995–2005[b]	1995–2005[b]	1995–2005[b]
54 Saint Kitts and Nevis	..	..	..	..	..	..	..	..	..	..	..
55 Tonga	47.5	126	63	..	..	..	..	..	..	..	..
56 Libyan Arab Jamahiriya	32.1	168	40	..	..	..	..	..	..	..	..
57 Antigua and Barbuda	..	..	..	3[c]	5[c]	7[c]	29[c]	87[c]	63[c]	..	..
58 Oman	22.7	149	28	5	7	14	11	80	82	..	..
59 Trinidad and Tobago	46.7	112	61	2	10	14	37	84	53	..	..
60 Romania	50.1	94	80	33	31	25	35	42	34	70	30
61 Saudi Arabia	17.6	118	22	1	5	1	24	98	71	..	..
62 Panama	50.8	131	64	4	22	9	22	86	56	..	..
63 Malaysia	46.5	105	57	11	16	27	35	62	49	..	..
64 Belarus	52.5	87	82	..	..	..	..	..	..	..	..
65 Mauritius	42.7	102	54	9	11	29	34	62	55	..	..
66 Bosnia and Herzegovina	58.3	97	86	..	..	..	..	..	..	..	..
67 Russian Federation	54.3	90	80	8	12	21	38	71	50	24	76
68 Albania	49.0	84	70	..	..	..	..	..	..	..	..
69 Macedonia (TFYR)	40.8	85	63	19	20	30	34	51	46	54	46
70 Brazil	56.7	127	71	16	25	13	27	71	48	..	..
MEDIUM HUMAN DEVELOPMENT											
71 Dominica	..	..	..	14	31	10	24	72	40	..	..
72 Saint Lucia	54.0	116	67	9	14	11	23	62	45	..	..
73 Kazakhstan	65.3	106	87	32	35	10	24	58	41	54	46
74 Venezuela (Bolivarian Republic of)	57.4	152	69	2	16	11	25	86	59	..	..
75 Colombia	61.3	135	76	8	32	16	21	76	48	..	..
76 Ukraine	49.6	86	79	17	21	21	38	62	41	50	50
77 Samoa	39.2	97	51	..	..	..	..	..	..	..	..
78 Thailand	65.6	87	81	41	44	19	22	41	34	..	..
79 Dominican Republic	46.4	127	57	2	23	15	24	83	53	..	..
80 Belize	43.3	139	52	6	37	12	19	83	44	..	..
81 China	68.8	94	83	..	..	..	..	..	..	..	..
82 Grenada	..	..	..	10	17	12	32	77	46	..	..
83 Armenia	47.9	67	79	..	..	..	..	..	..	38	63
84 Turkey	27.7	81	36	52	22	15	28	33	50	67	33
85 Suriname	33.6	92	52	2	8	1	22	97	64	..	..
86 Jordan	27.5	155	36	2	4	13	23	83	73	..	..
87 Peru	59.1	126	72	(.)	1	13	31	86	68	..	..
88 Lebanon	32.4	102	41	..	..	..	..	..	..	..	..
89 Ecuador	60.0	184	73	4	11	12	27	84	62	..	..
90 Philippines	54.7	115	66	25	45	12	17	64	39	..	..
91 Tunisia	28.6	138	38	..	..	..	..	..	..	..	..
92 Fiji	51.8	106	64	..	..	..	..	..	..	..	..
93 Saint Vincent and the Grenadines	55.3	124	68	8	20	8	27	72	46	..	..
94 Iran (Islamic Republic of)	38.6	180	52	34	23	28	31	37	46	..	..
95 Paraguay	65.1	126	77	20	39	10	19	70	42	..	..
96 Georgia	50.1	73	66	57	52	4	14	38	34	65	35
97 Guyana	43.5	120	53	16	34	20	24	61	42	..	..
98 Azerbaijan	60.2	95	82	37	41	9	15	54	44	..	..
99 Sri Lanka	34.9	77	45	40	32	35	40	25	29	..	..
100 Maldives	48.5	233	67	5	18	24	16	39	56	..	..
101 Jamaica	54.1	83	73	9	25	5	27	86	48	..	..
102 Cape Verde	34.0	81	45	..	..	..	..	..	..	..	..
103 El Salvador	47.3	93	62	3	30	22	25	75	45	..	..
104 Algeria	35.7	158	45	22	20	28	26	49	54	..	..
105 Viet Nam	72.2	98	92	60	56	14	21	26	23	..	..
106 Occupied Palestinian Territories	10.3	111	15	34	12	8	28	56	59	..	..

Human development indicators

TABLE 31

Gender inequality in economic activity

	Female economic activity (aged 15 and older)			Employment by economic activity [a] (%)						Contributing family workers (%)	
				Agriculture		Industry		Services			
HDI rank	Rate (%) 2005	Index (1990=100) 2005	As % of male rate 2005	Women 1995– 2005 [b]	Men 1995– 2005 [b]	Women 1995– 2005 [b]	Men 1995– 2005 [b]	Women 1995– 2005 [b]	Men 1995– 2005 [b]	Women 1995– 2005 [b]	Men 1995– 2005 [b]
107 Indonesia	51.0	101	60	45	43	15	20	40	37	..	..
108 Syrian Arab Republic	38.6	135	44	58	24	7	31	35	45	..	..
109 Turkmenistan	60.5	94	83	..	..	..	..	..	..	..	..
110 Nicaragua	35.7	100	41	10	43	17	19	52	32	..	..
111 Moldova	56.6	92	81	40	41	12	21	48	38	75	25
112 Egypt	20.1	76	27	39	28	6	23	55	49	..	..
113 Uzbekistan	56.6	95	78	..	..	..	..	..	..	..	..
114 Mongolia	53.9	97	66	38	43	14	19	49	39	..	..
115 Honduras	54.0	162	61	13	51	23	20	63	29	..	..
116 Kyrgyzstan	55.0	94	74	55	51	7	13	38	36	65	35
117 Bolivia	62.6	129	74	3	6	14	39	82	55	..	..
118 Guatemala	33.8	116	41	18	50	23	18	56	27	..	..
119 Gabon	61.4	98	75	..	..	..	..	..	..	..	..
120 Vanuatu	79.3	99	91	..	..	..	..	..	..	..	..
121 South Africa	45.9	85	58	7	13	14	33	79	54	..	..
122 Tajikistan	46.3	89	74	..	..	..	..	..	..	..	..
123 Sao Tome and Principe	29.8	83	40	..	..	..	..	..	..	..	..
124 Botswana	45.3	79	67	19	26	13	29	58	43	..	..
125 Namibia	46.6	96	74	29	33	7	17	63	49	..	..
126 Morocco	26.8	110	33	57	39	19	21	25	40	..	..
127 Equatorial Guinea	50.3	106	56	..	..	..	..	..	..	..	..
128 India	34.0	94	42	..	..	..	..	..	..	..	..
129 Solomon Islands	54.3	98	66	..	..	..	..	..	..	..	..
130 Lao People's Democratic Republic	54.0	101	67	89	81	3	4	8	14	..	..
131 Cambodia	74.4	96	93	75	72	10	7	15	20	..	..
132 Myanmar	68.2	99	79	..	..	..	..	..	..	..	..
133 Bhutan	46.7	134	58	..	..	..	..	..	..	..	..
134 Comoros	57.9	92	67	..	..	..	..	..	..	..	..
135 Ghana	70.3	92	94	50	60	15	14	36	27	..	..
136 Pakistan	32.7	117	39	65	38	16	22	20	40	..	..
137 Mauritania	54.4	98	65	..	..	..	..	..	..	..	..
138 Lesotho	45.7	81	63	45	66	13	17	31	17	..	..
139 Congo	56.4	98	65	..	..	..	..	..	..	..	..
140 Bangladesh	52.7	83	61	59	50	18	12	23	38	..	..
141 Swaziland	31.2	82	43	..	..	..	..	..	..	..	..
142 Nepal	49.9	104	64	..	..	..	..	..	..	..	..
143 Madagascar	78.9	100	92	79	77	6	7	15	16	..	..
144 Cameroon	51.7	92	65	68 [c]	53 [c]	4 [c]	14 [c]	23 [c]	26 [c]	..	..
145 Papua New Guinea	71.8	101	96	..	..	..	..	..	..	..	..
146 Haiti	55.6	97	67	37	63	6	15	57	23	..	..
147 Sudan	23.7	86	33	..	..	..	..	..	..	..	..
148 Kenya	69.1	93	78	16	20	10	23	75	57	..	..
149 Djibouti	52.9	94	64	(.) [c]	3 [c]	1 [c]	11 [c]	88 [c]	78 [c]	..	..
150 Timor-Leste	54.3	109	67	..	..	..	..	..	..	..	..
151 Zimbabwe	64.0	92	76	..	..	..	..	..	..	..	..
152 Togo	50.3	93	56	..	..	..	..	..	..	..	..
153 Yemen	29.7	108	39	88	43	3	14	9	43	..	..
154 Uganda	79.7	99	92	77	60	5	11	17	28	..	..
155 Gambia	59.1	94	69	..	..	..	..	..	..	..	..
LOW HUMAN DEVELOPMENT											
156 Senegal	56.3	92	69	..	..	..	..	..	..	..	..
157 Eritrea	58.1	95	64	..	..	..	..	..	..	..	..
158 Nigeria	45.4	95	53	2	4	11	30	87	67	..	..
159 Tanzania (United Republic of)	85.8	97	95	84	80	1	4	15	16	..	..

Human development indicators

TABLE 31

HDI rank	Female economic activity (aged 15 and older)			Employment by economic activity [a] (%)						Contributing family workers (%)	
	Rate (%) 2005	Index (1990=100) 2005	As % of male rate 2005	Agriculture		Industry		Services			
				Women 1995–2005[b]	Men 1995–2005[b]	Women 1995–2005[b]	Men 1995–2005[b]	Women 1995–2005[b]	Men 1995–2005[b]	Women 1995–2005[b]	Men 1995–2005[b]
160 Guinea	79.4	100	91	..	..	..	..	..	..	..	..
161 Rwanda	80.0	93	95	..	..	..	..	..	..	..	..
162 Angola	73.7	99	81	..	..	..	..	..	..	..	..
163 Benin	53.7	92	62	..	..	..	..	..	..	..	..
164 Malawi	85.4	100	95	..	..	..	..	..	..	..	..
165 Zambia	66.0	100	73	78	64	2	10	20	27	..	..
166 Côte d'Ivoire	38.8	89	44	..	..	..	..	..	..	..	..
167 Burundi	91.8	101	99	..	..	..	..	..	..	..	..
168 Congo (Democratic Republic of the)	61.2	101	68	..	..	..	..	..	..	..	..
169 Ethiopia	70.8	98	79	91[c]	94[c]	3[c]	3[c]	6[c]	3[c]	..	..
170 Chad	65.6	102	85	..	..	..	..	..	..	..	..
171 Central African Republic	70.3	99	79	..	..	..	..	..	..	..	..
172 Mozambique	84.5	96	102	..	..	..	..	..	..	..	..
173 Mali	72.5	100	87	..	..	..	..	..	..	..	..
174 Niger	71.3	101	75	..	..	..	..	..	..	..	..
175 Guinea-Bissau	61.0	105	66	..	..	..	..	..	..	..	..
176 Burkina Faso	77.6	101	87	..	..	..	..	..	..	..	..
177 Sierra Leone	56.1	105	60	..	..	..	..	..	..	..	..
Developing countries	52.4	101	64	..	..	..	..	..	..	..	..
Least developed countries	61.8	95	72	..	..	..	..	..	..	..	..
Arab States	26.7	110	34	..	..	..	..	..	..	..	..
East Asia and the Pacific	65.2	96	79	..	..	..	..	..	..	..	..
Latin America and the Caribbean	51.9	127	65	..	..	..	..	..	..	..	..
South Asia	36.2	99	44	..	..	..	..	..	..	..	..
Sub-Saharan Africa	62.6	96	73	..	..	..	..	..	..	..	..
Central and Eastern Europe and the CIS	52.4	89	79	..	..	..	..	..	..	..	..
OECD	50.3	105	72	..	..	..	..	..	..	..	..
High-income OECD	52.8	107	76	..	..	..	..	..	..	..	..
High human development	51.6	107	73	..	..	..	..	..	..	..	..
Medium human development	52.2	98	64	..	..	..	..	..	..	..	..
Low human development	63.4	97	72	..	..	..	..	..	..	..	..
High income	52.1	107	75	..	..	..	..	..	..	..	..
Middle income	57.0	101	72	..	..	..	..	..	..	..	..
Low income	45.7	96	55	..	..	..	..	..	..	..	..
World	52.5	101	67	..	..	..	..	..	..	..	..

NOTES

Because of limitations in the data, comparisons of labour statistics over time and across countries should be made with caution. For detailed notes on the data, see ILO 2005.

a. The percentage shares of employment by economic activity may not sum to 100 because of rounding or the omission of activities not classified.

b. Data refer to the most recent year available during the period specified.

c. Data refer to a year or period other than that specified.

SOURCES

Columns 1 and 4–9 : ILO 2005.

Columns 2, 3, 10 and 11: calculated on the basis of data on economically active rates from ILO 2005.

Human development indicators

TABLE 32

. . . and achieving equality for all women and men . . .

Gender, work and time allocation

		Total work in market and nonmarket activities (hours and minutes per day)		Market activities[a] (as % of total work time)		Cooking and cleaning[b] (hours and minutes per day)		Care of children[c]		Free time[d] (hours and minutes per day)		Personal care[e]	
HDI rank	Year	Women	Men	Women	Men	Women	Men	Women	Men	Women	Men	Women	Men
HIGH HUMAN DEVELOPMENT													
2 Norway	2000–01	7:13	7:23	41	61	2:14	0:52	0:34	0:17	6:08	6:23	10:18	9:59
3 Australia	1997	7:15	6:58	30	62	..	..	..	..	..	..	..	..
4 Canada f	2005	7:57	7:51	40	59	1:54	0:48	0:35 g	0:17 g	5:28	5:53	10:49	10:26
5 Ireland	2005	6:38	6:10	30	72	2:46	1:14	1:55 g	0:31 g	5:35	6:08	10:06	9:54
6 Sweden	2000–01	7:32	7:43	42	59	2:04	0:59	0:29	0:16	5:16	5:37	10:39	10:12
8 Japan h	1996	6:33	6:03	43	93	..	..	..	..	..	..	..	..
9 Netherlands h	1995	5:08	5:15	27	69	..	..	..	..	..	..	..	..
10 France	1998–99	7:01	6:27	33	59	3:04	0:48	0:28	0:09	3:52	4:26	11:57	11:46
11 Finland	1999–00	7:20	6:58	38	59	2:28	1:01	0:28	0:11	5:29	6:08	10:38	10:23
12 United States	2005	8:06	7:54	42	64	1:54	0:36	0:48 g	0:24 g	4:54	5:18	10:42	10:24
13 Spain	2002–03	7:54	6:51	30	71	3:22	0:37	0:30	0:12	4:34	5:34	11:05	11:11
16 United Kingdom	2000–01	7:41	7:32	35	62	2:34	0:59	0:33	0:12	5:11	5:44	10:43	10:22
17 Belgium	1999–00	6:35	6:04	29	54	2:57	0:55	0:35	0:19	4:40	5:12	11:12	10:55
19 New Zealand h	1999	7:00	6:57	32	60	..	..	..	..	..	..	..	..
20 Italy	2002–03	8:08	6:51	26	70	4:02	0:31	0:28	0:11	4:15	5:29	11:12	11:16
22 Germany	2001–02	7:00	6:49	30	55	2:32	0:52	0:26	0:10	5:35	6:02	11:02	10:44
26 Korea (Republic of)	2004	7:30	6:51	40	86	2:36	0:20	0:55	0:15	5:03	5:34	10:41	10:45
Rural f	2005	11:11	10:35	67	96	2:22	0:07	0:37 g	0:11 g	3:37	3:52	9:08	9:29
27 Slovenia	2000–01	8:22	7:24	35	57	3:21	0:54	0:29	0:12	4:40	5:43	10:32	10:30
29 Portugal f	1999	7:39	6:05	39	82	3:59	0:57	0:42 g	0:10 g	3:08	4:05	11:26	11:25
36 Hungary	1999–00	8:00	7:08	32	56	3:16	0:47	0:35	0:15	4:44	5:36	11:00	11:00
37 Poland	2003–04	7:55	7:25	31	59	3:13	1:02	0:39	0:16	4:33	5:23	11:03	10:44
43 Lithuania	2003–04	8:55	8:00	43	65	3:05	1:05	0:25	0:07	3:51	4:52	10:57	10:53
44 Estonia	1999–00	8:55	8:09	38	60	3:07	1:01	0:37	0:10	4:19	5:01	10:30	10:35
45 Latvia	2003–04	8:31	8:02	46	70	2:31	0:47	0:22	0:04	4:17	4:58	10:53	10:46
46 Uruguay i	2002	7:20	6:56	33	68	..	..	..	..	..	..	..	..
52 Mexico f	2002	8:10	6:25	23	78	4:43	0:39	1:01 g	0:21 g	2:37	3:01	9:56	9:43
65 Mauritius j	2003	6:33	6:09	30	80	3:33	0:30	0:44	0:13	4:34	5:09	11:49	11:35
MEDIUM AND LOW HUMAN DEVELOPMENT													
110 Nicaragua j	1998	6:29	6:08	28	74	3:31	0:31	1:01	0:17	5:05	5:05	10:48	10:42
Rural j	1998	6:33	6:40	36	73	3:49	0:21	1:00	0:11	5:05	5:18	11:00	10:42
Urban j	1998	6:30	5:30	18	76	3:16	0:43	1:01	0:24	5:52	5:56	10:42	10:36
114 Mongolia f	2000	9:02	8:16	49	76	3:49	1:45	0:45	0:16	2:54	3:39	10:29	10:40
Rural j	2000	10:35	9:52	48	80	4:46	1:46	0:43	0:12	2:18	2:51	10:20	10:31
Urban j	2000	7:41	6:49	51	70	3:00	1:44	0:47	0:19	3:25	4:23	10:38	10:47
121 South Africa	2000	6:52	6:01	38	76	3:06	1:00	0:39 g	0:04 g	4:08	4:53	12:11	11:58
128 India k	2000	7:37	6:31	35	92	..	..	..	..	..	..	..	..
143 Madagascar j	2001	7:14	7:03	50	80	2:51	0:17	0:31	0:08	1:45	2:15	13:09	13:04
Rural j	2001	7:30	7:40	53	78	2:52	0:14	0:31	0:07	1:24	1:54	13:18	13:13
Urban j	2001	6:36	5:37	44	86	2:49	0:22	0:31	0:11	2:35	3:05	12:47	12:43
163 Benin j	1998	8:03	5:36	59	80	2:49	0:27	0:45	0:05	1:32	3:22	12:05	11:59
Rural j	1998	8:20	5:50	61	81	2:50	0:22	0:50	0:05	1:51	3:26	11:52	11:55
Urban j	1998	7:23	5:02	53	78	2:46	0:37	0:35	0:04	1:58	3:16	12:13	12:06

NOTES

Comparisons between countries and areas must be made with caution. Unless otherwise noted, time use data in this table refer to an average day of the year for the total population aged 20 to 74. Travel time for each of the activities is included in the reported time for most of the countries, but exceptions may exist.

a. Refers to market-oriented production activities as defined by the 1993 revised UN System of National Accounts.

b. Includes the following activities: dishwashing, cleaning dwelling, laundry, ironing and other household upkeep.

c. Includes physical care of children, teaching, playing, etc. with children and other childcare.

d. Includes social life, entertainment, resting, doing sports, arts, computers, exposure to media, etc.

e. Includes sleep, eating and other personal care.

f. Data refer to age groups other than specified in the standard definition.

g. In addition to childcare, the value represented includes caring for adults with special needs or elderly persons, either in the home or elsewhere (e.g. help with personal care).

h. Harvey 2001.

i. Data refer to urban population only.

j. Data in columns 1-4 pertain to an age group different from the data in columns 5-12. In neither case is the reference population the same as in the standard definition.

k. UN 2002.

SOURCE
All columns: Time use 2007.

TABLE 33

... and achieving equality for all women and men

Women's political participation

		Year women received right [a]		Year first woman elected (E) or appointed (A) to parliament	Women in government at ministerial level (% of total) [b]	MDG Seats in parliament held by women (% of total) [c]		Upper house or senate
						Lower or single house		
HDI rank		To vote	To stand for election		2005	1990	2007	2007
HIGH HUMAN DEVELOPMENT								
1	Iceland	1915, 1920	1915, 1920	1922 E	27.3	20.6	31.7	—
2	Norway	1913	1907, 1913	1911 A	44.4	35.8	37.9	—
3	Australia	1902, 1962	1902, 1962	1943 E	20.0	6.1	24.7	35.5
4	Canada	1917, 1960	1920, 1960	1921 E	23.1	13.3	20.8	35.0
5	Ireland	1918, 1928	1918, 1928	1918 E	21.4	7.8	13.3	16.7
6	Sweden	1919, 1921	1919, 1921	1921 E	52.4	38.4	47.3	—
7	Switzerland	1971	1971	1971 E	14.3	14.0	25.0	23.9
8	Japan	1945, 1947	1945, 1947	1946 E	12.5	1.4	9.4	14.5
9	Netherlands	1919	1917	1918 E	36.0	21.3	36.7	34.7
10	France	1944	1944	1945 E	17.6	6.9	12.2	16.9
11	Finland	1906	1906	1907 E	47.1	31.5	42.0	—
12	United States	1920, 1965	1788 [d]	1917 E	14.3	6.6	16.3	16.0
13	Spain	1931	1931	1931 E	50.0	14.6	36.0	23.2
14	Denmark	1915	1915	1918 E	33.3	30.7	36.9	—
15	Austria	1918	1918	1919 E	35.3	11.5	32.2	27.4
16	United Kingdom	1918, 1928	1918, 1928	1918 E	28.6	6.3	19.7	18.9
17	Belgium	1919, 1948	1921	1921 A	21.4	8.5	34.7	38.0
18	Luxembourg	1919	1919	1919 E	14.3	13.3	23.3	—
19	New Zealand	1893	1919	1933 E	23.1	14.4	32.2	—
20	Italy	1945	1945	1946 E	8.3	12.9	17.3	13.7
21	Hong Kong, China (SAR)	..	..	..	..	..	..	..
22	Germany	1918	1918	1919 E	46.2	..	31.6	21.7
23	Israel	1948	1948	1949 E	16.7	6.7	14.2	—
24	Greece	1952	1952	1952 E	5.6	6.7	13.0	—
25	Singapore	1947	1947	1963 E	0.0	4.9	24.5	—
26	Korea (Republic of)	1948	1948	1948 E	5.6	2.0	13.4	—
27	Slovenia	1946	1946	1992 E [e]	6.3	..	12.2	7.5
28	Cyprus	1960	1960	1963 E	0.0	1.8	14.3	—
29	Portugal	1931, 1976	1931, 1976	1934 E	16.7	7.6	21.3	—
30	Brunei Darussalam	—	—	—	9.1	.. [f]	.. [f]	.. [f]
31	Barbados	1950	1950	1966 A	29.4	3.7	13.3	23.8
32	Czech Republic	1920	1920	1992 E [e]	11.1	..	15.5	14.8
33	Kuwait	2005	2005	2005 A	0.0	..	3.1 [g]	—
34	Malta	1947	1947	1966 E	15.4	2.9	9.2	—
35	Qatar	2003 [h]	..	..	7.7	..	0.0	—
36	Hungary	1918, 1945	1918, 1945	1920 E	11.8	20.7	10.4	—
37	Poland	1918	1918	1919 E	5.9	13.5	20.4	13.0
38	Argentina	1947	1947	1951 E	8.3	6.3	35.0	43.1
39	United Arab Emirates	—	—	—	5.6	0.0	22.5	—
40	Chile	1949	1949	1951 E	16.7	..	15.0	5.3
41	Bahrain	1973, 2002	1973, 2002	2002 A	8.7	..	2.5	25.0
42	Slovakia	1920	1920	1992 E [e]	0.0	..	19.3	—
43	Lithuania	1919	1919	1920 A	15.4	..	24.8	—
44	Estonia	1918	1918	1919 E	15.4	..	21.8	—
45	Latvia	1918	1918	..	23.5	..	19.0	—
46	Uruguay	1932	1932	1942 E	0.0	6.1	11.1	9.7
47	Croatia	1945	1945	1992 E [e]	33.3	..	21.7	—
48	Costa Rica	1949	1949	1953 E	25.0	10.5	38.6	—
49	Bahamas	1961, 1964	1961, 1964	1977 A	26.7	4.1	12.2	53.8
50	Seychelles	1948	1948	1976 E+A	12.5	16.0	23.5	—
51	Cuba	1934	1934	1940 E	16.2	33.9	36.0	—
52	Mexico	1947	1953	1952 A	9.4	12.0	22.6	17.2
53	Bulgaria	1937, 1945	1945	1945 E	23.8	21.0	22.1	—

TABLE 33

Women's political participation

		Year women received right [a]		Year first woman elected (E) or appointed (A) to parliament	Women in government at ministerial level (% of total) [b]	MDG Seats in parliament held by women (% of total) [c]		
						Lower or single house		Upper house or senate
HDI rank		To vote	To stand for election		2005	1990	2007	2007
54	Saint Kitts and Nevis	1951	1951	1984 E	0.0	6.7	0.0	—
55	Tonga	1960	1960	1993 E	..	0.0	3.3	—
56	Libyan Arab Jamahiriya	1964	1964	..	..	..	7.7	—
57	Antigua and Barbuda	1951	1951	1984 A	15.4	0.0	10.5	17.6
58	Oman	1994, 2003	1994, 2003	..	10.0	..	2.4	15.5
59	Trinidad and Tobago	1946	1946	1962 E+A	18.2	16.7	19.4	32.3
60	Romania	1929, 1946	1929, 1946	1946 E	12.5	34.4	11.2	9.5
61	Saudi Arabia	—	—	—	0.0	..	0.0	—
62	Panama	1941, 1946	1941, 1946	1946 E	14.3	7.5	16.7	—
63	Malaysia	1957	1957	1959 E	9.1	5.1	9.1	25.7
64	Belarus	1918	1919	1990 E [e]	10.0	..	29.1	31.0
65	Mauritius	1956	1956	1976 E	8.0	7.1	17.1	—
66	Bosnia and Herzegovina	1946	1946	1990 E [e]	11.1	..	14.3	13.3
67	Russian Federation	1918	1918	1993 E [e]	0.0	..	9.8	3.4
68	Albania	1920	1920	1945 E	5.3	28.8	7.1	—
69	Macedonia (TFYR)	1946	1946	1990 E [e]	16.7	..	28.3	—
70	Brazil	1932	1932	1933 E	11.4	5.3	8.8	12.3
MEDIUM HUMAN DEVELOPMENT								
71	Dominica	1951	1951	1980 E	0.0	10.0	12.9	—
72	Saint Lucia	1951	1951	1979 A	8.3	0.0	5.6 [i]	18.2
73	Kazakhstan	1924, 1993	1924, 1993	1990 E [e]	17.6	..	10.4	5.1
74	Venezuela (Bolivarian Republic of)	1946	1946	1948 E	13.6	10.0	18.6	—
75	Colombia	1954	1954	1954 A	35.7	4.5	8.4	11.8
76	Ukraine	1919	1919	1990 E [e]	5.6	..	8.7	—
77	Samoa	1948, 1990	1948, 1990	1976 A	7.7	0.0	6.1	—
78	Thailand	1932	1932	1948 A	7.7	2.8	8.7	—
79	Dominican Republic	1942	1942	1942 E	14.3	7.5	19.7	3.1
80	Belize	1954	1954	1984 E+A	6.3	0.0	6.7	25.0
81	China	1949	1949	1954 E	6.3	21.3	20.3	—
82	Grenada	1951	1951	1976 E+A	40.0	..	26.7	30.8
83	Armenia	1918	1918	1990 E [e]	0.0	35.6	9.2	—
84	Turkey	1930, 1934	1930, 1934	1935 A	4.3	1.3	4.4	—
85	Suriname	1948	1948	1975 E	11.8	7.8	25.5	—
86	Jordan	1974	1974	1989 A	10.7	0.0	5.5	12.7
87	Peru	1955	1955	1956 E	11.8	5.6	29.2	—
88	Lebanon	1952	1952	1991 A	6.9	0.0	4.7	—
89	Ecuador	1929	1929	1956 E	14.3	4.5	25.0	—
90	Philippines	1937	1937	1941 E	25.0	9.1	22.5	18.2
91	Tunisia	1959	1959	1959 E	7.1	4.3	22.8	13.4
92	Fiji	1963	1963	1970 A	9.1	.. [j]	.. [j]	.. [j]
93	Saint Vincent and the Grenadines	1951	1951	1979 E	20.0	9.5	18.2	—
94	Iran (Islamic Republic of)	1963	1963	1963 E+A	6.7	1.5	4.1	—
95	Paraguay	1961	1961	1963 E	30.8	5.6	10.0	8.9
96	Georgia	1918, 1921	1918, 1921	1992 E [e]	22.2	..	9.4	—
97	Guyana	1953	1945	1968 E	22.2	36.9	29.0	—
98	Azerbaijan	1918	1918	1990 E [e]	15.0	..	11.3	—
99	Sri Lanka	1931	1931	1947 E	10.3	4.9	4.9	—
100	Maldives	1932	1932	1979 E	11.8	6.3	12.0	—
101	Jamaica	1944	1944	1944 E	17.6	5.0	11.7	19.0
102	Cape Verde	1975	1975	1975 E	18.8	12.0	15.3	—
103	El Salvador	1939	1961	1961 E	35.3	11.7	16.7	—
104	Algeria	1962	1962	1962 A	10.5	2.4	7.2	3.1
105	Viet Nam	1946	1946	1976 E	11.5	17.7	25.8	—
106	Occupied Palestinian Territories	..	..	..	..	..	..	..

Human development indicators

TABLE 33

HDI rank	Year women received right[a] To vote	To stand for election	Year first woman elected (E) or appointed (A) to parliament	Women in government at ministerial level (% of total)[b] 2005	MDG Seats in parliament held by women (% of total)[c] Lower or single house 1990	2007	Upper house or senate 2007
107 Indonesia	1945, 2003	1945	1950 A	10.8	12.4	11.3	—
108 Syrian Arab Republic	1949, 1953	1953	1973 E	6.3	9.2	12.0	—
109 Turkmenistan	1927	1927	1990 E [e]	9.5	26.0	16.0	—
110 Nicaragua	1955	1955	1972 E	14.3	14.8	18.5	—
111 Moldova	1924, 1993	1924, 1993	1990 E	11.1	..	21.8	—
112 Egypt	1956	1956	1957 E	5.9	3.9	2.0	6.8
113 Uzbekistan	1938	1938	1990 E [e]	3.6	..	17.5	15.0
114 Mongolia	1924	1924	1951 E	5.9	24.9	6.6	—
115 Honduras	1955	1955	1957 E	14.3	10.2	23.4	—
116 Kyrgyzstan	1918	1918	1990 E [e]	12.5	..	0.0	—
117 Bolivia	1938, 1952	1938, 1952	1966 E	6.7	9.2	16.9	3.7
118 Guatemala	1946	1946, 1965	1956 E	25.0	7.0	8.2	—
119 Gabon	1956	1956	1961 E	11.8	13.3	12.5	15.4
120 Vanuatu	1975, 1980	1975, 1980	1987 E	8.3	4.3	3.8	—
121 South Africa	1930, 1994	1930, 1994	1933 E	41.4	2.8	32.8 [k]	33.3 [k]
122 Tajikistan	1924	1924	1990 E [e]	3.1	..	17.5	23.5
123 Sao Tome and Principe	1975	1975	1975 E	14.3	11.8	7.3	—
124 Botswana	1965	1965	1979 E	26.7	5.0	11.1	—
125 Namibia	1989	1989	1989 E	19.0	6.9	26.9	26.9
126 Morocco	1963	1963	1993 E	5.9	0.0	10.8	1.1
127 Equatorial Guinea	1963	1963	1968 E	4.5	13.3	18.0	—
128 India	1935, 1950	1935, 1950	1952 E	3.4	5.0	8.3	10.7
129 Solomon Islands	1974	1974	1993 E	0.0	0.0	0.0	—
130 Lao People's Democratic Republic	1958	1958	1958 E	0.0	6.3	25.2	—
131 Cambodia	1955	1955	1958 E	7.1	..	9.8	14.8
132 Myanmar	1935	1946	1947 E	..	..[l]	..[l]	..[l]
133 Bhutan	1953	1953	1975 E	0.0	2.0	2.7	—
134 Comoros	1956	1956	1993 E	..	0.0	3.0	—
135 Ghana	1954	1954	1960 A	11.8	..	10.9	—
136 Pakistan	1935, 1947	1935, 1947	1973 E [e]	5.6	10.1	21.3	17.0
137 Mauritania	1961	1961	1975 E	9.1	..	17.9	17.0
138 Lesotho	1965	1965	1965 A	27.8	..	23.5	30.3
139 Congo	1947, 1961	1963	1963 E	14.7	14.3	8.5	13.3
140 Bangladesh	1935, 1972	1935, 1972	1973 E	8.3	10.3	15.1 [m]	—
141 Swaziland	1968	1968	1972 E+A	13.3	3.6	10.8	30.0
142 Nepal	1951	1951	1952 A	7.4	6.1	17.3 [n]	—
143 Madagascar	1959	1959	1965 E	5.9	6.5	6.9	11.1
144 Cameroon	1946	1946	1960 E	11.1	14.4	8.9	—
145 Papua New Guinea	1964	1963	1977 E	..	0.0	0.9	—
146 Haiti	1957	1957	1961 E	25.0	..	4.1	13.3
147 Sudan	1964	1964	1964 E	2.6	..	17.8	4.0
148 Kenya	1919, 1963	1919, 1963	1969 E+A	10.3	1.1	7.3	—
149 Djibouti	1946	1986	2003 E	5.3	0.0	10.8	—
150 Timor-Leste	..	..	..	22.2	..	25.3 [o]	—
151 Zimbabwe	1919, 1957	1919, 1978	1980 E+A	14.7	11.0	16.7	34.8
152 Togo	1945	1945	1961 E	20.0	5.2	8.6	—
153 Yemen	1967, 1970	1967, 1970	1990 E [e]	2.9	4.1	0.3	1.8
154 Uganda	1962	1962	1962 A	23.4	12.2	29.8	—
155 Gambia	1960	1960	1982 E	20.0	7.8	9.4	—
LOW HUMAN DEVELOPMENT							
156 Senegal	1945	1945	1963 E	20.6	12.5	19.2	—
157 Eritrea	1955 [p]	1955 [p]	1994 E	17.6	..	22.0	—
158 Nigeria	1958	1958	..	10.0	..	6.4 [q]	7.3
159 Tanzania (United Republic of)	1959	1959	..	15.4	..	30.4	—

Human development indicators

TABLE 33 Women's political participation

HDI rank	Year women received right [a] To vote	To stand for election	Year first woman elected (E) or appointed (A) to parliament	Women in government at ministerial level (% of total) [b] 2005	MDG Seats in parliament held by women (% of total) [c] Lower or single house 1990	Lower or single house 2007	Upper house or senate 2007
160 Guinea	1958	1958	1963 E	15.4	..	19.3	—
161 Rwanda	1961	1961	1981 E	35.7	17.1	48.8	34.6
162 Angola	1975	1975	1980 E	5.7	14.5	15.0	—
163 Benin	1956	1956	1979 E	19.0	2.9	8.4	—
164 Malawi	1961	1961	1964 E	14.3	9.8	13.6	—
165 Zambia	1962	1962	1964 E+A	25.0	6.6	14.6	—
166 Côte d'Ivoire	1952	1952	1965 E	17.1	5.7	8.5	—
167 Burundi	1961	1961	1982 E	10.7	..	30.5	34.7
168 Congo (Democratic Republic of the)	1967	1970	1970 E	12.5	5.4	8.4	4.6
169 Ethiopia	1955	1955	1957 E	5.9	..	21.9	18.8
170 Chad	1958	1958	1962 E	11.5	..	6.5	—
171 Central African Republic	1986	1986	1987 E	10.0	3.8	10.5	—
172 Mozambique	1975	1975	1977 E	13.0	15.7	34.8	—
173 Mali	1956	1956	1959 E	18.5	..	10.2	—
174 Niger	1948	1948	1989 E	23.1	5.4	12.4	—
175 Guinea-Bissau	1977	1977	1972 A	37.5	20.0	14.0	—
176 Burkina Faso	1958	1958	1978 E	14.8	..	11.7	—
177 Sierra Leone	1961	1961	..	13.0	..	14.5	—
OTHERS							
Afghanistan	1963	1963	1965 E	10.0	3.7	27.3	22.5
Andorra	1970	1973	1993 E	33.3	..	28.6	—
Iraq	1980	1980	1980 E	18.8	10.8	25.5	—
Kiribati	1967	1967	1990 E	0.0	0.0	7.1	—
Korea (Democratic People's Rep)	1946	1946	1948 E	..	21.1	20.1	—
Liberia	1946	1946	..	13.6	..	12.5	16.7
Liechtenstein	1984	1984	1986 E	20.0	4.0	24.0	—
Marshall Islands	1979	1979	1991 E	0.0	..	3.0	—
Micronesia (Federated States of)	1979	1979	..	..	..	0.0	—
Monaco	1962	1962	1963 E	0.0	11.1	20.8	—
Montenegro	1946 [r]	1946 [r]	..	..	..	8.6	—
Nauru	1968	1968	1986 E	0.0	5.6	0.0	—
Palau	1979	1979	..	12.5	..	0.0	0.0
San Marino	1959	1973	1974 E	12.5	11.7	11.7	—
Serbia	1946 [r]	1946 [r]	..	..	..	20.4	—
Somalia	1956	1956	1979 E	..	4.0	8.2	—
Tuvalu	1967	1967	1989 E	0.0	7.7	0.0	—

NOTES

a. Data refer to the year in which the right to vote or stand for national election on a universal and equal basis was recognized. Where two years are shown, the first refers to the first partial recognition of the right to vote or stand for election. In some countries, women were granted the right to vote or stand at local elections before obtaining these rights for national elections. Data on local election rights are not included in this table.

b. Data are as of 1 January 2005. The total includes deputy prime ministers and ministers. Prime ministers who hold ministerial portfolios and vice-presidents and heads of ministerial level departments or agencies who exercise a ministerial function in the government structure are also included.

c. Data are as of 31 May 2007 unless otherwise specified. The percentage was calculated using as a reference the number of total seats filled in parliament at that time.

d. No information is available on the year all women received the right to stand for election. However, the constitution does not mention gender with regard to this right.

e. Refers to the year women were elected to the current parliamentary system.

f. Brunei Darussalam does not currently have a parliament.

g. No woman candidate was elected in the 2006 elections. One woman was appointed to the 16-member cabinet sworn in July 2006. A new cabinet sworn in March 2007 included two women. As cabinet ministers also sit in parliament, there are two women out of a total of 65 members.

h. According to the new constitution approved in 2003, women are granted suffrage. To date no legislative elections have been held.

i. No woman was elected in the 2006 elections. However one woman was appointed Speaker of the House and therefore became a member of the House.

j. Parliament has been dissolved or suspended for an indefinite period.

k. The figures on the distribution of seats do not include the 36 special rotating delegates appointed on an ad hoc basis, and all percentages given are therefore calculated on the basis of the 54 permanent seats.

l. The parliament elected in 1990 has never been convened nor authorized to sit, and many of its members were detained or forced into exile.

m. In 2004, the number of seats in parliament was raised from 300 to 345, with the addition of 45 reserved seats for women. These reserved seats were filled in September and October 2005, being allocated to political parties in proportion to their share of the national vote received in the 2001 election.

n. A transitional legislative parliament was established in January 2007. Elections for the Constituent Assembly will be held in 2007.

o. The purpose of the elections held on 30 August 2001 was to elect the members of the Constituent Assembly of Timor-Leste. This body became the National Parliament on 20 May 2002, the date on which the country became independent, without any new elections.

p. In November 1955, Eritrea was part of Ethiopia. The Constitution of sovereign Eritrea adopted on 23 May 1997 stipulates that "All Eritrean citizens, of eighteen years of age or more, shall have the right to vote."

q. Data are as of 31 May 2006.

r. Serbia and Montenegro separated into two independent states in June 2006. Women received the right to vote and to stand for elections in 1946, when Serbia and Montenegro were part of the former Yugoslavia.

SOURCES
Columns 1–3: IPU 2007b.
Column 4: IPU 2007a.
Column 5: UN 2007c, based on data from IPU.
Columns 6 and 7: IPU 2007c.

TABLE 34

Human and labour rights instruments

Status of major international human rights instruments

HDI rank	International Convention on the Prevention and Punishment of the Crime of Genocide 1948	International Convention on the Elimination of All Forms of Racial Discrimination 1965	International Covenant on Civil and Political Rights 1966	International Covenant on Economic, Social and Cultural Rights 1966	Convention on the Elimination of All Forms of Discrimination against Women 1979	Convention against Torture and Other Cruel, Inhuman or Degrading Treatment or Punishment 1984	Convention on the Rights of the Child 1989
HIGH HUMAN DEVELOPMENT							
1 Iceland	1949	1967	1979	1979	1985	1996	1992
2 Norway	1949	1970	1972	1972	1981	1986	1991
3 Australia	1949	1975	1980	1975	1983	1989	1990
4 Canada	1952	1970	1976	1976	1981	1987	1991
5 Ireland	1976	2000	1989	1989	1985	2002	1992
6 Sweden	1952	1971	1971	1971	1980	1986	1990
7 Switzerland	2000	1994	1992	1992	1997	1986	1997
8 Japan	..	1995	1979	1979	1985	1999	1994
9 Netherlands	1966	1971	1978	1978	1991	1988	..
10 France	1950	1971	1980	1980	1983	1986	1990
11 Finland	1959	1970	1975	1975	1986	1989	1991
12 United States	1988	1994	1992	1977	1980	1994	1995
13 Spain	1968	1968	1977	1977	1984	1987	1990
14 Denmark	1951	1971	1972	1972	1983	1987	1991
15 Austria	1958	1972	1978	1978	1982	l1987	1992
16 United Kingdom	1970	1969	1976	1976	1986	1988	1991
17 Belgium	1951	1975	1983	1983	1985	1999	1991
18 Luxembourg	1981	1978	1983	1983	1989	1987	1994
19 New Zealand	1978	1972	1978	1978	1985	1989	1993
20 Italy	1952	1976	1978	1978	1985	1989	1991
22 Germany	1954	1969	1973	1973	1985	1990	1992
23 Israel	1950	1979	1991	1991	1991	1991	1991
24 Greece	1954	1970	1997	1985	1983	1988	1993
25 Singapore	1995	..	..	..	1995	..	1995
26 Korea (Republic of)	1950	1978	1990	1990	1984	1995	1991
27 Slovenia	1992	1992	1992	1992	1992	1993	1992
28 Cyprus	1982	1967	1969	1969	1985	1991	1991
29 Portugal	1999	1982	1978	1978	1980	1989	1990
30 Brunei Darussalam	..	..	..	..	2006	..	1995
31 Barbados	1980	1972	1973	1973	1980	..	1990
32 Czech Republic	1993	1993	1993	1993	1993	1993	1993
33 Kuwait	1995	1968	1996	1996	1994	1996	1991
34 Malta	..	1971	1990	1990	1991	1990	1990
35 Qatar	..	1976	..	..	..	2000	1995
36 Hungary	1952	1967	1974	1974	1980	1987	1991
37 Poland	1950	1968	1977	1977	1980	1989	1991
38 Argentina	1956	1968	1986	1986	1985	1986	1990
39 United Arab Emirates	2005	1974	..	..	2004	..	1997
40 Chile	1953	1971	1972	1972	1989	1988	1990
41 Bahrain	1990	1990	2006	..	2002	1998	1992
42 Slovakia	1993	1993	1993	1993	1993	1993	1993
43 Lithuania	1996	1998	1991	1991	1994	1996	1992
44 Estonia	1991	1991	1991	1991	1991	1991	1991
45 Latvia	1992	1992	1992	1992	1992	1992	1992
46 Uruguay	1967	1968	1970	1970	1981	1986	1990
47 Croatia	1992	1992	1992	1992	1992	1992	1992
48 Costa Rica	1950	1967	1968	1968	1986	1993	1990
49 Bahamas	1975	1975	..	..	1993	..	1991
50 Seychelles	1992	1978	1992	1992	1992	1992	1990
51 Cuba	1953	1972	..	..	1980	1995	1991
52 Mexico	1952	1975	1981	1981	1981	1986	1990
53 Bulgaria	1950	1966	1970	1970	1982	1986	1991
54 Saint Kitts and Nevis	..	2006	..	..	1985	..	1990

Human development indicators

TABLE 34

Status of major international human rights instruments

HDI rank	International Convention on the Prevention and Punishment of the Crime of Genocide 1948	International Convention on the Elimination of All Forms of Racial Discrimination 1965	International Covenant on Civil and Political Rights 1966	International Covenant on Economic, Social and Cultural Rights 1966	Convention on the Elimination of All Forms of Discrimination against Women 1979	Convention against Torture and Other Cruel, Inhuman or Degrading Treatment or Punishment 1984	Convention on the Rights of the Child 1989
55 Tonga	1972	1972	..	..	..	..	1995
56 Libyan Arab Jamahiriya	1989	1968	1970	1970	1989	1989	1993
57 Antigua and Barbuda	1988	1988	..	..	1989	1993	1993
58 Oman	..	2003	..	..	2006	..	1996
59 Trinidad and Tobago	2002	1973	1978	1978	1990	..	1991
60 Romania	1950	1970	1974	1974	1982	1990	1990
61 Saudi Arabia	1950	1997	..	..	2000	1997	1996
62 Panama	1950	1967	1977	1977	1981	1987	1990
63 Malaysia	1994	..	..	..	1995	..	1995
64 Belarus	1954	1969	1973	1973 ·	1981	1987	1990
65 Mauritius	..	1972	1973	1973	1984	1992	1990
66 Bosnia and Herzegovina	1992	1993	1993	1993	1993	1993	1993
67 Russian Federation	1954	1969	1973	1973	1981	1987	1990
68 Albania	1955	1994	1991	1991	1994	1994	1992
69 Macedonia (TFYR)	1994	1994	1994	1994	1994	1994	1993
70 Brazil	1952	1968	1992	1992	1984	1989	1990
MEDIUM HUMAN DEVELOPMENT							
71 Dominica	..	..	1993	1993	1980	..	1991
72 Saint Lucia	..	1990	..	..	1982	..	1993
73 Kazakhstan	1998	1998	2006	2006	1998	1998	1994
74 Venezuela (Bolivarian Republic of)	1960	1967	1978	1978	1983	1991	1990
75 Colombia	1959	1981	1969	1969	1982	1987	1991
76 Ukraine	1954	1969	1973	1973	1981	1987	1991
77 Samoa	..	..	..	..	1992	..	1994
78 Thailand	..	2003	1996	1999	1985	..	1992
79 Dominican Republic	**1948**	1983	1978	1978	1982	**1985**	1991
80 Belize	1998	2001	1996	**2000**	1990	1986	1990
81 China	1983	1981	**1998**	2001	1980	1988	1992
82 Grenada	..	**1981**	1991	1991	1990	..	1990
83 Armenia	1993	1993	1993	1993	1993	1993	1993
84 Turkey	1950	2002	2003	2003	1985	1988	1995
85 Suriname	..	1984	1976	1976	1993	..	1993
86 Jordan	1950	1974	1975	1975	1992	1991	1991
87 Peru	1960	1971	1978	1978	1982	1988	1990
88 Lebanon	1953	1971	1972	1972	1997	2000	1991
89 Ecuador	1949	1966	1969	1969	1981	1988	1990
90 Philippines	1950	1967	1986	1974	1981	1986	1990
91 Tunisia	1956	1967	1969	1969	1985	1988	1992
92 Fiji	1973	1973	..	..	1995	..	1993
93 Saint Vincent and the Grenadines	1981	1981	1981	1981	1981	2001	1993
94 Iran (Islamic Republic of)	1956	1968	1975	1975	..	..	1994
95 Paraguay	2001	2003	1992	1992	1987	1990	1990
96 Georgia	1993	1999	1994	1994	1994	1994	1994
97 Guyana	..	1977	1977	1977	1980	1988	1991
98 Azerbaijan	1996	1996	1992	1992	1995	1996	1992
99 Sri Lanka	1950	1982	1980	1980	1981	1994	1991
100 Maldives	1984	1984	2006	2006	1993	2004	1991
101 Jamaica	1968	1971	1975	1975	1984	..	1991
102 Cape Verde	..	1979	1993	1993	1980	1992	1992
103 El Salvador	1950	1979	1979	1979	1981	1996	1990
104 Algeria	1963	1972	1989	1989	1996	1989	1993
105 Viet Nam	1981	1982	1982	1982	1982	..	1990
106 Occupied Palestinian Territories	..	..	..	..	..	..	..
107 Indonesia	..	1999	2006	2006	1984	1998	1990

TABLE 34

HDI rank	International Convention on the Prevention and Punishment of the Crime of Genocide 1948	International Convention on the Elimination of All Forms of Racial Discrimination 1965	International Covenant on Civil and Political Rights 1966	International Covenant on Economic, Social and Cultural Rights 1966	Convention on the Elimination of All Forms of Discrimination against Women 1979	Convention against Torture and Other Cruel, Inhuman or Degrading Treatment or Punishment 1984	Convention on the Rights of the Child 1989
108 Syrian Arab Republic	1955	1969	1969	1969	2003	2004	1993
109 Turkmenistan	..	1994	1997	1997	1997	1999	1993
110 Nicaragua	1952	1978	1980	1980	1981	2005	1990
111 Moldova	1993	1993	1993	1993	1994	1995	1993
112 Egypt	1952	1967	1982	1982	1981	1986	1990
113 Uzbekistan	1999	1995	1995	1995	1995	1995	1994
114 Mongolia	1967	1969	1974	1974	1981	2002	1990
115 Honduras	1952	2002	1997	1981	1983	1996	1990
116 Kyrgyzstan	1997	1997	1994	1994	1997	1997	1994
117 Bolivia	2005	1970	1982	1982	1990	1999	1990
118 Guatemala	1950	1983	1992	1988	1982	1990	1990
119 Gabon	1983	1980	1983	1983	1983	2000	1994
120 Vanuatu	..	..	..	..	1995	..	1993
121 South Africa	1998	1998	1998	1994	1995	1998	1995
122 Tajikistan	..	1995	1999	1999	1993	1995	1993
123 Sao Tome and Principe	..	2000	1995	..	2003	2000	1991
124 Botswana	..	1974	2000	..	1996	2000	1995
125 Namibia	1994	1982	1994	1994	1992	1994	1990
126 Morocco	1958	1970	1979	1979	1993	1993	1993
127 Equatorial Guinea	..	2002	1987	1987	1984	2002	1992
128 India	1959	1968	1979	1979	1993	1997	1992
129 Solomon Islands	..	1982	..	1982	2002	..	1995
130 Lao People's Democratic Republic	1950	1974	2000 a	2007	1981	..	1991
131 Cambodia	1950	1983	1992	1992	1992	1992	1992
132 Myanmar	1956	..	..	..	1997	..	1991
133 Bhutan	..	1973	..	..	1981	..	1990
134 Comoros	2004	2004	..	..	1994	2000	1993
135 Ghana	1958	1966	2000	2000	1986	2000	1990
136 Pakistan	1957	1966	..	2004	1996	..	1990
137 Mauritania	..	1988	2004	2004	2001	2004	1991
138 Lesotho	1974	1971	1992	1992	1995	2001	1992
139 Congo	..	1988	1983	1983	1982	2003	1993
140 Bangladesh	1998	1979	2000	1998	1984	1998	1990
141 Swaziland	..	1969	2004	2004	2004	2004	1995
142 Nepal	1969	1971	1991	1991	1991	1991	1990
143 Madagascar	..	1969	1971	1971	1989	2005	1991
144 Cameroon	..	1971	1984	1984	1994	1986	1993
145 Papua New Guinea	1982	1982	..	..	1995	..	1993
146 Haiti	1950	1972	1991	..	1981	..	1995
147 Sudan	2003	1977	1986	1986	..	1986	1990
148 Kenya	..	2001	1972	1972	1984	1997	1990
149 Djibouti	..	2006	2002	2002	1998	2002	1990
150 Timor-Leste	..	2003	2003	2003	2003	2003	2003
151 Zimbabwe	1991	1991	1991	1991	1991	..	1990
152 Togo	1984	1972	1984	1984	1983	1987	1990
153 Yemen	1987	1972	1987	1987	1984	1991	1991
154 Uganda	1995	1980	1995	1987	1985	1986	1990
155 Gambia	1978	1978	1979	1978	1993	1985	1990
LOW HUMAN DEVELOPMENT							
156 Senegal	1983	1972	1978	1978	1985	1986	1990
157 Eritrea	..	2001	2002	2001	1995	..	1994
158 Nigeria	..	1967	1993	1993	1985	2001	1991
159 Tanzania (United Republic of)	1984	1972	1976	1976	1985	..	1991
160 Guinea	2000	1977	1978	1978	1982	1989	1990

Human development indicators

TABLE 34

Status of major international human rights instruments

HDI rank	International Convention on the Prevention and Punishment of the Crime of Genocide 1948	International Convention on the Elimination of All Forms of Racial Discrimination 1965	International Covenant on Civil and Political Rights 1966	International Covenant on Economic, Social and Cultural Rights 1966	Convention on the Elimination of All Forms of Discrimination against Women 1979	Convention against Torture and Other Cruel, Inhuman or Degrading Treatment or Punishment 1984	Convention on the Rights of the Child 1989
161 Rwanda	1975	1975	1975	1975	1981	..	1991
162 Angola	..	..	1992	1992	1986	..	1990
163 Benin	..	2001	1992	1992	1992	1992	1990
164 Malawi	..	1996	1993	1993	1987	1996	1991
165 Zambia	..	1972	1984	1984	1985	1998	1991
166 Côte d'Ivoire	1995	1973	1992	1992	1995	1995	1991
167 Burundi	1997	1977	1990	1990	1992	1993	1990
168 Congo (Democratic Republic of the)	1962	1976	1976	1976	1986	1996	1990
169 Ethiopia	1949	1976	1993	1993	1981	1994	1991
170 Chad	..	1977	1995	1995	1995	1995	1990
171 Central African Republic	..	1971	1981	1981	1991	..	1992
172 Mozambique	1983	1983	1993	..	1997	1999	1994
173 Mali	1974	1974	1974	1974	1985	1999	1990
174 Niger	..	1967	1986	1986	1999	1998	1990
175 Guinea-Bissau	..	2000[a]	2000[a]	1992	1985	2000[a]	1990
176 Burkina Faso	1965	1974	1999	1999	1987	1999	1990
177 Sierra Leone	..	1967	1996	1996	1988	2001	1990
OTHERS[a]							
Afghanistan	1956	1983	1983	1983	2003	1987	1994
Andorra	2006	2006	2006	..	1997	2006	1996
Iraq	1959	1970	1971	1971	1986	..	1994
Kiribati	..	..		..	2004	..	1995
Democratic People's Republic of Korea	1989	..	1981	1981	2001	..	1990
Liberia	1950	1976	2004	2004	1984	2004	1993
Liechtenstein	1994	2000	1998	1998	1995	1990	1995
Marshall Islands	..	..		..	2006	..	1993
Monaco	1950	1995	1997	1997	2005	1991	1993
Montenegro[b]	2006	2006	2006	2006	2006	2006	2006
Nauru	..	2001	2001	..	..	2001[a]	1994
Palau	..	..		..	..	..	1995
San Marino	..	2002	1985	1985	2003	2006	1991
Serbia[b]	2001	2001	2001	2001	2001	2001	2001
Somalia	..	1975	1990	1990	..	1990	2002
Tuvalu	..	..		..	1999	..	1995
Total state parties[c]	**140**	**172**	**160**	**156**	**183**	**143**	**189**
Treaties signed, not yet ratified	**1**	**6**	**5**	**5**	**1**	**8**	**2**

NOTES

Data refer to year of ratification, accession or succession unless otherwise specified. All these stages have the same legal effects. **Bold** signifies signature not yet followed by ratification. Data are as of 1 July 2007.

a. Countries or areas, in addition to the 177 countries or areas included in the main indicator tables, that have signed at least one of the seven human rights instruments.

b. Following separation of Serbia and Montenegro into two independent states in June 2006, all treaty actions (ratification or signature) continue in force for the Republic of Serbia. As of 1 July 2007, the UN Secretary-General had not received notification from the Republic of Montenegro with regard to the treaties reported in this table, unless otherwise specified.

c. Refers to ratification, accession or succession.

SOURCE
Columns 1–7: UN 2007a.

TABLE 35

Human and labour rights instruments

Status of fundamental labour rights conventions

HDI rank	Freedom of association and collective bargaining		Elimination of forced and compulsory labour		Elimination of discrimination in respect of employment and occupation		Abolition of child labour	
	Convention 87 [a]	Convention 98 [b]	Convention 29 [c]	Convention 105 [d]	Convention 100 [e]	Convention 111 [f]	Convention 138 [g]	Convention 182 [h]
HIGH HUMAN DEVELOPMENT								
1 Iceland	1950	1952	1958	1960	1958	1963	1999	2000
2 Norway	1949	1955	1932	1958	1959	1959	1980	2000
3 Australia	1973	1973	1932	1960	1974	1973	..	2006
4 Canada	1972	..	..	1959	1972	1964	..	2000
5 Ireland	1955	1955	1931	1958	1974	1999	1978	1999
6 Sweden	1949	1950	1931	1958	1962	1962	1990	2001
7 Switzerland	1975	1999	1940	1958	1972	1961	1999	2000
8 Japan	1965	1953	1932	..	1967	..	2000	2001
9 Netherlands	1950	1993	1933	1959	1971	1973	1976	2002
10 France	1951	1951	1937	1969	1953	1981	1990	2001
11 Finland	1950	1951	1936	1960	1963	1970	1976	2000
12 United States	..	..	..	1991	..	..	..	1999
13 Spain	1977	1977	1932	1967	1967	1967	1977	2001
14 Denmark	1951	1955	1932	1958	1960	1960	1997	2000
15 Austria	1950	1951	1960	1958	1953	1973	2000	2001
16 United Kingdom	1949	1950	1931	1957	1971	1999	2000	2000
17 Belgium	1951	1953	1944	1961	1952	1977	1988	2002
18 Luxembourg	1958	1958	1964	1964	1967	2001	1977	2001
19 New Zealand	..	2003	1938	1968	1983	1983	..	2001
20 Italy	1958	1958	1934	1968	1956	1963	1981	2000
22 Germany	1957	1956	1956	1959	1956	1961	1976	2002
23 Israel	1957	1957	1955	1958	1965	1959	1979	2005
24 Greece	1962	1962	1952	1962	1975	1984	1986	2001
25 Singapore	..	1965	1965	[1965] [i]	2002	..	2005	2001
26 Korea (Republic of)	..	..	..	..	1997	1998	1999	2001
27 Slovenia	1992	1992	1992	1997	1992	1992	1992	2001
28 Cyprus	1966	1966	1960	1960	1987	1968	1997	2000
29 Portugal	1977	1964	1956	1959	1967	1959	1998	2000
30 Brunei Darussalam	..	..	..	..	..	..	..	..
31 Barbados	1967	1967	1967	1967	1974	1974	2000	2000
32 Czech Republic	1993	1993	1993	1996	1993	1993	2007	2001
33 Kuwait	1961	..	1968	1961	..	1966	1999	2000
34 Malta	1965	1965	1965	1965	1988	1968	1988	2001
35 Qatar	..	..	1998	2007	..	1976	2006	2000
36 Hungary	1957	1957	1956	1994	1956	1961	1998	2000
37 Poland	1957	1957	1958	1958	1954	1961	1978	2002
38 Argentina	1960	1956	1950	1960	1956	1968	1996	2001
39 United Arab Emirates	..	..	1982	1997	1997	2001	1998	2001
40 Chile	1999	1999	1933	1999	1971	1971	1999	2000
41 Bahrain	..	..	1981	1998	..	2000	..	2001
42 Slovakia	1993	1993	1993	1997	1993	1993	1997	1999
43 Lithuania	1994	1994	1994	1994	1994	1994	1998	2003
44 Estonia	1994	1994	1996	1996	1996	2005	2007	2001
45 Latvia	1992	1992	2006	1992	1992	1992	2006	2006
46 Uruguay	1954	1954	1995	1968	1989	1989	1977	2001
47 Croatia	1991	1991	1991	1997	1991	1991	1991	2001
48 Costa Rica	1960	1960	1960	1959	1960	1962	1976	2001
49 Bahamas	2001	1976	1976	1976	2001	2001	2001	2001
50 Seychelles	1978	1999	1978	1978	1999	1999	2000	1999
51 Cuba	1952	1952	1953	1958	1954	1965	1975	..
52 Mexico	1950	..	1934	1959	1952	1961	..	2000
53 Bulgaria	1959	1959	1932	1999	1955	1960	1980	2000
54 Saint Kitts and Nevis	2000	2000	2000	2000	2000	2000	2005	2000

Human development indicators

TABLE 35

Status of fundamental labour rights conventions

HDI rank	Freedom of association and collective bargaining		Elimination of forced and compulsory labour		Elimination of discrimination in respect of employment and occupation		Abolition of child labour	
	Convention 87 [a]	Convention 98 [b]	Convention 29 [c]	Convention 105 [d]	Convention 100 [e]	Convention 111 [f]	Convention 138 [g]	Convention 182 [h]
55 Tonga	..	..	..	..	..	..	..	..
56 Libyan Arab Jamahiriya	2000	1962	1961	1961	1962	1961	1975	2000
57 Antigua and Barbuda	1983	1983	1983	1983	2003	1983	1983	2002
58 Oman	..	..	1998	2005	..	..	2005	2001
59 Trinidad and Tobago	1963	1963	1963	1963	1997	1970	2004	2003
60 Romania	1957	1958	1957	1998	1957	1973	1975	2000
61 Saudi Arabia	..	..	1978	1978	1978	1978	..	2001
62 Panama	1958	1966	1966	1966	1958	1966	2000	2000
63 Malaysia	..	1961	1957	[1958] [j]	1997	..	1997	2000
64 Belarus	1956	1956	1956	1995	1956	1961	1979	2000
65 Mauritius	2005	1969	1969	1969	2002	2002	1990	2000
66 Bosnia and Herzegovina	1993	1993	1993	2000	1993	1993	1993	2001
67 Russian Federation	1956	1956	1956	1998	1956	1961	1979	2003
68 Albania	1957	1957	1957	1997	1957	1997	1998	2001
69 Macedonia (TFYR)	1991	1991	1991	2003	1991	1991	1991	2002
70 Brazil	..	1952	1957	1965	1957	1965	2001	2000
MEDIUM HUMAN DEVELOPMENT								
71 Dominica	1983	1983	1983	1983	1983	1983	1983	2001
72 Saint Lucia	1980	1980	1980	1980	1983	1983	..	2000
73 Kazakhstan	2000	2001	2001	2001	2001	1999	2001	2003
74 Venezuela (Bolivarian Republic of)	1982	1968	1944	1964	1982	1971	1987	2005
75 Colombia	1976	1976	1969	1963	1963	1969	2001	2005
76 Ukraine	1956	1956	1956	2000	1956	1961	1979	2000
77 Samoa	..	..	..	..	..	..	..	..
78 Thailand	..	..	1969	1969	1999	..	2004	2001
79 Dominican Republic	1956	1953	1956	1958	1953	1964	1999	2000
80 Belize	1983	1983	1983	1983	1999	1999	2000	2000
81 China	..	..	..	..	1990	2006	1999	2002
82 Grenada	1994	1979	1979	1979	1994	2003	2003	2003
83 Armenia	2006	2003	2004	2004	1994	1994	2006	2006
84 Turkey	1993	1952	1998	1961	1967	1967	1998	2001
85 Suriname	1976	1996	1976	1976	..	..	..	2006
86 Jordan	..	1968	1966	1958	1966	1963	1998	2000
87 Peru	1960	1964	1960	1960	1960	1970	2002	2002
88 Lebanon	..	1977	1977	1977	1977	1977	2003	2001
89 Ecuador	1967	1959	1954	1962	1957	1962	2000	2000
90 Philippines	1953	1953	2005	1960	1953	1960	1998	2000
91 Tunisia	1957	1957	1962	1959	1968	1959	1995	2000
92 Fiji	2002	1974	1974	1974	2002	2002	2003	2002
93 Saint Vincent and the Grenadines	2001	1998	1998	1998	2001	2001	2006	2001
94 Iran (Islamic Republic of)	..	..	1957	1959	1972	1964	..	2002
95 Paraguay	1962	1966	1967	1968	1964	1967	2004	2001
96 Georgia	1999	1993	1997	1996	1993	1993	1996	2002
97 Guyana	1967	1966	1966	1966	1975	1975	1998	2001
98 Azerbaijan	1992	1992	1992	2000	1992	1992	1992	2004
99 Sri Lanka	1995	1972	1950	2003	1993	1998	2000	2001
100 Maldives	..	..	..	..	..	..	..	..
101 Jamaica	1962	1962	1962	1962	1975	1975	2003	2003
102 Cape Verde	1999	1979	1979	1979	1979	1979	..	2001
103 El Salvador	2006	2006	1995	1958	2000	1995	1996	2000
104 Algeria	1962	1962	1962	1969	1962	1969	1984	2001
105 Viet Nam	..	..	2007	..	1997	1997	2003	2000
107 Indonesia	1998	1957	1950	1999	1958	1999	1999	2000
108 Syrian Arab Republic	1960	1957	1960	1958	1957	1960	2001	2003

TABLE 35

HDI rank	Freedom of association and collective bargaining		Elimination of forced and compulsory labour		Elimination of discrimination in respect of employment and occupation		Abolition of child labour	
	Convention 87[a]	Convention 98[b]	Convention 29[c]	Convention 105[d]	Convention 100[e]	Convention 111[f]	Convention 138[g]	Convention 182[h]
109 Turkmenistan	1997	1997	1997	1997	1997	1997	..	..
110 Nicaragua	1967	1967	1934	1967	1967	1967	1981	2000
111 Moldova	1996	1996	2000	1993	2000	1996	1999	2002
112 Egypt	1957	1954	1955	1958	1960	1960	1999	2002
113 Uzbekistan	..	1992	1992	1997	1992	1992	..	..
114 Mongolia	1969	1969	2005	2005	1969	1969	2002	2001
115 Honduras	1956	1956	1957	1958	1956	1960	1980	2001
116 Kyrgyzstan	1992	1992	1992	1999	1992	1992	1992	2004
117 Bolivia	1965	1973	2005	1990	1973	1977	1997	2003
118 Guatemala	1952	1952	1989	1959	1961	1960	1990	2001
119 Gabon	1960	1961	1960	1961	1961	1961	..	2001
120 Vanuatu	2006	2006	2006	2006	2006	2006	..	2006
121 South Africa	1996	1996	1997	1997	2000	1997	2000	2000
122 Tajikistan	1993	1993	1993	1999	1993	1993	1993	2005
123 Sao Tome and Principe	1992	1992	2005	2005	1982	1982	2005	2005
124 Botswana	1997	1997	1997	1997	1997	1997	1997	2000
125 Namibia	1995	1995	2000	2000	..	2001	2000	2000
126 Morocco	..	1957	1957	1966	1979	1963	2000	2001
127 Equatorial Guinea	2001	2001	2001	2001	1985	2001	1985	2001
128 India	..	..	1954	2000	1958	1960	..	..
129 Solomon Islands	..	..	1985	..	..	..	..	..
130 Lao People's Democratic Republic	..	..	1964	..	..	..	2005	2005
131 Cambodia	1999	1999	1969	1999	1999	1999	1999	2006
132 Myanmar	1955	..	1955	..	..	..	..	..
133 Bhutan	..	..	..	..	..	..	..	..
134 Comoros	1978	1978	1978	1978	1978	2004	2004	2004
135 Ghana	1965	1959	1957	1958	1968	1961	..	2000
136 Pakistan	1951	1952	1957	1960	2001	1961	2006	2001
137 Mauritania	1961	2001	1961	1997	2001	1963	2001	2001
138 Lesotho	1966	1966	1966	2001	1998	1998	2001	2001
139 Congo	1960	1999	1960	1999	1999	1999	1999	2002
140 Bangladesh	1972	1972	1972	1972	1998	1972	..	2001
141 Swaziland	1978	1978	1978	1979	1981	1981	2002	2002
142 Nepal	..	1996	2002	..	1976	1974	1997	2002
143 Madagascar	1960	1998	1960	2007	1962	1961	2000	2001
144 Cameroon	1960	1962	1960	1962	1970	1988	2001	2002
145 Papua New Guinea	2000	1976	1976	1976	2000	2000	2000	2000
146 Haiti	1979	1957	1958	1958	1958	1976	..	..
147 Sudan	..	1957	1957	1970	1970	1970	2002	2003
148 Kenya	..	1964	1964	1964	2001	2001	1979	2001
149 Djibouti	1978	1978	1978	1978	1978	2005	2005	2005
150 Timor-Leste	..	..	..	..	..	..	..	..
151 Zimbabwe	2003	1998	1998	1998	1989	1999	2000	2000
152 Togo	1960	1983	1960	1999	1983	1983	1984	2000
153 Yemen	1976	1969	1969	1969	1976	1969	2000	2000
154 Uganda	2005	1963	1963	1963	2005	2005	2003	2001
155 Gambia	2000	2000	2000	2000	2000	2000	2000	2001
LOW HUMAN DEVELOPMENT								
156 Senegal	1960	1961	1960	1961	1962	1967	1999	2000
157 Eritrea	2000	2000	2000	2000	2000	2000	2000	..
158 Nigeria	1960	1960	1960	1960	1974	2002	2002	2002
159 Tanzania (United Republic of)	2000	1962	1962	1962	2002	2002	1998	2001
160 Guinea	1959	1959	1959	1961	1967	1960	2003	2003
161 Rwanda	1988	1988	2001	1962	1980	1981	1981	2000

Human development indicators

TABLE 35

Status of fundamental labour rights conventions

HDI rank	Freedom of association and collective bargaining		Elimination of forced and compulsory labour		Elimination of discrimination in respect of employment and occupation		Abolition of child labour	
	Convention 87 [a]	Convention 98 [b]	Convention 29 [c]	Convention 105 [d]	Convention 100 [e]	Convention 111 [f]	Convention 138 [g]	Convention 182 [h]
162 Angola	2001	1976	1976	1976	1976	1976	2001	2001
163 Benin	1960	1968	1960	1961	1968	1961	2001	2001
164 Malawi	1999	1965	1999	1999	1965	1965	1999	1999
165 Zambia	1996	1996	1964	1965	1972	1979	1976	2001
166 Côte d'Ivoire	1960	1961	1960	1961	1961	1961	2003	2003
167 Burundi	1993	1997	1963	1963	1993	1993	2000	2002
168 Congo (Democratic Republic of the)	2001	1969	1960	2001	1969	2001	2001	2001
169 Ethiopia	1963	1963	2003	1999	1999	1966	1999	2003
170 Chad	1960	1961	1960	1961	1966	1966	2005	2000
171 Central African Republic	1960	1964	1960	1964	1964	1964	2000	2000
172 Mozambique	1996	1996	2003	1977	1977	1977	2003	2003
173 Mali	1960	1964	1960	1962	1968	1964	2002	2000
174 Niger	1961	1962	1961	1962	1966	1962	1978	2000
175 Guinea-Bissau	..	1977	1977	1977	1977	1977	..	..
176 Burkina Faso	1960	1962	1960	1997	1969	1962	1999	2001
177 Sierra Leone	1961	1961	1961	1961	1968	1966	..	..
OTHERS [k]								
Afghanistan	..	..	..	1963	1969	1969	..	..
Iraq	..	1962	1962	1959	1963	1959	1985	2001
Kiribati	2000	2000	2000	2000	..	..	..	..
Liberia	1962	1962	1931	1962	..	1959	..	2003
Montenegro	2006	2006	2006	2006	2006	2006	2006	2006
San Marino	1986	1986	1995	1995	1985	1986	1995	2000
Serbia	2000	2000	2000	2003	2000	2000	2000	2003
Somalia	..	..	1960	1961	..	1961	..	..
Total ratifications	**142**	**150**	**164**	**158**	**158**	**158**	**145**	**158**

NOTES

Table includes UN member states. Information is as of 1 July 2007. Years indicate the date of ratification.

a. Freedom of Association and Protection of the Right to Organize Convention (1948).
b. Right to Organize and Collective Bargaining Convention (1949).
c. Forced Labour Convention (1930).
d. Abolition of Forced Labour Convention (1957).
e. Equal Remuneration Convention (1951).
f. Discrimination (Employment and Occupation) Convention (1958).
g. Minimum Age Convention (1973).
h. Worst Forms of Child Labour Convention (1999).
i. Convention was denounced in 1979.
j. Convention was denounced in 1990.
k. Countries or areas, in addition to the 177 countries or areas included in the main indicator tables, that are members of ILO.

SOURCE
All columns: ILO 2007a.

TECHNICAL NOTE 1

Calculating the human development indices

The diagrams here summarize how the five human development indices used in the *Human Development Report* are constructed, highlighting both their similarities and their differences. The text on the following pages provides a detailed explanation.

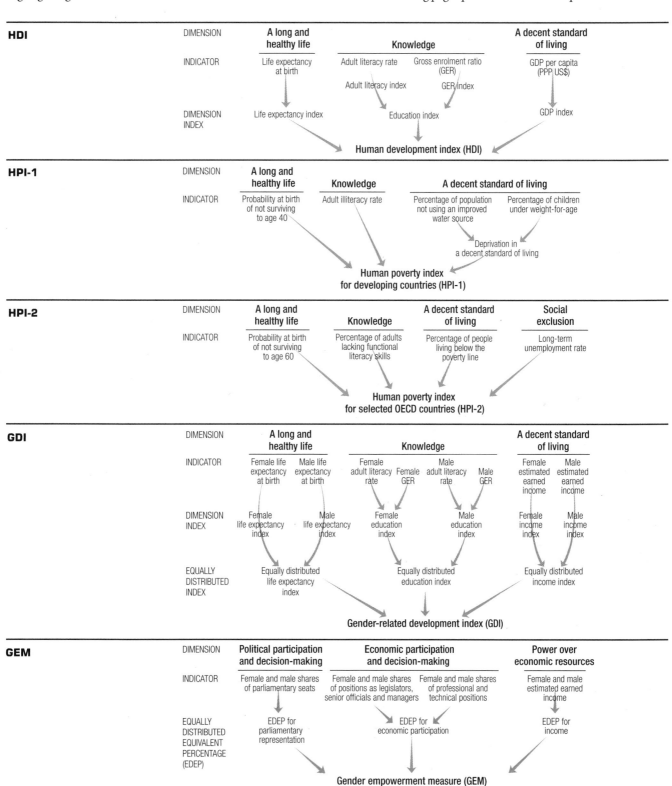

The human development index (HDI)

The HDI is a summary measure of human development. It measures the average achievements in a country in three basic dimensions of human development:

- A long and healthy life, as measured by life expectancy at birth.
- Knowledge, as measured by the adult literacy rate (with two-thirds weight) and the combined primary, secondary and tertiary gross enrolment ratio (with one-third weight).
- A decent standard of living, as measured by GDP per capita in purchasing power parity (PPP) terms in US dollars.

Before the HDI itself is calculated, an index needs to be created for each of these dimensions. To calculate these indices—the life expectancy, education and GDP indices—minimum and maximum values (goalposts) are chosen for each underlying indicator.

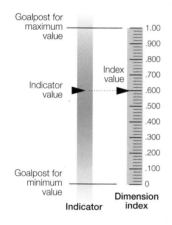

Performance in each dimension is expressed as a value between 0 and 1 by applying the following general formula:

$$\text{Dimension index} = \frac{\text{actual value} - \text{minimum value}}{\text{maximum value} - \text{minimum value}}$$

The HDI is then calculated as a simple average of the dimension indices. The box on the right illustrates the calculation of the HDI for a sample country.

Goalposts for calculating the HDI

Indicator	Maximum value	Minimum value
Life expectancy at birth (years)	85	25
Adult literacy rate (%)*	100	0
Combined gross enrolment ratio (%)	100	0
GDP per capita (PPP US$)	40,000	100

* The goalpost for calculating adult literacy implies the maximum literacy rate is 100%. In practice, the HDI is calculated using an upper bound of 99%.

Calculating the HDI

This illustration of the calculation of the HDI uses data for Turkey.

1. Calculating the life expectancy index
The life expectancy index measures the relative achievement of a country in life expectancy at birth. For Turkey, with a life expectancy of 71.4 years in 2005, the life expectancy index is 0.773.

$$\text{Life expectancy index} = \frac{71.4 - 25}{85 - 25} = \mathbf{0.773}$$

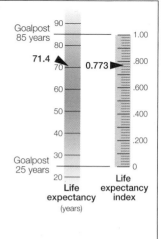

2. Calculating the education index
The education index measures a country's relative achievement in both adult literacy and combined primary, secondary and tertiary gross enrolment. First, an index for adult literacy and one for combined gross enrolment are calculated. Then these two indices are combined to create the education index, with two-thirds weight given to adult literacy and one-third weight to combined gross enrolment. For Turkey, with an adult literacy rate of 87.4% in 2005 and a combined gross enrolment ratio of 68.7% in 2005, the education index is 0.812.

$$\text{Adult literacy index} = \frac{87.4 - 0}{100 - 0} = 0.874$$

$$\text{Gross enrolment index} = \frac{68.7 - 0}{100 - 0} = 0.687$$

$$\text{Education index} = 2/3 \text{ (adult literacy index)} + 1/3 \text{ (gross enrolment index)}$$
$$= 2/3 \text{ (0.874)} + 1/3 \text{ (0.687)} = \mathbf{0.812}$$

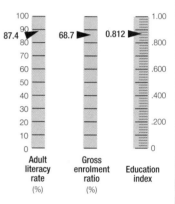

3. Calculating the GDP index
The GDP index is calculated using adjusted GDP per capita (PPP US$). In the HDI income serves as a surrogate for all the dimensions of human development not reflected in a long and healthy life and in knowledge. Income is adjusted because achieving a respectable level of human development does not require unlimited income. Accordingly, the logarithm of income is used. For Turkey, with a GDP per capita of 8,407 (PPP US$) in 2005, the GDP index is 0.740.

$$\text{GDP index} = \frac{\log (8,407) - \log (100)}{\log (40,000) - \log (100)} = \mathbf{0.740}$$

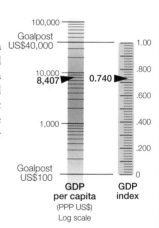

4. Calculating the HDI
Once the dimension indices have been calculated, determining the HDI is straightforward. It is a simple average of the three dimension indices.

$$\text{HDI} = 1/3 \text{ (life expectancy index)} + 1/3 \text{ (education index)}$$
$$+ 1/3 \text{ (GDP index)}$$
$$= 1/3 \text{ (0.773)} + 1/3 \text{ (0.812)} + 1/3 \text{ (0.740)} = \mathbf{0.775}$$

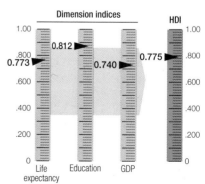

The human poverty index for developing countries (HPI-1)

While the HDI measures average achievement, the HPI-1 measures *deprivations* in the three basic dimensions of human development captured in the HDI:

- A long and healthy life—vulnerability to death at a relatively early age, as measured by the probability at birth of not surviving to age 40.
- Knowledge—exclusion from the world of reading and communications, as measured by the adult illiteracy rate.
- A decent standard of living—lack of access to overall economic provisioning, as measured by the unweighted average of two indicators, the percentage of the population not using an improved water source and the percentage of children under weight-for-age.

Calculating the HPI-1 is more straightforward than calculating the HDI. The indicators used to measure the deprivations are already normalized between 0 and 100 (because they are expressed as percentages), so there is no need to create dimension indices as for the HDI.

The human poverty index for selected OECD countries (HPI-2)

The HPI-2 measures deprivations in the same dimensions as the HPI-1 and also captures social exclusion. Thus it reflects deprivations in four dimensions:

- A long and healthy life—vulnerability to death at a relatively early age, as measured by the probability at birth of not surviving to age 60.
- Knowledge—exclusion from the world of reading and communications, as measured by the percentage of adults (ages 16–65) lacking functional literacy skills.
- A decent standard of living—as measured by the percentage of people living below the income poverty line (50% of the median adjusted household disposable income).
- Social exclusion—as measured by the rate of long-term unemployment (12 months or more).

Calculating the HPI-1

1. Measuring deprivation in a decent standard of living

An unweighted average of two indicators is used to measure deprivation in a decent standard of living.

$$\text{Unweighted average} = 1/2 \text{ (population not using an improved water source)} + 1/2 \text{ (children under weight-for-age)}$$

A sample calculation: Bolivia

Percentage of population not using an improved water source = 15%
Percentage of children under weight-for-age = 8%

$$\text{Unweighted average} = 1/2 \, (15) + 1/2 \, (8) = 11.3\%$$

2. Calculating the HPI-1

The formula used to calculate the HPI-1 is as follows:

$$\text{HPI-1} = [1/3 \, (P_1^{\alpha} + P_2^{\alpha} + P_3^{\alpha})]^{1/\alpha}$$

Where:

P_1 = Probability at birth of not surviving to age 40 (times 100)
P_2 = Adult illiteracy rate
P_3 = Unweighted average of population not using an improved water source and children under weight-for-age
α = 3

A sample calculation: Bolivia

P_1 = 15.5%
P_2 = 13.3%
P_3 = 11.3%

$$\text{HPI-1} = [1/3 \, (15.5^3 + 13.3^3 + 11.3^3)]^{1/3} = \textbf{13.6}$$

Calculating the HPI-2

The formula used to calculate the HPI-2 is as follows:

$$\text{HPI-2} = [1/4 \, (P_1^{\alpha} + P_2^{\alpha} + P_3^{\alpha} + P_4^{\alpha})]^{1/\alpha}$$

Where:

P_1 = Probability at birth of not surviving to age 60 (times 100)
P_2 = Percentage of adults lacking functional literacy skills
P_3 = Percentage of population below income poverty line (50% of median adjusted household disposable income)
P_4 = Rate of long-term unemployment (lasting 12 months or more)
α = 3

A sample calculation: Canada

P_1 = 8.1%
P_2 = 14.6%
P_3 = 11.4%
P_4 = 0.5%

$$\text{HPI-2} = [1/4 \, (8.1^3 + 14.6^3 + 11.4^3 + 0.5^3)]^{1/3} = \textbf{10.9}$$

Why α = 3 in calculating the HPI-1 and HPI-2

The value of α has an important impact on the value of the HPI. If $\alpha = 1$, the HPI is the average of its dimensions. As α rises, greater weight is given to the dimension in which there is the most deprivation. Thus as α increases towards infinity, the HPI will tend towards the value of the dimension in which deprivation is greatest (for Bolivia, the example used to calculate the HPI-1, would be 15.5, equal to the probability at birth of not surviving to age 40).

In this Report the value 3 is used to give additional but not overwhelming weight to areas of more acute deprivation. For a detailed analysis of the HPI's mathematical formulation, see Sudhir Anand and Amartya Sen's "Concepts of Human Development and Poverty: A Multidimensional Perspective" and the technical note in *Human Development Report 1997* (see the list of selected readings at the end of this technical note).

The gender-related development index (GDI)

While the HDI measures average achievement, the GDI adjusts the average achievement to reflect the *inequalities* between men and women in the following dimensions:

- A long and healthy life, as measured by life expectancy at birth.
- Knowledge, as measured by the adult literacy rate and the combined primary, secondary and tertiary gross enrolment ratio.
- A decent standard of living, as measured by estimated earned income (PPP US$).

The calculation of the GDI involves three steps. First, female and male indices in each dimension are calculated according to this general formula:

$$\text{Dimension index} = \frac{\text{actual value} - \text{minimum value}}{\text{maximum value} - \text{minimum value}}$$

Second, the female and male indices in each dimension are combined in a way that penalizes differences in achievement between men and women. The resulting index, referred to as the equally distributed index, is calculated according to this general formula:

$$\text{Equally distributed index}$$
$$= \{[\text{female population share (female index}^{1-\epsilon})]$$
$$+ [\text{male population share (male index}^{1-\epsilon})]\}^{1/1-\epsilon}$$

ϵ measures the aversion to inequality. In the GDI $\epsilon = 2$. Thus the general equation becomes:

$$\text{Equally distributed index}$$
$$= \{[\text{female population share (female index}^{-1})]$$
$$+ [\text{male population share (male index}^{-1})]\}^{-1}$$

which gives the harmonic mean of the female and male indices.

Third, the GDI is calculated by combining the three equally distributed indices in an unweighted average.

Goalposts for calculating the GDI

Indicator	Maximum value	Minimum value
Female life expectancy at birth (years)	87.5	27.5
Male life expectancy at birth (years)	82.5	22.5
Adult literacy rate (%)	100	0
Combined gross enrolment ratio (%)	100	0
Estimated earned income (PPP US$)	40,000	100

Note: The maximum and minimum values (goalposts) for life expectancy are 5 years higher for women to take into account their longer life expectancy. To preserve the relationship between female and male values of each indicator, scaled values are computed and used in place of figures where either the female or male value exceeds the threshold (in the case of Adult Literacy a practical threshold value of 99% is used). The scaling is achieved by multiplying the female and male values by the practical threshold value divided by the maximum reported value for either females or males.

Calculating the GDI

This illustration of the calculation of the GDI uses data for Botswana.

1. Calculating the equally distributed life expectancy index

The first step is to calculate separate indices for female and male achievements in life expectancy, using the general formula for dimension indices.

FEMALE
Life expectancy: 48.4 years

$$\text{Life expectancy index} = \frac{48.4 - 27.5}{87.5 - 27.5} = 0.348$$

MALE
Life expectancy: 47.6 years

$$\text{Life expectancy index} = \frac{47.6 - 22.5}{82.5 - 22.5} = 0.419$$

Next, the female and male indices are combined to create the equally distributed life expectancy index, using the general formula for equally distributed indices.

FEMALE
Population share: 0.504
Life expectancy index: 0.348

MALE
Population share: 0.496
Life expectancy index: 0.419

$$\text{Equally distributed life expectancy index} = \{[0.504\,(0.348^{-1})] + [0.496\,(0.419^{-1})]\}^{-1} = \mathbf{0.380}$$

2. Calculating the equally distributed education index

First, indices for the adult literacy rate and the combined primary, secondary and tertiary gross enrolment ratio are calculated separately for females and males. Calculating these indices is straightforward, since the indicators used are already normalized between 0 and 100.

FEMALE
Adult literacy rate: 81.8%
Adult literacy index: 0.818
Gross enrolment ratio: 70.1%
Gross enrolment index: 0.701

MALE
Adult literacy rate: 80.4%
Adult literacy index: 0.804
Gross enrolment ratio: 69.0%
Gross enrolment index: 0.690

Second, the education index, which gives two-thirds weight to the adult literacy index and one-third weight to the gross enrolment index, is computed separately for females and males.

$$\text{Education index} = 2/3\,(\text{adult literacy index}) + 1/3\,(\text{gross enrolment index})$$

$$\text{Female education index} = 2/3\,(0.818) + 1/3\,(0.701) = 0.779$$

$$\text{Male education index} = 2/3\,(0.804) + 1/3\,(0.690) = 0.766$$

Finally, the female and male education indices are combined to create the equally distributed education index.

FEMALE
Population share: 0.504
Education index: 0.779

MALE
Population share: 0.496
Education index: 0.766

$$\text{Equally distributed education index} = \{[0.504\,(0.779^{-1})] + [0.496\,(0.766^{-1})]\}^{-1} = \mathbf{0.773}$$

3. Calculating the equally distributed income index

First, female and male earned income (PPP US$) are estimated (for details on this calculation, see the addendum to this technical note). Then the income index is calculated for each gender. As with the HDI, income is adjusted by taking the logarithm of estimated earned income (PPP US$):

$$\text{Income index} = \frac{\log(\text{actual value}) - \log(\text{minimum value})}{\log(\text{maximum value}) - \log(\text{minimum value})}$$

FEMALE
Estimated earned income (PPP US$): 5,913

MALE
Estimated earned income (PPP US$): 19,094

$$\text{Income index} = \frac{\log(5{,}913) - \log(100)}{\log(40{,}000) - \log(100)} = 0.681$$

$$\text{Income index} = \frac{\log(19{,}094) - \log(100)}{\log(40{,}000) - \log(100)} = 0.877$$

Calculating the GDI continues on next page

Calculating the GDI (continued)

Second, the female and male income indices are combined to create the equally distributed income index :

FEMALE
Population share: 0.504
Income index: 0.681

MALE
Population share: 0.496
Income index: 0.877

$$\text{Equally distributed income index} = \{[0.504 \ (0.681^{-1})] + [0.496 \ (0.877^{-1})]\}^{-1} = 0.766$$

4. Calculating the GDI

Calculating the GDI is straightforward. It is simply the unweighted average of the three component indices—the equally distributed life expectancy index, the equally distributed education index and the equally distributed income index.

$$\text{GDI} = 1/3 \ (\text{life expectancy index}) + 1/3 \ (\text{education index}) + 1/3 \ (\text{income index})$$
$$= 1/3 \ (0.380) + 1/3 \ (0.773) + 1/3 \ (0.766) = 0.639$$

Why $\epsilon = 2$ in calculating the GDI

The value of ϵ is the size of the penalty for gender inequality. The larger the value, the more heavily a society is penalized for having inequalities.

If $\epsilon = 0$, gender inequality is not penalized (in this case the GDI would have the same value as the HDI). As ϵ increases towards infinity, more and more weight is given to the lesser-achieving group.

The value 2 is used in calculating the GDI (as well as the GEM). This value places a moderate penalty on gender inequality in achievement.

For a detailed analysis of the GDI's mathematical formulation, see Sudhir Anand and Amartya Sen's "Gender Inequality in Human Development: Theories and Measurement," Kalpana Bardhan and Stephan Klasen's "UNDP's Gender-Related Indices: A Critical Review" and the technical notes in *Human Development Report 1995* and *Human Development Report 1999* (see the list of selected readings at the end of this technical note).

The gender empowerment measure (GEM)

Focusing on women's opportunities rather than their capabilities, the GEM captures gender inequality in three key areas:

- Political participation and decision-making power, as measured by women's and men's percentage shares of parliamentary seats.
- Economic participation and decision-making power, as measured by two indicators—women's and men's percentage shares of positions as legislators, senior officials and managers and women's and men's percentage shares of professional and technical positions.
- Power over economic resources, as measured by women's and men's estimated earned income (PPP US$).

For each of these three dimensions, an equally distributed equivalent percentage (EDEP) is calculated, as a population-weighted average, according to the following general formula:

$$\text{EDEP} = \{[\text{female population share (female index}^{1-\epsilon})] + [\text{male population share (male index}^{1-\epsilon})]\}^{1/1-\epsilon}$$

ϵ measures the aversion to inequality. In the GEM (as in the GDI) $\epsilon = 2$, which places a moderate penalty on inequality. The formula is thus:

$$\text{EDEP} = \{[\text{female population share (female index}^{-1})] + [\text{male population share (male index}^{-1})]\}^{-1}$$

For political and economic participation and decision-making, the EDEP is then indexed by dividing it by 50. The rationale for this indexation is that in an ideal society, with equal empowerment of the sexes, the GEM variables would equal 50%—that is, women's share would equal men's share for each variable.

Where a male or female index value is zero, the EDEP according to the above formula is not defined. However, the limit of EDEP, when the index tends towards zero, is zero. Accordingly, in these cases the value of the EDEP is set to zero.

Finally, the GEM is calculated as a simple average of the three indexed EDEPs.

Calculating the GEM

This illustration of the calculation of the GEM uses data for the Russian Federation.

1. Calculating the EDEP for parliamentary representation

The EDEP for parliamentary representation measures the relative empowerment of women in terms of their political participation. The EDEP is calculated using the female and male shares of the population and female and male percentage shares of parliamentary seats according to the general formula.

FEMALE	MALE
Population share: 0.536	Population share: 0.464
Parliamentary share: 8.0%	Parliamentary share: 92.0%

$$\text{EDEP for parliamentary representation} = \{[0.536 \, (8.0^{-1})] + [0.464 \, (92.0^{-1})]\}^{-1} = 13.88$$

Then this initial EDEP is indexed to an ideal value of 50%.

$$\text{Indexed EDEP for parliamentary representation} = \frac{13.88}{50} = \mathbf{0.278}$$

2. Calculating the EDEP for economic participation

Using the general formula, an EDEP is calculated for women's and men's percentage shares of positions as legislators, senior officials and managers, and another for women's and men's percentage shares of professional and technical positions. The simple average of the two measures gives the EDEP for economic participation.

FEMALE	MALE
Population share: 0.536	Population share: 0.464
Percentage share of positions as legislators, senior officials and managers: 39.0%	Percentage share of positions as legislators, senior officials and managers: 61.0%
Percentage share of professional and technical positions: 64.7%	Percentage share of professional and technical positions: 35.3%

$$\text{EDEP for positions as legislators, senior officials and managers} = \{[0.536 \, (39.0^{-1})] + [0.464 \, (61.0^{-1})]\}^{-1} = 46.85$$

$$\text{Indexed EDEP for positions as legislators, senior officials and managers} = \frac{46.85}{50} = 0.937$$

$$\text{EDEP for professional and technical positions} = \{[0.536 \, (64.7^{-1})] + [0.464 \, (35.3^{-1})]\}^{-1} = 46.67$$

$$\text{Indexed EDEP for professional and technical positions} = \frac{46.67}{50} = 0.933$$

The two indexed EDEPs are averaged to create the EDEP for economic participation:

$$\text{EDEP for economic participation} = \frac{0.937 + 0.933}{2} = \mathbf{0.935}$$

3. Calculating the EDEP for income

Earned income (PPP US$) is estimated for women and men separately and then indexed to the scaled goalposts as was done for the GDI (for details, see the addendum to this technical note.). For the GEM, however, the income index is based on unadjusted values, not the logarithm of estimated earned income.

FEMALE	MALE
Population share: 0.536	Population share: 0.464
Estimated earned income (PPP US$): 8,476	Estimated earned income (PPP US$): 13,581

$$\text{Income index} = \frac{8,476 - 100}{40,000 - 100} = 0.210 \qquad \text{Income index} = \frac{13,581 - 100}{40,000 - 100} = 0.338$$

The female and male indices are then combined to create the equally distributed index:

$$\text{EDEP for income} = \{[0.536 \, (0.210^{-1})] + [0.464 \, (0.338^{-1})]\}^{-1} = \mathbf{0.255}$$

4. Calculating the GEM

Once the EDEP has been calculated for the three dimensions of the GEM, determining the GEM is straightforward. It is a simple average of the three EDEP indices.

$$\text{GEM} = \frac{0.278 + 0.935 + 0.255}{3} = \mathbf{0.489}$$

TECHNICAL NOTE 1 ADDENDUM
Female and male earned income

Despite the importance of having gender-disaggregated data on income, direct measures are unavailable. For this Report crude estimates of female and male earned income have therefore been derived.

Income can be seen in two ways: as a resource for consumption and as earnings by individuals. The use measure is difficult to disaggregate between men and women because they share resources within a family unit. By contrast, earnings are separable because different members of a family tend to have separate earned incomes.

The income measure used in the GDI and the GEM indicates a person's capacity to earn income. It is used in the GDI to capture the disparities between men and women in command over resources and in the GEM to capture women's economic independence. (For conceptual and methodological issues related to this approach, see Sudhir, Anand and Amartya Sen's "Gender Inequality in Human Development" and, in *Human Development Report 1995*, chapter 3 and *Technical notes 1 and 2*; see the list of selected readings at the end of this technical note.)

Female and male earned income (PPP US$) are estimated using the following data:

- Ratio of the female nonagricultural wage to the male nonagricultural wage.
- Male and female shares of the economically active population.
- Total female and male population.
- GDP per capita (PPP US$).

Key

W_f/W_m = ratio of female nonagricultural wage to male nonagricultural wage
EA_f = female share of economically active population
EA_m = male share of economically active population
S_f = female share of wage bill
Y = total GDP (PPP US$)
N_f = total female population
N_m = total male population
Y_f = estimated female earned income (PPP US$)
Y_m = estimated male earned income (PPP US$)

Note

Because of rounding, calculations carried out by hand may yield results that differ from those printed in the technical notes and indicator tables.

Estimating female and male earned income

This illustration of the estimation of female and male earned income uses 2005 data for Sweden.

1. Calculating total GDP (PPP US$)
Total GDP (PPP US$) is calculated by multiplying the total population by GDP per capita (PPP US$).

Total population: 9,024 (thousand)
GDP per capita (PPP US$): 32,525
Total GDP (PPP US$) = 9,024 (32,525) = 293,510,764 (thousand)

2. Calculating the female share of the wage bill
Because data on wages in rural areas and in the informal sector are rare, the Report has used nonagricultural wages and assumed that the ratio of female wages to male wages in the nonagricultural sector applies to the rest of the economy. The female share of the wage bill is calculated using the ratio of the female nonagricultural wage to the male nonagricultural wage and the female and male percentage shares of the economically active population. Where data on the wage ratio are not available, a value of 75% is used.

Ratio of female to male nonagricultural wage (W_f/W_m) = 0.907
Female percentage share of economically active population (EA_f) = 47.4%
Male percentage share of economically active population (EA_m) = 52.6%

$$\text{Female share of wage bill } (S_f) = \frac{W_f/W_m\,(EA_f)}{[W_f/W_m\,(EA_f)] + EA_m} = \frac{0.907\,(47.4)}{[0.907\,(47.4) + 52.6]} = 0.450$$

3. Calculating female and male earned income (PPP US$)
An assumption has to be made that the female share of the wage bill is equal to the female share of GDP.

Female share of wage bill (S_f) = 0.450
Total GDP (PPP US$) (Y) = 293,510,764 (thousand)
Female population (N_f) = 4,546 (thousand)

$$\text{Estimated female earned income (PPP US\$) } (Y_f) = \frac{S_f\,(Y)}{N_f} = \frac{0.450\,(293,510,764)}{4,546} = 29,044$$

Male population (N_m) = 4,478 (thousand)

$$\text{Estimated male earned income (PPP US\$) } (Y_m) = \frac{Y - S_f\,(Y)}{N_m} = \frac{293,510,764 - [0.450\,(293,510,764)]}{4,478} = 36,059$$

Selected readings

Anand, Sudhir, and Amartya Sen. 1994. "Human Development Index: Methodology and Measurement". Occasional Paper 12, United Nations Development Programme, Human Development Report Office, New York. *(HDI)*

——, **1995,** "Gender Inequality in Human Development Theories and Measurement." Occasional Paper 19, United Nations Development Programme, Human Development Report Office, New York. *(GDI, GEM)*

——, **1997,** "Concepts of Human Development and Poverty: A Multi-dimensional Perspective." In United Nations Development Programme, *Human Development Report 1997 Papers : Poverty and Human Development* New York. *(HPI-1, HPI-2)*

Bardhan, Kalpana, and Stephan Klasen, 1999. "UNDP's Gender-Related Indices. A Critical Review." *World Development* 27 (6): 985–1010 *(GDI, GEM)*

United Nations Development Programme, 1995. *Human Development Report 1995.* New York: Oxford University Press, Technical notes 1 and 2 and chapter 3. *(GDI, GEM)*

——, **1997,** *Human Development Report 1997.* New York: Oxford University Press. Technical note 1 and chapter 1. *(HPI-1, HPI-2)*

——, **1999,** *Human Development Report 1999.* New York: Oxford University Press. Technical note *(HDI, GDI)*

Klasen, Stephan. 2006. "UNDP's Gender-related Measures: Some Conceptual Problems and Possible Solutions." Journal of Human Development Alternative Economics in Action, 7 (2): 243 - 274.

Measuring the short and long-term effects of climate-related disasters

Human development is about expanding freedoms and capabilities. Yet, as explained in chapter 2, this process can be derailed by climate-related disasters. Besides their immediate costs in terms of lives lost and livelihoods disrupted, climate-related shocks carry substantial intrinsic costs that are likely to follow people throughout their lives, locking them into low human development traps. Climate change promises to raise these stakes for billions of vulnerable people.

To capture the extent of the threat to human development that is embedded in climate-related shocks, the short and long-term effects of being born in a disaster-affected area were measured. More specifically, some critical determinants of human development outcomes were examined for children under five years of age and adult women between the ages of 15 and 30, and those who were affected by a disaster were compared with those who were not.

Data

Data for the research were derived from Demographic and Health Surveys (DHS) and the international disasters database EM-DAT maintained by the University of Louvain.

Demographic and Health Surveys (DHS)

The DHS are household and community surveys administered by Macro International and partly financed by the United States Agency for International Development (USAID). These surveys collect information on a wide range of socio-economic variables at individual, household and community levels, and are usually conducted every five years to allow comparisons over time. DHS generally consist of a sample of 5,000–30,000 households but are not longitudinal in design. The survey design is representative at national, urban and rural levels.

Although their primary focus is on women aged 15–49, DHS also collect information on demographic indicators for all members of the household. For children under five years of age, these surveys also collect such monitoring and impact evaluation variables as health and nutrition indicators.

International disasters database EM-DAT

The EM-DAT is an international disasters database that presents core data on the occurrence of disasters worldwide from 1900 to the present. Disasters in EM-DAT are defined as: "a situation or event which overwhelms local capacity, necessitating a request to the national or international level for external assistance, or is recognized as such by a multilateral agency or at least by two sources, such as national, regional or international assistance groups and the media". For a disaster to be recorded in the database, it has to meet one or more of the following criteria:

- 10 or more people are killed;
- 100 people or more are reported affected;
- A state of emergency is declared;
- An international call for assistance is issued.

A key feature of this database is that it records both the date of occurrence of a disaster—relatively recent ones—its location, and the extent of its severity through the number of people affected, the number of casualties and the financial damage.[1]

Country selection criteria

For the purposes of this study, only countries where over 1,000,000 people were reported affected by a disaster were selected. For children under the age of five countries that had a DHS with a geographic positioning system (GPS) module two to three years following a disaster were selected. The selection of countries with GPS modules was necessary, especially for countries where some administrative districts were more affected than others. For adult women selection was limited to major disasters that had occurred during the 1970s and 1980s; with the requirement that the disaster in question occurred at least 15 years prior to the first DHS. See table for country coverage and sample characteristics.

Methodology

This approach borrows from impact evaluation techniques widely used in the social sciences. For children under the age of five, the outcome indicators used were: stunting (low height for age), wasting (low weight for height) and malnourishment (low weight for age). For adult women 15–30, the outcome indicator was educational outcome. In the absence of longitudinal data, a set of synthetic before and after cohorts were constructed and their outcomes compared using logit regressions with a difference-in-difference approach, controlling for individual, household and community characteristics.

To construct the cohorts, children and adult women in DHS were identified and their birth dates tracked. The subject's birth date and birth location were then crosschecked against the occurrence of a natural disaster as indicated in EM-DAT. The following groups were identified:

- Subjects born before a disaster in an area that was subsequently affected (born before, affected—group 1, affected).
- Subjects born before a disaster in an area that was not subsequently affected (born before, not affected—group 1, not affected).
- Subjects born during a disaster in an area that was affected (born during, affected—group 2, affected).

- Subjects born during a disaster in an area that was not affected (born during, not affected—group 2, not affected).

Using these different groups, the following model was estimated:

$$\hat{\phi} = \frac{1}{N}\sum_{i=1}^{n}[(y_{i2}^{a} - y_{i1}^{a}) - (y_{i2}^{na} - y_{i2}^{na})] \text{ where } y_i \text{ is the outcome}$$
in question for the ith person.[2]

At each step, a set of control variables were used to identify the effects of specific characteristics on children's nutritional outcomes. These included individual variables (the sex of the child, birth intervals and such maternal characteristics as mother's age and education) and community-level variables (e.g., urban/rural location). A regression analysis was then conducted to isolate the specific risks associated with being affected by a disaster.

For adults, if it is assumed that disasters are a deterministic process, then virtually every indicator including household socio-economic characteristics is determined by early exposure to a disaster, and is therefore endogenous. As a result, only variables that can reasonably be assumed exogenous, such as religion, were included.

Most of the results are shown and discussed in chapter 2 and in Fuentes and Seck 2007.

Notes

1 Guha-Sapir et al. 2004
2 Cameron and Trivedi 2005

Table	Country coverage and sample characteristics				
Country	**Year of survey**	**Sample size**	**Stunted (%)**	**Malnourished (%)**	**Wasted (%)**
Children					
Ethiopia	2005	9,861	43.4	37.8	11.1
Kenya	2003	5,949	32.5	20.2	6.7
Niger	1992	6,899	38.2	38.9	14.5
Adults	**Year of survey**	**Sample size**	**No education (%)**	**At least primary education (%)**	**At least secondary education (%)**
India	1998	90,303	35.3	50.5	33.6

Definitions of statistical terms

Antimalarial measures, fevers treated with antimalarial drugs The percentage of children under age five who were ill with fever in the two weeks before the survey and received antimalarial drugs.

Antimalarial measures, use of insecticide treated bednets The percentage of children under age five sleeping under insecticide trreated bednets.

Armed forces, total Strategic, land, naval, air, command, administrative and support forces. Also included are paramilitary forces such as the gendarmerie, customs service and border guard, if these are trained in military tactics.

Arms transfers, conventional Refers to the voluntary transfer by the supplier (and thus excludes captured weapons and weapons obtained through defectors) of weapons with a military purpose destined for the armed forces, paramilitary forces or intelligence agencies of another country. These include major conventional weapons or systems in six categories: ships, aircraft, missiles, artillery, armoured vehicles and guidance and radar systems (excluded are trucks, services, ammunition, small arms, support items, components and component technology and towed or naval artillery under 100-millimetre calibre).

Births attended by skilled health personnel The percentage of deliveries attended by personnel (including doctors, nurses and midwives) trained to give the necessary care, supervision and advice to women during pregnancy, labour and the post-partum period; to conduct deliveries on their own; and to care for newborns. Traditional birth attendants, trained or not, are not included in this category.

Birthweight, infants with low The percentage of infants with a birthweight of less than 2,500 grams.

Carbon dioxide emissions Anthropogenic (human originated) carbon dioxide emissions stemming from the burning of fossil fuels, gas flaring and the production of cement. Emissions are calculated from data on the consumption of solid, liquid and gaseous fuels; gas flaring; and the production of cement. Carbon dioxide can also be emitted by forest biomass through depletion of forest areas.

Carbon intensity of energy refers to the amount of carbon dioxide (CO_2) generated for every unit of energy used. It is the ratio of emitted CO_2 to energy use.

Carbon intensity of growth also known as the carbon intensity of the economy, refers to the amount of carbon dioxide generated by every dollar of growth in the world economy. It is the ratio of emitted CO_2 to GDP (in PPP terms).

Cellular subscribers Subscribers to an automatic public mobile telephone service that provides access to the public switched telephone network using cellular technology. Systems can be analogue or digital.

Children reaching grade 5 The percentage of children starting primary school who eventually attain grade 5. The estimates are based on the reconstructed cohort student flow method, which uses data on enrolment and repeaters for two consecutive school years in order to estimate the survival rates to successive grades of primary school.

Children under age five with diarrhoea receiving oral rehydration and continued feeding The percentage of children (aged 0–4) with diarrhoea in the two weeks preceding the survey who received either oral rehydration therapy (oral rehydration solutions or recommended homemade fluids) or increased fluids and continued feeding.

Condom use at last high-risk sex The percentage of men and women who have had sex with a nonmarital, noncohabiting partner in the last 12 months and who say they used a condom the last time they did so.

Consumer price index, average annual change in Reflects changes in the cost to the average consumer of acquiring a basket of goods and services that may be fixed or may change at specified intervals.

Contraceptive prevalence rate The percentage of women of reproductive age (15–49 years) who are using, or whose partners are using, any form of contraception, whether modern or traditional.

Contributing family worker Defined according to the 1993 International Classification by Status in Employment (ICSE) as a person who works without pay in an economic enterprise operated by a related person living in the same household.

Debt service, total The sum of principal repayments and interest actually paid in foreign currency, goods or services on long-term debt (having a maturity of more than one year), interest paid on short-term debt and repayments to the International Monetary Fund.

Earned income (PPP US$), estimated Derived on the basis of the ratio of the female nonagricultural wage to the male nonagricultural wage, the female and male shares of the economically active population, total female and male population and GDP per capita (in purchasing power parity terms in US dollars; see *PPP*). For details of this estimation, see *Technical note 1*.

Earned income, ratio of estimated female to male The ratio of estimated female earned income to estimated male earned income. See *Earned income (PPP US$), estimated*.

Education expenditure, current public Spending on goods and services that are consumed within the current year and that would need to be renewed the following year, including such expenditures as staff salaries and benefits, contracted or purchased services, books and teaching materials, welfare services, furniture and equipment, minor repairs, fuel, insurance, rents, telecommunications and travel.

Education expenditure, public Includes both capital expenditures (spending on construction, renovation, major repairs and purchases of heavy equipment or vehicles) and current expenditures. See *Education expenditure, current public*.

Education index One of the three indices on which the human development index is built. It is based on the adult literacy rate and the combined gross enrolment ratio for primary, secondary and tertiary schools. See *Literacy rate, adult*, and *enrolment ratio, gross combined, for primary, secondary and tertiary schools*. For details on how the index is calculated, see *Technical note 1*.

Education levels Categorized as pre-primary, primary, secondary, post-secondary and tertiary in accordance with the International Standard Classification of Education (ISCED). *Pre-primary education* (ISCED level 0) is the initial stage of organized instruction, designed primarily to introduce very young children to a school-type environment and to provide a bridge between home and school. *Primary education* (ISCED level 1) provides a sound basic education in reading, writing and mathematics along with an elementary understanding of other subjects such as history, geography, natural and social science, art, music and religion. *Secondary education* (ISCED levels 2 and 3) is generally designed to continue the basic programmes of the primary level but the instruction is typically more subject-focused, requiring more specialized teachers for each subject area. *Post-secondary (non-tertiary) education* (ISCED level 4) includes programmes which lie between upper secondary (ISCED 3) and tertiary education (ISCED 5 and 6) in an international context though typically are clearly within one or other level in the national context in different countries. ISCED 4 programmes are usually not significantly more advanced than ISCED 3 programmes but they serve to broaden the knowledge of students who have already completed an upper secondary programme. *Tertiary education* (ISCED levels 5 and 6) refers to programmes with an educational content that is more advanced than upper secondary or post-secondary education. The first stage of tertiary education (ISCED 5) is composed both of programmes of a theoretical nature (ISCED 5A) intended to provide access to advanced research programmes and professions with high skill requirements as well as programmes of a more practical, technical or occupationally specific nature (ISCED 5B). The second stage of tertiary education (ISCED 6) comprises programmes devoted to advanced study and original research, leading to the award of an advanced research qualification such as a doctorate.

Energy supply, primary refers to the supply of energy extracted or captured directly from natural resources such as crude oil, hard coal, natural gas, or are produced from primary commodities. Primary energy commodities may also be divided into fuels of fossil origin and renewable energy commodities. See fossil fuels and renewable energy.

Electricity consumption per capita Refers to gross production in per capita terms and includes consumption by station auxiliaries and any losses in transformers that are considered integral parts of the station. Also included is total electric energy produced by pumping installations without deduction of electric energy absorbed by pumping.

Electricity, people without access refers to the lack of access to electricity at the household level; that is the number of people who do not have electricity in their home. Access to electricity is comprised of electricity sold commercially, both on-grid and off-grid. It also includes self-generated electricity in those countries where access to electricity has been assessed through surveys by national administrations. This data does not capture unauthorised connections.

Electrification rates indicate the number of people with electricity access as a percentage of the total population.

Employment by economic activity Employment in industry, agriculture or services as defined according to the International Standard Industrial Classification (ISIC) system (revisions 2 and 3). *Industry* refers to mining and quarrying, manufacturing, construction and public utilities (gas, water and electricity). *Agriculture* refers to activities in agriculture, hunting, forestry and fishing. *Services* refer to wholesale and retail trade; restaurants and hotels; transport, storage and communications; finance, insurance, real estate and business services; and community, social and personal services.

Energy use, GDP per unit of The ratio of GDP (in 2000 PPP US$) to commercial energy use, measured in kilograms of oil equivalent. This indicator provides a measure of energy efficiency by showing comparable and consistent estimates of real GDP across countries relative to physical inputs (units of energy use). See *GDP (gross domestic product)* and *PPP (purchasing power parity)*. Differences in this ratio over time and across countries partly reflect structural changes in the economy, changes in energy efficiency of particular sectors, and differences in fuel mixes.

Enrolment ratio, gross The total number of pupils or students enrolled in a given level of education, regardless of age, expressed as a percentage of the population in the theoretical age group for the same level of education. For the tertiary level, the population used is the five-year age group following on from the secondary school leaving age. Gross enrolment ratios in excess of 100% indicate that there are pupils or students outside the theoretical age group who are enrolled in that level of education. See *Education levels*.

Enrolment ratio, gross combined, for primary, secondary and tertiary schools The number of students enrolled in primary, secondary and tertiary levels of education, regardless of age, as a percentage of the population of theoretical school age for the three levels. See *Education levels* and *Enrolment ratio, gross*.

Enrolment rate, net The number of pupils of the theoretical school-age group for a given level of education level who are enrolled in that level, expressed as a percentage of the total population in that age group. See *Education levels*.

Exports, high-technology Exports of products with a high intensity of research and development. They include high-technology products such as those used in aerospace, computers, pharmaceuticals, scientific instruments and electrical machinery.

Exports, manufactured Defined according to the Standard International Trade Classification to include exports of chemicals, basic manufactures, machinery and transport equipment and other miscellaneous manufactured goods.

Exports of goods and services The value of all goods and other market services provided to the rest of the world. Included is the value of merchandise, freight, insurance, transport, travel, royalties, licence fees and other services, such as communication, construction, financial, information, business, personal and government services. Excluded are labour and property income and transfer payments.

Exports, primary Defined according to the Standard International Trade Classification to include exports of food, agricultural raw materials, fuels and ores and metals.

Fertility rate, total The number of children that would be born to each woman if she were to live to the end of her child-bearing years and bear children at each age in accordance with prevailing age-specific fertility rates in a given year/period, for a given country, territory or geographical area.

Foreign direct investment, net inflows of Net inflows of investment to acquire a lasting management interest (10% or more of voting stock) in an enterprise operating in an economy other than that of the investor. It is the sum of equity capital, reinvestment of earnings, other long-term capital and short-term capital.

Forest area is land under natural or planted stands of trees, whether productive or not.

Fossil fuels are fuels taken from natural resources which were formed from biomass in the geological past. The main fossil fuels are coal, oil and natural gas. By extension, the term fossil is also applied to any secondary fuel manufactured from a fossil fuel. Fossil Fuels belong to the primary energy commodities group.

GDP (gross domestic product) The sum of value added by all resident producers in the economy plus any product taxes (less subsidies) not included in the valuation of output. It is calculated without making deductions for depreciation of fabricated capital assets or for depletion and degradation of natural resources. Value added is the net output of an industry after adding up all outputs and subtracting intermediate inputs.

GDP (US$) Gross domestic product converted to US dollars using the average official exchange rate reported by the International Monetary Fund. An alternative conversion factor is applied if the official exchange rate is judged to diverge by an exceptionally large margin from the rate effectively applied to transactions in foreign currencies and traded products. See *GDP (gross domestic product)*.

GDP index One of the three indices on which the human development index is built. It is based on gross domestic product per capita (in purchasing power parity terms in US dollars; see *PPP*). For details on how the index is calculated, see *Technical note 1*.

GDP per capita (PPP US$) Gross domestic product (in purchasing power parity terms in US dollars) divided by midyear population. See *GDP (gross domestic product), PPP (purchasing power parity)* and *Population, total*.

GDP per capita (US$) Gross domestic product in US dollar terms divided by midyear population. See *GDP (US$)* and *Population, total*.

GDP per capita annual growth rate Least squares annual growth rate, calculated from constant price GDP per capita in local currency units.

Gender empowerment measure (GEM) A composite index measuring gender inequality in three basic dimensions of empowerment—economic participation and decision-making, political participation, and decision-making and power over economic resources. For details on how the index is calculated, see *Technical note 1*.

Gender-related development index (GDI) A composite index measuring average achievement in the three basic dimensions captured in the human development index—a long and healthy life, knowledge and a decent standard of living—adjusted to account for inequalities between men and women. For details on how the index is calculated, see *Technical note 1*.

Gini index Measures the extent to which the distribution of income (or consumption) among individuals or households within a country deviates from a perfectly equal distribution. A Lorenz curve plots the cumulative percentages of total income received against the cumulative number of recipients, starting with the poorest individual or household. The Gini index measures the area between the Lorenz curve and a hypothetical line of absolute equality, expressed as a percentage of the maximum area under the line. A value of 0 represents absolute equality, a value of 100 absolute inequality.

GNI (gross national income) The sum of value added by all resident producers in the economy plus any product taxes (less subsidies) not included in the valuation of output plus net receipts of primary income (compensation of employees and property income) from abroad. Value added is the net output of an industry after adding up all outputs and subtracting intermediate inputs. Data are in current US dollars converted using the *World Bank Atlas* method.

Health expenditure per capita (PPP US$) The sum of public and private expenditure (in purchasing power parity terms in US dollars), divided by the mid-year population. Health expenditure includes the provision of health services (preventive and curative), family planning activities, nutrition activities and emergency aid designated for health, but excludes the provision of water and sanitation. See *Health expenditure, private; Health expenditure, public; Population, total;* and *PPP (purchasing power parity).*

Health expenditure, private Direct household (out of pocket) spending, private insurance, spending by non-profit institutions serving households and direct service payments by private corporations. Together with public health expenditure, it makes up total health expenditure. See *Health expenditure per capita (PPP US$)* and *Health expenditure, public.*

Health expenditure, public Current and capital spending from government (central and local) budgets, external borrowings and grants (including donations from international agencies and nongovernmental organizations) and social (or compulsory) health insurance funds. Together with private health expenditure, it makes up total health expenditure. See *Health expenditure per capita (PPP US$)* and *Health expenditure, private.*

HIV prevalence The percentage of people aged 15–49 years who are infected with HIV.

Human development index (HDI) A composite index measuring average achievement in three basic dimensions of human development—a long and healthy life, knowledge and a decent standard of living. For details on how the index is calculated, see *Technical note 1.*

Human poverty index for developing countries (HPI-1) A composite index measuring deprivations in the three basic dimensions captured in the human development index—a long and healthy life, knowledge and a decent standard of living. For details on how the index is calculated, see *Technical note 1.*

Human poverty index for selected high-income OECD countries (HPI-2) A composite index measuring deprivations in the three basic dimensions captured in the human development index—a long and healthy life, knowledge and a decent standard of living—and

also capturing social exclusion. For details on how the index is calculated, see *Technical note 1.*

Homicide, intentional Death deliberately inflicted on a person by another person, including infanticide.

Illiteracy rate, adult Calculated as 100 minus the adult literacy rate. See *Literacy rate, adult.*

Immunization, one-year-olds fully immunized against measles or tuberculosis One-year-olds injected with an antigen or a serum containing specific antibodies against measles or tuberculosis.

Imports of goods and services The value of all goods and other market services received from the rest of the world. Included is the value of merchandise, freight, insurance, transport, travel, royalties, licence fees and other services, such as communication, construction, financial, information, business, personal and government services. Excluded are labour and property income and transfer payments.

Income poverty line, population below The percentage of the population living below the specified poverty line:

- US$1 a day—at 1985 international prices (equivalent to US$1.08 at 1993 international prices), adjusted for purchasing power parity.

- US$2 a day—at 1985 international prices (equivalent to US$2.15 at 1993 international prices), adjusted for purchasing power parity.

- US$4 a day—at 1990 international prices, adjusted for purchasing power parity.

- US$11 a day (per person for a family of three)—at 1994 international prices, adjusted for purchasing power parity.

- National poverty line—the poverty line deemed appropriate for a country by its authorities. National estimates are based on population-weighted subgroup estimates from household surveys.

- 50% of median income—50% of the median adjusted disposable household income. See *PPP (purchasing power parity).*

Income or consumption, shares of The shares of income or consumption accruing to subgroups of population indicated by deciles or quintiles, based on national household surveys covering various years. Consumption surveys produce results showing lower levels of inequality between poor and rich than do income surveys, as poor people generally consume a greater share of their income. Because data come from surveys covering different years and using different methodologies, comparisons between countries must be made with caution.

Infant mortality rate See *Mortality rate, infant.*

Informal sector The informal sector, as defined by the International Expert Group on Informal Sector

Statistics (the Delhi Group) includes private unincorporated enterprises (excluding quasi-corporations), which produce at least some of their goods and services for sale or barter, have less than five paid employees, are not registered, and are engaged in nonagricultural activities (including professional or technical activities). Paid domestic employees are excluded from this category.

Informal sector, employment in, as a percentage of nonagricultural employment Refers to the ratio of total employment in the informal sector to total employment in all nonagricultural sectors. See *Informal sector*.

Internally displaced people People or groups of people who have been forced or obliged to flee or to leave their homes or places of habitual residence, in particular as a result of or to avoid the effects of armed conflict, situations of generalized violence, violations of human rights or natural or human-made disasters, and who have not crossed an internationally recognized state border.

Internet users People with access to the world-wide network.

Labour force All people employed (including people above a specified age who, during the reference period, were in paid employment, at work, self-employed or with a job but not at work) and unemployed (including people above a specified age who, during the reference period, were without work, currently available for work and actively seeking work).

Labour force participation rate A measure of the proportion of a country's working-age population that engages actively in the labour market, either by working or actively looking for work. It is calculated by expressing the number of persons in the labour force as a percentage of the working-age population. The working-age population is the population above 15 years of age (as used in this Report). See *Labour force*.

Labour force participation rate, female The number of women in the labour force expressed as a percentage of the female working-age population. See *Labour force participation rate* and *Labour force*.

Legislators, senior officials and managers, female Women's share of positions defined according to the International Standard Classification of Occupations (ISCO-88) to include legislators, senior government officials, traditional chiefs and heads of villages, senior officials of special-interest organizations, corporate managers, directors and chief executives, production and operations department managers and other department and general managers.

Life expectancy at birth The number of years a newborn infant would live if prevailing patterns of age-specific mortality rates at the time of birth were to stay the same throughout the child's life.

Life expectancy index One of the three indices on which the human development index is built. For details on how the index is calculated, see *Technical note 1*.

Literacy rate, adult The proportion of the adult population aged 15 years and older which is literate, expressed as a percentage of the corresponding population, total or for a given sex, in a given country, territory, or geographic area, at a specific point in time, usually mid-year. For statistical purposes, a person is literate who can, with understanding, both read and write a short simple statement on his/her everyday life.

Literacy rate, youth The percentage of people aged 15–24 years who can, with understanding, both read and write a short, simple statement related to their everyday life, see *Literacy rate, adult*.

Literacy skills, functional, people lacking The share of the population aged 16–65 years scoring at level 1 on the prose literacy scale of the International Adult Literacy Survey. Most tasks at this level require the reader to locate a piece of information in the text that is identical to or synonymous with the information given in the directive.

Market activities See *Time use, market activities*.

Medium-variant projection Population projections by the United Nations Population Division assuming medium-fertility path, normal mortality and normal international migration. Each assumption implies projected trends in fertility, mortality and net migration levels, depending on the specific demographic characteristics and relevant policies of each country or group of countries. In addition, for the countries highly affected by the HIV/AIDS epidemic, the impact of HIV/AIDS is included in the projection. The United Nations Population Division also publishes low-and high-variant projections. For more information, see http://esa.un.org/unpp/assumptions.html.

Military expenditure All expenditures of the defence ministry and other ministries on recruiting and training military personnel as well as on construction and purchase of military supplies and equipment. Military assistance is included in the expenditures of the donor country.

Mortality rate, infant The probability of dying between birth and exactly one year of age, expressed per 1,000 live births.

Mortality rate, under-five The probability of dying between birth and exactly five years of age, expressed per 1,000 live births.

Mortality ratio, maternal The quotient between the number of maternal deaths in a given year and the number of live births in that same year, expressed per 100,000 live births, for a given country, territory, or geographic area. Maternal death is defined as the death of a woman while pregnant or within the 42 days after termination of that pregnancy, regardless of the length and site of the pregnancy, due to any cause related to or aggravated by the pregnancy itself or its care, but not due to accidental or incidental causes.

Mortality ratio, maternal adjusted Maternal mortality ratio adjusted to account for well-documented problems of under reporting and misclassification of maternal deaths, as well as estimates for countries with no data. See *Mortality ratio, maternal*.

Mortality ratio, maternal reported Maternal mortality ratio as reported by national authorities. See *Mortality ratio, maternal*.

Nonmarket activities See *Time use, nonmarket activities*.

Official aid Grants or loans that meet the same standards as for official development assistance (ODA) except that recipient countries do not qualify as recipients of ODA. These countries are identified in part II of the Development Assistance Committee (DAC) list of recipient countries, which includes more advanced countries of Central and Eastern Europe, the countries of the former Soviet Union and certain advanced developing countries and territories. See *Official development assistance (ODA), net*.

Official development assistance (ODA), net Disbursements of loans made on concessional terms (net of repayments of principal) and grants by official agencies of the members of the Development Assistance Committee (DAC), by multilateral institutions and by non-DAC countries to promote economic development and welfare in countries and territories in part I of the DAC list of aid recipients. It includes loans with a grant element of at least 25% (calculated at a discount rate of 10%).

Official development assistance (ODA), per capita of donor country Official development assistance granted by a specific country divided by the country's total population. See *Official development assistance (ODA), net* and *population, total*.

Official development assistance (ODA) to basic social services ODA directed to basic social services, which include basic education (primary education, early childhood education and basic life skills for youth and adults), basic health (including basic health care, basic health infrastructure, basic nutrition, infectious disease control, health education and health personnel development) and population policies and programmes and reproductive health (population policy and administrative management; reproductive health care; family planning; control of sexually transmitted diseases, including HIV/AIDS; and personnel development for population and reproductive health). Aid to water supply and sanitation is included only if it has a poverty focus.

Official development assistance (ODA) to least developed countries See *Official development assistance (ODA), net* and country classifications for least developed countries.

Official development assistance (ODA), untied Bilateral ODA for which the associated goods and services may be fully and freely procured in substantially all countries and that is given by one country to another.

Patents granted to residents Refer to documents issued by a government office that describe an inven-

tion and create a legal situation in which the patented invention can normally be exploited (made, used, sold, imported) only by or with the authorization of the patentee. The protection of inventions is generally limited to 20 years from the filing date of the application for the grant of a patent.

Physicians Includes graduates of a faculty or school of medicine who are working in any medical field (including teaching, research and practice).

Population growth rate, annual Refers to the average annual exponential growth rate for the period indicated. See *Population, total*.

Population, total Refers to the de facto population in a country, area or region as of 1 July of the year indicated.

Population, urban Refers to the de facto population living in areas classified as urban according to the criteria used by each area or country. Data refer to 1 July of the year indicated. See *Population, total*.

PPP (purchasing power parity) A rate of exchange that accounts for price differences across countries, allowing international comparisons of real output and incomes. At the PPP US$ rate (as used in this Report), PPP US$1 has the same purchasing power in the domestic economy as US$1 has in the United States.

Private flows, other A category combining non-debt-creating portfolio equity investment flows (the sum of country funds, depository receipts and direct purchases of shares by foreign investors), portfolio debt flows (bond issues purchased by foreign investors) and bank and trade-related lending (commercial bank lending and other commercial credits).

Probability at birth of not surviving to a specified age Calculated as 100 minus the probability (expressed as a percentage) of surviving to a specified age for a given cohort. See *Probability at birth of surviving to a specified age*.

Probability at birth of surviving to a specified age The probability of a newborn infant surviving to a specified age if subject to prevailing patterns of age-specific mortality rates, expressed as a percentage.

Professional and technical workers, female Women's share of positions defined according to the International Standard Classification of Occupations (ISCO-88) to include physical, mathematical and engineering science professionals (and associate professionals), life science and health professionals (and associate professionals), teaching professionals (and associate professionals) and other professionals and associate professionals.

Refugees People who have fled their country because of a well-founded fear of persecution for reasons of their race, religion, nationality, political opinion or membership in a particular social group and who cannot or do not want to return. *Country of asylum* is the country in which a refugee has filed a claim of asylum but has not yet received a decision or is otherwise registered as an asylum seeker. *Country of origin* refers to the claimant's nationality or country of citizenship.

Renewable energy Energy derived from natural processes that are constantly replenished. Among the forms of renewable energy are deriving directly or indirectly from the sun, or from heat generated deep within the earth. Renewable energy includes energy generated from solar, wind, biomass, geothermal, hydropower and ocean resources and some waste. Renewable energy commodities belong to the primary energy commodities group.

Research and development (R&D) expenditures Current and capital expenditures (including overhead) on creative, systematic activity intended to increase the stock of knowledge. Included are fundamental and applied research and experimental development work leading to new devices, products or processes.

Researchers in R&D People trained to work in any field of science who are engaged in professional research and development activity. Most such jobs require the completion of tertiary education.

Royalties and licence fees, receipts of Receipts by residents from nonresidents for the authorized use of intangible, nonproduced, nonfinancial assets and proprietary rights (such as patents, trademarks, copyrights, franchises and industrial processes) and for the use, through licensing agreements, of produced originals of prototypes (such as films and manuscripts). Data are based on the balance of payments.

Sanitation facilities, improved, population using The percentage of the population with access to adequate excreta disposal facilities, such as a connection to a sewer or septic tank system, a pour-flush latrine, a simple pit latrine or a ventilated improved pit latrine. An excreta disposal system is considered adequate if it is private or shared (but not public) and if it can effectively prevent human, animal and insect contact with excreta.

Science, maths and engineering, tertiary students in The share of tertiary students enrolled in natural sciences; engineering; mathematics and computer sciences; architecture and town planning; transport and communications; trade, craft and industrial programmes; and agriculture, forestry and fisheries. See *Education levels.*

Seats in parliament held by women Refers to seats held by women in a lower or single house or an upper house or senate, where relevant.

Smoking, prevalence among adults of The percentage of men and women who smoke cigarettes.

Telephone mainlines Telephone lines connecting a customer's equipment to the public switched telephone network.

Terms of trade The ratio of the export price index to the import price index measured relative to a base year. A value of more than 100 means that the price of exports has risen relative to the price of imports.

Time use, market activities Time spent on activities such as employment in establishments, primary production not in establishments, services for income and other production of goods not in establishments as defined according to the 1993 revised UN System of National Accounts. See *Time use, nonmarket activities* and *Time use, work time, total.*

Time use, nonmarket activities Time spent on activities such as household maintenance (cleaning, laundry and meal preparation and cleanup), management and shopping for own household; care for children, the sick, the elderly and the disabled in own household; and community services, as defined according to the 1993 revised UN System of National Accounts. See *Time use, market activities* and *Time use, work time, total.*

Time use, work time, total Time spent on market and nonmarket activities as defined according to the 1993 revised UN System of National Accounts. See *Time use, market activities* and *Time use, nonmarket activities.*

Treaties, ratification of After signing a treaty, a country must ratify it, often with the approval of its legislature. Such process implies not only an expression of interest as indicated by the signature, but also the transformation of the treaty's principles and obligations into national law.

Tuberculosis cases, prevalence The total number of tuberculosis cases reported to the World Health Organization. A tuberculosis case is defined as a patient in whom tuberculosis has been bacteriologically confirmed or diagnosed by a clinician.

Tuberculosis cases cured under DOTS The percentage of estimated new infectious tuberculosis cases cured under DOTS, the internationally recommended tuberculosis control strategy.

Tuberculosis cases detected under DOTS The percentage of estimated new infectious tuberculosis cases detected (diagnosed in a given period) under DOTS, the internationally recommended tuberculosis control strategy.

Under-five mortality rate See *Mortality rate, under-five.*

Under height for age, children under age five Includes moderate stunting (defined as between two and three standard deviations below the median height-for-age of the reference population), and severe stunting (defined as more than three standard deviations below the median height-for-age of the reference population).

Under weight for age, children under age five Includes moderate underweight (defined as between two and three standard deviations below the median weight-for-age of the reference population), and severe underweight (defined as more than three standard deviations below the median weight-for-age of the reference population).

Undernourished people People whose food intake is chronically insufficient to meet their minimum energy requirements.

Unemployment Refers to all people above a specified age who are not in paid employment or self-employed, but are available for work and have taken specific steps to seek paid employment or self-employment.

Unemployment, long-term Unemployment lasting 12 months or longer. See *Unemployment*.

Unemployment rate The unemployed divided by the labour force (those employed plus the unemployed). See *Unemployment* and *Labour force*.

Unemployment rate, youth Refers to the unemployment rate between the ages of 15 or 16 and 24, depending on the national definition. See *Unemployment* and *Unemployment rate*.

Water source, improved, population not using Calculated as 100 minus the percentage of the population using an improved water source. Unimproved sources include vendors, bottled water, tanker trucks and unprotected wells and springs. See *Water source, improved, population using*.

Water source, improved, population using The share of the population with reasonable access to any of the following types of water supply for drinking: household connections, public standpipes, boreholes, protected dug wells, protected springs and rainwater collection. *Reasonable access* is defined as the availability of at least 20 litres a person per day from a source within one kilometre of the user's dwelling.

Women in government at ministerial level Includes deputy prime ministers and ministers. Prime ministers were included when they held ministerial portfolios. Vice-presidents and heads of ministerial-level departments or agencies were also included when exercising a ministerial function in the government structure.

Work time, total See *Time use, work time, total*.

Statistical references

Amnesty International. 2007. "Facts and Statistics on the Death Penalty." [http://www.amnesty.org/]. Accessed June 2007.

Cameron, A. Colin and Pravin K. Trivedi. 2005. Microeconometrics: Methods and Applications, Cambridge University Press.

CDIAC (Carbon Dioxide Information Analysis Center). 2007. Correspondence on carbon dioxide emissions. July. Oak Ridge.

Charmes, Jacques and Uma Rani. 2007. "An overview of size and contribution of informal sector in the total economy: A comparison across countries". Paris. l'Institut de Recherche pour le Développement.

FAO (Food and Agriculture Organization). 2006. Global Forest Resources Assessment 2005. Rome. FAO.

———. 2007a. FAOSTAT Database. [http://faostat.fao.org/]. Accessed May 2007.

———. 2007b. "Forest Resources Assessment". Correspondence on carbon stocks in forests; extract from database. August. Rome.

Fuentes, Ricardo and Papa Seck. 2007."The short- and long-term human development effects of climate-related shocks: some empirical evidence."

Guha-Sapir, Debarati, David Hargitt, Philippe Hoyois. 2004. Thirty years of Natural Disasters 1974–2003: the numbers. Presses universitaires de Louvain, Louvain-la-Neuve. Brussels, Belgium.

Harvey, Andrew S. 2001. "National Time Use Data on Market and Non-Market Work by Both Women and Men." Background paper for UNDP, Human Development Report 2001. United Nations Development Programme, Human Development Report Office, New York.

Heston, Alan, Robert Summers, and Bettina Aten. 2001. Correspondence on data from the Penn World Table Version 6.0. University of Pennsylvania, Center for International Comparisons of Production, Income and Prices. [http://pwt.econ.upenn.edu/]. March. Philadelphia.

———. 2006. "Penn World Table Version 6.2." University of Pennsylvania, Center for International Comparisons of Production, Income and Prices, Philadelphia. [http://pwt.econ.upenn.edu/]. Accessed June 2007.

ICPS (International Centre for Prison Studies). 2007. World Prison Population List. Seventh Edition. King's College London. London.

IDMC (Internally Displaced Monitoring Centre). 2007. "Global Statistics." [http://www.internal-displacement.org/]. Accessed April 2007.

IEA (International Energy Agency). 2002. World Energy Outlook 2002. Paris. IEA Publication Service.

———. 2006. World Energy Outlook 2006. Paris. IEA Publication Service.

———. 2007. Energy Balances for OECD and non-OECD countries Vol 2007, release 01 Database. Paris. IEA Energy Statistics and Balances. Accessed August 2007.

IISS (International Institute for Strategic Studies). 2007. Military Balance 2006–2007. London: Routledge, Taylor and Francis Group.

ILO (International Labour Organization). 2005. Key Indicators of the Labour Market. Fourth Edition. Geneva. CD-ROM. Geneva. [www.ilo.org/kilm/]. Accessed July 2006.

———. 2007a. International Labour Standards (ILOEX) Database. [http://www.ilo.org/ilolex/]. Accessed July 2007.

———. 2007b. LABORSTA Database. Geneva. [http://laborsta.ilo.org]. Accessed June 2007.

ILO (International Labour Organization) Bureau of Statistics. 2007. Correspondence on informal sector data. June. Geneva.

IPU (Inter-Parliamentary Union). 2007a. Correspondence on women in government at the ministerial level. June. Geneva.

———. 2007b. Correspondence on year women received the right to vote and year women first stand for election and year first woman was elected or appointed to parliament. June. Geneva.

———. 2007c. Parline Database. [www.ipu.org]. Accessed June 2007.

LIS (Luxembourg Income Studies). 2007. "Relative Poverty Rates for the Total Population, Children and the Elderly." Luxembourg. [http://www.lisproject.org/]. Accessed May 2007.

Macro International. 2007a. Correspondence on household data. May 2007. Calverton, MD.

———. 2007b. Demographic and Health Surveys (DHS) reports. Calverton, MD. [http://www.measuredhs.com/]. Accessed June 2007.

OECD (Organisation for Economic Co-operation and Development). 2007. OECD Main Economic Indicators. Paris. [http://www.oecd.org/statsportal]. Accessed July 2007.

OECD (Organisation for Economic Co-operation and Development) and Statistics Canada. 2000. Literacy in the Information age. Final Report on the International Adult Literacy Survey. OECD Publishing. Paris.

———. 2005. Learning a Living by Earning Skills: First Results of the Adult Literacy and Life Skills Survey. OECD Publishing. Paris.

OECD-DAC (Organisation for Economic Co-operation and Development, Development Assistance Committee). 2007a. OECD Journal on Development: Development Co-operation Report 2006. OECD Publishing. Paris.

———. 2007b. Correspondence on official development assistance disbursed. May. Paris.

Ruoen, Ren, and Chen Kai. 1995. "China's GDP in U.S. Dollars Based on Purchasing Power Parity." Policy Research Working Paper 1415. World Bank, Washington, D.C.

SIPRI (Stockholm International Peace Research Institute). 2007a. Correspondence on arms transfers. March. Stockholm.

———. 2007b. Correspondence on military expenditures. March. Stockholm.

———. 2007c. SIPRI Yearbook: Armaments, Disarmaments and International Security. Oxford, U.K.: Oxford University Press.

Smeeding, Timothy M. 1997. "Financial Poverty in Developed Countries: The Evidence from the Luxembourg Income Study." Background paper for UNDP, Human Development Report 1997. United Nations Development Programme, Human Development Report Office, New York.

Smeeding, Timothy M., Lee Rainwater, and Gary Burtless. 2000. "United States Poverty in a Cross-National Context." In Sheldon H. Danziger and Robert H. Haveman, eds., Understanding Poverty. New York: Russell Sage Foundation; and Cambridge, MA: Harvard University Press.

Statec. 2006. Correspondence on gross enrolment ratio for
Luxembourg. May. Luxembourg.

Time use. 2007. Correspondence with time use professionals: Debbie
Budlender (Community Agency for Social Enquiry) for South Africa
based on "A Survey of Time Use"; Jacques Charmes (Institut
de recherche pour le développement) for Benin, Nicaragua,
Madagascar, Mauritius and Uruguay based on country specific
time use surveys 1998–2002; Choi Yoon Ji (Rural Development
Administration of the Republic of Korea) for Rural Republic of Korea;
Jamie Spinney (St. Mary's University), Marcel Bechard (Statistics
Canada) and Isabelle Marchand (Statistics Canada) for Canada
based on "Canadian Time Use Survey 2005"; Marcela Eternod and
Elsa Contreras (INEGI) for Mexico based on "Encuesta Nacional
sobre Uso del Tiempo 2002"; Elsa Fontainha (ISEG - Technical
University of Lisbon) for Portugal based on "INE, Inquérito à
Ocupação do Tempo, 1999"; Rachel Krantz-Kent (Bureau of Labor
Statistics) for the United Sates based on "American Time Use
Survey 2005"; Fran McGinnity (Economic and Social Research
Institute) for Ireland based on "Irish National Time Use Survey
2005"; Iiris Niemi (Statistics Finland) for Belgium, Finland, France,
Estonia, Germany, Hungary, Italy, Latvia, Lithuania, Norway, Poland,
Slovenia, Spain, Sweden, United Kingdom based on Harmonized
European Time Use Surveys 1998–2004; Andries van den Broek
(Social and Culture Planning Office of The Netherlands) for the
Netherlands based on "Trends in Time"; Jayoung Yoon (University
of Massachusetts) for Republic of Korea based on "Korean Time Use
Survey 2004."

UN (United Nations). 2002. Correspondence on time use surveys.
Department of Economic and Social Affairs. Statistics Division.
February. New York.

———. 2006a. Millennium Development Goals Indicators Database.
Department of Economic and Social Affairs, Statistics Division. New
York. [http://mdgs.un.org]. Accessed May 2007.

———. 2006b. World Urbanization Prospects: The 2005 Revision.
Database. Department of Economic and Social Affairs, Population
Division. New York.

———. 2007a. Multilateral Treaties Deposited with the Secretary-
General. New York. [http://untreaty.un.org]. Accessed June 2007.

———. 2007b. Correspondence on electricity consumption.
Department of Economic and Social Affairs, Statistics Division.
March. New York.

———. 2007c. Correspondence on the Millennium Development Goals
Indicators. Department of Economic and Social Affairs, Statistics
Division. July. New York.

———. 2007d. The 2004 Energy Statistics Yearbook. Department of
Economic and Social Affairs, Statistics Division. New York.

———. 2007e. World Population Prospects 1950–2050: The 2006
Revision. Database. Department of Economic and Social Affairs,
Population Division. New York. Accessed July 2007.

UNAIDS (Joint United Nations Programme on HIV/AIDS). 2006.
Correspondence on HIV prevalence. May 2006. Geneva.

UNDP (United Nations Development Programme). 2006. The Path
out of Poverty. National Human Development Report for Timor-
Leste. Dili.

———. 2007. Social Inclusion in BiH. National Human Development
Report for Bosnia and Herzegovina. Sarajevo.

UNESCO (United Nations Educational, Scientific and Cultural
Organization). 1997. "International Standard Classification of
Education 1997." Paris. [http://www.uis.unesco.org/TEMPLATE/
pdf/isced/ISCED_A.pdf]. Accessed August 2007.

UNESCO (United Nations Educational, Scientific and Cultural
Organization) Institute for Statistics. 1999. Statistical yearbook.
Montreal.

———. 2003. Correspondence on adult and youth literacy rates.
March. Montreal.

———. 2006. Correspondence on students in science, engineering,
manufacturing and construction. April. Montreal.

———. 2007a. Correspondence on adult and youth literacy rates.
May. Montreal.

———. 2007b. Correspondence on education expenditure data.
April. Montreal.

———. 2007c. Correspondence on gross and net enrolment ratios,
children reaching grade 5 and tertiary education. April. Montreal.

UNHCR (United Nations High Commission for Refugees). 2007.
Correspondence on refugees by country of asylum and country of
origin. May. Geneva.

UNICEF (United Nations Children's Fund). 2004. State of the World's
Children 2005. New York.

———. 2005. State of the World's Children 2006. New York.

———. 2006. State of the World's Children 2007. New York.

———. 2007a. Correspondence on maternal mortality. New York.
August 2007.

———. 2007b. Multiple Indicator Cluster Surveys (MICS) reports. New
York. [http://www.childinfo.org]. Accessed June 2007.

UNODC (United Nations Office on Drugs and Crime). 2007.
Correspondence on "The Ninth United Nations Survey on Crime
Trends and the Operations of the Criminal Justice Systems".May
Vienna.

WHO (World Health Organization). 2007a. Core Health Indicators
2007 Database. Geneva. [http://www.who.int/whosis/database/].
Accessed July 2007.

———. 2007b. Global Tuberculosis Control: WHO Report 2007.
Geneva.[http://www.who.int/tb/publications/global_report/2007/
en/index.html]. Accessed July 2007.

WIPO (World Intellectual Property Organization). 2007. "Patents
Granted by Office (1985–2005)." Geneva. [http://wipo.int/ipstats/
en/statistics/]. Accessed May 2007.

World Bank. 2006. World Development Indicators 2006. CD-ROM.
Washington, D.C.

———. 2007a. Povcalnet. Washington, D.C.. [http://iresearch.
worldbank.org/]. Accessed May 2007.

———. 2007b. World Development Indicators 2007. CD-ROM.
Washington, D.C.

Classification of countries

Countries in the human development aggregates

High human development
(HDI 0.800 and above)

Albania	Poland
Antigua and Barbuda	Portugal
Argentina	Qatar
Australia	Romania
Austria	Russian Federation
Bahamas	Saint Kitts and Nevis
Bahrain	Saudi Arabia
Barbados	Seychelles
Belarus	Singapore
Belgium	Slovakia
Bosnia and Herzegovina	Slovenia
Brazil	Spain
Brunei Darussalam	Sweden
Bulgaria	Switzerland
Canada	Tonga
Chile	Trinidad and Tobago
Costa Rica	United Arab Emirates
Croatia	United Kingdom
Cuba	United States
Cyprus	Uruguay
Czech Republic	(70 countries or areas)
Denmark	
Estonia	
Finland	
France	
Germany	
Greece	
Hong Kong, China (SAR)	
Hungary	
Iceland	
Ireland	
Israel	
Italy	
Japan	
Korea (Republic of)	
Kuwait	
Latvia	
Libyan Arab Jamahiriya	
Lithuania	
Luxembourg	
Macedonia (TFYR)	
Malaysia	
Malta	
Mauritius	
Mexico	
Netherlands	
New Zealand	
Norway	
Oman	
Panama	

Medium human development
(HDI 0.500–0.799)

Algeria	Myanmar
Armenia	Namibia
Azerbaijan	Nepal
Bangladesh	Nicaragua
Belize	Occupied Palestinian
Bhutan	Territories
Bolivia	Pakistan
Botswana	Papua New Guinea
Cambodia	Paraguay
Cameroon	Peru
Cape Verde	Philippines
China	Saint Lucia
Colombia	Saint Vincent and the
Comoros	Grenadines
Congo	Samoa
Djibouti	Sao Tome and Principe
Dominica	Solomon Islands
Dominican Republic	South Africa
Ecuador	Sri Lanka
Egypt	Sudan
El Salvador	Suriname
Equatorial Guinea	Swaziland
Fiji	Syrian Arab Republic
Gabon	Tajikistan
Gambia	Thailand
Georgia	Timor-Leste
Ghana	Togo
Grenada	Tunisia
Guatemala	Turkey
Guyana	Turkmenistan
Haiti	Uganda
Honduras	Ukraine
India	Uzbekistan
Indonesia	Vanuatu
Iran (Islamic Republic of)	Venezuela (Bolivarian
Jamaica	Republic of)
Jordan	Viet Nam
Kazakhstan	Yemen
Kenya	Zimbabwe
Kyrgyzstan	(85 countries or areas)
Lao People's Democratic	
Republic	
Lebanon	
Lesotho	
Madagascar	
Maldives	
Mauritania	
Moldova	
Mongolia	
Morocco	

Low human development
(HDI below 0.500)

Angola
Benin
Burkina Faso
Burundi
Central African Republic
Chad
Congo (Democratic Republic
  of the)
Côte d'Ivoire
Eritrea
Ethiopia
Guinea
Guinea-Bissau
Malawi
Mali
Mozambique
Niger
Nigeria
Rwanda
Senegal
Sierra Leone
Tanzania (United Republic of)
Zambia
(22 countries or areas)

Note: The following UN member countries are not included in the human development aggregates because the HDI cannot be computed for them: Afghanistan, Andorra, Iraq, Kiribati, Korea (Democratic People's Republic of), Liberia, Liechtenstein, Marshall Islands, Micronesia (Federated States of), Monaco, Montenegro, Nauru, Palau, San Marino, Serbia, Somalia and Tuvalu.

Countries in the income aggregates

High income
(GNI per capita of US$10,726 or more in 2005)

Andorra	United Arab Emirates
Antigua and Barbuda	United Kingdom
Aruba	United States
Australia	United States Virgin Islands
Austria	(55 countries or areas)
Bahamas	
Bahrain	
Belgium	
Bermuda	
Brunei Darussalam	
Canada	
Cayman Islands	
Cyprus	
Denmark	
Faeroe Islands	
Finland	
France	
French Polynesia	
Germany	
Greece	
Greenland	
Guam	
Hong Kong, China (SAR)	
Iceland	
Ireland	
Isle of Man	
Israel	
Italy	
Japan	
Korea (Republic of)	
Kuwait	
Liechtenstein	
Luxembourg	
Macao, China (SAR)	
Malta	
Monaco	
Netherlands	
Netherlands Antilles	
New Caledonia	
New Zealand	
Norway	
Portugal	
Puerto Rico	
Qatar	
San Marino	
Saudi Arabia	
Singapore	
Slovenia	
Spain	
Sweden	
Switzerland	

Middle income
(GNI per capita of US$876–US$10,725 in 2005)

Albania	Libya Arab Jamahiriya
Algeria	Lithuania
American Samoa	Macedonia (TFYR)
Angola	Malaysia
Argentina	Maldives
Armenia	Marshall Islands
Azerbaijan	Mauritius
Barbados	Mexico
Belarus	Micronesia (Federated
Belize	States of)
Bolivia	Moldova
Bosnia and Herzegovina	Montenegro
Botswana	Morocco
Brazil	Namibia
Bulgaria	Nicaragua
Cameroon	Northern Mariana Islands
Cape Verde	Occupied Palestinian
Chile	Territories
China	Oman
Colombia	Palau
Congo	Panama
Costa Rica	Paraguay
Croatia	Peru
Cuba	Philippines
Czech Republic	Poland
Djibouti	Romania
Dominica	Russian Federation
Dominican Republic	Saint Kitts and Nevis
Ecuador	Saint Lucia
Egypt	Saint Vincent and the
El Salvador	Grenadines
Equatorial Guinea	Samoa
Estonia	Serbia
Fiji	Seychelles
Gabon	Slovakia
Georgia	South Africa
Grenada	Sri Lanka
Guatemala	Suriname
Guyana	Swaziland
Honduras	Syrian Arab Republic
Hungary	Thailand
Indonesia	Tonga
Iran (Islamic Republic of)	Tunisia
Iraq	Turkey
Jamaica	Turkmenistan
Jordan	Ukraine
Kazakhstan	Uruguay
Kiribati	Vanuatu
Latvia	Venezuela (Bolivarian
Lebanon	Republic of)
Lesotho	(97 countries or areas)

Low income
(GNI per capita of US$875 or less in 2005)

Afghanistan	Uganda
Bangladesh	Uzbekistan
Benin	Viet Nam
Bhutan	Yemen
Burkina Faso	Zambia
Burundi	Zimbabwe
Cambodia	(54 countries or areas)
Central African Republic	
Chad	
Comoros	
Congo (Democratic Republic	
of the)	
Côte d'Ivoire	
Eritrea	
Ethiopia	
Gambia	
Ghana	
Guinea	
Guinea-Bissau	
Haiti	
India	
Kenya	
Korea (Democratic People's	
Republic of)	
Kyrgyzstan	
Lao People's Democratic	
Republic	
Liberia	
Madagascar	
Malawi	
Mali	
Mauritania	
Mongolia	
Mozambique	
Myanmar	
Nepal	
Niger	
Nigeria	
Pakistan	
Papua New Guinea	
Rwanda	
Sao Tome and Principe	
Senegal	
Sierra Leone	
Solomon Islands	
Somalia	
Sudan	
Tajikistan	
Tanzania (United Republic of)	
Timor-Leste	
Togo	

Note: Income aggregates use World Bank classification (effective 1 July 2006) based on gross national income (GNI) per capita. They include the following countries or areas that are not UN member states and therefore not included in the HDI tables: high income - Aruba, Bermuda, Cayman Islands, Faeroe Islands, French Polynesia, Greenland, Guam, Isle of Man, Macao, China (SAR), Netherlands Antilles, New Caledonia, Puerto Rico and United States Virgin Islands; middle income - American Samoa. These countries or areas are included in the aggregates by income level. UN member countries Nauru and Tuvalu are not included because of lack of data.

Countries in the major world aggregates

Developing countries

Afghanistan
Algeria
Angola
Antigua and Barbuda
Argentina
Bahamas
Bahrain
Bangladesh
Barbados
Belize
Benin
Bhutan
Bolivia
Botswana
Brazil
Brunei Darussalam
Burkina Faso
Burundi
Cambodia
Cameroon
Cape Verde
Central African Republic
Chad
Chile
China
Colombia
Comoros
Congo
Congo (Dem. Rep. of the)
Costa Rica
Côte d'Ivoire
Cuba
Cyprus
Djibouti
Dominica
Dominican Republic
Ecuador
Egypt
El Salvador
Equatorial Guinea
Eritrea
Ethiopia
Fiji
Gabon
Gambia
Ghana
Grenada
Guatemala
Guinea
Guinea-Bissau

Guyana
Haiti
Honduras
Hong Kong, China (SAR)
India
Indonesia
Iran (Islamic Republic of)
Iraq
Jamaica
Jordan
Kenya
Kiribati
Korea (Democratic People's
    Republic of)
Korea (Republic of)
Kuwait
Lao People's Democratic
    Republic
Lebanon
Lesotho
Liberia
Libya
Madagascar
Malawi
Malaysia
Maldives
Mali
Marshall Islands
Mauritania
Mauritius
Mexico
Micronesia (Federated
    States of)
Mongolia
Morocco
Mozambique
Myanmar
Namibia
Nauru
Nepal
Nicaragua
Niger
Nigeria
Occupied Palestinian
    Territories
Oman
Pakistan
Palau
Panama
Papua New Guinea

Paraguay
Peru
Philippines
Qatar
Rwanda
Saint Kitts and Nevis
Saint Lucia
Saint Vincent and the
    Grenadines
Samoa
Sao Tome and Principe
Saudi Arabia
Senegal
Seychelles
Sierra Leone
Singapore
Solomon Islands
Somalia
South Africa
Sri Lanka
Sudan
Suriname
Swaziland
Syrian Arab Republic
Tanzania (United Republic of)
Thailand
Timor-Leste
Togo
Tonga
Trinidad and Tobago
Tunisia
Turkey
Tuvalu
Uganda
United Arab Emirates
Uruguay
Vanuatu
Venezuela (Bolivarian
    Republic of)
Viet Nam
Yemen
Zambia
Zimbabwe
(137 countries or areas)

Least developed
countries[a]

Afghanistan
Angola
Bangladesh

Benin
Bhutan
Burkina Faso
Burundi
Cambodia
Cape Verde
Central African Republic
Chad
Comoros
Congo (Democratic Republic
    of the)
Djibouti
Equatorial Guinea
Eritrea
Ethiopia
Gambia
Guinea
Guinea-Bissau
Haiti
Kiribati
Lao People's Democratic
    Republic
Lesotho
Liberia
Madagascar
Malawi
Maldives
Mali
Mauritania
Mozambique
Myanmar
Nepal
Niger
Rwanda
Samoa
Sao Tome and Principe
Senegal
Sierra Leone
Solomon Islands
Somalia
Sudan
Tanzania (United Republic of)
Timor-Leste
Togo
Tuvalu
Uganda
Vanuatu
Yemen
Zambia
(50 countries or areas)

Central and Eastern
Europe and the
Commonwealth
of Independent
States (CIS)

Albania
Armenia
Azerbaijan
Belarus
Bosnia and Herzegovina
Bulgaria
Croatia
Czech Republic
Estonia
Georgia
Hungary
Kazakhstan
Kyrgyzstan
Latvia
Lithuania
Macedonia (TFYR)
Moldova
Montenegro
Poland
Romania
Russian Federation
Serbia
Slovakia
Slovenia
Tajikistan
Turkmenistan
Ukraine
Uzbekistan
(28 countries or areas)

Organisation for
Economic
Co-operation
and Development
(OECD)

Australia
Austria
Belgium
Canada
Czech Republic
Denmark
Finland
France
Germany
Greece
Hungary

Iceland
Ireland
Italy
Japan
Korea (Republic of)
Luxembourg
Mexico
Netherlands
New Zealand
Norway
Poland
Portugal
Slovakia
Spain
Sweden
Switzerland
Turkey
United Kingdom
United States
(30 countries or areas)

High-income OECD
countries

Australia
Austria
Belgium
Canada
Denmark
Finland
France
Germany
Greece
Iceland
Ireland
Italy
Japan
Korea (Republic of)
Luxembourg
Netherlands
New Zealand
Norway
Portugal
Spain
Sweden
Switzerland
United Kingdom
United States
(24 countries or areas)

a UN classification based on UN-OHRLLS 2007.

Developing countries in the regional aggregates

Arab States
Algeria
Bahrain
Djibouti
Egypt
Iraq
Jordan
Kuwait
Lebanon
Libya
Morocco
Occupied Palestinian
  Territories
Oman
Qatar
Saudi Arabia
Somalia
Sudan
Syrian Arab Republic
Tunisia
United Arab Emirates
Yemen
(20 countries or areas)

East Asia and the Pacific
Brunei Darussalam
Cambodia
China
Fiji
Hong Kong, China (SAR)
Indonesia
Kiribati
Korea (Democratic People's
  Republic of)
Korea (Republic of)
Lao People's Democratic
  Republic
Malaysia
Marshall Islands
Micronesia (Federated
  States of)
Mongolia
Myanmar
Nauru
Palau
Papua New Guinea
Philippines
Samoa
Singapore
Solomon Islands
Thailand
Timor-Leste
Tonga
Tuvalu
Vanuatu
Viet Nam
(28 countries or areas)

South Asia
Afghanistan
Bangladesh
Bhutan
India
Iran (Islamic Republic of)
Maldives
Nepal
Pakistan
Sri Lanka
(9 countries or areas)

Latin America and Caribbean
Antigua and Barbuda
Argentina
Bahamas
Barbados
Belize
Bolivia
Brazil
Chile
Colombia
Costa Rica
Cuba
Dominica
Dominican Republic
Ecuador
El Salvador
Grenada
Guatemala
Guyana
Haiti
Honduras
Jamaica
Mexico
Nicaragua
Panama
Paraguay
Peru
Saint Kitts and Nevis
Saint Lucia
Saint Vincent and the
  Grenadines
Suriname
Trinidad and Tobago
Uruguay
Venezuela (Bolivarian
  Republic of)
(33 countries or areas)

Southern Europe
Cyprus
Turkey
(2 countries or areas)

Sub-Saharan Africa
Angola
Benin
Botswana
Burkina Faso
Burundi
Cameroon
Cape Verde
Central African Republic
Chad
Comoros
Congo
Congo (Democratic Republic
  of the)
Côte d'Ivoire
Equatorial Guinea
Eritrea
Ethiopia
Gabon
Gambia
Ghana
Guinea
Guinea-Bissau
Kenya
Lesotho
Liberia
Madagascar
Malawi
Mali
Mauritania
Mauritius
Mozambique
Namibia
Niger
Nigeria
Rwanda
Sao Tome and Principe
Senegal
Seychelles
Sierra Leone
South Africa
Swaziland
Tanzania (United Republic of)
Togo
Uganda
Zambia
Zimbabwe
(45 countries or areas)

Index to indicators

Index to Millennium Development Goal indicators in the HDR indicator tables

Goals and targets from the Millennium Declaration*	Indicators for monitoring progress	Indicator tables
Goal 1: Eradicate extreme poverty and hunger		
Target 1: Halve, between 1990 and 2015, the proportion of people whose income is less than one dollar a day	1. Proportion of population below one dollar (PPP) a day 2. Poverty gap ratio [incidence x depth of poverty] 3. Share of poorest quintile in national consumption	3 15
Target 2: Halve, between 1990 and 2015, the proportion of people who suffer from hunger	4. Prevalence of underweight children under-five years of age 5. Proportion of population below minimum level of dietary energy consumption	3, 7 1a[a], 7[a]
Goal 2: Achieve universal primary education		
Target 3: Ensure that, by 2015, children everywhere, boys and girls alike, will be able to complete a full course of primary schooling	6. Net enrolment ratio in primary education 7. Proportion of pupils starting grade 1 who reach grade 5 8. Literacy rate of 15–24 year-olds	1a, 12 12 12
Goal 3: Promote gender equality and empower women		
Target 4: Eliminate gender disparity in primary and secondary education, preferably by 2005, and in all levels of education not later than 2015	9. Ratio of girls to boys in primary, secondary and tertiary education 10. Ratio of literate women to men, 15–24 years old 11. Share of women in wage employment in the non-agricultural sector 12. Proportion of seats held by women in national parliament	30 [b] 30 31 [c] 29, 33 [d]
Goal 4: Reduce child mortality		
Target 5: Reduce by two-thirds, between 1990 and 2015, the under-five mortality rate	13. Under-five mortality rate 14. Infant mortality rate 15. Proportion of 1 year-old children immunised against measles	1a, 10 10 6
Goal 5: Improve maternal health		
Target 6: Reduce by three-quarters, between 1990 and 2015, the maternal mortality ratio	16. Maternal mortality ratio 17. Proportion of births attended by skilled health personnel	10 6
Goal 6: Combat HIV/AIDS, malaria and other diseases		
Target 7: Have halted by 2015 and begun to reverse the spread of HIV/AIDS	18. HIV prevalence among pregnant women aged 15–24 years 19. Condom use rate of the contraceptive prevalence rate 19a. Condom use at last high-risk sex 19b. Proportion of population aged 15–24 years with comprehensive correct knowledge of HIV/AIDS 19c. Contraceptive prevalence rate 20. Ratio of school attendance of orphans to school attendance of non-orphans aged 10–14 years	1a [e], 9 [e] 9 6
Target 8: Have halted by 2015 and begun to reverse the incidence of malaria and other major diseases	21. Prevalence and death rates associated with malaria 22. Proportion of population in malaria-risk areas using effective malaria prevention and treatment measures 23. Prevalence and death rates associated with tuberculosis 24. Proportion of tuberculosis cases detected and cured under directly observed treatment short course (DOTS)	 9 [f] 9 [g] 9
Goal 7: Ensure environmental sustainability		
Target 9: Integrate the principles of sustainable development into country policies and programmes and reverse the loss of environmental resources	25. Proportion of land area covered by forest 26. Ratio of area protected to maintain biological diversity to surface area 27. Energy use (kg oil equivalent) per US$1 GDP (PPP) 28. Carbon dioxide emissions per capita and consumption of ozone-depleting CFCs (ODP tons) 29. Proportion of population using solid fuels	22 22 [h] 24 [i]
Target 10: Halve, by 2015, the proportion of people without sustainable access to safe drinking water and basic sanitation	30. Proportion of population with sustainable access to an improved water source, urban and rural 31. Proportion of population with access to improved sanitation, urban and rural	1a, 7 , 3 [j] 7

Goals and targets from the Millennium Declaration*	Indicators for monitoring progress	Indicator tables
Target 11: By 2020, to have achieved a significant improvement in the lives of at least 100 million slum dwellers	32. Proportion of households with access to secure tenure.	

Goal 8: Develop a global partnership for development

Target 12: Develop further an open, rule-based, predictable, non-discriminatory trading and financial system. Includes a commitment to good governance, development and poverty reduction—both nationally and internationally	Some of the indicators listed below are monitored separately for the least developed countries (LDCs), Africa, landlocked developing countries and small island developing States.	

Official development assistance (ODA)

Target 13: Address the special needs of the least developed countries Includes: tariff and quota free access for the least developed countries' exports; enhanced programme of debt relief for heavily indebted poor countries (HIPC) and cancellation of official bilateral debt; and more generous ODA for countries committed to poverty reduction	33. Net ODA, total and to the least developed countries, as percentage of OECD/DAC donors' gross national income	17 k
	34. Proportion of total bilateral, sector-allocable ODA of OECD/DAC donors to basic social services (basic education, primary health care, nutrition, safe water and sanitation)	17
	35. Proportion of bilateral official development assistance of OECD/DAC donors that is untied	17
Target 14: Address the special needs of landlocked developing countries and small island developing States (through the Programme of Action for the Sustainable Development of Small Island Developing States and the outcome of the twenty-second special session of the General Assembly)	36. ODA received in landlocked developing countries as a proportion of their gross national incomes	18 l
	37. ODA received in small island developing States as a proportion of their gross national incomes	18 l

Market access

Target 15: Deal comprehensively with the debt problems of developing countries through national and international measures in order to make debt sustainable in the long term	38. Proportion of total developed country imports (by value and excluding arms) from developing countries and least developed countries, admitted free of duty	
	39. Average tariffs imposed by developed countries on agricultural products and textiles and clothing from developing countries	
	40. Agricultural support estimate for OECD countries as a percentage of their gross domestic product	
	41. Proportion of ODA provided to help build trade capacity	
	Debt sustainability	
	42. Total number of countries that have reached their HIPC decision points and number that have reached their HIPC completion points (cumulative)	
	43. Debt relief committed under HIPC initiative	
	44. Debt service as a percentage of exports of goods and services	18
Target 16: In cooperation with developing countries, develop and implement strategies for decent and productive work for youth.	45. Unemployment rate of young people aged 15–24 years, each sex and total	
Target 17: In cooperation with pharmaceutical companies, provide access to affordable essential drugs in developing countries	46. Proportion of population with access to affordable essential drugs on a sustainable basis	
Target 18: In cooperation with the private sector, make available the benefits of new technologies, especially information and communications	47. Telephone lines and cellular subscribers per 100 people	13 m
	48a. Personal computers in use per 100 people	
	48b. Internet users per 100 people	13 m

* The Millennium Development Goals and targets come from the Millennium Declaration, signed by 189 countries, including 147 heads of State and Government, in September 2000 (http://www.un.org/millennium/declaration/ares552e.htm). The goals and targets are interrelated and should be seen as a whole. They represent a partnership between the developed countries and the developing countries "to create an environment – at the national and global levels alike – which is conducive to development and the elimination of poverty".

a Tables 1a and 7 present this indicator as undernourished people as a percentage of total population.
b Table presents female (net or gross) enrolment ratio as a percentage of male ratio for primary, secondary and tertiary education levels separately.
c Table includes data on female employment by economic activity.
d Table 33 presents a breakdown of the percentage of lower and upper house seats held by women.
e Tables 1a and 9 present HIV prevalence among people ages 15–49.
f Table includes data on children under five using insecticide-treated bed nets, and children under five with fever treated with antimalarial drugs.
g Table presents tuberculosis prevalence rates. Data on death rates are not included.
h Table shows data as GDP per unit of energy use in 2000 PPP US$ per kg of oil equivalent.
i Table shows data on carbon dioxide emissions per capita. Data on consumption of ozone depleting CFCs are not included.
j Tables 1a and 7 present this indicator as the percentage of people with access to an improved drinking water source, and Table 3 includes data on people without access to an improved drinking water source.
k Table includes data on official development assistance (ODA) to least developed countries as a percentage of total ODA.
l Table includes data on received ODA by all recipient countries as percentage of GDP.
m Data on telephone mainlines, cellular subscribers and internet users expressed in 'per 1,000 people'.